Intermediate Algebra

Concepts and Applications

FOURTH EDITION

Intermediate Algebra
Concepts and Applications

FOURTH EDITION

MARVIN L. BITTINGER
Indiana University–Purdue University at Indianapolis

MERVIN L. KEEDY
Purdue University

DAVID ELLENBOGEN
St. Michael's College

ADDISON-WESLEY PUBLISHING COMPANY
Reading, Massachusetts • Menlo Park, California • New York
Don Mills, Ontario • Wokingham, England • Amsterdam
Bonn • Sydney • Singapore • Tokyo • Madrid
San Juan • Milan • Paris

Sponsoring Editor	Melissa Acuña
Managing Editor	Karen Guardino
Production Supervisor	Jack Casteel
Design, Editorial, and Production Services	Quadrata, Inc.
Illustrator	Scientific Illustrators and Leo Harrington
Manufacturing Supervisor	Roy Logan
Cover Designer	Geri Davis, Quadrata, Inc., and Hannus Design Associates
Cover Photograph	Jurgen Vogt, The Image Bank

Photo Credits

1, ©91, Joe Towers, The Stock Market **45,** Bob Daemmrich, Stock, Boston **55,** NASA **63,** Stock Photos Unlimited, The Image Bank **80,** Alexander Tsiaras, Stock, Boston **113,** ©Richard Megna, 1991, Fundamental Photographs **173,** Keith Wood, ©Tony Stone Worldwide **179,** AP/Wide World Photos **206,** Comstock **217,** Benn Mitchell, The Image Bank **220,** ©Simon Wilkinson, The Image Bank **275,** ©Lou Jones, The Image Bank **335,** ©Mark E. Gibson, The Stock Market **397,** Bettman/UPI **455,** AP/Wide World Photos **471,** Sotheby's Inc. **522,** AP/Wide World Photos **529,** The Finest Image Photography **565,** Grant Faint, The Image Bank Portrait photograph on page 1 courtesy of USAir

Reprinted with corrections, September 1994.

Library of Congress Cataloging-in-Publication Data

Bittinger, Marvin L.
   Intermediate algebra : concepts and applications / Marvin L.
Bittinger, Mervin L. Keedy, David Ellenbogen. — 4th ed.
     p.   cm
   Includes index.
   ISBN 0-201-53786-9
   1. Algebra.   I. Keedy, Mervin Laverne.   II. Ellenbogen, David.
III. Title.
QA154.2.B54   1994
512.9 — dc20

93-5668
CIP

4 5 6 7 8 9 10—DOW—969594

For Kit

An extraordinary mother, grandmother, and advocate for the educationally disadvantaged

D.J.E.

Appropriate for a one-term course in intermediate algebra, this text is intended for students who have completed a first course in algebra. It is the second of two texts in an algebra series that also includes *Elementary Algebra: Concepts and Applications,* Fourth Edition, by Bittinger/Keedy/Ellenbogen. *Intermediate Algebra: Concepts and Applications,* Fourth Edition, is a significant revision of the Third Edition with respect to design, contents, pedagogy, and an expanded supplements package. This series is designed to prepare students for any mathematics course at the college algebra level.

APPROACH

Our approach, which has been developed over many years, is designed to help today's students both learn and retain mathematical concepts. Our goal in preparing this revision was to address the major challenges for teachers of developmental mathematics courses that we have seen emerging during the early 1990s. The first challenge is to prepare students of developmental mathematics to make the transition from "skills-oriented" elementary and intermediate algebra courses to the more "concept-oriented" presentation of college algebra or other college-level mathematics courses. The second is to teach these same students critical-thinking skills: to reason mathematically, to communicate mathematically, and to solve mathematical problems. The third challenge is to reduce the amount of content overlap between elementary algebra and intermediate algebra texts.

Following are some aspects of the approach that we have used in this revision to help meet the challenges we all face teaching developmental mathematics.

PROBLEM SOLVING

One distinguishing feature of our approach is our treatment of and emphasis on problem solving. We use problem solving and applications to motivate the material wherever possible, and we include real-life applications and problem-solving techniques throughout the text. We feel that problem solving encourages students to think about how mathematics can be used. It also challenges students and helps to prepare them for more difficult material in later courses.

- In Chapter 1, we introduce the five-step process for solving problems: (1) Familiarize, (2) Translate, (3) Carry out, (4) Check, and (5) State the answer. These steps are used throughout the text whenever we encounter a problem-solving situation. Repeated use of this algorithm gives students a sense that they have a starting point for any type of problem they encounter, and frees them to focus on the mathematics necessary to successfully translate the problem situation. (See pages 26–33, 265, and 449.)

FUNCTIONS AND GRAPHING

To retain skills and to apply them at a more conceptual level in later courses, students must have an intuitive understanding of the material. A visual interpretation of mathematical concepts can provide this type of understanding to those students with a visual, rather than symbolic, orientation. In addition, familiarity and practice with functions and graphing techniques make students more comfortable with this essential tool when they move on to later courses.

- We introduce functions and graphing in Chapter 2, which is substantially earlier than in many intermediate algebra texts. We then present functions and graphs throughout the text to help students develop an intuitive understanding of different types of equations and their solutions. For instance, examples of polynomial and rational functions are introduced along with polynomial and rational expressions and equations in Chapters 5 and 6. (See pages 64–71 and 72–80.)

- The inclusion of *Technology Connections*, an optional feature that allows students to use a graphing calculator or computer to help visualize concepts, provides additional opportunities for students to see the usefulness of functions and graphing. (See pages 69, 262, and 556.)

CONTENT

Many intermediate algebra texts contain a substantial review of elementary algebra topics. This lulls students into complacency and does not allow instructors sufficient time to cover the intermediate algebra topics necessary to prepare students for later courses.

- By introducing graphing and functions in Chapter 2, we present students with ''intermediate algebra'' topics almost immediately. These topics are then used throughout the text to give students familiarity and practice with concepts that will be critical to them at the college level.

- Systems of equations are introduced in Chapter 3 to provide students with a valuable problem-solving tool. Students can then translate problem situations into systems of equations throughout the remainder of the text. This approach provides a useful alternative to always translating problems into equations in which only one variable is used.

PEDAGOGY

Skill Maintenance Exercises and *Cumulative Reviews.*   Retention of skills is critical to the future success of our students. In nearly all exercise sets, we include carefully chosen exercises that review skills and concepts from preceding chapters of the text. Each chapter test includes Skill Maintenance Exercises selected from the three or four text sections that are identified at the beginning of each chapter. After every three chapters, and at the end of the text, we have also included a Cumulative Review, which reviews skills and concepts from all preceding chapters of the text. (See pages 164, 170, 171, and 205.)

Synthesis Exercises. Each exercise set ends with a set of synthesis exercises. These problems can offer opportunities for students to synthesize skills and concepts from earlier sections with the present material, or can provide students with deeper insights into the current topic. Synthesis exercises are generally more challenging than those in the main body of the exercise set. (See pages 95, 121, 122, and 315.)

Verbalization Skills. Wherever appropriate throughout the text, we have discussed how mathematical terms are used in language. The Summary and Review sections emphasize key terms and important properties and formulas. In addition, thinking and writing exercises are included in the Synthesis Exercises. These encourage students to verbalize mathematical concepts, leading to better understanding. (See pages 390 and 406.)

WHAT'S NEW IN THE FOURTH EDITION?

We have rewritten many key topics in response to user and reviewer feedback and have made significant improvements in design and pedagogy. Detailed information about the content changes is available in the form of a Conversion Guide. Please ask your local Addison-Wesley sales representative for more information. Following is a list of the major changes in this revision.

New Design

- The new design is more open and readable. Pedagogical use of color makes it easier to see where exercises, explanations, and examples begin and end.
- The entire art program is new for this edition. We have ensured the accuracy of the graphical art through the use of computer-generated graphs. Color in the graphical art is used pedagogically and precisely to help the student visualize the mathematics. (See pages 475 and 546.)

Technology Connections

- These features integrate technology, increase the understanding of concepts through visualization, encourage exploration, and motivate discovery learning. Optional Technology Connection exercises occur in many exercise sets. (See pages 277 and 546.)

Writing Exercises

- Nearly every set of Synthesis Exercises begins with two writing exercises. These exercises are usually not as difficult as other synthesis exercises, but require written answers that aid in student comprehension, critical thinking, and conceptualization. Because some instructors may collect answers to writing exercises, and because more than one answer may be correct, answers to writing exercises are not listed at the back of the text. (See pages 19, 406, and 550.)

Content Changes. A variety of content changes have been made. Some of the more significant changes are listed below.

- Rational exponents are now presented early in Chapter 7. Doing so has enabled us to use rational exponents in our subsequent work with radical notation. (See pages 343–346.)
- Although our fear of students performing ''illegal'' cancellations is as acute as ever, we now use canceling as a way to simplify rational expressions. We do so in recognition of the fact that we use canceling when working on our own. Whenever canceling is used, we point out that we are effectively ''removing'' a factor of 1. (See pages 280–285.)
- In response to numerous requests, we have included a new section on ''Geometric Applications,'' in which the important properties of 30°–60°–90° and 45°–45°–90° triangles are developed and used. (See pages 343–346.)
- Because so many students remain convinced that they ''cannot do word problems,'' we have made increased use of guessing as a means of familiarizing oneself with a problem-solving situation. By checking to see if a guess is correct, students can more easily discover an algebraic translation of the problem. (See pages 114 and 115.)
- Throughout the text, we have included a variety of new applications that appeal to a large cross section of the student population. By emphasizing applications that students and faculty find interesting, we hope that we have made the text enjoyable to use. (See pages 164, 327, and 479.)

SUPPLEMENTS FOR THE INSTRUCTOR

INSTRUCTOR'S SOLUTIONS MANUAL
by Judith A. Penna

This supplement contains worked-out solutions to all exercises in the text.

INSTRUCTOR'S RESOURCE GUIDE
by Donna DeSpain

This supplement contains the following:

- Extra practice problems for challenging topics in the text
- Black-line masters of grids and number lines for transparency masters or test preparation
- Videotape index and section cross references to the tutorial software packages available with this text
- Conversion guide from the Third Edition to the Fourth Edition

PRINTED TEST BANK
by Donna DeSpain

This supplement contains the following:

- Six alternative test forms for each chapter and six final examinations

- Two multiple-choice versions of each chapter test

All test forms have been completely rewritten.

COMPUTERIZED TESTING
Omnitest II (for IBM and Macintosh).
This computerized test bank allows you to create up to 99 versions of a customized test with just a few keystrokes, and allows the option of choosing items by chapter, section, or objective. It contains over 400 multiple-choice and open-ended algorithms. You may enter your own test items, edit existing items, and define the level of difficulty of problems.

SUPPLEMENTS FOR THE STUDENT

STUDENT'S SOLUTIONS MANUAL
by Judith A. Penna

This manual contains completely worked-out solutions with step-by-step annotations for all the odd-numbered exercises in the text, and answers for all even-numbered exercises in the text.

VIDEOTAPES

Developed especially for the Bittinger/Keedy/Ellenbogen texts, these videotapes feature an engaging team of lecturers presenting material from each section of the text in an interactive format that includes a group of students. The lecturers' presentation also incorporates slides, sophisticated computer-generated graphics, and a white board to support an approach that emphasizes visualization and problem solving.

TUTORIAL SOFTWARE

THE MATHLAB✛ (IBM and Macintosh). This software combines a unique combination of drill and practice modules with an interactive and easy-to-use graphing tool. The drill and practice segments feature feedback for wrong answers and detailed record keeping. The graphing tool allows students to graph and explore a wide variety of two-dimensional functions.

Algebra Problem Solver (IBM). After selecting a topic and an exercise type, students can enter their own exercises or request an exercise from the computer. In each case, the student is given detailed, annotated, step-by-step solutions.

ACKNOWLEDGMENTS

We wish to express our appreciation to the many people who helped with the development of this book. Barbara Johnson and Laurie A. Hurley deserve special thanks for their many fine suggestions. Their proofreadings of the text, in spite of almost endless time pressure, contributed immeasurably to the accuracy and readability of the text. Judy Penna also merits special thanks for her preparation of the *Student's Solution Manual,* the *Instructor's Solution Manual,* and the indexes. Judy's work is always performed with a thoroughness that amounts to another proofreading of the book and for that we are grateful. We are also indebted to Stuart Ball for his expert guidance in preparing the Technology Connections and the associated artwork.

This book's sponsoring editor, Melissa Acuña, performed admirably in coordinating the many intricacies of this project; George and Brian Morris of Scientific Illustrators generated a remarkable set of graphs and illustrations that are both precise and easily understood; and Leo Harrington drew the many fine sketches that enhance our exercises and examples. Geri Davis and Martha Morong of Quadrata, Inc., provided design, editorial, and production services second to none, ensuring that every last detail has been taken care of. To all of these people, we offer our deepest thanks.

In addition, we thank the following professors for their thoughtful reviews and insightful comments.

Dick J. Clark, *Portland Community College*
Linda Cook, *University of South Carolina — Spartanburg*
Linda Crabtree, *Longview Community College*
Jeannine Dawson, *Henry Ford Community College*
Lynette Goff, *Glendale Community College*
Jeff Koleno, *Lorain County Community College*
Linda Kyle, *Tarrant County Community College*
Peter Lindstrom, *North Lake College*
Karen Norwood, *North Carolina State University*
Wing Park, *College of Lake County*
Joseph Parker, *University of South Carolina*
William Radulovich, *Florida Community College — Jacksonville*
Minnie Shuler, *Gulf Coast Community College*
Lawrence Small, *Los Angeles Pierce College*
Larry Symrski, *Henry Ford Community College*
Bruce F. Teague, *Santa Fe Community College*
Richard Twaddle, *Anoka–Ramsey Community College*
Frances Ventola, *Brookdale Community College*
Joyce Vilseck, *Texas A&I University*
Elizabeth Whitener, *University of South Carolina — Spartanburg*
Deborah A. Zopf, *Henry Ford Community College*

Finally, a special thank you to all those who so generously agreed to discuss their professional uses for mathematics in our chapter openers. These dedicated people, none of whom we knew prior to writing this text, all share a desire to make math more meaningful to students. We cannot imagine a finer set of role models.

M.L.B.
M.L.K.
D.J.E.

C O N T E N T S

Cumulative Review: Chapters 1–3   170

Intermediate Algebra

Concepts and Applications

FOURTH EDITION

CHAPTER 1

Algebra and Problem Solving

AN APPLICATION

A commercial jet, flying from Chicago to Los Angeles, has been instructed to climb from its present altitude of 8000 ft to a cruising altitude of 29,000 ft. If the plane ascends at a rate of 3500 ft per minute, how long will it take to reach the cruising altitude?

This problem appears as Exercise 9 in Section 1.4.

Carole Perry
COMMERCIAL COPILOT

"A good working knowledge of math is essential to the success of each flight. I must be able to calculate quickly and accurately without a calculator."

T he principal theme of this text is problem solving in algebra. An overall strategy for solving problems is presented in Section 1.4. Additional and increasing emphasis on problem solving appears throughout the book. This chapter begins with a short review of algebraic symbolism and properties of numbers. As you will see, the manipulations of algebra, such as simplifying expressions and solving equations, are based on the properties of numbers.

1.1

The Beginnings of Algebra

Algebraic Expressions and Their Use • **Translating to Algebraic Expressions** • **Evaluating Algebraic Expressions** • **Solutions to Equations** • **Sets of Numbers** • **Set Notation** • **Notation for Rational Numbers**

This section is intended to introduce some of the basic concepts of algebra. We will study the use of algebraic expressions in problem solving and some of the types of numbers needed for problem solving.

Algebraic Expressions and Their Use

In arithmetic, you worked with expressions like

$$42 + 58, \qquad 9 \times 12, \qquad 17 - 5, \quad \text{and} \quad \frac{5}{7}.$$

In algebra, we will work with expressions like

$$42 + x, \qquad l \cdot w, \qquad 17 - t, \quad \text{and} \quad \frac{d}{y}.$$

Sometimes a letter can stand for various numbers. In that case, we call the letter a **variable.** Sometimes a letter can stand for just one number. In that case, we call the letter a **constant.** Let $b =$ your date of birth. Then b is a constant. Let $a =$ your age. Then a is a variable since a changes from year to year.

An **algebraic expression** consists of variables, numbers, and operation signs. Thus all of the expressions above are examples of algebraic expressions.

Algebraic expressions frequently arise in problem-solving situations. For example, consider the chart at the top of page 3. Suppose we want to determine how many Americans were uninsured in 1992. We might use algebra to translate the problem into an equation, with x representing the number of uninsured Americans.

Number of Americans with health insurance	plus	Number of uninsured Americans	is	Total number of Americans
	↓		↓	
212,000,000	+	x	=	249,000,000

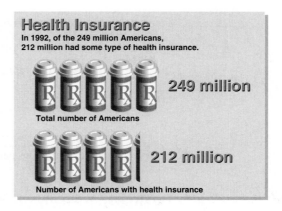

Note that we have an algebraic expression on the left. To find the number x, we can subtract 212,000,000 on both sides of the equation:

$$x = 249,000,000 - 212,000,000.$$

Then we carry out the subtraction and obtain the answer: In 1992, there were 37,000,000 uninsured Americans.

In arithmetic, you probably would do this subtraction right away without considering an equation. In algebra, however, you will find most problems difficult without first solving an equation.

Translating to Algebraic Expressions

In algebra, we translate problems to equations. To do this, we need to know what words translate to certain operation symbols:

KEY WORDS

Addition	Subtraction	Multiplication	Division
add	subtract	multiply	divide
sum	difference	product	divided by
plus	minus	times	quotient
increased by	less than	twice	ratio
more than	decreased by	of	

Phrase	Algebraic Expression
Five more than some number	$n + 5$ or $5 + n$
Half of a number	$\frac{1}{2}t$ or $\frac{t}{2}$
Five more than three times some number	$3p + 5$
The difference of two numbers	$x - y$
Six less than the product of two numbers	$mn - 6$
Seventy-six percent of some number	$76\%z$ or $0.76z$

Note that expressions like mn represent products and can also be written as $m \cdot n$, $m \times n$, or $(m)(n)$.

EXAMPLE 1

Translate to an algebraic expression.

Five less than forty-five percent of the quotient of two numbers

Solution   We let m and n represent the two numbers.

$$(0.45)\ \frac{m}{n}\ -\ 5$$

Five less than forty-five percent of the quotient of two numbers

Evaluating Algebraic Expressions

When we replace a variable by a number, we say that we are **substituting** for the variable. This process is called **evaluating the expression.**

EXAMPLE 2

Evaluate the expression $3xy + z$ when $x = 2$, $y = 5$, and $z = 7$.

Solution   We substitute and carry out the multiplication and addition:

$$3xy + z = 3 \cdot 2 \cdot 5 + 7$$
$$= 30 + 7$$
$$= 37.$$

EXAMPLE 3

The area of a triangular sail with a base of length b and a height of length h is $\frac{1}{2} \cdot b \cdot h$. Find the area when b is 8 m and h is 6.4 m.

Solution   We substitute 8 for b and 6.4 for h and carry out the multiplication:
$$\tfrac{1}{2} \cdot b \cdot h = \tfrac{1}{2} \cdot 8 \cdot 6.4 = 25.6 \text{ square meters (sq m)}.$$

Solutions to Equations

The use of the symbol $=$ in the work above indicates that the symbols on either side of the equals sign represent, or name, the same number. An **equation** is a number sentence with the verb $=$. At the beginning of this section, we saw that $212{,}000{,}000 + x = 249{,}000{,}000$ is a true equation when x is replaced by $37{,}000{,}000$.

Solution to an Equation

A replacement or substitution that makes an equation true is called a *solution*. Some equations may have more than one solution and some may have no solution. When we have found all the solutions, we say that we have *solved* the equation.

EXAMPLE 4

Determine whether 5.6 is a solution of the equation $3x = 16.8$.

Solution   We substitute 5.6 for the variable and see whether we get a true sentence.

$$3x = 16.8$$

$$3(5.6) \; ? \; 16.8 \qquad \text{Substituting 5.6 for } x \text{ and then}$$
$$\qquad\qquad\qquad\quad \text{simplifying the left side}$$
$$16.8 \mid 16.8 \quad \text{TRUE}$$

The number 5.6 is a solution. In fact, it is the *only* solution.   ❑

Sets of Numbers

When solving equations, or evaluating algebraic expressions, we often need to concern ourselves with the *type* of numbers used. For example, if we are solving for the optimal number of seats for a lecture hall, a fractional solution must be rounded up or down, since of course it makes no sense to discuss a fractional part of a seat. Three frequently used sets of numbers are listed below.

Three Important Sets

Natural Numbers

Those numbers used for counting: $\{1, 2, 3, \ldots\}$

Whole Numbers

The set of natural numbers with 0 included: $\{0, 1, 2, 3, \ldots\}$

Integers

The set of all whole numbers and their opposites:

$$\{\ldots, -4, -3, -2, -1, 0, 1, 2, 3, 4, \ldots\}$$

The dots mean that the pattern continues in the direction indicated.

The integers correspond to the points on a number line as follows:

$$-7 \;\; -6 \;\; -5 \;\; -4 \;\; -3 \;\; -2 \;\; -1 \;\; 0 \;\; 1 \;\; 2 \;\; 3 \;\; 4 \;\; 5 \;\; 6 \;\; 7$$

To fill in the rest of the points on our number line, we will need to describe two more sets of numbers. To do so, we must first discuss set notation.

Set Notation

The set containing the numbers -2, 1, and 3 can be written $\{-2, 1, 3\}$. This method of writing a set is known as **roster notation.** We used the roster notation for the sets listed above.

We can also name a set by specifying conditions under which a number is in the set. The following symbolism is read as indicated, and the notation is known as **set-builder notation:**

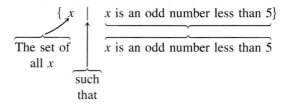

$$\{ \, x \mid x \text{ is an odd number less than } 5\}$$

The set of all x — x is an odd number less than 5

such that

EXAMPLE 5

Using both roster notation and set-builder notation, name the set consisting of the first four even natural numbers.

Solution

Using roster notation: $\{2, 4, 6, 8\}$

Using set-builder notation: $\{n \mid n \text{ is an even number between 1 and 9}\}$ ❑

The Greek letter epsilon, $\in$, is used to indicate that an element belongs to a set. Thus if $A = \{2, 4, 6, 8\}$, we might write $4 \in A$ to indicate that 4 *is an element of* A. We might also write $5 \notin A$ to indicate that 5 *is not an element of* A.

EXAMPLE 6

Classify the statement $8 \in \{x \mid x \text{ is an even number}\}$ as true or false.

Solution   Since 8 *is* an element of the set of all even numbers, the statement is true. In other words, because 8 is even, it belongs to the set.   ❑

Using set-builder notation, we can now describe the set of all *rational numbers*.

Rational Numbers

Numbers that can be expressed as an integer divided by a nonzero integer are called *rational numbers:*

$$\left\{ \frac{p}{q} \mid p \text{ is an integer, } q \text{ is an integer, and } q \neq 0 \right\}.$$

Notation for Rational Numbers

Rational numbers can be written using fractional or decimal notation. *Fractional notation* uses symbolism like the following:

$$\frac{5}{8}, \quad \frac{12}{-7}, \quad \frac{-17}{15}, \quad -\frac{9}{7}, \quad \frac{39}{1}, \quad \frac{0}{6}.$$

In *decimal notation,* rational numbers either *terminate* or *repeat.*

EXAMPLE 7

When written in decimal form, does each of the following numbers terminate or repeat? **(a)** $\frac{5}{8}$; **(b)** $\frac{6}{11}$.

Solution

a) Since $\frac{5}{8}$ means $5 \div 8$, we perform long division to find that $\frac{5}{8} = 0.625$. Thus we can write $\frac{5}{8}$ as a terminating decimal.

b) Using long division, we find that $6 \div 11 = 0.5454 \ldots$, so we can write $\frac{6}{11}$ as a repeating decimal. Repeating decimal notation can be abbreviated by writing a bar over the repeating part—in this case, $0.\overline{54}$. ❑

The set of all rational numbers does not fill up the number line. Numbers like π, $\sqrt{2}$, and $\sqrt{15}$, which will be used later in this text, can be only approximated by rational numbers. As decimals, these numbers are nonterminating and nonrepeating. Numbers like π, $\sqrt{2}$, and $\sqrt{15}$ are said to be **irrational.**

The set of all irrational numbers, combined with the set of all rational numbers, gives us the set of all **real numbers.**

Real Numbers

Numbers that are either rational or irrational are called *real* numbers:

$\{x| \ x \text{ is rational or } x \text{ is irrational}\}.$

The set of all real numbers *does* fill up the number line.

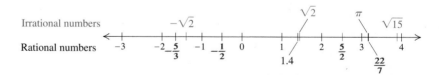

To help see that $\sqrt{2}$ is a very "real" point on the number line, you may recall from geometry that when a right triangle has two legs of length 1 unit, the remaining side has length $\sqrt{2}$ units. In this manner, we can "measure" a value for $\sqrt{2}$.

The following figure shows the relationships among various kinds of numbers.

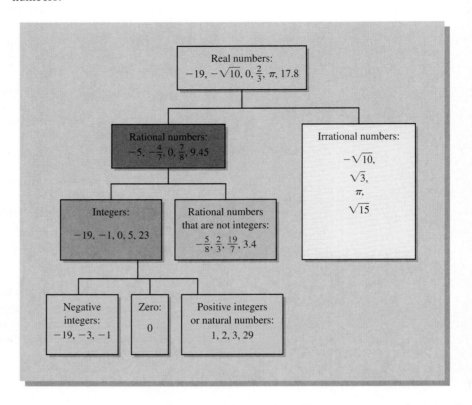

When *all* the members of a set are found in a second set, we say that the first set is a **subset** of the second set. Thus if $A = \{1, 2, 3\}$ and $B = \{1, 2, 3, 4, 5\}$, we can write $A \subseteq B$ to indicate that *A is a subset of B*. We can see from the preceding diagram that if $\mathbb{Z}$ represents the set of all integers and $\mathbb{R}$ represents the set of all real numbers, then $\mathbb{Z} \subseteq \mathbb{R}$. Similar statements may be made using other sets in the diagram.

EXERCISE SET │ 1.1

Translate the phrase to mathematical language.

1. Four less than some number

2. Six more than some number

3. Twice a number

4. Five times a number

5. Thirty-two percent of some number

6. Forty-seven percent of some number

7. Seven more than half of a number

8. Eight less than twice a number

9. Four less than nineteen percent of some number

10. Three more than eighty-two percent of some number

11. Five more than the difference of two numbers

12. Six less than the sum of two numbers

13. Four less than the product of two numbers

14. Ten more than the product of two numbers

15. One more than thirty-five percent of some number

16. Two less than eight percent of some number

Evaluate the given expression using the values provided.

17. $2x + y$ when $x = 3$ and $y = 7$

18. $3a - b$ when $a = 7$ and $b = 5$

19. $7abc$ when $a = 2$, $b = 1$, and $c = 3$

20. $2xyz$ when $x = 5$, $y = 2$, and $z = 1$

21. $8mn - p$ when $m = 1$, $n = 2$, and $p = 9$

22. $7rs + q$ when $r = 3$, $s = 2$, and $q = 5$

23. $5ab \div c$ when $a = 4$, $b = 2$, and $c = 10$

24. $4xy \div z$ when $x = 6$, $y = 3$, and $z = 12$

25. $2ab - a$ when $a = 5$ and $b = 3$

26. $3xy + y$ when $x = 2$ and $y = 3$

27. $pqr \div q$ when $p = 2$, $q = 3$, and $r = 2$

28. $abc \div a$ when $a = 5$, $b = 3$, and $c = 1$

Determine whether the given value is a solution of the equation.

29. $3; x - 2 = 1$

30. $4; x + 3 = 7$

31. $7; 4 + a = 13$

32. $6; 9 - y = 2$

33. $6; 13 - y = 7$

34. $7; 8 + a = 13$

35. $5; 2x + 3 = 13$

36. $4; 2n - 3 = 5$

37. $0.4; 8n - 1.7 = 2.5$

38. $0.5; 3x + 2.9 = 4.4$

39. $\frac{17}{3}; 3x - 4 = 13$

40. $\frac{9}{4}; 4x + 2 = 11$

Use roster notation to name the set.

41. The set of all vowels in the alphabet

42. The set of all the days of the week

43. The set of all even natural numbers

44. The set of all odd natural numbers

45. The set of all natural numbers that are multiples of 5

46. The set of all natural numbers that are multiples of 3

Use set-builder notation to name the set.

47. The set of all odd numbers between 10 and 30

48. The set of all multiples of 4 between 22 and 45

49. $\{0, 1, 2, 3, 4\}$

50. $\{-3, -2, -1, 0, 1, 2\}$

51. The set of all multiples of 5 between 7 and 79

52. The set of all even numbers between 9 and 99

Classify each statement as true or false. The following sets are used:

$\mathbb{N}$ = the set of natural numbers
$\mathbb{W}$ = the set of whole numbers
$\mathbb{Z}$ = the set of integers
$\mathbb{Q}$ = the set of rational numbers
$\mathbb{R}$ = the set of real numbers

53. $7 \in \mathbb{N}$

54. $3.4 \in \mathbb{N}$

55. $5.1 \in \mathbb{Z}$

56. $9.\overline{32} \in \mathbb{R}$

57. $-7.\overline{4} \in \mathbb{Q}$

58. $3.9 \in \mathbb{Q}$

59. $\sqrt{7} \in \mathbb{R}$

60. $\sqrt{8} \in \mathbb{R}$

61. $7.1 \notin \mathbb{W}$

62. $-3 \notin \mathbb{W}$

63. $\mathbb{N} \subseteq \mathbb{Z}$

64. $\mathbb{N} \subseteq \mathbb{R}$

65. $\mathbb{N} \subseteq \mathbb{W}$

66. $\mathbb{Z} \subseteq \mathbb{W}$

67. $\mathbb{W} \subseteq \mathbb{N}$

68. $\mathbb{Z} \subseteq \mathbb{Q}$

69. $\mathbb{Q} \subseteq \mathbb{R}$

70. $\mathbb{N} \subseteq \mathbb{Q}$

To the student and the instructor: The *synthesis exercises* found at the end of every exercise set will challenge students to combine concepts or skills studied in that section or in preceding parts of the text. *Writing exercises,* denoted by the ◈ icon, are meant to be answered with one or more English sentences.

Synthesis

71. ◈ Explain the difference between rational numbers and irrational numbers.

72. ◈ What advantage does set-builder notation have over roster notation?

Translate to mathematical language.

73. Three times the sum of two numbers

74. Half of the difference of two numbers

75. The quotient of the difference of two numbers and their sum

76. The product of the sum of two numbers and their difference

Use roster notation to name the set.

77. The set of all integers that are whole numbers but not natural numbers

78. The set of all integers that are not whole numbers

Use set-builder notation to name the set.

79. The set of all solutions of the equation $2x + 6 = 2(x + 3)$

80. The set of all solutions of the equation $3x - 6 = 3(x - 2)$

81. Draw a right triangle that could be used to "measure" $\sqrt{13}$ units.

1.2

Operations and Properties of Real Numbers

Absolute Value   •   Inequalities   •   Addition   •   Subtraction and Opposites   •   Multiplication   •   Division   •   Multiplication and Division by 1   •   The Commutative and Associative Laws   •   The Distributive Law

In this section, we review how real numbers are added, subtracted, multiplied, and divided. We also study some important rules for the manipulation of algebraic expressions.

The following discussions of absolute value and inequalities will help us when we begin to add real numbers.

Absolute Value

It is convenient to have a notation that represents a number's distance from zero on the number line.

Absolute Value

> We write $|a|$, read "the absolute value of a," to represent the number of units that a is from zero.

EXAMPLE 1

Find the absolute value: **(a)** $|-3|$; **(b)** $|7.2|$; **(c)** $|0|$.

Solution

a) $|-3| = 3$ -3 is 3 units from 0.
b) $|7.2| = 7.2$ 7.2 is 7.2 units from 0.
c) $|0| = 0$ 0 is 0 units from itself.   ❑

Note that whereas the absolute value of a nonnegative number is the number itself, the absolute value of a negative number is positive.

Inequalities

We need a notation to indicate how any two real numbers compare with each other. For any two numbers on the number line, the one to the left is said to be less than, or smaller than, the one to the right. The symbol $<$ means "is less than" and the symbol $>$ means "is greater than." The symbol $\leq$ means "is less than or equal to" and the symbol $\geq$ means "is greater than or equal to." These symbols are used to form **inequalities.**

In the following figure, note that although $|-3| > |-1|$, we have $-3 < -1$ since -3 is to the left of -1.

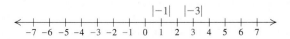

EXAMPLE 2

Write out the meaning of the inequality and determine whether it is a true statement: **(a)** $-7 < -2$; **(b)** $4 > -1$; **(c)** $-3 \geqslant -2$; **(d)** $5 \leqslant 6$; **(e)** $6 \leqslant 6$.

Solution

	Inequality	*Meaning*
a)	$-7 < -2$	-7 is less than -2, a true statement since -7 is to the left of -2.
b)	$4 > -1$	4 is greater than -1, a true statement.
c)	$-3 \geqslant -2$	-3 is greater than or equal to -2, a false statement since -3 is to the left of -2.
d)	$5 \leqslant 6$	5 is less than or equal to 6. Because $5 < 6$ is true, $5 \leqslant 6$ is true.
e)	$6 \leqslant 6$	6 is less than or equal to 6. Because $6 = 6$ is true, $6 \leqslant 6$ is true. ❏

Addition

We are now ready to review the method for adding any two real numbers.

Rules for Addition of Real Numbers

1. *Positive numbers:* Add the numbers. The result is positive.
2. *Negative numbers:* Add absolute values. Make the answer negative.
3. *A negative and a positive number:* Subtract the smaller absolute value from the larger one. Then:
   a) If the positive number has the larger absolute value, make the answer positive.
   b) If the negative number has the larger absolute value, make the answer negative.
   c) If the numbers have the same absolute value, the answer is 0.
4. *One number is zero:* The sum is the other number.

EXAMPLE 3

Add: **(a)** $-9 + (-5)$; **(b)** $-3.2 + 9.7$; **(c)** $-\frac{3}{4} + \frac{1}{3}$.

Solution

a) $-9 + (-5)$ We add the absolute values, getting 14. The answer is *negative*, -14.

b) $-3.2 + 9.7$ The absolute values are 3.2 and 9.7. Subtract 3.2 from 9.7 to get 6.5. The larger absolute value came from the positive number, so the answer is *positive*, 6.5.

c) $-\frac{3}{4} + \frac{1}{3} = -\frac{9}{12} + \frac{4}{12}$ The absolute values are $\frac{9}{12}$ and $\frac{4}{12}$. Subtract to get $\frac{5}{12}$. The larger absolute value came from the negative number, so the answer is *negative*, $-\frac{5}{12}$. ❏

Subtraction and Opposites

When numbers like 7 and -7 are added, the result is 0. Such numbers are called **opposites,** or **additive inverses,** of one another.

The Law of Opposites

For any two numbers a and $-a$,

$$a + (-a) = 0.$$

(When opposites are added, their sum is 0.)

EXAMPLE 4

Find the opposite: **(a)** -17.5; **(b)** $\frac{4}{5}$; **(c)** 0.

Solution

a) The opposite of -17.5 is 17.5 because $-17.5 + 17.5 = 0$.
b) The opposite of $\frac{4}{5}$ is $-\frac{4}{5}$ because $\frac{4}{5} + \left(-\frac{4}{5}\right) = 0$.
c) The opposite of 0 is 0 because $0 + 0 = 0$. ❏

To name the opposite, we use the symbol "$-$" and read the symbolism $-a$ as "the opposite of a."

Note that $-a$ does not necessarily denote a negative number. In fact, when a is *negative*, $-a$ is *positive*.

EXAMPLE 5

Find $-x$ for the following: **(a)** $x = -2$; **(b)** $x = \frac{3}{4}$.

Solution

a) If $x = -2$, then $-x = -(-2) = 2$.   The opposite of -2 is 2.
b) If $x = \frac{3}{4}$, then $-x = -\frac{3}{4}$.   The opposite of $\frac{3}{4}$ is $-\frac{3}{4}$. ❏

We can now give a more formal definition of absolute value.

Absolute Value

$$|x| = \begin{cases} x & \text{if } x \geq 0, \\ -x & \text{if } x < 0 \end{cases}$$

(The absolute value of x is x if x is nonnegative. The absolute value of x is its opposite if x is negative.)

A negative number is said to have a negative "sign" and a positive number a positive "sign." To subtract, we can add an opposite. Thus we sometimes say that we "change the sign of the number being subtracted and then add."

EXAMPLE 6

Subtract: **(a)** $5 - 9$; **(b)** $-1.2 - (-3.7)$; **(c)** $-\frac{4}{5} - \frac{2}{3}$.

Solution

a) $5 - 9 = 5 + (-9)$ Change the sign and add.
$\quad\quad\quad = -4$

b) $-1.2 - (-3.7) = -1.2 + 3.7$
$\quad\quad\quad\quad\quad\quad = 2.5$

c) $-\frac{4}{5} - \frac{2}{3} = -\frac{4}{5} + \left(-\frac{2}{3}\right)$
$\quad\quad\quad\quad = -\frac{12}{15} + \left(-\frac{10}{15}\right)$ Finding a common denominator
$\quad\quad\quad\quad = -\frac{22}{15}$ ❏

Multiplication

Multiplication of real numbers can be regarded as repeated addition or as repeated subtraction that begins at 0. For example,

$$3 \cdot (-2) = 0 + (-2) + (-2) + (-2) = -6$$

and

$$(-2)(-5) = 0 - (-5) - (-5) = 5 + 5 = 10.$$

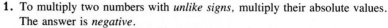

1. To multiply two numbers with *unlike signs*, multiply their absolute values. The answer is *negative*.
2. To multiply two numbers having the *same sign*, multiply their absolute values. The answer is *positive*.

Thus we have $(-3)9 = -27$ and $\left(-\frac{2}{3}\right)\left(-\frac{3}{8}\right) = \frac{1}{4}$.

Division

To divide, we use the definition of division. The quotient $a \div b$ (also denoted a/b) is that number c, if it exists, such that $c \cdot b = a$. For example, $10 \div (-2) = -5$ since $(-5)(-2) = 10$; $(-12) \div 3 = -4$ since $(-4)3 = -12$; and $-18 \div (-6) = 3$ since $3(-6) = -18$. We obtain the following rules for division, which are just like the rules for multiplication.

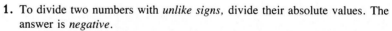

1. To divide two numbers with *unlike signs*, divide their absolute values. The answer is *negative*.
2. To divide two numbers having the *same sign*, divide their absolute values. The answer is *positive*.

Thus we have

$$\frac{-45}{-15} = 3 \quad \text{and} \quad \frac{20}{-4} = -5.$$

Note that since

$$\frac{-8}{2} = \frac{8}{-2} = -\frac{8}{2} = -4,$$

we have the following generalization.

For any number a and any nonzero number b,

$$\frac{-a}{b} = \frac{a}{-b} = -\frac{a}{b}.$$

Recall that

$$\frac{a}{b} = \frac{a}{1} \cdot \frac{1}{b} = a \cdot \frac{1}{b}.$$

That is, if we prefer, we can multiply by $1/b$ rather than divide by b. Provided that b is not 0, the number $1/b$ is called the **reciprocal,** or **multiplicative inverse,** of b.

The Law of Reciprocals

For any two numbers a and $1/a$ $(a \neq 0)$,

$$a \cdot \frac{1}{a} = 1.$$

(When reciprocals are multiplied, their product is 1.)

EXAMPLE 7

Find the reciprocal: **(a)** $\frac{7}{8}$; **(b)** $-\frac{3}{4}$; **(c)** -8.

Solution

a) The reciprocal of $\frac{7}{8}$ is $\frac{8}{7}$ because $\frac{7}{8} \cdot \frac{8}{7} = 1$.

b) The reciprocal of $-\frac{3}{4}$ is $-\frac{4}{3}$.

c) The reciprocal of -8 is $\frac{1}{-8}$ or $-\frac{1}{8}$.   ❑

To divide, we can multiply by a reciprocal. We sometimes say that we "invert and multiply."

EXAMPLE 8

Divide: **(a)** $-\frac{1}{4} \div \frac{3}{5}$; **(b)** $\frac{2}{3} \div \left(-\frac{4}{9}\right)$.

Solution

a) $-\frac{1}{4} \div \frac{3}{5} = -\frac{1}{4} \cdot \frac{5}{3}$ "Inverting" the $\frac{3}{5}$ and changing the division to a multiplication

$\qquad\qquad = -\frac{5}{12}$

b) $\frac{2}{3} \div \left(-\frac{4}{9}\right) = \frac{2}{3} \cdot \left(-\frac{9}{4}\right) = -\frac{18}{12}$,   or $-\frac{3}{2}$ ❑

Thus far, we have avoided dividing by 0 or, what is equivalent, having a denominator of 0. There is a reason for this. Suppose we were to divide 5 by 0. Then our answer would have to be some number such that when we multiplied it by 0, we got 5. But any number times 0 is 0. Thus we cannot divide 5 or any other nonzero number by 0.

What if we divide 0 by 0? In this case, our solution would need to be some number such that when we multiplied it by 0, we got 0. But then *any* number would work as a solution to $0 \div 0$. This could lead to contradictions so we agree to exclude division of 0 by 0 also.

Division by Zero

We never divide by 0. If asked to divide a nonzero number by 0, we say that the answer is *undefined*. If asked to divide 0 by 0, we say that the answer is *indeterminate*.

Multiplication and Division by 1

Whereas division by 0 is problematic, division or multiplication by 1 is easy to perform. The result is always the original number. We can use this fact with fractional notation to help find **equivalent expressions.**

Equivalent Expressions ▬▬▬▬▬ Two expressions that have the same value for all possible replacements are called *equivalent expressions*.

EXAMPLE 9

Use multiplication by 1 to find an expression equivalent to $x/5$ with a denominator of 15.

Solution   We multiply by 1, using $\frac{3}{3}$ as a name for 1:

$$\frac{x}{5} \cdot \frac{3}{3} = \frac{x \cdot 3}{15} = \frac{3x}{15}.$$

For any replacement of x, say $x = 2$, $x/5$ and $3x/15$ have the same value.   ❑

When we "simplify," we form equivalent expressions.

EXAMPLE 10

Use multiplication by 1 to simplify $-\dfrac{20x}{12x}.$

Solution

$$-\frac{20x}{12x} = -\frac{5 \cdot 4x}{3 \cdot 4x} \qquad \text{We factored the numerator and the denominator, after identifying (mentally) a common factor of } 4x.$$

$$= -\frac{5}{3} \cdot \frac{4x}{4x} \qquad \text{Next we factored the fractional expression, with one factor equal to 1.}$$

$$= -\frac{5}{3} \qquad \text{Finally, we leave off the factor of 1.}$$

Note that for all replacements except 0, $-\dfrac{20x}{12x} = -\dfrac{5}{3}.$ ❑

The Commutative and Associative Laws

In Example 9, you may have noticed that we made use of the fact that $x \cdot 3$ is the same as $3x$. This concept, that the result for multiplication is not dependent on the order in which the numbers are multiplied, is known as the *commutative law for multiplication*. A similar result holds for addition.

The Commutative Laws ▬▬▬▬▬ *Addition.*   For any real numbers a and b,

$$a + b = b + a.$$

Multiplication.   For any real numbers a and b,

$$a \cdot b = b \cdot a.$$

The commutative laws are often used to formulate equivalent expressions. Another pair of tools that can produce equivalent expressions, the *associative laws*, help us change groupings.

The Associative Laws

Addition.   For any real numbers a, b, and c,

$$a + (b + c) = (a + b) + c.$$

Multiplication.   For any real numbers a, b, and c,

$$a \cdot (b \cdot c) = (a \cdot b) \cdot c.$$

EXAMPLE 11

Write an expression equivalent to $3x + 4y$, using the commutative law of addition.

Solution

$$3x + 4y = 4y + 3x \qquad \text{Using the commutative law to change order}$$

The expressions $3x + 4y$ and $4y + 3x$ are equivalent. They name the same number for all replacements of x and y.   ❏

EXAMPLE 12

Write an expression equivalent to $(3x + 7y) + 9z$, using the associative law of addition.

Solution

$$(3x + 7y) + 9z = 3x + (7y + 9z)$$

The expressions $(3x + 7y) + 9z$ and $3x + (7y + 9z)$ are equivalent. They name the same number for all replacements of x, y, and z.   ❏

EXAMPLE 13

Write an expression equivalent to

$$\frac{5}{x} \cdot (2yz),$$

using the commutative and associative laws of multiplication.

Solution

$$\frac{5}{x} \cdot (2yz) = \left(\frac{5}{x} \cdot 2y \right) \cdot z \qquad \text{Using the associative law}$$

$$= \left(2y \cdot \frac{5}{x} \right) \cdot z \qquad \text{Using the commutative law in the parentheses} \qquad ❏$$

Although you probably don't think about it, you use the associative law of multiplication when you multiply a term such as $8x$ by a number such as 5, to get $40x$:

$$5(8x) = (5 \cdot 8)x = 40x.$$

The Distributive Law

Let's look at another way of forming equivalent expressions. We say that multiplication is distributive over addition, meaning that when we multiply a number by a sum, we can either add first or multiply first. Multiplication is also distributive over subtraction. If we regard subtraction as addition of an opposite, the following law holds for both addition and subtraction.

The Distributive Law

For any numbers a, b, and c,
$$a(b + c) = ab + ac.$$

EXAMPLE 14

Obtain an expression equivalent to $5x(y + 4)$ by multiplying.

Solution   We use the distributive law to get

$$5x(y + 4) = 5xy + 5x \cdot 4 \qquad \text{Using the distributive law}$$
$$= 5xy + 5 \cdot 4 \cdot x \qquad \text{Using the commutative law of multiplication}$$
$$= 5xy + 20x. \qquad \text{Simplifying}$$

The expressions $5x(y + 4)$ and $5xy + 20x$ are equivalent. They name the same number for all replacements of x and y. ❑

When we reverse the steps in Example 14, we say that we are **factoring** an expression.

EXAMPLE 15

Obtain an expression equivalent to $3xy - 6x$ by factoring.

Solution   We use the distributive law to get

$$3xy - 6x = 3x(y - 2).$$ ❑

In Example 15, since the product of $3x$ and $y - 2$ is $3xy - 6x$, we say that $3x$ and $y - 2$ are **factors** of $3xy - 6x$. Thus the word "factor" can act as a noun as well as a verb.

EXERCISE SET | 1.2

Find the absolute value.

1. $|-7|$ **2.** $|-9|$ **3.** $|9|$

4. $|12|$ **5.** $|-6.2|$ **6.** $|-7.9|$

7. $|0|$ **8.** $|3\frac{3}{4}|$ **9.** $|1\frac{7}{8}|$

10. $|0.91|$ **11.** $|-4.21|$ **12.** $|-5.309|$

Write the meaning of the inequality, and determine whether it is a true statement.

13. $-9 \leqslant -1$ **14.** $-1 \leqslant -5$

15. $-7 > 1$ **16.** $7 \geqslant -2$

17. $3 \geqslant -5$ **18.** $9 \leqslant 9$

19. $-9 < -4$ **20.** $7 \geqslant -8$

21. $-4 \geq -4$

22. $2 < 2$

23. $-5 < -5$

24. $-2 > -12$

Add.

25. $5 + 12$

26. $9 + 7$

27. $-4 + (-7)$

28. $-8 + (-3)$

29. $-5.9 + 2.7$

30. $-1.9 + 7.3$

31. $\frac{2}{7} + \left(-\frac{3}{5}\right)$

32. $\frac{3}{8} + \left(-\frac{2}{5}\right)$

33. $-4.9 + (-3.6)$

34. $-2.1 + (-7.5)$

35. $-\frac{1}{9} + \frac{2}{3}$

36. $-\frac{1}{2} + \frac{4}{5}$

37. $0 + (-4.5)$

38. $-3.19 + 0$

39. $-7.24 + 7.24$

40. $-9.46 + 9.46$

41. $15.9 + (-22.3)$

42. $21.7 + (-28.3)$

Find the opposite, or additive inverse.

43. 7.29 **44.** 5.43 **45.** -4.8

46. -8.1 **47.** 0 **48.** $-2\frac{3}{4}$

49. $-6\frac{1}{3}$ **50.** $4\frac{1}{5}$

Find $-x$ for each of the following.

51. $x = 7$

52. $x = 3$

53. $x = -2.7$

54. $x = -1.9$

55. $x = 1.79$

56. $x = 3.14$

57. $x = 0$

58. $x = -1$

59. $x = -0.03$

60. $x = -1.09$

Subtract.

61. $9 - 7$

62. $8 - 3$

63. $4 - 9$

64. $3 - 10$

65. $-6 - (-10)$

66. $-3 - (-9)$

67. $-4 - 13$

68. $-7 - 8$

69. $2.7 - 5.8$

70. $3.7 - 4.2$

71. $-\frac{3}{5} - \frac{1}{2}$

72. $-\frac{2}{3} - \frac{1}{5}$

73. $-3.9 - (-6.8)$

74. $-5.4 - (-4.3)$

75. $0 - (-7.9)$

76. $0 - 5.3$

Multiply.

77. $(-4)7$

78. $(-5)9$

79. $(-3)(-8)$

80. $(-7)(-8)$

81. $(4.2)(-5)$

82. $(3.5)(-8)$

83. $(-7.2)(1)$

84. $1(5.9)$

85. $(-17.45) \cdot 0$

86. 15.2×0

87. $(-3.2) \times (-1.7)$

88. $(1.9) \cdot (4.3)$

Divide.

89. $\dfrac{-10}{-2}$

90. $\dfrac{-15}{-3}$

91. $\dfrac{-100}{20}$

92. $\dfrac{-50}{5}$

93. $\dfrac{73}{-1}$

94. $\dfrac{-62}{1}$

95. $\dfrac{0}{-7}$

96. $\dfrac{0}{-11}$

97. $\dfrac{-42}{-6}$

98. $\dfrac{-48}{-6}$

Find the reciprocal, or multiplicative inverse.

99. 5 **100.** 3 **101.** -9

102. -7 **103.** $\frac{2}{3}$ **104.** $\frac{4}{7}$

105. $-\frac{3}{11}$ **106.** $-\frac{7}{3}$

Divide.

107. $\frac{2}{3} \div \frac{4}{5}$

108. $\frac{2}{7} \div \frac{6}{5}$

109. $-\frac{3}{5} \div \frac{1}{2}$

110. $\left(-\frac{4}{7}\right) \div \frac{1}{3}$

111. $\left(-\frac{2}{9}\right) \div \left(-\frac{3}{4}\right)$

112. $\left(-\frac{2}{11}\right) \div \frac{4}{7}$

113. $\left(-\frac{3}{8}\right) \div 1$

114. $\left(-\frac{2}{7}\right) \div (-1)$

115. $\frac{7}{3} \div (-1)$

116. $-\frac{5}{4} \div 1$

Use multiplication by 1 to find an expression satisfying the given conditions.

117. Equivalent to $3x/7$ with a denominator of 35

118. Equivalent to $2a/5$ with a denominator of 40

119. Equivalent to $-3a/8$ with a denominator of 32

120. Equivalent to $-4x/3$ with a denominator of 33

Simplify.

121. $\dfrac{50x}{5}$

122. $\dfrac{75x}{3}$

123. $\dfrac{56a}{7}$

124. $\dfrac{72a}{9}$

Write an equivalent expression using a commutative law.

125. $5y + 2x$ **126.** $4y + x$

127. $2x - 3y$ **128.** $5y - 4x$

129. $\dfrac{x}{2} \cdot \dfrac{y}{3}$ **130.** $3 \cdot \dfrac{y}{4}$

Write an equivalent expression using an associative law.

131. $2 \cdot (8x)$ **132.** $x \cdot (3y)$

133. $x + (2y + 5)$ **134.** $(3y + 4) + 10$

Obtain an equivalent expression by multiplying.

135. $3(a + 1)$ **136.** $8(x + 1)$

137. $4(x - y)$ **138.** $9(a - b)$

139. $-5(2a + 3b)$ **140.** $-2(3c + 5d)$

141. $2a(b - c + d)$ **142.** $5x(y - z + w)$

Obtain an equivalent expression by factoring.

143. $8x + 8y$ **144.** $7a + 7b$

145. $9p - 9$ **146.** $12x - 12$

147. $7x - 21$ **148.** $6y - 36$

149. $2x - 2y + 2z$ **150.** $3x + 3y - 3z$

Skill Maintenance _____

To the student and the instructor: Exercises included for Skill Maintenance review skills previously studied in the text. You can expect such exercises in almost every exercise set and, beginning in Chapter 2, all Chapter Reviews and Tests.

151. Translate to an algebraic expression:

Five less than seventy percent of a number.

152. Translate to an algebraic expression:

Two more than half a number.

153. Evaluate $7xy - z$ given $x = 2$, $y = 1$, and $z = 20$.

154. Evaluate $5ab + c$ given $a = 2$, $b = 3$, and $c = -7$.

Synthesis _____

155. ◈   What is the difference between the associative law of multiplication and the distributive law?

156. ◈   Explain in your own words why 7/0 is undefined.

Use multiplication by 1 to simplify the following. Assume nonzero denominators.

157. $\dfrac{2x + 10}{3x + 15}$ **158.** $\dfrac{4x + 6}{6x + 9}$

159. $\dfrac{6x - 12}{3x + 6}$ **160.** $\dfrac{5x - 10}{10x + 15}$

1.3

Solving Equations

Equivalent Equations • The Addition and Multiplication Principles • Collecting Like Terms • Grouping Symbols • Special Cases • Solution Sets

Solving equations is essential for problem solving in algebra. In this section, we will review and practice the solving of simple equations.

Equivalent Equations

We saw in Section 1.1 that the equation $212,000,000 + x = 249,000,000$ has $37,000,000$ as a solution. Although this problem may seem simple to you, it is important for us to understand how such an equation is solved using the principles of algebra.

Equivalent Equations

Two equations are said to be *equivalent* if they have the same solution(s).

EXAMPLE 1

Determine whether $2x = 6$ and $10x = 30$ are equivalent equations.

Solution   When x is replaced by 3, both equations are true, and for any other replacement, both equations are false. Thus the equations are equivalent.   ❑

EXAMPLE 2

Determine whether $4 + x = 9$ and $x = 5$ are equivalent equations.

Solution When x is replaced by 5, both equations are true. For other replacements, both equations are false. Thus the equations are equivalent. ❏

EXAMPLE 3

Determine whether $3x = 4x$ and $3/x = 4/x$ are equivalent equations.

Solution When x is replaced by 0, neither $3/x$ nor $4/x$ is defined, so 0 is *not* a solution of $3/x = 4/x$. Since 0 *is* a solution of $3x = 4x$, we conclude that the equations are not equivalent. ❏

The Addition and Multiplication Principles

Suppose that a and b represent the *same number* and we add a number c to a. We will get the same answer if we add c to b, because a and b are the same number. The same is true if we multiply both a and b by c.

The Addition and Multiplication Principles for Equations

For any real numbers a, b, and c,

a) if $a = b$, then $a + c = b + c$; and **b)** if $a = b$, then $a \cdot c = b \cdot c$.

EXAMPLE 4

Solve: $y - 4.7 = 13.9$.

Solution

$$y - 4.7 = 13.9$$
$$y - 4.7 + 4.7 = 13.9 + 4.7 \qquad \text{Using the addition principle; adding 4.7}$$
$$y + 0 = 13.9 + 4.7 \qquad \text{The law of opposites}$$
$$y = 18.6$$

Check: $\quad\dfrac{y - 4.7 = 13.9}{18.6 - 4.7 \;?\; 13.9}$ $\qquad$ Substituting 18.6 for y and calculating
$\qquad\qquad\qquad$ $13.9 \mid 13.9$ TRUE

The solution is 18.6. ❏

In Example 4, why did we add 4.7 on both sides? Because we wanted the variable y alone on one side of the equation. When we added 4.7, we got $y + 0$, which left y alone on the left. The other side showed us what calculations to make to find the equation $y = 18.6$ from which the solution, 18.6, could be easily read.

EXAMPLE 5

Solve: $4x = 9$.

Solution

$$4x = 9$$
$$\tfrac{1}{4} \cdot 4x = \tfrac{1}{4} \cdot 9 \qquad \text{Using the multiplication principle,}$$
$$\qquad\qquad\qquad \text{we multiply by } \tfrac{1}{4}, \text{ the reciprocal of 4.}$$
$$1x = \tfrac{9}{4} \qquad \text{The law of reciprocals; simplifying}$$
$$x = \tfrac{9}{4}$$

The check is left to the student. The solution is $\tfrac{9}{4}$. ❏

In Example 5, why did we choose to multiply by $\frac{1}{4}$? Because we wanted x alone on one side of the equation. We multiplied by the reciprocal of 4. Then we got $1x$, which simplified to x. This eliminated the 4 on the left.

Collecting Like Terms

In the expression $5xy + 20x$, the parts separated by the plus sign are called **terms.** Thus, $5xy$ and $20x$ are terms in $5xy + 20x$. When terms have variable factors that are exactly the same, we refer to the terms as **like** or **similar** terms. Thus, $7xy$ and $3xy$ are like terms, but $2ab$ and $6ac$ are not. We often simplify expressions by using the distributive law to **collect** or **combine like terms.**

EXAMPLE 6

Collect like terms: $3x + 4x$.

Solution

$$3x + 4x = (3 + 4)x \qquad \text{Using the distributive law}$$
$$\text{(in reverse), or factoring}$$
$$= 7x \qquad\qquad\qquad\qquad\qquad\qquad \square$$

Grouping Symbols

In mathematics, parentheses ( ), brackets [ ], or braces { } are used to indicate groupings. We have seen that the distributive law enables us to "remove" parentheses. When grouping symbols contain other grouping symbols, we work from the inside out when simplifying.

EXAMPLE 7

Simplify the expression $3x + 2[4 + 5(x + 2y)]$.

Solution

$$3x + 2[4 + 5(x + 2y)] = 3x + 2[4 + 5x + 10y] \qquad \text{Using the distributive law}$$
$$= 3x + 8 + 10x + 20y \qquad \text{Using the distributive law}$$
$$= 13x + 8 + 20y \qquad\quad \text{Collecting like terms} \quad \square$$

When we multiply a number by -1, we get its opposite, or additive inverse. For example,

$$-1 \cdot 8 = -8 \qquad \text{(the opposite of 8)}.$$

Reading from right to left, we have $-8 = -1 \cdot 8$, and in general, $-x = -1 \cdot x$. We can use this fact along with the distributive law to remove parentheses that are preceded by a negative sign or subtraction.

EXAMPLE 8

Obtain an expression equivalent to $-(a - b)$, using the property of -1.

Solution

$$-(a - b) = -1 \cdot (a - b) \qquad \text{Replacing } - \text{ by } -1$$
$$= -1 \cdot a - (-1) \cdot b \qquad \text{Using the distributive law}$$
$$= -a - (-b) \qquad\qquad \text{Replacing } -1 \cdot a \text{ by } -a \text{ and } (-1) \cdot b \text{ by } -b$$
$$= -a + b, \quad \text{or} \quad b - a$$

The expressions $-(a - b)$ and $b - a$ are equivalent. They name the same number for all replacements of a and b.   ❏

Example 8 illustrates something worth remembering, because it gives us a useful shortcut — that is,

The opposite, or additive inverse, of $a - b$ is $b - a$.

EXAMPLE 9

Simplify the expression $3[x + 2(x + y)] - (5x + y)$.

Solution

$$3[x + 2(x + y)] - (5x + y) = 3[x + 2x + 2y] - 5x - y \qquad \text{Using the distributive law twice}$$

$$= 3[3x + 2y] - 5x - y \qquad \text{Collecting like terms}$$

$$= 9x + 6y - 5x - y \qquad \text{Using the distributive law}$$

$$= 4x + 5y \qquad \text{Collecting like terms}$$

❏

In Examples 7–9, we were able to obtain simplified expressions that were equivalent to the original expressions. We now use this ability in conjunction with the addition and multiplication principles to solve equations containing parentheses.

EXAMPLE 10

Solve: $8x - 3(x + 4) = 4[x + 2(3 - x)]$.

Solution

$$8x - 3(x + 4) = 4[x + 2(3 - x)]$$

$$8x - 3x - 12 = 4[x + 6 - 2x] \qquad \text{Using the distributive law}$$

$$5x - 12 = 4[-x + 6] \qquad \text{Collecting like terms}$$

$$5x - 12 = -4x + 24 \qquad \text{Using the distributive law}$$

$$4x + 5x - 12 = 4x - 4x + 24 \qquad \text{Using the addition principle; adding } 4x$$

$$9x - 12 = 24 \qquad \text{Collecting like terms}$$

$$9x - 12 + 12 = 24 + 12 \qquad \text{Using the addition principle; adding 12}$$

$$9x = 36 \qquad \text{Simplifying}$$

$$\tfrac{1}{9} \cdot 9x = \tfrac{1}{9} \cdot 36 \qquad \text{Using the multiplication principle, we multiply by } \tfrac{1}{9}, \text{ the reciprocal of 9.}$$

$$x = 4 \qquad \text{The law of reciprocals; simplifying}$$

Check:

$$\begin{array}{c|c} \multicolumn{2}{c}{8x - 3(x + 4) = 4[x + 2(3 - x)]} \\ \hline 8 \cdot 4 - 3(4 + 4) \ ? & 4[4 + 2(3 - 4)] \\ 32 - 3(8) & 4[4 + 2(-1)] \\ 32 - 24 & 4[4 - 2] \\ 8 & 4 \cdot 2 \\ 8 & 8 \qquad \text{TRUE} \end{array}$$

The solution is 4.   ❏

You may wonder whether all of the preceding steps must be performed in a specific order. Often there is more than one ''valid'' sequence of steps. For instance, in Example 10, we could have ''distributed'' the 4 earlier.

Don't rush to do steps in your head. As long as you are careful in your use of the principles we've studied, the number of steps in your solution should be of little concern. What *is* important is that each step produces a simpler, yet equivalent, equation. The goal is always to get the variable alone on one side.

Special Cases

Sometimes we encounter equations that have *infinitely many solutions,* like $x + 3 = 3 + x$; other times we encounter equations that have *no solutions,* like $3 + x = 4 + x$. The rules of algebra can help us to recognize either one of these situations.

EXAMPLE 11

Solve: $x + 3 = 3 + x$.

Solution

$$x + 3 = 3 + x$$
$$x + 3 + (-x) = 3 + x + (-x) \qquad \text{Using the addition principle}$$
$$3 = 3 \qquad \text{Simplifying}$$

Since the equation $3 = 3$ is true regardless of our choice of x, and because $x + 3 = 3 + x$ is equivalent to $3 = 3$, we see that $x + 3 = 3 + x$ is true for any choice of x. All real numbers are solutions. If it troubles you that the solution set to $3 = 3$ is all real numbers, think of $3 = 3$ as $3 + 0 \cdot x = 3 + 0 \cdot x$. All real numbers are solutions. ❏

EXAMPLE 12

Solve: $3x - 5 = 3(x - 2) + 4$.

Solution

$$3x - 5 = 3(x - 2) + 4$$
$$3x - 5 = 3x - 6 + 4 \qquad \text{Using the distributive law}$$
$$3x - 5 = 3x - 2$$
$$-3x + 3x - 5 = -3x + 3x - 2 \qquad \text{Using the addition principle}$$
$$-5 = -2$$

Since our original equation is equivalent to the equation $-5 = -2$, which is false for any choice of x, there is no solution to this problem. There is no choice of x that will solve the original equation. ❏

Solution Sets

We will sometimes refer to the set of solutions, or **solution set,** of a particular problem. Thus the solution set for Example 10 is $\{4\}$. The solution set for Example 12 is the set containing *no* elements, denoted by $\{ \ \}$ or $\varnothing$, and referred to as the **empty set.** The solution set in Example 11 can be written simply as $\mathbb{R}$, the set of all real numbers.

EXERCISE SET | 1.3

Determine whether the two equations in each pair are equivalent.

1. $3x = 12$ and $2x = 8$

2. $5x = 20$ and $15x = 60$

3. $2x - 1 = -7$ and $x = -3$

4. $x + 2 = -5$ and $x = -7$

5. $x + 5 = 11$ and $3x = 18$

6. $x - 3 = 7$ and $3x = 24$

7. $13 - x = 4$ and $2x = 20$

8. $3x - 4 = 8$ and $3x = 12$

9. $5x = 2x$ and $\dfrac{4}{x} = 3$

10. $6 = 2x$ and $5 = \dfrac{2}{3 - x}$

Solve. Don't forget to check.

11. $x - 5.2 = 9.4$ **12.** $y + 4.3 = 11.2$

13. $9y = 72$ **14.** $7x = 63$

15. $4x - 12 = 60$ **16.** $4x - 6 = 70$

17. $5y + 3 = 28$ **18.** $7t + 11 = 74$

19. $2y - 11 = 37$ **20.** $3x - 13 = 29$

21. $-4x - 7 = -35$ **22.** $-9y + 8 = -91$

Obtain an equivalent expression by collecting like terms. Use the distributive law.

23. $4a + 5a$ **24.** $9x + 3x$

25. $8b - 11b$ **26.** $9c - 12c$

27. $14y + y$ **28.** $13x + x$

29. $12a - a$ **30.** $15x - x$

31. $t - 9t$ **32.** $x - 6x$

33. $5x - 3x + 8x$ **34.** $3x - 11x + 2x$

35. $5x - 8y + 3x$ **36.** $9a - 10b + 4a$

37. $7c + 8d - 5c + 2d$ **38.** $12a + 3b - 5a + 6b$

39. $4x - 7 + 18x + 25$ **40.** $13p + 5 - 4p + 7$

41. $13x + 14y - 11x - 47y$

42. $17a + 17b - 12a - 38b$

Solve. Don't forget to check.

43. $5x + 2x = 56$ **44.** $3x + 7x = 120$

45. $9y - 7y = 42$ **46.** $8t - 3t = 65$

47. $-6y - 10y = -32$ **48.** $-9y - 5y = 28$

49. $7y - 1 = 23 - 5y$

50. $15x + 20 = 8x - 22$

51. $5 - 4a = a - 13$

52. $8 - 5x = x - 16$

53. $3m - 7 = -7 - 4m - m$

54. $5x - 8 = -8 + 3x - x$

55. $5r - 2 + 3r = 2r + 6 - 4r$

56. $5m - 17 - 2m = 6m - 1 - m$

57. $\frac{1}{4} + \frac{3}{8}y = \frac{3}{4}$

58. $\frac{1}{5} + \frac{3}{10}x = \frac{4}{5}$

59. $-\frac{5}{2}x + \frac{1}{2} = -18$

60. $0.9y - 0.7 = 4.2$

61. $0.8t - 0.3t = 6.5$

62. $1.4x + 5.02 = 0.4x$

Obtain an equivalent but simpler expression, by removing parentheses and simplifying.

63. $a - (2a + 5)$ **64.** $x - (5x + 9)$

65. $4m - (3m - 1)$ **66.** $5a - (4a - 3)$

67. $3d - 7 - (5 - 2d)$ **68.** $8x - 9 - (7 - 5x)$

69. $-2(x + 3) - 5(x - 4)$

70. $-9(y + 7) - 6(y - 3)$

71. $5x - 7(2x - 3)$

72. $8y - 4(5y - 6)$

73. $9a - [7 - 5(7a - 3)]$

74. $12b - [9 - 7(5b - 6)]$

75. $5\{-2 + 3[4 - 2(3 + 5)]\}$

76. $7\{-7 + 8[5 - 3(4 + 6)]\}$

77. $2y + \{7[3(2y - 5) - (8y + 7)] + 9\}$

78. $7b - \{6[4(3b - 7) - (9b + 10)] + 11\}$

Solve. Don't forget to check.

79. $2(x + 6) = 8x$ **80.** $3(y + 5) = 8y$

81. $80 = 10(3t + 2)$ **82.** $27 = 9(5y - 2)$

83. $180(n - 2) = 900$ **84.** $210(x - 3) = 840$

85. $5y - (2y - 10) = 25$

86. $8x - (3x - 5) = 40$

87. $0.7(3x + 6) = 1.1 - (x + 2)$

88. $0.9(2x + 8) = 20 - (x + 5)$

89. $\frac{1}{8}(16y + 8) - 17 = -\frac{1}{4}(8y - 16)$

90. $\frac{1}{6}(12t + 48) - 20 = -\frac{1}{8}(24t - 144)$

91. $a + (a - 3) = (a + 2) - (a + 1)$

92. $0.8 - 4(b - 1) = 0.2 + 3(4 - b)$

93. $5[2 + 3(x - 1)] = 4$

94. $3[t - 4(t + 7)] = -3$

95. $5 + 2(x - 3) = 2[5 - 4(x + 2)]$

96. $3[2 - 4(x - 1)] = 3 - 4(x + 2)$

97. $2\{9 - 3[2x - 4(x + 1)]\} = 5(2x + 8)$

98. $3\{7 - 2[3x + 4(x - 1)]\} = 7(6 - 3x)$

Solve and find the solution set.

99. $4x - 2x - 2 = 2x$

100. $2x + 4 + x = 4 + 3x$

101. $2 + 9x = 3(3x + 1) - 1$

102. $4 + 7x = 7(x + 1)$

103. $-8x + 5 = 14 - 8x$

104. $-8x + 5 = 5 - 8x$

105. $2\{9 - 3[-2x - 4]\} = 12x + 42$

106. $3\{7 - 2[7x - 4]\} = -42x + 45$

Skill Maintenance

107. Write the set consisting of the positive integers less than 10, using both roster notation and set-builder notation.

108. Write the set consisting of the negative integers

greater than -9, using both roster notation and set-builder notation.

Perform the indicated operation.

109. $-7 - (-5.3)$

110. $-9 - (-3.7)$

111. $(-9)(-6)$

112. $(-4)(-9)$

113. $(-12) \div (-3)$

114. $(-15) \div (-5)$

Synthesis

115. ◈ Explain the difference between equivalent expressions and equivalent equations.

116. ◈ Explain how the distributive and commutative laws can be used to rewrite $3x + 6y + 4x + 2y$ as $7x + 8y$.

Solve and check. The symbol ▦ designates exercises designed to be solved with a calculator.

117. ▦ $0.0008x = 0.00000564$

118. ▦ $43.008z = 1.201135$

119. ▦ $4.23x - 17.898 = -1.65x - 42.454$

120. ▦ $-0.00458y + 1.7787 = 13.002y - 1.005$

121. $x - \{3x - [2x - (5x - (7x - 1))]\} = x + 7$

122. $3x - \{5x - [7x - (4x - (3x + 1))]\} = 3x + 5$

123. $17 - 3\{5 + 2[x - 2]\} + 4\{x - 3(x + 7)\}$
$= 9\{x + 3[2 + 3(4 - x)]\}$

124. $23 - 2\{4 + 3[x - 1]\} + 5\{x - 2(x + 3)\}$
$= 7\{x - 2[5 - (2x + 3)]\}$

1.4

Problem Solving

The Five-Step Strategy • **The First Step: Familiarization** •
The Other Steps for Problem Solving • **Problem Solving**

We now begin to study and practice the "art" of problem solving. Although we are interested mainly in the use of algebra to solve problems, much of what we say here applies to solving all kinds of problems.

What do we mean by a *problem*? Perhaps you've already used algebra to solve some "real-world" problems. What procedure did you use? Was there anything in your approach that could be used to solve problems of a more general nature? These are some questions that we will answer in this section.

In this text, we do not restrict the use of the word "problem" to computational situations involving arithmetic or algebra, such as $589 + 437 = a$ or $3x + 5x = 9$.

We mean instead some question to which we wish to find an answer. Perhaps this can best be illustrated with some sample problems:

1. How can I schedule my classes so that all of them are in the morning?
2. What is the fastest way to get from New York City to Washington, D.C.?
3. Can I eat 3000 calories a day and still lose weight?
4. An airplane traveling at a speed of 210 mph in still air encounters a head wind of 46 mph. How long will it take for the plane to travel 369 mi into the head wind?

Although these problems are all different, there are some similarities. We cannot give rules for problem solving, but there is a general *strategy* that can be used. Some problems can be solved using algebra and some cannot, but the overall strategy can be used in any case.

The Five-Step Strategy

Since you have already studied some algebra, you have had some experience with problem solving. The following steps make up a strategy that you may already have used. They constitute a good strategy for problem solving in general.

Five Steps for Problem Solving with Algebra

1. *Familiarize* yourself with the problem situation.
2. *Translate* to mathematical language.
3. *Carry out* some mathematical manipulation.
4. *Check* your possible answer in the original problem.
5. *State* the answer clearly.

The First Step: Familiarization

Of the five steps, probably the most important is the first: becoming familiar with the problem situation. Here are some hints for familiarization.

The First Step in Problem Solving with Algebra

Familiarize yourself with a problem situation.

1. If a problem is given in words, read it carefully.
2. Reread the problem, perhaps aloud. Try to verbalize the problem to yourself.
3. List the information given and the question to be answered. Choose a variable or variables to represent any unknown(s) and clearly state what each variable represents. Be descriptive! For example, let l = length in meters, d = distance in miles, and so on.
4. Find further information. Look up a formula at the back of the book or in a reference book. Talk to a reference librarian or an expert in the field.
5. Make a table of the information given and the information you have collected. Look for patterns that may help in the translation to an equation.
6. Make a drawing and label it with known information. Also indicate unknown information, using specific units if given.
7. Guess or estimate the answer.

EXAMPLE 1

How might you familiarize yourself with the situation of Problem 1: "How can I schedule my classes so that all of them are in the morning?"

Solution   Clearly you will need to find further information in order to solve this problem. You might:

a) List all courses that you are *required* to take.
b) Get a schedule of course offerings and study it.
c) Talk to counselors.

When the information is known, it might be wise to make a table or chart to assist with your schedule selection.   ❑

EXAMPLE 2

How might you familiarize yourself with the situation given in Problem 4: "How long will it take for the plane to travel 369 mi into the head wind?"

Solution   First read the question *very* carefully. This may even involve speaking aloud. You may need to reread the problem one or more times to fully understand what information you are given and what information is required. A sketch is often helpful.

In this case, a table can be constructed to clearly list the relevant information.

210 mph

Plane

46-mph wind

Speed of Plane in Still Air	210 mph
Speed of Head Wind	46 mph
Speed of Plane in Head Wind	?
Distance to Be Traveled	369 mi
Time Required	?

As a next step in the familiarization process, we should determine, possibly with the aid of outside references, what relationships exist between the various quantities in the problem. With some effort it can be learned that a head wind's speed should be subtracted from the plane's speed in still air if we are to determine the speed of the plane in the head wind. We might consult a physics book to note that

Distance = Speed × Time.

We rewrite part of the table, letting t represent the time, in hours, required for the plane to fly 369 mi into the head wind.

Speed of Plane in Head Wind	$210 - 46 = 164$ mph
Distance to Be Traveled	369 mi
Time Required	t

At this point, we could attempt to guess an answer. Suppose that the plane flew into the head wind for 2 hr. Then the plane will have traveled

Speed × Time = Distance.

$$164 \times 2 = 328 \text{ mi}$$

Since $328 \neq 369$, we conclude that our guess is wrong. However, an examination of how we checked our guess gives added insight into the problem. We notice that a better guess, when multiplied by 164, would yield a number closer to 369.   □

The Other Steps for Problem Solving

The second step in problem solving is to translate the situation to mathematical language. In algebra, this usually consists of forming an equation.

The Second Step in Problem Solving with Algebra

> Translate the situation of the problem to mathematical language. In some cases, translation can be done by writing an algebraic expression, but most problems in this text can be solved by translating to an equation.

In the third step of our process, we work with the results of the first two steps. Often this will require us to use the algebra that we have studied.

The Third Step in Problem Solving with Algebra

> Carry out some mathematical manipulation. If you have translated to an equation, this means to solve the equation.

To properly complete the problem-solving process, we should always **check** our solution and then **state** the solution in a clear and precise manner. Normally our check consists of returning to the original problem and determining whether all its conditions have been satisfied. If our answer checks, we write a complete English sentence to state what the solution is. Our five steps are listed again. Try to apply them regularly in your work in mathematics.

Five Steps for Problem Solving with Algebra

> 1. *Familiarize* yourself with the problem situation.
> 2. *Translate* to mathematical language.
> 3. *Carry out* some mathematical manipulation.
> 4. *Check* your possible answer in the original problem.
> 5. *State* the answer clearly.

Problem Solving

At this point, our study of algebra is still in a beginning stage. Thus we have few algebraic tools with which to work problems. As the number of tools in our algebraic ''tool box'' increases, so will the level of difficulty of the problems to be solved. For now our problems may seem simple; however, to gain practice with the problem-solving process, you should attempt to proceed through all five steps. Later some steps can be skipped or shortened.

EXAMPLE 3

Elka paid $1187.20 for a computer. If the price paid included a 6% sales tax, what was the price of the computer itself?

Solution

1. **FAMILIARIZE.** Familiarize yourself with the problem. We note that the tax is calculated from, and then added to, the computer's price. We let

C = the computer's price.

Let's guess that the computer's price was $1000. To check the guess, we calculate the amount of tax, $(0.06)(\$1000) = \$60$, and add it to $1000:

$$(0.06)(\$1000) + \$1000 = \$60 + \$1000$$
$$= \$1060.$$

Our guess was wrong, but it was useful. The manner in which we manipulated our guess will guide us in the next step.

2. **TRANSLATE.** Translate the problem to mathematical language. Our guess leads us to the following translation:

6% of the computer's price	plus	the computer's price	is	the price with sales tax.
$(0.06)C$	$+$	C	$=$	$\$1187.20$

3. **CARRY OUT.** Carry out the algebraic manipulation:

$$0.06C + 1C = 1187.20$$
$$1.06C = 1187.20 \qquad \text{Collecting like terms}$$
$$\frac{1}{1.06} \cdot 1.06C = \frac{1}{1.06} \cdot 1187.20 \qquad \text{Using the multiplication principle}$$
$$C = 1120.$$

4. **CHECK.** Check the answer in the original problem. To do this, we calculate the amount of tax, $(0.06)(\$1120) = \$67.20$, and add it to $1120:

$$(0.06)(\$1120) + \$1120 = \$67.20 + \$1120$$
$$= \$1187.20.$$

We see that $1120 checks in the original problem.

5. **STATE.** State the answer clearly. The computer itself cost $1120.00.   ❏

EXAMPLE 4

A piece of wood molding 100 in. long is to be cut into two pieces, and those pieces are each to be cut into the shape of a square frame. The length of a side of one square is to be $1\frac{1}{2}$ times the length of a side of the other. How should the wood be cut?

Solution

1. **FAMILIARIZE.** We note that the *perimeter* (distance around) of each square is four times the length of a side. Furthermore, if s is used to represent the length of a side of the smaller square, then $\left(1\frac{1}{2}\right)s$ will represent the length of a side of the larger square. Finally, note that the two perimeters must add up to 100 in.

Perimeter of a square $= 4 \cdot$ *length of a side*

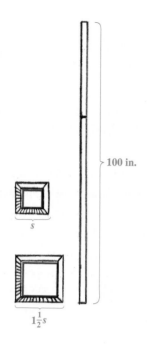

100 in.

s

$1\frac{1}{2}s$

2. TRANSLATE. Rewording the problem can help us translate:

Rewording:	The perimeter of one square	plus	the perimeter of the other	is	100 in.
Translating:	$4s$	$+$	$4\left(1\frac{1}{2}s\right)$	$=$	100

3. CARRY OUT. We solve the equation:

$$4s + 4\left(1\frac{1}{2}s\right) = 100$$

$$4s + 6s = 100 \qquad \text{Simplifying}$$

$$10s = 100 \qquad \text{Collecting like terms}$$

$$s = \frac{1}{10} \cdot 100 \qquad \text{Multiplying by } \frac{1}{10} \text{ on both sides}$$

$$s = 10. \qquad \text{Simplifying}$$

4. CHECK. If 10 is the length of the smaller side, then $\left(1\frac{1}{2}\right)(10) = 15$ is the length of the larger side. The two perimeters would then be

$$4 \cdot 10 = 40 \quad \text{and} \quad 4 \cdot 15 = 60.$$

Since $40 + 60 = 100$, our answer checks.

5. STATE. The wood should be cut into two pieces, one 40 in. long and the other 60 in. long. Each piece should then be quartered to form the frames. ❑

EXAMPLE 5

Three numbers are such that the second is 6 less than 3 times the first and the third is 2 more than $\frac{2}{3}$ the first. The sum of the three numbers is 150. Find the largest of the three numbers.

Solution   We proceed according to the five-step process.

1. FAMILIARIZE. Three numbers are involved, and we want to find the largest. We list the information in a table, letting x represent the first number.

First Number	x
Second Number	6 less than 3 times the first
Third Number	2 more than $\frac{2}{3}$ the first

$$\text{First} + \text{Second} + \text{Third} = 150$$

Try to make and check a guess at this point. We will proceed to the next step.

2. TRANSLATE. We can now name the second and third numbers by using x. (We often say that we name them "in terms of x.") We'll go back to the table and add another column:

First Number	x	x
Second Number	6 less than 3 times the first	$3x - 6$
Third Number	2 more than $\frac{2}{3}$ the first	$\frac{2}{3}x + 2$

We know that the sum is 150. Substituting, we obtain an equation:

First	$+$	Second	$+$	Third	$=$	150.
x	$+$	$(3x - 6)$	$+$	$\left(\frac{2}{3}x + 2\right)$	$=$	150

3. CARRY OUT. We solve the equation:

$$x + 3x - 6 + \tfrac{2}{3}x + 2 = 150 \qquad \text{Leaving off unnecessary parentheses}$$

$$\left(4 + \tfrac{2}{3}\right)x - 4 = 150 \qquad \text{Collecting like terms}$$

$$\tfrac{14}{3}x - 4 = 150$$

$$\tfrac{14}{3}x = 154 \qquad \text{Adding 4 on both sides}$$

$$x = \tfrac{3}{14} \cdot 154 \qquad \text{Multiplying on both sides by } \tfrac{3}{14}$$

$$x = 33. \qquad \text{Remember: } x \text{ represents the first number.}$$

We could check to see whether 33 is a solution of the equation, but we can skip that step because we will check later in the original problem.

Going back to the table, we can find the other two numbers:

Second: $3x - 6 = 3 \cdot 33 - 6 = 93$;

Third: $\tfrac{2}{3}x + 2 = \tfrac{2}{3} \cdot 33 + 2 = 24.$

4. CHECK. We go back to the original problem. We have three numbers: 33, 93, and 24. Is the second number 6 less than 3 times the first?

$$3 \times 33 - 6 = 99 - 6 = 93$$

The answer is *yes*.

Is the third number 2 more than $\tfrac{2}{3}$ the first?

$$\tfrac{2}{3} \times 33 + 2 = 22 + 2 = 24$$

The answer is *yes*.

Is the sum of the three numbers 150?

$$33 + 93 + 24 = 150$$

The answer is *yes*. The numbers do check.

5. STATE. The problem asks us to find the largest number, so the answer is: ''The largest number is 93.'' ❑

Note in Example 5 that although the equation $x = 33$ enabled us to find the largest number, 93, the number 33 was *not* the solution to the problem. By carefully labeling our variable in the first step of problem solving, we may avoid the temptation of thinking that our variable always represents the solution to the problem.

EXERCISE SET | 1.4 |

For each problem, familiarize yourself with the situation. Then translate to mathematical language. You need not actually solve the problem; just carry out the first two steps.

1. The sum of two numbers is 81. One of the numbers is 9 more than the other. What are the numbers?

2. The sum of two numbers is 95. One of the numbers is 11 more than the other. What are the numbers?

3. Deirdre swims at a speed of 5 km/h in still water. The current in a river is moving at 3.2 km/h. How long will it take Deirdre to swim 2.7 km upriver?

4. Francis swims at a speed of 4 km/h in still water.

The current in a river is moving at 1.5 km/h. How long will it take Francis to swim 3.75 km upriver?

5. A paddleboat moves at a rate of 12 km/h in still water. How long will it take the boat to travel 35 km downriver if the river's current moves at a rate of 3 km/h?

River's current
3 km/h

6. A paddleboat moves at a rate of 14 km/h in still water. How long will it take the boat to travel 56 km downriver if the river's current moves at a rate of 7 km/h?

7. The degree measures of the angles in a triangle are three consecutive integers. Find the measures of the angles.

8. The degree measures of the angles in a triangle are three consecutive even integers. Find the measures of the angles.

9. A commercial jet has been instructed to climb from its present altitude of 8000 ft to a cruising altitude of 29,000 ft. If the plane ascends at a rate of 3500 ft/min, how long will it take to reach the cruising altitude?

10. A piece of wire 10 m long is to be cut into two pieces, one of them $\frac{2}{3}$ as long as the other. How should the wire be cut?

11. One angle of a triangle is three times as great as a second angle. The third angle measures 12° less than twice the second angle. Find the measures of the angles.

12. One angle of a triangle is four times as great as a second angle. The third angle measures 5° more than twice the second angle. Find the measures of the angles.

13. Find three consecutive odd integers such that the sum of the first, two times the second, and three times the third is 70.

14. Find two consecutive even integers such that two times the first plus three times the second is 76.

15. A piece of wire 100 cm long is to be cut into two pieces, each to be bent to make a square. The length of a side of one square is to be 2 cm greater

than the length of a side of the other. How should the wire be cut?

16. A piece of wire 100 cm long is to be cut into two pieces, and those pieces are each to be bent to make a square. The area of one square is to be 144 cm^2 greater than that of the other. How should the wire be cut?

17. Three numbers are such that the second is six less than three times the first, and the third is two more than two thirds of the second. The sum of the three numbers is 172. Find the largest number.

18. Whitney's appliance store is having a sale on 13 TV sets. They are displayed in order of increasing price from left to right. The price of each set differs by $20 from either set next to it. For the price of the set at the extreme right, a customer can buy both the second and seventh sets. What is the price of the least expensive set?

19. A student's scores on five tests are 93, 89, 72, 80, and 96. What must the score be on the next test so that the average will be 88?

20. The changes in population of a city for three consecutive years are, respectively, 20% increase, 30% increase, and 20% decrease. What is the percent of total change for those three years?

Solve the problem. Carry out all five problem-solving steps.

21. The product of two numbers is 12.3. If one of the factors is 3, find the other number.

22. The quotient of two numbers is 0.75. If the divisor is 50, find the other number.

23. The number 38.2 is less than some number by 12.1. What is the number?

24. The number 173.5 is greater than a certain number by 16.8. What is the number?

25. The number 128 is 0.4 of what number?

26. The number 456 is $\frac{1}{3}$ of what number?

27. One number is greater than another by 12. The sum of the numbers is 114. What is the larger number?

28. One number is less than another by 65. The sum of the numbers is 92. What is the smaller number?

29. One number is twice another number. The sum of the numbers is 495. What are the numbers?

30. One number is five times another. The sum of the numbers is 472. What are the numbers?

31. Solve the problem of Exercise 4.

32. Solve the problem of Exercise 3.

33. Solve the problem of Exercise 14.

34. Solve the problem of Exercise 13.

35. Solve the problem of Exercise 10.

36. Solve the problem of Exercise 9.

37. Solve the problem of Exercise 11.

38. Solve the problem of Exercise 15.

Skill Maintenance

Solve.

39. $3[2x - (5 + 4x)] = 5 - 7x$

40. $5[3x - (2 + 4x)] = 9 - 4x$

41. $4 + 2(a - 7) = 2(a - 5)$

42. $5 + 3(a - 2) = 3(4 + a)$

Synthesis

43. ◈ Write a problem for a classmate to solve. Devise the problem so that the solution is "The material should be cut into two pieces, one 30 cm long and the other 45 cm long."

44. ◈ Write a problem for a classmate to solve. Devise the problem so that the solution is "The first angle is 40°, the second angle is 50°, and the third angle is 90°."

45. The height and sides of a triangle are four consecutive integers. The height is the first integer, and the base is the fourth integer. The perimeter of the triangle is 42 in. Find the area of the triangle.

46. Blanche's salary is reduced $n\%$ during a period of financial difficulty. By what percent would her salary need to be raised in order to bring it back to where it was before the reduction?

47. Brookdale's population grew 8% in 1991, 10% in 1992, and 11% in 1993. Find the population of the city at the start of 1991, if the population is 1,582,416 at the end of 1993.

48. Tico's scores on four tests are 83, 91, 78, and 81. How many points above the average must Tico score on the next test in order to raise the average 2 points?

1.5

Exponential Notation

Whole Numbers as Exponents • **The Zero Exponent** • **Negative Integers as Exponents** • **Raising Powers to Powers** • **Raising a Product or a Quotient to a Power** • **Order of Operations** • **Calculators**

In this section, we review the definitions for integer exponents, study rules for manipulating integer exponents, and conclude with a review of the order in which operations are performed when making calculations.

Whole Numbers as Exponents

Just as $3 \cdot 4$ is shorthand notation for $4 + 4 + 4$ (repeated addition), notation such as 5^2 is shorthand notation for $5 \cdot 5$ (repeated multiplication). Thus the following definition is very important.

Exponential Notation

A symbol a^n, where n is an integer greater than 1, means

$$\underbrace{a \cdot a \cdot a \cdot \cdots \cdot a \cdot a}_{n \text{ factors.}}$$

In a^n, a is called the *base* and n is the *exponent*, or *power*.
The symbol a^1 means a.
We read a^n as "a raised to the nth power" or simply "a to the nth."

We read s^2 as "s-squared" and x^3 as "x-cubed." This terminology comes from the fact that the area of a square of side s is $s \cdot s = s^2$ and the volume of a cube of side x is $x \cdot x \cdot x = x^3$.

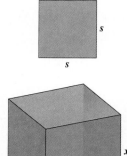

EXAMPLE 1

Write an expression equivalent to $x^3 \cdot x^4$.

Solution By the definition,

$$x^3 \cdot x^4 = \underbrace{x \cdot x \cdot x}_{3 \text{ factors}} \cdot \underbrace{x \cdot x \cdot x \cdot x}_{4 \text{ factors}}$$

$$= \underbrace{x \cdot x \cdot x \cdot x \cdot x \cdot x \cdot x}_{7 \text{ factors}}$$

$$= x^7.$$

The result of Example 1 can be generalized.

Multiplying with Like Bases: The Product Rule

For any number a and any positive integers m and n,

$$a^m \cdot a^n = a^{m+n}.$$

(When multiplying with exponential notation, if the bases are the same, keep the base and add the exponents.)

EXAMPLE 2

Multiply and simplify: **(a)** $m^5 \cdot m^7$; **(b)** $(5a^2b^3)(3a^4b^5)$.

Solution

a) $m^5 \cdot m^7 = m^{5+7} = m^{12}$

b) $(5a^2b^3)(3a^4b^5) = 5 \cdot 3 \cdot a^2 \cdot a^4 \cdot b^3 \cdot b^5$ Using the associative and commutative laws

$\qquad = 15a^{2+4}b^{3+5}$ Multiplying; using the product rule

$\qquad = 15a^6b^8$ ◻

We next simplify a quotient:

$$\frac{x^8}{x^3} = \frac{x \cdot x \cdot x \cdot x \cdot x \cdot x \cdot x \cdot x}{x \cdot x \cdot x}$$ Using the definition of exponential notation

$$= \frac{x \cdot x \cdot x}{x \cdot x \cdot x} \cdot x \cdot x \cdot x \cdot x \cdot x$$

$$= x \cdot x \cdot x \cdot x \cdot x$$ Removing a factor of 1

$$= x^5.$$

Again, we can generalize our result.

Dividing with Like Bases: The Quotient Rule

For any nonzero number a and any positive integers m and n, $m > n$,

$$\frac{a^m}{a^n} = a^{m-n}.$$

(When dividing with exponential notation, if the bases are the same, keep the base and subtract the exponent of the denominator from the exponent of the numerator.)

EXAMPLE 3

Divide and simplify: **(a)** $\dfrac{r^9}{r^3}$; **(b)** $\dfrac{10x^{11}y^5}{2x^4y^3}$.

Solution

a) $\dfrac{r^9}{r^3} = r^{9-3} = r^6$ Using the quotient rule

b) $\dfrac{10x^{11}y^5}{2x^4y^3} = 5 \cdot x^{11-4} \cdot y^{5-3}$ Dividing; using the quotient rule

$\qquad = 5x^7y^2$ ◻

The Zero Exponent

Suppose now that the bases in the numerator and the denominator are both raised to the same power:

$$\frac{x^5}{x^5} = 1 \quad \text{or} \quad \frac{8^3}{8^3} = 1.$$

These results follow from the fact that any (nonzero) expression, divided by itself, is equal to 1. On the other hand, were we to subtract exponents, we would obtain

$$\frac{x^5}{x^5} = x^{5\,-\,5} = x^0 \quad \text{or} \quad \frac{8^3}{8^3} = 8^{3\,-\,3} = 8^0.$$

Thus, in order to continue subtracting exponents when dividing with like bases, we must have $x^0 = 1$ and $8^0 = 1$.

The Zero Exponent

> For any real number a, $a \neq 0$,
>
> $$a^0 = 1.$$
>
> (Any nonzero number raised to the zero power is 1.)

Defining the zero exponent in this manner preserves the following pattern:

$4^3 = 4 \cdot 4 \cdot 4,$

$4^2 = 4 \cdot 4,$ Dividing by 4 on both sides

$4^1 = 4,$ Dividing by 4 on both sides

$4^0 = 1.$ Dividing by 4 on both sides

EXAMPLE 4

Evaluate x^0 when $x = -7.25$.

Solution   When $x = -7.25$, $x^0 = (-7.25)^0 = 1$.   ❑

Negative Integers as Exponents

We can use the quotient rule to lead us to a definition for negative integer exponents. Consider $5^3/5^7$ and first simplify it using procedures we have learned for working with fractions:

$$\frac{5^3}{5^7} = \frac{5 \cdot 5 \cdot 5}{5 \cdot 5 \cdot 5 \cdot 5 \cdot 5 \cdot 5 \cdot 5} = \frac{5 \cdot 5 \cdot 5 \cdot 1}{5 \cdot 5 \cdot 5 \cdot 5 \cdot 5 \cdot 5 \cdot 5}$$

$$= \frac{5 \cdot 5 \cdot 5}{5 \cdot 5 \cdot 5} \cdot \frac{1}{5 \cdot 5 \cdot 5 \cdot 5}$$

$$= \frac{1}{5^4}.$$

Now suppose we apply the rule for dividing powers with the same bases. Then

$$\frac{5^3}{5^7} = 5^{3\,-\,7}$$

$$= 5^{-4}.$$

From these two expressions for $5^3/5^7$, it follows that

$$5^{-4} = \frac{1}{5^4}.$$

This leads to our definition of negative exponents.

Negative Exponents

For any real number a that is nonzero and any integer n,

$$a^{-n} = \frac{1}{a^n}.$$

(The numbers a^{-n} and a^n are reciprocals.)

EXAMPLE 5

Express using positive exponents and then simplify:

a) 3^{-2} **b)** $5x^{-4}y^3$ **c)** $\dfrac{1}{5^{-2}}$

Solution

a) $3^{-2} = \dfrac{1}{3^2} = \dfrac{1}{9}$

b) $5x^{-4}y^3 = 5\left(\dfrac{1}{x^4}\right)y^3 = \dfrac{5y^3}{x^4}$

c) $\dfrac{1}{5^{-2}} = \dfrac{1}{\dfrac{1}{5^2}} = 1 \cdot \dfrac{5^2}{1} = 25$ □

Example 5(c) reveals that when a factor of the numerator or the denominator is raised to any power, the factor can be moved to the other side of the fraction bar provided the sign of the exponent is changed. Thus, for example,

$$\frac{a^{-2}b^3}{c^{-4}} = \frac{c^4}{a^2 b^{-3}}.$$

The rules for multiplying and dividing powers with like bases still hold when exponents are zero or negative.

EXAMPLE 6

Simplify: **(a)** $7^{-3} \cdot 7^8$; **(b)** $\dfrac{b^{-5}}{b^{-4}}$.

Solution

a) $7^{-3} \cdot 7^8 = 7^{-3+8}$ ⎽⎽⎽⎽ Adding exponents
$= 7^5$

b) $\dfrac{b^{-5}}{b^{-4}} = b^{-5-(-4)} = b^{-1}$ ⎽⎽⎽⎽ Subtracting exponents

$= \dfrac{1}{b}$ □

Raising Powers to Powers

Next, consider an expression like $(3^4)^2$. In this case, we are raising 3^4 to the second power:

$$(3^4)^2 = (3^4)(3^4)$$
$$= (3 \cdot 3 \cdot 3 \cdot 3)(3 \cdot 3 \cdot 3 \cdot 3)$$
$$= 3 \cdot 3 \cdot 3 \cdot 3 \cdot 3 \cdot 3 \cdot 3 \cdot 3 \qquad \text{Using the associative law}$$
$$= 3^8.$$

Note that in this case, we could have multiplied the exponents:

$$(3^4)^2 = 3^{4 \cdot 2}$$
$$= 3^8.$$

Likewise, $(y^8)^3 = (y^8)(y^8)(y^8) = y^{24}$. Once again, we get the same result if we multiply the exponents:

$$(y^8)^3 = y^{8 \cdot 3}$$
$$= y^{24}.$$

Raising a Power to a Power: The Power Rule

For any real number a and any integers m and n,

$$(a^m)^n = a^{mn}.$$

(To raise a power to a power, multiply the exponents.)

EXAMPLE 7

Simplify: **(a)** $(3^5)^4$; **(b)** $(y^{-5})^7$; **(c)** $(a^{-3})^{-7}$.

Solution

a) $(3^5)^4 = 3^{5 \cdot 4} = 3^{20}$

b) $(y^{-5})^7 = y^{-5 \cdot 7} = y^{-35}$

c) $(a^{-3})^{-7} = a^{(-3)(-7)} = a^{21}$ ❑

Raising a Product or a Quotient to a Power

When an expression inside parentheses is raised to a power, the inside expression is the base. Let us compare $2a^3$ and $(2a)^3$.

$$2a^3 = 2 \cdot a \cdot a \cdot a; \qquad\qquad (2a)^3 = (2a)(2a)(2a)$$
$$= (2 \cdot 2 \cdot 2)(a \cdot a \cdot a)$$
$$= 2^3 a^3$$
$$= 8a^3$$

We see that $2a^3$ and $(2a)^3$ are *not* equivalent. We also see that we can evaluate the term $(2a)^3$ by raising each factor to the power 3. This leads us to the following rule.

Raising a Product to a Power

For any real numbers a and b and any integer n (provided $ab \neq 0$ when $n \leq 0$),

$$(ab)^n = a^n b^n.$$

(To raise a product to the nth power, raise each factor to the nth power.)

EXAMPLE 8

Simplify: **(a)** $(-2x)^3$; **(b)** $(-2x^3y^{-1})^{-4}$.

Solution

a) $(-2x)^3 = (-2)^3 \cdot x^3$ Raising each factor to the third power

$$= -8x^3$$

b) $(-2x^3y^{-1})^{-4} = (-2)^{-4}(x^3)^{-4}(y^{-1})^{-4}$ Raising each factor to the negative fourth power

$$= \frac{1}{(-2)^4} \cdot x^{-12}y^4$$ Multiplying powers

$$= \frac{y^4}{16x^{12}}$$

There is a similar rule for raising a quotient to a power.

Raising a Quotient to a Power

For any real numbers a and b for which a/b, a^n, and b^n exist,

$$\left(\frac{a}{b}\right)^n = \frac{a^n}{b^n}.$$

(To raise a quotient to a power, raise the numerator to the power and divide by the denominator to the power.)

EXAMPLE 9

Simplify: **(a)** $\left(\dfrac{x^2}{3}\right)^4$; **(b)** $\left(\dfrac{y^2z^3}{5}\right)^{-3}$.

Solution

a) $\left(\dfrac{x^2}{3}\right)^4 = \dfrac{(x^2)^4}{3^4} = \dfrac{x^8}{81}$

b) $\left(\dfrac{y^2z^3}{5}\right)^{-3} = \dfrac{(y^2z^3)^{-3}}{5^{-3}}$

$$= \frac{5^3}{(y^2z^3)^3}$$ Moving factors to the other side of the fraction bar and changing the -3 to 3

$$= \frac{125}{y^6z^9}$$

The following is a summary of the definitions and rules for exponents that we have considered in this section.

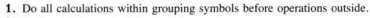

**Definitions and
Rules for Exponents**

For any integers m and n (assuming 0 is not raised to a nonpositive power):

Zero as an exponent:	$a^0 = 1$
Negative integers as exponents:	$a^{-n} = \dfrac{1}{a^n}$
Multiplying with like bases:	$a^m \cdot a^n = a^{m+n}$
Dividing with like bases:	$\dfrac{a^m}{a^n} = a^{m-n},\ a \neq 0$
Raising a product to a power:	$(ab)^n = a^n b^n$
Raising a power to a power:	$(a^m)^n = a^{mn}$
Raising a quotient to a power:	$\left(\dfrac{a}{b}\right)^n = \dfrac{a^n}{b^n},\ b \neq 0$

Order of Operations

What does $1 + 2 \cdot 5^2$ mean? If we add 1 and 2 and multiply by 5^2, or 25, we get 75. If we multiply 2 times 25 and add 1, we get 51. Clearly, both results cannot be correct. To help us determine which procedure to use, mathematicians have adopted the following conventions.

**Rules for Order
of Operations**

1. Do all calculations within grouping symbols before operations outside.
2. Evaluate all exponential expressions.
3. Do all multiplications and divisions in order from left to right.
4. Do all additions and subtractions in order from left to right.

EXAMPLE 10

Calculate: $5 + 2(1 - 4)^2$.

Solution

$$
\begin{aligned}
5 + 2(1 - 4)^2 &= 5 + 2(-3)^2 & \text{Working within parentheses first} \\
&= 5 + 2(9) & \text{Simplifying } (-3)^2 \\
&= 5 + 18 & \text{Multiplying} \\
&= 23 & \text{Adding} \qquad \square
\end{aligned}
$$

It is important to remember that division should be performed before multiplication if, after calculations within grouping symbols are made and exponential expressions are simplified, the division is encountered first when reading from left to right. Similarly, if subtraction appears before addition, we should subtract before adding.

EXAMPLE 11

Calculate: $7 - (-1 + 3) + 6 \div 2 \cdot (-3)^2$.

Solution

$$7 - (-1 + 3) + 6 \div 2 \cdot (-3)^2 = 7 - 2 + 6 \div 2 \cdot 9$$

Working within parentheses; simplifying $(-3)^2$

$$= 7 - 2 + 3 \cdot 9$$ Dividing

$$= 7 - 2 + 27$$ Multiplying

$$= 5 + 27$$ Subtracting

$$= 32$$ Adding   ❏

In addition to the usual grouping symbols — parentheses, brackets, and braces — a fraction bar or absolute value symbol may indicate groupings.

EXAMPLE 12

Calculate: $\dfrac{12|7 - 9| + 4 \cdot 5}{(-3)^4 + 2^3}$.

Solution   We do the calculations in the numerator and in the denominator and divide the results.

$$\frac{12|7 - 9| + 4 \cdot 5}{(-3)^4 + 2^3} = \frac{12|-2| + 4 \cdot 5}{81 + 8}$$

$$= \frac{12(2) + 20}{89}$$

$$= \frac{44}{89}$$ Multiplying and adding   ❏

Calculators

Most of today's calculators make use of the order of operations that we have adopted. These calculators are said to possess *algebraic logic*. Because many of these calculators do not have parentheses, we must be especially careful when calculating products and quotients and when raising an expression to a power. If you find that your calculator works differently, consult your instructor or an owner's manual.

EXAMPLE 13

Write the sequence of keys that should be pressed to calculate each of the following.

a) $\dfrac{5 + 3}{2}$

b) $(7 + 2 \cdot 6 - 9)4$

Solution

a) $\dfrac{5 + 3}{2}$ means ⑤ ⊞ ③ ⊟ ⊡ ② ⊟ .

b) $(7 + 2 \cdot 6 - 9)4$ means ⑦ ⊞ ② ⊠ ⑥ ⊟ ⑨ ⊟ ⊠ ④ ⊟ .   ❏

Note that had we not pressed the first equals sign in Example 13(a), only the 3 would have been divided by 2. Similarly, had we not pressed the first equals sign in Example 13(b), only the 9 would have been multiplied by 4. A similar use of the equals sign is made when raising a bracketed expression to a power.

EXAMPLE 14

Write the sequence of keys that should be pressed to calculate $(5 + 2 \cdot 3)^4$.

Solution

$(5 + 2 \cdot 3)^4$ means   .

Various symbols are used
to denote this key.   ❑

EXERCISE SET | 1.5

Multiply and simplify. Leave the answer in exponential notation.

1. $3^2 \cdot 3^5$

2. $2^3 \cdot 2^8$

3. $5^6 \cdot 5^3$

4. $6^2 \cdot 6^6$

5. $a^3 \cdot a^0$

6. $x^0 \cdot x^5$

7. $5x^4 \cdot 3x^2$

8. $4a^3 \cdot 2a^7$

9. $(-3m^4)(-7m^9)$

10. $(-2a^5)(7a^4)$

11. $(x^3y^4)(x^7y^6z^0)$

12. $(m^6n^5)(m^4n^7p^0)$

Divide and simplify.

13. $\dfrac{a^9}{a^3}$

14. $\dfrac{x^{12}}{x^3}$

15. $\dfrac{8x^7}{4x^4}$

16. $\dfrac{20a^{20}}{5a^4}$

17. $\dfrac{m^7n^9}{m^2n^5}$

18. $\dfrac{m^{12}n^9}{m^4n^6}$

19. $\dfrac{35x^8y^5}{7x^2y}$

20. $\dfrac{45x^7y^8}{5xy^2}$

21. $\dfrac{-49a^5b^{12}}{7a^2b^2}$

22. $\dfrac{-42a^2b^7}{7ab^4}$

23. $\dfrac{18x^6y^7z^9}{-6x^2yz^3}$

24. $\dfrac{28x^8y^{10}z^{12}}{-7x^4y^2z}$

Write an equivalent expression without negative exponents.

25. 6^{-3}

26. 8^{-4}

27. 9^{-5}

28. 16^{-2}

29. $(-11)^{-1}$

30. $(-4)^{-3}$

31. $(5x)^{-3}$

32. $(4xy)^{-5}$

33. x^2y^{-3}

34. $2a^2b^{-5}$

35. x^2y^{-2}

36. $a^2b^{-3}c^4d^{-5}$

37. $\dfrac{x^3}{y^{-2}}$

38. $\dfrac{y^4z^3}{x^{-1}}$

39. $\dfrac{y^{-5}}{x^2}$

40. $\dfrac{z^{-4}}{3x^5}$

41. $\dfrac{y^{-5}}{x^{-3}}$

42. $\dfrac{(7x)^{-4}}{y^{-7}}$

43. $\dfrac{x^{-2}y^7}{z^{-4}}$

44. $\dfrac{y^4x^{-3}}{x^{-2}}$

Write an equivalent expression with negative exponents.

45. $\dfrac{1}{3^4}$

46. $\dfrac{1}{9^2}$

47. $\dfrac{1}{(-16)^2}$

48. $\dfrac{1}{(-8)^6}$

49. 6^4

50. 8^5

51. $6x^2$

52. $-4y^5$

53. $\dfrac{1}{(5y)^3}$

54. $\dfrac{1}{(5x)^5}$

55. $\dfrac{1}{3y^4}$

56. $\dfrac{1}{4b^3}$

Simplify. When negative exponents appear in the answer, write a second answer using only positive exponents.

57. $8^{-6} \cdot 8^2$

58. $9^{-5} \cdot 9^3$

59. $8^{-2} \cdot 8^{-4}$

60. $9^{-1} \cdot 9^{-6}$

61. $b^2 \cdot b^{-5}$

62. $a^4 \cdot a^{-3}$

63. $a^{-3} \cdot a^4 \cdot a^2$

64. $x^{-8} \cdot x^5 \cdot x^3$

65. $(14m^2n^3)(-2m^3n^2)$

66. $(6x^5y^{-2})(-3x^2y^3)$

67. $(-2x^{-3})(7x^{-8})$

68. $(6x^{-4}y^3)(-4x^{-8}y^{-2})$

69. $5^{a+1} \cdot 5^{2a-1}$

70. $7^{a+b} \cdot 7^{a+1}$

71. $\dfrac{4^3}{4^{-2}}$

72. $\dfrac{5^8}{5^{-3}}$

73. $\dfrac{10^{-3}}{10^6}$

74. $\dfrac{12^{-4}}{12^8}$

75. $\dfrac{9^{-4}}{9^{-6}}$ **76.** $\dfrac{2^{-7}}{2^{-5}}$

77. $\dfrac{a^3}{a^{-2}}$ **78.** $\dfrac{y^4}{y^{-5}}$

79. $\dfrac{9a^2}{3ab^3}$ **80.** $\dfrac{24a^5b^3}{-8a^4b}$

81. $\dfrac{-24x^6y^7}{18x^{-3}y^9}$ **82.** $\dfrac{14a^4b^{-3}}{-8a^8b^{-5}}$

83. $\dfrac{-18x^{-2}y^3}{-12x^{-5}y^5}$ **84.** $\dfrac{-14a^{14}b^{-5}}{-18a^{-2}b^{-10}}$

85. $\dfrac{10^{2a}}{10^a}$ **86.** $\dfrac{11^b}{11^{3b}}$

87. $(4^3)^2$ **88.** $(5^4)^5$

89. $(8^4)^{-3}$ **90.** $(9^3)^{-4}$

91. $(6^{-4})^{-3}$ **92.** $(7^{-8})^{-5}$

93. $(3x^2y^2)^3$ **94.** $(2a^3b^4)^5$

95. $(-2x^3y^{-4})^{-2}$ **96.** $(-3a^2b^{-5})^{-3}$

97. $(-6a^{-2}b^3c)^{-2}$ **98.** $(-8x^{-4}y^5z^2)^{-4}$

99. $\dfrac{(5a^3b)^2}{10a^2b}$ **100.** $\dfrac{(3x^3y^4)^3}{6xy^3}$

101. $\left(\dfrac{2x^3y^{-2}}{3y^{-3}}\right)^3$ **102.** $\left(\dfrac{-4x^4y^{-2}}{5x^{-1}y^4}\right)^{-4}$

103. $\left(\dfrac{30x^5y^{-7}}{6x^{-2}y^{-6}}\right)^0$ **104.** $\left(\dfrac{3a^{-2}b^5}{9a^{-4}b^0}\right)^{-2}$

105. $\left(\dfrac{5x^0y^{-7}}{2x^{-2}y^4}\right)^{-2}$ **106.** $\left(\dfrac{4a^3b^{-9}}{2a^{-2}b^5}\right)^0$

Calculate using the rules for order of operations.

107. $5 + 2 \cdot 3^2$ **108.** $9 - 3 \cdot 2^2$

109. $12 - (9 - 3 \cdot 2^3)$ **110.** $19 - (4 + 2 \cdot 3^2)$

111. $\dfrac{5 \cdot 2 - 4^2}{27 - 2^4}$ **112.** $\dfrac{7 \cdot 3 - 5^2}{9 + 4 \cdot 2}$

113. $\dfrac{3^4 - (5 - 3)^4}{1 - 2^3}$

114. $\dfrac{4^3 - (7 - 4)^2}{3^2 - 7}$

115. $5^3 - [2(4^2 - 3^2 - 6)]^3$

116. $7^2 - [3(5^2 - 4^2 - 7)]^2$

117. $|2^2 - 7|^3 + 1$

118. $|-2 - 3| \cdot 4^2 - 1$

119. $30 - (-5)^2 + 15 \div (-3) \cdot 2$

120. $55 - (-9 + 2)^2 + 18 \div 6 \cdot (-2)$

121. $12 - (7 - 5) + 4 \div 3 \cdot 2^3$

122. $15 - (3 - 8) + 5 \div 10 \cdot 3^2$

List the sequence of keys that should be pushed on a calculator in order to compute each of the following. Assume algebraic logic.

123. $\dfrac{9 + 3 \cdot 4 + 1}{2}$ **124.** $\dfrac{7 - 2 \cdot 3 + 5}{4}$

125. $8 - 2 \cdot 3 + \dfrac{4}{15}$ **126.** $12 + 3 \cdot 2 - \dfrac{7}{12}$

127. $\dfrac{9 - 2^8 + 4}{3}$ **128.** $\dfrac{15 - 3^7 + 8}{14}$

129. $\left(7 - \dfrac{2}{3}\right)^4 - \dfrac{5}{8}$ **130.** $\left(9 - \dfrac{4}{7}\right)^5 + \dfrac{3}{11}$

Skill Maintenance

131. Evaluate $x + (2xy - z)^2$ when $x = 3$, $y = -4$, and $z = -20$.

132. Evaluate $a - (bc - 3a)^2$ when $a = -4$, $b = 5$, and $c = -2$.

133. Find three consecutive odd integers whose sum is 183.

134. Find three consecutive even integers whose sum is 114.

Synthesis

135. ◈ Explain why $(-1)^n = 1$ for any even number n.

136. ◈ Explain why $(-17)^{-8}$ is positive.

Simplify. Assume that all variables represent nonzero integers.

137. $\dfrac{9a^{x-2}}{3a^{2x+2}}$

138. $\dfrac{-12x^{a+1}}{4x^{2-a}}$

139. $[7y(7 - 8)^{-2} - 8y(8 - 7)^{-2}]^{(-2)^2}$

140. $\{[(8^{-a})^{-2}]^b\}^{-c} \cdot [(8^0)^a]^c$

141. $(3^{a+2})^a$ **142.** $(12^{3-a})^{2b}$

143. $\dfrac{4x^{2a+3}y^{2b-1}}{2x^{a+1}y^{b+1}}$ **144.** $\dfrac{25x^{a+b}y^{b-a}}{-5x^{a-b}y^{b+a}}$

145. $\dfrac{(2^{-2})^a \cdot (2^b)^{-a}}{(2^{-2})^{-b}(2^b)^{-2a}}$ **146.** $\dfrac{-28x^{b+5}y^{4+c}}{7x^{b-5}y^{c-4}}$

147. $\dfrac{3^{q+3} - 3^2(3^q)}{3(3^{q+4})}$

148. $\left[\left(\dfrac{a^{-2c}}{b^{7c}}\right)^{-3}\left(\dfrac{a^{4c}}{b^{-3c}}\right)^2\right]^{-a}$

1.6

Geometry, Formulas, and Problem Solving

Solving Formulas • Formulas in Translating

A **formula** is an equation that uses letters to represent a relationship between two or more quantities. For example, in Section 1.4 we made use of the formula $P = 4s$, where P represents the perimeter of a square and s the length of a side. Other formulas that you may recall from geometry are $A = \pi r^2$ (for the area A of a circle of radius r), $C = \pi d$ (for the circumference C of a circle of diameter d), and $A = b \cdot h$ (for the area A of a parallelogram of height h and base length b).* A more complete list of geometric formulas appears on page 606.

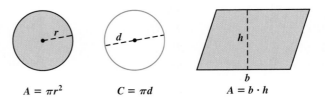

$$A = \pi r^2 \qquad C = \pi d \qquad A = b \cdot h$$

Solving Formulas

Suppose that we know the area and the width of a rectangular room and want to find the length. To do so, we might begin with the equation $A = l \cdot w$, which expresses a rectangle's area A in terms of its length l and its width w. We then "solve" for l.

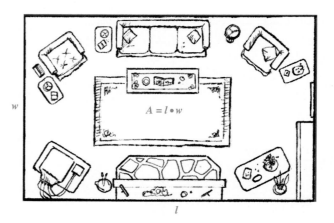

$$A = l \bullet w$$

EXAMPLE 1

Solve the formula $A = l \cdot w$ for l.

---

*The Greek letter π, read "pi," is *approximately* 3.14159265358979323846264. In this text, we will use 3.14 to approximate π unless otherwise noted.

Solution

$$A = l \cdot w \qquad \text{We want this letter alone.}$$

$$A \cdot \frac{1}{w} = l \cdot w \cdot \frac{1}{w} \qquad \text{Multiplying by } \frac{1}{w}$$

$$\frac{A}{w} = l \qquad \text{Simplifying} \qquad \square$$

Thus to find the length of a rectangular room, we can divide the area of the room by its width. Were we to do this calculation for a number of different rectangular rooms, the formula $l = A/w$ would be most useful.

When we solve a formula, we do the same things that we do to solve any equation. The idea is to get a certain letter alone on one side of the equals sign.

EXAMPLE 2

The formula $I = Prt$ is used to determine the amount of simple interest I, earned on P dollars, when invested for t years at an interest rate r. Solve the above formula for t.

Solution

$$I = Prt \qquad \text{We want this letter alone.}$$

$$\frac{1}{Pr} \cdot I = \frac{1}{Pr} \cdot Prt \qquad \text{Multiplying by } \frac{1}{Pr}$$

$$\frac{I}{Pr} = t \qquad \text{Simplifying} \qquad \square$$

EXAMPLE 3

A trapezoid is a geometric shape with four sides, two of which, the bases, are parallel to each other. The formula for calculating the area A of a trapezoid with bases b_1 and b_2 (read ''b sub one'' and ''b sub two'') and height h is given by

$$A = \frac{h}{2}(b_1 + b_2),$$

where the *subscripts* 1 and 2 are used to distinguish the two bases from each other. Solve for b_1.

Solution

$$A = \frac{h}{2}(b_1 + b_2)$$

$$\frac{2}{h} \cdot A = \frac{2}{h} \cdot \frac{h}{2}(b_1 + b_2) \qquad \text{Multiplying by } \frac{2}{h}$$

$$\frac{2A}{h} = b_1 + b_2 \qquad \text{Simplifying. The right side is cleared of fractions.}$$

$$\frac{2A}{h} - b_2 = b_1 \qquad \text{Adding } -b_2 \qquad \square$$

The similarities between solving formulas and solving equations can be seen below. In (a), we solve as we did before; in (b), we do not carry out all calculations; and in (c), we cannot carry out all calculations because the numbers are unknown.

a) $9 = \frac{3}{2}(x + 5)$ **b)** $9 = \frac{3}{2}(x + 5)$ **c)** $A = \frac{h}{2}(b_1 + b_2)$

$\frac{2}{3} \cdot 9 = \frac{2}{3} \cdot \frac{3}{2}(x + 5)$ $\frac{2}{3} \cdot 9 = \frac{2}{3} \cdot \frac{3}{2}(x + 5)$ $\frac{2}{h} \cdot A = \frac{2}{h} \cdot \frac{h}{2}(b_1 + b_2)$

$6 = x + 5$ $\frac{2 \cdot 9}{3} = x + 5$ $\frac{2A}{h} = b_1 + b_2$

$1 = x$

$\frac{2 \cdot 9}{3} - 5 = x$ $\frac{2A}{h} - b_2 = b_1$

EXAMPLE 4

The formula $A = P + Prt$ tells how much a principal P, in dollars, will be worth in t years when invested at simple interest at a rate r. Solve the formula for P.

Solution

$A = P + Prt$ We want this letter alone.

$A = P(1 + rt)$ Factoring (or using the distributive law)

$A \cdot \frac{1}{1 + rt} = P(1 + rt) \cdot \frac{1}{1 + rt}$ Multiplying by $\frac{1}{1 + rt}$

$\frac{A}{1 + rt} = P$ Simplifying   ❏

Note in Example 4 that the factoring enabled us to write P once rather than twice. This is comparable to collecting like terms when solving an equation like $16 = x + 7x$.

You may find the following summary useful.

To solve a formula for a given letter:

1. Multiply on both sides to clear fractions or decimals, if that is needed.
2. Collect like terms on each side where convenient.
3. Get all terms with the letter to be solved for on one side of the equation and all other terms on the other side, using the addition principle.
4. Collect like terms again, if necessary. This may require factoring.
5. Solve for the letter in question, using the multiplication principle.

Formulas in Translating

The next example illustrates the use of formulas in translating problem situations to mathematical language and in carrying out the mathematical manipulations (steps 2 and 3 of the problem-solving process).

EXAMPLE 5

Chris suspects that a ''gold'' medal is not really gold. The density of gold is 7.72 grams per cubic centimeter (g/cm^3) and the medal is 0.5 cm thick with a radius of 3 cm. If the medal really is gold, how much should it weigh?

Solution

1. **FAMILIARIZE.** In a science book, we can find a formula for density and learn that an object's density depends on its mass and volume. A formula for the volume of a right circular cylinder can be found in Table 2 at the back of the book.

2. **TRANSLATE.** We need to use two formulas:

$$D = \frac{m}{V},$$

where D is the density, m is the mass, and V is the volume of a given material, and

$$V = \pi r^2 h,$$

where V is the volume, r is the radius, and h is the height of a right circular cylinder. Since we are interested in mass, we solve the first formula for m:

$$D = \frac{m}{V}$$

$$V \cdot D = V \cdot \frac{m}{V} \qquad \text{Multiplying by } V$$

$$V \cdot D = m. \qquad \text{Simplifying}$$

3. **CARRY OUT.** This is a two-step problem. We first substitute into the formula $V = \pi r^2 h$:

$$V = \pi r^2 h$$
$$= \pi (3)^2 (0.5) \qquad \text{Substituting}$$
$$= 14.13. \qquad \text{Using 3.14 for } \pi$$

Finally we substitute into the formula $V \cdot D = m$:

$$V \cdot D = m$$
$$(14.13)(7.72) = m \qquad \text{Substituting}$$
$$109.0836 = m.$$

4. **CHECK.** We leave the check to the student.

5. **STATE.** The solution is that the medal, if it is actually gold, should have a mass of 109.0836 grams. ❏

EXERCISE SET | 1.6 |

Solve.

1. $A = lw$, for w
2. $F = ma$, for a
3. $W = EI$, for I (an electricity formula)
4. $W = EI$, for E
5. $d = rt$, for r (a distance formula)
6. $d = rt$, for t
7. $V = lwh$, for l (a volume formula)
8. $I = Prt$, for r

9. $E = mc^2$, for m (a relativity formula)

10. $E = mc^2$, for c^2

11. $P = 2l + 2w$, for l (a perimeter formula)

12. $P = 2l + 2w$, for w

13. $c^2 = a^2 + b^2$, for a^2 (a geometry formula)

14. $c^2 = a^2 + b^2$, for b^2

15. $A = \pi r^2$, for r^2 (an area formula)

16. $A = \pi r^2$, for π

17. $W = \frac{11}{2}(h - 40)$, for h

18. $C = \frac{5}{9}(F - 32)$, for F (a temperature formula)

19. $V = \frac{4}{3}\pi r^3$, for r^3 (a volume formula)

20. $V = \frac{4}{3}\pi r^3$, for π

21. $A = \frac{h}{2}(b_1 + b_2)$, for h (an area formula)

22. $A = \frac{h}{2}(b_1 + b_2)$, for b_2

23. $F = \frac{mv^2}{r}$, for m (a physics formula)

24. $F = \frac{mv^2}{r}$, for v^2

25. $A = \frac{q_1 + q_2 + q_3}{n}$, for n

(*Hint:* First clear the fraction.)

26. $r = \frac{s + t}{d}$, for d

27. $v = \frac{d_2 - d_1}{t}$, for t (a physics formula)

28. $v = \frac{s_2 - s_1}{m}$, for m 29. $v = \frac{d_2 - d_1}{t}$, for d_1

30. $v = \frac{s_2 - s_1}{m}$, for s_1 31. $r = m + mnp$, for m

32. $p = x - xyz$, for x 33. $y = ab - ac^2$, for a

34. $d = mn - mp^3$, for m

Problem Solving

35. The area of a parallelogram is 72 cm². The height of the figure is 6 cm. How long is the base?

36. The area of a parallelogram is 78 cm². The base of the figure is 13 cm. What is the height?

Ultrasonic images of 29-week-old fetuses can be used to predict birth weight. One formula, developed by Thurnau,* is $P = 9.337da - 299$; a second formula, developed by Weiner,* is $P = 94.593c + 34.227a - 2134.616$. For both formulas, P represents the predicted birth weight in grams, d represents the diameter of the fetal head in centimeters, c represents the circumference of the fetal head in centimeters, and a represents the circumference of the fetal abdomen in centimeters.

37. Use Thurnau's formula to estimate the diameter of a fetus' head at 29 weeks when the birthweight is 1614 g and the circumference of the fetal abdomen is 24.1 cm.

38. Use Weiner's formula to estimate the circumference of a fetus' head at 29 weeks when the birthweight is 1277 g and the circumference of the fetal abdomen is 23.4 cm.

39. You are going to buy certificates of deposit at a bank. They pay 7% simple interest. You want your money to earn $110 in a year. How much must you invest?

40. You have $250 to invest for 6 months, and you expect it to earn at least $8 in that time. What rate of simple interest will your money have to earn?

41. A garden is being constructed in the shape of a trapezoid. The dimensions are as shown in the figure. The unknown dimension is to be such that the area of the garden is 90 ft². Find that unknown dimension.

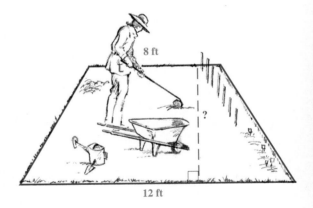

12 ft

42. A rectangular garden is being constructed. There is 76 ft of fencing available, so the perimeter must be 76 ft at most. The width of the garden is to be 13 ft. What should the length be, in order to use just 76 ft of fence? (See the formula in Exercise 11.)

43. Bok Lum Chan is going to invest $1600 at simple interest at 9%. How long will it take for the investment to be worth $2608?

*Thurnau, G. R., R. K. Tamura, R. E. Sabbagha, et al. *Am J Obstet Gynecol* 1983; **145**:557.

*Weiner, C. P., R. E. Sabbagha, N. Vaisrub, et al. *Obstet Gynecol* 1985; **65**:812.

44. You are going to invest $950 at simple interest at 7%. How long will it take for your investment to be worth $1349?

In an effort to minimize waiting time for patients at a doctor's office without increasing a physician's idle time, Michael Goiten of Massachusetts General Hospital has developed a model. Goiten suggests that the interval time I, in minutes, between scheduled appointments be related to the total number of minutes T that a physician spends with patients in a day and the number of scheduled appointments N according to the formula $I = 1.08(T/N)$.*

45. A doctor determines that she has a total of 8 hr a day to see patients. If she insists on an interval time of 15 min, according to Goiten's model, how many appointments should she make in one day?

46. A doctor insists on an interval time of 20 min and must be able to see 25 appointments a day. According to Goiten's model, how many hours a day should the doctor be prepared to spend with patients?

47. The density of iron is 7.5 g/cm³. A boat's anchor occupies a volume of 1230 cm³. Assuming that the anchor is pure iron, determine its mass.

48. The density of copper is 8.93 g/cm³. A solid copper bar occupies a volume of 40.185 cm³. Determine its mass.

Skill Maintenance

49. What percent of 5800 is 4176?

50. Simplify: $-5a + 9b - (3a - 4b)$.

51. Subtract: $-72.5 - (-14.06)$.

52. Simplify: $\dfrac{45x}{15x}$.

---

*New England Journal of Medicine, 30 August 1990, pp. 604–608.

Synthesis

53. ◈ Is every rectangle a trapezoid? Explain why or why not.

54. ◈ Which would you expect to have the greater density, and why: cork or steel?

55. ▣ The density of copper is 8.93 g/cm³. The mass of a roll of pennies is 177.6 g. If the diameter of a penny is 1.85 cm, how tall is a roll of pennies?

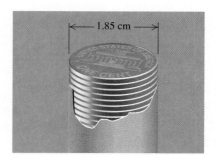

56. ▣ The density of copper is 8.93 g/cm³. How long must a copper wire be if it is 1 cm thick and has a mass of 4280 g?

57. In Example 2, we solved the formula $I = Prt$ for t. Now use it to find how much time is needed in order to earn $6 on $200 at 12% simple interest.

58. In Exercise 8, the formula $I = Prt$ was solved for r. Now use it to find the rate of interest required for a principal of $120 to earn $6.60 in half a year.

Solve.

59. $s = v_i t + \frac{1}{2}at^2$, for a

60. $A = 4lw + w^2$, for l

61. $\dfrac{P_1 V_1}{T_1} = \dfrac{P_2 V_2}{T_2}$, for V_1

62. $\dfrac{P_1 V_1}{T_1} = \dfrac{P_2 V_2}{T_2}$, for T_2

63. $x = \dfrac{a}{b + c}$, for c

64. $m = \dfrac{(d/e)}{(e/f)}$, for d

1.7

Scientific Notation

Conversions • **Significant Digits and Rounding** • **Scientific Notation in Problem Solving**

There are many kinds of symbolism, or *notation,* for numbers. You are already familiar with fractional notation, decimal notation, and percent notation. We now study **scientific notation,** so named because of its usefulness in work with the very large and very small numbers that often occur in the study of science. It is very important for problem solving and is also useful in estimating.

The following are examples of scientific notation:

7.2×10^5 means   720,000;

3.4×10^{-6} means   0.0000034;

4.89×10^{-3} means   0.00489.

Scientific Notation

> *Scientific notation* for a number is an expression of the type $N \times 10^n$, where $1 \leq N < 10$, N is in decimal notation, and n is an integer.

Conversions

Note that $10^b \cdot 10^{-b} = 10^{b + (-b)} = 10^0 = 1$. We can convert to scientific notation by multiplying by 1, choosing a name like $10^b \cdot 10^{-b}$ for the number 1.

EXAMPLE 1

It has been estimated that in the year 2000 the world population will be 6,251,000,000. Write scientific notation for this number.

Solution   We want to move the decimal point nine places — to between the 6 and the 2 — so we choose $10^{-9} \times 10^9$ as a name for 1. Then we multiply.

$$6,251,000,000 = 6,251,000,000 \times 10^{-9} \times 10^9 \qquad \text{Multiplying by 1}$$
$$= (6,251,000,000 \times 10^{-9}) \times 10^9 \qquad \text{Using the associative law}$$
$$= 6.251 \times 10^9 \qquad \text{The } 10^{-9} \text{ moved the decimal point 9 places to the left. We now have scientific notation.} \qquad \square$$

EXAMPLE 2

The mass of a million atoms of chlorine is approximately

0.0000000000000589 gram.

Write scientific notation for this number.

Solution   We want to move the decimal point 14 places to the right. We choose $10^{14} \times 10^{-14}$ as a name for 1. Then we multiply.

$$0.0000000000000589 = 0.0000000000000589 \times 10^{14} \times 10^{-14} \qquad \text{Multiplying by 1}$$
$$= 5.89 \times 10^{-14} \qquad \text{The } 10^{14} \text{ moved the decimal point 14 places to the right and we have scientific notation.} \qquad \square$$

Try to make conversions to scientific notation mentally as often as possible. In doing so, remember that negative powers of 10 are used for small numbers and positive powers of 10 are used for large numbers.

EXAMPLE 3

Convert mentally to scientific notation: **(a)** 82,500,000; **(b)** 0.0000091.

Solution

a) $82,500,000 = 8.25 \times 10^7$ *Think:* Multiplying 8.25 by 10^7 moves the decimal point seven places to the right.

b) $0.0000091 = 9.1 \times 10^{-6}$ *Think:* Multiplying 9.1 by 10^{-6} moves the decimal point six places to the left. $\square$

EXAMPLE 4

Convert mentally to decimal notation: **(a)** 4.371×10^7; **(b)** 1.73×10^{-5}.

Solution

a) $4.371 \times 10^7 = 43,710,000$ Moving the decimal point 7 places to the right

b) $1.73 \times 10^{-5} = 0.0000173$ Moving the decimal point 5 places to the left   ❑

Significant Digits and Rounding

In the world of science, it is important to know just how accurate a measurement is. Clearly the measurement 5.12 cm is more precise than the measurement 5.1 cm. We say that the number 5.12 has three **significant digits** whereas 5.1 has only two significant digits. When two or more measurements are added, subtracted, multiplied, or divided, the result is only as accurate as the *least* precise measurement used in the computation. Thus scientists have agreed on the following conventions.

1. The sum or difference of two numbers should be rounded off so that it has the same number of significant digits to the right of the decimal as the number in the calculation with the fewest significant digits to the right of the decimal.

    For example,

$$135.4 \text{ cm} + 50.28 \text{ cm} = 185.68 \text{ cm}$$

1 digit   2 digits

should be rounded off to

1 digit

$$185.7 \text{ cm}.$$

2. The product or quotient of two numbers should be rounded off so that it contains the same number of significant digits as the number in the calculation with the fewest significant digits.

    For example,

$$2.1 \text{ cm} \times 6.45 \text{ cm} = 13.545 \text{ cm}^2$$

2 digits   3 digits

should be rounded off to

2 digits

$$14 \text{ cm}^2.$$

For future work in this text with measurements and scientific notation, results will be rounded off according to the above conventions.

EXAMPLE 5

Multiply and write scientific notation for the answer:

$$(7.2 \times 10^5)(4.3 \times 10^9).$$

Solution

$$\begin{aligned}
(7.2 \times 10^5)(4.3 \times 10^9) &= (7.2 \times 4.3)(10^5 \times 10^9) && \text{Using the commutative and associative laws} \\
&= 30.96 \times 10^{14} && \text{Adding exponents} \\
&= 31 \times 10^{14} && \text{Rounding to 2 significant digits}
\end{aligned}$$

To find scientific notation for the result, we convert 31 to scientific notation and simplify:

$$31 \times 10^{14} = (3.1 \times 10^1) \times 10^{14} = 3.1 \times 10^{15}. \qquad \square$$

EXAMPLE 6

Divide and write scientific notation for the answer:

$$\frac{3.48 \times 10^{-7}}{4.64 \times 10^6}.$$

Solution

$$\frac{3.48 \times 10^{-7}}{4.64 \times 10^6} = \frac{3.48}{4.64} \times \frac{10^{-7}}{10^6} \qquad \text{Factoring. Our answer must have 3 significant digits.}$$

$$= 0.75 \times 10^{-13} \qquad \text{Subtracting exponents; simplifying}$$

$$= (7.5 \times 10^{-1}) \times 10^{-13} \qquad \text{Converting 0.75 to scientific notation}$$

$$= 7.50 \times 10^{-14} \qquad \text{Adding exponents. We write 7.50 to indicate 3 significant digits.} \quad \square$$

Scientific Notation in Problem Solving

The following examples show how scientific notation can be useful in problem solving.

EXAMPLE 7

Alpha Centauri is the star — apart from the sun — that is closest to earth. Its distance from earth is approximately 2.4×10^{13} mi. How many light years is it from earth to Alpha Centauri?

Solution

1. **FAMILIARIZE.** From an astronomy text, we learn that light travels about 5.88×10^{12} mi in one year. Thus 1 light year $= 5.88 \times 10^{12}$ mi. We will let y represent the number of light years from earth to Alpha Centauri. Let's guess that the answer is 3 light years. Then the distance in miles would be

$$(5.88 \times 10^{12}) \cdot 3 = 17.64 \times 10^{12}$$
$$= 1.764 \times 10^{13}.$$

1 light year $= 5.88 \times 10^{12}$ mi

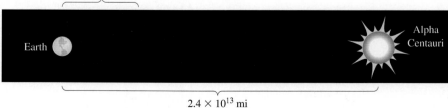

Earth

Alpha Centauri

2.4×10^{13} mi

Although our guess is not correct, we have gained some insight into the problem.

2. **TRANSLATE.** Note that the distance to Alpha Centauri is y light years, or $(5.88 \times 10^{12})y$ mi. We also are told that the distance is 2.4×10^{13} mi. Since the quantities $(5.88 \times 10^{12})y$ and 2.4×10^{13} both represent the number of

miles to Alpha Centauri, we form the equation

$$(5.88 \times 10^{12})y = 2.4 \times 10^{13}.$$

3. CARRY OUT. We solve the equation:

$$(5.88 \times 10^{12})y = 2.4 \times 10^{13}$$

$$\frac{1}{5.88 \times 10^{12}} (5.88 \times 10^{12})y = \frac{1}{5.88 \times 10^{12}} \times 2.4 \times 10^{13}$$ Multiplying by $1/(5.88 \times 10^{12})$ on both sides

$$y = \frac{2.4 \times 10^{13}}{5.88 \times 10^{12}}$$ Simplifying

$$y = \frac{2.4}{5.88} \times \frac{10^{13}}{10^{12}}$$ Factoring. Our answer must have 2 significant digits.

$$y \approx 0.41 \times 10 = 4.1.$$

4. CHECK. Since light travels 5.88×10^{12} mi in one year, in 4.1 yr it will travel $4.1 \times 5.88 \times 10^{12} = 2.4108 \times 10^{13}$ mi, which is approximately the distance from the earth to Alpha Centauri.

5. STATE. The distance to Alpha Centauri from the earth is approximately 4.1 light years.   ❏

EXAMPLE 8

A certain kind of wire will be used to construct 350 km of transmission line. The wire has a diameter of 1.2 cm. What is the volume of wire needed for 350 km of transmission line?

Solution

1. FAMILIARIZE. Drawing a picture, we see that we have a cylinder (a very *long* one). Its length is 350 km and the base has a diameter of 1.2 cm.

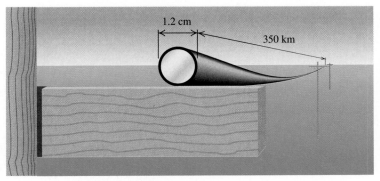

Recall that the formula for the volume of a cylinder is

$$V = \pi r^2 h,$$

where r is the radius of the base and h is the height (in this case, the length of the wire).

2. TRANSLATE. We will use the volume formula, but it is important to make the units the same. Let's put everything into meters:

Length: 350 km = 350,000 m,   or 3.5×10^5 m;

Diameter: 1.2 cm = 0.012 m,   or 1.2×10^{-2} m.

The radius, which we will need in the formula, is half the diameter:

Radius: 0.6×10^{-2} m, or 6×10^{-3} m.

We now substitute into the formula:

$$V = \pi(6 \times 10^{-3})^2(3.5 \times 10^5).$$

3. CARRY OUT. We do the calculation, using 3.14 for π.

$$
\begin{aligned}
V &= 3.14 \times (6 \times 10^{-3})^2(3.5 \times 10^5) \\
&= 3.14 \times 6^2 \times 10^{-6} \times 3.5 \times 10^5 \\
&= (3.14 \times 6^2 \times 3.5) \times (10^{-6} \times 10^5) \\
&= 395.64 \times 10^{-1} \\
&\approx 4.0 \times 10^1, \quad \text{or} \quad 40.
\end{aligned}
$$

Rounding to 2 significant digits. The decimal point emphasizes that we didn't round to the nearest ten.

4. CHECK. In this case, about all we can do is recheck the calculations and the translation.

5. STATE. The answer is that the volume of wire is about 40. m^3 (cubic meters). ❑

EXERCISE SET | 1.7

Convert to scientific notation.

1. 47,000,000,000   **2.** 2,600,000,000,000

3. 863,000,000,000,000,000

4. 957,000,000,000,000,000

5. 0.000000016   **6.** 0.000000263

7. 0.00000000007   **8.** 0.00000000009

9. 407,000,000,000   **10.** 3,090,000,000,000

11. 0.000000603   **12.** 0.00000000802

13. 492,700,000,000   **14.** 953,400,000,000

Convert to decimal notation.

15. 4×10^{-4} **16.** 5×10^{-5}

17. 6.73×10^8 **18.** 9.24×10^7

19. 8.923×10^{-10} **20.** 7.034×10^{-2}

21. 9.03×10^{10} **22.** 1.01×10^{12}

23. 4.037×10^{-8} **24.** 3.007×10^{-9}

25. 8.007×10^{12} **26.** 9.001×10^{10}

Simplify and write scientific notation for the answer. Use the correct number of significant digits.

27. $(2.3 \times 10^6)(4.2 \times 10^{-11})$

28. $(6.5 \times 10^3)(5.2 \times 10^{-8})$

29. $(2.34 \times 10^{-8})(5.7 \times 10^{-4})$

30. $(3.26 \times 10^{-6})(8.2 \times 10^{-6})$

31. $(3.2 \times 10^6)(2.6 \times 10^4)$

32. $(3.11 \times 10^3)(1.01 \times 10^{13})$

33. $(3.01 \times 10^{-5})(6.5 \times 10^7)$

34. $(4.08 \times 10^{-10})(7.7 \times 10^5)$

35. $(5.01 \times 10^{-7})(3.02 \times 10^{-6})$

36. $(7.04 \times 10^{-9})(9.01 \times 10^{-7})$

37. $\dfrac{8.5 \times 10^8}{3.4 \times 10^5}$ **38.** $\dfrac{5.1 \times 10^6}{3.4 \times 10^3}$

39. $\dfrac{4.0 \times 10^{-6}}{8.0 \times 10^{-3}}$ **40.** $\dfrac{7.5 \times 10^{-9}}{2.5 \times 10^{-4}}$

41. $\dfrac{12.6 \times 10^8}{4.2 \times 10^{-3}}$ **42.** $\dfrac{3.2 \times 10^{-7}}{8.0 \times 10^8}$

43. $\dfrac{2.42 \times 10^5}{1.21 \times 10^{-5}}$ **44.** $\dfrac{9.36 \times 10^{-11}}{3.12 \times 10^{11}}$

45. $\dfrac{4.7 \times 10^{-9}}{2.35 \times 10^7}$ **46.** $\dfrac{6.12 \times 10^{19}}{3.06 \times 10^{-7}}$

47. $\dfrac{1.05 \times 10^{-6}}{4.2 \times 10^{-7}}$ **48.** $\dfrac{1.1 \times 10^{-8}}{8.8 \times 10^{-5}}$

49. $\dfrac{(6.1 \times 10^4)(7.2 \times 10^{-6})}{9.8 \times 10^{-4}}$

50. $\dfrac{(8.05 \times 10^{-11})(5.9 \times 10^7)}{3.1 \times 10^{14}}$

51. $\dfrac{780,000,000 \times 0.00071}{0.000005}$

52. $\dfrac{830,000,000 \times 0.12}{3,100,000}$

53. $\dfrac{43,000,000 \times 0.095}{63,000}$

54. $\dfrac{0.0073 \times 0.84}{0.000006}$

Problem Solving

55. The diameter of the Milky Way galaxy is approximately 5.88×10^{17} mi. How many light years is it from one end of the galaxy to the other?

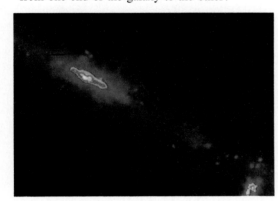

56. The brightest star in the night sky, Sirius, is about 4.704×10^{13} mi from the earth. How many light years is it from the earth to Sirius?

Named in tribute to Anders Ångström, a Swedish physicist who measured light waves, 1 Å (read "one Angstrom") equals 10^{-10} meters. One parsec is about 3.26 light years, and one light year equals 9.46×10^{15} meters.

57. How many Angstroms are in one parsec?

58. How many kilometers are in one parsec?

For Exercises 59 and 60, approximate the average distance from the earth to the sun by 1.50×10^{11} meters (which is one astronomical unit, AU).

59. Determine the volume of a cylindrical sunbeam that is 3 Å in diameter.

60. Determine the volume of a cylindrical sunbeam that is 5 Å in diameter.

61. A certain thin plastic sheet is used in many applications of building and landscaping. The sheet is 1 m wide and 30 m long. The thickness of the sheet is 0.8 mm. Find the volume of a plastic sheet.

62. If a star 5.9×10^{14} mi from earth were to explode today, its light would not reach us for 100 years. How far does light travel in 13 weeks?

63. The average distance of the earth from the sun is about 9.3×10^7 miles. About how far does the earth travel in a yearly orbit about the sun?

64. A *mil* is one thousandth of a dollar. The taxation rate in a certain school district is 5.0 mils for every dollar of assessed valuation. The assessed valuation for the district is 13.4 million dollars. How much tax revenue will be raised?

65. Subtract: $-\frac{5}{6} - \left(-\frac{3}{4}\right)$.

66. Multiply: $(-7.2)(-4.3)$.

67. Simplify: $-2(x - 3) - 3(4 - x)$.

68. Solve: $4(3x - 7) + 9 = 2$.

69. ◈ Write a problem for which the solution is "The area is 6.4×10^{11} m^2."

70. ◈ Explain why 1 m = 100 cm, but 1 m$^2 \neq 100$ cm^2.

71. Compare $8 \cdot 10^{-90}$ and $9 \cdot 10^{-91}$. Which is the larger value? How much larger? Write scientific notation for the difference.

72. Write the reciprocal of 8.00×10^{-23} in scientific notation.

73. ▦ Evaluate: $(4096)^{0.05}(4096)^{0.2}$.

74. What is the unit's digit in 513^{128}?

75. ▦ A grain of sand is placed on the first square of a chessboard, two grains on the second square, four grains on the third, eight on the fourth, sixteen on the fifth, and so on. Use scientific notation to approximate the number of grains of sand required for the last square. (*Hint:* Use the fact that $2^{10} \approx 10^3$.)

SUMMARY AND REVIEW | 1

KEY TERMS

Variable, p. 2
Constant, p. 2
Algebraic expression, p. 2
Substituting, p. 4
Evaluate, p. 4
Equation, p. 5
Solution, p. 5
Natural numbers, p. 5
Whole numbers, p. 5
Integers, p. 5
Roster notation, p. 6
Set-builder notation, p. 6
Element, p. 6
Rational numbers, p. 6
Fractional notation, p. 7

Decimal notation, p. 7
Irrational numbers, p. 7
Real numbers, p. 7
Subset, p. 8
Absolute value, p. 10
Inequality, p. 10
Opposite, p. 11
Additive inverse, p. 11
Reciprocal, p. 14
Multiplicative inverse, p. 14
Indeterminate, p. 14
Equivalent expressions, p. 15
Factor, p. 17
Equivalent equations, p. 19

Collect like terms, p. 21
Grouping symbols, p. 21
Solution set, p. 23
Empty set, p. 23
Perimeter, p. 29
Exponent, p. 34
Base, p. 34
Order of operations, p. 40
Formula, p. 44
Area, p. 44
Scientific notation, p. 49
Significant digits, p. 51

IMPORTANT PROPERTIES AND FORMULAS

Area of a rectangle: $A = lw$
Area of a square: $A = s^2$
Area of a parallelogram: $A = bh$

Area of a trapezoid: $A = \dfrac{h}{2}(b_1 + b_2)$

Area of a triangle: $A = \frac{1}{2}bh$
Area of a circle: $A = \pi r^2$
Circumference of a circle: $C = \pi d$
Volume of a cube: $V = s^3$
Volume of a right circular cylinder: $V = \pi r^2 h$
Perimeter of a square: $P = 4s$
Distance traveled: $d = rt$
Simple interest: $I = Prt$

Rules for Addition

1. *Positive numbers:* Add the numbers. The result is positive.
2. *Negative numbers:* Add absolute values. Make the answer negative.
3. *A negative and a positive number:* Subtract the smaller absolute value from the larger one. Then:
   a) If the positive number has the larger absolute value, make the answer positive.

b) If the negative number has the larger absolute value, make the answer negative.
c) If the numbers have the same absolute value, the answer is 0.
4. *One number is zero:* The sum is the other number.

_____

Rules for Multiplication

1. To multiply two numbers with *unlike signs*, multiply their absolute values. The answer is *negative*.
2. To multiply two numbers with the *same sign*, multiply their absolute values. The answer is *positive*.

_____

Rules for Division

1. To divide two numbers with *unlike signs*, divide their absolute values. The answer is *negative*.
2. To divide two numbers with the *same sign*, divide their absolute values. The answer is *positive*.

_____

The law of opposites: $a + (-a) = 0$

The law of reciprocals: $a \cdot \dfrac{1}{a} = 1,\ a \neq 0$

_____

Absolute value: $|x| = \begin{cases} x, & \text{if } x \geq 0, \\ -x, & \text{if } x < 0 \end{cases}$

_____

For any number a and any nonzero number b,

$$\frac{-a}{b} = \frac{a}{-b} = -\frac{a}{b}.$$

_____

Commutative laws: $a + b = b + a,$
$\qquad\qquad\qquad\quad ab = ba$

Associative laws: $a + (b + c) = (a + b) + c,$
$\qquad\qquad\qquad\quad a(bc) = (ab)c$

Distributive law: $a(b + c) = ab + ac$

_____

The addition principle for equations: If $a = b$, then $a + c = b + c$.
The multiplication principle for equations: If $a = b$, then $a \cdot c = b \cdot c$.

Five Steps for Problem Solving in Algebra

1. *Familiarize* yourself with the problem situation.
2. *Translate* to mathematical language.
3. *Carry out* some mathematical manipulation.
4. *Check* your possible answer in the original problem.
5. *State* the answer clearly.

Definitions and Rules for Exponents

For any integers m and n (assuming 0 is not raised to a nonpositive power):

Zero as an exponent: $a^0 = 1$

Negative integers as exponents: $a^{-n} = \dfrac{1}{a^n}$

Multiplying with like bases: $a^m \cdot a^n = a^{m+n}$ (Product Rule)

Dividing with like bases: $\dfrac{a^m}{a^n} = a^{m-n}$; $a \neq 0$ (Quotient Rule)

Raising a product to a power: $(ab)^n = a^n b^n$

Raising a power to a power: $(a^m)^n = a^{mn}$ (Power Rule)

Raising a quotient to a power: $\left(\dfrac{a}{b}\right)^n = \dfrac{a^n}{b^n}$; $b \neq 0$

Rules for Order of Operations

1. Do all calculations within grouping symbols before operations outside.
2. Evaluate all exponential expressions.
3. Do all multiplications and divisions in order from left to right.
4. Do all additions and subtractions in order from left to right.

To solve a formula for a given letter, identify the letter, and:

1. Multiply on both sides to clear fractions or decimals, if that is needed.
2. Collect like terms on each side where convenient.
3. Get all terms with the letter to be solved for on one side of the equation and all other terms on the other side, using the addition principle.
4. Collect like terms again, if necessary. This may require factoring.
5. Solve for the letter in question, using the multiplication principle.

Scientific notation for a number is an expression of the type $N \times 10^n$, where $1 \leq N < 10$, N is in decimal notation, and n is an integer.

REVIEW EXERCISES

The following review exercises are for practice. Answers are at the back of the book. If you miss an exercise, restudy the section indicated alongside the answer.

1. Translate to an algebraic expression: Five less than the quotient of two numbers.

2. Evaluate the expression $5xy - z$ if $x = -2$, $y = 3$, and $z = -5$.

3. Name the set consisting of the first six even natural numbers using both roster notation and set-builder notation.

4. Determine whether $\frac{1}{3}$ is a solution of the equation $3x + 2 = 3$.

Find the absolute value.

5. $|-7.3|$ 6. $|4.09|$ 7. $|0|$

Add.

8. $-9.4 + (-3.7)$ 9. $\left(-\frac{4}{5}\right) + \left(\frac{1}{7}\right)$
10. $\left(-\frac{1}{3}\right) + \frac{4}{5}$

11. Find $-a$ if $a = -4.01$.

Subtract.

12. $-7.9 - 3.6$ 13. $-\frac{2}{3} - \left(-\frac{1}{2}\right)$
14. $12.5 - 17.9$

Multiply.

15. $(-2.1)(-3)$ 16. $\left(-\frac{2}{3}\right)\left(\frac{5}{8}\right)$
17. $(1.2)(-4)$

Divide.

18. $\dfrac{-15}{-3}$ 19. $\dfrac{72.8}{-8}$ 20. $-7 \div \dfrac{4}{3}$

Use a commutative law to write an equivalent expression.

21. $5 + a$ 22. $7y$ 23. $5x + y$

Use an associative law to write an equivalent expression.

24. $(4 + a) + b$ 25. $(xy)7$

26. Obtain an expression that is equivalent to $7mn + 14m$ by factoring.

27. Obtain an expression that is equivalent to $3x + 7 - 2x + 5$ by collecting like terms.

28. Simplify: $7x - 4[2x + 3(5 - 4x)]$.

Solve. If the solution set is $\mathbb{R}$ or $\varnothing$, state so.

29. $x - 4.9 = 1.7$

30. $\frac{2}{3}a = 9$

31. $-9x + 4(2x - 3) = 5(2x - 3) + 7$

32. $4 + 3x = 2 + 3x + 1$

33. $3(x - 4) + 2 = x + 2(x - 5)$

34. Translate to an equation. Do not solve. 13 less than twice a number is 21.

35. One number is less than another number by 17. The sum of the numbers is 115. What is the smaller number?

36. One angle of a triangle is three times as great as a second angle. The third angle is twice as great as the second angle. Find the measures of the angles.

37. Multiply and simplify: $(5a^2b^7)(-2a^3b)$.

38. Divide and simplify: $\dfrac{12x^3y^8}{3x^2y^2}$.

39. Evaluate a^0 when $a = 17.2$.

Simplify. When negative exponents appear in the answer, write a second answer using only positive exponents.

40. $3^{-4} \cdot 3^7$ 41. $(5a^2)^3$

42. $(-2a^{-3}b^2)^{-3}$ 43. $\left(\dfrac{x^2y^3}{z^4}\right)^{-2}$

44. $\left(\dfrac{2a^{-2}b}{4a^3b^{-3}}\right)^4$

Simplify.

45. $\dfrac{7(5 - 2 \cdot 3) - 3^2}{4^2 - 3^2}$

46. $1 - (2 - 5)^2 + 5 \div 10 \cdot 4^2$

47. Solve for m: $P = m/S$.

48. Solve for x: $c = mx - rx$.

49. The volume of a film canister is 62.8 cm^3. If the canister is 5 cm tall, determine its radius.

50. One *parsec* (a unit that is used in astronomy) is $30,860,000,000,000$ km. Write scientific notation for this number.

51. Convert 0.000000103 to scientific notation.

Simplify and write scientific notation for the answer. Use the correct number of significant digits.

52. $(8.7 \times 10^{-9}) \times (4.3 \times 10^{15})$

53. $\dfrac{1.2 \times 10^{-12}}{6.1 \times 10^{-7}}$

54. A sheet of plastic has a thickness of 0.00015 mm. The sheet is 1.2 m wide and 79 m long. Use scientific notation to find the volume of the sheet of plastic.

Synthesis

55. ◈   Describe a method that could be used to write equations that have no solution.

56. ◈   Explain how you might determine whether a letter in an expression is a variable or a constant.

57. If the smell of gasoline is detectable at 3 parts per billion, what percent of the air is occupied by the gasoline?

58. Evaluate $a + b(c - a^2)^0 + (abc)^{-1}$ when $a = 2$, $b = -3$, and $c = -4$.

59. What's a better deal: a 13-in. diameter pizza for \$5 or a 17-in. diameter pizza for \$8? Explain.

60. The surface area of a cube is 486 cm². Find the volume of the cube.

61. Solve for z:
$$m = \frac{x}{y - z}.$$

62. Simplify:
$$\frac{(3^{-2})^a \cdot (3^b)^{-2a}}{(3^{-2})^b \cdot (9^{-b})^{-3a}}.$$

63. Each of a student's test scores is three times as important as each of a student's quiz scores. If after 4 quizzes a student's average is 82.5, what score is needed on a test in order to raise the average to 85?

64. Fill in the following blank so as to assure an infinite number of solutions.
$$5x - 7(x + 3) - 4 = 2(7 - x) + \_\_\_\_$$

65. Fill in the following blank so as to assure that no solution exists.
$$20 - 7[3(2x + 4) - 10] = 9 - 2(x - 5) + \_\_\_\_$$

66. Use the commutative law for addition once and the distributive law twice to show that
$$a2 + cb + cd + ad = a(d + 2) + c(b + d).$$

67. Find an irrational number between $\frac{1}{2}$ and $\frac{3}{4}$.

CHAPTER TEST   1

1. A student's scores on five tests are 94, 80, 76, 91, and 75. What must the score be on the sixth test so that the average will be 85?

2. One number is 27.4 less than another number. The sum of the numbers is 83.6. What are the numbers?

Add.

3. $-25 + (-16)$

4. $-10.5 + 6.8$

5. $6.21 + (-8.32)$

Subtract.

6. $29.5 - 43.7$

7. $-17.8 - 25.4$

8. $-\dfrac{7}{3} - \left(-\dfrac{3}{4}\right)$

Multiply.

9. $-6.4(5.3)$

10. $8.4(-10.1)$

11. $-\dfrac{2}{7}\left(-\dfrac{5}{14}\right)$

Divide.

12. $\dfrac{-42.6}{-7.1}$

13. $\dfrac{-9.1}{-13}$

14. $\dfrac{2}{5} \div \left(-\dfrac{3}{10}\right)$

Collect like terms.

15. $4y - 15y + 19y$

16. $35a + 16b - 24a - 43b$

Multiply.

17. $-11(3c - 4d)$ **18.** $4x(-6y - 7z)$

Solve.

19. $13x - 7 = 41x + 49$ **20.** $9y - (5y - 3) = 33$

21. Find three consecutive odd integers such that the sum of four times the first, three times the second, and two times the third is 167.

Simplify. When negative exponents appear in the answer, write a second answer using only positive exponents.

22. $-5(x - 4) - 3(x + 7)$

23. $6b - [7 - 2(9b - 1)]$

24. $(12x^{-4}y^{-7})(-6x^{-6}y)$

25. $5^{-1} + 2 \cdot 3^{-1}$

26. $(-6x^2y^{-4})^{-2}$

27. $\left(\dfrac{2x^3y^{-6}}{-4y^{-2}}\right)^2$

28. $(5x^3y)^0$

29. $5 + (1 - 3)^2 - 7 \div 2^2 \cdot 6$

Simplify and write scientific notation for the answer. Use the correct number of significant digits.

30. $(2.9 \times 10^8)(6.1 \times 10^{-4})$

31. $(9.05 \times 10^{-3})(2.22 \times 10^{-5})$

32. $\dfrac{3.6 \times 10^7}{7.2 \times 10^{-3}}$ **33.** $\dfrac{1.8 \times 10^{-4}}{4.8 \times 10^{-7}}$

review

Solve.

34. The average distance from the planet Venus to the sun is 6.7×10^7 mi. About how far does Venus travel in one orbit around the sun?

35. $\dfrac{P_1 V_1}{T_1} = \dfrac{P_2 V_2}{T_2}$, for V_2

Synthesis

Simplify.

36. $(4x^{3a}y^{b+1})^{2c}$ **37.** $\dfrac{-27a^{x+1}}{3a^{x-2}}$

38. $\dfrac{(-16x^{x-1}y^{y-2})(2x^{x+1}y^{y+1})}{(-7x^{x+2}y^{y+2})(8x^{x-2}y^{y-1})}$

CHAPTER 2

Graphs, Functions, and Linear Equations

AN APPLICATION

A gift shop experiencing constant growth totalled $250,000 in sales in 1988 and $285,000 in 1993. Use a graph that displays the shop's total sales as a function of time to predict total sales for the year 1997.

This problem appears as Exercise 43 in Section 2.2.

Terry Palatino
GIFTSHOP OWNER

"I can't see owning a business without having a strong background in math. Inventory, payroll, taxes, balance sheets and sales projections all involve math. The sales clerks and managers who have a solid math background learn faster and make fewer mistakes."

Graphs are important because they allow us to *see* relationships. For example, a graph of an equation in two variables helps us see how those two variables are related. Graphs are useful in problem solving, and in other ways as well. In this chapter, you will take a close look at graphs of equations.

A certain kind of relationship between two variables is known as a *function*. Functions are very important in mathematics in general, and in problem solving in particular. You will learn what we mean by a function and then begin to use functions to solve problems.

In addition to the material in this chapter, the review and test for Chapter 2 include material from Sections 1.2, 1.3, 1.5, and 1.7.

2.1

Graphs

Points and Ordered Pairs   •   Quadrants   •   Solutions of Equations   •   Nonlinear Equations

It has often been said that a picture is worth a thousand words. As we turn our attention to the study of graphs, we discover that in mathematics this is quite literally the case. Graphs are a compact means of displaying information and provide a pictorial way to solve problems.

Points and Ordered Pairs

On a number line, each point corresponds to a number. On a plane, each point corresponds to a pair of numbers. The idea of using two perpendicular number lines, called **axes,** to identify points in a plane is commonly attributed to the great French mathematician and philosopher René Descartes (1596–1650). Because the variable x is most frequently represented on the horizontal axis and the variable y is most commonly represented on the vertical axis, we often refer to the **x, y coordinate system.** In honor of Descartes, this representation is also called the **Cartesian coordinate system.**

Plotting Points

Notice on the figure at right that (2, 3) and (3, 2) give different points. These pairs of numbers are called **ordered pairs** because the order in which the numbers are listed is important. The ordered pair (0, 0) is called the **origin.**

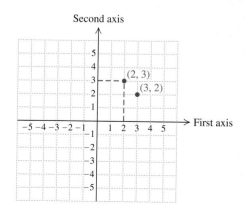

EXAMPLE 1

Plot the points $(-4, 3)$, $(-5, -3)$, $(0, 4)$, and $(2.5, 0)$.

Solution   To plot $(-4, 3)$, we note that the first number, -4, tells us the distance in the first, or horizontal, direction. We go 4 units *left*. The second number tells us the distance in the second, or vertical, direction. We go 3 units *up*. The point $(-4, 3)$ is then marked, or "plotted."

The points $(-5, -3)$, $(0, 4)$, and $(2.5, 0)$ are also plotted below.

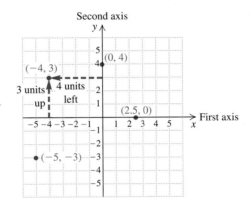

The numbers in an ordered pair are called **coordinates.** In $(-4, 3)$, the *first coordinate* is -4 and the *second coordinate** is 3.

Quadrants

The axes divide the plane into four regions called **quadrants,** as shown here.

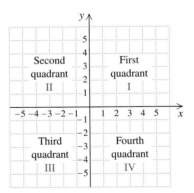

In region I (the *first* quadrant), both coordinates of a point are positive. In region II (the *second* quadrant), the first coordinate is negative and the second coordinate is positive. In the third quadrant, both coordinates are negative, and in the fourth quadrant, the first coordinate is positive while the second coordinate is negative.

Points with one or more 0's as coordinates, such as $(0, -6)$, $(4, 0)$, and $(0, 0)$, are on axes and *not* in quadrants.

---

*The first coordinate is sometimes called the **abscissa** and the second coordinate is called the **ordinate.**

Solutions of Equations

If an equation has two variables, its solutions are pairs of numbers. To determine which part of the pair is listed first, we usually begin with the variable that occurs first alphabetically.

EXAMPLE 2

Determine whether the pairs $(4, 2)$, $(-1, -4)$, and $(2, 5)$ are solutions of the equation $y = 3x - 1$.

Solution   To determine whether each pair is a solution, we replace x by the first coordinate and y by the second coordinate. When the replacements make the equation true, we say that the ordered pair is a solution.

$$\begin{array}{c|c}
\multicolumn{2}{l}{y = 3x - 1} \\
\hline
2 \ ? & 3(4) - 1 \\
 & 12 - 1 \\
2 & 11
\end{array}
\qquad
\begin{array}{c|c}
\multicolumn{2}{l}{y = 3x - 1} \\
\hline
-4 \ ? & 3(-1) - 1 \\
 & -3 - 1 \\
-4 & -4
\end{array}
\qquad
\begin{array}{c|c}
\multicolumn{2}{l}{y = 3x - 1} \\
\hline
5 \ ? & 3(2) - 1 \\
 & 6 - 1 \\
5 & 5
\end{array}$$

Since $2 = 11$ is *false*, the pair $(4, 2)$ is *not* a solution.

Since $-4 = -4$ is *true*, the pair $(-1, -4)$ *is* a solution.

Since $5 = 5$ is *true*, the pair $(2, 5)$ *is* a solution.   ❑

In fact, there are infinitely many solutions of the equation $y = 3x - 1$. Rather than attempt to list all these solutions, we will use a graph as a convenient representation of such a large set of solutions. Thus to *graph* an equation means to make a drawing that represents its solutions.

EXAMPLE 3

Graph the equation $y = x$.

Solution   We label the horizontal axis as the x-axis and the vertical axis as the y-axis.

Next, we find some ordered pairs that are solutions of the equation. In this case, it is easy. Here are a few pairs that satisfy the equation $y = x$:

$$(0, 0), \qquad (1, 1), \qquad (5, 5), \qquad (-1, -1), \qquad (-6, -6).$$

Now we plot these points. We can see that if we were to plot a million solutions, the dots that we drew would merge into a solid line. Observing the pattern, we can draw the line with a ruler. The line is the graph of the equation $y = x$. We label the line $y = x$.

Note that the coordinates of *any* point on the line—for example, $(2.5, 2.5)$—satisfy the equation $y = x$.

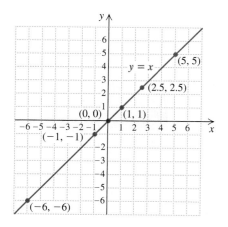

❑

EXAMPLE 4

Graph the equation $y = 2x$.

Solution   We find some ordered pairs that are solutions. This time we list the pairs in a table. To find an ordered pair, we can choose *any* number for x and then determine y. For example, if we choose 3 for x, then $y = 2 \cdot 3 = 6$ (substituting into the equation $y = 2x$). We make some negative choices for x, as well as some positive ones. If a number takes us off the graph paper, we generally do not use it. Next, we plot these points. If we plotted *many* such points, they would appear to make a solid line. We draw the line with a ruler and label it $y = 2x$.

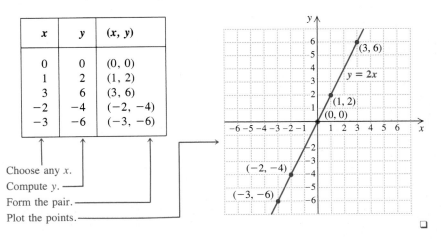

x	y	(x, y)
0	0	$(0, 0)$
1	2	$(1, 2)$
3	6	$(3, 6)$
-2	-4	$(-2, -4)$
-3	-6	$(-3, -6)$

Choose any x.
Compute y.
Form the pair.
Plot the points.

EXAMPLE 5

Graph the equation $y = -\frac{1}{2}x$.

Solution   To find an ordered pair, we choose any convenient number for x and then determine y. For example, if we choose 4 for x, we get $y = \left(-\frac{1}{2}\right)(4)$, or -2. When we choose -6 for x, we get $y = \left(-\frac{1}{2}\right)(-6)$, or 3. We find several ordered pairs, plot them, and draw the line.

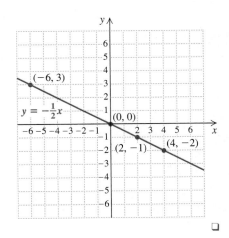

x	y	(x, y)
4	-2	$(4, -2)$
-6	3	$(-6, 3)$
0	0	$(0, 0)$
2	-1	$(2, -1)$

As you can see, the graphs in Examples 3–5 are straight lines. We will refer to any equation whose graph is a straight line as a **linear equation.** Linear equations will be discussed in more detail in Sections 2.3–2.5.

Nonlinear Equations

There are many equations whose graphs are not straight lines. Let's look at some of these **nonlinear equations.**

EXAMPLE 6

Graph: $y = |x|$.

Solution   We select numbers for x and find the corresponding values for y. For example, if we choose -1 for x, we get $y = |-1| = 1$. Several ordered pairs are listed in the table below.

x	y
-3	3
-2	2
-1	1
0	0
1	1
2	2
3	3

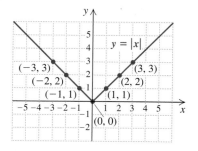

We plot these points, noting that the absolute value of a positive number is the same as the absolute value of its opposite. Thus the x-values 3 and -3 both are paired with the y-value 3. Note that the graph is V-shaped, centered at the origin.   ❏

EXAMPLE 7

Graph: $y = 1/x$.

Solution   We select x-values and find the corresponding y-values. The table lists the ordered pairs $\left(2, \frac{1}{2}\right)$, $\left(-2, -\frac{1}{2}\right)$, $\left(\frac{1}{2}, 2\right)$, and so on.

x	y
3	$\frac{1}{3}$
2	$\frac{1}{2}$
1	1
$\frac{1}{2}$	2
$-\frac{1}{2}$	-2
-1	-1
-2	$-\frac{1}{2}$
-3	$-\frac{1}{3}$

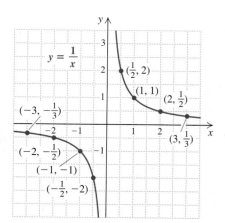

We plot these points, noting that each first coordinate is paired with its reciprocal. Since 1/0 is undefined, we cannot use 0 as a first coordinate. Thus there are two "branches" to this graph—one on either side of the y-axis. Note that for x-values far to the right or far to the left of 0, the graph approaches, but does not touch, the x-axis.   ❏

EXAMPLE 8

Graph: $y = x^2 - 5$.

Solution We select numbers for x and find the corresponding values for y. For example, if we choose -2 for x, we get $y = (-2)^2 - 5 = 4 - 5 = -1$. The table lists several ordered pairs.

x	y
0	-5
-1	-4
1	-4
-2	-1
2	-1
-3	4
3	4

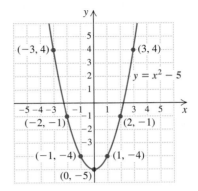

Next, we plot these points. Note that as the absolute value of x increases, $x^2 - 5$ also increases. Thus the graph is a curve that rises on either side of the y-axis, as shown in the figure. ❑

 Calculators can be used to speed our search for ordered pairs, especially when we are uncertain about the shape of a graph. Very complicated graphs are often drawn with the aid of a computer or a graphing calculator. As you may have discovered, determining just a few ordered pairs that solve an equation can be quite time-consuming. Computers can calculate large numbers of ordered pairs in very little time (and without complaining!) when properly programmed.

TECHNOLOGY CONNECTION

Beginning in this chapter, we include in some sections and exercise sets activities that utilize graphing calculators or computer graphing software. Such calculators and software will be referred to simply as *graphers*. Most activities will use only basic features common to virtually all graphers. All activities are presented in a generic form—check your user's manual or ask your instructor for the exact procedures.

One feature common to all graphers is the *window*. This refers to the rectangular portion of the screen in which a graph appears. Windows are described by four numbers, [L, R, B, T], that represent the *L*eft and *R*ight endpoints of the x-axis and the *B*ottom and *T*op endpoints of the y-axis. A Range key is sometimes used to set these dimensions.

The primary use for graphers is graphing equations. For example, let's graph the equation $y = -4x + 3$. Selecting the window $[-10, 10, -10, 10]$ results in the graph shown at right.

Once you have graphed an equation on a grapher, you can investigate some of its points by using the Trace feature

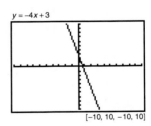

$y = -4x + 3$

$[-10, 10, -10, 10]$

that most graphers offer. Usually, a Trace key is pressed to access this feature. Once it is pressed, a cursor (often blinking) appears somewhere on the graph while its x- and y-coordinates are shown elsewhere on the screen. These coordinates will change as the cursor moves along the graph.

Graph each of the following equations using a $[-10, 10, -10, 10]$ window. Then use the Trace function to find the coordinates of at least three points on the graph.

TC1. $y = 5x - 3$

TC2. $y = x^2 - 4x + 3$

TC3. $y = (x + 4)^2$

EXERCISE SET | 2.1

Plot the points. Label each point with the indicated letter.

1. $A(5, 3)$, $B(2, 4)$, $C(0, 2)$, $D(0, -6)$, $E(3, 0)$, $F(-2, 0)$, $G(1, -3)$, $H(-5, 3)$, $J(-4, 4)$

2. $A(3, 5)$, $B(1, 5)$, $C(0, 4)$, $D(0, -4)$, $E(5, 0)$, $F(-5, 0)$, $G(1, -5)$, $H(-7, 4)$, $J(-5, 5)$

3. $A(3, 0)$, $B(4, 2)$, $C(5, 4)$, $D(6, 6)$, $E(3, -4)$, $F(3, -3)$, $G(3, -2)$, $H(3, -1)$

4. $A(1, 1)$, $B(2, 3)$, $C(3, 5)$, $D(4, 7)$, $E(-2, 1)$, $F(-2, 2)$, $G(-2, 3)$, $H(-2, 4)$, $J(-2, 5)$, $K(-2, 6)$

5. Plot the points $M(2, 3)$, $N(5, -3)$, and $P(-2, -3)$. Draw $\overline{MN}$, $\overline{NP}$, and $\overline{MP}$. ($\overline{MN}$ means the line segment from M to N). What kind of geometric figure is formed? What is its area?

6. Plot the points $Q(-4, 3)$, $R(5, 3)$, $S(2, -1)$, and $T(-7, -1)$. Draw $\overline{QR}$, $\overline{RS}$, $\overline{ST}$, and $\overline{TQ}$. What kind of figure is formed? What is its area?

In which quadrant is the point located?

7. $(-3, -5)$
8. $(2, 17)$
9. $(-6, 1)$
10. $(4, -8)$
11. $(3, \frac{1}{2})$
12. $(-1, -8)$
13. $(7, -0.2)$
14. $(-4, 31)$

Determine whether the ordered pair is a solution of the equation.

15. $(1, -1)$; $y = 2x - 3$
16. $(2, 5)$; $y = 3x - 1$
17. $(3, 4)$; $3s + t = 4$
18. $(2, 3)$; $2p + q = 5$
19. $(3, 5)$; $4x - y = 7$
20. $(2, 7)$; $5x - y = 3$
21. $(0, \frac{3}{5})$; $2a + 5b = 3$
22. $(0, \frac{3}{2})$; $3f + 4g = 6$
23. $(2, -1)$; $4r + 3s = 5$
24. $(2, -4)$; $5w + 2z = 2$
25. $(3, 2)$; $3x - 2y = -4$
26. $(1, 2)$; $2x - 5y = -6$
27. $(-1, 3)$; $y = 3x^2$
28. $(2, 4)$; $2r^2 - s = 5$
29. $(2, 3)$; $5s^2 - t = 7$
30. $(2, 3)$; $y = x^3 - 5$

Graph.

31. $y = -2x$
32. $y = -\frac{1}{2}x$
33. $y = x + 3$
34. $y = x - 2$
35. $y = 3x - 2$
36. $y = -4x + 1$
37. $y = -2x + 3$
38. $y = -3x + 1$

39. $y = \frac{2}{3}x + 1$
40. $y = \frac{1}{3}x + 2$
41. $y = -\frac{3}{2}x + 1$
42. $y = -\frac{2}{3}x - 2$
43. $y = \frac{3}{4}x + 1$
44. $y = x^2$
45. $y = -x^2$
46. $y = x^2 + 2$
47. $y = x^2 - 2$
48. $x = y^2 + 2$
49. $y = |x| + 2$
50. $y = -|x|$
51. $y = 3 - x^2$
52. $y = x^3 - 2$
53. $y = -\dfrac{1}{x}$
54. $y = \dfrac{3}{x}$

Skill Maintenance

55. A garden is being constructed in the shape of a triangle. One side is 12 ft long. How tall should the triangle be in order to make the area of the garden 156 ft²?

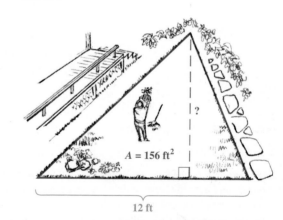

$A = 156$ ft²

12 ft

56. What rate of interest is required in order for a principal of $320 to earn $17.60 in half a year?

57. Subtract: $-3.9 - (-2.5)$.

58. Simplify: $\dfrac{-3 - 7}{2 - 4}$.

Synthesis

59. ◈ Using the equation $y = |x|$, explain why it is "dangerous" to draw a graph after plotting just two points.

60. ◈ Without making a drawing, explain why the graph of the equation $y = x - 30$ passes through three quadrants.

61. ◈ Graph $y = 6x$, $y = 3x$, $y = \frac{1}{2}x$, $y = -6x$, $y = -3x$, and $y = -\frac{1}{2}x$ using the same set of axes, and compare the slants of the lines. Describe the pattern that relates the slant of the line to the multiplier of x.

62. If $(-10, -2)$, $(-3, 4)$, and $(6, 4)$ are the coordinates of three consecutive vertices of a parallelogram, what are the coordinates of the fourth vertex?

63. Which of the following equations have $\left(-\frac{1}{3}, \frac{1}{4}\right)$ as a solution?

a) $-\frac{3}{2}x - 3y = -\frac{1}{4}$
b) $8y - 15x = \frac{7}{2}$
c) $0.16y = -0.09x + 0.1$
d) $2(-y + 2) - \frac{1}{4}(3x - 1) = 4$

▨ Graph the equation after plotting at least 10 points.

64. $y = x^3 + 3x^2 + 3x + 1$; use values of x from -3 to 1

65. $y = 1/(x - 2)$; use values of x from -1 to 5

66. $y = 1/x^2$; use values of x from -3 to 3

67. $y = -1/x^2$; use values of x from -3 to 3

68. If $(-1, 1)$ and $(4, -4)$ are the endpoints of a diagonal of a square, what are the coordinates of the other two vertices of the square?

69. ◈ Using the same set of axes, graph $y = 2x$, $y = 2x - 3$, and $y = 2x + 3$. Describe the pattern that relates each line to the number that is added to $2x$.

▨ In Exercises 70 and 71, use a grapher to draw the graph of each equation. For each equation, select a window that shows the curvature of the graph.

70. **a)** $y = -12.4x + 7.8$
b) $y = -3.5x^2 + 6x - 8$
c) $y = (x - 3.4)^3 + 5.6$

71. **a)** $y = 2.3x^4 + 3.4x^2 + 1.2x - 4$
b) $y = 12.3x - 3.5$
c) $y = 3(x + 2.3)^2 + 2.3$

2.2

Functions

Notation for Functions • **Functions and Graphs** •
The Vertical-Line Test

We now develop the idea of a *function*—one of the most important concepts in mathematics. In much the same way that the ordered pairs of Section 2.1 formed correspondences between first coordinates and second coordinates, a function is a correspondence from one set to another. For example:

To each person in a class	there corresponds	his or her mother.
To each item in a store	there corresponds	its price.
To each real number	there corresponds	the cube of that number.

In each example, the first set is called the **domain.** The second set is called the **range.** Given a member of the domain, there is *just one* member of the range to which it corresponds. This kind of correspondence is called a **function.**

EXAMPLE 1

Determine whether the correspondence is a function.

a) $a \longrightarrow 4$
     $b \longrightarrow 0$
     c

b)   San Francisco $\longrightarrow$ Giants
     New York $\longrightarrow$ Mets
     Cleveland $\longrightarrow$ Browns

Solution

a) The correspondence *is* a function because each member of the domain corresponds to just one member of the range.
b) The correspondence *is not* a function because a member of the domain (New York) corresponds to more than one member of the range. ❑

Function

A *function* is a correspondence between a first set, called the *domain,* and a second set, called the *range,* such that each member of the domain corresponds to *exactly one* member of the range.

EXAMPLE 2

Determine whether the correspondence is a function.

Domain	Correspondence	Range
a) A family	Each person's weight	A set of positive numbers
b) The natural numbers	Each number's square	A set of natural numbers
c) The set of all states	Each state's members of the U.S. Senate	A set of U.S. senators

Solution

a) The correspondence *is* a function, because each person has *only one* weight.
b) The correspondence *is* a function, because each natural number has *only one* square.
c) The correspondence *is not* a function, because each state has two U.S. senators. ❑

When a correspondence between two sets is not a function, it is still an example of a **relation.**

Relation

A *relation* is a correspondence between a first set, called the *domain,* and a second set, called the *range,* such that each member of the domain corresponds to *at least one* member of the range.

Thus, although the correspondences of Examples 1 and 2 are not all functions, they *are* all relations. A function is a special type of relation — one in which each member of the domain is paired with *exactly one* member of the range.

Notation for Functions

To understand function notation, it helps to imagine a "function machine." Think of putting a member of the domain (an *input*) into the machine. The machine knows the correspondence and gives you a member of the range (the *output*).

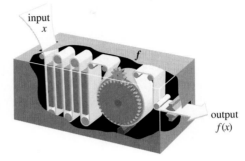

The function has been named f. We call the input x, and its output $f(x)$. This is read "f of x," or "f at x" or "the value of f at x." Note that $f(x)$ does *not* mean "f times x."

Most functions are described by equations. For example, $f(x) = 2x + 3$ describes the function that takes an input x, multiplies it by 2, and then adds 3.

$$f(x) \;=\; \underset{\text{Double}}{2x} \;\; \underset{\text{Add 3}}{+\,3}$$
Input

To find the output $f(4)$, we take the input 4, double it, and add 3 to get 11. That is, we substitute 4 into the formula for $f(x)$:

$$f(4) = 2 \cdot 4 + 3$$
$$= 11.$$

Sometimes, instead of writing $f(x) = 2x + 3$, we might write $y = 2x + 3$, where it is understood that the value of y, the *dependent variable*, is calculated after first choosing a value for x, the *independent variable*. To understand why $f(x)$ notation is so useful, consider two equivalent statements:

a) If $f(x) = 2x + 3$, then $f(4) = 11$.
b) If $y = 2x + 3$, then the value of y is 11 when x is 4.

The notation used in part (a) is far more concise.

EXAMPLE 3

Find the indicated function value.

a) $f(5)$, for $f(x) = 3x + 2$ **b)** $g(3)$, for $g(z) = 5z^2 - 4$
c) $A(-2)$, for $A(r) = 3r^2 + 2r$ **d)** $F(a + 1)$, for $F(x) = 3x + 2$

Solution

a) $f(5) = 3 \cdot 5 + 2 = 17$
b) $g(3) = 5(3)^2 - 4 = 41$
c) $A(-2) = 3(-2)^2 + 2(-2) = 8$
d) $F(a + 1) = 3(a + 1) + 2 = 3a + 3 + 2 = 3a + 5$

When all of a function's inputs share one output value, we say that we have a *constant function*. Thus, for the constant function $h(x) = 7$, we have $h(3) = 7$ and $h(5) = 7$.

Note that whether we write $f(x) = 3x + 2$, or $f(t) = 3t + 2$, or $f(\Box) = 3\Box + 2$, we still have $f(5) = 17$. Thus the independent variable can be thought of as a *dummy variable*. The letter chosen for the dummy variable is not as important as the algebraic manipulations to which it is subjected.

Although you have probably already used functions that are described by formulas, you may not have seen function notation in those applications. For example, the formula for finding the area A of a circle with radius r is

$$A = \pi r^2.$$

We say that the area A is *a function of r*. To emphasize that fact, we often write

$$A(r) = \pi r^2.$$

Thus, to find the area of a circle with a radius of 12 cm, we might write

$$A(12) = \pi (12)^2 \qquad \text{Substituting for } r$$
$$= \pi \times 144$$
$$\approx 452 \text{ cm}^2. \qquad \text{Using 3.14 for } \pi$$

Functions and Graphs

Functions are often described by graphs. To use a graph in problem solving, we note that each point on the graph represents a pair of values — one from the horizontal axis (the domain) and one from the vertical axis (the range). In the following example, we first draw a graph and then estimate a function value from it.

EXAMPLE 4

According to the Federal Center for Disease Control, there were 309 newly reported cases of AIDS in the United States in 1981, 4436 cases in 1984, 21,114 cases in 1987, and 43,339 cases in 1990. Estimate the number of newly reported cases for the years 1983 and 1988.

Solution

1. and **2.** FAMILIARIZE and TRANSLATE. The given information enables us to plot four points on a graph. We let the horizontal axis represent the year and the vertical axis the number of reported cases.

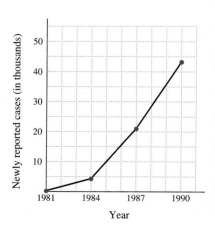

3. CARRY OUT. To estimate the reported number of cases in 1983, we locate the point that is directly above the year 1983. After doing so, we estimate its second coordinate by moving horizontally from the point to the vertical axis.

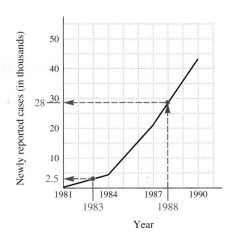

We can estimate from the graph that the function value is 2.5 thousand. Following a similar procedure, we estimate that 1988 is paired with 28 thousand.

4. CHECK. Although a precise check is impossible, note that 2.5 thousand is between 309 and 4436. Similarly, 28 thousand is between 21,114 and 43,339. Thus both answers are at least plausible.

5. STATE. There were about 2500 newly reported cases of AIDS in 1983 and about 28,000 newly reported cases in 1988.   ❑

The Vertical-Line Test

When we are graphing functions, the domain consists of all values on the horizontal axis that serve as a first coordinate for some point on the graph. The range consists of all values on the vertical axis that serve as a second coordinate for some point on the graph. If any value on the horizontal axis is the first coordinate of more than one point on the graph, the graph cannot represent a function (otherwise one member of the domain would correspond to more than one member of the range). This observation is the basis of the *vertical-line test*.

The Vertical-Line Test

A graph is that of a function provided it is not possible to draw a vertical line that intersects the graph more than once.

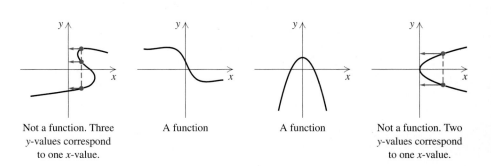

Not a function. Three *y*-values correspond to one *x*-value.   A function   A function   Not a function. Two *y*-values correspond to one *x*-value.

EXERCISE SET | 2.2

Determine whether the correspondence is a function.

1. $a \longrightarrow P$
$b \longrightarrow Q$
$c \longrightarrow R$
$d \longrightarrow S$
$ \longrightarrow T$

2. $1 \longrightarrow a$
$2 \longrightarrow b$
$3 \longrightarrow c$
$4 \longrightarrow d$
5

3.

Firm	Number of Partners
Brown & Jones	850
Smith & Hawkens	850
Hernandez & Rowle	1900
Ciani & Ross	1270

4.

Firm	Number of Female Partners
Brown & Jones	38
Smith & Hawkens	44
Hernandez & Rowle	38
Ciani & Ross	27
Morong & Davis	54

5.
cat — dog
fish — worm
dog — cat
tiger — fish
teacher — student

6. $A \longrightarrow a$
$B \longrightarrow b$
$C \longrightarrow c$
$D \longrightarrow d$
$ \longrightarrow e$

7.

Pitcher	Favorite Pitch
Viola	Slider
Hammaker	Fastball
Worrell	Curveball
Morris	

8.

Site	Year
Lake Placid	1980
Oslo	1976
Squaw Valley	1960
Innsbruck	1952
	1932

9. $a \longrightarrow M$
$b \longrightarrow N$
$c \longrightarrow O$
$d \longrightarrow P$
e

10.
Jack — John
Kate — James
Marnie — Katherine
Jim — Margaret
Peggy

Determine whether each of the following is a function. Identify any relations that are not functions.

	Domain	Correspondence	Range
11.	A math class	Each person's seat number	A set of numbers
12.	A set of numbers	Square each number and then add 4.	A set of numbers
13.	A set of shapes	Find the area of each shape.	A set of numbers
14.	A family	Each person's eye color	A set of colors
15.	The people in a town	Each person's aunt	A set of females
16.	A set of avenues	Find an intersecting road.	A set of cross streets

Find the function values.

17. $g(x) = x + 1$
    a) $g(0)$ **b)** $g(-4)$ **c)** $g(-7)$
    d) $g(8)$ **e)** $g(a + 2)$

18. $h(x) = x - 4$
    a) $h(4)$ **b)** $h(8)$ **c)** $h(-3)$
    d) $h(-4)$ **e)** $h(a - 1)$

19. $f(n) = 5n^2 + 4$
    a) $f(0)$ **b)** $f(-1)$ **c)** $f(3)$
    d) $f(t)$ **e)** $f(2a)$

20. $g(n) = 3n^2 - 2$
    a) $g(0)$ **b)** $g(-1)$ **c)** $g(3)$
    d) $g(t)$ **e)** $g(2a)$

21. $g(r) = 3r^2 + 2r - 1$
    a) $g(2)$ **b)** $g(3)$ **c)** $g(-3)$
    d) $g(1)$ **e)** $g(3r)$

22. $h(r) = 4r^2 - r + 2$

  a) $h(3)$ **b)** $h(0)$ **c)** $h(-1)$
  d) $h(-2)$ **e)** $h(3r)$

23. $f(x) = \dfrac{x - 3}{2x - 5}$

  a) $f(0)$ **b)** $f(4)$ **c)** $f(-1)$
  d) $f(3)$ **e)** $f(x + 2)$

24. $s(x) = \dfrac{3x - 4}{2x + 5}$

  a) $s(10)$ **b)** $s(2)$ **c)** $s\left(-\frac{5}{2}\right)$
  d) $s(-1)$ **e)** $s(x + 3)$

The function A described by $A(s) = s^2 \dfrac{\sqrt{3}}{4}$ gives the area of an equilateral triangle with side s.

s

25. Find the area when a side measures 4 cm.

26. Find the area when a side measures 6 in.

The function V described by $V(r) = 4\pi r^2$ gives the surface area of a sphere with radius r.

27. Find the area when the radius is 3 in.

28. Find the area when the radius is 5 cm.

The function F described by $F(C) = \frac{9}{5}C + 32$ gives the Fahrenheit temperature corresponding to the Celsius temperature C.

29. Find the Fahrenheit temperature equivalent to $-10°C$.

30. Find the Fahrenheit temperature equivalent to $5°C$.

The function H described by $H(x) = 2.75x + 71.48$ can be used to predict the height, in centimeters, of a woman whose *humerus* (the bone from the elbow to the shoulder) is x cm long. Predict the height of a woman whose humerus is the length given.

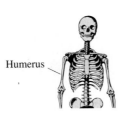

Humerus

31. 32 cm   **32.** 35 cm

For Exercises 33 and 34, use the following graph, which shows the annual heart attack rate per 10,000 men as a function of blood cholesterol level.*

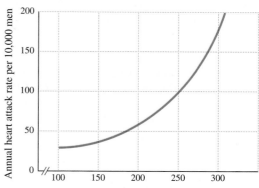

Blood cholesterol (in milligrams per deciliter)

33. Approximate the annual heart attack rate per 10,000 men for those whose blood cholesterol level is 225 mg/dl.

34. Approximate the annual heart attack rate per 10,000 men for those whose blood cholesterol level is 275 mg/dl.

For Exercises 35 and 36, use the following graph, which shows the number of baseball bats sold as a function of time.†

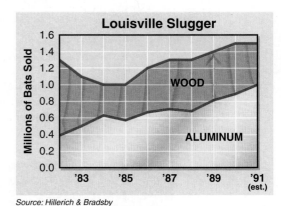

Source: Hillerich & Bradsby

35. Approximate the number of wood bats sold in 1989.

36. Approximate the number of aluminum bats sold in 1987.

---

*Copyright 1989, CSPI. Adapted from *Nutrition Action Healthletter* (1875 Connecticut Avenue, N.W., Suite 300, Washington, DC 20009-5728. $20.00 for 10 issues).
†*The New York Times*, 7/7/91. Copyright © 1991 by The New York Times Company. Reprinted with permission.

The following table can be used to predict the number of drinks required for a person of a specified weight to be legally intoxicated (blood alcohol level of 0.08 or above) in Vermont. One 12-oz glass of beer, a 5-oz glass of wine, or a cocktail containing 1 oz of a distilled liquor all count as one drink. Assume that all drinks are consumed within one hour.

Input, Body Weight (in Pounds)	Output, Number of Drinks
100	2.5
160	4
180	4.5
200	5

37. Use the table above to draw a graph and to estimate the number of drinks that a 140-lb person would have to drink to be considered intoxicated.

38. Use the graph from Exercise 37 to estimate the number of drinks a 120-lb person would have to drink to be considered intoxicated.

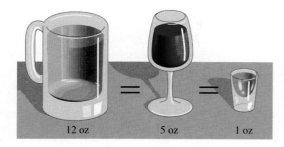

12 oz   5 oz   1 oz

Jamaal buys a video cassette recorder on which there is a revolution counter. There is also a booklet with a table that relates the counter reading and the time for which the tape has run.

Counter Reading	Time of Tape (in Hours)
000	0
300	1
500	2
675	3
800	4

39. Use the data in the table to draw a graph of the time that a tape has run as a function of the counter reading and then estimate the time elapsed when the counter has reached 600.

40. Use the graph from Exercise 39 to estimate the time elapsed when the counter has reached 200.

A city experiencing rapid growth recorded the following dates and populations.

Input, Year	Output, Population (in Tens of Thousands)
1985	5.8
1987	6
1989	7
1991	10

41. Use the data in the table to draw a graph of the population as a function of time. Then estimate what the population was in 1988.

42. Use the graph in Exercise 41 to predict the city's population in the year 1993.

43. A gift shop experiencing constant growth totalled $250,000 in sales in 1988 and $285,000 in 1993. Use a graph that displays the store's total sales as a function of time to predict total sales for the year 1997.

44. Use the graph in Exercise 43 to estimate what the total sales were in 1991.

Determine whether each of the following is the graph of a function.

45.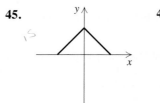

46.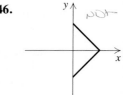

47.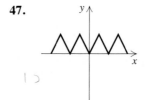

48.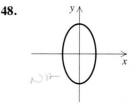

49.

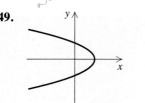

50.

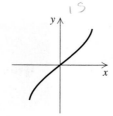

51.

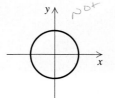

52.

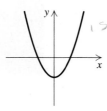

Skill Maintenance

53. Find three consecutive even integers such that the sum of the first, two times the second, and three times the third is 124.

54. The changes in the salary of a vice president of a corporation for three consecutive years are, respectively, a 10% increase, a 15% increase, and a 5% decrease. What is the percent of total change for those three years?

55. The surface area of a rectangular solid of length l, width w, and height h is given by $S = 2lh + 2lw + 2wh$. Solve for l.

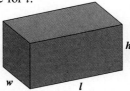

56. For what value is the expression

$$(5x - 2)/(3x - 5)$$

undefined?

Synthesis

Researchers at Yale University have suggested that the following graphs* may represent three different aspects of love.

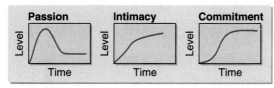

---

*From "A Triangular Theory of Love," by R. J. Sternberg, 1986, *Psychological Review*, **93**(2), 119–135. Copyright 1986 by the American Psychological Association, Inc. Reprinted by permission.

57. ◈ In what unit would you measure time if the horizontal length of each graph were ten units? Why?

58. ◈ Do you agree with the researchers that these graphs should be shaped as they are? Why or why not?

For each function, find the indicated function values.

59. $f(x) = 4.3x^2 - 1.4x$

    a) $f(1.034)$ **b)** $f(-3.441)$
    c) $f(27.35)$ **d)** $f(-16.31)$

60. $g(x) = 2.2x^3 + 3.5$

    a) $g(17.3)$ **b)** $g(-64.2)$
    c) $g(0.095)$ **d)** $g(-6.33)$

61. Suppose that a function g is such that $g(-1) = -7$ and $g(3) = 8$. Find a formula for g if $g(x)$ is of the form $g(x) = mx + b$, where m and b are constants.

62. Suppose that for some function f, $f(x - 1) = 5x$. What is $f(6)$?

63. ◈ Does the following chart constitute a function? Why or why not?

APPROXIMATE ENERGY EXPENDITURE BY A
150-POUND PERSON IN VARIOUS ACTIVITIES

Activity	Calories per Hour
Lying down or sleeping	80
Sitting	100
Driving an automobile	120
Standing	140
Domestic work	180
Walking, $2\frac{1}{2}$ mph	210
Bicycling, $5\frac{1}{2}$ mph	210
Gardening	220
Golf; lawn mowing, power mower	250
Bowling	270
Walking, $3\frac{3}{4}$ mph	300
Swimming, $\frac{1}{4}$ mph	300
Square dancing, volleyball, roller skating	350
Wood chopping or sawing	400
Tennis	420
Skiing, 10 mph	600
Squash and handball	600
Bicycling, 13 mph	660
Running, 10 mph	900

Source: Based on material prepared by Robert E. Johnson, M.D., Ph.D., and colleagues, University of Illinois.

For Exercises 64–67, use the following graph of a woman's "stress test." This graph shows the size of a pregnant woman's contractions as a function of time.

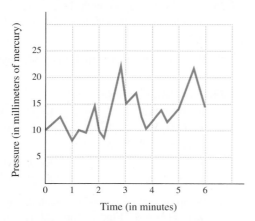

64. How large is the largest contraction that occurred during the test?

65. At what time during the test did the largest contraction occur?

66. ◈   On the basis of the information provided, how large a contraction would you expect 60 seconds from the end of the test? Why?

67. What is the frequency of the woman's largest contractions?

68. The *greatest integer function* $f(x) = [\![x]\!]$ is defined as follows: $[\![x]\!]$ is the greatest integer that is less than or equal to x. For example, if $x = 3.74$, then $[\![x]\!] = 3$; and if $x = -0.98$, then $[\![x]\!] = -1$. Graph the greatest integer function for values of x such that $-5 \leq x \leq 5$. (The notation $f(x) = \text{INT}[x]$ is used in many computer programs for the greatest integer function.)

2.3

Graphs of Linear Functions

Graphs of Equations of the Type $y = mx + b$ **or** $f(x) = mx + b$ •
Graphs of Equations of the Type $y = b$ **or** $x = a$ •   **Recognizing Linear Equations**

Different functions have different graphs. In this section, we examine functions whose graphs are straight lines. Such functions and their graphs are called *linear*.

Graphs of Equations of the Type $y = mx + b$ or $f(x) = mx + b$

Examples 3–5 in Section 2.1 showed that for any number m, the graph of $y = mx$ is a straight line passing through the origin. What will happen if we add a number b on the right side to get the equation $y = mx + b$?

EXAMPLE 1

Graph $y = 2x$ and $y = 2x + 3$, using the same set of axes.

Solution   We first make a table of solutions of both equations.

x	y y = 2x	y y = 2x + 3
0	0	3
1	2	5
−1	−2	1
2	4	7
−2	−4	−1

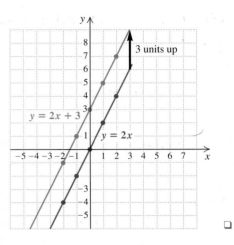

We then plot these points. Drawing a blue line for $y = 2x + 3$ and a red line for $y = 2x$, we see that the graph of $y = 2x + 3$ can be obtained by moving, or *translating*, the graph of $y = 2x$ three units up. The lines are parallel.

EXAMPLE 2

Graph $f(x) = \frac{1}{3}x$ and $g(x) = \frac{1}{3}x - 2$, using the same set of axes.

Solution   We first make a table of solutions of both equations. By choosing multiples of 3, we avoid fractions.

x	f(x) $f(x) = \frac{1}{3}x$	g(x) $g(x) = \frac{1}{3}x - 2$
0	0	−2
3	1	−1
−3	−1	−3
6	2	0

TECHNOLOGY
CONNECTION

A grapher can be used to explore the effect of b when graphing $y = mx + b$. To see this, begin with the graph of $y = x$. On the same set of axes, graph the lines $y = x + 3$ and $y = x - 4$. How do these lines differ from $y = x$? What do you think the line $y = x - 5$ will look like? Try drawing lines like $y = x + \frac{1}{4}$ and $y = x + (-3.2)$ and describe what happens to the graph of $y = x$ when a number b is added.

We then plot these points. Drawing a blue line for $g(x) = \frac{1}{3}x - 2$ and a red line for $f(x) = \frac{1}{3}x$, we see that the graph of $g(x) = \frac{1}{3}x - 2$ looks just like the graph of $f(x) = \frac{1}{3}x$ but is shifted, or translated, 2 units down.

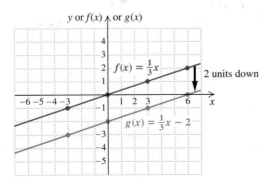

Intercepts and Slopes

Note that the graph of $y = 2x + 3$ passed through the point $(0, 3)$ and the graph of $g(x) = \frac{1}{3}x - 2$ passed through the point $(0, -2)$. In general, the graph of $y = mx + b$ is a line parallel to $y = mx$, passing through the point $(0, b)$. The point $(0, b)$ is called the **y-intercept.** To save time, we sometimes simply refer to the number b as the y-intercept.

EXAMPLE 3

For each equation, find the y-intercept: **(a)** $y = -5x + 4$; **(b)** $f(x) = 5.3x - 12$.

Solution

a) The y-intercept is $(0, 4)$, or simply 4.
b) The y-intercept is $(0, -12)$, or simply -12. In function notation, the y-intercept is $f(0)$, or -12. ❑

In examining the graphs in Examples 1 and 2, note that the slant of the red line appears to be the same as the slant of the blue line. This leads us to suspect that it is the number m, in the equation $y = mx + b$, that is responsible for the slant of the line. In Section 2.4, we will prove that this is indeed the case, but for now we simply state that the number m is called the *slope* of the line $y = mx + b$. Thus the slope of the lines in Example 1 is 2 and the slope of the lines in Example 2 is $\frac{1}{3}$. Note that $2 = \frac{2}{1}$ and that from any point on either line in Example 1, if we go *up* 2 units and *to the right* 1 unit (or *down* 2 units and *to the left* 1 unit), we return to the line. Similarly, in Example 2, if we go *up* 1 unit and *to the right* 3 units (or *down* 1 unit and *to the left* 3 units), we return to the line. In Example 2, we could also have gone up $\frac{1}{3}$ of a unit and to the right 1 unit and still returned to the line.

EXAMPLE 4

Determine the slope of the line given by $y = \frac{2}{3}x + 4$, and draw its graph.

Solution   Here $m = \frac{2}{3}$, so the slope is $\frac{2}{3}$. This means that from *any* point on the graph, we can find a second point by simply going *up* 2 units and *to the right* 3 units. Since the y-intercept, $(0, 4)$, is a point that is known to be on the graph, we calculate that $(0 + 3, 4 + 2)$, or $(3, 6)$, is also on the graph. Knowing two points, we can draw the graph.

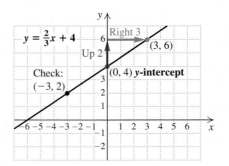

Important: As a check, we choose some other value for x, say -3, and determine y (in this case, 2). We plot that point and see whether it is on the line. If it is not, there has been some error. ❑

In Example 4, the slope is $\frac{2}{3}$. We can find some other names for $\frac{2}{3}$ by multiplying by 1 and then find other points on the graph.

$$\frac{2}{3} = \frac{2}{3} \cdot \frac{2}{2} = \frac{4}{6}$$ ← Thus we can locate another point by going *up* 4 units and *to the right* 6 units from any point on the graph.

$$\frac{2}{3} = \frac{2}{3} \cdot \frac{3}{3} = \frac{6}{9}$$ ← Thus we can locate a point by going *up* 6 units and *to the right* 9 units from any point on the graph.

When numbers are negative, we reverse directions.

$$\frac{2}{3} = \frac{2}{3} \cdot \frac{-1}{-1} = \frac{-2}{-3}$$ ← Thus we can locate another point by going *down* 2 units and then *to the left* 3 units from any point on the graph.

If the slope of a line is negative, it slants downward from left to right.

Graph: $y = -\frac{1}{2}x + 5$.

Solution The y-intercept is $(0, 5)$. The slope is $-\frac{1}{2}$, or $\frac{-1}{2}$. From the y-intercept, we go *down* 1 unit and *to the right* 2 units. That gives us the point $(2, 4)$. We can now draw the graph.

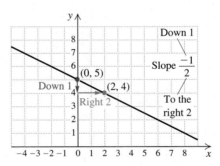

As a new type of check, we rename the slope and find another point:

$$\frac{-1}{2} = \frac{-1}{2} \cdot \frac{-3}{-3} = \frac{3}{-6}.$$

Thus we can go *up* 3 units and then *to the left* 6 units. This gives the point $(-6, 8)$. Since $(-6, 8)$ is on the line, we have a check.

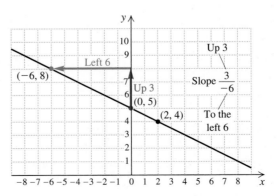

We also could have checked by choosing a new number for x and finding y. In either case, we find a third point not already known to be on the graph. If the third point does not line up with the other two, a mistake has been made.

Because an equation of the form $y = mx + b$ enables us to make use of the slope and the y-intercept when graphing, it is called the *slope–intercept* form of a linear equation.

**The Slope–
Intercept Equation**

Any equation $y = mx + b$ has a graph that is a straight line. It goes through the y-intercept $(0, b)$ and has slope m. Any equation of the form $y = mx + b$ is said to be a *slope–intercept equation*.

EXAMPLE 6

Determine an equation for a line with slope $-\frac{2}{3}$ and y-intercept $(0, 4)$.

Solution   We use the slope–intercept form, $y = mx + b$:

$$y = -\tfrac{2}{3}x + 4. \qquad \text{Substituting } -\tfrac{2}{3} \text{ for } m \text{ and } 4 \text{ for } b$$

□

EXAMPLE 7

Determine the slope and the y-intercept for the line given by $5x - 4y = 8$.

Solution   We convert to a slope–intercept equation:

$$
\begin{aligned}
5x - 4y &= 8 \\
-4y &= -5x + 8 && \text{Adding } -5x \\
y &= -\tfrac{1}{4}(-5x + 8) && \text{Multiplying by } -\tfrac{1}{4} \\
y &= \tfrac{5}{4}x - 2. && \text{Using the distributive law}
\end{aligned}
$$

Because we have an equation $y = mx + b$, we know that the slope is $\frac{5}{4}$ and the y-intercept is $(0, -2)$.

□

Linear Functions

Consider the function $f(x) = 3x + 2$. If we think of it as $y = 3x + 2$, we realize that the graph is a line and that we already know how to graph it. Because the function's domain is not specified, we will assume the domain to be all numbers that could work as inputs. Thus the domain of f is the set of all real numbers.

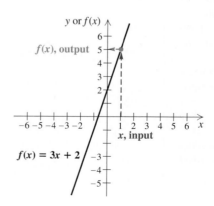

EXAMPLE 8

A firm uses the function $C(x) = \frac{3}{5}x + 2$ to calculate the cost in dollars, $C(x)$, of shipping x ounces of chocolates.

a) Using the horizontal axis for values of x and the vertical axis for values of $C(x)$, graph the equation $C(x) = \frac{3}{5}x + 2$.

b) Use the graph to estimate the cost of shipping $6\frac{1}{2}$ oz of chocolates. Compare this value to the one obtained from the function formula.

Solution

a) We locate the y-intercept $(0, 2)$ and from there count *up* 3 units and *to the right* 5 units. That gives us the point $(5, 5)$. We can now draw the line. As a check, we note that the pair $(10, 8)$ also satisfies the equation.

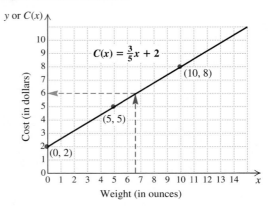

b) To estimate the cost of shipping $6\frac{1}{2}$ oz, we determine the coordinate on the vertical (cost) axis that appears to be paired with the coordinate $6\frac{1}{2}$ on the horizontal (weight) axis. We do so by drawing a vertical line segment from $6\frac{1}{2}$ up to the line and then a horizontal segment from the line over to the cost axis.

We estimate that it costs approximately \$6.00 to ship $6\frac{1}{2}$ oz of chocolates. An exact calculation is made by substitution into the function formula:

$$C\left(6\tfrac{1}{2}\right) = C\left(\tfrac{13}{2}\right) = \tfrac{3}{5} \cdot \tfrac{13}{2} + 2 = \$5.90. \qquad \square$$

Graphs of Equations of the Type $y = b$ or $x = a$

Sometimes an equation has a graph that is parallel to one of the axes. Such an equation will have a missing variable.

EXAMPLE 9

Graph $y = 3$. Determine whether the graph is that of a function.

Solution   Because x is missing, we know that no matter which value of x we choose, y must be 3. Thus the pairs $(-1, 3)$, $(0, 3)$, and $(2, 3)$ all satisfy the equation. The graph is a line parallel to the x-axis.

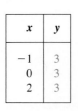

x	y
-1	3
0	3
2	3

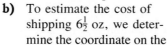

Note that $y = 3 = 0x + 3$, so the line here has a slope of 0. Because $0 = 0/2$ (any nonzero number could be used in place of 2), we can draw the graph by going up 0 units and to the right 2 units from the y-intercept of $(0, 3)$.

Since the graph satisfies the vertical-line test, the graph is that of a function. $\quad \square$

EXAMPLE 10

Graph $x = -2$. Determine whether the graph is that of a function.

Solution   With y missing, no matter which value of y we choose, x must be -2. Thus the pairs $(-2, 3)$, $(-2, 0)$, and $(-2, -4)$ all satisfy the equation. The graph is a line parallel to the y-axis.

x	y
-2	3
-2	0
-2	-4

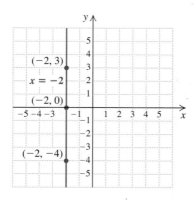

Because y is absent from this equation, it is impossible to write it in the form $y = mx + b$.

Recalling our definition of a function, we note that since the first coordinate, -2, is paired with more than one second coordinate, the graph is *not* that of a function. The graph *fails* the vertical-line test.   ❑

Which Equations Are Linear?

Let us now consider an equation of the form $Ax + By = C$, where A, B, and C are real numbers. Suppose that A and B are both nonzero and solve for y:

$$Ax + By = C$$
$$By = -Ax + C \qquad \text{Adding } -Ax \text{ on both sides}$$
$$y = -\frac{A}{B}x + \frac{C}{B}. \qquad \text{Dividing by } B \text{ on both sides}$$

Since the last equation is a slope–intercept equation, we see that $Ax + By = C$ is a linear equation when $A \neq 0$ and $B \neq 0$.

Suppose next that A or B (but not both) is 0. If A is 0, then $By = C$ and $y = C/B$. If B is 0, then $Ax = C$ and $x = C/A$. In the first case, the graph is a horizontal line, and in the second case, the line is vertical. Thus, $Ax + By = C$ is a linear equation when A or B (but not both) is 0. We have now justified the following result.

The Standard
Form of a Linear
Equation

> Any equation of the form $Ax + By = C$, where A, B, and C are real numbers and A and B are not both 0, is linear.
>
> Any equation of the form $Ax + By = C$ is said to be a linear equation in *standard form*.

EXAMPLE 11

Determine whether the equation $y = x^2 - 5$ is linear.

Solution   We attempt to put the equation in standard form:

$$y = x^2 - 5$$
$$-x^2 + y = -5. \quad \text{Adding } -x^2 \text{ on both sides}$$

This last equation is not linear because it has an x^2-term. From Example 8 in Section 2.1, we see that the graph is not a straight line.   ❏

EXERCISE SET | 2.3

Determine the slope and the y-intercept.

1. $y = 4x + 5$

2. $y = 5x + 3$

3. $f(x) = -2x - 6$

4. $g(x) = -5x + 7$

5. $y = -\frac{3}{8}x - 0.2$

6. $y = \frac{15}{7}x + 2.2$

7. $g(x) = 0.5x - 9$

8. $f(x) = -3.1x + 5$

9. $y = 7$

10. $y = -2$

Give an equation for a line that has the given slope and y-intercept.

11. Slope $\frac{2}{3}$, y-intercept -7

12. Slope $-\frac{3}{4}$, y-intercept 5

13. Slope -4, y-intercept 2

14. Slope 2, y-intercept -1

15. Slope $-\frac{7}{9}$, y-intercept 3

16. Slope $-\frac{4}{11}$, y-intercept 9

17. Slope 5, y-intercept $\frac{1}{2}$

18. Slope 6, y-intercept $\frac{2}{3}$

Determine the slope and the y-intercept. Then draw a graph. Be sure to use a third point as a check.

19. $y = \frac{5}{2}x + 1$

20. $y = \frac{2}{5}x + 4$

21. $f(x) = -\frac{5}{2}x + 4$

22. $f(x) = -\frac{2}{5}x + 3$

23. $2x - y = 5$

24. $2x + y = 4$

25. $f(x) = \frac{1}{3}x + 6$

26. $f(x) = -3x + 6$

27. $7y + 2x = 7$

28. $4y + 20 = x$

29. $f(x) = -0.25x + 2$

30. $f(x) = 1.5x - 3$

31. $4x - 5y = 10$

32. $5x + 4y = 4$

33. $f(x) = \frac{5}{4}x - 2$

34. $f(x) = \frac{4}{3}x + 2$

35. $12 - 4f(x) = 3x$

36. $15 + 5f(x) = -2x$

37. $f(x) = 4$

38. $f(x) = -1$

39. *Sales of cotton goods.*   The function $f(t) = 2.6t + 17.8$ can be used to estimate the yearly sales of cotton goods, in billions of dollars, as a function of time, in years, measured from 1975.

a) Using the horizontal axis for values of t and the vertical axis for values of $f(t)$, graph the equation $f(t) = 2.6t + 17.8$.

b) Use the graph to estimate the total sales of cotton goods in 1981. Compare this value with one obtained from the function formula.

40. *Cost of a movie ticket.*   The average price $P(t)$, in dollars, of a movie ticket can be estimated by the function

$$P(t) = 0.1522t + 4.29,$$

where t is the number of years since 1990.

a) Using the horizontal axis for values of t and the vertical axis for values of $P(t)$, graph the equation $P(t) = 0.1522t + 4.29$.

b) Use the graph to estimate the average price of a movie ticket in the year 2000. Compare this value with one obtained from the function formula.

41. *Cost of a taxi ride.*   The cost, in dollars, of a taxi ride in Pelham is given by $C(m) = 0.75m + 2$, where m is the number of miles traveled.

a) Graph $C(m) = 0.75m + 2$.

b) Use the graph to estimate the cost of a $5\frac{1}{2}$-mi taxi ride.

42. *Cost of renting a car.*   The cost, in dollars, of renting a car for a day is given by $C(m) = 0.3m + 10$, where m is the number of miles driven.

a) Graph $C(m) = 0.3m + 10$.

b) Use the graph to estimate the cost of driving the rental car 40 mi.

43. *Natural gas demand.* The demand, in quadrillions of joules, for natural gas is approximated by $D(t) = \frac{1}{5}t + 20$, where t is the number of years after 1960.

a) Graph the equation $D(t) = \frac{1}{5}t + 20$.
b) Use the graph to predict the demand for gas in 1995.

44. *Cricket chirps per minute.* The number of cricket chirps per minute is given by $N(t) = 7.2t - 32$, where t is the temperature in degrees Celsius.

a) Graph the equation $N(t) = 7.2t - 32$.
b) Use the graph to predict the number of cricket chirps per minute when it is 5°C.

45. *Cost of a telephone call.* The cost, in dollars, of a long distance telephone call is given by $C(m) = 0.80m + 1$, where m is the length of the call in minutes.

a) Graph $C(m) = 0.80m + 1$.
b) Use the graph to approximate the cost of a 4-min call.
c) Use the graph to determine how long a phone call can be made for $7.40.

46. *Life expectancy of American women.* The life expectancy of American women t years after 1950 is given by $A(t) = \frac{3}{20}t + 72$.

a) Graph $A(t) = \frac{3}{20}t + 72$.
b) Use the graph to predict the life expectancy of American women in the year 2000.
c) Use the graph to determine the year in which the life expectancy of American women reached 78 yr.

Graph.

47. $3y = 9$

48. $6g(x) + 24 = 0$

49. $3x = -15$

50. $2x = 10$

51. $4g(x) + 3x = 12 + 3x$

52. $6x - 4y + 12 = -4y$

Determine whether the equation is linear. Find the slope of any nonvertical lines.

53. $3x + 5f(x) + 15 = 0$ **54.** $5x - 3f(x) = 15$

55. $3x - 12 = 0$ **56.** $16 + 4y = 0$

57. $2x + 4g(x) = 19$ **58.** $3g(x) = 5x^2 + 4$

59. $5x - 4xy = 12$ **60.** $3y = 7xy - 5$

61. $\frac{3y}{4x} = 5y + 2$ **62.** $6y - \frac{4}{y} = 0$

63. $f(x) = x^3$ **64.** $f(x) = \frac{1}{2}x^2$

Skill Maintenance _____

65. Simplify: $9\{2x - 3[5x + 2(-3x + y^0 - 2)]\}$.

66. Solve: $3[7x - 2(4 + 5x)] = 5(x + 2) - 7$.

Synthesis _____

67. ◈ Wind friction, or *air resistance,* increases with speed. Here are some measurements made in a wind tunnel. Plot the data and explain why a linear function does or does not give an approximate fit.

Velocity, in Kilometers per Hour	Force of Resistance, in Newtons
10	3
21	4.2
34	6.2
40	7.1
45	15.1
52	29.0

68. ◈ Wind chill is a measure of how cold the wind makes you feel. Below are some measurements of wind chill for a 15-mph breeze.* Without plotting the data, explain why a linear function will not give an approximate fit.

Temperature	15-mph Wind Chill
30°	9°
25°	2°
20°	-5°
15°	-11°
10°	-18°
5°	-25°
0°	-31°

_____

Source: National Oceanic & Atmospheric Administration, as reported in the *Burlington Free Press,* 17 January 1992.

In Exercises 69–72, assume that r, p, and s are constants and that x and y are variables. Determine the slope and the y-intercept.

69. $ry = -5x + p$ **70.** $px + 2 = 4y - 9r$

71. $rx + py = s$ **72.** $rx + py = s - ry$

In Exercises 73–76, assume that r, p, and s are constants and that x and y are variables. Determine whether each equation is linear.

73. $rx + 3y = p - s$ **74.** $py = sx - ry + 2$

75. $r^2x = py + 5$ **76.** $\dfrac{x}{r} - py = 17$

77. The graph of a linear function passes through the points $(0, 3.1)$ and $(2, 7.8)$. Write an equation for the function.

78. The graph of a function passes through the points $(1, 3)$ and $(4, -2)$. Write an equation for the function.

79. 〰 Graph the equations

$$y_1 = 1.4x + 2, \qquad y_2 = 0.6x + 2,$$
$$y_3 = 1.4x + 5, \quad \text{and} \quad y_4 = 0.6x + 5$$

using a grapher. If possible, use the *simultaneous* mode so that you cannot tell which equation is being graphed first. Then decide which line corresponds to each equation.

2.4

Another Look at Linear Graphs

Graphing Using Intercepts • **Finding the Slope** • **Zero Slope and Lines with Undefined Slope**

Graphing Using Intercepts

Once we have determined that an equation is linear, we can sometimes draw its graph very quickly by plotting the *y-intercept* and the *x-intercept*. These are the points at which the graph crosses the y-axis and the x-axis, respectively. Note that to find the y-intercept, we replace x with 0 and solve for y. To find the x-intercept, we replace y with 0 and solve for x. Once the intercepts are plotted, we can draw a straight line through those points.

Locating Intercepts

> The x-intercept is $(a, 0)$. To find a, let $y = 0$ and solve the original equation for x.
>
> The y-intercept is $(0, b)$. To find b, let $x = 0$ and solve the original equation for y.

EXAMPLE 1

Graph the equation $3x + 2y = 12$ by using intercepts.

Solution This equation is linear. *To find the y-intercept, we let $x = 0$ and solve for y:*

$$3 \cdot 0 + 2y = 12$$
$$2y = 12$$
$$y = 6.$$

The *y*-intercept is (0, 6).

To find the x-intercept, we let y = 0 and solve for x:

$$3x + 2 \cdot 0 = 12$$
$$3x = 12$$
$$x = 4.$$

The *x*-intercept is (4, 0).

We plot the two intercepts and draw the line. A third point could be calculated and used as a check.

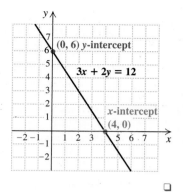

❏

Finding the Slope

Knowing the coordinates of *any* two points on a line, we can calculate the slope of the line. To do this, we divide the vertical distance between the points, the *rise,* by the horizontal distance, the *run.* That is, "slope = rise/run." Equivalently, we often say "slope = change in y/change in x."

In the following definition, (x_1, y_1) and (x_2, y_2) — read "*x* sub-one, *y* sub-one and *x* sub-two, *y* sub-two" — represent two different points on a line. The letter *m* is traditionally used for slope and has its roots in the French verb *monter,* to climb.

Slope

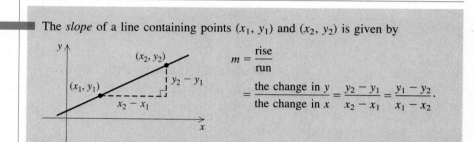

The *slope* of a line containing points (x_1, y_1) and (x_2, y_2) is given by

$$m = \frac{\text{rise}}{\text{run}}$$

$$= \frac{\text{the change in } y}{\text{the change in } x} = \frac{y_2 - y_1}{x_2 - x_1} = \frac{y_1 - y_2}{x_1 - x_2}.$$

EXAMPLE 2

Find the slope of the line shown in the following graph.

Solution We can choose any two points on the line. Using the points (1, 3) and (5, 6) gives us

$$\text{Slope} = \frac{\text{rise}}{\text{run}} = \frac{\text{change in } y}{\text{change in } x} = \frac{6 - 3}{5 - 1}$$

$$= \frac{3}{4}.$$

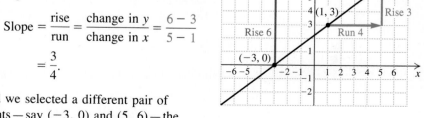

Had we selected a different pair of points — say $(-3, 0)$ and $(5, 6)$ — the calculations would still yield $\frac{3}{4}$:

$$\text{Slope} = \frac{\text{rise}}{\text{run}} = \frac{\text{change in } y}{\text{change in } x} = \frac{6 - 0}{5 - (-3)} = \frac{6}{8} = \frac{3}{4}.$$

❏

EXAMPLE 3

Graph the line containing the points $(1, -3)$ and $(-2, 2)$, and find the slope.

Solution   The graph reveals that from $(1, -3)$ to $(-2, 2)$, the change in y, or the rise, is $2 - (-3)$, or 5. The change in x, or the run, is $-2 - 1$, or -3.

$$\text{Slope} = \frac{5}{-3} = -\frac{5}{3}$$

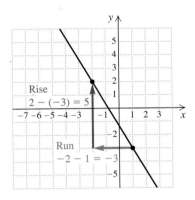

Suppose that, in Example 3, we subtracted coordinates in the opposite order. For the change in y, we would get $-3 - 2$, or -5. For the change in x, we would get $1 - (-2)$, or 3. This gives us

$$\text{Slope} = \frac{-5}{3} = -\frac{5}{3}.$$

This is the same answer that we got before. It does not matter which point is considered (x_1, y_1) and which is considered (x_2, y_2) so long as we subtract coordinates in the same order in both the numerator and the denominator.

EXAMPLE 4

A racer bikes 3 km in 4 min and 9 km in 12 min. Assuming that the racer maintains a steady pace, determine the rate of travel.

Solution

1. FAMILIARIZE.   We might approach this problem in one of two ways. First, we might note that $d = r \cdot t$ is a formula that expresses distance in terms of rate and time. Second, we might use a graph to give us a ''picture'' of the problem.

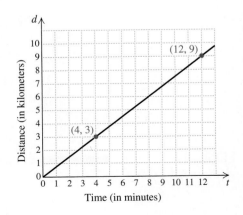

2. TRANSLATE. Using the first approach, we could substitute either 4 for t and 3 for d, or 12 for t and 9 for d. Thus we could form the equation

$$3 \text{ km} = r \cdot 4 \text{ min} \quad \text{or} \quad 9 \text{ km} = r \cdot 12 \text{ min}.$$

Using the second approach, we could form the equation

$$\text{Rate} = \text{change in distance/change in time}.$$

Thus we could form the equation

$$r = (9 \text{ km} - 3 \text{ km})/(12 \text{ min} - 4 \text{ min}).$$

Note that this expression for r is the slope of the line drawn.

3. CARRY OUT. Using the first approach, we solve either

$$3 = r \cdot 4 \quad \text{or} \quad 9 = r \cdot 12$$

to obtain $r = \frac{3}{4}$ km/min.

Using the second approach, we calculate

$$r = \frac{9 - 3}{12 - 4}$$

$$= \frac{6}{8} = \frac{3}{4} \text{ km/min}.$$

4. CHECK. If the rate is $\frac{3}{4}$ km/min, in 4 min the racer has biked $\frac{3}{4} \cdot 4 = 3$ km, and in 12 min, $\frac{3}{4} \cdot 12 = 9$ km. Our answer checks. The fact that both approaches produced the same answer serves as another check.

5. STATE. The racer bikes at a rate of $\frac{3}{4}$ km/min. ☐

In general, *slope can be considered a rate of change*—that is, the ratio of vertical change to horizontal change. Thus a slope of $\frac{3}{4}$ can be interpreted as a rise of $\frac{3}{4}$ unit for each horizontal change of 1 unit, since $\frac{3}{4}$ is equivalent to $\frac{3}{4}/1$.

Zero Slope and Lines with Undefined Slope

If two different points have the same second coordinate, what about the slope of the line joining them? Since, in this case, $y_2 = y_1$, we have

$$\frac{y_2 - y_1}{x_2 - x_1} = \frac{0}{x_2 - x_1} = 0.$$

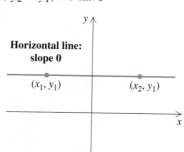

Slope of a Horizontal Line

Every horizontal line has a slope of 0.

Suppose that two different points are on a vertical line. Then they have the same first coordinate. In this case, we have $x_2 = x_1$, so

$$\frac{y_2 - y_1}{x_2 - x_1} = \frac{y_2 - y_1}{0}.$$

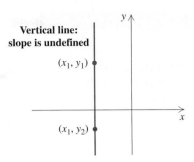

Vertical line: slope is undefined

(x_1, y_1)

(x_1, y_2)

Since we cannot divide by 0, this is undefined.

Slope of a Vertical Line

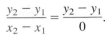

The slope of a vertical line is undefined.

EXAMPLE 5

Find the slope of each given line. If the slope is undefined, state this.

a) $3y + 2 = 8$ **b)** $2x = 10$

Solution

a) We solve for y:

$$3y + 2 = 8$$
$$3y = 6$$
$$y = 2. \qquad \text{The graph of } y = 2 \text{ is a horizontal line.}$$

Since $3y + 2 = 8$ is equivalent to $y = 2$, the slope of the line $3y + 2 = 8$ is 0.

b) When y does not appear, we solve for x:

$$2x = 10$$
$$x = 5. \qquad \text{The graph of } x = 5 \text{ is a vertical line.}$$

Since $2x = 10$ is equivalent to $x = 5$, the slope of the line $2x = 10$ is undefined.   ❑

We can now prove that in the equation $y = mx + b$, m does indeed represent the slope of the line.

Proof.   Observe that $(0, b)$ and $(1, m + b)$ are solutions to any equation of the form $y = mx + b$. Using the slope formula, we see that the slope of the line is given by

$$\text{Slope} = \frac{y_2 - y_1}{x_2 - x_1} = \frac{(m + b) - b}{1 - 0} = m.$$ ❑

EXERCISE SET | 2.4

Find the intercepts. Then graph by using the intercepts and a third point as a check.

1. $x - 2 = y$ **2.** $x - 4 = y$

3. $3x - 1 = y$ **4.** $3x - 4 = y$

5. $5x - 4y = 20$

6. $3x + 5y = 15$

7. $y = -5 - 5x$

8. $y = -2 - 2x$

9. $5y = -15 + 3x$

10. $7x = 3y - 21$

11. $6x - 7 + 3y = 9x - 2y + 8$

12. $7x - 8 + 4y = 8y + 4x + 4$

13. $1.4y - 3.5x = -9.8$

14. $3.6x - 2.1y = 22.68$

15. $5x + 2y = 7$

16. $3x - 4y = 10$

For each pair of points, find the slope of the line containing them. If the slope is undefined, state so.

17. $(6, 9)$ and $(4, 5)$

18. $(8, 7)$ and $(2, -1)$

19. $(3, 8)$ and $(9, -4)$

20. $(17, -12)$ and $(-9, -15)$

21. $(-8, -7)$ and $(-9, -12)$

22. $(14, 3)$ and $(2, 12)$

23. $(-16.3, 12.4)$ and $(-5.2, 8.7)$

24. $(14.4, -7.8)$ and $(-12.5, -17.6)$

25. $(3.2, -12.8)$ and $(3.2, 2.4)$

26. $(-1.5, 7.6)$ and $(-1.5, 8.8)$

27. $(7, 3.4)$ and $(-1, 3.4)$

28. $(-3, 4.2)$ and $(5.1, 4.2)$

29. $(0, 9.1)$ and $(9.1, 0)$

30. $(4.3, 0)$ and $(0, -4.3)$

31. *Running rate.* An Olympic marathoner passes the 5-km point of a race after 30 min and reaches the 25-km point after 2.5 hr. Find the speed of the marathoner.

32. *Skiing rate.* A cross-country skier reaches the 3-km mark of a race in 15 min and the 12-km mark 45 min later. Find the speed of the skier.

12 km

3 km

33. *Rate of production.* At the beginning of a production run, 4.5 tons of sugar had already been refined. Six hours later, the total amount of refined sugar reached 8.1 tons. Calculate the rate of production.

34. *Work rate.* As a painter begins work, one fourth of a house has already been painted. Eight hours later, the house is two-thirds done. Calculate the painter's work rate.

35. *Rate of descent.* A plane descends to sea level from 12,000 ft after being airborne for $1\frac{1}{2}$ hr. The entire flight time is 2 hr and 10 min. Determine the plane's average rate of descent.

36. *Rate of descent.*   A climber leaves a 6200-ft peak at 1:40 P.M. and reaches the base (elevation 950 ft) at 6:10 P.M. Find the climber's average rate of descent.

37. *Sales rate of Avon Products, Co.*   In 1985, the total sales of Avon Products, Co., were about $2.0 million. In 1987, they were about $2.8 million. Determine the rate at which their sales were increasing.

38. In 1985, the number of U.S. visitors overseas was about 7.5 million. In 1988, the number grew to 11.6 million. Determine the rate at which the number of U.S. visitors overseas was growing.

For each equation, find the slope. If the slope is undefined, state so.

39. $5x - 6 = 15$

40. $5y - 12 = 3x$

41. $3x = 12 + y$

42. $-12 = 4x - 7$

43. $5y = 6$

44. $19 = -6y$

45. $5x - 7y = 30$

46. $2x - 3y = 18$

47. $12 - 4x = 9 + x$

48. $15 + 7x = 3x - 5$

49. $2y - 4 = 35 + x$

50. $2x - 17 + y = 0$

51. $3y + x = 3y + 2$

52. $x - 4y = 12 - 4y$

53. $4y + 8x = 6$

54. $5y + 6x = -3$

55. $y - 6 = 14$

56. $3y - 5 = 8$

57. $3y - 2x = 5 + 9y - 2x$

58. $17y + 4x + 3 = 7 + 4x$

59. $7x - 3y = -2x + 1$

60. $9x - 4y = 3x + 5$

Skill Maintenance

61. The fare for a taxi ride from Johnson Street to Elm Street is $5.20. If the rate of the taxi is $1.00 for the first $\frac{1}{2}$ mile and 30¢ for each additional $\frac{1}{4}$ mile, how far is it from Johnson Street to Elm Street?

62. The formula
$$f = \frac{F(c - v_0)}{c - v_s}$$
is used when studying the physics of sound. Solve for F.

Synthesis

63. ◈   Belly Up, Inc., is losing $1.5 million per year while Spinning Wheels, Inc., is losing $170 an hour. Which firm would you rather own and why?

64. ◈   Rosie Picshure claims that her firm's profits continue to go up, but the rate of increase is going down.

  a) Sketch a graph that might represent her firm's profits as a function of time.

  b) Explain why the graph can go up while the rate of increase goes down.

65. A line contains the points $(-100, 4)$ and $(0, 0)$. List four more points of the line.

66. Give an equation, in standard form, for the line whose x-intercept is 5 and whose y-intercept is -4.

67. Find the x-intercept of $y = mx + b$, assuming $m \neq 0$.

68. Determine a so that the slope of the line through this pair of points has the value m.
$$(-2, 3a), (4, -a); \quad m = -\tfrac{5}{12}$$

69. Find the slope of the line that contains the pair of points.

  a) $(5b, -6c), (b, -c)$

  b) $(b, d), (b, d + e)$

  c) $(c + f, a + d), (c - f, -a - d)$

70. Suppose that two linear equations have the same y-intercept but that equation A has an x-intercept that is half the x-intercept of equation B. How do the slopes compare?

2.5

Other Equations of Lines

Point–Slope Equations   •   **Parallel and Perpendicular Lines**

Specifying a line's slope and one point through which the line passes enables us to draw the line. In this section, we study how this same information can be used to produce an *equation* of the line.

Point–Slope Equations

EXAMPLE 1

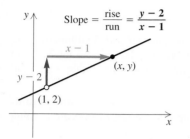

Find an equation of a line with slope $\frac{3}{4}$ that passes through the point $(1, 2)$.

Solution Because any other point (x, y) on the line satisfies the slope formula, we must have

$$\frac{y - 2}{x - 1} = \frac{3}{4}.$$

Thus any pair (x, y) that is a solution of the preceding equation lies on the graph. Unfortunately, the point $(1, 2)$ must be deleted from the graph of

$$\frac{y - 2}{x - 1} = \frac{3}{4}$$

since substitution would give 0/0. We can avoid this difficulty as follows:

$$\frac{y - 2}{x - 1} = \frac{3}{4}$$

$$(x - 1) \cdot \frac{y - 2}{x - 1} = \frac{3}{4}(x - 1) \qquad \text{Multiplying on both sides by } x - 1$$

$$y - 2 = \frac{3}{4}(x - 1). \qquad \text{Simplifying}$$

The pair $(1, 2)$ *is* a solution of this last equation. ❑

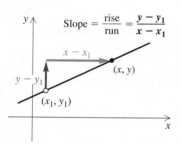

As a generalization of Example 1, let's now suppose that a line of slope m passes through the point (x_1, y_1). For any other point (x, y) to lie on this line, we must have

$$\frac{y - y_1}{x - x_1} = m. \qquad \textbf{(1)}$$

Multiplying by $x - x_1$ gives us the equation we are looking for.

Point–Slope Equation

> The *point–slope equation* of a line is
>
> $$y - y_1 = m(x - x_1).$$

Since Equation (1) is not true when x is x_1 and y is y_1, its graph excludes the point (x_1, y_1). The graph of the point–slope equation includes this point.

Because the slope of a vertical line is undefined, vertical lines do not have point–slope equations. All other lines do.

EXAMPLE 2

Find and graph an equation of the line containing the point $(5, -1)$ with slope $-\frac{1}{2}$.

Solution We substitute in the equation $y - y_1 = m(x - x_1)$:

$$y - y_1 = m(x - x_1)$$

$$y - (-1) = -\tfrac{1}{2}(x - 5). \qquad \text{Substituting}$$

We now graph this point–slope equation by plotting $(5, -1)$, counting off a slope of $\frac{1}{-2}$, and drawing a line.

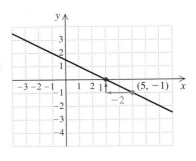

EXAMPLE 3

Find an equation of the line passing through the points $(1, 4)$ and $(3, 7)$. Write the equation using function notation.

Solution We first determine the slope of the line and then use the point–slope equation. Note that

$$m = \frac{7 - 4}{3 - 1} = \frac{3}{2}.$$

Since the line passes through $(1, 4)$, we have

$y - 4 = \frac{3}{2}(x - 1)$	Substituting
$y - 4 = \frac{3}{2}x - \frac{3}{2}$	Using the distributive law
$y = \frac{3}{2}x + \frac{5}{2}$	Adding 4 on both sides
$f(x) = \frac{3}{2}x + \frac{5}{2}.$	Using function notation

You can confirm that substituting $(3, 7)$ instead of $(1, 4)$ in $y - y_1 = m(x - x_1)$ will yield the same expression for $f(x)$.

EXAMPLE 4

A car dealership's sales figures indicate that 160 new cars were sold in 1986 and 200 new cars were sold in 1990. Assuming constant growth since 1985, how many new cars can the dealership expect to sell in 1997?

Solution

1. FAMILIARIZE. We form the pairs $(1, 160)$ and $(5, 200)$ and plot these data points, choosing suitable scales on the two axes. A constant growth rate means that a linear relationship exists between years and number of cars sold. We will let n represent the number of cars sold and t the number of years since 1985.

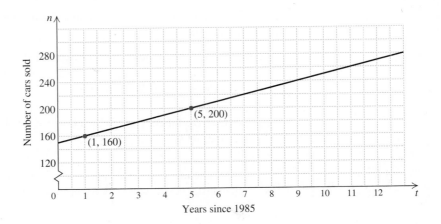

2. TRANSLATE. We seek an equation relating n and t. To accomplish this, we find the slope of the line and use the point–slope equation:

$$m = \frac{200 - 160}{5 - 1} = \frac{40}{4} = 10.$$

Thus,

$$n - 160 = 10(t - 1) \qquad \text{Writing point–slope form}$$
$$n - 160 = 10t - 10 \qquad \text{Using the distributive law}$$
$$n = 10t + 150. \qquad \text{Adding 160 on both sides}$$

3. CARRY OUT. Using function notation, we have

$$n(t) = 10t + 150.$$

To predict sales in 1997, we find

$$n(12) = 10 \cdot 12 + 150 \qquad \text{1997 is 12 years from 1985.}$$
$$= 270.$$

4. CHECK. To check, we observe that 270 seems reasonable since $(12, 270)$ appears to lie on the graph.

5. STATE. The dealership can expect to sell about 270 cars in 1997. ❏

Because of the different scales used on the axes of Example 4, the line is not as steep as the equation $n(t) = 10t + 150$ would lead us to expect.

Parallel and Perpendicular Lines

If two lines are vertical, they are parallel. How can we tell whether nonvertical lines are parallel? The answer is simple: We look at their slopes (see Examples 1 and 2 in Section 2.3).

Two nonvertical lines are parallel if they have the same slope.

EXAMPLE 5

Determine whether the line passing through the points $(1, 7)$ and $(4, -2)$ is parallel to the line $f(x) = -3x + 4.2$.

Solution   The slope of the line passing through $(1, 7)$ and $(4, -2)$ is given by

$$m = \frac{7 - (-2)}{1 - 4} = \frac{9}{-3} = -3.$$

Because the function given by $f(x) = -3x + 4.2$ also has a slope of -3, we conclude that the lines are parallel. ❏

If one line is vertical and another is horizontal, they are perpendicular. There are other instances in which two lines are perpendicular.

Consider a line $\overleftrightarrow{RS}$, as shown in the graph, with slope a/b. Then think of rotating the figure 90° to get a line $\overleftrightarrow{R'S'}$ perpendicular to $\overleftrightarrow{RS}$. For the new line, the rise and the run are interchanged, but the run is now negative. Thus the slope of the new line is $-b/a$. Let us multiply the slopes:

$$\frac{a}{b}\left(-\frac{b}{a}\right) = -1.$$

This is the condition under which lines are perpendicular.

Slope and Perpendicular Lines

Two lines are perpendicular if the product of their slopes is -1. (If one line has slope m, the slope of a line perpendicular to it is $-1/m$. That is, we take the reciprocal and change the sign.) Lines are also perpendicular if one of them is vertical and the other is horizontal.

EXAMPLE 6

Consider the line given by the equation $8y = 7x - 24$.

a) Find an equation for a parallel line passing through $(-1, 2)$.
b) Find an equation for a perpendicular line passing through $(-1, 2)$.

Solution   To find the slope of the line $8y = 7x - 24$, we solve for y to obtain slope–intercept form:

$$8y = 7x - 24$$
$$y = \tfrac{7}{8}x - 3. \qquad \text{Multiplying by } \tfrac{1}{8} \text{ on both sides}$$

The slope is $\tfrac{7}{8}$.

a) Using the point–slope equation, we have

$$y - 2 = \tfrac{7}{8}[x - (-1)] \qquad \text{Substituting } \tfrac{7}{8} \text{ for the slope and } (-1, 2) \text{ for the point}$$
$$y = \tfrac{7}{8}x + \tfrac{23}{8}.$$

b) The slope of a perpendicular line is given by the opposite of the reciprocal of $\tfrac{7}{8}$, $-\tfrac{8}{7}$. The point–slope equation yields

$$y - 2 = -\tfrac{8}{7}[x - (-1)] \qquad \text{Substituting } -\tfrac{8}{7} \text{ for the slope and } (-1, 2) \text{ for the point}$$
$$y = -\tfrac{8}{7}x + \tfrac{6}{7}.$$

EXERCISE SET | 2.5

Find an equation in point–slope form of the line having the specified slope and containing the point indicated. Then graph the line.

1. $m = 4$, $(3, 2)$ **2.** $m = 5$, $(5, 4)$

3. $m = -2$, $(4, 7)$ **4.** $m = -3$, $(7, 3)$

5. $m = 3$, $(-2, -4)$ **6.** $m = 1$, $(-5, -7)$

7. $m = -2$, $(8, 0)$ **8.** $m = -3$, $(-2, 0)$

9. $m = 0$, $(0, -7)$ **10.** $m = 0$, $(0, 4)$

11. $m = \frac{3}{4}$, $(5, -1)$ **12.** $m = \frac{2}{5}$, $(-3, 7)$

Find an equation of the line having the specified slope and containing the indicated point. Write your final answer in slope–intercept form.

13. $m = 5$, $(2, -3)$ **14.** $m = -4$, $(-1, 5)$

15. $m = -\frac{2}{3}$, $(4, -7)$ **16.** $m = \frac{3}{7}$, $(-2, 1)$

17. $m = -0.6$, $(-3, -4)$

18. $m = -3.1$, $(5, -2)$

Find a function for the line containing the pair of points.

19. $(1, 4)$ and $(5, 6)$ **20.** $(2, 6)$ and $(4, 1)$

21. $(-1, -1)$ and $(2, 2)$ **22.** $(-3, -3)$ and $(6, 6)$

23. $(-2, 0)$ and $(0, 5)$ **24.** $(6, 0)$ and $(0, -3)$

25. $(3, 5)$ and $(-5, 3)$ **26.** $(4, 6)$ and $(-6, 4)$

27. $(0, 0)$ and $(5, 2)$ **28.** $(0, 0)$ and $(7, 3)$

29. $(-4, -7)$ and $(-2, -1)$ **30.** $(-2, -3)$ and $(-4, -6)$

31. *Records in the 400-meter run.* In 1930, the record for the 400-m run was 46.8 sec. In 1970, it was 43.8 sec. Let R represent the record in the 400-m run and t the number of years since 1930.

  a) Find a linear function $R(t)$ that fits the data.

  b) Use the function of part (a) to predict the record in 1995; in 1998.

  c) When will the record be 40 sec?

32. *Records in the 1500-meter run.* In 1930, the record for the 1500-m run was 3.85 min. In 1950, it was 3.70 min. Let R represent the record in the 1500-m run and t the number of years since 1930.

  a) Find a linear function $R(t)$ that fits the data.

  b) Use the function of part (a) to predict the record in 1998; in 2002.

  c) When will the record be 3.3 min?

33. *PAC contributions.* In 1986, Political Action Committees (PACs) contributed \$132.7 million to congressional candidates. In 1990, the figure rose to \$150 million. Let A represent the amount of PAC contributions and t the number of years since 1986.*

  a) Find a linear function $A(t)$ that fits the data.

  b) Use the function of part (a) to predict the amount of PAC contributions in 1998.

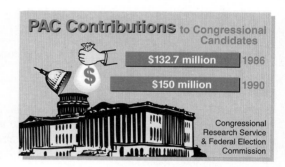

34. *Consumer demand.* Suppose that 6.5 million lb of coffee are sold when the price is \$6 per pound, and 4.0 million lb are sold at \$7 per pound.

  a) Find a linear function that expresses the amount of coffee sold as a function of the price per pound.

  b) Use the function of part (a) to predict the consumer demand should the price drop to \$4 per pound.

35. *Recycling.* In 1980, Americans recycled 14.5 million tons of garbage. In 1988, the figure grew to 23.5 million tons.† Let N represent the number of tons recycled and t the number of years since 1980.

  a) Find a linear function $N(t)$ that fits the data.

  b) Use the function of part (a) to predict the amount recycled in 1997.

---

*Source: Congressional Research Service and Federal Election Commission.

†Source: Franklin Associates and Gannett News Service, reported in the *Burlington Free Press,* 22 January 1992.

36. *Seller's supply.* Suppose that suppliers are willing to sell 5.0 million lb of coffee at a price of $6 per pound and 7.0 million lb at $7 per pound.

a) Find a linear function that expresses the amount suppliers are willing to sell as a function of the price per pound.

b) Use the function of part (a) to predict how much suppliers would be willing to sell at a price of $4 per pound.

37. *Advertising.* An automobile dealer discovers that when $1000 is spent on radio advertising, weekly sales increase by $101,000. When $1250 is spent on radio advertising, sales increase by $126,000. Suppose that the sales increase is a linear function of the amount of radio advertising. Calculate the increase in sales when the amount spent on radio advertising is $1500; when it is $2000.

38. *Items in a supermarket.* In 1981, the average supermarket contained 12,877 items. In 1985, that figure had grown to 17,459 items. Assuming constant growth, predict the average number of items to be found in a supermarket in the year 1997.

39. *Life expectancy of males in the United States.* In 1950, the life expectancy of males was 65 years. In 1970, it was 68 years. Let E represent life expectancy and t the number of years since 1950.

a) Find a linear function $E(t)$ that fits the data.

b) Use the function of part (a) to predict the life expectancy of males in the year 2000.

40. *Pressure at sea depth.* The pressure 100 ft beneath the ocean's surface is approximately 4 atm (atmospheres), whereas at a depth of 200 ft, the pressure is about 7 atm.

a) Find a linear function that expresses pressure as a function of depth.

b) Use the function of part (a) to determine the pressure at a depth of 690 ft.

Without graphing, tell whether the graphs of each pair of equations are parallel.

41. $x + 6 = y,$
$y - x = -2$

42. $2x - 7 = y,$
$y - 2x = 8$

43. $y + 3 = 5x,$
$3x - y = -2$

44. $y + 8 = -6x,$
$-2x + y = 5$

45. $y = 3x + 9,$
$2y = 6x - 2$

46. $y = -7x - 9,$
$-3y = 21x + 7$

Write an equation of the line containing the specified point and parallel to the indicated line.

47. $(3, 7),\ x + 2y = 6$

48. $(0, 3),\ 3x - y = 7$

49. $(2, -1),\ 5x - 7y = 8$

50. $(-4, -5),\ 2x + y = -3$

51. $(-6, 2),\ 3x - 9y = 2$

52. $(-7, 0),\ 5x + 2y = 6$

53. $(-3, -2),\ 3x + 2y = -7$

54. $(-4, 3),\ 6x - 5y = 4$

Without graphing, tell whether the graphs of each pair of equations are perpendicular.

55. $y = 4x - 5,$
$4y = 8 - x$

56. $2x - 5y = -3,$
$2x + 5y = 4$

57. $x + 2y = 5,$
$2x + 4y = 8$

58. $y = -x + 7,$
$y = x + 3$

Write an equation of the line containing the specified point and perpendicular to the indicated line.

59. $(2, 5),\ 2x + y = -3$

60. $(4, 0),\ x - 3y = 0$

61. $(3, -2),\ 3x + 4y = 5$

62. $(-3, -5),\ 5x - 2y = 4$

63. $(0, 9),\ 2x + 5y = 7$

64. $(-3, -4),\ -3x + 6y = 2$

65. $(-4, -7),\ 3x - 5y = 6$

66. $(-4, 5),\ 7x - 2y = 1$

Skill Maintenance

67. The price of a radio, including 5% sales tax, is $36.75. Find the price of the radio before the tax was added.

68. A basketball team increases its score by 7 points in each of three consecutive games. If the team scored a total of 228 points in all three games, what was its score in the first game?

69. 15% of what number is 12.4?

70. Subtract: $-\frac{7}{4} - \left(-\frac{2}{3}\right)$.

Synthesis

71. ◈ The total number of reported cases of AIDS in the United States has risen from 372 in 1981 to 100,000 in 1989 and 200,000 in 1992. Can a linear function be used to express the number of cases as a function of the number of years since 1981? Why or why not?

72. ◈ A firm offers its entering employees a starting salary with a guaranteed 7% increase each year. Can a linear function be used to express the yearly salary as a function of the number of years an employee has worked? Why or why not?

For Exercises 73 and 74, assume that a linear equation fits the situation.

73. Water freezes at 32° Fahrenheit and at 0° Celsius. Water boils at 212°F and at 100°C. What Celsius temperature corresponds to a room temperature of 70°F?

74. The value of a copying machine is $5200 when it is purchased. After 2 years, its value is $4225. Find its value after 8 years.

75. For a linear function f, $f(-1) = 3$ and $f(2) = 4$.

   a) Find an equation for f.
   b) Find $f(3)$.
   c) Find a such that $f(a) = 100$.

76. For a linear function g, $g(3) = -5$ and $g(7) = -1$.

   a) Find an equation for g.
   b) Find $g(-2)$.
   c) Find a such that $g(a) = 75$.

77. Find the value of k so that the graph of $5y - kx = 7$ and the line containing the points $(7, -3)$ and $(-2, 5)$ are parallel.

78. Find the value of k so that the graph of $7y - kx = 9$ and the line containing the points $(2, -1)$ and $(-4, 5)$ are perpendicular.

2.6

The Algebra of Functions

The Domain of a Sum, Difference, Product, or Quotient of Two Functions • Domains and Graphs

Let's now return to the idea of a function as a machine. Suppose that a is in the domain of two functions, f and g. The input a is paired with $f(a)$ by f and with $g(a)$ by g. The outputs can then be added to get $f(a) + g(a)$.

EXAMPLE 1

Let $f(x) = x + 5$ and $g(x) = x^2$. Find $f(2) + g(2)$.

Solution   We visualize two function machines. Because 2 is in the domain of each function, we can compute $f(2)$ and $g(2)$.

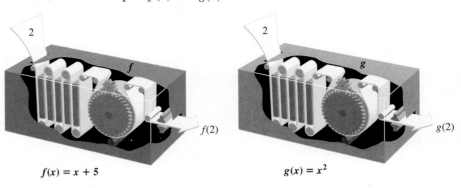

$$f(x) = x + 5 \qquad\qquad g(x) = x^2$$

Since

$$f(2) = 2 + 5 = 7 \quad \text{and} \quad g(2) = 2^2 = 4,$$

we have

$$f(2) + g(2) = 7 + 4 = 11.$$

In Example 1, suppose that we were to write $f(x) + g(x)$ as $(x + 5) + x^2$ so that $f(x) + g(x) = x^2 + x + 5$. This could then be regarded as a "new" function. The notation $(f + g)(x)$ is generally used to denote a function formed in this manner. Similar notations exist for subtraction, multiplication, and division of functions.

The Algebra of Functions

If f and g are functions and x is in the domain of both functions, then:

1. $(f + g)(x) = f(x) + g(x)$;
2. $(f - g)(x) = f(x) - g(x)$;
3. $(f \cdot g)(x) = f(x) \cdot g(x)$;
4. $(f/g)(x) = f(x)/g(x)$, provided $g(x) \neq 0$.

EXAMPLE 2

For $f(x) = x^2 - 1$ and $g(x) = x + 2$, find the following.

a) $(f + g)(3)$

b) $(f - g)(x)$ and $(f - g)(-1)$

c) $(f/g)(x)$ and $(f/g)(-4)$

d) $(f \cdot g)(3)$

Solution

a) Since $f(3) = 3^2 - 1 = 8$ and $g(3) = 3 + 2 = 5$, we have

$$(f + g)(3) = f(3) + g(3)$$
$$= 8 + 5 \qquad \text{Substituting}$$
$$= 13.$$

Alternatively, we could first find $(f + g)(x)$:

$$(f + g)(x) = f(x) + g(x)$$
$$= x^2 - 1 + x + 2$$
$$= x^2 + x + 1. \qquad \text{Combining like terms}$$

Thus,

$$(f + g)(3) = 3^2 + 3 + 1 = 13.$$

b) We have

$$(f - g)(x) = f(x) - g(x)$$
$$= x^2 - 1 - (x + 2) \qquad \text{Substituting}$$
$$= x^2 - x - 3. \qquad \text{Removing parentheses and combining like terms}$$

Thus

$$(f - g)(-1) = (-1)^2 - (-1) - 3$$
$$= -1. \qquad \text{Simplifying}$$

c) We have

$$(f/g)(x) = f(x)/g(x)$$
$$= \frac{x^2 - 1}{x + 2}.$$

Thus,

$$(f/g)(-4) = \frac{(-4)^2 - 1}{-4 + 2}$$ Substituting

$$= \frac{15}{-2}$$

$$= -7.5.$$

d) Using our work in part (a), we have

$$(f \cdot g)(3) = f(3) \cdot g(3)$$

$$= 8 \cdot 5$$

$$= 40.$$

It is also possible to compute $(f \cdot g)(3)$ by first multiplying $x^2 - 1$ and $x + 2$ using methods we will discuss in Chapter 5. ❑

Although it is usually difficult to visualize the product or quotient of two or more functions, sums and differences *can* be visualized. In the following graph,* the total number of airline passengers, $F(t)$, in the New York area is regarded as a function of time. The number of passengers using Kennedy Airport is denoted by $k(t)$, the number of passengers using LaGuardia Airport is $l(t)$, and the number of passengers using Newark Airport is $n(t)$. Although separate graphs for k, l, and n have not been drawn, we can see that

$$F(t) = k(t) + l(t) + n(t).$$

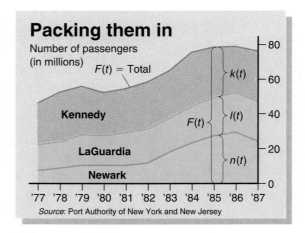

In the next graph, the functions *are* graphed separately before being added. Here all braces extend to the horizontal axis.

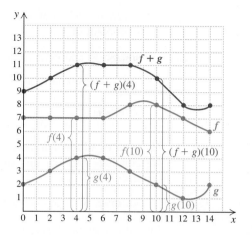

The Domain of a Sum, Difference, Product, or Quotient of Two Functions

It makes sense that in order to find $(f + g)(a)$, $(f - g)(a)$, $(f \cdot g)(a)$, or $(f/g)(a)$, we must first be able to find $f(a)$ and $g(a)$. Thus we need to determine whether a is in the domains of f and g.

EXAMPLE 3

Let

$$f(x) = \frac{5}{x} \quad \text{and} \quad g(x) = \frac{2x - 6}{x + 1}.$$

Find the domain of $f + g$, the domain of $f - g$, and the domain of $f \cdot g$.

Solution   Note that because division by 0 is undefined, we have

Domain of $f = \{x \mid x \text{ is a real number and } x \neq 0\}$

and

Domain of $g = \{x \mid x \text{ is a real number and } x \neq -1\}$.

In order to find $f(a) + g(a)$, $f(a) - g(a)$, or $f(a) \cdot g(a)$, we must know that a is in *both* of the above domains. Thus,

Domain of $f + g =$ Domain of $f - g =$ Domain of $f \cdot g$
$= \{x \mid x \text{ is a real number and } x \neq 0 \text{ and } x \neq -1\}.$ ❑

Suppose that in Example 2(c) we needed to find $(f/g)(-2)$. Finding $f(-2)$ and $g(-2)$ poses no problem:

$$f(-2) = (-2)^2 - 1 = 3 \quad \text{and} \quad g(-2) = -2 + 2 = 0;$$

but then

$$(f/g)(-2) = f(-2)/g(-2)$$
$$= 3/0.$$

Thus, $(f/g)(-2)$ is undefined. Although -2 is in the domain of both f and g, it is not in the domain of f/g.

EXAMPLE 4

Let $F(x) = x^3$ and $G(x) = 4x - 3$. Find the domain of F/G.

Solution   The domain of F and G is all real numbers. However, the domain of

$$(F/G)(x) = \frac{x^3}{4x - 3}$$

is the set of all real numbers such that $4x - 3 \neq 0$. Because $4x - 3 = 0$ when x is $\frac{3}{4}$, we conclude that the domain of F/G is the set of all real numbers except $\frac{3}{4}$. In set-builder notation,

Domain of $F/G = \{x \mid x$ is a real number and $x \neq \frac{3}{4}\}$.   ❏

EXAMPLE 5

Find the domain of f/g, if

$$f(x) = \frac{3}{x - 4} \quad \text{and} \quad g(x) = \frac{6}{x + 2}.$$

Solution   Since the domain of $f = \{x \mid x$ is a real number and $x \neq 4\}$ and the domain of $g = \{x \mid x$ is a real number and $x \neq -2\}$, we conclude that the domain of f/g is the set of all real numbers except -2, 4, and any x-values for which $g(x) = 0$. Because $6/(x + 2)$ is never 0, there is no x-value such that $g(x) = 0$. Thus,

Domain of $f/g = \{x \mid x$ is a real number and $x \neq 4$ and $x \neq -2\}$.   ❏

Comment: It is tempting to write

$$(f/g)(x) = \frac{\dfrac{3}{x - 4}}{\dfrac{6}{x + 2}} = \frac{3}{x - 4} \div \frac{6}{x + 2}$$

$$= \frac{3}{x - 4} \cdot \frac{x + 2}{6}$$

$$= \frac{x + 2}{2(x - 4)},$$

in which case the domain of f/g would exclude only 4. Because of the fact that $(f/g)(x)$ is defined as $f(x)/g(x)$, we can use only x-values that are in the domains of *both* functions. Thus we stipulate

$$(f/g)(x) = \frac{x + 2}{2(x - 4)}, \qquad \text{provided } x \neq -2.$$

EXAMPLE 6

Find the domain of p/q, if

$$p(x) = \frac{5}{x} \quad \text{and} \quad q(x) = \frac{2x - 6}{x + 1}.$$

Solution   We have

Domain of $p = \{x \mid x$ is a real number and $x \neq 0\}$,

Domain of $q = \{x \mid x$ is a real number and $x \neq -1\}$.

Since $q(x) = 0$ when $2x - 6 = 0$, we have $q(x) = 0$ when x is 3. We conclude that

Domain of $p/q = \{x \mid x$ is a real number and $x \neq 0$, $x \neq -1$, and $x \neq 3\}$. ❑

Determining the Domain

To find the domain of a sum, difference, product, or quotient of two functions:

1. Determine the domain of each function individually.
2. The domain of the sum, difference, or product is the set of all values common to both domains.
3. The domain of the quotient is the set of all values common to both domains, excluding any value that would lead to division by 0.

Domains and Graphs

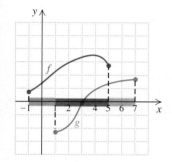

There is an interesting visual interpretation to the algebra of functions. Consider f and g as sketched in the figure.

Note that

Domain of $f = \{x \mid -1 \leqslant x \leqslant 5\}$

and

Domain of $g = \{x \mid 1 \leqslant x \leqslant 7\}$

can be regarded as the *projections,* or "shadows," of f and g on the x-axis. Thus for $f + g, f - g$, or $f \cdot g$, the domain is $\{x \mid 1 \leqslant x \leqslant 5\}$, the values common to the domains of f and g. Since $g(3) = 0$,

Domain of $f/g = \{x \mid 1 \leqslant x \leqslant 5$ and $x \neq 3\}$.

EXERCISE SET | 2.6

Let $f(x) = -3x + 1$ and $g(x) = x^2 + 2$. Find the following.

1. $f(1) + g(1)$ **2.** $f(2) + g(2)$

3. $f(-1) + g(-1)$ **4.** $f(-2) + g(-2)$

5. $f(-7) - g(-7)$ **6.** $f(-5) - g(-5)$

7. $f(5) - g(5)$ **8.** $f(4) - g(4)$

9. $f(2) \cdot g(2)$ **10.** $f(3) \cdot g(3)$

11. $f(-3) \cdot g(-3)$ **12.** $f(-4) \cdot g(-4)$

13. $f(0)/g(0)$ **14.** $f(1)/g(1)$

15. $f(-3)/g(-3)$ **16.** $f(-4)/g(-4)$

Let $F(x) = x^2 - 3$ and $G(x) = 4 - x$. Find the following.

17. $(F + G)(-3)$ **18.** $(F + G)(-2)$

19. $(F + G)(x)$ **20.** $(F + G)(a)$

21. $(F - G)(-4)$ **22.** $(F - G)(-5)$

23. $(F \cdot G)(2)$ **24.** $(F \cdot G)(3)$

25. $(F \cdot G)(-3)$ **26.** $(F \cdot G)(-4)$

27. $(F/G)(0)$ **28.** $(F/G)(1)$

29. $(F/G)(-2)$ **30.** $(F/G)(-1)$

For each pair of functions f and g, determine the domain of the sum, difference, and product of the two functions.

31. $f(x) = x^2$,
$g(x) = 3x - 4$

32. $f(x) = 5x - 1$,
$g(x) = 2x^2$

33. $f(x) = \dfrac{1}{x - 2}$,
$g(x) = 4x^3$

34. $f(x) = 3x^2$,
$g(x) = \dfrac{1}{x - 4}$

35. $f(x) = \dfrac{2}{x}$,
$g(x) = x^2 - 4$

36. $f(x) = x^3 + 1$,
$g(x) = \dfrac{5}{x}$

37. $f(x) = 4x + \dfrac{2}{x - 1}$,
$g(x) = 3x^3$

38. $f(x) = 9 - x^2$,
$g(x) = \dfrac{3}{x - 5} + 2x$

39. $f(x) = \dfrac{3}{x - 2}$,
$g(x) = \dfrac{5}{4 - x}$

40. $f(x) = \dfrac{5}{x - 3}$,
$g(x) = \dfrac{1}{x - 2}$

41. $f(x) = \dfrac{3}{x + 2}$,
$g(x) = \dfrac{x}{x - 4}$

42. $f(x) = \dfrac{2x}{3 - x}$,
$g(x) = \dfrac{4}{x - 5}$

For each pair of functions f and g, determine the domain of f/g.

43. $f(x) = x^4$,
$g(x) = x - 3$

44. $f(x) = 2x^3$,
$g(x) = 5 - x$

45. $f(x) = 3x - 2$,
$g(x) = 2x - 8$

46. $f(x) = 5 + x$,
$g(x) = 6 - 2x$

47. $f(x) = \dfrac{3}{x - 4}$,
$g(x) = 5 - x$

48. $f(x) = \dfrac{1}{2 - x}$,
$g(x) = 7 - x$

49. $f(x) = \dfrac{2x}{x + 1}$,
$g(x) = 2x + 5$

50. $f(x) = \dfrac{7x}{x - 2}$,
$g(x) = 3x + 7$

51. $f(x) = \dfrac{x - 1}{x - 4}$,
$g(x) = 3x^2$

52. $f(x) = \dfrac{x + 2}{x + 3}$,
$g(x) = 4x^3$

53. $f(x) = 3x^2$,
$g(x) = \dfrac{x - 1}{x - 4}$

54. $f(x) = 4x^3$,
$g(x) = \dfrac{x + 2}{x + 3}$

55. $f(x) = \dfrac{x - 1}{x - 2}$,
$g(x) = \dfrac{x - 3}{x - 4}$

56. $f(x) = \dfrac{x + 4}{x + 3}$,
$g(x) = \dfrac{x + 2}{x + 1}$

For Exercises 57–60, consider the functions F and G as shown at the top of the next column.

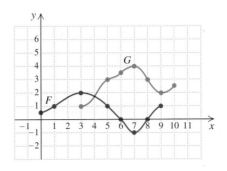

57. Find the domain of F, the domain of G, the domain of $F + G$, and the domain of F/G.

58. Find the domain of $F - G$, the domain of $F \cdot G$, and the domain of G/F.

59. Graph $F + G$.   **60.** Graph $G - F$.

Skill Maintenance

61. Solve: $5x - 7 = 0$.

62. Evaluate $3x - y^2$ if $x = -4$ and $y = 5$.

63. A student's average after 4 tests is 78.5. What score is needed on the fifth test in order to raise the average to 80?

64. Write scientific notation for 0.00000631.

Synthesis

In the graph* that follows, $W(t)$ represents the number of gallons of whole milk, $L(t)$ the number of gallons of lowfat milk, and $S(t)$ the number of gallons of skim milk consumed by the average American in a year.

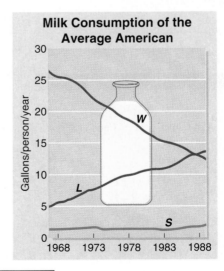

Milk Consumption of the Average American

---

*Copyright 1990, CSPI. Adapted from Nutrition Action Health-letter (1875 Connecticut Avenue, N.W., Suite 300, Washington, DC 20009-5728. $20.00 for 10 issues).

65. ◈ Explain in words what $(W - L)(t)$ represents and what it would mean to have $(W - L)(t) < 0$.

66. ◈ Consider $(W + L + S)(t)$ and explain why you feel that total milk consumption per person has or has not changed over the years 1968–1988.

67. Find the domain of p/q, if

$$p(x) = \begin{cases} 2x, & \text{if } x > 1, \\ x^2, & \text{if } x < 1 \end{cases}$$

and

$$q(x) = \frac{x - 3}{x - 2}.$$

68. Find the domain of m/n, if

$$m(x) = 3x \text{ for } -1 < x < 5$$

and

$$n(x) = 2x - 3.$$

69. Find the domains of $f + g$, $f - g$, $f \cdot g$, and f/g, if

$$f = \{(-2, 1), (-1, 2), (0, 3), (1, 4), (2, 5)\}$$

and

$$g = \{(-4, 4), (-3, 3), (-2, 4), (-1, 0), (0, 5), (1, 6)\}.$$

70. For f and g as defined in Exercise 69, find $(f + g)(-2)$, $(f \cdot g)(0)$, and $(f/g)(1)$.

71. Find the domain of F/G, if

$$F(x) = \frac{1}{x - 4} \quad \text{and} \quad G(x) = \frac{x^2 - 4}{x - 3}.$$

72. Find the domain of f/g, if

$$f(x) = \frac{3x}{2x + 5} \quad \text{and} \quad g(x) = \frac{x^4 - 1}{3x + 9}.$$

73. Write equations for two functions f and g such that the domain of $f + g$ is

$$\{x \mid x \text{ is a real number and } x \neq -2 \text{ and } x \neq 5\}.$$

74. Sketch the graph of two functions f and g such that the domain of f/g is

$$\{x \mid -2 \leqslant x \leqslant 3 \text{ and } x \neq 1\}.$$

75. Use the graph of the number of airline passengers that appears after Example 2 to estimate each of the following.

a) $(k + l + n)(1983)$
b) $(l + n)(1986)$
c) $(l + k)(1986)$

SUMMARY AND REVIEW | 2

KEY TERMS			
Axes, p. 64	Domain, p. 71	Dummy variable, p. 74
x, y-coordinate system, p. 64	Range, p. 71	Linear function, p. 80
Cartesian coordinate system, p. 64	Function, p. 71	y-intercept, p. 82
Ordered pair, p. 64	Relation, p. 72	Slope, p. 82
Origin, p. 64	Input, p. 73	x-intercept, p. 89
Coordinates, p. 65	Output, p. 73	Rise, p. 90
Quadrant, p. 65	Dependent variable, p. 73	Run, p. 90
Graph, p. 66	Independent variable, p. 73	Zero slope, p. 92
Linear equation, p. 67	Constant function, p. 74	Undefined slope, p. 92
Nonlinear equation, p. 68		

IMPORTANT PROPERTIES AND FORMULAS

The Vertical-Line Test

A graph is that of a function provided it is not possible to draw a vertical line that intersects the graph more than once.

The x-intercept is $(a, 0)$. To find a, let $y = 0$ and solve the original equation for x.
The y-intercept is $(0, b)$. To find b, let $x = 0$ and solve the original equation for y.

$$\text{Slope} = m = \frac{\text{rise}}{\text{run}} = \frac{\text{change in } y}{\text{change in } x} = \frac{y_2 - y_1}{x_2 - x_1}$$

Every horizontal line has a slope of 0.
The slope of a vertical line is undefined.

The slope–intercept equation of a line is $y = mx + b$.
The point–slope equation of a line is $y - y_1 = m(x - x_1)$.
The standard form of a linear equation is $Ax + By = C$.

Parallel lines: slopes equal, y-intercepts different.
Perpendicular lines: product of slopes $= -1$.

The Algebra of Functions

1. $(f + g)(x) = f(x) + g(x)$
2. $(f - g)(x) = f(x) - g(x)$
3. $(f \cdot g)(x) = f(x) \cdot g(x)$
4. $(f/g)(x) = f(x)/g(x)$, provided $g(x) \neq 0$

To find the domain of a sum, difference, product, or quotient of two functions:
1. Determine the domain of each function.
2. The domain of the sum, difference, or product is the set of all values common to both domains.
3. The domain of the quotient is the set of all values common to both domains, excluding any value that would lead to division by 0.

REVIEW EXERCISES

This chapter's review and test include Skill Mainte-nance exercises from Sections 1.2, 1.3, 1.5, and 1.7.

Determine whether the ordered pair is a solution.

1. $(3, 7)$, $4p - q = 5$

2. $(-2, 4)$, $x - 3y = 10$

3. $\left(0, -\frac{1}{2}\right)$, $3a + 4b = 2$

4. $(8, -2)$, $3c - 2d = 28$

Graph.

5. $y = -3x + 2$ 6. $y = -x^2 + 1$

7. $8x + 32 = 0$

Find the slope and the y-intercept.

8. $g(x) = -4x - 9$ 9. $-6y + 2x = 7$

Determine whether each of these is a linear equation.

10. $2x - 7 = 0$ 11. $3x - 8f(x) = 7$

12. $2a + 7b^2 = 3$ 13. $2p - \dfrac{7}{q} = 1$

14. Graph using intercepts: $-2x + 4y = 7$.

Find the slope of the line. If the slope is undefined, state so.

15. Containing the points $(4, 5)$ and $(-3, 1)$

16. Containing $(-16.4, 2.8)$ and $(-16.4, 3.5)$

17. Find an equation in point–slope form of the line with slope -2 and containing $(-3, 4)$.

18. Use function notation to write an equation for the line containing $(2, 5)$ and $(-4, -3)$.

Determine whether the lines are parallel, perpen-dicular, or neither.

19. $y + 5 = -x$,   20. $3x - 5 = 7y$,
    $x - y = 2$ $7y - 3x = 7$

21. $4y + x = 3$,
    $2x + 8y = 5$

Find an equation of the line.

22. Containing the point $(2, -5)$ and parallel to the line $3x - 5y = 9$

23. Containing the point $(2, -5)$ and perpendicular to the line $3x - 5y = 9$

Let $g(x) = 2x - 5$ and $h(x) = 3x + 7$. Find the fol-lowing.

24. $g(0)$ 25. $h(-5)$

26. $(g \cdot h)(4)$ 27. $(g - h)(-2)$

28. $(g/h)(-1)$ 29. $g(a + b)$

30. The domain of $g + h$ and $g \cdot h$

31. The domain of h/g

Skill Maintenance

32. Simplify: $-\frac{2}{3} - \left(-\frac{4}{5}\right)$.

33. Simplify: $(5a^3b)^2$.

34. Solve: $3(x - 2) + x = 5(x + 4)$.

35. Use scientific notation to write the number of feet in 1000 miles.

Synthesis

36. ◈   Explain why every function is a relation, but not every relation is a function.

37. ◈   Explain why the slope of a vertical line is undefined whereas the slope of a horizontal line is 0.

38. Find the y-intercept of the function given by
$$f(x) + 3 = 0.17x^2 + (5 - 2x)^x - 7.$$

39. Determine the value of a so that the lines $3x - 4y = 12$ and $ax + 6y = -9$ are parallel.

CHAPTER TEST | 2

Determine whether the ordered pair is a solution of the equation.

1. $(2, 5)$, $3y - 4z = -14$

2. $(-6, 8)$, $2s - t = -4$

3. $(0, -5)$, $x - 4y = -20$

4. $(1, -4)$, $-2p + 5q = 18$

Graph.

5. $y = -5x + 4$ **6.** $y = -2x^2 + 3$

Find the slope and y-intercept.

7. $y = 3x - 5$ **8.** $-3y + 4x = 9$

9. Graph: $3x - 18 = 0$.

10. Which of these are linear equations?

   a) $8x - 7 = 0$
   b) $4b - 9a^2 = 2$
   c) $2x - 5y = 3$

11. Graph using intercepts: $-5x + 2y = -12$.

Find the slope of the line containing the following points. If the slope is undefined, state so.

12. $(-2, -2)$ and $(6, 3)$

13. $(-3.1, 5.2)$ and $(-4.4, 5.2)$

14. Which of these equations has zero slope and which has an undefined slope?

   a) $2y = 7$
   b) $3x - 4y = 6x - 4y$

15. Find an equation in point–slope form of the line with slope 4 and containing $(-2, -4)$.

16. Use function notation to write an equation for the line containing $(3, -1)$ and $(4, -2)$.

Determine without graphing whether the pair of lines is parallel, perpendicular, or neither.

17. $4y + 2 = 3x$,   **18.** $y = -2x + 5$,
     $-3x + 4y = -12$ $2y - x = 6$

Find an equation of the line.

19. Containing $(-3, 2)$ and parallel to the line $2x - 5y = 8$

20. Containing $(-3, 2)$ and perpendicular to the line $2x - 5y = 8$

21. Find the following function values, given that $g(x) = -3x - 4$ and $h(x) = x^2 + 1$.

   a) $g(0)$ **b)** $h(-2)$
   c) $(g/h)(2)$ **d)** $(g - h)(-3)$

22. *The cost of renting a car.* If you rent a car for one day and drive it 250 miles, the cost is $100. If you drive it 300 miles, the cost is $115. Let $C(m)$ represent the cost, in dollars, of driving m miles.

   a) Find a linear function that fits the data.
   b) Use the function to find how much it will cost to rent the car for one day and drive it 500 miles.

Skill Maintenance _____

23. Solve: $3(2x - 4) = 5x - (12 - x)$.

24. Write the number 0.000528 in scientific notation.

25. Simplify: $\dfrac{3(2 - 4 \cdot 3)^2 - 2 \cdot 15}{\cdot\, 5^2 - 4^2}$.

26. Subtract: $19.7 - 26.5$.

Synthesis _____

27. The function $f(t) = 5 + 15t$ can be used to determine a bicycle racer's position, in miles from the starting line, measured in hours since passing the 5-mile mark.

   a) How far from the start will the racer be 1 hour and 40 minutes after passing the 5-mile mark?
   b) Assuming a constant rate, how fast is the racer traveling?

28. The graph of the function $f(x) = mx + b$ contains the points $(r, 3)$ and $(7, s)$. Express s in terms of r if the graph is parallel to the line $3x - 2y = 7$.

CHAPTER 3

Systems of Equations and Problem Solving

"Arctic Antifreeze" is 18% alcohol. "Frost-No-More" is 10% alcohol. How many liters of each should be mixed together in order to get 20 L of a mixture that is 15% alcohol?

This problem can be solved using a *system of equations.* It appears as Exercise 19 in Section 3.3.

James C. Letton
CHEMIST

"I learned very early that studies in science require a solid math background. Today, I often use systems of equations to determine the rate of change in the concentration of reactants, intermediates, and products during a chemical reaction."

T
he most difficult part of solving a problem in algebra is almost always translating the problem situation to mathematical language. Once an equation is translated, the rest is usually straightforward. In this chapter, we study *systems of equations* and how to solve them using graphing, substitution, and elimination. One of the great advantages of using a system of equations is that many problem situations then become easier to translate to mathematical language.

Systems of equations have extensive application to many fields, such as psychology, sociology, business, education, engineering, and science. Systems of inequalities are also useful in a branch of mathematics called *linear programming*. This chapter includes a brief introduction to linear programming as well as a study of *matrices*, which can also be used to solve systems of equations.

In addition to the material in this chapter, the review and test for Chapter 3 include material from Sections 1.3, 1.4, 1.6, and 2.5.

3.1

Systems of Equations in Two Variables

Translating • **Systems of Equations and Solutions** • **Solving Systems Graphically**

Translating

Let us see how more than one equation can be used when translating a problem to mathematical language. Many problems involving more than one unknown can be more easily solved by translating to two or more equations than by translating to a single equation.

EXAMPLE 1

Translate the following problem situation to mathematical language, using two equations.

> Two angles are complementary. One angle is 12° less than three times the other. Find the measures of the angles.

Solution

1. **FAMILIARIZE.** Recall that we may sometimes have to look up some kind of information when problem solving. If you did not know the definition of complementary angles, you might look it up in a geometry book. It turns out that two angles are complementary if the sum of their measures is 90°. Suppose we make a guess about the answers. The measures 30° and 60° have a sum that is 90°. One angle is supposed to be 12° less than three times the other. Thus, $3 \cdot 30° - 12° = 90° - 12° = 78°$, which is not the same as the second part of our guess, 60°. Although our guess was wrong, the guessing has familiarized us with the problem. We decide to let $x =$ the measure of one angle and $y =$ the measure of the other angle.

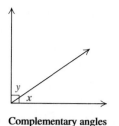

Complementary angles

2. TRANSLATE. There are two statements in the problem. The fact that the angles are complementary can be reworded and translated as follows:

Two angles are complementary.

Rewording: <u>The sum of the measures</u> is 90°.

Translating: $x + y$ $= 90$

The second statement of the problem can be translated directly, again using x and y:

Wording: <u>One angle</u> is <u>12° less than three times the other.</u>

Translating: y $=$ $3x - 12$

The second statement could have been translated as $x = 3y - 12$, which would also have been correct. The problem has been translated to a pair, or **system of equations:**

$$x + y = 90,$$
$$y = 3x - 12.$$ ❏

EXAMPLE 2

In one day, Glovers, Inc., sold 20 pairs of gloves. Cloth gloves sold for $24.95 per pair and pigskin gloves sold for $37.50 per pair. The company took in $687.25. Write a system of equations that could be used to find how many of each kind were sold.

Solution

1. FAMILIARIZE. To familiarize ourselves with this problem, let's make a guess:

Glovers sold 12 pairs of cloth gloves and

11 pairs of pigskin gloves.

Does the guess check? Does it make sense?
 The total number of pairs of gloves sold was supposed to be 20, so our guess cannot be right. Let's try another guess:

12 pairs of cloth gloves and

8 pairs of pigskin gloves.

Now the total number sold is 20, so our guess is right in that respect.
 How much money would then have been taken in? Since cloth gloves sold for $24.95, Glovers would have taken in

12($24.95)

from the cloth gloves. They would have taken in

8($37.50)

from the pigskin gloves. This makes the total received

12($24.95) + 8($37.50) = $299.40 + $300.00 = $599.40.

Our guess is not the answer to the problem, because the total, according to the problem, was $687.25. Since $599.40 is smaller than $687.25, it seems reasonable that more of the expensive gloves were sold than we had guessed. We could now adjust our guess accordingly. Instead, let's work toward an algebraic approach that avoids guessing.

2. TRANSLATE. We let c = the number of pairs of cloth gloves sold and p = the number of pairs of pigskin gloves sold. The information can be organized in a table, which will help with the translating.

Kind of Glove	Cloth	Pigskin	Total
Number Sold	c	p	20
Price	$24.95	$37.50	
Amount Taken In	24.95c	37.50p	687.25

$\longrightarrow c + p = 20$

$\longrightarrow$ $24.95c + 37.50p = 687.25$

The first row of the table and the first sentence of the problem indicate that a total of 20 pairs of gloves were sold:

$$c + p = 20.$$

Since each pair of cloth gloves cost $24.95 and c pairs were sold, $24.95c$ represents the amount taken in from the sale of cloth gloves. Similarly, $37.50p$ represents the amount taken in from the sale of p pairs of pigskin gloves. From the third row of the table and the third sentence of the problem, we get the second equation:

$$24.95c + 37.50p = 687.25.$$

Multiplying both sides by 100, we clear the decimals. Thus we have the translation, a system of equations:

$$c + p = 20,$$
$$2495c + 3750p = 68,725.$$ ❑

Systems of Equations and Solutions

A *solution* of a system of equations in two variables is an ordered pair of numbers that makes *both* equations true.

EXAMPLE 3

Determine whether $(-4, 7)$ is a solution of the system

$$x + y = 3,$$
$$5x - y = -27.$$

Solution   We use alphabetical order of the variables. Thus we replace x by -4 and y by 7:

$$\frac{x + y = 3}{-4 + 7 \;?\; 3}$$
$$3 \mid 3 \quad \text{TRUE}$$

$$\frac{5x - y = -27}{5(-4) - (7) \;?\; -27}$$
$$-20 - 7$$
$$-27 \mid -27 \quad \text{TRUE}$$

The pair $(-4, 7)$ makes both equations true, so it is a solution of the system. We sometimes describe such a solution by saying, in this case, that $x = -4$ and $y = 7$. Set notation can also be used to list the solution as $\{(-4, 7)\}$. ❑

Solving Systems Graphically — SKIP

Recall that the graph of an equation is a drawing that represents its solution set. If the graph of an equation is a line, then every point on that line corresponds to an ordered pair that is a solution of the equation. If we graph a *system* of two linear equations, any point at which the lines intersect is a solution of *both* equations. Such a point is a solution of the system.

EXAMPLE 4

Solve the system graphically.

a) $y - x = 1,$
$y + x = 3$

b) $y = -3x + 5,$
$y = -3x - 2$

c) $3y - 2x = 6,$
$-12y + 8x = -24$

Solution

a) We draw the graphs of each equation using any method studied in Chapter 2. All ordered pairs from line L_1 are solutions of the first equation. All ordered pairs from line L_2 are solutions of the second equation. The point of intersection has coordinates that make *both* equations true. The solution seems to be the point $(1, 2)$. Graphing is not perfectly accurate, so solving by graphing may get only approximate answers. Since the pair $(1, 2)$ does check, it is the solution.

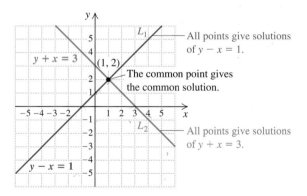

Check: $\begin{array}{c} y - x = 1 \\ \hline 2 - 1 \;?\; 1 \\ 1 \;\vert\; 1 \;\; \text{TRUE} \end{array}$ $\qquad$ $\begin{array}{c} y + x = 3 \\ \hline 2 + 1 \;?\; 3 \\ 3 \;\vert\; 3 \;\; \text{TRUE} \end{array}$

b) We graph the equations. The lines have the same slope, -3, and different y-intercepts, so they are parallel. There is no point at which they cross, so the system has no solution.

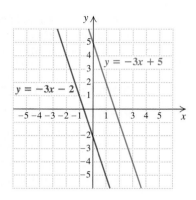

No matter what point we try, it will *not* check in *both* equations. There is no solution. The solution set is thus the empty set, denoted $\varnothing$ or $\{\ \}$.

c) We graph the equations and see that the graphs are the same. Thus any solution of one of the equations is a solution of the other. Each equation has an infinite number of solutions, some of which are listed on the graph. Each of these is also a solution of the other equation. We check one solution, $(0, 2)$, which is the y-intercept of each equation.

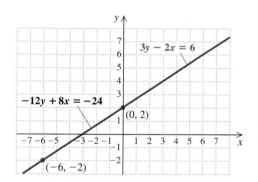

$$
\begin{array}{c|c}
3y - 2x = 6 \\ \hline
3(2) - 2(0) \ ? \ 6 \\
6 - 0 \\
6 \ | \ 6 \ \text{TRUE}
\end{array}
\qquad
\begin{array}{c|c}
-12y + 8x = -24 \\ \hline
-12(2) + 8(0) \ ? \ -24 \\
-24 + 0 \\
-24 \ | \ -24 \ \text{TRUE}
\end{array}
$$

You can check that $(-6, -2)$ is another solution of both equations. In fact, any pair that is a solution of one equation is a solution of the other equation as well. Thus the solution set is $\{(x, y) | 3y - 2x = 6\}$ or, in words, "the set of all pairs (x, y) for which $3y - 2x = 6$." In place of $3y - 2x = 6$, we could have used $-12y + 8x = -24$ since the two equations are equivalent.   ❏

Example 4 illustrates that when we graph a system of two linear equations in two variables, one of the following three outcomes will occur.

1. The lines have one point in common, and that point is the only solution of the system (see part (a)).

2. The lines are parallel, with no point in common, and the system has no solution (see part (b)). This system is called **inconsistent.**

3. The lines coincide, sharing the same graph. Since every solution of one equation is a solution of the other, the system has an infinite number of solutions (see part (c)). The equations are said to be **dependent.**

Any system of equations that has at least one solution is said to be **consistent.** The systems of Examples 4(a) and 4(c) are both consistent.

Graphing is helpful when solving systems because it allows us to "see" the solution, and in many practical situations, an estimate made graphically is quite satisfactory. However, graphing has the disadvantage of not yielding exact answers when fractional or decimal solutions are involved. In Section 3.2, we will develop two algebraic methods of solving systems. Both methods produce exact answers.

TECHNOLOGY
CONNECTION

A grapher can be used to solve systems of equations, especially if it has a Zoom feature that allows a small portion of the graph to be magnified. Since most graphers require equations to have y alone on one side, the first step is to solve for y in each equation. After graphing both equations, use the Zoom feature to magnify the intersection (shrink the window's dimensions if your grapher lacks the Zoom feature). By doing this repeatedly and using the Trace feature, you can determine the coordinates of the point of intersection to the desired accuracy.

A grapher is especially useful when the equations in a system include fractions or decimals or when the coordinates of the intersection are not integers. To illustrate, let's solve the following system:

$$2.34x + 5.71y = -12.45,$$
$$-9.08x + 14.45y = 9.45.$$

After solving each equation for y, enter them into the grapher. The dimensions of the "standard" window for most graphers are $[-10, 10, -10, 10]$. In such a window, these two lines should appear as shown here:

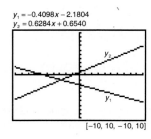

Now use the Zoom and Trace features to magnify the region in which the intersection appears. (Changing the window dimensions to $[-3, -2, -2, 0]$ gives a similar result.) You will now see the following:

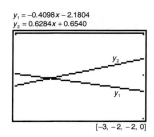

By "zooming" and "tracing" over and over, we can determine that, to the nearest hundredth, the coordinates of the intersection are $(-2.73, -1.06)$.

Use a grapher to find the solution to the system. Round the x- and y-coordinates to the nearest hundredth.

TC1. $y = -4.56x + 12.95,$
$\quad\quad y = 7.88x - 6.77$

TC2. $y = 123.52x + 89.32,$
$\quad\quad y = -89.22x + 33.76$

TC3. $1.76x + 8.21y = 12.22,$
$\quad\quad 6.02x - 3.22y = 9.18$

TC4. $-9.25x - 12.94y = -3.88,$
$\quad\quad 21.83x + 16.33y = 13.69$

EXERCISE SET | 3.1

Translate the problem situation to a system of equations. Do not attempt to solve, but save for later use.

1. The sum of two numbers is -42. The first number minus the second number is 52. What are the numbers?

2. The difference between two numbers is 11. Twice the smaller plus three times the larger is 123. What are the numbers?

3. Windy City Mufflers sold 40 scarves. White ones sold for $4.95 and printed ones sold for $7.95. In all, $282 worth of scarves were sold. How many of each kind were sold?

4. Twin City Pens sold 45 pens, one kind at $8.50 and another kind at $9.75. In all, $398.75 was taken in. How many of each kind were sold?

5. Two angles are complementary. The sum of the measures of the first angle and half the second angle is $64°$. Find the measures of the angles.

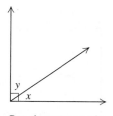

Complementary angles

6. Two angles are supplementary. One angle is $3°$ less than twice the other. Find the measures of the angles.

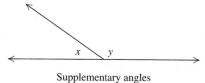

Supplementary angles

7. At a barbecue, there were 250 dinners served. A child's plate cost $3.50 and an adult's plate cost $7.00. If the total amount of money collected for dinners was $1347.50, how many of each type of plate was served?

8. Mary Monroe scored 18 times during one basketball game. She scored a total of 30 points, two for each field goal and one for each free throw. How many field goals did she make? How many free throws?

9. The perimeter of a standard tennis court when used for doubles play is 228 ft. The width is 42 ft less than the length. Find the dimensions.

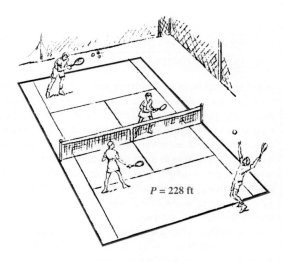

$P = 228$ ft

10. The perimeter of a standard basketball court is 288 ft. The length is 44 ft longer than the width. Find the dimensions.

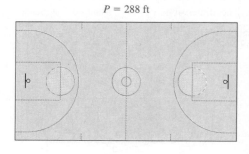

$P = 288$ ft

11. The Central College Cougars made 40 field goals in a recent basketball game, some 2-pointers and

some 3-pointers. Altogether the 40 baskets count-
ed for 89 points. How many of each type of field
goal was made?

12. Acme Video rents general interest films for $3.00
and children's films for $1.50. In one day, a total
of $213 was taken in from the rental of 77 videos.
How many of each type of video was rented?

13. A lumber company can convert logs into either
lumber or plywood. In a given day, the mill turns
out a total of 400 units of lumber and plywood. It
makes a profit of $20 on a unit of lumber and $30
on a unit of plywood. How many of each unit must
be produced and sold in order to make a profit of
$11,000?

14. Hockey teams receive 2 points when they win and
1 point when they tie. One season, the Wildcats
won a championship with 60 points. They won 9
more games than they tied. How many wins and
how many ties did the Wildcats have?

15. A disc jockey must play 12 commercial spots during one hour of a radio show. Each commercial is
either 30 sec or 60 sec long. If the total commercial
time during that hour is 10 min, how many 30-sec
commercials were played that hour? How many
60-sec commercials?

16. An airplane has a total of 152 seats. The number of
coach-class seats is five more than six times the
number of first-class seats. How many of each type
of seat are there on the plane?

Determine whether the ordered pair is a solution of the
given system of equations. Remember to use alphabetical order of variables.

17. $(1, 2)$; $4x - y = 2$,
$\qquad 10x - 3y = 4$

18. $(-1, -2)$; $2x + y = -4$,
$\qquad x - y = 1$

19. $(2, 5)$; $y = 3x - 1$,
$\qquad 2x + y = 4$

20. $(-1, -2)$; $x + 3y = -7$,
$\qquad 3x - 2y = 12$

21. $(1, 5)$; $x + y = 6$,
$\qquad y = 2x + 3$

22. $(5, 2)$; $a + b = 7$,
$\qquad 2a - 8 = b$

23. $(2, -7)$; $3a + b = -1$,
$\qquad 2a - 3b = -8$

24. $(2, 1)$; $3p + 2q = 5$,
$\qquad 4p + 5q = 2$

25. $(3, 1)$; $3x + 4y = 13$,
$\qquad 5x - 4y = 11$

26. $(4, -2)$; $-3x - 2y = -8$,
$\qquad y = 2x - 5$

Solve the system graphically. Be sure to check.

27. $x + y = 4$,
$\quad x - y = 2$

28. $x - y = 3$,
$\quad x + y = 5$

29. $2x - y = 4$,
$\quad 5x - y = 13$

30. $3x + y = 5$,
$\quad x - 2y = 4$

31. $4x - y = 9$,
$\quad x - 3y = 16$

32. $4y = x + 8$,
$\quad 3x - 2y = 6$

33. $a = 1 + b$,
$\quad b = -2a + 5$

34. $x = y - 1$,
$\quad 2x = 3y$

35. $2u + v = 3$,
$\quad 2u = v + 7$

36. $2b + a = 11$,
$\quad a - b = 5$

37. $y = -\frac{1}{3}x - 1$,
$\quad 4x - 3y = 18$

38. $y = -\frac{1}{4}x + 1$,
$\quad 2y = x - 4$

39. $6x - 2y = 2$,
$\quad 9x - 3y = 1$

40. $y - x = 5$,
$\quad 2x - 2y = 10$

41. $x = 4$,
$\quad y = -5$

42. $x = -3$,
$\quad y = 2$

43. $y = -x - 1$,
$\quad 4x - 3y = 24$

44. $2a + b = 4$,
$\quad b = 4a + 1$

45. $2x - 3y = 6$,
$\quad 3y - 2x = -6$

46. $y = 3 - x$,
$\quad 2x + 2y = 6$

47. For the systems in the odd-numbered exercises
27–45, which are consistent?

48. For the systems in the even-numbered exercises
28–46, which are consistent?

49. For the systems in the odd-numbered exercises
27–45, which are dependent?

50. For the systems in the even-numbered exercises
28–46, which are dependent?

Skill Maintenance _____

Solve.

51. $3x + 4 = x - 2$ **52.** $\frac{3}{5}x + 2 = \frac{2}{5}x - 5$

53. $4x - 5x = 8x - 9 + 11x$

54. Solve $Q = \frac{1}{4}(a - b)$ for b.

Synthesis _____

55. ◈   Write a problem for a classmate to solve that
can be translated into a system of two equations.
Devise the problem so that the solution is ''Lucy
sold 5 apple pies and 7 cherry pies.''

56. ◈   Write a problem for a classmate to solve that requires writing a system of two equations. Devise the problem so that the solution is "The Lakers made 6 three-point baskets and 31 two-point baskets."

For Exercises 57–59, consider the following graph.*

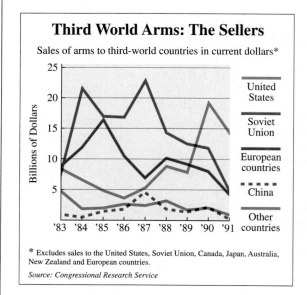

Third World Arms: The Sellers

Sales of arms to third-world countries in current dollars*

United States

Soviet Union

European countries

China

Other countries

* Excludes sales to the United States, Soviet Union, Canada, Japan, Australia, New Zealand and European countries.

Source: Congressional Research Service

57. In what years did U.S. arms sales to third-world countries exceed Soviet arms sales to third-world countries?

58. What was the last year in which European arms sales to third-world countries exceeded U.S. arms sales to third-world countries?

59. Determine the most recent year in which Soviet arms sales to third-world countries exceeded the combined sales of the United States, Europe, and China.

60. The solution of the following system is $(4, -5)$. Find A and B.

$$Ax - 6y = 13,$$
$$x - By = -8.$$

61. Write a system of equations for which:

a) $(5, 1)$ is a solution,
b) there is no solution, and
c) there are infinitely many solutions.

62. A system of linear equations has $(1, -1)$ and $(-2, 3)$ as solutions. Determine:

a) a third point that is a solution, and
b) how many solutions there are.

Translate to a system of equations. Do not solve.

63. Burl is twice as old as his son. Ten years ago, Burl was three times as old as his son. How old are they now?

64. Lou and Juanita have been mathematics professors at a state university. Together, they have 46 years of service. Two years ago, Lou had taught 2.5 times as many years as Juanita. How long has each taught at the university?

65. A piece of posterboard has a perimeter of 156 in. If you cut 6 in. off the width, the length becomes four times the width. What are the dimensions of the original piece of posterboard?

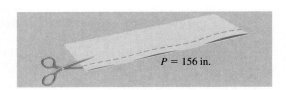

$P = 156$ in.

66. The Mudhens have played 104 baseball games. Before winning their last eight games in a row, they had won three times as many games as they had lost. What is their present won–loss record?

Solve graphically.

67. $y = |x|,$
$x + 4y = 15$

68. $x - y = 0,$
$y = x^2$

◢◣   In Exercises 69–72, use a grapher to solve the system of linear equations for x and y. Round all coordinates to the nearest hundredth.

69. $y = 8.23x + 2.11,$
$y = -9.11x - 4.66$

70. $y = -3.44x - 7.72,$
$y = 4.19x - 8.22$

71. $14.12x + 7.32y = 2.98,$
$21.88x - 6.45y = -7.22$

72. $5.22x - 8.21y = -10.21,$
$-12.67x + 10.34y = 12.84$

3.2

Solving by Substitution or Elimination

The Substitution Method • **The Elimination Method** •
Comparing Methods

The Substitution Method

One nongraphical method for solving systems of equations, the *substitution method*, relies on having one variable isolated.

EXAMPLE 1

Solve the system

$$x + y = 4, \quad (1)$$
$$x = y + 1. \quad (2)$$

Solution Equation (2) says that x and $y + 1$ name the same number. Thus we can substitute $y + 1$ for x in Equation (1):

$$x + y = 4 \qquad \text{Equation (1)}$$
$$(y + 1) + y = 4. \qquad \text{Substituting } y + 1 \text{ for } x$$

We solve this last equation, using methods learned earlier:

$$(y + 1) + y = 4$$
$$2y + 1 = 4 \qquad \text{Removing parentheses and collecting like terms}$$
$$2y = 3 \qquad \text{Subtracting 1 on both sides}$$
$$y = \tfrac{3}{2}. \qquad \text{Dividing by 2}$$

We now return to the original pair of equations and substitute $\tfrac{3}{2}$ for y in either equation so that we can solve for x. Calculations will be easier if we choose Equation (2):

$$x = y + 1 \qquad \text{Equation (2)}$$
$$x = \tfrac{3}{2} + 1 \qquad \text{Substituting } \tfrac{3}{2} \text{ for } y$$
$$x = \tfrac{3}{2} + \tfrac{2}{2} = \tfrac{5}{2}.$$

We obtain the ordered pair $\left(\tfrac{5}{2}, \tfrac{3}{2}\right)$. We check to be sure that it is a solution.

Check:

$$
\begin{array}{c|c}
x + y = 4 & x = y + 1 \\
\hline
\tfrac{5}{2} + \tfrac{3}{2} \; ? \; 4 & \tfrac{5}{2} \; ? \; \tfrac{3}{2} + 1 \\
\tfrac{8}{2} & \tfrac{3}{2} + \tfrac{2}{2} \\
4 \;\big|\; 4 \;\; \text{TRUE} & \tfrac{5}{2} \;\big|\; \tfrac{5}{2} \;\; \text{TRUE}
\end{array}
$$

Since $\left(\tfrac{5}{2}, \tfrac{3}{2}\right)$ checks, it is the solution. ❑

The solution to Example 1 would have been difficult to find graphically because it involves fractions.

If neither equation in a system has a variable alone on one side, we first isolate a variable in one equation and then substitute.

EXAMPLE 2

Solve the system

$$2x + y = 6, \qquad (1)$$
$$3x + 4y = 4. \qquad (2)$$

Solution   First we solve one equation for one variable. To isolate y, we can add $-2x$ to both sides of Equation (1) to get

$$y = 6 - 2x. \qquad (3)$$

Then we substitute $6 - 2x$ for y in Equation (2) and solve for x:

$$3x + 4(6 - 2x) = 4 \qquad \text{Substituting } 6 - 2x \text{ for } y. \text{ Use parentheses!}$$
$$3x + 24 - 8x = 4 \qquad \text{Multiplying to remove parentheses}$$
$$3x - 8x = 4 - 24$$
$$-5x = -20$$
$$x = 4.$$

Now we can substitute in either Equation (1), (2), or (3). It is easiest to use Equation (3) since it is already solved for y:

$$y = 6 - 2x = 6 - 2(4) = 6 - 8 = -2.$$

The pair $(4, -2)$ appears to be the solution.

Check:

$2x + y = 6$		$3x + 4y = 4$	
$2(4) + (-2)$? 6		$3(4) + 4(-2)$? 4	
$8 - 2$		$12 - 8$	
6	6 TRUE	4	4 TRUE

Since $(4, -2)$ checks, it is the solution.   ❑

Some systems have no solution, as we saw graphically in Section 3.1. How do we recognize such systems if we are solving using an algebraic method?

EXAMPLE 3

Solve the system

$$y = -3x + 5,$$
$$y = -3x - 2.$$

Solution   We solved this problem graphically in Example 4(b) of Section 3.1, and found that the lines are parallel and the system has no solution. Let us now try to solve the system using the substitution method. We substitute $-3x - 2$ for y in the first equation:

$$-3x - 2 = -3x + 5 \qquad \text{Substituting } -3x - 2 \text{ for } y$$
$$-2 = 5. \qquad \text{Adding } 3x$$

When we add $3x$ to get the x-terms on one side, the x-terms drop out and we end up with a *false* equation. When solving algebraically yields a false equation, the system has no solution.   ❑

The Elimination Method

The *elimination method* for solving systems of equations makes use of the *addition principle:* If $a = b$, then $a + c = b + c$. Consider the following system:

$$2x - 3y = 0, \quad (1)$$
$$-4x + 3y = -1. \quad (2)$$

The key to the advantage of the elimination method for solving this system involves the $-3y$ in one equation and the $3y$ in the other. These terms are opposites. If we add the terms on the left side of the equations, these terms will add to 0, and in effect, the variable y will be "eliminated."

To use the addition principle for equations, note that according to Equation (2), $-4x + 3y$ and -1 are the same number. Thus we can use a vertical form and add $-4x + 3y$ to the left side of Equation (1) and -1 to the right side:

$$
\begin{array}{ll}
2x - 3y = 0 & (1) \\
\underline{-4x + 3y = -1} & (2) \\
-2x + 0y = -1. & \text{Adding}
\end{array}
$$

We have eliminated the variable y, which is why we call this the *elimination method.* We now have an equation with just one variable, x, which we solve for:

$$-2x = -1$$
$$x = \tfrac{1}{2}.$$

Next we substitute $\tfrac{1}{2}$ for x in Equation (1) and solve for y:

$$
\begin{array}{ll}
2 \cdot \tfrac{1}{2} - 3y = 0 & \text{Substituting} \\
\left.\begin{array}{l} 1 - 3y = 0 \\ y = \tfrac{1}{3}. \end{array}\right\} & \text{Solving for } y
\end{array}
$$

Check:

$$
\begin{array}{c|c}
2x - 3y = 0 & -4x + 3y = -1 \\
\hline
2\left(\tfrac{1}{2}\right) - 3\left(\tfrac{1}{3}\right) \ ? \ 0 & -4\left(\tfrac{1}{2}\right) + 3\left(\tfrac{1}{3}\right) \ ? \ -1 \\
1 - 1 & -2 + 1 \\
0 \ \big| \ 0 \ \text{TRUE} & -1 \ \big| \ -1 \ \text{TRUE}
\end{array}
$$

Since $\left(\tfrac{1}{2}, \tfrac{1}{3}\right)$ checks, it is the solution.

In order to eliminate a variable, we sometimes must multiply before adding.

| EXAMPLE 4 |

Solve the system

$$3x + 3y = 15, \quad (1)$$
$$2x + 6y = 22. \quad (2)$$

Solution If we add, we will not eliminate a variable. However, if the $3y$ in Equation (1) were $-6y$, we would. So we multiply by -2 on both sides of the first equation:

$$
\begin{array}{ll}
-6x - 6y = -30 & \text{Multiplying by } -2 \text{ on both sides of Equation (1)} \\
\underline{2x + 6y = 22} & \\
-4x + 0 = -8 & \text{Adding} \\
x = 2. & \text{Solving for } x
\end{array}
$$

Then

$$2 \cdot 2 + 6y = 22 \qquad \text{Substituting 2 for } x \text{ in Equation (2)}$$
$$4 + 6y = 22$$
$$y = 3. \qquad \text{Solving for } y$$

We obtain (2, 3), or $x = 2$, $y = 3$. This checks, so it is the solution. ❑

Sometimes we must multiply twice in order to make two terms become opposites.

EXAMPLE 5

Solve the system

$$2x + 3y = 17, \qquad (1)$$
$$5x + 7y = 29. \qquad (2)$$

Solution   We have

$$2x + 3y = 17, \longrightarrow \begin{array}{l}\text{Multiplying Equation} \\ \text{(1) by 5}\end{array} \longrightarrow \quad 10x + 15y = 85$$
$$5x + 7y = 29 \longrightarrow \begin{array}{l}\text{Multiplying Equation} \\ \text{(2) by } -2\end{array} \longrightarrow \quad \underline{-10x - 14y = -58}$$
$$0 + \quad y = 27 \qquad \text{Adding}$$
$$y = 27.$$

We then find that

$$2x + 3 \cdot 27 = 17 \qquad \text{Substituting 27 for } y \text{ in Equation (1)}$$
$$2x + 81 = 17$$
$$2x = -64$$
$$x = -32. \qquad \text{Solving for } x$$

Check:

$$\frac{2x + 3y = 17}{2(-32) + 3(27) \;?\; 17}$$
$$-64 + 81 \;\big|$$
$$17 \;\big|\; 17 \;\text{TRUE}$$

$$\frac{5x + 7y = 29}{5(-32) + 7(27) \;?\; 29}$$
$$-160 + 189 \;\big|$$
$$29 \;\big|\; 29 \;\text{TRUE}$$

We obtain (−32, 27), or $x = -32$, $y = 27$, as the solution. ❑

EXAMPLE 6

Solve the system

$$3y - 2x = 6, \qquad (1)$$
$$-12y + 8x = -24. \qquad (2)$$

Solution   We graphed this system in Example 4(c) of Section 3.1, and found that the lines coincided and the system has an infinite number of solutions. Suppose we try to solve this system using the elimination method:

$$12y - 8x = 24 \qquad \text{Multiplying Equation (1) by 4}$$
$$\underline{-12y + 8x = -24}$$
$$0 = 0. \qquad \text{Adding, we obtain a true equation.}$$

Note that we have eliminated both variables and what remains is a true equation. If a pair solves Equation (1), then it will also solve Equation (2). The equations are dependent and the solution set contains infinitely many pairs: $\{(x, y)|3y - 2x = 6\}$. ❑

When solving a system of two linear equations in two variables:

1. If a false equation is obtained, such as $0 = 7$, then the system has no solution. The system is inconsistent and the equations are independent.*
2. If a true equation is obtained, such as $0 = 0$, then the system has an infinite number of solutions. The system is consistent and the equations are dependent.

Should decimals or fractions appear, it often helps to *clear* before solving.

EXAMPLE 7

Solve the system

$$0.2x + 0.3y = 1.7,$$
$$\tfrac{1}{7}x + \tfrac{1}{5}y = \tfrac{29}{35}.$$

Solution   We have

$$0.2x + 0.3y = 1.7, \qquad \longrightarrow \text{Multiplying by 10} \longrightarrow \qquad 2x + 3y = 17$$
$$\tfrac{1}{7}x + \tfrac{1}{5}y = \tfrac{29}{35} \qquad \longrightarrow \text{Multiplying by 35} \longrightarrow \qquad 5x + 7y = 29.$$

We multiplied by 10 to clear the decimals. Multiplication by 35, the least common multiple of the denominators 7, 5, and 35, clears the fractions. The problem is now identical to the one solved in Example 5. The solution is $(-32, 27)$, or $x = -32$, $y = 27$. ❑

Comparing Methods

The following table is a summary that compares the graphical, substitution, and elimination methods for solving systems of equations.

Method	Strengths	Weaknesses
Graphical	Can "see" solutions.	Inexact when solutions involve numbers that are not integers.
Substitution	Yields exact solutions. Easy to use when a variable is alone on one side.	Introduces extensive computations with fractions when solving more complicated systems. Cannot "see" solutions quickly.
Elimination	Yields exact solutions. Easy to use when fractions or decimals appear in the system.	Cannot "see" solutions quickly.

*Consistency and dependency are discussed in detail in Section 3.4.

When deciding which method to use, consider this table and directions from your instructor. The situation is analogous to having a piece of wood to cut and three saws with which to cut it. The right saw for the job depends on the wood, the result you want, and how you want to go about it.

EXERCISE SET | 3.2

Solve using the substitution method.

1. $3x + 5y = 3,$
$x = 8 - 4y$

2. $2x - 3y = 13,$
$y = 5 - 4x$

3. $9x - 2y = 3,$
$3x - 6 = y$

4. $x = 3y - 3,$
$x + 2y = 9$

5. $5m + n = 8,$
$3m - 4n = 14$

6. $4x + y = 1,$
$x - 2y = 16$

7. $4x + 12y = 4,$
$-5x + y = 11$

8. $-3b + a = 7,$
$5a + 6b = 14$

9. $3x - y = 1,$
$2x + 2y = 2$

10. $5p + 7q = 1,$
$4p - 2q = 16$

11. $3x - y = 7,$
$2x + 2y = 5$

12. $5x + 3y = 4,$
$x - 4y = 3$

13. $x + 2y = 6,$
$x = 4 - 2y$

14. $y - 2x = 1,$
$2x - 3 = y$

15. $x - 3 = y,$
$2x - 2y = 6$

16. $3y = x - 2,$
$x = 2 + 3y$

Solve using the elimination method.

17. $x + 3y = 7,$
$-x + 4y = 7$

18. $x + y = 9,$
$2x - y = -3$

19. $2x + y = 6,$
$x - y = 3$

20. $x - 2y = 6,$
$-x + 3y = -4$

21. $9x + 3y = -3,$
$2x - 3y = -8$

22. $6x - 3y = 18,$
$6x + 3y = -12$

23. $5x + 3y = 19,$
$2x - 5y = 11$

24. $3x + 2y = 3,$
$9x - 8y = -2$

25. $5r - 3s = 24,$
$3r + 5s = 28$

26. $5x - 7y = -16,$
$2x + 8y = 26$

27. $0.3x - 0.2y = 4,$
$0.2x + 0.3y = 1$

28. $0.7x - 0.3y = 0.5,$
$-0.4x + 0.7y = 1.3$

29. $\frac{1}{2}x + \frac{1}{3}y = 4,$
$\frac{1}{4}x + \frac{1}{3}y = 3$

30. $\frac{2}{3}x + \frac{1}{7}y = -11,$
$\frac{1}{7}x - \frac{1}{3}y = -10$

31. $\frac{2}{5}x + \frac{1}{2}y = 2,$
$\frac{1}{2}x - \frac{1}{6}y = 3$

32. $\frac{1}{3}x + \frac{1}{5}y = 7,$
$\frac{1}{6}x - \frac{2}{5}y = -4$

33. $2x + 3y = 1,$
$4x + 6y = 2$

34. $3x - 2y = 1,$
$-6x + 4y = -2$

35. $2x - 4y = 5,$
$2x - 4y = 6$

36. $3x - 5y = -2,$
$5y - 3x = 7$

Solve.

37. $5x - 9y = 7,$
$7y - 3x = -5$

38. $a - 2b = 16,$
$b + 3 = 3a$

39. $3(a - b) = 15,$
$4a = b + 1$

40. $10x + y = 306,$
$10y + x = 90$

41. $x - \frac{1}{10}y = 100,$
$y - \frac{1}{10}x = -100$

42. $\frac{1}{8}x + \frac{3}{5}y = \frac{19}{2},$
$-\frac{3}{10}x - \frac{7}{20}y = -1$

43. $0.05x + 0.25y = 22,$
$0.15x + 0.05y = 24$

44. $1.3x - 0.2y = 12,$
$0.4x + 17y = 89$

Skill Maintenance

45. What rate of interest would have to be charged in order for a principal of \$320 to earn \$17.60 in $\frac{1}{2}$ year?

46. Simplify: $-9(y + 7) - 6(y - 4)$.

Synthesis

47. ◈ Write a system of linear equations that would be most easily solved using the substitution method. Explain why substitution would be easier to use than the elimination method.

48. ◈ Write a system of linear equations that would be most easily solved using the elimination method. Explain why elimination would be easier to use than the substitution method.

Solve.

49. ▦ $3.5x - 2.1y = 106.2,$
$4.1x + 16.7y = -106.28$

50. $\dfrac{x + y}{2} - \dfrac{x - y}{5} = 1,$

$\dfrac{x - y}{2} + \dfrac{x + y}{6} = -2$

51. Solve for x and y in terms of a and b:

$$5x + 2y = a,$$
$$x - y = b.$$

52. Determine a and b for which $(-4, -3)$ will be a solution of the system

$$ax + by = -26,$$
$$bx - ay = 7.$$

53. The points $(0, -3)$ and $\left(-\frac{3}{2}, 6\right)$ are two of the solutions of the equation $px - qy = -1$. Find p and q.

54. For $f(x) = mx + b$, two solutions are $(1, 2)$ and $(-3, 4)$. Find m and b.

Each of the following is a system of equations that is *not* linear. But each is *reducible to linear*, because an appropriate substitution (say, u for $1/x$ and v for $1/y$) yields a linear system. Solve for the new variable and then solve for the original variable.

55. $\dfrac{1}{x} - \dfrac{3}{y} = 2,$

$\dfrac{6}{x} + \dfrac{5}{y} = -34$

56. $\dfrac{2}{x} + \dfrac{1}{y} = 0,$

$\dfrac{5}{x} + \dfrac{2}{y} = -5$

3.3

Problem Solving Using Systems of Two Equations in Two Variables

Problems Involving Money • Problems Involving Mixtures •
Problems Involving Motion

You are in a much better position to solve problems now that systems of equations can be used. Using systems often makes the translating step easier.

EXAMPLE 1

Two angles are complementary. One angle is 12° less than three times the other. Find the measures of the angles.

Solution   The **Familiarize** and **Translate** steps have been done in Example 1 of Section 3.1. The resulting system of equations is

$$x + y = 90,$$
$$y = 3x - 12,$$

where $x =$ the measure of the first angle and $y =$ the measure of the other angle.

3. CARRY OUT.   We solve the system of equations. What method should we use?

Since we have a variable alone on one side, let's use the substitution method:

$$\begin{aligned}
x + y &= 90 \\
x + (3x - 12) &= 90 \qquad \text{Substituting } 3x - 12 \text{ for } y \\
4x - 12 &= 90 \qquad \text{Collecting like terms} \\
4x &= 102 \\
x &= 25.5.
\end{aligned}$$

We return to the second equation and substitute 25.5 for x and compute y:

$$y = 3x - 12 = 3(25.5) - 12 = 64.5.$$

The angles seem to be 25.5° and 64.5°.

4. CHECK. The sum of the angles is 90°, so they are complementary. Also, three times the 25.5° angle minus 12°, is 64.5°, the measure of the second angle. Thus the measures of the angles check.

5. STATE. The measure of one angle is 25.5°, and the measure of the other is 64.5°.

❏

EXAMPLE 2

In one day, Glovers, Inc., sold 20 pairs of gloves. Cloth gloves sold for $24.95 per pair and pigskin gloves sold for $37.50 per pair. The company took in $687.25. How many of each kind were sold?

Solution   The **Familiarize** and **Translate** steps were done in Example 2 of Section 3.1.

3. CARRY OUT. We are to solve the system of equations

$$c + p = 20, \qquad (1)$$
$$2495c + 3750p = 68{,}725, \qquad (2)$$

where $c = $ the number of pairs of cloth gloves sold and $p = $ the number of pairs of pigskin gloves sold. What method should we use? We could try graphing, but the large numbers would make that cumbersome. Since no variable appears alone and the equations are in the form $Ax + By = C$, let's try the elimination method. We eliminate c by multiplying Equation (1) by -2495 and adding it to Equation (2):

$$-2495c - 2495p = -49{,}900 \qquad \text{Multiplying Equation (1) by } -2495$$
$$\underline{2495c + 3750p = 68{,}725}$$
$$1255p = 18{,}825 \qquad \text{Adding}$$
$$p = 15. \qquad \text{Solving for } p$$

To find c, we substitute 15 for p in Equation (1) and solve for c:

$$c + p = 20 \qquad \text{Equation (1)}$$
$$c + 15 = 20 \qquad \text{Substituting 15 for } p$$
$$c = 5. \qquad \text{Solving for } c$$

We obtain (5, 15), or $c = 5$, $p = 15$.

4. CHECK. We check in the original problem. Remember that c is the number of cloth gloves and p is the number of pigskin gloves. Thus:

Number of gloves: $\qquad c + p = 5 + 15 = 20$
Money from cloth gloves: $\qquad \$24.95c = 24.95 \times 5 = \$124.75$
Money from pigskin gloves: $\qquad \$37.50p = 37.50 \times 15 = \underline{\$562.50}$
$\qquad \qquad \qquad \qquad \qquad \qquad \qquad \text{Total} = \$687.25$

The numbers check.

5. STATE. Glovers sold 5 pairs of cloth gloves and 15 pairs of pigskin gloves.

❏

EXAMPLE 3

Yardbirds Gardening, Inc., carries two brands of solutions containing weedkiller and water. "Gently Green" is 5% weedkiller and "Sun Saver" is 15% weedkiller. Yardbirds Gardening needs to combine the two types of solutions to make 100 L of a solution that is 12% weedkiller. How much of each brand should be used?

Solution

1. FAMILIARIZE. Suppose that 40 L of Gently Green and 60 L of Sun Saver are mixed. The resulting mixture will be the right size, 100 L, but will it be the right strength? To find out, note that 40 L of Gently Green would contribute $0.05(40) = 2$ L of weedkiller to the mixture. Since the Sun Saver is 15% weedkiller, 60 L would contribute $0.15(60) = 9$ L of weedkiller to the mixture. Altogether, 40 L of Gently Green and 60 L of Sun Saver would make 100 L of a mixture that has $2 + 9 = 11$ L of weedkiller. Since this would mean that the final mixture is 11% weedkiller, our guess of 40 L and 60 L is incorrect. Still, the process of checking our guess has familiarized us with the problem.

2. TRANSLATE. Let g = the number of liters of Gently Green and s = the number of liters of Sun Saver. The information can be organized in a table.

	Gently Green	Sun Saver	Mixture	
Number of Liters	g	s	100	$\longrightarrow g + s = 100$
Percent of Weedkiller	5%	15%	12%	
Amount of Weedkiller	$0.05g$	$0.15s$	0.12×100, or 12 liters	$\longrightarrow 0.05g + 0.15s = 12$

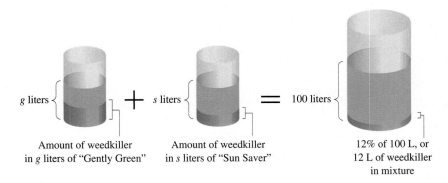

| Amount of weedkiller in g liters of "Gently Green" | Amount of weedkiller in s liters of "Sun Saver" | 12% of 100 L, or 12 L of weedkiller in mixture |

If we add g and s in the first row, we get one equation. It represents the total amount of mixture: $g + s = 100$.

If we add the amounts of weedkiller listed in the third row, we get a second equation. This equation represents the amount of weedkiller in the mixture: $0.05g + 0.15s = 12$.

After clearing decimals, we have the problem translated to the system

$$g + s = 100, \quad (1)$$
$$5g + 15s = 1200. \quad (2)$$

3. CARRY OUT. We use the elimination method to solve the system:

$$-5g - 5s = -500 \qquad \text{Multiplying Equation (1) by } -5$$
$$\underline{5g + 15s = 1200}$$
$$10s = 700 \qquad \text{Adding}$$
$$s = 70; \qquad \text{Solving for } s$$
$$g + 70 = 100 \qquad \text{Substituting into Equation (1)}$$
$$g = 30. \qquad \text{Solving for } g$$

4. CHECK. Remember, g is the number of liters of Gently Green and s is the number of liters of Sun Saver.

Total number of
liters of mixture: $\qquad\qquad g + s = 30 + 70 = 100$

Total amount
of weedkiller: $\quad 5\% \times 30 + 15\% \times 70 = 1.5 + 10.5 = 12$

Percentage of
weedkiller in $\qquad \dfrac{\text{Total amount of weedkiller}}{\text{Total number of liters of mixture}} = \dfrac{12}{100} = 12\%$
mixture:

The numbers do check in the original problem.

5. STATE. Yardbird Gardening should mix 30 L of Gently Green with 70 L of Sun Saver. □

EXAMPLE 4

Two investments are made totaling $4800. Part of the money is invested at 8% and the rest at 9%. In the first year, the total yield is $412 in simple interest. Find the amount invested at each rate of interest.

Solution

1. FAMILIARIZE. As in Example 3, we can begin with a guess. If $1000 was invested at 8% and $3800 was invested at 9%, the two investments would total $4800. The interest would then be 8%($1000), or $80, and 9%($3800), or $342, for a total of $422 in interest. Our guess was wrong, but the manner in which we checked the guess has familiarized us with the problem.

2. TRANSLATE. Let $x =$ the amount of money invested at 8% and $y =$ the amount of money invested at 9%. We can then organize the information in a table. Each column in the table comes from the formula for simple interest: *Interest = Principal · Rate · Time*.

	First Investment	Second Investment	Total	
Principal	x	y	$4800	⟶ $x + y = \$4800$
Rate of Interest	8%	9%		
Time	1 yr	1 yr		
Interest	$0.08x$	$0.09y$	$412	⟶ $0.08x + 0.09y = \$412$

The total of the amounts invested is found in the first row. This gives us one equation:

$x + y = 4800.$

Look at the last row. The interest, or *yield*, totals $412. This gives us a second equation:

$8\%x + 9\%y = 412,$ or $0.08x + 0.09y = 412.$

After we multiply on both sides to clear the decimals, we have

$8x + 9y = 41{,}200.$

3. CARRY OUT. The following system can be solved by elimination or substitution:

$$x + y = 4800,$$
$$8x + 9y = 41{,}200.$$

We find that $x = 2000$ and $y = 2800$.

4. CHECK. The sum is $2000 + $2800, or $4800. The interest from $2000 at 8% for 1 yr is 8%($2000), or $160. The interest from $2800 at 9% for 1 yr is 9%($2800), or $252. The total interest is $160 + $252, or $412, so the numbers check.

5. STATE. An amount of $2000 was invested at 8% and $2800 at 9%. ❑

EXAMPLE 5

The ground floor of the John Hancock building in Chicago is a rectangle whose perimeter is 860 ft. The length is 100 ft more than the width. Find the length and the width.

$100 + w$

Solution

1. FAMILIARIZE. We make a drawing and label it. We recall, or look up, the definition of *perimeter:* $P = 2l + 2w$. We let l = the length of the ground floor of the building and w = the width. Guesses could be made now, but this time we proceed to the next step.

2. TRANSLATE.   We translate as follows:

The perimeter is 860.

$$2l + 2w \quad = 860$$

We translate the second statement:

The length is 100 ft more than the width.

$$l \quad = \quad w + 100$$

This gives us the system of equations

$$2l + 2w = 860, \qquad (1)$$
$$l = w + 100. \qquad (2)$$

3. CARRY OUT.   In this case, it is probably easier to use the substitution method:

$2(w + 100) + 2w = 860$	Substituting $w + 100$ for l in Equation (1)
$2w + 200 + 2w = 860$	Multiplying to remove parentheses
$4w + 200 = 860$	Collecting like terms
$4w = 660$	Adding -200
$w = 165.$	Multiplying by $\frac{1}{4}$

Then we substitute 165 for w in Equation (2):

$$l = 165 + 100 = 265.$$

4. CHECK.   The length is 100 ft more than the width, and the perimeter is $2(265 \text{ ft}) + 2(165 \text{ ft})$, or 860 ft. The numbers check in the original problem.

5. STATE.   The length is 265 ft and the width is 165 ft.   ❑

Problems Involving Motion

When a problem deals with distance, speed (rate), and time, we need to recall the following.

If r represents rate, t represents time, and d represents distance, then:

$$d = rt, \qquad r = \frac{d}{t}, \qquad \text{and} \qquad t = \frac{d}{r}.$$

You should remember at least one of these equations and obtain the others by using algebraic manipulations as needed in a problem situation.

EXAMPLE 6

A freight train leaves Ames traveling east at a speed of 60 km/h. Two hours later, a passenger train leaves Ames traveling in the same direction on a parallel track at 90 km/h. At what point will the passenger train catch up to the freight train?

Solution

1. FAMILIARIZE. To familiarize ourselves with the problem, we make a guess — say, 180 km — and check to see if it is correct. A freight train, traveling 60 km/h, would reach a point 180 km from Ames in $\frac{180}{60} = 3$ hr. The passenger train, traveling 90 km/h, would cover 180 km in $\frac{180}{90} = 2$ hr. Since the problem states that the freight train has been running for two hours before the passenger train departs, and since $\frac{180}{60} = 3$ hr is *not* two hours more than $\frac{180}{90} = 2$ hr, we see that our guess of 180 km is incorrect. Although our guess was wrong, we now see that the time that the trains are running is an unknown as well as the point at which they meet up. Let $t =$ the time the freight train is running, $t - 2 =$ the time the passenger train is running, and $d =$ the distance at which the trains meet. Then make a sketch.

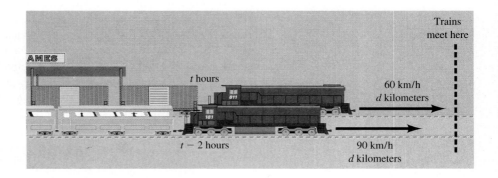

2. TRANSLATE. We can organize the information in a chart. Each row is determined by the formula $d = rt$.

$$d \;=\; r \;\cdot\; t$$

	Distance	Rate	Time
Freight Train	d	60	t	⟶ $d = 60t,$
Passenger Train	d	90	$t - 2$	⟶ $d = 90(t - 2)$

Using $d = rt$ in each row of the table, we get an equation. Thus we have a system of equations:

$$d = 60t, \qquad (1)$$
$$d = 90(t - 2). \qquad (2)$$

3. CARRY OUT. We solve the system:

$$60t = 90(t - 2) \qquad \text{Substituting } 60t \text{ for } d \text{ in Equation (2)}$$
$$60t = 90t - 180$$
$$-30t = -180$$
$$t = 6.$$

The time for the freight train is 6 hr, which means that the time for the passenger train is $6 - 2$, or 4 hr. For $t = 6$, we have $d = 60 \cdot 6 = 360$ km.

4. CHECK. At 60 km/h, the freight train will travel $60 \cdot 6$, or 360 km, in 6 hr. At 90 km/h, the passenger train will travel $90 \cdot (6 - 2) = 360$ km in 4 hr. Remember that it is distance, not time, that the problem asked for. The numbers check.

5. STATE. The trains will meet at a point 360 km east of Ames.   ❏

EXAMPLE 7

A motorboat took 4 hr to make a trip downstream with a 6-mph current. The return trip against the same current took 5 hr. Find the speed of the boat in still water.

Solution

1. FAMILIARIZE. We visualize the problem mentally and then make a sketch. Observe that the current *slows* the boat traveling upstream and *speeds up* the boat traveling downstream. Since the distances traveled each way are the same and the times are known, we could make a guess of, say, 40 mph for the speed of the boat in still water. The boat would then go $40 - 6 = 34$ mph upstream and would travel $34 \cdot 5 = 170$ mi against the current. The boat would go $40 + 6 = 46$ mph downstream, and would travel $46 \cdot 4 = 184$ mi downstream. Since $170 \neq 184$, our guess of 40 mph is incorrect. Rather than guess again, we decide to let $r =$ the speed, in miles per hour, of the boat in still water, $r - 6 =$ the boat's speed going upstream, and $r + 6 =$ the boat's speed going downstream. We also let $d =$ the distance traveled, in miles.

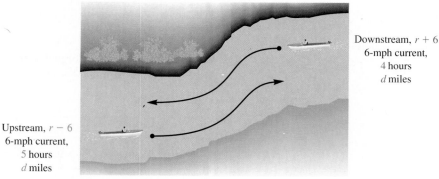

Downstream, $r + 6$
6-mph current,
4 hours
d miles

Upstream, $r - 6$
6-mph current,
5 hours
d miles

2. TRANSLATE. The information can be organized in a chart. The distances traveled are the same, so we use the formula *Distance = Rate* (or *Speed*) $\cdot$ *Time*. Each row of the chart gives an equation.

	Distance	Rate	Time	
Downstream	d	$r + 6$	4	$\longrightarrow d = (r + 6)4$
Upstream	d	$r - 6$	5	$\longrightarrow d = (r - 6)5$

The two equations constitute a system:

$$d = (r + 6)4, \quad (1)$$
$$d = (r - 6)5. \quad (2)$$

3. CARRY OUT. We solve the system:

$(r - 6)5 = (r + 6)4$ Substituting $(r - 6)5$ for d in Equation (1)

$5r - 30 = 4r + 24$ Using the distributive law

$r = 54.$ Solving for r

4. CHECK. When $r = 54$, $r + 6 = 60$, and $60 \cdot 4 = 240$, the distance. When $r = 54$, $r - 6 = 48$, and $48 \cdot 5 = 240$. In both cases, we get the same distance.

5. STATE. The speed of the boat in still water is 54 mph.   ❑

Tips for Solving Motion Problems

1. Draw a diagram using an arrow or arrows to represent distance and the direction of each object in motion.
2. Organize the information in a chart.
3. Look for as many things as you can that are the same, so you can write equations.
4. Translating to a system of equations eases the solution of many motion problems.
5. When checking, be sure that you have solved for what the problem asked for.

EXERCISE SET | 3.3

1.–16. For Exercises 1–16, solve Exercises 1–16 from Exercise Set 3.1.

17. Soybean meal is 16% protein and corn meal is 9% protein. How many pounds of each should be mixed together in order to get a 350-lb mixture that is 12% protein?

18. Lucinda has one solution that is 25% acid and a second that is 50% acid. How many liters of each should be mixed together in order to get 10 L of a solution that is 40% acid?

19. "Arctic Antifreeze" is 18% alcohol and "Frost-No-More" is 10% alcohol. How many liters of each should be mixed together in order to get 20 L of a mixture that is 15% alcohol?

20. "Orange-Thirst" is 15% orange juice and "Quen-cho" is 5% orange juice. How many liters of each should be mixed together in order to get 10 L of a mixture that is 10% orange juice?

21. Two investments are made totaling $15,000. For a certain year, these investments yield $1432 in simple interest. Part of the $15,000 is invested at 9% and part at 10%. Find the amount invested at each rate.

22. Two investments are made totaling $8800. For a certain year, these investments yield $1326 in simple interest. Part of the $8800 is invested at 14% and part at 16%. Find the amount invested at each rate.

23. $27,000 is invested, part of it at 10% and part of it at 12%. The total yield at simple interest for one year is $2990. How much was invested at each rate?

24. $1150 is invested, part of it at 12% and part of it at 11%. The total yield at simple interest for one year is $133.75. How much was invested at each rate?

25. At a club play, 117 tickets were sold. Adults' tickets cost $1.25 each and children's tickets cost $0.75 each. In all, $129.75 was taken in. How many of each kind of ticket were sold?

26. I.Q. Bean's sold 30 sweatshirts. White ones cost $9.95 and yellow ones cost $10.50. In all, $310.60

worth of sweatshirts were sold. How many of each color were sold?

27. Paula is 12 years older than her brother Bob. Four years from now, Bob will be $\frac{2}{3}$ as old as Paula. How old are they now?

28. Carlos is 8 years older than his sister Maria. Four years ago, Maria was $\frac{2}{3}$ as old as Carlos. How old are they now?

29. The perimeter of a lot is 190 m. The width is one fourth of the length. Find the dimensions.

30. The perimeter of a rectangular field is 194 yd. The length is two more than four times the width. Find the dimensions.

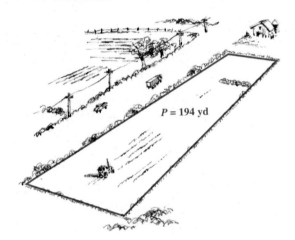

$P = 194$ yd

31. A customer goes to a bank and gets change for a $50 bill consisting of all $5 bills and $1 bills. There are 22 bills in all. How many of each kind are there?

32. Cecilia makes a $9.25 purchase at the bookstore with a $20 bill. The store has no bills and gives her the change in quarters and fifty-cent pieces. There are 30 coins in all. How many of each kind are there?

33. A train leaves Danville Junction and travels north at a speed of 75 km/h. Two hours later, a second train leaves on a parallel track and travels north at 125 km/h. How far from the station will they meet?

34. Two cars leave Denver traveling in opposite directions. One car travels at a speed of 80 km/h and the other at 96 km/h. In how many hours will they be 528 km apart?

35. Two motorcycles travel toward each other from Chicago and Indianapolis, which are about 350 km apart, at rates of 110 km/h and 90 km/h. They started at the same time. In how many hours will they meet?

36. Two planes travel toward each other from cities that are 780 km apart at rates of 190 km/h and 200 km/h. They started at the same time. In how many hours will they meet?

37. A motorboat took 3 hr to make a trip downstream with a 6-mph current. The return trip against the same current took 5 hr. Find the speed of the boat in still water.

38. A canoeist paddled for 4 hr with a 6-km/h current to reach a campsite. The return trip against the same current took 10 hr. Find the speed of the canoe in still water.

Skill Maintenance

39. Simplify: $-3(x - 7) - 2[x - (4 + 3x)]$.

40. Solve: $3(x - 5) = 7(x - 6)$.

41. Write an equation of a line containing $(2, -5)$ with slope $-\frac{3}{4}$.

42. Graph: $y = -\frac{1}{2}x + 5$.

Synthesis

43.–46. For Exercises 43–46, solve Exercises 63–66 from Exercise Set 3.1.

47. The radiator in Michelle's car contains 16 L of antifreeze and water. This mixture is 30% antifreeze. How much of this mixture should she drain and replace with pure antifreeze so that there will be a mixture of 50% antifreeze?

48. Natalie jogs and walks to school each day. She averages 4 km/h walking and 8 km/h jogging. The distance from home to school is 6 km and she makes the trip in 1 hr. How far does she jog in a trip?

49. The ten's digit of a two-digit positive integer is two more than three times the unit's digit. If the digits are interchanged, the new number is thirteen less than half the given number. Find the given integer. (*Hint:* Let $x =$ the ten's-place digit and $y =$ the unit's-place digit; then $10x + y$ is the number.)

50. A limited edition of a book published by a historical society was offered for sale to its membership. The cost was one book for $12 or two books for $20. The society sold 880 books, and the total amount of money taken in was $9840. How many members ordered two books?

51. A train leaves Union Station for Central Station, 216 km away, at 9 A.M. One hour later, a train leaves Central Station for Union Station. They meet at noon. If the second train had started at 9 A.M. and the first train at 10:30 A.M., they would still have met at noon. Find the speed of each train.

52. Dianne Osborne's station wagon gets 18 miles per gallon (mpg) in city driving and 24 mpg in highway driving. The car is driven 465 mi on 23 gal of gasoline. How many miles were driven in the city and how many were driven on the highway?

53. Phil and Phyllis are siblings. Phyllis has twice as many brothers as she has sisters. Phil has the same number of brothers as sisters. How many girls and how many boys are in the family?

54. ◈ Write three or four study tips of your own for use by someone beginning this exercise set.

3.4

Systems of Equations in Three Variables

Identifying Solutions • Solving Systems in Three Variables • Dependency, Inconsistency, and Geometric Considerations

Some problem situations easily translate to two equations. Others more naturally call for a translation to three or more equations. In this section, we consider how to solve systems of three linear equations. Later, we will use such systems in problem-solving situations.

Identifying Solutions

A **linear equation in three variables** is an equation equivalent to one in the form $Ax + By + Cz = D$, where A, B, C, and D are real numbers. We will refer to the form $Ax + By + Cz = D$ as *standard form* for a linear equation in three variables.

A solution of a system of three equations in three variables is an ordered triple (p, q, r) that makes *all three* equations true.

EXAMPLE 1

Determine whether $\left(\frac{3}{2}, -4, 3\right)$ is a solution of the system

$$4x - 2y - 3z = 5,$$
$$-8x - y + z = -5,$$
$$2x + y + 2z = 5.$$

Solution   We substitute $\left(\frac{3}{2}, -4, 3\right)$ into the three equations, using alphabetical order.

$$\frac{4x - 2y - 3z = 5}{4 \cdot \frac{3}{2} - 2(-4) - 3 \cdot 3 \ ? \ 5}$$
$$6 + 8 - 9$$
$$5 \mid 5 \ \text{TRUE}$$

$$\frac{-8x - y + z = -5}{-8 \cdot \frac{3}{2} - (-4) + 3 \ ? \ -5}$$
$$-12 + 4 + 3$$
$$-5 \mid -5 \ \text{TRUE}$$

$$\frac{2x + y + 2z = 5}{2 \cdot \frac{3}{2} + (-4) + 2 \cdot 3 \ ? \ 5}$$
$$3 - 4 + 6$$
$$5 \mid 5 \ \text{TRUE}$$

The triple makes all three equations true, so it is a solution.   ❏

Solving Systems in Three Variables

Graphical methods for solving linear equations in three variables are unsatisfactory, because a three-dimensional coordinate system is required and the graph of a linear equation in three variables is a plane. The substitution method can be used in any situation, but it is not as useful unless one or more of the equations has only two variables. Fortunately, we can use the elimination method to eliminate a variable and obtain a system of two equations in two variables.

EXAMPLE 2

Solve the following system of equations:

$$
\begin{aligned}
x + y + z &= 4, &\quad (1) \\
x - 2y - z &= 1, &\quad (2) \\
2x - y - 2z &= -1. &\quad (3)
\end{aligned}
$$

Solution   We first use *any* two of the three equations to get an equation in two variables. Let's use Equations (1) and (2) and add to eliminate z:

$$
\begin{array}{ll}
x + y + z = 4 & (1) \\
\underline{x - 2y - z = 1} & (2) \\
2x - y \phantom{{}-z} = 5. & (4) \qquad \text{Adding}
\end{array}
$$

Next we use a different pair of equations and eliminate the *same variable* we did above. Let's use Equations (1) and (3) to again eliminate z. Be careful here! A common error is to eliminate a different variable in this step.

$$
\begin{array}{l}
x + y + z = 4, \\
2x - y - 2z = -1
\end{array}
\xrightarrow[\text{(1) by 2}]{\text{Multiplying Equation}}
\begin{array}{l}
2x + 2y + 2z = 8 \\
\underline{2x - y - 2z = -1} \\
4x + y \phantom{{}- 2z} = 7 \qquad (5)
\end{array}
$$

Now we solve the resulting system of Equations (4) and (5). That solution will give us two of the numbers.

$$
\begin{array}{ll}
2x - y = 5 & (4) \\
\underline{4x + y = 7} & (5) \\
6x \phantom{{}- y} = 12 & \text{Adding} \\
x = 2
\end{array}
$$

Note that we now have two equations in two variables. Had we eliminated different variables above, this would not be the case.

We can use either Equation (4) or (5) to find y. We choose Equation (5):

$$
\begin{array}{ll}
4x + y = 7 & (5) \\
4 \cdot 2 + y = 7 & \text{Substituting 2 for } x \text{ in Equation (5)} \\
8 + y = 7 \\
y = -1.
\end{array}
$$

We have $x = 2$ and $y = -1$. To find the value for z, we use any of the original three equations and substitute to find the third number, z. Let's use Equation (1) and substitute our two numbers in it:

$$
\begin{array}{ll}
x + y + z = 4 & (1) \\
2 + (-1) + z = 4 & \text{Substituting 2 for } x \text{ and } -1 \text{ for } y \\
1 + z = 4 \\
z = 3.
\end{array}
$$

We have obtained the triple $(2, -1, 3)$. We now check in *all three* equations:

$$\frac{x + y + z = 4}{2 + (-1) + 3 \; ? \; 4}$$
$$4 \mid 4 \quad \text{TRUE}$$

$$\frac{x - 2y - z = 1}{2 - 2(-1) - 3 \; ? \; 1}$$
$$1 \mid 1 \quad \text{TRUE}$$

$$\frac{2x - y - 2z = -1}{2 \cdot 2 - (-1) - 2 \cdot 3 \; ? \; -1}$$
$$-1 \mid -1 \quad \text{TRUE}$$

The solution is $(2, -1, 3)$. ❑

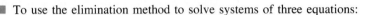

To use the elimination method to solve systems of three equations:

1. Write all equations in the standard form $Ax + By + Cz = D$.
2. Clear any decimals or fractions.
3. Choose a variable to eliminate. Then use any two of the three equations to get an equation in two variables.
4. Next use a different pair of equations and get another equation *in the same two variables*. That is, eliminate the same variable that you did in step (3).
5. Solve the resulting system (pair) of equations. That will give two of the numbers.
6. Then use any of the original three equations to find the third number.

EXAMPLE 3

Solve the system

$$4x - 2y - 3z = 5, \qquad (1)$$
$$-8x - y + z = -5, \qquad (2)$$
$$2x + y + 2z = 5. \qquad (3)$$

Solution

1., 2. The equations are already in standard form and there are no fractions or decimals.

3. We must choose a variable to eliminate. We decide on y because the y-terms are opposites of each other in Equations (2) and (3). We add:

$$-8x - y + z = -5 \qquad (2)$$
$$\underline{2x + y + 2z = \quad 5} \qquad (3)$$
$$-6x \qquad + 3z = \quad 0. \qquad (4) \qquad \text{Adding}$$

4. We use another pair of equations to get an equation in the same two variables, x and z. That is, we eliminate the same variable, y, that we did in step (3). We use Equations (1) and (3) and eliminate y:

$$4x - 2y - 3z = 5,$$
$$2x + y + 2z = 5 \longrightarrow \begin{array}{c} \text{Multiplying Equation} \\ \text{(3) by 2} \end{array} \longrightarrow$$

$$4x - 2y - 3z = \quad 5$$
$$\underline{4x + 2y + 4z = 10}$$
$$8x \qquad + z = 15. \qquad (5)$$

5. Now we solve the resulting system of Equations (4) and (5). That will give us two of the numbers.

$$-6x + 3z = 0,$$
$$8x + z = 15 \longrightarrow \begin{array}{c}\text{Multiplying Equation}\\ \text{(5) by } -3\end{array} \longrightarrow$$

$$\begin{array}{r} -6x + 3z = 0 \\ -24x - 3z = -45 \\ \hline -30x \quad\quad = -45 \\ x = \frac{-45}{-30} = \frac{3}{2} \end{array}$$

We use Equation (5) to find z:

$$8x + z = 15$$
$$8 \cdot \tfrac{3}{2} + z = 15 \qquad \text{Substituting } \tfrac{3}{2} \text{ for } x$$
$$12 + z = 15$$
$$z = 3.$$

6. Then we use any of the original equations and substitute to find the third number, y. We choose Equation (3):

$$2x + y + 2z = 5$$
$$2 \cdot \tfrac{3}{2} + y + 2 \cdot 3 = 5 \qquad \text{Substituting } \tfrac{3}{2} \text{ for } x \text{ and } 3 \text{ for } z$$
$$3 + y + 6 = 5$$
$$y + 9 = 5$$
$$y = -4.$$

The solution is $\left(\tfrac{3}{2}, -4, 3\right)$. The check is in Example 1.   ❑

Sometimes, certain variables are missing at the outset.

EXAMPLE 4

Solve the system

$$\begin{array}{rl} x + y + z = 180, & (1) \\ x \quad\quad - z = -70, & (2) \\ 2y - z = 0. & (3) \end{array}$$

Solution

1., 2. The equations are already in standard form with no fractions or decimals.

3., 4. Observe that there is no y in Equation (2). Thus, at the outset, we already have y eliminated from one equation. We need another equation with y eliminated. Let's use Equations (1) and (3):

$$\begin{array}{rl} x + y + z = 180, & \\ 2y - z = 0 & \end{array} \rightarrow \begin{array}{c}\text{Multiplying Equation}\\ \text{(1) by } -2\end{array} \rightarrow$$

$$\begin{array}{r} -2x - 2y - 2z = -360 \\ 2y - z = 0 \\ \hline -2x \quad - 3z = -360. \quad (4) \end{array}$$

5., 6. Now we solve the resulting system of Equations (2) and (4):

$$\begin{array}{rl} x - z = -70, & \\ -2x - 3z = -360 & \end{array} \rightarrow \begin{array}{c}\text{Multiplying Equation}\\ \text{(2) by } 2\end{array} \rightarrow$$

$$\begin{array}{r} 2x - 2z = -140 \\ -2x - 3z = -360 \\ \hline -5z = -500 \\ z = 100. \end{array}$$

Continuing as in previous examples, we get the solution $(30, 50, 100)$.   ❑

Dependency, Inconsistency, and Geometric Considerations

Each equation in Examples 2, 3, and 4 has a graph that is a plane in three dimensions. The solution in each case is a point common to all the planes of the system. Since it is possible for three planes to have infinitely many points in common or no points at all in common, we need to extend the notions of dependent equations and inconsistent systems to systems of equations in three variables.

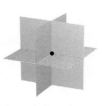  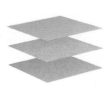 

| One solution: planes intersecting in exactly one point. | The planes intersect along a common line. There are infinitely many points common to the three planes. | Three parallel planes. There is no common point of intersection. | Planes intersect two at a time, but there is no point common to all three. |

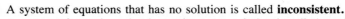

A system of equations that has no solution is called **inconsistent.**
A system of equations that has at least one solution is called **consistent.**

EXAMPLE 5

Solve:

$$y + 3z = 4, \qquad (1)$$
$$-x - y + 2z = 0, \qquad (2)$$
$$x + 2y + z = 1. \qquad (3)$$

Solution   The variable x is missing in Equation (1). Thus we can use Equations (2) and (3) and add to eliminate x. We get

$$-x - y + 2z = 0 \qquad (2)$$
$$\underline{x + 2y + z = 1} \qquad (3)$$
$$y + 3z = 1. \qquad (4) \qquad \text{Adding}$$

Equations (1) and (4) are both in y and z. We multiply Equation (1) by -1 and add:

$$y + 3z = 4, \quad \rightarrow \text{Multiplying Equation} \rightarrow \quad -y - 3z = -4$$
$$y + 3z = 1 \qquad \text{(1) by } -1 \qquad \qquad \underline{y + 3z = \quad 1}$$
$$0 = -3. \qquad \text{Adding}$$

Since we end up with a *false* equation, we know that the system has no solution. It is *inconsistent.* The solution set is $\varnothing$.   ❑

We now extend our definition of the word dependent from its meaning in Section 3.1.

> If a system of n linear equations is equivalent to a system of fewer than n of them, we say that the equations are *dependent*. If such is not the case, we call the equations *independent*.

EXAMPLE 6

Solve:

$$2x + y + z = 3, \quad (1)$$
$$x - 2y - z = 1, \quad (2)$$
$$3x + 4y + 3z = 5. \quad (3)$$

Solution   We use Equations (1) and (2) and add to eliminate z:

$$2x + y + z = 3$$
$$\underline{x - 2y - z = 1}$$
$$3x - y \quad\;\; = 4. \quad (4)$$

Next we use Equations (2) and (3) to eliminate z again:

$$x - 2y - z = 1, \quad \xrightarrow[\text{(2) by 3}]{\text{Multiplying Equation}} \quad 3x - 6y - 3z = 3$$
$$3x + 4y + 3z = 5 \qquad\qquad\qquad\qquad \underline{3x + 4y + 3z = 5}$$
$$\qquad\qquad\qquad\qquad\qquad\qquad\qquad 6x - 2y \quad\;\; = 8. \quad (5)$$

We now try to solve the resulting system of Equations (4) and (5):

$$3x - y = 4, \quad \xrightarrow[\text{(4) by } -2]{\text{Multiplying Equation}} \quad -6x + 2y = -8$$
$$6x - 2y = 8 \qquad\qquad\qquad\qquad\quad \underline{6x - 2y = \quad 8}$$
$$\qquad\qquad\qquad\qquad\qquad\qquad\qquad 0 = 0. \quad (6)$$

Equation (6) indicates that Equations (1), (2), and (3) are *dependent*. To see that the original system of three equations is equivalent to a system of two equations, note that two times Equation (1), minus Equation (2), is Equation (3). Thus, removing Equation (3) from the system does not affect the solution of the system. In writing an answer to this problem, we simply state that the equations are dependent.   ❏

The observant student might have noticed that when dependent equations appeared in Section 3.1, the solution sets were always infinite in size and were written in set-builder notation. That is, in Section 3.1, all systems of dependent equations were *consistent*. This is not always the case for systems of three equations. The following figure illustrates some possibilities geometrically.

The planes intersect along a common line. The equations are dependent and the system is consistent. There is an infinite number of solutions.

The planes coincide. The equations are dependent and the system is consistent. There is an infinite number of solutions.

Two planes coincide. The third plane is parallel. The equations are dependent and the system is inconsistent. There is no solution.

EXERCISE SET | 3.4

1. Determine whether $(1, -2, 3)$ is a solution of the system

$$x + y + z = 2,$$
$$x - 2y - z = 2,$$
$$3x + 2y + z = 2.$$

2. Determine whether $(2, -1, -2)$ is a solution of the system

$$x + y - 2z = 5,$$
$$2x - y - z = 7,$$
$$-x - 2y + 3z = 6.$$

Solve.

3. $x + y + z = 6,$
$2x - y + 3z = 9,$
$-x + 2y + 2z = 9$

4. $2x - y + z = 10,$
$4x + 2y - 3z = 10,$
$x - 3y + 2z = 8$

5. $2x - y - 3z = -1,$
$2x - y + z = -9,$
$x + 2y - 4z = 17$

6. $x - y + z = 6,$
$2x + 3y + 2z = 2,$
$3x + 5y + 4z = 4$

7. $2x - 3y + z = 5,$
$x + 3y + 8z = 22,$
$3x - y + 2z = 12$

8. $6x - 4y + 5z = 31,$
$5x + 2y + 2z = 13,$
$x + y + z = 2$

9. $3a - 2b + 7c = 13,$
$a + 8b - 6c = -47,$
$7a - 9b - 9c = -3$

10. $x + y + z = 0,$
$2x + 3y + 2z = -3,$
$-x + 2y - 3z = -1$

11. $2x + 3y + z = 17,$
$x - 3y + 2z = -8,$
$5x - 2y + 3z = 5$

12. $2x + y - 3z = -4,$
$4x - 2y + z = 9,$
$3x + 5y - 2z = 5$

13. $2x + y + z = -2,$
$2x - y + 3z = 6,$
$3x - 5y + 4z = 7$

14. $2x + y + 2z = 11,$
$3x + 2y + 2z = 8,$
$x + 4y + 3z = 0$

15. $x - y + z = 4,$
$5x + 2y - 3z = 2,$
$4x + 3y - 4z = -2$

16. $-2x + 8y + 2z = 4,$
$x + 6y + 3z = 4,$
$3x - 2y + z = 0$

17. $4x - y - z = 4,$
$2x + y + z = -1,$
$6x - 3y - 2z = 3$

18. $a + 2b + c = 1,$
$7a + 3b - c = -2,$
$a + 5b + 3c = 2$

19. $2r + 3s + 12t = 4,$
$4r - 6s + 6t = 1,$
$r + s + t = 1$

20. $10x + 6y + z = 7,$
$5x - 9y - 2z = 3,$
$15x - 12y + 2z = -5$

21. $4a + 9b = 8,$
$8a + 6c = -1,$
$6b + 6c = -1$

22. $3p + 2r = 11,$
$q - 7r = 4,$
$p - 6q = 1$

23. $x + y + z = 57,$
$-2x + y = 3,$
$x - z = 6$

24. $x + y + z = 105,$
$10y - z = 11,$
$2x - 3y = 7$

25. $2a - 3b = 2,$
$7a + 4c = \frac{3}{4},$
$2c - 3b = 1$

26. $a - 3c = 6,$
$b + 2c = 2,$
$7a - 3b - 5c = 14$

27. $x + y + z = 180,$
$y = 2 + 3x,$
$z = 80 + x$

28. $l + m = 7,$
$3m + 2n = 9,$
$4l + n = 5$

29. $x + z = 0,$
$x + y + 2z = 3,$
$y + z = 2$

30. $x + y = 0,$
$x + z = 1,$
$2x + y + z = 2$

31. $x + y + z = 1,$
$-x + 2y + z = 2,$
$2x - y = -1$

32. $y + z = 1,$
$x + y + z = 1,$
$x + 2y + 2z = 2$

Skill Maintenance

33. Solve $F = \frac{1}{2}t(c - d)$ for c.

34. Solve $F = \frac{1}{2}t(c - d)$ for d.

Synthesis

35. ◈ Suggest a procedure that could be used to solve a system of four equations in four variables.

36. ◈ Explain, in your own words, what it means for the equations of a system to be dependent.

Solve.

37. $\dfrac{x + 2}{3} - \dfrac{y + 4}{2} + \dfrac{z + 1}{6} = 0,$

$\dfrac{x - 4}{3} + \dfrac{y + 1}{4} - \dfrac{z - 2}{2} = -1,$

$\dfrac{x + 1}{2} + \dfrac{y}{2} + \dfrac{z - 1}{4} = \dfrac{3}{4}$

38. $0.2x + 0.3y + 1.1z = 1.6,$
$0.5x - 0.2y + 0.4z = 0.7,$
$-1.2x + y - 0.7z = -0.9$

39. $w + x + y + z = 2,$
$w + 2x + 2y + 4z = 1,$
$w - x + y + z = 6,$
$w - 3x - y + z = 2$

40. $w + x - y + z = 0,$
$w - 2x - 2y - z = -5,$
$w - 3x - y + z = 4,$
$2w - x - y + 3z = 7$

For Exercises 41 and 42, let u represent $1/x$, v represent $1/y$, and w represent $1/z$. Solve for u, v, and w, and then solve for x, y, and z.

41. $\dfrac{2}{x} - \dfrac{1}{y} - \dfrac{3}{z} = -1,$

$\dfrac{2}{x} - \dfrac{1}{y} + \dfrac{1}{z} = -9,$

$\dfrac{1}{x} + \dfrac{2}{y} - \dfrac{4}{z} = 17$

42. $\dfrac{2}{x} + \dfrac{2}{y} - \dfrac{3}{z} = 3,$

$\dfrac{1}{x} - \dfrac{2}{y} - \dfrac{3}{z} = 9,$

$\dfrac{7}{x} - \dfrac{2}{y} + \dfrac{9}{z} = -39$

Determine k so that the system is dependent.

43. $x - 3y + 2z = 1,$
$2x + y - z = 3,$
$9x - 6y + 3z = k$

44. $5x - 6y + kz = -5,$
$x + 3y - 2z = 2,$
$2x - y + 4z = -1$

In each case, three solutions of an equation are given. Find the equation.

45. $Ax + By + Cz = 12;$
$\left(1, \frac{3}{4}, 3\right), \left(\frac{4}{3}, 1, 2\right),$ and $(2, 1, 1)$

46. $z = b - mx - ny;$
$(1, 1, 2), (3, 2, -6),$ and $\left(\frac{3}{2}, 1, 1\right)$

3.5

Problem Solving Using Systems of Three Equations

Applications of Three Equations in Three Unknowns

Solving systems of three or more equations is important in many applications. Systems of equations occur very often in such fields as social science, business, natural science, and engineering.

EXAMPLE 1

The sum of three numbers is 4. The first number minus twice the second, minus the third is 1. Twice the first number minus the second, minus twice the third is -1. Find the numbers.

Solution

1. **FAMILIARIZE.** In this case, there are three statements in the problem. The translation looks as though it can be made directly from these statements once we decide what letters to assign to the unknown numbers.

2. **TRANSLATE.** Let's call the three numbers x, y, and z. Then we can translate directly, from the words of the problem, as follows.

The sum of the three numbers is 4.

$$x + y + z = 4$$

The first number minus twice the second minus the third is 1.

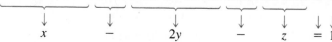

$$x - 2y - z = 1$$

Twice the first number minus the second minus twice the third is -1.

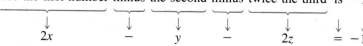

$$2x - y - 2z = -1$$

We now have a system of three equations:

$$x + y + z = 4,$$
$$x - 2y - z = 1,$$
$$2x - y - 2z = -1.$$

3. CARRY OUT.   We need to solve the system of equations. Note that we have already solved the system as Example 2 in Section 3.4. We obtained $(2, -1, 3)$ for the solution.

4. CHECK.   We go to the original problem. The first statement says that the sum of the three numbers is 4. That checks. The second statement says that the first number minus twice the second, minus the third is 1. We calculate: $2 - 2(-1) - 3 = 1$. That checks. We leave the check of the third statement to the student.

5. STATE.   The three numbers are 2, -1, and 3.   ❏

EXAMPLE 2

In a triangle, the largest angle is 70° greater than the smallest angle. The largest angle is also twice as large as the remaining angle. Find the measure of each angle.

Solution

1. FAMILIARIZE.   The first thing we do is make a drawing, or a sketch.

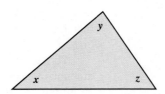

Since we don't know the size of any angle, we use x, y, and z for the measures of the angles. Recall that the measures of the angles of a triangle add up to 180°.

2. TRANSLATE.   This geometric fact about triangles gives us one equation:

$$x + y + z = 180.$$

There are two statements in the problem that we can translate almost directly.

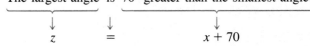

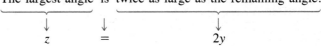

We now have a system of three equations:

$$
\begin{array}{lll}
x + y + z = 180, & & x + y + z = 180, \\
x + 70 = z, & \text{or} \quad x \quad - z = -70, & \\
2y = z; & 2y - z = 0. &
\end{array}
$$

Converting to standard form

3. **CARRY OUT.** The system was solved in Example 4 of Section 3.4. The solution is (30, 50, 100).

4. **CHECK.** The sum of the numbers is 180, so that checks.

The largest angle measures 100° and the smallest measures 30°. The largest angle is 70° greater than the smallest.

The remaining angle measures 50°. The largest angle measures 100°, so it is twice as large. We do have an answer to the problem.

5. **STATE.** The measures of the angles of the triangle are 30°, 50°, and 100°.

❏

EXAMPLE 3

Americans are becoming very conscious of their cholesterol levels. Recent studies indicate that a child should ingest no more than 300 mg of cholesterol per day. By eating 1 egg, 1 cupcake, and 1 slice of pizza, a child would ingest 302 mg of cholesterol. If the child eats 2 cupcakes and 3 slices of pizza, he or she ingests 65 mg of cholesterol. By eating 2 eggs and 1 cupcake, the child consumes 567 mg of cholesterol. How much cholesterol is in each item?

Solution

1. **FAMILIARIZE.** After we have read the problem a few times, it becomes clear that an egg contains considerably more cholesterol than the other foods. Let's guess that one egg contains 200 mg of cholesterol and one cupcake contains 50 mg. Because of the third sentence in the problem, it would then follow that a slice of pizza contains 52 mg of cholesterol since $200 + 50 + 52 = 302$.

To see if our guess satisfies the other statements in the problem, we find the amount of cholesterol that 2 cupcakes and 3 slices of pizza would contain: $2 \cdot 50 + 3 \cdot 52 = 256$. Since this does not match the 65 mg listed in the fourth sentence of the problem, we know that our guess is incorrect. Rather than guess again, we examine how we checked our guess. We decide to let e, c, and s = the number of milligrams of cholesterol in an egg, a cupcake, and a slice of pizza, respectively.

2. **TRANSLATE.** By rewording some of the sentences in the problem, we can translate it into three equations.

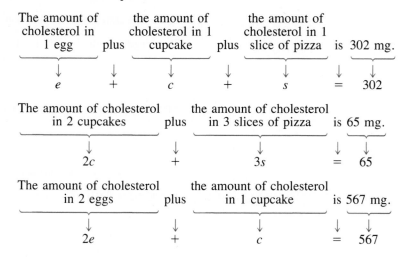

We now have a system of three equations:

$$e + c + \ s = 302,$$
$$2c + 3s = 65,$$
$$2e + c \quad\ = 567.$$

3. CARRY OUT.   We solve and get $e = 274$, $c = 19$, $s = 9$, or $(274, 19, 9)$.

4. CHECK.   The sum of 274, 19, and 9 is 302 so the total cholesterol in 1 egg, 1 cupcake, and 1 slice of pizza checks. Two cupcakes and three slices of pizza would contain $2 \cdot 19 + 3 \cdot 9 = 65$ mg, so that checks. Finally, two eggs and one cupcake would contain $2 \cdot 274 + 19 = 567$ mg of cholesterol. The answer checks.

5. STATE.   An egg contains 274 mg of cholesterol, a cupcake contains 19 mg of cholesterol, and a slice of pizza contains 9 mg of cholesterol.   ❑

EXERCISE SET | 3.5

Solve.

1. The sum of three numbers is 105. The third is 11 less than 10 times the second. Twice the first is 7 more than 3 times the second. Find the numbers.

2. The sum of three numbers is 57. The second is 3 more than the first. The third is 6 more than the first. Find the numbers.

3. The sum of three numbers is 5. The first number minus the second plus the third is 1. The first minus the third is 3 more than the second. Find the numbers.

4. The sum of three numbers is 26. Twice the first minus the second is 2 less than the third. The third is the second minus three times the first. Find the numbers.

5. In triangle ABC, the measure of angle B is 2° more than three times the measure of angle A. The measure of angle C is 8° more than the measure of angle A. Find the angle measures.

6. In triangle ABC, the measure of angle B is three times the measure of angle A. The measure of angle C is 30° greater than the measure of angle A. Find the angle measures.

7. In triangle ABC, the measure of angle B is twice the measure of angle A. The measure of angle C is 80° more than that of angle A. Find the angle measures.

8. In triangle ABC, the measure of angle B is three times that of angle A. The measure of angle C is 20° more than that of angle A. Find the angle measures.

9. In a recent year, companies spent a total of $84.8 billion on newspaper, television, and radio ads. The total amount spent on television and radio ads was only $2.6 billion more than the amount spent on newspaper ads alone. The amount spent on newspaper ads was $5.1 billion more than what was spent on television ads. How much was spent on each form of advertising? (*Hint:* Let the variables represent numbers of billions of dollars.)

10. A recent basic model of a particular automobile had a cost of $12,685. The basic model with the added features of automatic transmission and power door locks was $14,070. The basic model with air conditioning (AC) and power door locks was $13,580. The basic model with AC and automatic transmission was $13,925. What was the individual cost of each of the three options?

11. A dietician in a hospital prepares meals under the guidance of a physician. Suppose that for a particular patient a physician prescribes a meal to have 800 Calories, 55 g of protein, and 220 mg of vitamin C. The dietician prepares the meal using steak (each 3-oz serving contains 300 Cal, 20 g of protein, and no vitamin C), baked potatoes (one baked

potato contains 100 Cal, 5 g of protein, and 20 mg of vitamin C), and broccoli (one 156-g serving contains 50 Cal, 5 g of protein, and 100 mg of vitamin C). How many servings of each food are needed in order to satisfy the physician's requirements? (*Hint:* Let s = the number of servings of steak, p = the number of baked potatoes, and b = the number of servings of broccoli. Find an equation for the total number of calories, the total amount of protein, and the total amount of vitamin C.)

12. Repeat Exercise 11 but replace the broccoli with asparagus, for which one 180-g serving contains 50 calories, 5 g of protein, and 44 mg of vitamin C. Which meal would you prefer eating?

13. In the United States, the highest incidence of fraternal twin births occurs among Orientals, then African-Americans, and then Caucasians. Out of every 15,400 births, the total number of fraternal twin births for all three is 739, where there are 185 more for Orientals than African-Americans and 231 more for Orientals than Caucasians. How many births of fraternal twins are there for each race out of every 15,400 births?

14. The sum of the average number of times a man, a woman, and a one-year-old child cry each month is 71.7. A one-year-old cries 46.4 more times than a man. The average number of times a one-year-old cries per month is 28.3 more than the average number of times combined that a man and a woman cry. What is the average number of times per month that each cries?

15. In a factory there are three polishing machines, A, B, and C. When all three of them are working, 5700 lenses can be polished in one week. When only A and B are working, 3400 lenses can be polished in one week. When only B and C are working, 4200 lenses can be polished in one week. How many lenses can be polished in a week by each machine?

16. Sawmills A, B, and C can produce 7400 board-feet of lumber per day. Mills A and B together can produce 4700 board-feet per day, while mills B and C together can produce 5200 board-feet per day. How many board-feet can each mill produce by itself?

17. When three pumps, A, B, and C, are running together, they can pump 3700 gal/hr. When only A and B are running, 2200 gal/hr can be pumped. When only A and C are running, 2400 gal/hr can be pumped. What is the pumping capacity of each pump?

18. Pat, Chris, and Jean can weld 37 linear feet per hour when working together. Pat and Chris together can weld 22 linear feet per hour, while Pat and Jean can weld 25 linear feet per hour. How many linear feet per hour can each weld alone?

19. One year an investment of $80,000 was made by a business club. The investment was split into three parts and lasted for one year. The first part of the investment earned 8% interest, the second 6%, and the third 9%. Total interest from the investments was $6300. The interest from the first investment was four times the interest from the second. Find the amounts of the three parts of the investment.

20. Find the year in which the first U.S. transcontinental railroad was completed. The following are some facts about the number. The sum of the digits in the year is 24. The one's digit is one more than the hundred's digit. Both the ten's and the one's digits are multiples of three.

Skill Maintenance

21. Solve for x: $3(5 - x) + 7 = 5(x + 3) - 9$.

22. Compute: $(5 - 3^2 \div 2) \cdot (-4)^2$.

23. Simplify: $\dfrac{(a^2 b^3)^5}{a^7 b^{16}}$.

24. Give a slope–intercept equation for a line with slope $-\frac{3}{5}$ and y-intercept $(0, -7)$.

Synthesis

25. Tammy's age is the sum of the ages of Carmen and Dennis. Carmen's age is two more than the sum of the ages of Dennis and Mark. Dennis's age is four times Mark's age. The sum of all four ages is 42. How old is Tammy?

26. Find a three-digit positive integer such that the sum of all three digits is 14, the ten's digit is two more than the one's digit, and if the digits are reversed, the number is unchanged.

27. Find the sum of the angle measures at the tips of the star in this figure.

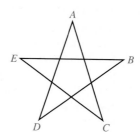

28. A theater audience of 100 people consists of adults, students, and children. The ticket prices are $10 for adults, $3 for students, and 50¢ for children. The total amount of money taken in is $100. How many adults, students, and children are in attendance? Does there seem to be some information missing? Do some more careful reasoning.

29. Hal gives Tom as many raffle tickets as Tom has and Gary as many as Gary has. In like manner, Tom then gives Hal and Gary as many tickets as each then has. Similarly, Gary gives Hal and Tom as many tickets as each then has. If each finally has 40 tickets, with how many tickets does Tom begin?

3.6

Elimination Using Matrices

Definition of a Matrix • Row-Equivalent Operations

In solving systems of equations, we perform computations with the constants. The variables play no important role until the end. Thus we can simplify writing a system by omitting the variables. For example, the system

$$3x + 4y = 5,$$
$$x - 2y = 1$$

simplifies to

$$\begin{array}{ccc} 3 & 4 & 5 \\ 1 & -2 & 1 \end{array}$$

if we leave off the variables, omit the operation of addition, and omit the equals signs.

In this example, we have written a rectangular array of numbers. Such an array is called a **matrix** (plural, **matrices**). We ordinarily write brackets around matrices. The following are matrices:

$$\begin{bmatrix} 4 & 1 & 3 & 5 \\ 1 & 0 & 1 & 2 \\ 6 & 3 & -2 & 0 \end{bmatrix}, \quad \begin{bmatrix} 6 & 2 & 1 & 4 & 7 \\ 1 & 2 & 1 & 3 & 1 \\ 4 & 0 & -2 & 0 & -3 \end{bmatrix}, \quad \begin{bmatrix} 1 & 2 \\ 145 & 0 \\ -7 & 9 \\ 8 & 1 \\ 0 & 0 \end{bmatrix}.$$

The **rows** of a matrix are horizontal, and the **columns** are vertical.

$$\begin{bmatrix} 5 & -2 & 2 \\ 1 & 0 & 1 \\ 0 & 1 & 2 \end{bmatrix}$$

⟵ row 1
⟵ row 2
⟵ row 3

↑ column 1   ↑ column 2   ↑ column 3

Let us now use matrices to solve systems of linear equations.

EXAMPLE 1

Solve the system

$$5x - 4y = -1,$$
$$-2x + 3y = 2.$$

Solution   We write a matrix using only the constants, keeping in mind that x corresponds to the first column and y to the second. A dashed line separates the coefficients from the constants at the end of each equation:

$$\begin{bmatrix} 5 & -4 & \vdots & -1 \\ -2 & 3 & \vdots & 2 \end{bmatrix}.$$ The individual numbers are called *elements* or *entries*.

Our goal is to transform this matrix into one of the form

$$\begin{bmatrix} a & b & \vdots & c \\ 0 & d & \vdots & e \end{bmatrix}.$$

The variables can then be reinserted to form equations from which we can complete the solution.

We do calculations that are similar to those that we would do if we wrote the entire equations. The first step, if possible, is to multiply and/or interchange the rows so that each number in the first column below the first number is a multiple of that number. In this case, we do so by multiplying Row 2 by 5. This corresponds to multiplying the second equation by 5.

$$\begin{bmatrix} 5 & -4 & \vdots & -1 \\ -10 & 15 & \vdots & 10 \end{bmatrix}$$ New Row 2 = 5(Row 2)

Next, we multiply the first row by 2 and add the result to the second row. This corresponds to multiplying the first equation by 2 and adding the result to the second equation. You should write out these computations if necessary — we perform them mentally.

$$\begin{bmatrix} 5 & -4 & \vdots & -1 \\ 0 & 7 & \vdots & 8 \end{bmatrix}$$ New Row 2 = 2(Row 1) + (Row 2)

If we now reinsert the variables, we have

$$5x - 4y = -1, \quad (1)$$
$$7y = 8. \quad (2)$$

We can now proceed as before, solving Equation (2) for y:

$$7y = 8 \quad (2)$$
$$y = \tfrac{8}{7}.$$

Next we substitute $\tfrac{8}{7}$ for y back in Equation (1). This procedure is called *back-substitution*.

$$5x - 4y = -1 \quad (1)$$
$$5x - 4 \cdot \tfrac{8}{7} = -1 \quad \text{Substituting } \tfrac{8}{7} \text{ for } y \text{ in Equation (1)}$$
$$x = \tfrac{5}{7} \quad \text{Solving for } x$$

The solution is $\left(\tfrac{5}{7}, \tfrac{8}{7}\right)$.   ❏

EXAMPLE 2   Solve the system

$$2x - y + 4z = -3,$$
$$x \qquad - 4z = 5,$$
$$6x - y + 2z = 10.$$

Solution   We first write a matrix, using only the constants. Where there are missing terms, we must write 0's:

$$\begin{bmatrix} 2 & -1 & 4 & \vdots & -3 \\ 1 & 0 & -4 & \vdots & 5 \\ 6 & -1 & 2 & \vdots & 10 \end{bmatrix} \begin{matrix} (P1) \\ (P2) \\ (P3) \end{matrix}$$

(P1), (P2), and (P3) designate the equations that are in the first, second, and third position, respectively.

Our goal is to find an equivalent matrix of the form

$$\begin{bmatrix} a & b & c & \vdots & d \\ 0 & e & f & \vdots & g \\ 0 & 0 & h & \vdots & i \end{bmatrix}.$$

A matrix of this form can be rewritten as a system of equations from which a solution can be found easily.

The first step, if possible, is to interchange the rows so that each number in the first column below the first number is a multiple of that number. In this case, we do so by interchanging Rows 1 and 2:

$$\begin{bmatrix} 1 & 0 & -4 & \vdots & 5 \\ 2 & -1 & 4 & \vdots & -3 \\ 6 & -1 & 2 & \vdots & 10 \end{bmatrix}$$ This corresponds to interchanging the first two equations.

Next, we multiply the first row by -2 and add it to the second row:

$$\begin{bmatrix} 1 & 0 & -4 & \vdots & 5 \\ 0 & -1 & 12 & \vdots & -13 \\ 6 & -1 & 2 & \vdots & 10 \end{bmatrix}.$$ This corresponds to multiplying new equation (P1) by -2 and adding it to new equation (P2). We perform the calculations mentally.

Now we multiply the first row by -6 and add it to the third row:

$$\begin{bmatrix} 1 & 0 & -4 & \vdots & 5 \\ 0 & -1 & 12 & \vdots & -13 \\ 0 & -1 & 26 & \vdots & -20 \end{bmatrix}.$$ This corresponds to multiplying equation (P1) by -6 and adding it to equation (P3).

Next, we multiply Row 2 by -1 and add it to the third row:

$$\begin{bmatrix} 1 & 0 & -4 & \vdots & 5 \\ 0 & -1 & 12 & \vdots & -13 \\ 0 & 0 & 14 & \vdots & -7 \end{bmatrix}.$$ This corresponds to multiplying equation (P2) by -1 and adding it to equation (P3).

Reinserting the variables gives us

$$\begin{aligned} x \qquad - 4z &= 5, & (P1) \\ - y + 12z &= -13, & (P2) \\ 14z &= -7. & (P3) \end{aligned}$$

We now solve (P3) for z and get $z = -\frac{1}{2}$. Next we back-substitute $-\frac{1}{2}$ for z in (P2) and solve for y: $-y + 12\left(-\frac{1}{2}\right) = -13$, so $y = 7$. Since there is no y-term in (P1), we need only substitute $-\frac{1}{2}$ for z in (P1) and solve for x: $x - 4\left(-\frac{1}{2}\right) = 5$, so $x = 3$. The solution is $\left(3, 7, -\frac{1}{2}\right)$.   ❑

All the operations used in the preceding example correspond to operations with the equations and produce equivalent systems of equations. We call the matrices **row-equivalent** and the operations that produce them **row-equivalent operations.**

Row-Equivalent Operations

Each of the following row-equivalent operations produces an equivalent matrix:

a) Interchanging any two rows.
b) Multiplying each element of a row by the same nonzero number.
c) Multiplying each element of a row by a nonzero number and adding the result to another row.

The best overall method for solving systems of equations is by row-equivalent matrices; even computers are programmed to use them. Matrices are part of a branch of mathematics known as linear algebra. They are also studied in more detail in many courses in finite mathematics.

EXERCISE SET | 3.6

Solve using matrices.

1. $4x + 2y = 11,$
$3x - y = 2$

2. $3x - 3y = 11,$
$9x - 2y = 5$

3. $x + 4y = 8,$
$3x + 5y = 3$

4. $x + 4y = 5,$
$-3x + 2y = 13$

5. $5x - 3y = -2,$
$4x + 2y = 5$

6. $3x + 4y = 7,$
$-5x + 2y = 10$

7. $4x - y - 3z = 1,$
$8x + y - z = 5,$
$2x + y + 2z = 5$

8. $3x + 2y + 2z = 3,$
$x + 2y - z = 5,$
$2x - 4y + z = 0$

9. $p + q + r = 1,$
$p - 2q - 3r = 3,$
$4p + 5q + 6r = 4$

10. $x + 2y - 3z = 9,$
$2x - y + 2z = -8,$
$3x - y - 4z = 3$

11. $3p + 2r = 11,$
$q - 7r = 4,$
$p - 6q = 1$

12. $4a + 9b = 8,$
$8a + 6c = -1,$
$6b + 6c = -1$

13. $2x + 2y - 2z - 2w = -10,$
$w + y + z + x = -5,$
$x - y + 4z + 3w = -2,$
$w - 2y + 2z + 3x = -6$

14. $-w - 3y + z + 2x = -8,$
$x + y - z - w = -4,$
$w + y + z + x = 22,$
$x - y - z - w = -14$

Problem Solving

15. A collection of 34 coins consists of dimes and nickels. The total value is $1.90. How many dimes and how many nickels are there?

16. A collection of 43 coins consists of dimes and quarters. The total value is $7.60. How many dimes and how many quarters are there?

17. A grocer has two kinds of granola. One is worth $4.05 per pound and the other is worth $2.70 per pound. The grocer wants to blend the two granolas to get a 15-lb mixture worth $3.15 per pound. How much of each kind of granola should be used?

18. A grocer mixes candy worth $0.80 per pound with nuts worth $0.70 per pound to get a 20-lb mixture worth $0.77 per pound. How many pounds of candy and how many pounds of nuts should be used?

19. Elena receives $212 per year in simple interest from three investments totaling $2500. Part is invested at 7%, part at 8%, and part at 9%. There is $1100 more invested at 9% than at 8%. Find the amount invested at each rate.

20. Miguel receives $306 per year in simple interest from three investments totaling $3200. Part is invested at 8%, part at 9%, and part at 10%. There is $1900 more invested at 10% than at 9%. Find the amount invested at each rate.

Skill Maintenance

Solve.

21. $0.1x - 12 = 3.6x - 2.34 - 4.9x$

22. $180 = 2x - 11x$

23. $4(9 - x) - 6(8 - 3x) = 5(3x + 4)$

24. $5b = c - ab$, for b

Synthesis

25. ◈ Explain how you can recognize a dependent system when solving with matrices.

26. ◈ Explain how you can recognize an inconsistent system when solving with matrices.

27. The sum of the digits in a four-digit number is 10. Twice the sum of the thousand's digit and the ten's digit is one less than the sum of the other two digits. The ten's digit is twice the thousand's digit. The one's digit equals the sum of the thousand's digit and the hundred's digit. Find the four-digit number.

28. Solve for x and y:

$$ax + by = c,$$
$$dx + ey = f.$$

3.7

Determinants and Cramer's Rule

Determinants of 2 × 2 Matrices • Cramer's Rule: 2 × 2 Systems • Cramer's Rule: 3 × 3 Systems

Determinants of 2 × 2 Matrices

When a matrix has m rows and n columns, it is called an "m by n" matrix. Thus its *dimensions* are denoted by $m \times n$. If a matrix has the same number of rows and columns, it is called a **square matrix.** With every square matrix is associated a number called its **determinant,** defined as follows for 2×2 matrices.

The determinant of the matrix $\begin{bmatrix} a & c \\ b & d \end{bmatrix}$ is denoted $\begin{vmatrix} a & c \\ b & d \end{vmatrix}$,

and is defined as follows:

$$\begin{vmatrix} a & c \\ b & d \end{vmatrix} = ad - bc.$$

EXAMPLE 1

Evaluate: $\begin{vmatrix} 2 & -5 \\ 6 & 7 \end{vmatrix}$.

Solution We multiply and subtract as follows:

$$\begin{vmatrix} 2 & -5 \\ 6 & 7 \end{vmatrix} = 2 \cdot 7 - 6 \cdot (-5) = 14 + 30 = 44.$$ ❑

Cramer's Rule: 2 × 2 Systems

One of the many uses for determinants is in solving systems of linear equations in which the number of variables is the same as the number of equations and the constants are not all 0. Let us consider a system of two equations:

$$a_1x + b_1y = c_1,$$
$$a_2x + b_2y = c_2.$$

Using the elimination method, we can solve to obtain

$$x = \frac{c_1b_2 - c_2b_1}{a_1b_2 - a_2b_1}, \qquad y = \frac{a_1c_2 - a_2c_1}{a_1b_2 - a_2b_1}.$$

The numerators and the denominators of the expressions for x and y can be written as determinants.

Cramer's Rule: 2 × 2 Systems

The solution of the system

$$a_1x + b_1y = c_1,$$
$$a_2x + b_2y = c_2,$$

if it is unique, is given by

$$x = \frac{\begin{vmatrix} c_1 & b_1 \\ c_2 & b_2 \end{vmatrix}}{\begin{vmatrix} a_1 & b_1 \\ a_2 & b_2 \end{vmatrix}}, \qquad y = \frac{\begin{vmatrix} a_1 & c_1 \\ a_2 & c_2 \end{vmatrix}}{\begin{vmatrix} a_1 & b_1 \\ a_2 & b_2 \end{vmatrix}}.$$

The equations above make sense only if the determinant in the denominator is not 0. If the denominator *is* 0, then one of two things happens.

1. If the denominator is 0 and the other two determinants in the numerators are also 0, then the equations in the system are dependent.
2. If the denominator is 0 and at least one of the other determinants in the numerators is not 0, then the system is inconsistent.

To use Cramer's rule, we find the three determinants and compute x and y as shown above. Note that the denominator in both cases contains a_1, a_2, b_1, and b_2 in the same position as in the original equations. For x, the numerator is obtained by replacing a_1 and a_2 by c_1 and c_2. For y, the numerator is obtained by replacing b_1 and b_2 by c_1 and c_2.

EXAMPLE 2

Solve using Cramer's rule:

$$2x + 5y = 7,$$
$$5x - 2y = -3.$$

Solution   We have

$$x = \frac{\begin{vmatrix} 7 & 5 \\ -3 & -2 \end{vmatrix}}{\begin{vmatrix} 2 & 5 \\ 5 & -2 \end{vmatrix}} \qquad \text{Using Cramer's rule}$$

$$= \frac{7(-2) - (-3)5}{2(-2) - 5 \cdot 5} = -\frac{1}{29},$$

and

$$y = \dfrac{\begin{vmatrix} 2 & 7 \\ 5 & -3 \end{vmatrix}}{\begin{vmatrix} 2 & 5 \\ 5 & -2 \end{vmatrix}} \qquad \text{Using Cramer's rule}$$

$$= \frac{2(-3) - 5 \cdot 7}{-29} = \frac{41}{29}.$$

The solution is $\left(-\frac{1}{29}, \frac{41}{29}\right)$. ❏

Cramer's Rule: 3 × 3 Systems

A similar method has been developed for solving systems of three equations in three unknowns. Before stating the rule, though, we must develop some terminology.

The *determinant* of a three-by-three matrix is defined as follows:

$$\begin{vmatrix} a_1 & b_1 & c_1 \\ a_2 & b_2 & c_2 \\ a_3 & b_3 & c_3 \end{vmatrix} = a_1 \overset{\text{Subtract.}}{\begin{vmatrix} b_2 & c_2 \\ b_3 & c_3 \end{vmatrix}} - a_2 \begin{vmatrix} b_1 & c_1 \\ b_3 & c_3 \end{vmatrix} \overset{\text{Add.}}{+} a_3 \begin{vmatrix} b_1 & c_1 \\ b_2 & c_2 \end{vmatrix}$$

Note that the a's come from the first column.

Note too that the second-order determinants above can be obtained by crossing out the row and column in which the a occurs.

For a_1:
$$\begin{vmatrix} a_1 & b_1 & c_1 \\ a_2 & b_2 & c_2 \\ a_3 & b_3 & c_3 \end{vmatrix}$$

For a_2:
$$\begin{vmatrix} a_1 & b_1 & c_1 \\ a_2 & b_2 & c_2 \\ a_3 & b_3 & c_3 \end{vmatrix}$$

For a_3:
$$\begin{vmatrix} a_1 & b_1 & c_1 \\ a_2 & b_2 & c_2 \\ a_3 & b_3 & c_3 \end{vmatrix}$$

EXAMPLE 3

Evaluate:

$$\begin{vmatrix} -1 & 0 & 1 \\ -5 & 1 & -1 \\ 4 & 8 & 1 \end{vmatrix}.$$

Solution   We have

$$\begin{vmatrix} -1 & 0 & 1 \\ -5 & 1 & -1 \\ 4 & 8 & 1 \end{vmatrix} = -1 \begin{vmatrix} 1 & -1 \\ 8 & 1 \end{vmatrix} \overset{\text{Subtract.}}{-} (-5) \begin{vmatrix} 0 & 1 \\ 8 & 1 \end{vmatrix} \overset{\text{Add.}}{+} 4 \begin{vmatrix} 0 & 1 \\ 1 & -1 \end{vmatrix}$$

$$= -1(1 + 8) + 5(0 - 8) + 4(0 - 1) \qquad \text{Evaluating the three determinants}$$

$$= -9 - 40 - 4 = -53. \qquad ❏$$

Cramer's Rule:
3 × 3 Systems

The solution of the system

$$a_1x + b_1y + c_1z = d_1,$$
$$a_2x + b_2y + c_2z = d_2,$$
$$a_3x + b_3y + c_3z = d_3$$

is found by considering the following determinants:

$$D = \begin{vmatrix} a_1 & b_1 & c_1 \\ a_2 & b_2 & c_2 \\ a_3 & b_3 & c_3 \end{vmatrix}, \quad D_x = \begin{vmatrix} d_1 & b_1 & c_1 \\ d_2 & b_2 & c_2 \\ d_3 & b_3 & c_3 \end{vmatrix},$$

$$D_y = \begin{vmatrix} a_1 & d_1 & c_1 \\ a_2 & d_2 & c_2 \\ a_3 & d_3 & c_3 \end{vmatrix}, \quad D_z = \begin{vmatrix} a_1 & b_1 & d_1 \\ a_2 & b_2 & d_2 \\ a_3 & b_3 & d_3 \end{vmatrix}.$$

The solution, if it is unique, is given by

$$x = \frac{D_x}{D}, \quad y = \frac{D_y}{D}, \quad z = \frac{D_z}{D}.$$

EXAMPLE 4

Solve using Cramer's rule:

$$x - 3y + 7z = 13,$$
$$x + y + z = 1,$$
$$x - 2y + 3z = 4.$$

Solution   We compute D, D_x, D_y and D_z:

$$D = \begin{vmatrix} 1 & -3 & 7 \\ 1 & 1 & 1 \\ 1 & -2 & 3 \end{vmatrix} = -10; \quad D_x = \begin{vmatrix} 13 & -3 & 7 \\ 1 & 1 & 1 \\ 4 & -2 & 3 \end{vmatrix} = 20;$$

$$D_y = \begin{vmatrix} 1 & 13 & 7 \\ 1 & 1 & 1 \\ 1 & 4 & 3 \end{vmatrix} = -6; \quad D_z = \begin{vmatrix} 1 & -3 & 13 \\ 1 & 1 & 1 \\ 1 & -2 & 4 \end{vmatrix} = -24.$$

Then

$$x = \frac{D_x}{D} = \frac{20}{-10} = -2;$$

$$y = \frac{D_y}{D} = \frac{-6}{-10} = \frac{3}{5};$$

$$z = \frac{D_z}{D} = \frac{-24}{-10} = \frac{12}{5}.$$

The solution is $\left(-2, \frac{3}{5}, \frac{12}{5}\right)$.

In Example 4, we would not have needed to evaluate D_z. Once we found x and y, we could have substituted them into one of the equations to find z. In practice, it is faster to use determinants to find only two of the numbers; then we find the third by substitution into an equation.

In using Cramer's rule, we divide by D. If D should be 0, however, we could not do so. If $D = 0$ and at least one of the other determinants is not 0, then the system is inconsistent. If $D = 0$ and all the other determinants are also 0, then the equations in the system are dependent.

EXERCISE SET | 3.7

Evaluate.

1. $\begin{vmatrix} 2 & 7 \\ 1 & 5 \end{vmatrix}$

2. $\begin{vmatrix} 3 & 2 \\ 2 & -3 \end{vmatrix}$

3. $\begin{vmatrix} 6 & -9 \\ 2 & 3 \end{vmatrix}$

4. $\begin{vmatrix} 3 & 2 \\ -7 & 5 \end{vmatrix}$

5. $\begin{vmatrix} 0 & 2 & 0 \\ 3 & -1 & 1 \\ 1 & -2 & 2 \end{vmatrix}$

6. $\begin{vmatrix} 3 & 0 & -2 \\ 5 & 1 & 2 \\ 2 & 0 & -1 \end{vmatrix}$

7. $\begin{vmatrix} -1 & -2 & -3 \\ 3 & 4 & 2 \\ 0 & 1 & 2 \end{vmatrix}$

8. $\begin{vmatrix} 1 & 2 & 2 \\ 2 & 1 & 0 \\ 3 & 3 & 1 \end{vmatrix}$

9. $\begin{vmatrix} 3 & 2 & -2 \\ -2 & 1 & 4 \\ -4 & -3 & 3 \end{vmatrix}$

10. $\begin{vmatrix} 2 & -1 & 1 \\ 1 & 2 & -1 \\ 3 & 4 & -3 \end{vmatrix}$

Solve using Cramer's rule.

11. $3x - 4y = 6,$
$5x + 9y = 10$

12. $5x + 8y = 1,$
$3x + 7y = 5$

13. $-2x + 4y = 3,$
$3x - 7y = 1$

14. $5x - 4y = -3,$
$7x + 2y = 6$

15. $3x + 2y - z = 4,$
$3x - 2y + z = 5,$
$4x - 5y - z = -1$

16. $3x - y + 2z = 1,$
$x - y + 2z = 3,$
$-2x + 3y + z = 1$

17. $2x - 3y + 5z = 27,$
$x + 2y - z = -4,$
$5x - y + 4z = 27$

18. $x - y + 2z = -3,$
$x + 2y + 3z = 4,$
$2x + y + z = -3$

19. $r - 2s + 3t = 6,$
$2r - s - t = -3,$
$r + s + t = 6$

20. $a - 3c = 6,$
$b + 2c = 2,$
$7a - 3b - 5c = 14$

Skill Maintenance

Solve.

21. $0.5x - 2.34 + 2.4x = 7.8x - 9$

22. $5x + 7x = -144$

23. A piece of wire 32.8 ft long is to be cut into two pieces, and those pieces are each to be bent to make a square. The length of a side of one square is to be 2.2 ft greater than the length of a side of the other. How should the wire be cut?

Synthesis

Solve.

24. $\begin{vmatrix} y & -2 \\ 4 & 3 \end{vmatrix} = 44$

25. $\begin{vmatrix} 2 & x & -1 \\ -1 & 3 & 2 \\ -2 & 1 & 1 \end{vmatrix} = -12$

26. $\begin{vmatrix} m+1 & -2 \\ m-2 & 1 \end{vmatrix} = 27$

27. Show that an equation of the line through (x_1, y_1) and (x_2, y_2) can be written

$$\begin{vmatrix} x & y & 1 \\ x_1 & y_1 & 1 \\ x_2 & y_2 & 1 \end{vmatrix} = 0.$$

28. ◈ Cramer's rule states that whenever the equations $a_1x + b_1y = c_1$ and $a_2x + b_2y = c_2$ are dependent, we have

$$\begin{vmatrix} a_1 & b_1 \\ a_2 & b_2 \end{vmatrix} = 0.$$

Explain why this occurs.

3.8

Business and Economic Applications

Break-Even Analysis   •   Supply and Demand

Break-Even Analysis

When a company manufactures x units of a product, it invests money. This is **total cost** and can be thought of as a function C, where $C(x)$ is the total cost of producing x units. When the company sells x units of the product, it takes in money. This is **total revenue** and can be thought of as a function R, where $R(x)$ is the total revenue from the sale of x units. **Total profit** is the money taken in less the money spent, or total revenue minus total cost. Total profit from the production and sale of x units is a function P given by

$$\text{Profit} = \text{Revenue} - \text{Cost}, \quad \text{or} \quad P(x) = R(x) - C(x).$$

If $R(x)$ is greater than $C(x)$, then the company makes money. If $C(x)$ is greater than $R(x)$, then the company has a loss. If $R(x) = C(x)$, then the company breaks even.

There are two kinds of costs. First, there are costs like rent, insurance, machinery, and so on. These costs, which must be paid whether a product is produced or not, are called *fixed costs*. When a product is being produced, there are costs for labor, materials, marketing, and so on. These are called *variable costs*. They vary according to the amount being produced. Adding these together, we find the *total cost* of producing a product.

EXAMPLE 1

Ergs, Inc., is planning to make a new kind of radio. During the first year, fixed costs will be $90,000, and it will cost $15 to produce each radio (variable costs). Each radio will sell for $26.

a)   Find the total cost $C(x)$ of producing x radios.
b)   Find the total revenue $R(x)$ from the sale of x radios.
c)   Find the total profit $P(x)$ from the production and sale of x radios.
d)   What profit or loss will the company realize from the production and sale of 3000 radios? of 14,000 radios?
e)   Graph the total-cost, total-revenue, and total-profit functions using the same set of axes. Determine the break-even point.

Solution

a)   Total cost is given by

$$C(x) = (\text{Fixed costs}) \text{ plus } (\text{Variable costs}),$$
$$\text{or} \quad C(x) = \quad 90{,}000 \quad + \quad 15x,$$

where x is the number of radios produced.

b)   Total revenue is given by

$$R(x) = 26x. \qquad \text{\$26 times the number of radios sold. We are assuming that all radios produced are sold.}$$

c) Total profit is given by

$$P(x) = R(x) - C(x)$$
$$= 26x - (90,000 + 15x)$$
$$= 11x - 90,000.$$

d) Total profit is found by

$$P(3000) = 11 \cdot 3000 - 90,000 = -\$57,000$$

when 3000 radios are produced and sold, and

$$P(14,000) = 11 \cdot 14,000 - 90,000 = \$64,000$$

when 14,000 radios are produced and sold. Thus the company loses money from the production and sale of 3000 radios, but makes money from the production and sale of 14,000 radios.

e) The graphs of each of the three functions are shown below:

$$R(x) = 26x, \qquad (1)$$
$$C(x) = 90,000 + 15x, \qquad (2)$$
$$P(x) = 11x - 90,000. \qquad (3)$$

$R(x)$, $C(x)$, and $P(x)$ are all in dollars.

Equation (1) has a graph that goes through the origin and has a slope of 26. Equation (2) has an intercept on the $-axis of 90,000 and has a slope of 15. Equation (3) has an intercept on the $-axis of $-90,000$ and has a slope of 11. It is shown by the dashed line. The color dashed line shows a "negative" profit, which is a loss. (That is what is known as "being in the red.") The black dashed line shows a "positive" profit, or gain. (That is what is known as "being in the black.")

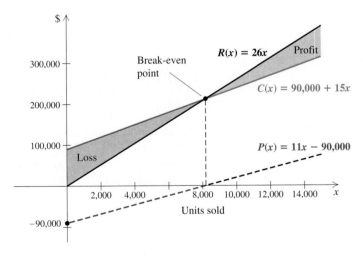

Profits occur where the revenue is greater than the cost. Losses occur where the revenue is less than the cost. The **break-even point** occurs where the graphs of R and C cross. Thus to find the break-even point, we solve a system:

$$R(x) = 26x,$$
$$C(x) = 90,000 + 15x.$$

Since both revenue and cost are in *dollars* and they are equal at the break-even point, the system can be rewritten as

$$d = 26x, \qquad (1)$$
$$d = 90{,}000 + 15x \qquad (2)$$

and solved using substitution:

$$26x = 90{,}000 + 15x \qquad \text{Substituting } 26x \text{ for } d \text{ in Equation (2)}$$
$$11x = 90{,}000$$
$$x \approx 8181.8.$$

The firm will break even if it produces and sells about 8182 radios (8181 will yield a tiny loss and 8182 a tiny gain), and takes in $R(8182) = 26 \cdot 8182 = \$212{,}732$ in revenue. Note that the x-coordinate of the break-even point can also be found by solving $P(x) = 0$.   ❏

Supply and Demand

Demand Varies with Price.   As the price of coffee varied over a period of years, the amount sold varied. The table and graph both show that the amount *demanded* by consumers goes down as the price goes up. As price goes down, demand goes up.

DEMAND FUNCTION, D

Price, p, per Kilogram	Quantity, $D(p)$, in Millions of Kilograms
$ 8.00	25
9.00	20
10.00	15
11.00	10
12.00	5

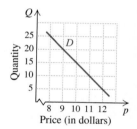

Supply Varies with Price.   As the price of coffee varied, we see that the amount available varied. The table and graph show that the seller *supplies* less as the price goes down, but is willing to supply more as the price goes up.

SUPPLY FUNCTION, S

Price, p, per Kilogram	Quantity, $S(p)$, in Millions of Kilograms
$ 9.00	5
9.50	10
10.00	15
10.50	20
11.00	25

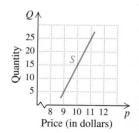

The Equilibrium Point.   Let us look at the above graphs together. We see that as price increases, demand decreases. As price increases, supply increases. The point of intersection is called the *equilibrium point*. At that price, the amount that the seller will supply is the same amount that the consumer will buy. The situation is analogous to a buyer and a seller negotiating the price of an item. The equilibrium point, or selling price, is what they finally agree on.

Any ordered pair of coordinates from the graph is (price, quantity), because the horizontal axis is the price axis and the vertical axis is the quantity axis. If D is the demand function for a product and S is the supply function, then the equilibrium point is where demand equals supply:

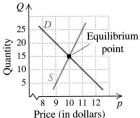

$$D(p) = S(p).$$

EXAMPLE 2

Find the equilibrium point for the demand and supply functions given:

$$D(p) = 1000 - 60p, \quad (1)$$
$$S(p) = \ \ 200 + \ 4p. \quad (2)$$

Solution Since both demand and supply are *quantities* and they are equal at the equilibrium point, the system can be rewritten as

$$q = 1000 - 60p, \quad (1)$$
$$q = \ \ 200 + \ 4p. \quad (2)$$

We substitute $200 + 4p$ for q in Equation (1) and solve:

$$200 + \ \ 4p = 1000 - 60p$$
$$200 + 64p = 1000 \qquad \text{Adding } 60p$$
$$64p = 800 \qquad \text{Adding } -200$$
$$p = \tfrac{800}{64} = 12.5.$$

Thus the equilibrium price is $12.50 per unit.

To find the equilibrium quantity, we substitute $12.50 into either $D(p)$ or $S(p)$. We use $S(p)$:

$$S(12.5) = 200 + 4(12.5) = 200 + 50 = 250.$$

Thus the equilibrium quantity is 250 units, and the equilibrium point is ($12.50, 250). ❑

EXERCISE SET | 3.8

For each of the following pairs of total-cost and total-revenue functions, find (a) the total-profit function and (b) the break-even point.

1. $C(x) = 45x + 600,000$;
$R(x) = 65x$

2. $C(x) = 25x + 360,000$;
$R(x) = 70x$

3. $C(x) = 10x + 120,000$;
$R(x) = 60x$

4. $C(x) = 30x + 49,500$;
$R(x) = 85x$

5. $C(x) = 20x + 10,000$;
$R(x) = 100x$

6. $C(x) = 40x + 22,500;$
$R(x) = 85x$

7. $C(x) = 15x + 75,000;$
$R(x) = 55x$

8. $C(x) = 22x + 16,000;$
$R(x) = 40x$

9. $C(x) = 50x + 195,000;$
$R(x) = 125x$

10. $C(x) = 34x + 928,000;$
$R(x) = 128x$

Find the equilibrium point for each of the following pairs of demand and supply functions.

11. $D(p) = 2000 - 60p,$
$S(p) = 460 + 94p$

12. $D(p) = 1000 - 10p,$
$S(p) = 250 + 5p$

13. $D(p) = 760 - 13p,$
$S(p) = 430 + 2p$

14. $D(p) = 800 - 43p,$
$S(p) = 210 + 16p$

15. $D(p) = 7500 - 25p,$
$S(p) = 6000 + 5p$

16. $D(p) = 8800 - 30p,$
$S(p) = 7000 + 15p$

17. $D(p) = 1600 - 53p,$
$S(p) = 320 + 75p$

18. $D(p) = 5500 - 40p,$
$S(p) = 1000 + 85p$

Solve.

19. Sky View Electronics is planning to introduce a new line of computers. For the first year, the fixed costs for setting up the production line are $125,000. The variable costs for producing each computer are $750. The revenue from each computer is $1050.

a) Find the total cost $C(x)$ of producing x computers.

b) Find the total revenue $R(x)$ from the sale of x computers.

c) Find the total profit $P(x)$ from the production and sale of x computers.

d) What profit or loss will the company realize from the production and sale of 400 computers? of 700 computers?

e) Find the break-even point.

20. City Lights, Inc., is planning a new type of lamp. For the first year, the fixed costs for setting up the production line are $22,500. The variable costs for producing each lamp are estimated to be $40. The revenue from each lamp is to be $85.

a) Find the total cost $C(x)$ of producing x lamps.

b) Find the total revenue $R(x)$ from the sale of x lamps.

c) Find the total profit $P(x)$ from the production and sale of x lamps.

d) What profit or loss will the company realize from the production and sale of 3000 lamps? of 400 lamps?

e) Find the break-even point.

21. Sarducci's is planning a new line of sport coats. For the first year, the fixed costs for setting up the production line are $10,000. The variable costs for producing each coat are $20. The revenue from each coat is to be $100.

a) Find the total cost $C(x)$ of producing x coats.

b) Find the total revenue $R(x)$ from the sale of x coats.

c) Find the total profit $P(x)$ from the production and sale of x coats.

d) What profit or loss will the company realize from the production and sale of 2000 coats? of 50 coats?

e) Find the break-even point.

22. Martina's Custom Printing is planning on adding painter's caps to its product line. For the first year, the fixed costs for setting up the production line are $16,400. The variable costs for producing a dozen caps are $6.00. The revenue on each dozen caps will be $18.00.

a) Find the total cost $C(x)$ of producing x dozen caps.

b) Find the total revenue $R(x)$ from the sale of x dozen caps.

c) Find the total profit $P(x)$ from the production and sale of x dozen caps.

d) What profit or loss will the company realize from the production and sale of 3000 dozen caps? of 1000 dozen caps?

e) Find the break-even point.

Skill Maintenance

23. Graph: $y - 3 = \frac{2}{5}(x - 1)$.

24. When two consecutive integers are added and then doubled, the result is three less than five times the smaller number. Find both numbers.

25. Solve: $9x = 5x - \{3(2x - 7) - 4\}$.

26. Solve $v - st = rw$ for t.

Synthesis ——————————————————

27. Bing Boing Hobbies is willing to produce 100 yo-yo's at $2.00 each and 500 yo-yo's at $8.00 each. Research indicates that the public will buy 500 yo-yo's at $1.00 each and 100 yo-yo's at $9.00 each. Find the equilibrium point.

28. Fidelity Speakers, Inc., has fixed costs of $15,400 and variable costs of $100 for each pair of speakers produced. If the speakers sell for $250 a pair, how many pairs of speakers need to be produced (and sold) in order to have enough profit to cover the fixed costs of two new facilities? Assume that all fixed costs are identical.

29. ◈   Variable costs and fixed costs are often compared to the slope and the *y*-intercept, respectively, of an equation for a line. Explain why you feel this analogy is or is not valid.

30. ◈   In this section, we examined supply and demand functions for coffee. Does it seem realistic to you for the graph of *D* to have a constant slope? Why or why not?

◤◢   In Exercises 31 and 32, use a grapher to solve each of the problems.

31. a) The Number Cruncher Computer Corporation is planning a new line of computers, each of which will sell for $970. The fixed costs in setting up the production line are $1,235,580 and the variable costs for each computer are $697. What is the break-even point? (Round to the nearest whole number.)

b) The marketing department at Number Cruncher is not sure that the $970 price is the best price. They have determined that the demand function for the new computers will be $D(p) = -304.5p + 374,580$ and the supply function will be $S(p) = 788.7p - 576,504$. To the nearest dollar, what price *p* would result in equilibrium between supply and demand?

32. a) Puppy Love, Inc., will soon begin producing a new line of puppy food. The marketing department predicts that the demand function will be $D(p) = -14.97p + 987.35$ and the supply function will be $S(p) = 98.55p - 5.13$. To the nearest cent, what price per unit should be charged in order to have equilibrium between supply and demand?

b) The production of the puppy food involves $87,985 in fixed costs and $5.15 per unit in variable costs. If the price per unit is the value you found in part (a), how many units must be sold in order to break even?

SUMMARY AND REVIEW | 3

IMPORTANT PROPERTIES AND FORMULAS

To use the elimination method to solve systems of three equations:

1. Write all equations in the standard form $Ax + By + Cz = D$.
2. Clear any decimals or fractions.
3. Choose a variable to eliminate. Then use any two of the three equations to get an equation in two variables.
4. Next use a different pair of equations and get another equation *in the same two variables*. That is, eliminate the same variable that you did in step (3).
5. Solve the resulting system (pair) of equations. That will give two of the numbers.
6. Then use any of the original three equations to find the third number.

Row-Equivalent Operations

Each of the following row-equivalent operations produces an equivalent matrix:

a) Interchanging any two rows.
b) Multiplying each element of a row by the same nonzero number.
c) Multiplying each element of a row by a nonzero number and adding the result to another row.

Determinant of a 2 × 2 Matrix

$$\begin{vmatrix} a & c \\ b & d \end{vmatrix} = ad - bc$$

Determinant of a 3 × 3 Matrix

$$\begin{vmatrix} a_1 & b_1 & c_1 \\ a_2 & b_2 & c_2 \\ a_3 & b_3 & c_3 \end{vmatrix} = a_1 \begin{vmatrix} b_2 & c_2 \\ b_3 & c_3 \end{vmatrix} - a_2 \begin{vmatrix} b_1 & c_1 \\ b_3 & c_3 \end{vmatrix} + a_3 \begin{vmatrix} b_1 & c_1 \\ b_2 & c_2 \end{vmatrix}$$

Cramer's Rule: 2 × 2 Systems

The solution of the system
$$a_1x + b_1y = c_1,$$
$$a_2x + b_2y = c_2,$$
if it is unique, is given by

$$x = \frac{\begin{vmatrix} c_1 & b_1 \\ c_2 & b_2 \end{vmatrix}}{\begin{vmatrix} a_1 & b_1 \\ a_2 & b_2 \end{vmatrix}}, \qquad y = \frac{\begin{vmatrix} a_1 & c_1 \\ a_2 & c_2 \end{vmatrix}}{\begin{vmatrix} a_1 & b_1 \\ a_2 & b_2 \end{vmatrix}}.$$

Cramer's Rule: 3 × 3 Systems

The solution of the system
$$a_1x + b_1y + c_1z = d_1,$$
$$a_2x + b_2y + c_2z = d_2,$$
$$a_3x + b_3y + c_3z = d_3$$
is found by considering the following determinants:

$$D = \begin{vmatrix} a_1 & b_1 & c_1 \\ a_2 & b_2 & c_2 \\ a_3 & b_3 & c_3 \end{vmatrix}, \qquad D_x = \begin{vmatrix} d_1 & b_1 & c_1 \\ d_2 & b_2 & c_2 \\ d_3 & b_3 & c_3 \end{vmatrix},$$

$$D_y = \begin{vmatrix} a_1 & d_1 & c_1 \\ a_2 & d_2 & c_2 \\ a_3 & d_3 & c_3 \end{vmatrix}, \qquad D_z = \begin{vmatrix} a_1 & b_1 & d_1 \\ a_2 & b_2 & d_2 \\ a_3 & b_3 & d_3 \end{vmatrix}.$$

The solution, if it is unique, is given by

$$x = \frac{D_x}{D}, \qquad y = \frac{D_y}{D}, \qquad z = \frac{D_z}{D}.$$

REVIEW EXERCISES

This chapter's review and test include Skill Maintenance exercises from Sections 1.3, 1.4, 1.6, and 2.5.

Solve graphically.   *practice* *w/ Mrs H*

1. $4x - y = 10,$
$2x + 3y = 12$

2. $y = 3x + 7,$
$3x + 2y = -4$

Solve using the substitution method.

3. $7x - 4y = 6,$
$y - 3x = -2$

4. $y = x + 2,$
$y - x = 8$ *DO*

5. $9x - 6y = 2,$
$x = 4y + 5$

Solve using the elimination method.

6. $8x - 2y = 10,$
$-4y - 3x = -17$

7. $4x - 7y = 18,$
$9x + 14y = 40$

8. $3x - 5y = -4,$
$5x - 3y = 4$ *DO*

Solve.

9. Sean has $37 to spend. He can spend all the money for two compact discs and a cassette, or he can buy one CD and two cassettes and have $5.00 left over. What is the price of a CD? What is the price of a cassette?

10. A train leaves Watsonville at noon traveling north at a speed of 44 mph. One hour later, another train, going 55 mph, travels north on a parallel track. How many hours will the second train travel before it overtakes the first train?

11. Cleanse-O is 30% alcohol. Tingle is 50% alcohol. How much of each should be mixed in order to obtain 40 L of a solution that is 45% alcohol?

Solve.

12. $x + 2y + z = 10,$
$2x - y + z = 8,$
$3x + y + 4z = 2$

13. $3x + 2y + z = 3,$
$6x - 4y - 2z = -34,$
$-x + 3y - 3z = 14$

14. $2x - 5y - 2z = -4,$
$7x + 2y - 5z = -6,$
$-2x + 3y + 2z = 4$

15. $-5x + 5y = -6,$
$2x - 2y = 4$

16. $3x + y = 2,$
$x + 3y + z = 0,$
$x + z = 2$

17. $3x + 4y = 6,$
$1.5x - 3 = -2y$

18. $x + y + 2z = 1,$
$x - y + z = 1,$
$x + 2y + z = 2$

Solve.

19. In triangle ABC, the measure of angle A is four times the measure of angle C, and the measure of angle B is 45° more than the measure of angle C. What are the measures of the angles of the triangle?

20. Find the three-digit number in which the sum of the digits is 11, the ten's digit is three less than the sum of the hundred's and one's digits, and the one's digit is five less than the hundred's digit.

21. Lynn has $194 in her purse, consisting of $20, $5, and $1 bills. The number of $1 bills is one less than the total number of $20 and $5 bills. If she has 39 bills in her purse, how many of each denomination does she have?

Solve using matrices. Show your work.

22. $3x + 4y = -13,$
$5x + 6y = 8$

23. $3x - y + z = -1,$
$2x + 3y + z = 4,$
$5x + 4y + 2z = 5$

Evaluate.

24. $\begin{vmatrix} -2 & 4 \\ -3 & 5 \end{vmatrix}$ *DO*

25. $\begin{vmatrix} 2 & 3 & 0 \\ 1 & 4 & -2 \\ 2 & -1 & 5 \end{vmatrix}$

Solve using Cramer's rule. Show your work.

26. $2x + 3y = 6,$
$x - 4y = 14$ *DO*

27. $2x + y + z = -2,$
$2x - y + 3z = 6,$
$3x - 5y + 4z = 7$

28. Find the equilibrium point for the demand and supply functions

$$D(p) = 60 + 7p \quad \text{and} \quad S(p) = 120 - 13p.$$

29. Kregel Furniture is planning to produce a new type of bed. For the first year, the fixed costs for setting up the production line are $35,000. The variable costs for producing each bed are $175. The revenue from each bed is $225.

  a) Find the total cost $C(x)$ of producing x beds.
  b) Find the total revenue $R(x)$ from the sale of x beds.
  c) Find the total profit $P(x)$ from the production and sale of x beds.
  d) What profit or loss will the company realize

from the production and sale of 1200 beds? of 500 beds?
e) Find the break-even point.

30. Solve: $4x - 5x + 8 = -9x + 2x$.

31. Solve $Q = at - 4t$ for t.

32. Write an equation of the line containing $(3, 5)$ and perpendicular to the line $3x - 7y = 4$.

33. A basketball team increases its score by seven points in each of three consecutive games. In those three games, the team scored a total of 228 points. What was their score in each game?

34. ◈   How would you go about solving a problem-solving situation that involved four variables?

35. ◈   Explain how a system of equations can be both dependent and inconsistent.

36. Solve graphically:
$$y = x + 2,$$
$$y = x^2 + 2.$$

37. The graph of $f(x) = ax^2 + bx + c$ contains the points $(-2, 3)$, $(1, 1)$, and $(0, 3)$. Find a, b, and c and give a formula for the function.

CHAPTER TEST   3

Solve using the substitution method.

1. $x + 3y = -8,$
$4x - 3y = 23$

2. $2x + 4y = -6,$
$y = 3x - 9$

Solve using the elimination method.

3. $4x - 6y = 3,$
$6x - 4y = -3$

4. $4y + 2x = 18,$
$3x + 6y = 26$

5. The perimeter of a rectangle is 96. The length of the rectangle is 6 less than twice the width. Find the dimensions of the rectangle.

6. Chicken-To-Go sold 132 chicken dinners during one day. Two-piece dinners sold for $4.50, and three-piece dinners sold for $5.50. The restaurant's total receipts that day from chicken dinners were $656. How many of each size dinner did they sell?

Solve.

7. $-3x + y - 2z = 8,$
$-x + 2y - z = 5,$
$2x + y + z = -3$

8. $6x + 2y - 4z = 15,$
$-3x - 4y + 2z = -6,$
$4x - 6y + 3z = 8$

9. $2x + 2y = 0,$
$4x + 4z = 4,$
$2x + y + z = 2$

10. $3x + 3z = 0,$
$2x + 2y = 2,$
$3y + 3z = 3$

Solve using matrices.

11. $7x - 8y = 10,$
$9x + 5y = -2$

12. $x + 3y - 3z = 12,$
$3x - y + 4z = 0,$
$-x + 2y - z = 1$

Evaluate.

13. $\begin{vmatrix} 4 & -2 \\ 3 & 7 \end{vmatrix}$

14. $\begin{vmatrix} 3 & 4 & 2 \\ 2 & -5 & 4 \\ 4 & 5 & -3 \end{vmatrix}$

15. Solve using Cramer's rule:
$$8x - 3y = 5,$$
$$2x + 6y = 3.$$

16. An electrician, a carpenter, and a plumber are hired to work on a house. The electrician earns $21 per hour, the carpenter earns $19.50 per hour, and the plumber earns $24 per hour. The first day on the job, they worked a total of 21.5 hr and earned a total of $469.50. If the plumber worked two more hours than the carpenter did, how many hours did the electrician work?

17. Find the equilibrium point for the demand and supply functions
$$D(p) = 79 - 8p \quad \text{and} \quad S(p) = 37 + 6p.$$

18. A sporting goods manufacturer is planning a new type of tennis racket. For the first year, the fixed costs for setting up production lines are $40,000. The variable costs for producing each racket are $45. The sales department predicts that 1500 rackets can be sold during the first year. The revenue from each racket is $80.

   a) Find the total cost $C(x)$ of producing x tennis rackets.

   b) Find the total revenue $R(x)$ from the sale of x tennis rackets.

   c) Find the total profit $P(x)$.

   d) What profit or loss will the company realize if the expected sales of 1500 rackets occur?

   e) Find the break-even point.

Skill Maintenance

19. The price of a radio, including 5% sales tax, is $36.75. Find the price of the radio before the tax was added.

20. Solve $P = 4a - 3b$ for a.

21. Solve: $-3x - 5 + 6x = 8x - 14$.

22. Find an equation of the line passing through the points $(5, -1)$ and $(-3, 4)$.

Synthesis

23. The graph of the function $f(x) = mx + b$ contains the points $(-1, 3)$ and $(-2, -4)$. Find m and b.

24. At a county fair, an adult's ticket sold for $5.50, a senior citizen's ticket sold for $4.00, and a child's ticket sold for $1.50. On the opening day, the number of tickets sold to children and senior citizens was 30 more than the number of tickets sold to adults. The number of tickets sold to senior citizens was 6 more than four times the number of tickets sold to children. Total receipts from the ticket sales were $14,967. How many of each type of ticket were sold?

CUMULATIVE REVIEW 1–3

Solve.

1. $-14.3 + 29.17 = x$

2. $x + 9.4 = -12.6$

3. $3.9(-11) = x$

4. $-2.4x = -48$

5. $4x + 7 = -14$

6. $-3 + 5x = 2x + 15$

7. $3n - (4n - 2) = 7$

8. $6y - 5(3y - 4) = 10$

9. $14 + 2c = -3(c + 4) - 6$

10. $5x - [4 - 2(6x - 1)] = 12$

Simplify.

11. $x^4 \cdot x^{-6} \cdot x^{13}$

12. $(4x^{-3}y^2)(-10x^4y^{-7})$

13. $(6x^2y^3)^2(-2x^0y^4)^3$

14. $\dfrac{y^4}{y^{-6}}$

15. $\dfrac{-10a^7b^{-11}}{25a^{-4}b^{22}}$

16. $\left(\dfrac{3x^4y^{-2}}{4x^{-5}}\right)^4$

Multiply.

17. $(2.6 \times 10^4)(4.3 \times 10^{-5})$

18. $(1.95 \times 10^{-3})(5.73 \times 10^8)$

Divide.

19. $\dfrac{3.2 \times 10^{-3}}{8.0 \times 10^{-6}}$

20. $\dfrac{2.42 \times 10^5}{6.05 \times 10^{-2}}$

21. Solve $A = \frac{1}{2}h(b + t)$ for b.

Determine whether the ordered pair is a solution of the equation.

22. $(-3, 4)$; $5a - 2b = -23$

23. $(2, -1)$; $-x - 5y = 4$

Graph.

24. $y = -2x + 3$

25. $y = x^2 - 1$

26. $4x + 16 = 0$

27. $-3x + 2y = 6$

28. Find the slope and the y-intercept of the line with equation $-4y + 9x = 12$.

29. Find the slope, if it exists, of the line containing the points $(2, 7)$ and $(-1, 3)$.

30. Find an equation of the line with slope -3 and containing the point $(2, -11)$.

31. Find an equation of the line containing the points $(-6, 3)$ and $(4, 2)$.

32. Determine whether the lines are parallel or perpendicular:

$$2x = 4y + 7,$$
$$x - 2y = 5.$$

33. Find an equation of the line containing the point $(2, 1)$ and perpendicular to the line $x - 2y = 5$.

Given $g(x) = 4x - 3$ and $h(x) = -2x^2 + 1$, find the following function values.

34. $h(4)$

35. $-g(0)$

36. $3g(-3) - 2h(-1)$

37. $g(a) - h(2a)$

Solve.

38. $3x + y = 4,$
$6x - y = 5$

39. $4x + 4y = 4,$
$5x - 3y = -19$

40. $6x - 10y = -22,$
$-11x - 15y = 27$

41. $x + y + z = -5,$
$2x + 3y - 2z = 8,$
$x - y + 4z = -21$

42. $2x + 5y - 3z = -11,$
$-5x + 3y - 2z = -7,$
$3x - 2y + 5z = 12$

Evaluate.

43. $\begin{vmatrix} 2 & -3 \\ 4 & 1 \end{vmatrix}$

44. $\begin{vmatrix} 1 & 0 & 1 \\ -1 & 2 & 1 \\ 2 & 1 & 3 \end{vmatrix}$

45. The sum of two numbers is 26. Three times the smaller plus twice the larger is 60. Find the numbers.

46. Soakem is 34% salt and the rest water. Rinsem is 61% salt and the rest water. How many ounces of each would be needed to obtain 120 oz of a mixture that is 50% salt?

47. Find three consecutive odd numbers such that the sum of 4 times the first number and 5 times the third number is 47.

48. Belinda's scores on four tests are 83, 92, 100, and 85. What must the score be on the fifth test so that the average will be 90?

49. The perimeter of a rectangle is 32 cm. If 5 times the width equals 3 times the length, what are the dimensions of the rectangle?

50. There are four more nickels than dimes in a piggy bank. The total amount of money in the bank is $2.45. How many of each type of coin are in the bank?

51. One month a family spent $680 for electricity, rent, and telephone. The electric bill was $\frac{1}{4}$ of the rent and the rent was $400 more than the phone bill. How much was the electric bill?

52. A hockey team played 64 games one season. It won 15 more games than it tied and lost 10 more games than it won. How many games did it win? lose? tie?

53. The sum of three numbers is 71. The first number is 12 more than twice the third number. The first number minus the second number is −47. What are the numbers?

Synthesis

54. Simplify: $(6x^{a+2}y^{b+2})(-2x^{a-2}y^{y+1})$.

55. An automotive dealer discovers that when $1000 is spent on radio advertising, weekly sales increase by $101,000. When $1250 is spent on radio advertising, weekly sales increase by $126,000. Assuming that sales increase according to a linear equation, by what amount would sales increase when $1500 is spent on radio advertising?

56. Two solutions of the equation $y = mx + b$ are $(5, -3)$ and $(-4, 2)$. Find m and b.

CHAPTER | 4

Inequalities and Linear Programming

AN APPLICATION

An insurance company offers two types of medical coverage. With plan A, the employee pays the first $100 of medical bills and the insurance company pays 80% of the rest. With plan B, the employee pays the first $250 of medical bills and the insurer pays 90% of the rest. For what amount of medical bills would plan B save an employee money?

This problem appears as Exercise 71 in Section 4.1.

Mike Mezo
PRESIDENT,
USWA LOCAL 1010

"Successful contract negotiations often rely on a solid understanding of math. A group insurance package is typically an important part of our contracts, and failure to procure the most beneficial plan could cost our members thousands of dollars."

nequalities are mathematical sentences containing symbols such as < (is less than). Principles similar to those used for solving equations enable us to solve inequalities and the problems that translate to inequalities. In this section, we will develop a procedure for solving systems of inequalities.

In addition to material from this chapter, the review and test for Chapter 4 include material from Sections 1.3, 2.3, 3.2, and 3.3.

4.1

Inequalities and Problem Solving

Solving Inequalities • Interval Notation • The Addition Principle • The Multiplication Principle • Using the Principles Together • Problem Solving

Inequalities

We can extend our equation-solving skills to the solving of inequalities. An **inequality** is any sentence having one of the verbs $<$, $>$, $\leq$, or $\geq$ (see Section 1.2)—for example,

$$-2 < a, \qquad x > 4, \qquad x + 3 \leq 6, \quad \text{and} \quad 16 - 7y \geq 10y - 4.$$

Some replacements for the variable in an inequality make it true, and some make it false. A replacement that makes it true is called a **solution.** The set of all solutions is called the **solution set.** When we have found the set of all solutions of an inequality, we say that we have **solved** the inequality.

EXAMPLE 1

Determine whether the given number is a solution of the inequality.

a) $x + 3 < 6, \quad 5$ **b)** $2x - 3 > -5, \quad 1$

Solution

a) We substitute and get $5 + 3 < 6$, or $8 < 6$, a false sentence. Therefore, 5 is not a solution.

b) We substitute and get $2 \cdot 1 - 3 > -5$, or $-1 > -5$, a true sentence. Therefore, 1 is a solution. ❑

A *graph* of an inequality is a drawing that represents its solutions. An inequality in one variable can be graphed on a number line. Inequalities in two variables can be graphed on a coordinate plane, and will be considered later in this chapter.

EXAMPLE 2

Graph $x < 4$ on a number line.

Solution   The solutions are all real numbers less than 4, so we shade all numbers less than 4. Since 4 is not a solution, we use an open circle at 4.

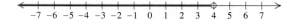

The solution set can be written using *set-builder notation* (see Chapter 1):

$\{x|\ x < 4\}.$

This is read

"The set of all x such that x is less than 4."   ❑

Interval Notation

Another way to list the solutions of an inequality in one variable is to use **interval notation.** Pay special attention to the manner in which parentheses, ( ), and brackets, [ ], are used.

If a and b are real numbers such that $a < b$, we define the **open interval (a, b)** as the set of all numbers x for which $a < x < b$. Thus,

$(a,\ b) = \{x|\ a < x < b\}.$

Its graph excludes the endpoints:

Be careful not to confuse the *interval* (a, b) with the *ordered pair* (a, b). The context of the discussion usually makes the meaning clear.

The **closed interval $[a, b]$** is defined as the set of all numbers x for which $a \le x \le b$. Thus,

$[a,\ b] = \{x|\ a \le x \le b\}.$

Its graph includes the endpoints:

There are two kinds of **half-open intervals** defined as follows:

1. $(a,\ b] = \{x|\ a < x \le b\}.$ This is open on the left. Its graph is as follows:

2. $[a,\ b) = \{x|\ a \le x < b\}.$ This is open on the right. Its graph is as follows:

We use the symbols ∞ and $-\infty$ to represent positive and negative infinity, respectively. Thus the notation (a, ∞) represents the set of all real numbers greater than a, and $(-\infty, a)$ represents the set of all real numbers less than a.

The notations $[a, \infty)$ and $(-\infty, a]$ are used when we want to include the endpoint a.

EXAMPLE 3

Graph $y \geq -2$ on a number line and write the solution set using both set-builder and interval notations.

Solution   Using set-builder notation, we write the solution set as $\{y \mid y \geq -2\}$.
Using interval notation, we write the solution set as $[-2, \infty)$.
To graph the solution, we shade all numbers to the right of -2 and use a solid endpoint to indicate that -2 is also a solution.

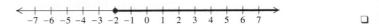

The Addition Principle

The addition principle for inequalities is similar to the addition principle for equations. Both principles are used for forming equivalent sentences.

The Addition
Principle for
Inequalities

For any real numbers a, b, and c:

if $a < b$, then $a + c < b + c$;   if $a > b$, then $a + c > b + c$;
if $a \leq b$, then $a + c \leq b + c$;   if $a \geq b$, then $a + c \geq b + c$.

As with equations, we try to get the variable alone on one side in order to determine solutions easily.

EXAMPLE 4

Solve and graph: **(a)** $x + 5 > 3$; **(b)** $4x - 1 \geq 5x - 2$.

Solution

a)
$$x + 5 > 3$$
$$x + 5 + (-5) > 3 + (-5) \qquad \text{Using the addition principle, add } -5.$$
$$x > -2$$

The solution set is $\{x \mid x > -2\}$, or $(-2, \infty)$, and the graph of the inequality is as follows:

We cannot check all the solutions of an inequality by substitution, because there are too many of them. A partial check could be done by substituting a number greater than -2, say -1, into the original inequality:

$$\frac{x + 5 > 3}{-1 + 5 \; ? \; 3}$$
$$4 \mid 3 \quad \text{TRUE}$$

Since $4 > 3$ is true, -1 is a solution. Any number greater than -2 is a solution.

b)
$$4x - 1 \geqslant 5x - 2$$

$$4x - 1 + 2 \geqslant 5x - 2 + 2 \qquad \text{Adding 2}$$

$$4x + 1 \geqslant 5x \qquad \text{Simplifying}$$

$$4x + 1 - 4x \geqslant 5x - 4x \qquad \text{Adding } -4x$$

$$1 \geqslant x \qquad \text{Simplifying}$$

We know that $1 \geqslant x$ has the same meaning as $x \leqslant 1$. Thus any number less than or equal to 1 is a solution. We can express the solution set as $\{x \mid 1 \geqslant x\}$ or as $\{x \mid x \leqslant 1\}$. The latter is probably used most often. Using interval notation, we write the solution set as $(-\infty, 1]$. The graph is as follows:

The Multiplication Principle

The multiplication principle for inequalities is somewhat different from the principle for equations.

Consider this true inequality:

$$4 < 9.$$

If we multiply both numbers by 2, we get another true inequality:

$$4 \cdot 2 < 9 \cdot 2, \quad \text{or} \quad 8 < 18.$$

If we multiply both numbers by -2, we get a false inequality:

$$\text{FALSE} \rightarrow \quad 4(-2) < 9(-2), \quad \text{or} \quad -8 < -18. \qquad \leftarrow \text{FALSE}$$

This is because negation reverses relative position on the number line. However, if we now reverse the inequality symbol, we get a true inequality:

$$-8 > -18.$$

⌐——— The $<$ symbol has been reversed!

The Multiplication Principle for Inequalities

For any real numbers a and b, and for any *positive* number c,

if $a < b$, then $ac < bc$; and if $a > b$, then $ac > bc$.

For any real numbers a and b, and for any *negative* number c,

if $a < b$, then $ac > bc$; and if $a > b$, then $ac < bc$.

Similar statements hold for $\leqslant$ and $\geqslant$.

The important thing to remember is that if you multiply by a negative number, you must reverse the inequality symbol.

When we solve an inequality using the multiplication principle, we can multiply on both sides by any number except zero.

EXAMPLE 5

Solve and graph: **(a)** $3y < \frac{3}{4}$; **(b)** $-5x \geqslant -80$.

Solution

a) $3y < \frac{3}{4}$

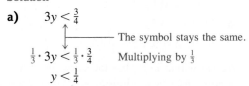

$\frac{1}{3} \cdot 3y < \frac{1}{3} \cdot \frac{3}{4}$ Multiplying by $\frac{1}{3}$

$y < \frac{1}{4}$

Any number less than $\frac{1}{4}$ is a solution. The solution set is $\left\{y \mid y < \frac{1}{4}\right\}$, or $\left(-\infty, \frac{1}{4}\right)$. The graph is as follows:

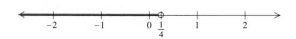

b) $-5x \geqslant -80$

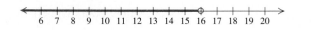

$-\frac{1}{5} \cdot (-5x) \leqslant -\frac{1}{5} \cdot (-80)$ Multiplying by $-\frac{1}{5}$

$x \leqslant 16$

The solution set is $\{x \mid x \leqslant 16\}$, or $(-\infty, 16]$. The graph is as follows:

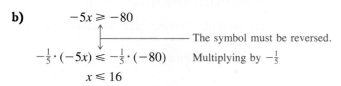   ❑

Using the Principles Together

We use the addition and multiplication principles together in solving inequalities in much the same way as in solving equations.

EXAMPLE 6

Solve: **(a)** $16 - 7y \geqslant 10y - 4$; **(b)** $-3(x + 8) - 5x > 4x - 9$.

Solution

a) $16 - 7y \geqslant 10y - 4$

$-16 + 16 - 7y \geqslant -16 + 10y - 4$ Adding -16

$-7y \geqslant 10y - 20$

$-10y + (-7y) \geqslant -10y + 10y - 20$ Adding $-10y$

$-17y \geqslant -20$

$-\frac{1}{17} \cdot (-17y) \leqslant -\frac{1}{17} \cdot (-20)$ The symbol must be reversed.

$y \leqslant \frac{20}{17}$ Multiplying by $-\frac{1}{17}$

The solution set is $\left\{y \mid y \leqslant \frac{20}{17}\right\}$, or $\left(-\infty, \frac{20}{17}\right]$.

b) $-3(x + 8) - 5x > 4x - 9$

$\quad\quad -3x - 24 - 5x > 4x - 9$ Using the distributive law

$\quad\quad\quad\quad -24 - 8x > 4x - 9$

$\quad -24 - 8x + 8x > 4x - 9 + 8x$ Adding $8x$ on both sides

$\quad\quad\quad\quad\quad -24 > 12x - 9$

$\quad\quad\quad -24 + 9 > 12x - 9 + 9$ Adding 9

$\quad\quad\quad\quad\quad\quad -15 > 12x$

                The symbol stays the same.

$\quad\quad\quad\quad\quad\quad -\frac{5}{4} > x$ Multiplying by $\frac{1}{12}$ and simplifying

The solution set is $\left\{x \mid -\frac{5}{4} > x\right\}$, or $\left\{x \mid x < -\frac{5}{4}\right\}$, or $\left(-\infty, -\frac{5}{4}\right)$.

Problem Solving

EXAMPLE 7

Records in the women's 100-m dash.   Florence Griffith Joyner set a world record of 10.49 sec in the women's 100-m dash in the 1988 Olympics. The formula

$$R = -0.0433t + 10.49$$

can be used to predict the world record in the women's 100-m dash t years after 1988. For example, to find the record in 2008, we would subtract 1988, to get $t = 20$. Determine (in terms of an inequality) those years for which the world record will be less than 10.0 sec.

Solution

1. **FAMILIARIZE.** We already have a formula. To become more familiar with it, we might make a substitution for t. Suppose we want to know the record after 30 years, in the year 2018. We substitute 30 for t:

$$R = -0.0433t + 10.49$$
$$= -0.0433(30) + 10.49 = 9.191 \text{ sec.}$$

2. **TRANSLATE.** The record R is to be *less than* 10.0 sec. Thus we have the inequality

$$R < 10.0.$$

We substitute $-0.0433t + 10.49$ for R to find the times t that satisfy the inequality:

$$-0.0433t + 10.49 < 10.0.$$

3. **CARRY OUT.**   We solve the inequality:

$$-0.0433t + 10.49 < 10.0$$
$$-0.0433t < -0.49 \qquad \text{Adding } -10.49$$
$$t > 11.32. \qquad \text{Dividing by } -0.0433 \text{ or multiplying}$$

by $-\frac{1}{0.0433}$ and rounding. You might use a calculator.

4. **CHECK.**   We can check by substituting a value for t greater than 11.32. We did that in the familiarization step.

5. **STATE.**   The record will be less than 10.0 for those races occurring more than 11.32 years after 1988, which is approximately $\{x \mid x \geq 1999\}$. ❏

EXAMPLE 8

On your new job, you can be paid in one of two ways:

Plan A: A salary of $600 per month, plus a commission of 4% of sales;

Plan B: A salary of $800 per month, plus a commission of 6% of sales in excess of $10,000.

For what amount of sales is plan A better than plan B, if we assume that sales are always more than $10,000?

Solution

1. **FAMILIARIZE.**   Listing the given information in a table will be helpful.

Plan A: Monthly Income	Plan B: Monthly Income
$600 salary 4% of sales *Total:* $600 + 4% of sales	$800 salary 6% of sales over $10,000 *Total:* $800 + 6% of sales over $10,000

Next, suppose that you were to sell a certain amount — say, $12,000 — in one month. Which plan would be better? Under plan A, you would earn $600 plus 4% of $12,000, or

$$600 + 0.04(12,000) = \$1080.$$

Since with plan B commissions are paid only on sales in excess of $10,000, you would earn $800 plus 6% of $(12,000 - 10,000)$, or

$$800 + 0.06(12,000 - 10,000) = \$920.$$

This shows that for monthly sales of $12,000, plan A is better. Similar calculations will show that for sales of $30,000 a month, plan B is better. To determine *all* values for which plan A earns more money, we must solve an inequality that is based on the above calculations.

2. **TRANSLATE.**   We let S represent the amount of monthly sales. Examining the calculations in the *Familiarize* step, we see that monthly income from plan A

is $600 + 0.04S$ and from plan B is $800 + 0.06 (S - 10,000)$. We want to find all values of S for which

Income from plan A	is greater than	income from plan B
↓	↓	↓
$600 + 0.04S$	$>$	$800 + 0.06(S - 10,000).$

3. CARRY OUT. We solve the inequality:

$$600 + 0.04S > 800 + 0.06(S - 10,000)$$

$$600 + 0.04S > 800 + 0.06S - 600 \qquad \text{Using the distributive law}$$

$$600 + 0.04S > 200 + 0.06S \qquad \text{Collecting like terms}$$

$$400 > 0.02S \qquad \text{Subtracting both 200 and } 0.04S$$

$$20,000 > S. \qquad \text{Dividing by 0.02}$$

4. CHECK. For $S = 20,000$, the income from plan A is

$$600 + 4\% \cdot 20,000, \quad \text{or} \quad \$1400.$$

The income from plan B is

$$800 + 6\% \cdot 10,000, \quad \text{or} \quad \$1400.$$

In the *Familiarize* step, we saw that for sales of $12,000, plan A pays more. Since $20,000 > 12,000$, we have a partial check. We cannot check all possible values of S so we will stop here.

5. STATE. For monthly sales of less than $20,000, plan A is better. ❏

EXERCISE SET | 4.1

Determine whether the given numbers are solutions of the inequality.

1. $x - 2 \geqslant 6$; $-4, 0, 4, 8$

2. $3x + 5 \leqslant -10$; $-5, -10, 0, 27$

3. $t - 8 > 2t - 3$; $0, -8, -9, -3$

4. $5y - 7 < 5 - y$; $2, -3, 0, 3$

Graph the inequality, and write the solution set using both set-builder and interval notation.

5. $x > 4$ **6.** $y < 5$

7. $t \leqslant 6$ **8.** $x \geqslant -4$

9. $y < -3$ **10.** $t > -2$

11. $x \geqslant -6$ **12.** $x \leqslant -5$

Solve. Then graph.

13. $x + 8 > 3$ **14.** $x + 5 > 2$

15. $y + 3 < 9$

16. $y + 4 < 10$

17. $a + 9 \leqslant -12$

18. $a + 7 \leqslant -13$

19. $t + 14 \geqslant 9$

20. $x - 9 \leqslant 10$

21. $y - 8 > -14$

22. $y - 9 > -18$

23. $x - 11 \leqslant -2$

24. $y - 18 \leqslant -4$

25. $8x \geqslant 24$

26. $9t < -81$

27. $0.3x < -18$

28. $0.5x < 25$

29. $-9x \geqslant -8.1$

30. $-8y \leqslant 3.2$

31. $-\frac{3}{4}x \geqslant -\frac{5}{8}$

32. $-\frac{5}{6}y \leqslant -\frac{3}{4}$

33. $2x + 7 < 19$

34. $5y + 13 > 28$

35. $5y + 2y \leqslant -21$

36. $-9x + 3x \geqslant -24$

37. $2y - 7 < 5y - 9$

38. $8x - 9 < 3x - 11$

39. $0.4x + 5 \leqslant 1.2x - 4$

40. $0.2y + 1 > 2.4y - 10$

41. $3x - \frac{1}{8} \leq \frac{3}{8} + 2x$

42. $2x - 3 < \frac{13}{4}x + 10 - 4.25x$

Solve.

43. $4(3y - 2) \geq 9(2y + 5)$

44. $4m + 5 \geq 14(m - 2)$

45. $3(2 - 5x) + 2x < 2(4 + 2x)$

46. $2(0.5 - 3y) + y > (4y - 0.2)8$

47. $5[3m - (m + 4)] > -2(m - 4)$

48. $[8x - 3(3x + 2)] - 5 \geq 3(x + 4) - 2x$

49. $3(r - 6) + 2 > 4(r + 2) - 21$

50. $5(t + 3) + 9 < 3(t - 2) + 6$

51. $19 - (2x + 3) \leq 2(x + 3) + x$

52. $13 - (2c + 2) \geq 2(c + 2) + 3c$

53. $\frac{1}{4}(8y + 4) - 17 < -\frac{1}{2}(4y - 8)$

54. $\frac{1}{3}(6x + 24) - 20 > -\frac{1}{4}(12x - 72)$

55. $2[4 - 2(3 - x)] - 1 \geq 4[2(4x - 3) + 7] - 25$

56. $5[3(7 - t) - 4(8 + 2t)] - 20 \leq -6[2(6 + 3t) - 4]$

Solve.

57. You are taking a history course in which there are 4 tests. You have scores of 89, 92, and 95 on the first three tests. You must score a total of at least 360 in order to get an A. What scores on the last test will give you an A?

58. You are taking a science course in which there are 5 tests, each worth 100 points. You have scores of 94, 90, and 89 on the first three. You must score a total of at least 450 in order to get an A. What scores on your fourth test will keep you eligible for an A?

59. A car rents for $30 per day plus 20¢ per mile. You are on a daily budget of $96. What mileages will allow you to stay within the budget?

60. A car can be rented for $35 per day with unlimited mileage, or for $28 per day plus 19¢ per mile. For what daily mileages would the unlimited mileage plan save you money?

61. A long-distance telephone call using Down East Calling costs 20 cents for the first minute and 16 cents for each additional minute. The same call, placed on Long Call Systems, costs 19 cents for the first minute and 18 cents for each additional minute. For what length phone calls is Down East Calling less expensive?

62. Musclebound Movers charges $85 plus $40 an hour to move households across town. Lug-a-lot Movers charges $60 an hour for cross-town

moves. For what lengths of time is Lug-a-lot the more expensive mover?

63. On your new job, you can be paid in one of two ways:

> *Plan A:* A salary of $500 per month, plus a commission of 4% of gross sales;
> *Plan B:* A salary of $750 per month plus a commission of 5% of gross sales over $8000.

For what gross sales is plan B better than plan A, assuming that gross sales are always more than $8000?

64. On your new job, you can be paid in one of two ways:

> *Plan A:* A salary of $25,000 per year;
> *Plan B:* A salary of $1500 per month plus a commission of 6% on gross sales.

For what gross sales is plan A better than plan B?

65. A painter can be paid in one of two ways:

> *Plan A:* $500 plus $6.00 per hour;
> *Plan B:* Straight $11.00 per hour.

Suppose that the job takes n hours. For what values of n is plan A better for the painter?

66. A mason can be paid in one of two ways:

> *Plan A:* $300 plus $9.00 per hour;
> *Plan B:* Straight $12.50 per hour.

Suppose that the job takes n hours. For what values of n is plan B better for the mason?

67. You are going to invest $25,000, part at 7% and part at 8%. What is the most that can be invested at 7% in order to make at least $1800 interest per year?

68. You are going to invest $20,000, part at 6% and part at 8%. What is the most that can be invested at 6% in order to make at least $1500 interest per year?

69. In planning for a college dance, you find that one band will play for $250 plus 50% of the total ticket sales. Another band will play for a flat fee of $550. In order for the first band to produce more profit for the school than the other band, what is the highest price you can charge per ticket, assuming that 300 people will attend?

70. A bank offers two checking account plans. Plan A charges a base service charge of $2.00 per month plus 15¢ per check. Plan B charges a base service charge of $4.00 per month plus 9¢ per check. For what numbers of checks per month will plan B be better than plan A?

71. An insurance company offers two plans. With plan A, the employee pays the first $100 of medical bills and the insurance company pays 80% of the rest. With plan B, the employee pays the first $250 of medical bills and the insurance company pays 90% of the rest. For what amount of medical bills will plan B save the employee money?

72. You can spend $3.50 at the laundromat washing your clothes, or you can have them do the laundry for 40¢ per pound. For what weights of clothes will it save you money to wash your clothes yourself?

73. The formula

$$C = \tfrac{5}{9}(F - 32)$$

can be used to convert Fahrenheit temperatures F to Celsius temperatures C.

a) Gold is solid at Celsius temperatures less than 1063° C. Find the Fahrenheit temperatures for which gold is solid.

b) Silver is solid at Celsius temperatures less than 960.8° C. Find the Fahrenheit temperatures for which silver is solid.

74. The percentage of the total active military duty force that is women has been steadily increasing. The number N of women in the active duty force t years since 1971 is approximated by

$$N = 12{,}197.8t + 44{,}000.$$

a) How many women were in the military in 1971 ($t = 0$)? in 1981 ($t = 10$)? in 1990 ($t = 19$)?

b) For what years will the number of women be at least 250,000?

75. Ergs, Inc., is planning to make a new kind of radio. Fixed costs will be $90,000, and variable costs will be $15 for the production of each radio. The total-cost function is

$$C(x) = 90{,}000 + 15x.$$

The company makes $26 in revenue for each radio sold. The total-revenue function is

$$R(x) = 26x.$$

(See Section 3.8.)

a) When $R(x) < C(x)$, the company loses money. Find the values of x for which the company loses money.

b) When $R(x) > C(x)$, the company makes a profit. Find the values of x for which the company makes a profit.

76. The demand and supply functions for a certain product are given by

$$D(p) = 2000 - 60p \quad \text{and}$$
$$S(p) = 460 + 94p.$$

(See Section 3.8.)

a) Find those values of p for which demand exceeds supply.

b) Find those values of p for which demand is less than supply.

Skill Maintenance

77. Graph: $5y - 10 = 2x$.

78. Solve: $-3x + 5 = 11$.

79. Simplify: $|-16|$. **80.** Simplify: $-|-4|$.

Synthesis

81. ◈ Explain in your own words why the inequality symbol must be reversed when both sides of an inequality are multiplied by a negative number.

82. ◈ Presto photocopiers cost $510 and Exact Image photocopiers cost $590. Write a problem that involves the cost of the copiers, the cost per page of photocopies, and the number of copies for which the Presto machine is the more expensive machine to own.

Solve. Assume that a, b, c, d, and m are positive constants.

83. $3ax + 2x \geq 5ax - 4$; assume $a > 1$

84. $6by - 4y \leq 7by + 10$

85. $a(by - 2) \geq b(2y + 5)$; assume $a > 2$

86. $c(6x - 4) < d(3 + 2x)$; assume $3c > d$

87. $c(2 - 5x) + dx > m(4 + 2x)$; assume $5c + 2m < d$

88. $a(3 - 4x) + cx < d(5x + 2)$; assume $c > 4a + 5d$

Determine whether the statement is true or false. If false, give an example that shows this.

89. For any real numbers a, b, c, and d, if $a < b$ and $c < d$, then $a - c < b - d$.

90. For all real numbers x and y, if $x < y$, then $x^2 < y^2$.

91. ◈ Determine whether the inequalities

$$x < 3 \quad \text{and} \quad x + \frac{1}{x} < 3 + \frac{1}{x}$$

are equivalent. Give reasons to support your answer.

92. ◈ Determine whether the inequalities

$$x < 3 \quad \text{and} \quad 0 \cdot x < 0 \cdot 3$$

are equivalent. Give reasons to support your answer.

Solve.

93. $x + 5 \leq 5 + x$ **94.** $x + 8 < 3 + x$

95. $x^2 > 0$ **96.** $x^2 + 1 > 0$

4.2

Compound Inequalities

Intersections of Sets and Conjunctions of Sentences   •
Unions of Sets and Disjunctions of Sentences

We now consider **compound inequalities,** that is, sentences formed by two or more inequalities, joined with the word *and* or the word *or*.

Intersections of Sets and Conjunctions of Sentences

The **intersection** of two sets A and B is the set of all members that are common to both A and B. We denote the intersection of sets A and B as

$A \cap B$.

The intersection of two sets is often pictured as follows:

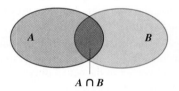

$A \cap B$

EXAMPLE 1

Find the intersection: $\{1, 2, 3, 4, 5\} \cap \{-2, -1, 0, 1, 2, 3\}$.

Solution   The numbers 1, 2, and 3 are common to both sets, so the intersection is $\{1, 2, 3\}$.   ❏

When two or more sentences are joined by the word *and* to make a compound sentence, the new sentence is called a **conjunction** of the sentences. The following is a conjunction of inequalities:

$-2 < x$ *and* $x < 1$.

For a conjunction to be true, all the individual sentences must be true. The solution set of a conjunction is the intersection of the solution sets of the individual sentences. Let us consider the conjunction

$-2 < x$ *and* $x < 1$.

The graphs of each separate sentence are shown below, and the intersection is the last graph. We use both set-builder and interval notations.

$\{x | -2 < x\}$

$-7\ -6\ -5\ -4\ -3\ -2\ -1\ \ 0\ \ 1\ \ 2\ \ 3\ \ 4\ \ 5\ \ 6\ \ 7$ $(-2, \infty)$

$\{x | x < 1\}$

$-7\ -6\ -5\ -4\ -3\ -2\ -1\ \ 0\ \ 1\ \ 2\ \ 3\ \ 4\ \ 5\ \ 6\ \ 7$ $(-\infty, 1)$

$\{x | -2 < x\} \cap \{x | x < 1\}$
$= \{x | -2 < x \text{ and } x < 1\}$

$-7\ -6\ -5\ -4\ -3\ -2\ -1\ \ 0\ \ 1\ \ 2\ \ 3\ \ 4\ \ 5\ \ 6\ \ 7$ $(-2, 1)$

The conjunction $-2 < x$ *and* $x < 1$ can be abbreviated by $-2 < x < 1$. Thus the interval $(-2, 1)$ can be represented as $\{x \mid -2 < x < 1\}$, the set of all numbers that are simultaneously greater than -2 *and* less than 1.

For any real numbers a and b with $a < b$:

The conjunction $a < x$ *and* $x < b$ can be abbreviated as $a < x < b$.

The conjunction $b > x$ *and* $x > a$ can be abbreviated as $b > x > a$.

EXAMPLE 2

Solve and graph: $-1 \le 2x + 5 < 13$.

Solution   This inequality is an abbreviation for the following conjunction:

$$-1 \le 2x + 5 \quad and \quad 2x + 5 < 13.$$

The word *and* corresponds to set *intersection*. The solution set is thus the intersection of the solution set of $-1 \le 2x + 5$ and the solution set of $2x + 5 < 13$:

$$\{x \mid -1 \le 2x + 5\} \cap \{x \mid 2x + 5 < 13\}.$$

Method 1. We write the conjunction with the word *and:*

$$
\begin{array}{lll}
-1 \le 2x + 5 & and & 2x + 5 < 13 \\
-6 \le 2x & and & 2x < 8 \\
-3 \le x & and & x < 4.
\end{array}
$$

We now abbreviate the answer:

$$-3 \le x < 4.$$

The solution set is $\{x \mid -3 \le x < 4\}$, or, in interval notation, $[-3, 4)$.

Method 2. Using Method 1, we did the same thing to each inequality. We can shorten the writing as follows:

$$
\begin{array}{ll}
-1 \le 2x + 5 < 13 & \\
-1 - 5 \le 2x + 5 - 5 < 13 - 5 & \text{Subtracting 5} \\
-6 \le 2x < 8 & \\
-3 \le x < 4. & \text{Dividing by 2}
\end{array}
$$

The solution set is $\{x \mid -3 \le x < 4\}$, or $[-3, 4)$.

The graph is the intersection of the individual graphs.

$\{x \mid -3 \le x\}$ $[-3, \infty)$

$\{x \mid x < 4\}$ $(-\infty, 4)$

$\{x \mid -3 \le x\} \cap \{x \mid x < 4\}$
$= \{x \mid -3 \le x < 4\}$ $[-3, 4)$

EXAMPLE 3

Solve and graph: $2x - 5 \geqslant -3$ *and* $5x + 2 \geqslant 17$.

Solution   We solve each inequality separately:

$$2x - 5 \geqslant -3 \quad and \quad 5x + 2 \geqslant 17$$
$$2x \geqslant 2 \quad and \quad 5x \geqslant 15$$
$$x \geqslant 1 \quad and \quad x \geqslant 3.$$

The solution set is the intersection of the solution sets of the individual inequalities.

$\{x \mid x \geqslant 1\}$  $[1, \infty)$

$\{x \mid x \geqslant 3\}$  $[3, \infty)$

$\{x \mid x \geqslant 1\} \cap \{x \mid x \geqslant 3\}$
$= \{x \mid x \geqslant 3\}$  $[3, \infty)$

The numbers common to both sets are those that are greater than or equal to 3. Thus the solution set is $\{x \mid x \geqslant 3\}$, or, in interval notation, $[3, \infty)$.   ❑

Intersection

> The word "and" corresponds to "intersection" and to the symbol "∩". For a number to be a solution of the conjunction, it must be in *both* solution sets.

If sets have no common members, we say that their intersection is the empty set, ∅. The following two sets have an empty intersection:

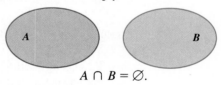

$A \cap B = \varnothing.$

EXAMPLE 4

Solve and graph: $2x - 3 > 1$ *and* $3x - 1 < 2$.

Solution   We solve each inequality separately:

$$2x - 3 > 1 \quad and \quad 3x - 1 < 2$$
$$2x > 4 \quad and \quad 3x < 3$$
$$x > 2 \quad and \quad x < 1.$$

The solution set is the intersection of the individual inequalities.

$\{x \mid x > 2\}$ 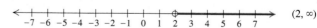 $(2, \infty)$

$\{x \mid x < 1\}$  $(-\infty, 1)$

$\{x \mid x > 2\} \cap \{x \mid x < 1\}$
$= \{x \mid x > 2 \text{ and } x < 1\} = \varnothing$ $\varnothing$

Since no number is both greater than 2 and less than 1, the solution set is the empty set, $\varnothing$.   ❑

Unions of Sets and Disjunctions of Sentences

The **union** of two sets A and B is formed by putting the sets together. We denote the union of sets A and B as

$A \cup B$.

The union of two sets is often pictured as shown below.

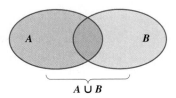

$A \cup B$

EXAMPLE 5

Find the union: $\{2, 3, 4\} \cup \{3, 5, 7\}$.

Solution   The numbers in either or both sets are 2, 3, 4, 5, and 7, so the union is $\{2, 3, 4, 5, 7\}$.   ❑

When two or more sentences are joined by the word *or* to make a compound sentence, the new sentence is called a **disjunction** of the sentences. Here are three examples:

$x < -3$ *or* $x > 3$;

y is an odd number   *or* y is a prime number;

$x < 0$ *or* $x = 0$ *or* $x > 0$.

For a disjunction to be true, at least one of the individual sentences must be true. The solution set of a disjunction is the union of the individual solution sets. Consider the disjunction

$x < -3$ *or* $x > 3$.

The graphs of each separate sentence are shown below, and the union is the last graph. Again, we use both set-builder and interval notations.

$\{x | x < -3\}$ [number line graph: open circle at -3, arrow left; points labeled $-6\ -5\ -4\ -3\ -2\ -1\ 0\ 1\ 2\ 3\ 4\ 5\ 6$] $(-\infty, -3)$

$\{x | x > 3\}$ [number line graph: open circle at 3, arrow right; points labeled $-6\ -5\ -4\ -3\ -2\ -1\ 0\ 1\ 2\ 3\ 4\ 5\ 6$] $(3, \infty)$

$\{x | x < -3\} \cup \{x | x > 3\}$
$= \{x | x < -3 \text{ or } x > 3\}$ [number line graph: open circle at -3 arrow left and open circle at 3 arrow right; points labeled $-6\ -5\ -4\ -3\ -2\ -1\ 0\ 1\ 2\ 3\ 4\ 5\ 6$] $(-\infty, -3) \cup (3, \infty)$

The interval notation $(-\infty, -3) \cup (3, \infty)$ cannot be simplified. The solution set of $x < -3 \text{ or } x > 3$ is most commonly written $\{x | x < -3 \text{ or } x > 3\}$

Union

> The word "or" corresponds to "union" and to the symbol "∪". In order for a number to be a solution of the disjunction, it must be in *at least one* of the solution sets.

EXAMPLE 6

Solve and graph: $7 + 2x < 1$ *or* $13 - 5x \leq 3$.

Solution   We solve each inequality separately:

$$7 + 2x < 1 \quad or \quad 13 - 5x \leq 3$$
$$2x < -6 \quad or \quad -5x \leq -10$$
$$x < -3 \quad or \quad x \geq 2.$$

To find the solution set of the disjunction, we consider the individual graphs. We graph $x < -3$. We also graph $x \geq 2$. Then we take the union of the two graphs:

$\{x | x < -3\}$ $(-\infty, -3)$

$\{x | x \geq 2\}$ $[2, \infty)$

$\{x | x < -3 \text{ or } x \geq 2\}$ $(-\infty, -3) \cup [2, \infty)$

The solution set is $\{x | x < -3 \text{ or } x \geq 2\}$, or, in interval notation, $(-\infty, -3) \cup [2, \infty)$.

CAUTION! A compound inequality like

$$x < -3 \quad or \quad x \geq 2,$$

as in Example 6, cannot be abbreviated to one like $2 \leq x < -3$ because to do so would be to say that x is *simultaneously* less than -3 and greater than or equal to 2. When the word *or* appears, you must keep that word.

EXAMPLE 7

Solve: $-2x - 5 < -2$ *or* $x - 3 < -10$.

Solution   We solve the individual inequalities separately, retaining the word *or:*

$$-2x - 5 < -2 \quad or \quad x - 3 < -10$$
$$-2x < 3 \quad or \quad x < -7$$

Reverse the symbol.

$$x > -\tfrac{3}{2} \quad or \quad x < -7.$$

Keep the word "or."

The solution set consists of all numbers that are less than -7 or greater than $-\frac{3}{2}$. We write the solution set as $\left\{x | x < -7 \text{ or } x > -\frac{3}{2}\right\}$, or, in interval notation, $(-\infty, -7) \cup \left(-\frac{3}{2}, \infty\right)$.

EXAMPLE 8

Solve: $3x - 11 < 4$ *or* $4x + 9 \geq 1$.

Solution   We solve the individual inequalities separately, retaining the word *or*.

$$3x - 11 < 4 \quad or \quad 4x + 9 \geq 1$$
$$3x < 15 \quad or \quad 4x \geq -8$$
$$x < 5 \quad or \quad x \geq -2$$

Keep the word "or."

To find the solution set, we look at the individual graphs.

$\{x \mid x < 5\}$ 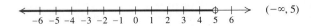 $(-\infty, 5)$

$\{x \mid x \geq -2\}$  $[-2, \infty)$

$\{x \mid x < 5\} \cup \{x \mid x \geq -2\}$
$= \{x \mid x < 5 \text{ } or \text{ } x \geq -2\}$  $(-\infty, \infty) = \mathbb{R}$

Since *all* numbers are less than 5 or greater than or equal to -2, the two sets fill up the entire number line. Thus the solution set is $\mathbb{R}$, the set of all real numbers.   ❑

EXERCISE SET | 4.2

Find the intersection.

1. $\{5, 6, 7, 8\} \cap \{4, 6, 8, 10\}$
2. $\{9, 10, 27\} \cap \{8, 10, 38\}$
3. $\{2, 4, 6, 8\} \cap \{1, 3, 5\}$
4. $\{-4, -2, 0\} \cap \{-1, 0, 1\}$
5. $\{1, 2, 3, 4\} \cap \{1, 2, 3, 4\}$
6. $\{8, 9, 10\} \cap \varnothing$

Graph and write interval notation.

7. $1 < x < 6$ **8.** $0 \leq y \leq 3$
9. $-7 \leq y \leq -3$ **10.** $-9 \leq x < -5$
11. $-4 \leq -x < 3$
12. $x > -8$ *and* $x < -3$
13. $6 > -x \geq -2$ **14.** $x > -4$ *and* $x < 2$
15. $5 > x \geq -2$ **16.** $3 > x \geq 0$
17. $x < 5$ *and* $x \geq 1$ **18.** $x \geq -2$ *and* $x < 2$

Solve.

19. $-2 < x + 2 < 8$ **20.** $-1 < x + 1 \leq 6$

21. $1 < 2y + 5 \leq 9$
22. $3 \leq 5x + 3 \leq 8$
23. $-10 \leq 3x - 5 \leq -1$
24. $-18 \leq -2x - 7 < 0$
25. $2 < x + 3 \leq 9$ **26.** $-6 \leq x + 1 < 9$
27. $-6 \leq 2x - 3 < 6$ **28.** $4 > -3m - 7 \geq 2$
29. $-\frac{1}{2} < \frac{1}{4}x - 3 \leq \frac{1}{2}$ **30.** $-\frac{2}{3} \leq 4 - \frac{1}{4}x < \frac{2}{3}$

Find the union.

31. $\{4, 5, 6, 7, 8\} \cup \{1, 4, 6, 11\}$
32. $\{8, 9, 27\} \cup \{2, 8, 27\}$
33. $\{2, 4, 6, 8\} \cup \{1, 3, 5\}$
34. $\{8, 9, 10\} \cup \varnothing$
35. $\{4, 8, 11\} \cup \varnothing$
36. $\varnothing \cup \varnothing$

Graph.

37. $x < -1$ *or* $x > 2$ **38.** $x < -2$ *or* $x > 0$
39. $x \leq -3$ *or* $x > 1$ **40.** $x \leq -1$ *or* $x > 3$

Solve.

41. $x + 7 < -2 \ or \ x + 7 > 2$

42. $x + 9 < -4 \ or \ x + 9 > 4$

43. $2x - 8 \leqslant -3 \ or \ x - 8 \geqslant 3$

44. $x + 7 \leqslant -2 \ or \ 3x - 7 \geqslant 2$

45. $7x + 4 \geqslant -17 \ or \ 6x + 5 \geqslant -7$

46. $4x - 4 < -8 \ or \ 4x - 4 < 12$

47. $7 > -4x + 5 \ or \ 10 \leqslant -4x + 5$

48. $6 > 2x - 1 \ or \ -4 \leqslant 2x - 1$

49. $3x - 7 > -10 \ or \ 5x + 2 \leqslant 22$

50. $3x + 2 < 2 \ or \ 4 - 2x < 14$

51. $-2x - 2 < -6 \ or \ -2x - 2 > 6$

52. $-3m - 7 < -5 \ or \ -3m - 7 > 5$

53. $\frac{2}{3}x - 14 < -\frac{5}{6} \ or \ \frac{2}{3}x - 14 > \frac{5}{6}$

54. $\frac{1}{4} - 3x \leqslant -3.7 \ or \ \frac{1}{4} - 5x \geqslant 4.8$

55. $\dfrac{2x - 5}{6} \leqslant -3 \ or \ \dfrac{2x - 5}{6} \geqslant 4$

56. $\dfrac{7 - 3x}{5} < -4 \ or \ \dfrac{7 - 3x}{5} > 4$

57. $5x - 7 \leqslant 13 \ or \ 2x - 1 \geqslant -7$

58. $5x + 4 \leqslant 14 \ or \ 7 - 2x \geqslant 9$

Skill Maintenance _____

Solve.

59. $2x - 3y = 7,$
     $3x + 2y = -10$

60. $3x - 9(x + 4) = 20(3x + 7)$

61. $5(2x + 3) = 3(x - 4)$

62. Graph: $3x - 4y = -12.$

Synthesis _____

63. ◈ Explain how the use of the word *or* in a compound inequality differs from the use of the word *or* in everyday English. (*Hint:* Consider the expression and/or.)

64. ◈ Explain why the conjunction $3 < x \ and \ x < 5$ can be rewritten as $3 < x < 5$, but the disjunction $3 < x \ or \ x < 5$ cannot be rewritten as $3 < x < 5$.

65. *Temperatures of liquids.* The formula
$$C = \tfrac{5}{9}(F - 32)$$
can be used to convert Fahrenheit temperatures F to Celsius temperatures C.

a) Gold is liquid for Celsius temperatures C such that $1063° \leqslant C < 2660°$. Find a similar such inequality for the corresponding Fahrenheit temperatures.

b) Silver is liquid for Celsius temperatures C such that $960.8° \leqslant C < 2180°$. Find a similar such inequality for the corresponding Fahrenheit temperatures.

66. We say that x is *between a and b* if $a < x < b$. Find all the numbers on a number line from which you can subtract 3, and still be between -8 and 8.

67. *Converting dress sizes.* The function
$$f(x) = 2(x + 10)$$
can be used to convert dress sizes x in the United States to dress sizes $f(x)$ in Italy. For what dress sizes in the United States will dress sizes in Italy be between 32 and 46?

68. *Pressure at sea depth.* The function
$$P(d) = 1 + \frac{d}{33}$$
gives the pressure, in atmospheres (atm), at a depth of d feet in the sea. For what depths d is the pressure at least 1 atm and at most 7 atm?

69. *Women in the military ranks.* The percentage of the total active military duty force that is women has been steadily increasing. The number N of women in the active duty force t years since 1971 can be predicted by

$$N = 12,197.8t + 44,000.$$

For what years will the number of women always be at least 50,000 and at most 250,000? (Measure from the end of 1971 and make sure that $N \not< 50,000$ and $N \not> 250,000$ at any point in the time span.)

70. *Records in the women's 100-m dash.* Florence Griffith Joyner set a world record of 10.49 sec in the women's 100-m dash in 1988. The formula

$$R = -0.0433t + 10.49$$

can be used to predict the world record in the women's 100-m dash t years after 1988. Predict (in terms of an inequality) those years for which the world record was between 11.5 and 10.8 sec. (Measure from the end of 1988.)

Solve and graph.

71. $4a - 2 \leqslant a + 1 \leqslant 3a + 4$

72. $4m - 8 > 6m + 5 \quad or \quad 5m - 8 < -2$

73. $x - 10 < 5x + 6 \leqslant x + 10$

74. $2[5(3 - y) - 2(y - 2)] > y + 4$

75. $3x < 4 - 5x < 5 + 3x$

76. $(x + 6)(x - 4) > (x + 1)(x - 3)$

Determine whether the sentence is true or false for all real numbers a, b, and c.

77. If $b > c$, then $b \not\leqslant c$.

78. If $-b < -a$, then $a < b$.

79. If $c \neq a$, then $a < c$.

80. If $a < c$ and $c < b$, then $b \not> a$.

81. If $a < c$ and $b < c$, then $a < b$.

82. If $-a < c$ and $-c > b$, then $a < b$.

Solve.

83. $[4x - 2 < 8 \ or \ 3(x - 1) < -2] \quad and$
$-2 \leqslant 5x \leqslant 10$

84. $-2 \leqslant 4m + 3 < 7 \quad and$
$[m - 5 \geqslant 4 \ or \ 3 - m > 12]$

4.3

Absolute-Value Equations and Inequalities

Equations with Absolute Value • **Inequalities with Absolute Value**

Equations with Absolute Value

Recall from Section 1.2 the definition of absolute value.

Absolute Value

> The absolute value of x, denoted $|x|$, is defined as
>
> $$|x| = \begin{cases} x, & \text{if } x \geqslant 0, \\ -x, & \text{if } x < 0. \end{cases}$$

In words, the definition states that the absolute value of a nonnegative number is the number itself, and the absolute value of a negative number is the opposite of the number.

Since distance is always nonnegative, we can think of a number's absolute value as its distance from zero on a number line.

EXAMPLE 1

Find the solution set: **(a)** $|x| = 4$; **(b)** $|x| = 0$; **(c)** $|x| = -7$.

Solution

a) We interpret $|x| = 4$ to mean that the number x is 4 units from zero on a number line. There are two such numbers, 4 and -4. Thus the solution set is $\{-4, 4\}$.

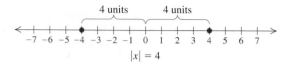

$|x| = 4$

b) We interpret $|x| = 0$ to mean that x is 0 units from zero on a number line. The only number that satisfies this is zero itself. Thus the solution set is $\{0\}$.

c) Since distance is always nonnegative, it doesn't make sense to talk about a number that is -7 units from zero. Remember that the absolute value of a number is never negative. Thus there is no solution; the solution set is $\varnothing$.

◻

Example 1 leads us to the following principle for solving equations with absolute value.

The Absolute-Value Principle for Equations

For any positive number p and any algebraic expression X:

The solutions of $|X| = p$ are those numbers that satisfy $X = -p$ or $X = p$.

If $|X| = 0$, then $X = 0$. If p is negative, then $|X| = p$ has no solution.

EXAMPLE 2

Find the solution set: **(a)** $|2x + 5| = 13$; **(b)** $|4 - 7x| = -8$.

Solution

a) We use the absolute-value principle, replacing X by $2x + 5$ and p by 13:

$$|X| = p$$
$$|2x + 5| = 13$$
$$2x + 5 = -13 \quad or \quad 2x + 5 = 13 \qquad \text{Using the absolute-value principle}$$
$$2x = -18 \quad or \qquad 2x = 8$$
$$x = -9 \quad or \qquad x = 4.$$

Check: For -9:   For 4:

$$\begin{array}{c}|2x + 5| = 13 \\ \hline |2(-9) + 5| \; ? \; 13 \\ |-18 + 5| \\ |-13| \\ 13 \end{array} \quad 13 \text{ TRUE}$$

$$\begin{array}{c}|2x + 5| = 13 \\ \hline |2 \cdot 4 + 5| \; ? \; 13 \\ |8 + 5| \\ |13| \\ 13 \end{array} \quad 13 \text{ TRUE}$$

The number $2x + 5$ is 13 units from 0 if x is replaced by -9 or 4. The solution set is $\{-9, 4\}$.

b)   The absolute-value principle reminds us that absolute value is always nonnegative. The equation $|4 - 7x| = -8$ has no solution. The solution set is $\varnothing$.   ❑

The absolute-value principle can be used together with the addition and multiplication principles to solve many types of equations with absolute value.

EXAMPLE 3

Solve: $2|x + 3| + 1 = 15$.

Solution   We first isolate $|x + 3|$. Then we use the absolute-value principle:

$$2|x + 3| + 1 = 15$$
$$2|x + 3| = 14 \qquad \text{Adding } -1 \text{ on both sides}$$
$$|x + 3| = 7 \qquad \text{Multiplying by } \tfrac{1}{2} \text{ on both sides}$$
$$x + 3 = -7 \quad or \quad x + 3 = 7 \qquad \text{Replacing } X \text{ by } x + 3 \text{ and } p \text{ by } 7 \text{ in}$$
$$\text{the absolute-value principle}$$
$$x = -10 \quad or \qquad x = 4.$$

We leave the check to the student. The solutions are -10 and 4. The solution set is $\{-10, 4\}$.   ❑

EXAMPLE 4

Solve: $|x - 2| = 3$.

Solution   This equation can be solved in two different ways.

Method 1.   This approach is helpful in calculus. The expressions $|a - b|$ and $|b - a|$ can be used to represent the *distance between* a and b on the number line. For example, the distance between 7 and 8 is given by $|8 - 7|$ or $|7 - 8|$. From this viewpoint, the equation $|x - 2| = 3$ states that the distance between x and 2 is 3 units. We draw a number line and locate those numbers that are 3 units from 2.

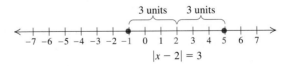

The solutions of $|x - 2| = 3$ are -1 and 5.

Method 2.   Using the absolute-value principle, we can interpret the equation as stating that the number $x - 2$ is 3 units from zero. Thus we replace X by $x - 2$ and p by 3:

$$|X| = p$$
$$|x - 2| = 3$$
$$x - 2 = -3 \quad or \quad x - 2 = 3 \qquad \text{Using the absolute-value principle}$$
$$x = -1 \quad or \qquad x = 5.$$

The check consists of observing that both methods gave the same solutions. The solution set is $\{-1, 5\}$.   ❑

Sometimes an equation has two absolute-value expressions. Consider $|a| = |b|$. This means that a and b are the same distance from zero.

If a and b are the same distance from zero, then either they are the same number or they are opposites.

EXAMPLE 5

Solve: $|2x - 3| = |x + 5|$.

Solution   Either $2x - 3 = x + 5$ or $2x - 3 = -(x + 5)$. We solve each equation separately:

$$2x - 3 = x + 5 \quad or \quad 2x - 3 = -(x + 5)$$
$$x - 3 = 5 \quad or \quad 2x - 3 = -x - 5$$
$$x = 8 \quad or \quad 3x - 3 = -5$$
$$3x = -2$$
$$x = -\tfrac{2}{3}.$$

We leave the check to the student. The solutions are 8 and $-\tfrac{2}{3}$. The solution set is $\left\{8, -\tfrac{2}{3}\right\}$.   ❑

Inequalities with Absolute Value

Our methods for solving equations with absolute value can be extended for solving inequalities.

EXAMPLE 6

Solve $|x| < 4$. Then graph.

Solution   The solutions of $|x| < 4$ are those numbers whose *distance from zero is less than* 4. By substituting or by looking at the number line, we can see that numbers like $-3, -2, -1, -\tfrac{1}{2}, -\tfrac{1}{4}, 0, \tfrac{1}{4}, \tfrac{1}{2}, 1, 2,$ and 3 are all solutions. In fact, the solutions are all the numbers between -4 and 4. The solution set is $\{x \mid -4 < x < 4\}$ or, in interval notation, $(-4, 4)$. The graph is as follows:

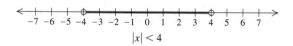

$$|x| < 4$$

❑

EXAMPLE 7

Solve $|x| \geq 4$. Then graph.

Solution   The solutions of $|x| \geq 4$ are those numbers whose *distance from zero is greater than or equal to* 4—in other words, those numbers x such that $x \leq -4$ or $4 \leq x$. The solution set is $\{x \mid x \leq -4 \ or \ x \geq 4\}$, or, in interval notation, $(-\infty, -4] \cup [4, \infty)$. We check with numbers like $-4.1, -5, 4.1,$ and 5. Note also that -3.9 and 3.9 are *not* solutions. The graph is as follows:

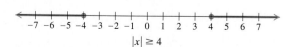

$$|x| \geq 4$$

❑

Examples 1, 6, and 7 illustrate three types of problems in which absolute-value signs appear. The following is a general principle for solving such problems.

Principles for Solving Absolute-Value Problems

For any positive number p and any expression X:

a) The solutions of $|X| = p$ are those numbers that satisfy $X = -p$ or $X = p$.

b) The solutions of $|X| < p$ are those numbers that satisfy $-p < X < p$.

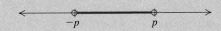

c) The solutions of $|X| > p$ are those numbers that satisfy $X < -p$ or $p < X$.

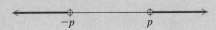

Of course, if p is negative, any value of X will satisfy the inequality $|X| > p$ since absolute value is never negative. By the same reasoning, $|X| < p$ has no solution when p is not positive. Thus the inequality $|2x - 7| > -3$ is true for any real number x, and the inequality $|2x - 7| < -3$ has no solution.

Note that an inequality of the form $|X| < p$ corresponds to a *con*junction, whereas an inequality of the form $|X| > p$ corresponds to a *dis*junction.

EXAMPLE 8

Solve $|3x - 2| < 4$. Then graph.

Solution   We use part (b) of the principles listed above. In this case, X is $3x - 2$ and p is 4:

$$|X| < p$$
$$|3x - 2| < 4 \qquad \text{Replacing } X \text{ by } 3x - 2 \text{ and } p \text{ by } 4$$
$$-4 < 3x - 2 < 4 \qquad \text{The number } 3x - 2 \text{ must be within 4 units of zero.}$$
$$-2 < \quad 3x \quad < 6 \qquad \text{Adding 2}$$
$$-\tfrac{2}{3} < \quad x \quad < 2. \qquad \text{Multiplying by } \tfrac{1}{3}$$

The solution set is $\left\{x \mid -\tfrac{2}{3} < x < 2\right\}$, or, in interval notation, $\left(-\tfrac{2}{3}, 2\right)$. The graph is as follows:

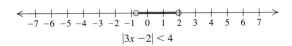

EXAMPLE 9

Solve $|4x + 2| \geqslant 6$. Then graph.

Solution   We use part (c) of the principles listed above. In this case, X is $4x + 2$ and p is 6:

$$|X| \geqslant p$$

$$|4x + 2| \geqslant 6 \qquad \text{Replacing } X \text{ by } 4x + 2 \text{ and } p \text{ by } 6$$

$$4x + 2 \leqslant -6 \quad or \quad 6 \leqslant 4x + 2 \qquad \text{The number } 4x + 2 \text{ must be at least 6 units from zero.}$$

$$4x \leqslant -8 \quad or \quad 4 \leqslant 4x \qquad \text{Adding } -2$$

$$x \leqslant -2 \quad or \quad 1 \leqslant x. \qquad \text{Multiplying by } \tfrac{1}{4}$$

The solution set is $\{x \mid x \leqslant -2 \text{ or } x \geqslant 1\}$, or, in interval notation, $(-\infty, -2] \cup [1, \infty)$. The graph is as follows:

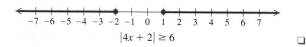

$$|4x + 2| \geqslant 6$$

TECHNOLOGY
CONNECTION

To solve an inequality like $|4x + 2| \geqslant 6$ with a grapher, simply graph the equation $y_1 = |4x + 2|$ using the ABS key or by typing ABS for the absolute-value function. On the same set of axes, graph $y_2 = 6$. Using the Trace and Zoom features, locate those points on the graph of $y_1 = |4x + 2|$ that are *on or above* the line $y = 6$. The x-values of those points solve the inequality. How can the same graph be used to solve the inequality $|4x + 2| < 6$ or the equation $|4x + 2| = 6$? Try using this procedure to solve Example 8 on your grapher.

EXERCISE SET | 4.3

Solve.

1. $|x| = 3$ **2.** $|x| = 5$

3. $|x| = -3$ **4.** $|x| = -5$

5. $|p| = 0$ **6.** $|y| = 8.6$

7. $|t| = 5.5$ **8.** $|m| = 0$

9. $|x - 3| = 12$ **10.** $|3x - 2| = 6$

11. $|2x - 3| = 4$ **12.** $|5x + 2| = 3$

13. $|2y - 7| = 10$ **14.** $|3y - 4| = 8$

15. $|3x - 10| = -8$ **16.** $|7x - 2| = -9$

17. $|x| + 7 = 18$ **18.** $|x| - 2 = 6.3$

19. $|5x| - 3 = 37$ **20.** $|2y| - 5 = 13$

21. $5|q| - 2 = 9$ **22.** $7|z| + 2 = 16$

23. $\left|\dfrac{2x - 1}{3}\right| = 5$ **24.** $\left|\dfrac{4 - 5x}{6}\right| = 7$

25. $|m + 5| + 9 = 16$ **26.** $|t - 7| + 3 = 4$

27. $\left|\dfrac{1 - 2x}{3}\right| = 1$ **28.** $\left|\dfrac{3x - 2}{5}\right| = 2$

29. $5 - 2|3x - 4| = -5$ **30.** $3|2x - 5| - 7 = -1$

Each pair of numbers represents two points on a number line. Find the distance between the points.

31. 13, 17   **32.** 9, 15

33. 25, 14   **34.** 32, 17

35. -9, 24   **36.** -18, -37

37. -8, -42 **38.** -9, -36

Solve.

39. $|3x + 4| = |x - 7|$ **40.** $|2x - 8| = |x + 3|$

41. $|x + 5| = |x - 2|$ **42.** $|x - 7| = |x + 8|$

43. $|2a + 4| = |3a - 1|$

44. $|5p + 7| = |4p + 3|$

45. $|y - 3| = |3 - y|$ **46.** $|m - 7| = |7 - m|$

47. $|5 - p| = |p + 8|$ **48.** $|8 - q| = |q + 19|$

49. $\left|\dfrac{2x - 3}{6}\right| = \left|\dfrac{4 - 5x}{8}\right|$

50. $\left|\dfrac{6 - 8x}{5}\right| = \left|\dfrac{7 + 3x}{2}\right|$

51. $\left|\tfrac{1}{2}x - 5\right| = \left|\tfrac{1}{4}x + 3\right|$ **52.** $\left|2 - \tfrac{2}{3}x\right| = \left|4 + \tfrac{7}{8}x\right|$

Solve and graph.

53. $|x| < 3$ **54.** $|x| \leq 5$

55. $|x| \geq 2$ **56.** $|y| > 8$

57. $|t| \geq 5.5$ **58.** $|m| > 0$

59. $|x - 3| < 1$ **60.** $|x - 2| < 6$

61. $|x + 2| \leq 5$ **62.** $|x + 4| \leq 1$

63. $|x - 3| > 1$ **64.** $|x - 2| > 6$

65. $|2x - 3| \leq 4$ **66.** $|5x + 2| \leq 3$

67. $|2y - 7| > -1$ **68.** $|3y - 4| > 8$

69. $|4x - 9| \geq 14$ **70.** $|9y - 1| \geq -3$

71. $|y - 3| < 12$ **72.** $|p - 2| < 3$

73. $|2x + 3| \leq 4$ **74.** $|5x - 2| \leq 3$

75. $|4 - 3y| > 8$ **76.** $|7 - 2y| < -6$

77. $|9 - 4x| \leq 14$ **78.** $|2 - 9p| \geq 17$

79. $|3 - 4x| < -5$ **80.** $|-5 - 7x| \leq 30$

81. $7 + |2x - 1| > 16$ **82.** $5 + |3x + 2| > 19$

83. $\left| \dfrac{x - 7}{3} \right| < 4$ **84.** $\left| \dfrac{x + 5}{4} \right| \leq 2$

85. $\left| \dfrac{2 - 5x}{4} \right| \geq \dfrac{2}{3}$ **86.** $\left| \dfrac{1 + 3x}{5} \right| > \dfrac{7}{8}$

87. $|m + 5| + 9 \leq 16$ **88.** $|t - 7| + 3 \geq 4$

89. $|g + 7| + 13 > 9$

90. $2|2x - 7| + 11 > 10$

91. $\left| \dfrac{2x - 1}{3} \right| \leq 1$ **92.** $\left| \dfrac{3x - 2}{5} \right| \leq 2$

Skill Maintenance

93. The perimeter of a rectangular field is 628 m. The length of the field is 6 m greater than the width. Find the area of the field.

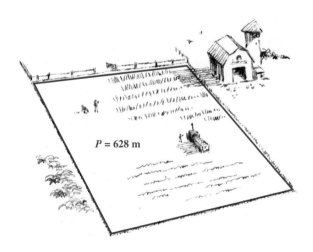

$P = 628$ m

94. At a barbecue, there were 250 dinners served. The cost of a dinner was $1.50 each for children and $4.00 each for adults. The total amount of money collected was $705. How many of each type of plate was served?

Synthesis

95. ◈   Explain why the inequality $|x + 7| < 1$ can be interpreted as "the distance between x and -7 is less than 1."

96. ◈   Explain why the inequality $|x + 5| \geq 2$ can be interpreted as "the number x is at least 2 units from -5."

97. From the definition of absolute value, $|x| = x$ only when $x \geq 0$. Thus, $|x + 3| = x + 3$ only when $x + 3 \geq 0$, which means that $x \geq -3$. Solve $|2x - 5| = 2x - 5$ using this same argument.

Solve.

98. $1 - |\frac{1}{4}x + 8| = \frac{3}{4}$ **99.** $|x + 5| = x + 5$

100. $|x - 1| = x - 1$ **101.** $|7x - 2| = x + 4$

102. $|3.7x - \frac{4}{9}| > -2$ **103.** $|5.2x - \frac{6}{7}| \leq -8$

104. $|\frac{5}{9} + 3x| < -\frac{1}{6}$ **105.** $|x + 5| > x$

106. $2 \leq |x - 1| \leq 5$

Find an equivalent inequality with absolute value.

107. $-3 < x < 3$

108. $-5 \leq y \leq 5$

109. $x \leq -6 \; or \; 6 \leq x$

110. $x < -4 \; or \; 4 < x$

111. $x < -8 \; or \; 2 < x$

112. $-5 < x < 1$

113. Pipe is being constructed so that it has a length of 5 ft with a tolerance of $\frac{1}{8}$ in. This means that the length of the pipe can be at most 5 ft plus $\frac{1}{8}$ in. and at least 5 ft minus $\frac{1}{8}$ in. Suppose that $p =$ the length of such a pipe. Find an inequality with absolute value whose solutions are all the possible lengths p.

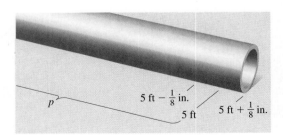

p $5 \text{ ft} - \frac{1}{8}$ in.

$5 \text{ ft} + \frac{1}{8}$ in.

5 ft

114. A weighted spring is bouncing up and down so that its distance d above the ground satisfies the inequality $|d - 6\text{ ft}| \leq \frac{1}{2}\text{ ft}$ (see the figure at right). Find all possible distances d.

115. 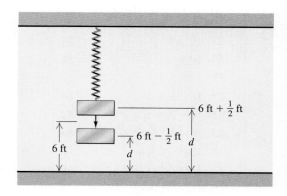   Use a grapher to check your solutions to Exercises 1, 9, 13, 41, 53, 63, 87, 99, and 105.

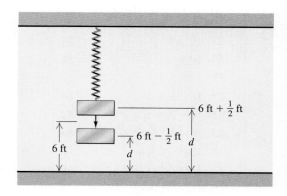

4.4

Inequalities in Two Variables

Solutions of Inequalities in Two Variables • Systems of Linear Inequalities

In Section 4.1, we graphed inequalities in one variable on a number line. Now we graph inequalities in two variables on a plane.

Solutions of Inequalities in Two Variables

The solutions of inequalities in two variables are ordered pairs.

EXAMPLE 1

Determine whether $(-3, 2)$ and $(6, -7)$ are solutions of the inequality $5x - 4y > 13$.

Solution   Below, on the left, we replace x by -3 and y by 2. On the right, we replace x by 6 and y by -7.

$$
\begin{array}{c|c}
5x - 4y > 13 \\
\hline
5(-3) - 4 \cdot 2 \;?\; 13 \\
-15 - 8 \\
\quad\quad -23 \;\bigm|\; 13 \quad \text{FALSE}
\end{array}
\qquad
\begin{array}{c|c}
5x - 4y > 13 \\
\hline
5(6) - 4(-7) \;?\; 13 \\
30 + 28 \\
\quad\quad 58 \;\bigm|\; 13 \quad \text{TRUE}
\end{array}
$$

Since $-23 > 13$ is false, $(-3, 2)$ is not a solution.

Since $58 > 13$ is true, $(6, -7)$ is a solution.

We now consider graphs of inequalities in two variables.

EXAMPLE 2

Graph: $y < x$.

Solution   We first graph the line $y = x$. Every solution of $y = x$ is an ordered pair like (3, 3). The first and second coordinates are the same. The graph of $y = x$ is shown on the left below. We draw it dashed because these points are *not* solutions of $y < x$.

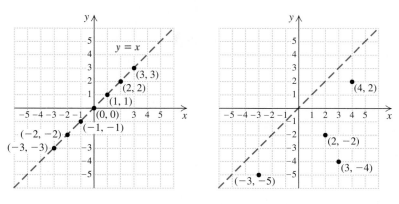

Notice that in the graph on the right each of the ordered pairs on the half-plane below $y = x$ contains a y-coordinate that is less than the x-coordinate. Thus all the pairs shown represent solutions of $y < x$. We can check a pair, (4, 2), as follows:

$$\frac{y < x}{2 \mid 4 \quad \text{TRUE}}$$

It turns out that *any* point on the same side of $y = x$ as (4, 2) is also a solution. Thus, if one point in a half-plane is a solution, then all points in that half-plane are solutions. In this text, we will usually indicate this by color shading. We shade the half-plane below $y = x$.

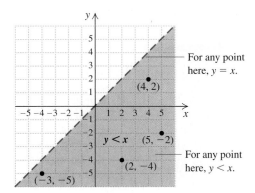

EXAMPLE 3

Graph: $8x + 3y \geq 24$.

Solution   First we sketch the line $8x + 3y = 24$. Since the inequality sign is $\geq$, points on the line $8x + 3y = 24$ are also in the graph of $8x + 3y \geq 24$, so we draw a solid line. This indicates that all points on the line are solutions. The rest of the solutions are either in the half-plane above the line or the half-plane below the line. To determine which, we select a point that is not on the line and determine whether it is a solution of $8x + 3y \geqslant 24$. We try $(-3, 4)$ as a test point:

$$\frac{8x + 3y \geqslant 24}{\begin{array}{c|c} 8(-3) + 3 \cdot 4 \ ? \ 24 \\ -24 + 12 \\ -12 \end{array} \ \begin{array}{c} \\ \\ 24 \ \ \text{FALSE} \end{array}}$$

We see that $-12 \geqslant 24$ is *false*. Since $(-3, 4)$ is not a solution, no point in the half-plane containing $(-3, 4)$ is a solution. Thus the points in the other half-plane are solutions. We shade that half-plane and obtain the graph:

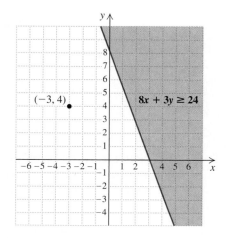

A **linear inequality** is one that we can get from a linear equation by changing the equals sign to an inequality sign. Every linear equation has a graph that is a straight line. The graph of a linear inequality is a half-plane, sometimes including the line along the edge. That line is the graph of what we call a *related equation*. We graph linear inequalities as follows.

> To graph an inequality in two variables:
>
> 1. Replace the inequality sign with an equals sign and graph this related equation. If the inequality symbol is $<$ or $>$, draw the line dashed. If the inequality symbol is $\leqslant$ or $\geqslant$, draw the line solid.
> 2. The graph consists of a half-plane, either above or below or to the left or right of the line, and, if the line is solid, the line as well. To determine which half-plane to shade, choose a point not on the line as a test point. Substitute to find whether that point is a solution. If so, shade the half-plane containing that point. If not, shade the other half-plane.

EXAMPLE 4

Graph: $6x - 2y < 12$.

Solution We first graph the line $6x - 2y = 12$. The intercepts are $(0, -6)$ and $(2, 0)$. The point $(3, 3)$ is also on the line. This line forms the boundary of the solutions of the inequality. Since the inequality symbol is $<$, points on the line are not solutions of the inequality and we draw a dashed line with open circles at $(0, -6)$

and (2, 0). To determine which half-plane to shade, we test a point *not* on the line. The point (0, 0) is easy to substitute:

$$\frac{6x - 2y < 12}{6 \cdot 0 - 2 \cdot 0 \ ? \ 12}$$
$$\begin{array}{c|c} 0 - 0 & \\ 0 & 12 \quad \text{TRUE} \end{array}$$

Since the inequality $0 < 12$ is *true,* the point (0, 0) is a solution; each point in the half-plane containing (0, 0) is a solution. Thus each point in the other half-plane is *not* a solution. The graph is shown below.

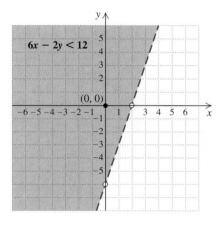

EXAMPLE 5

Graph $x > -3$ on a plane.

Solution   There is a missing variable in this inequality. If we graph the inequality on a line, its graph is as follows:

However, we can also write this inequality as $x + 0y > -3$ and consider graphing it in the plane. We use the same technique that we have used with the other examples. We first graph the related equation $x = -3$ in the plane. We then draw the boundary with a dashed line, and use some test point, say, (2, 5):

$$\frac{x + 0y > -3}{2 + 0 \cdot 5 \ ? \ -3}$$
$$\begin{array}{c|c} 2 & -3 \quad \text{TRUE} \end{array}$$

Since (2, 5) is a solution, all points in the half-plane containing (2, 5) are solutions. We shade that half-plane. Note that the solutions of $x > -3$ are all pairs with first coordinates greater than -3.

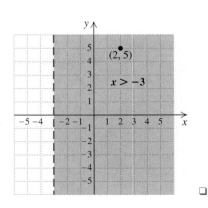

EXAMPLE 6

We can graph an inequality like $y < 1.2x + 3.49$ on a grapher by using the Shade feature, if it has one. On many calculators, this feature will shade regions only *between* two curves. Thus you may need to enter a second equation, like $y = -100$ or $y = 100$, that you know will bound a shaded region from above or below, out of sight of the $[-10, 10, -10, 10]$ window. To shade $y < 1.2x + 3.49$ in the $[-10, 10, -10, 10]$ window, enter $y_1 = 1.2x + 3.49$ as the upper curve and $y_2 = -100$ as the lower curve. You should see a graph similar to this:

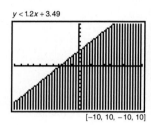

$y < 1.2x + 3.49$

$[-10, 10, -10, 10]$

Use a grapher to draw the graphs of the following inequalities. Note that for most graphers we must first solve for y if it is not already isolated. Note too that most graphers draw only solid (not dashed) lines.

TC1. $y > x + 3.5$ **TC2.** $7y \leqslant 2x + 5$

TC3. $8x - 2y < 11$ **TC4.** $11x + 13y + 4 \geqslant 0$

Graph $y \leqslant 4$ on a plane.

Solution We first graph $y = 4$ using a solid line to indicate that all points on the line are solutions. We then use $(2, -3)$ as a test point and substitute:

$$\frac{0x + y \leqslant 4}{0 \cdot 2 + (-3) \ ? \ 4}$$
$$-3 \mid 4 \quad \text{TRUE}$$

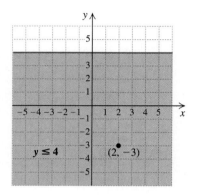

We see that $(2, -3)$ is a solution, so all points in the half-plane containing $(2, -3)$ are solutions. Note that this half-plane consists of all ordered pairs whose y-coordinates are less than or equal to 4.   ❑

Systems of Linear Inequalities

To graph a system of equations, we graph the individual equations and then find the intersection of the individual graphs. We do the same thing for a system of inequalities, that is, we graph each inequality and find the intersection of the individual graphs.

EXAMPLE 7

Graph the system

$$x + y \leqslant 4,$$
$$x - y < 4.$$

Solution To graph the inequality $x + y \leqslant 4$, we graph $x + y = 4$ using a solid line. We then consider $(0, 0)$ as a test point and find that it is a solution, so we shade all points in that region red. The arrows near the ends of the line also indicate the half-plane that contains solutions for each inequality.

Next, we graph $x - y < 4$. We graph $x - y = 4$ using a dashed line and consider $(0, 0)$ as a test point. Again, $(0, 0)$ is a solution, so we shade that side of the line using blue shading. The solution set of the system is the region that is shaded purple and part of the line $x + y = 4$.

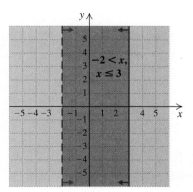

EXAMPLE 8

Graph: $-2 < x \leq 3$.

Solution

This is a system of inequalities:

$$-2 < x,$$
$$x \leq 3.$$

We graph the equation $-2 = x$, and see that the graph of the first inequality is the half-plane to the right of the line $-2 = x$. It is shaded red.

We graph the second inequality, starting with the line $x = 3$, and find that its graph is the line and also the half-plane to the left of it. It is shaded blue.

The solution set of the system is the region that is the intersection of the individual graphs. Since it is shaded both blue and red, it appears to be purple in the following graph.

A system of inequalities may have a graph that consists of a polygon and its interior. In the next section, we will need to find the vertices of such a graph.

EXAMPLE 9

Graph the system of inequalities. Find the coordinates of any vertices formed.

$$6x - 2y \leq 12, \quad \textbf{(1)}$$
$$y - 3 \leq 0, \quad \textbf{(2)}$$
$$x + y \geq 0. \quad \textbf{(3)}$$

Solution   We graph the lines

$$6x - 2y = 12,$$
$$y - 3 = 0,$$

and $x + y = 0$

using solid lines. The regions for each inequality are indicated by the arrows near the ends of the lines. We note where the regions overlap and shade the region of solutions using purple.

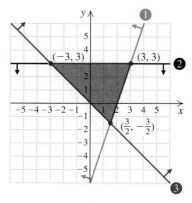

To find the vertices, we solve three different systems of equations. The system of related equations from inequalities (1) and (2) is

$$6x - 2y = 12,$$
$$y - 3 = 0. \qquad \text{Solving, we obtain the vertex } (3, 3).$$

The system of related equations from inequalities (1) and (3) is

$$6x - 2y = 12,$$
$$x + y = 0. \qquad \text{Solving, we obtain the vertex } \left(\tfrac{3}{2}, -\tfrac{3}{2}\right).$$

The system of related equations from inequalities (2) and (3) is

$$y - 3 = 0,$$
$$x + y = 0. \qquad \text{Solving, we obtain the vertex } (-3, 3).$$

❑

EXERCISE SET | 4.4

Determine whether the ordered pair is a solution of the given inequality.

1. $(-4, 2)$; $2x + y < -5$

2. $(3, -6)$; $4x + 2y \geq 0$

3. $(8, 14)$; $2y - 3x > 5$

4. $(7, 20)$; $3x - y > -1$

Graph on a plane.

5. $y > 2x$

6. $y < 3x$

7. $y < x + 1$

8. $y \leq x - 3$

9. $y > x - 2$

10. $y \geq x + 4$

11. $x + y < 4$

12. $x - y \geq 5$

13. $3x + 4y \leq 12$

14. $2x + 3y < 6$

15. $2y - 3x > 6$

16. $2y - x \leq 4$

17. $3x - 2 \leq 5x + y$

18. $2x - 2y \geq 8 + 2y$

19. $x < -4$

20. $y \geq 2$

21. $y > -2$

22. $x \leq 5$

23. $-4 < y < -1$

24. $-2 < y < 3$

25. $-3 \leq x \leq 3$

26. $-4 \leq x \leq 4$

27. $0 \leq x \leq 5$

28. $0 \leq y \leq 3$

Graph the system of inequalities. Find the coordinates of any vertices formed.

29. $y < x,$
$y > -x + 3$

30. $y > x,$
$y < -x + 1$

31. $y \geq x,$
$y \leq -x + 4$

32. $y \geq x,$
$y \leq -x + 2$

33. $y \geq -2,$
$x \geq 1$

34. $y \leq -2,$
$x \geq 2$

35. $x < 3,$
$y > -3x + 2$

36. $x > -2,$
$y < -2x + 3$

37. $y \geq -2,$
$y \geq x + 3$

38. $y \leq 4,$
$y \geq -x + 2$

39. $x + y < 1,$
$x - y < 2$

40. $x + y \leq 3,$
$x - y \leq 4$

41. $y - 2x \geq 1,$
$y - 2x \leq 3$

42. $y + 3x > 0,$
$y + 3x < 2$

43. $2y - x \leq 2,$
$y - 3x \geq -1$

44. $y \leq 2x + 1,$
$y \geq -2x + 1,$
$x \leq 2$

45. $x - y \leq 2,$
$x + 2y \geq 8,$
$y \leq 4$

46. $x + 2y \leq 12,$
$2x + y \leq 12,$
$x \geq 0,$
$y \geq 0$

47. $4y - 3x \geq -12,$
$4y + 3x \geq -36,$
$y \leq 0,$
$x \leq 0$

48. $8x + 5y \leq 40,$
$x + 2y \leq 8,$
$x \geq 0,$
$y \geq 0$

49. $3x + 4y \geq 12,$
$5x + 6y \leq 30,$
$1 \leq x \leq 3$

50. $y - x \geq 1,$
$y - x \leq 3,$
$2 \leq x \leq 5$

Skill Maintenance _____

51. One side of a square is 5 less than a side of an equilateral triangle. If the perimeter of the square is the same as the perimeter of the triangle, what is the length of a side of the square? of a side of the triangle?

Solve.

52. $4y - 3x = 8,$
$2x + 5y = -1$

53. $5(3x - 4) = -2(x + 5)$

54. $4(3x + 4) = 2 - x$

Synthesis _____

55. ◈ Do all systems of linear inequalities have solutions? Why or why not?

56. ◈ Explain how a system of linear inequalities could have a solution set containing exactly one pair.

Graph.

57. $x + y \geq 5,$
$x + y \leq -3$

58. $x + y \leq 8,$
$x + y \leq -2$

59. $x - 2y \leq 0,$
$-2x + y \leq 2,$
$x \leq 2,$
$y \leq 2,$
$x + y \leq 4$

60. $x + y \geq 1,$
$-x + y \geq 2,$
$x \leq 4,$
$y \geq 0,$
$y \leq 4,$
$x \leq 2$

61. *Widths of a basketball floor.* Sizes of basketball floors vary due to building sizes and other constraints such as cost. The length L is to be at most 94 ft and the width W is to be at most 50 ft. Graph a system of inequalities that describes the possible dimensions of a basketball floor.

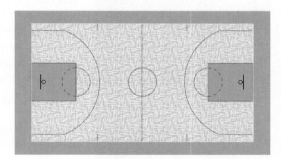

62. *Hockey wins and losses.* A hockey team determines that it needs at least 60 points for the season in order to make the playoffs. A win w is worth 2 points and a tie t is worth 1 point. Graph a system of inequalities that describes the situation.

63. *Elevators.* Many elevators have a capacity of 1 metric ton (1000 kg). Suppose that c children, each weighing 35 kg, and a adults, each 75 kg, are on an elevator. Graph a system of inequalities that indicates when the elevator is overloaded.

⚏ Exercises 64 and 65 can be worked only if your grapher has a Shade feature.

64. Use a grapher to graph the inequality.

a) $3x + 6y > 2$ **b)** $x - 5y \leq 10$
c) $13x - 25y + 10 \leq 0$ **d)** $2x + 5y > 0$

65. Use a grapher to check your answers to Exercises 29–43. If it is available, use the Trace feature to determine the point(s) of intersection.

4.5

Problem Solving Using Linear Programming

Objective Functions and Constraints • Linear Programming

There are many problems in real life in which we want to find a greatest value (a maximum) or a least value (a minimum). For example, if you are in business, you would like to know how to make the *most* profit. Or you might like to know how to make your expenses the *least* possible. Some such problems can be solved using systems of inequalities.

Objective Functions and Constraints

Often a quantity we wish to maximize depends on two or more other quantities. For instance, a gardener's profits P might depend on the number of shrubs s and the number of trees t that are planted. If the gardener makes a \$5 profit from each shrub and a \$9 profit from each tree, the total profit is given by the **objective function**

$$P = 5s + 9t.$$

Thus the gardener might be tempted to simply plant lots of trees since they yield the greater profit. This would be a good idea were it not for the fact that the number of trees and shrubs the gardener plants — and thus the total profit — is subject to the demands, or **constraints**, of the situation. For example, the gardener might be required to plant no more than a total of 10 plants. Thus the objective function would be subject to the *constraint*

$$s + t \leq 10.$$

He or she might also be required to plant at least 3 shrubs. This would subject the objective function to a *second* constraint:

$$s \geq 3.$$

Finally, the gardener might be told to spend no more than \$350 on the plants. If the shrubs cost \$20 each and the trees cost \$50 each, the objective function is subject to a *third* constraint:

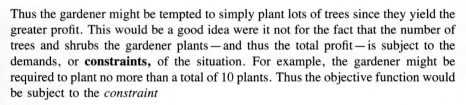

The cost of the shrubs plus the cost of the trees cannot exceed \$350.

$$20s \qquad + \qquad 50t \qquad \leq \qquad 350$$

In short, the gardener wishes to maximize the objective function

$$P = 5s + 9t$$

subject to the constraints

$$s + t \leq 10,$$
$$s \geq 3,$$
$$20s + 50t \leq 350,$$
$$s \geq 0,$$
$$t \geq 0.$$

← Because the number of trees and shrubs cannot be negative

Note that the constraints listed above form a system of linear inequalities that can be graphed.

Linear Programming

The problem facing the gardener is "How many shrubs and trees should be planted, subject to the constraints listed, in order to maximize profit?" To solve such a problem, we use an important result from a branch of mathematics known as **linear programming.**

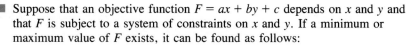

Suppose that an objective function $F = ax + by + c$ depends on x and y and that F is subject to a system of constraints on x and y. If a minimum or maximum value of F exists, it can be found as follows:

1. Graph the system of inequalities and find the vertices.
2. Find the value of the objective function at each vertex. The largest and the smallest of those values are the maximum and the minimum of the function, respectively.

This theorem was proven during World War II, when linear programming was developed to deal with the complicated process of shipping troops and supplies to Europe.

EXAMPLE 1

Solve the gardener's problem discussed above.

Solution   We are asked to maximize $P = 5s + 9t$, subject to

$$s + t \leqslant 10,$$
$$s \geqslant 3,$$
$$20s + 50t \leqslant 350,$$
$$s \geqslant 0,$$
$$t \geqslant 0.$$

We graph the system, using the techniques of Section 4.4. The portion of the graph shaded represents all pairs that satisfy the constraints. It is sometimes called the *feasible region.*

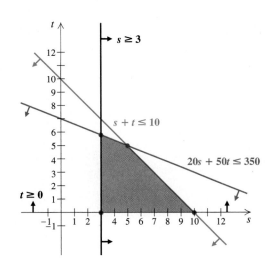

According to the linear programming theorem, P is maximized at one of the vertices of the shaded region. To determine the coordinates of the vertices, we solve the following systems:

$$20s + 50t = 350,$$
$$s = 3;$$

$$s + t = 10,$$
$$20s + 50t = 350;$$

$$s + t = 10,$$
$$t = 0;$$

$$t = 0,$$
$$s = 3.$$

The solutions of the systems are $(3, 5.8)$, $(5, 5)$, $(10, 0)$, and $(3, 0)$, respectively. We now find the value of P at each of these points:

Vertex (s, t)	Profit $P = 5s + 9t$	
$(3, 5.8)$	$5(3) + 9(5.8) = 67.2$	
$(5, 5)$	$5(5) + 9(5) = 70$	← Maximum
$(10, 0)$	$5(10) + 9(0) = 50$	
$(3, 0)$	$5(3) + 9(0) = 15$	← Minimum

The largest value of P occurs at $(5, 5)$. Thus profit will be maximized at $70 if the gardener plants 5 shrubs and 5 trees. Although we were not asked to do so, we have also shown that profit will be minimized at $15 if 3 shrubs and 0 trees are planted.   ❑

EXAMPLE 2

You are taking a test in which multiple-choice questions are worth 10 points each and short-answer questions are worth 15 points each. It takes you 3 min to answer each multiple-choice question and 6 min for each short-answer question. The total time allowed is 60 min, and you are not allowed to answer more than 16 questions. Assuming that all your answers are correct, how many items of each type should you answer in order to get the best score?

Solution

1. FAMILIARIZE.   Tabulating information will help us to see the picture.

Type	Number of Points for Each	Time Required for Each	Number Answered
Multiple-choice	10	3 min	x
Short-answer	15	6 min	y
Total time: 60 min			
Total number of items: 16 or fewer			

Note that we have used x to represent the number of multiple-choice questions and y to represent the number of short-answer questions that are answered.

2. TRANSLATE. In this case, it will help to extend the table.

Type	Number of Points for Each	Time Required for Each	Number Answered	Total Time for Type	Total Points for Type
Multiple-choice	10	3 min	x	$3x$	$10x$
Short-answer	15	6 min	y	$6y$	$15y$
Total			$x + y \leqslant 16$	$3x + 6y \leqslant 60$	$10x + 15y$

↑ Because no more than 16 items can be answered

↑ Because the time cannot be more than 60 min

↑ This is the total score on the test.

Suppose that the total score on the test is T. We write T as the objective function in terms of x and y:

$$T = 10x + 15y.$$

We wish to maximize T subject to these facts (constraints) about x and y:

$$x + y \leqslant 16,$$
$$3x + 6y \leqslant 60,$$
$$\left.\begin{array}{l} x \geqslant 0, \\ y \geqslant 0. \end{array}\right\} \leftarrow \text{Because the number of items answered cannot be negative}$$

3. CARRY OUT. The mathematical manipulation consists of graphing the system and evaluating T at each vertex. The graph is as follows:

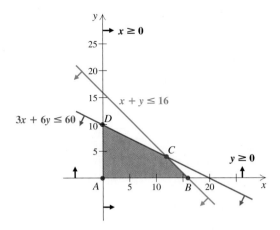

We need to find the coordinates of each vertex. Note that each vertex is the intersection of two lines. Thus we can find each one by solving a system of two linear equations. The coordinates of point A are obviously $(0, 0)$. To find the coordinates of point C, we solve the system

$$3x + 6y = 60, \quad \textbf{(1)}$$
$$x + y = 16, \quad \textbf{(2)}$$

as follows:

$$3x + 6y = 60$$
$$\underline{-3x - 3y = -48} \qquad \text{Multiplying both sides of Equation (2) by } -3$$
$$3y = 12 \qquad \text{Adding}$$
$$y = 4.$$

Then we find that $x = 12$. Thus the coordinates of vertex C are $(12, 4)$.

Continuing to find the coordinates of the vertices and computing the test score for each ordered pair, we obtain the following:

Vertex (x, y)	Score $T = 10x + 15y$
A (0, 0)	0
B (16, 0)	160
C (12, 4)	180
D (0, 10)	150

The greatest score in the table is 180, obtained when 12 multiple-choice and 4 short-answer questions are answered.

4. CHECK. We can go back to the original conditions of the problem and calculate the scores, using the ordered pairs in the table above. We can also check our work, the algebra and the arithmetic. In this case, there is no further checking that we can do without an undue amount of work.

5. STATE. The answer is that in order to maximize your score, you should answer 12 multiple-choice questions and 4 short-answer questions. ❏

EXERCISE SET 4.5

Find the maximum and the minimum values of the objective function and the values of x and y for which they occur.

1. $F = 4x + 28y$,
   subject to
   $5x + 3y \leq 34$,
   $3x + 5y \leq 30$,
   $x \geq 0$,
   $y \geq 0$

2. $G = 14x + 16y$,
   subject to
   $3x + 2y \leq 12$,
   $7x + 5y \leq 29$,
   $x \geq 0$,
   $y \geq 0$

3. $P = 16x - 2y + 40$,
   subject to
   $6x + 8y \leq 48$,
   $0 \leq y \leq 4$,
   $0 \leq x \leq 7$

4. $Q = 24x - 3y + 52$,
   subject to
   $5x + 4y \leq 20$,
   $0 \leq y \leq 4$,
   $0 \leq x \leq 3$

5. $F = 5x + 2y + 3$,
   subject to
   $y \leq 2x + 1$,
   $x \leq 5$,
   $y \geq 1$

6. $G = 2y - 3x$,
   subject to
   $y \leq 2x + 1$,
   $y \geq -2x + 1$,
   $x \leq 2$

Problem Solving

7. The Hockeypuck Biscuit Factory makes two types of biscuits, Biscuit Jumbos and Mitimite Biscuits. The oven can cook at most 200 biscuits per hour. Jumbos each require 2 oz of flour, Mitimites require 1 oz of flour, and there is at most 300 oz of flour available. The income from Jumbos is $0.10 and from Mitimites is $0.08. How many of each

type of biscuit should be made in order to maximize income? What is the maximum income?

8. Roschelle owns a car and a moped. She has at most 12 gal of gasoline to be used between the car and the moped. The car's tank holds at most 10 gal and the moped's 3 gal. The mileage for the car is 20 mpg and for the moped is 100 mpg. How many gallons of gasoline should each vehicle use if Roschelle wants to travel as far as possible? What is the maximum number of miles?

9. You are about to take a test that contains matching questions worth 10 points each and essay questions worth 25 points each. You must do at least 3 matching questions, but time restricts doing more than 12. You must do at least 4 essays, but time restricts doing more than 15. You can do no more than 20 questions in total. How many of each type of question must you do to maximize your score? What is this maximum score?

10. You are about to take a test that contains short-answer questions worth 4 points each and word problems worth 7 points each. You must do at least 5 short-answer questions, but time restricts doing more than 10. You must do at least 3 word problems, but time restricts doing more than 10. You can do no more than 18 questions in total. How many of each type of question must you do to maximize your score? What is this maximum score?

11. Yawaka manufactures motorcycles and bicycles. To stay in business, it must produce at least 10 motorcycles each month, but it does not have the facilities to produce more than 60 motorcycles or more than 120 bicycles. The total production of motorcycles and bicycles cannot exceed 160. The profit on a motorcycle is $134 and on a bicycle, $20. Find the number of each that should be manufactured in order to maximize profit.

12. Bernie's snack bar sells hamburgers and hot dogs during football games. To stay in business, it must sell at least 10 hamburgers but cannot cook more than 40. It must also sell at least 30 hot dogs but cannot cook more than 70. It cannot cook more than 90 sandwiches altogether. The profit is $0.33 on a hamburger and $0.21 on a hot dog. How many of each kind of sandwich should it sell in order to make the maximum profit?

13. Johnson Lumber can convert logs into either lumber or plywood. In a given week, the mill can turn out 400 units of production, of which 100 units of lumber and 150 units of plywood are required by regular customers. The profit on a unit of lumber is $20 and on a unit of plywood is $30. How many

units of each type should the mill produce in order to maximize the profit?

14. Thano's farm consists of 240 acres of cropland. Thano wishes to plant this acreage in corn or oats. Profit per acre in corn production is $40 and in oats, $30. An additional restriction is that the total number of hours of labor during the production period is 320. Each acre of land in corn production uses 2 hr of labor during the production period, while production of oats requires 1 hr per acre. Determine how the land should be divided between corn and oats in order to give maximum profit.

15. Rosa is planning to invest up to $40,000 in corporate or municipal bonds, or both. The least she is allowed to invest in corporate bonds is $6000, and she does not want to invest more than $22,000 in corporate bonds. She also does not want to invest more than $30,000 in municipal bonds. The interest on corporate bonds is 8% and on municipal bonds is $7\frac{1}{2}$%. This is simple interest for one year. How much should she invest in each type of bond to earn the most interest? What is the maximum income?

16. Jamaal is planning to invest up to $22,000 in City Bank or State Bank, or both. He wants to invest at least $2000 but no more than $14,000 in City Bank. State Bank does not insure more than a $15,000 investment, so he will invest no more than that in State Bank. The interest in City Bank is 6% and in State Bank, $6\frac{1}{2}$%. This is simple interest for one year. How much should he invest in each bank to earn the most interest? What is the maximum income?

17. A pipe tobacco company has 3000 lb of English tobacco, 2000 lb of Virginia tobacco, and 500 lb of Latakia tobacco. To make one batch of Smello tobacco, it takes 12 lb of English tobacco and 4 lb of Latakia. To make one batch of Roppo tobacco, it takes 8 lb of English and 8 lb of Virginia tobacco. The profit is $10.56 per batch for Smello and $6.40 for Roppo. How many batches of each kind of tobacco should be made to yield maximum profit? What is the maximum profit? (*Hint:* Organize the information in a table.)

18. It takes a tailoring firm 2 hr of cutting and 4 hr of sewing to make a knit suit. To make a worsted suit, it takes 4 hr of cutting and 2 hr of sewing. At most 20 hr per day are available for cutting and at most 16 hr per day are available for sewing. The profit on a knit suit is $34 and on a worsted suit is $31. How many of each kind of suit should be made to maximize profit?

Synthesis

19. An airline with two types of airplanes, P-1 and P-2, has contracted with a tour group to provide accommodations for a minimum of 2000 first-class, 1500 tourist-class, and 2400 economy-class passengers. Airplane P-1 costs $12,000 per mile to operate and can accommodate 40 first-class, 40 tourist-class, and 120 economy-class passengers, whereas airplane P-2 costs $10,000 per mile to operate and can accommodate 80 first-class, 30 tourist-class, and 40 economy-class passengers. How many of each type of airplane should be used to minimize the operating cost?

P-1: $12,000 per mile

P-2: $10,000 per mile

20. A new airplane P-3 becomes available, having an operating cost of $15,000 per mile and accommodating 40 first-class, 40 tourist-class, and 80 economy-class passengers. If airplane P-1 of Exercise 19 were replaced by airplane P-3, how many P-2's and how many P-3's would be needed in order to minimize the operating cost?

21. Guy's Home Furnishings produces chairs and sofas. The chairs require 20 ft of wood, 1 lb of foam rubber, and 2 sq yd of material. The sofas require 100 ft of wood, 50 lb of foam rubber, and 20 sq yd of material. Guy has in stock 1900 ft of wood, 500 lb of foam rubber, and 240 sq yd of material. The chairs can be sold for $20 each and the sofas for $300 each. How many of each should be produced in order to maximize income?

SUMMARY AND REVIEW | 4

IMPORTANT PROPERTIES AND FORMULAS

The Addition Principle for Inequalities

Adding the same number on both sides of an inequality forms an equivalent inequality.

The Multiplication Principle for Inequalities

Multiplying on both sides of a true inequality by a *positive* number produces another true inequality. If we multiply by a *negative* number on both sides, the inequality symbol must be reversed to produce another true inequality.

Set intersection: $A \cap B = \{x \mid x$ is in A and x is in $B\}$
Set union: $A \cup B = \{x \mid x$ is in A or in B, or both$\}$

For any real numbers a and b with $a < b$, "$a < x$ *and* $x < b$" is equivalent to "$a < x < b$." Intersection corresponds to "and"; union corresponds to "or."

$|a| = a$ if $a \geqslant 0$; $|a| = -a$ if $a < 0$.

The Absolute-Value Principles for Equations and Inequalities

For any positive number p and any algebraic expression X:

a) The solutions of $|X| = p$ are those numbers that satisfy $X = -p$ or $X = p$.
b) The solutions of $|X| < p$ are those numbers that satisfy $-p < X < p$.
c) The solutions of $|X| > p$ are those numbers that satisfy $X < -p$ or $p < X$.

If $|X| = 0$, then $X = 0$. If p is negative, then $|X| = p$ has no solution.

Suppose that an objective function $F = ax + by + c$ depends on x and y and that F is subject to a system of constraints on x and y. If a minimum or maximum value of F exists, it can be found as follows:

1. Graph the system of inequalities and find the vertices.
2. Find the value of the objective function at each vertex. The largest and the smallest of those values are the maximum and the minimum of the function, respectively.

REVIEW EXERCISES

This chapter's review and test include Skill Maintenance exercises from Sections 1.3, 2.3, 3.2, and 3.3.

Find the solution set. Graph.

1. $x \leq -4$ **2.** $x + 5 > 6$

3. $a + 7 \leq -14$ **4.** $y - 5 \geq -12$

5. $4y > -15$ **6.** $-0.3y < 9$

7. $-6x - 5 < 13$ **8.** $4y + 3 < -6y - 9$

9. $-\frac{1}{2}x - \frac{1}{4} > \frac{1}{2} - \frac{1}{4}x$

10. $0.3y - 7 < 2.6y + 15$

11. $-2(x - 5) \geq 6(x + 7) - 12$

Solve.

12. You are taking a biology course in which there will be 5 tests. You have scores of 91, 93, 86, and 88 on the first four. You must score a total of 450 in order to get an A. What scores on the last test will give you an A?

13. You are going to invest $30,000, part at 13% and part at 15%. What is the most that can be invested at 13% in order to make at least $4300 interest per year?

14. Find the intersection:

$$\{1, 2, 5, 6, 9\} \cap \{1, 3, 5, 9\}.$$

15. Find the union:

$$\{1, 2, 5, 6, 9\} \cup \{1, 3, 5, 9\}.$$

Solve.

16. $-4 < x + 3 \leq 5$

17. $-15 < -4x - 5 < 0$

18. $3x < -9$ or $-5x < -5$

19. $2x + 5 < -17$ or $-4x + 10 \leq 34$

20. $x + 5 < -6$ or $x + 5 > 6$

21. $2x + 7 \leq -5$ or $x + 7 \geq 15$

22. $|x| = 6$ **23.** $|x| < 0$

24. $|x| \geq 3.5$ **25.** $|x - 2| = 7$

26. $|2x + 5| < 12$ **27.** $|3x - 4| \geq 15$

28. $|2x + 5| = |x - 9|$ **29.** $|5x + 6| = -8$

30. $\left| \dfrac{x + 4}{8} \right| \leq 1$ **31.** $2|x - 5| - 7 > 3$

Graph. Find the coordinates of any vertices formed.

32. $y \geq -3,$
$\quad x \geq 2$

33. $x + 3y > -1,$
$\quad x + 3y < 4$

34. $x - 3y \leq 3,$
$\quad x + 3y \geq 9,$
$\quad y \leq 6$

35. Find the maximum and the minimum values of

$$F = 3x + y + 4$$

subject to

$$y \leq 2x + 1,$$
$$x \leq 7,$$
$$y \geq 3.$$

36. LaKenya wants to invest $60,000 in mutual funds and municipal bonds. She does not want to invest more than 50%, or less than 20%, of her money in mutual funds. The minimum investment for municipal bonds is $10,000, and they are guaranteed only up to $40,000, so she will not invest more than $40,000 in municipal bonds. The mutual funds should produce a return of 10%, and the municipal bonds a return of 12%. How much should she invest in each in order to maximize her income? What is her maximum income?

Skill Maintenance _____

Solve.

37. $5x - 4y = -10,$
$\quad 4x + 2y = 5$

38. $3(x + 4) = 2(x - 5)$

39. Graph: $y = -2x - 6$.

40. The perimeter of a rectangular field is 786 ft. The length is 9 ft longer than the width. Find the area of the field.

Synthesis _____

41. ◈   Explain in your own words why $|x| = p$ has two solutions when p is positive and no solution when p is negative.

42. ◈   Explain why the graph of the solution of a system of linear inequalities is the intersection, not the union, of the individual graphs.

43. Solve: $|2x + 5| \leq |x + 3|$.

44. Determine whether this is true or false: If $x < 3$, then $x^2 < 9$. If false, give an example showing why.

45. Just-For-Fun manufactures marbles with a 1.1-cm diameter and a ± 0.03-cm manufacturing tolerance, or allowable variation in diameter. Write the tolerance as an inequality with absolute value.

CHAPTER TEST | 4

Solve.

1. $x - 2 < 12$

2. $-0.6y < 30$

3. $-4y - 3 \geq 5$

4. $3a - 5 \leq -2a + 6$

5. $-5y - 1 > -9y + 3$

6. $4(5 - x) < 2x + 5$

7. $-8(2x + 3) + 6(4 - 5x) \geq 2(1 - 7x) - 4(4 + 6x)$

8. You can rent a car for either $40 per day with unlimited mileage or $30 per day with an extra charge of 15¢ a mile. For what numbers of miles traveled would the unlimited mileage plan save you money?

9. Agnaldo is taking an intermediate algebra course in which four tests are to be given. To get an A, a student must average at least 90 on the four tests. Agnaldo got scores of 89, 92, and 86 on the first three tests. What scores on the last test will allow him to get an A?

10. Find the intersection:

$$\{1, 3, 5, 7, 9\} \cap \{3, 5, 11, 13\}.$$

11. Find the union:

$$\{1, 3, 5, 7, 9\} \cup \{3, 5, 11, 13\}.$$

Solve.

12. $-3 < x - 2 < 4$

13. $-11 \leq -5x - 2 < 0$

14. $-3x > 12$ *or* $4x > -10$

15. $x - 7 \geq -5$ *or* $x - 7 \leq -10$

16. $3x - 2 < 7$ *or* $x - 2 > 4$

17. $-\frac{1}{3} \leq \frac{1}{6}x - 1 < \frac{1}{4}$

18. $|x| = 9$

19. $|x| > 3$

20. $|4x - 1| < 4.5$

21. $|-5x - 3| \geq 10$

22. $|x + 10| = |x - 12|$

23. $|2 - 5x| = -10$

24. $\left|\dfrac{6 - x}{7}\right| \leq 15$

Graph. Find the coordinates of any vertices formed.

25. $x + y \geq 3,$
$x - y \geq 5$

26. $2y - x \geq -7,$
$2y + 3x \leq 15,$
$y \leq 0,$
$x \leq 0$

27. Find the maximum and the minimum values of

$$F = 5x + 3y$$

subject to

$x + y \leq 15,$
$1 \leq x \leq 6,$
$0 \leq y \leq 12.$

28. You are about to take a test that contains questions of type A worth 7 points each and type B worth 12 points each. The total number of questions worked must be at least 8. If you know that type-A questions take 10 min and type-B questions take 8 min and that the maximum time for the test is 80 min, how many of each type of question should you answer in order to maximize your score? What is this maximum score?

Skill Maintenance _____

Solve.

29. $4(x - 2) - 3(2x + 7) = 4$

30. $3x + 5y = 8,$
$7x + 9y = -11$

31. Graph: $f(x) = \frac{3}{5}x - 3$.

32. A disc jockey must play 15 commercial spots during one hour of a radio show. Each commercial is either 30 sec or 60 sec long. The total commercial time during that hour is 10 min. How many of each type of commercial were played?

Synthesis _____

Solve.

33. $|3x - 4| \leq -3$

34. $7x < 8 - 3x < 6 + 7x$

CHAPTER 5

Polynomials and Polynomial Functions

AN APPLICATION

Gentamicin is an antibiotic frequently used in veterinary medicine. The concentration, in micrograms per milliliter (mcg/ml), of Gentamicin in a horse's bloodstream can be approximated by a polynomial function.

In Section 5.1, we will find the concentration of Gentamicin in a horse's bloodstream 2 hr after injection.

Karen B. Anderson
VETERINARIAN

"A career in veterinary medicine relies heavily on science and thus on mathematics. Dosages are nearly always based on calculations involving the animal's weight. Being able to make quick calculations using the correct formulas can save an animal's life."

A polynomial is a type of algebraic expression that contains one or more terms. We define polynomials more specifically in this chapter and learn how to manipulate them so that we can use polynomials and polynomial functions in problem solving.

In addition to material from this chapter, the review and test for Chapter 5 include material from Sections 3.4, 3.5, 4.1, and 4.3.

5.1

Introduction to Polynomials and Polynomial Functions

Polynomial Expressions •	**Polynomial Functions** •
Adding Polynomials •	**Opposites and Subtraction**

In this section, we introduce a type of algebraic expression known as a *polynomial*. We then develop some vocabulary, study the addition and subtraction of polynomials, and evaluate *polynomial functions*.

Polynomial Expressions

In Chapter 1, we introduced algebraic expressions like

$$5x^2, \quad \frac{3}{x^2+5}, \quad 7a^3b^4, \quad 3x^{-2}, \quad 6x^2+3x+1, \quad -9, \quad \text{and} \quad 5-2x.$$

Of the expressions listed, $5x^2$, $7a^3b^4$, $3x^{-2}$, and -9 are examples of *terms*. A **term** is simply a number or a product of a number and a variable or variables raised to a power.

Whenever the variables in a term are raised to whole-number powers, we say that the term is a **monomial.** Of the expressions listed, $5x^2$, $7a^3b^4$, and -9 are monomials.

A **polynomial** is a monomial or a sum of monomials. Of the expressions listed, $5x^2$, $7a^3b^4$, $6x^2+3x+1$, -9, and $5-2x$ are polynomials. In fact, with the exception of $7a^3b^4$, these are all polynomials *in one variable*. The expression $7a^3b^4$ is a *polynomial in two variables*. Note that $5-2x$ is the sum of 5 and $-2x$.

A number like 6 in the term $6x^2$ is called the **coefficient** of that term. Thus the coefficient of the monomial $7a^3b^4$ is 7, and the coefficient of $3x^{-2}$ is 3.

The **degree of a term** is the sum of the exponents of the variables, if there are variables. The degree of a constant term is 0, except when the constant term is 0. Mathematicians agree that the polynomial 0 has no degree. This is because we can express 0 as $0 = 0x^5 = 0x^8$, and so on, using any exponent we wish. The **degree of a polynomial** is the same as the degree of its term of highest degree.

The **leading term** of a polynomial is the term of highest degree. Its coefficient is called the **leading coefficient.**

EXAMPLE 1

For each polynomial given, find the degree of each term, the degree of the polynomial, the leading term, and the leading coefficient.

a) $2x^3 + 8x^2 - 17x - 3$
b) $6x^2 + 8x^2y^3 - 17xy - 24xy^2z^4 + 2y + 3$

Solution

	(a)				(b)				
Term	$2x^3$	$8x^2$	$-17x$	-3	$6x^2$	$8x^2y^3$	$-17xy$	$-24xy^2z^4$	$2y$	3
Degree	3	2	1	0	2	5	2	7	1	0
Degree of polynomial	3				7					
Leading term	$2x^3$				$-24xy^2z^4$					
Leading coefficient	2				-24					

A polynomial of degree 0 or 1 is called **linear**. A polynomial in one variable is said to be **quadratic** if it is of degree 2 and **cubic** if it is of degree 3.

The following are some names for certain kinds of polynomials.

Type	Definition	Examples
Monomial	A polynomial of one term	$4,\ -3p,\ 5x^2,\ -7a^2b^3,\ 0,\ xyz$
Binomial	A polynomial of two terms	$2x + 7,\ a - 3b,\ 5x^2 + 7y^3$
Trinomial	A polynomial of three terms	$x^2 - 7x + 12,\ 4a^2 + 2ab + b^2$

We generally arrange polynomials in one variable so that the exponents *decrease* from left to right, which is **descending order.** Sometimes they may be written so that the exponents *increase* from left to right, which is **ascending order.** Generally, if an exercise is written in one kind of order, we write the answer in that same order.

EXAMPLE 2

Arrange in ascending order: $12 + 2x^3 - 7x + x^2$.

Solution

$$12 + 2x^3 - 7x + x^2 = 12 - 7x + x^2 + 2x^3$$

Polynomials in several variables can be arranged with respect to the powers of one of the variables.

EXAMPLE 3

Arrange in descending powers of x: $y^4 + 2 - 5x^2 + 3x^3y + 7xy^2$.

Solution

$$y^4 + 2 - 5x^2 + 3x^3y + 7xy^2 = 3x^3y - 5x^2 + 7xy^2 + y^4 + 2$$

Polynomial Functions

A polynomial function is a function like

$$P(x) = 5x^7 + 3x^5 - 4x^2 - 5,$$

where the expression on the right is a polynomial. To evaluate a polynomial function, we substitute a number for each occurrence of the variable, as we did in Chapter 2.

EXAMPLE 4	For the polynomial function $P(x) = -x^2 + 4x - 1$, find the following: **(a)** $P(2)$; **(b)** $P(10)$; **(c)** $P(-10)$.

Solution

a) $P(2) = -2^2 + 4(2) - 1 = -4 + 8 - 1 = 3$ We square the input before taking its opposite.

b) $P(10) = -10^2 + 4(10) - 1 = -100 + 40 - 1 = -61$

c) $P(-10) = -(-10)^2 + 4(-10) - 1 = -100 - 40 - 1 = -141$ ❑

The next two examples are problem solving in nature, but involve only the evaluation of a polynomial function. For that reason, we do not apply all five problem-solving steps.

EXAMPLE 5	**Games in a sports league.** In a sports league of n teams in which each team plays every other team twice, the total number of games to be played is given by the polynomial function

$$P(n) = n^2 - n.$$

A women's softball league has 10 teams. What is the total number of games to be played?

Solution   We evaluate the function for $n = 10$:

$$P(10) = 10^2 - 10 = 100 - 10 = 90.$$

The league plays 90 games.   ❑

EXAMPLE 6	**Veterinary medicine.** Gentamicin is an antibiotic frequently used by veterinarians. The concentration, in micrograms per milliliter (mcg/ml), of Gentamicin in a horse's bloodstream t hours after injection can be approximated by the polynomial function

$$C(t) = -0.005t^4 + 0.003t^3 + 0.35t^2 + 0.5t.$$

What is the concentration 2 hr after injection?

Solution   We evaluate the function for $t = 2$:

$$C(2) = -0.005(2)^4 + 0.003(2)^3 + 0.35(2)^2 + 0.5(2)$$

$\qquad\quad = -0.005(16) + 0.003(8) + 0.35(4) + 0.5(2)$ We carry out the calculation using the rules for order of operations.

$\qquad\quad = -0.08 + 0.024 + 1.4 + 1$

$\qquad\quad = -0.08 + 2.424$

$\qquad\quad = 2.344.$

The concentration after 2 hr is 2.344 mcg/ml.   ❑

Adding Polynomials

Recall from Section 1.3 that if two terms of a polynomial have the same variable(s) raised to the same power(s), they are **similar,** or **like, terms** and can be "combined" or "collected."

EXAMPLE 7

Collect like terms.

a) $3x^2 - 4y + 2x^2$
b) $9x^3 + 5x - 4x^2 - 2x^3 + 5x^2$
c) $3x^2y + 5xy^2 - 3x^2y - xy^2$

Solution

a) $3x^2 - 4y + 2x^2 = 3x^2 + 2x^2 - 4y$ Rearranging using the commutative law for addition

$\qquad\qquad\qquad\quad = (3 + 2)x^2 - 4y$ Using the distributive law

$\qquad\qquad\qquad\quad = 5x^2 - 4y$

b) $9x^3 + 5x - 4x^2 - 2x^3 + 5x^2 = 7x^3 + x^2 + 5x$ We normally perform the middle steps mentally and write just the answer.

c) $3x^2y + 5xy^2 - 3x^2y - xy^2 = 4xy^2$

The sum of two polynomials can be found by writing a plus sign between them and then collecting like terms. Ordinarily, this can be done mentally.

EXAMPLE 8

Add: $(-3x^3 + 2x - 4) + (4x^3 + 3x^2 + 2)$.

Solution

$\qquad (-3x^3 + 2x - 4) + (4x^3 + 3x^2 + 2) = x^3 + 3x^2 + 2x - 2$

Using columns is often helpful. To do so, we write the polynomials one under the other, listing like terms under one another and leaving spaces for missing terms. Let us do the addition in Example 8 using columns.

$$
\begin{array}{rrrr}
-3x^3 & & +\ 2x & -\ 4 \\
4x^3 & +\ 3x^2 & & +\ 2 \\
\hline
x^3 & +\ 3x^2 & +\ 2x & -\ 2
\end{array}
$$

EXAMPLE 9

Add: $4ax^2 + 4bx - 5$ and $-6ax^2 + 5bx + 8$.

Solution

$$
\begin{array}{r}
4ax^2 + 4bx - 5 \\
-6ax^2 + 5bx + 8 \\
\hline
-2ax^2 + 9bx + 3
\end{array}
$$

Although using columns is helpful for complicated examples, you should attempt, for the sake of working faster, to write only the answer whenever you can.

EXAMPLE 10

Add: $13x^3y + 3x^2y - 5y$ and $x^3y + 4x^2y - 3xy$.

Solution

$\qquad (13x^3y + 3x^2y - 5y) + (x^3y + 4x^2y - 3xy) = 14x^3y + 7x^2y - 3xy - 5y$

Opposites and Subtraction

If the sum of two polynomials is 0, the polynomials are *opposites,* or *additive inverses,* of each other. For example,

$$(3x^2 - 5x + 2) + (-3x^2 + 5x - 2) = 0,$$

so the opposite of $(3x^2 - 5x + 2)$ is $(-3x^2 + 5x - 2)$. We can say the same thing using algebraic symbolism, as follows:

The opposite of $(3x^2 - 5x + 2)$ is $(-3x^2 + 5x - 2)$.

$$-\ (3x^2 - 5x + 2) = -3x^2 + 5x - 2$$

To form the opposite of a polynomial, we can think of distributing the ''$-$'' sign, or multiplying each term of the polynomial by -1, and removing the parentheses. The effect is to change the sign of each term in the polynomial.

> The *opposite* of a polynomial P can be symbolized by $-P$ or by replacing each term with its opposite. The two expressions for the opposite are equivalent.

EXAMPLE 11

Write two equivalent expressions for the opposite of

$$7xy^2 - 6xy - 4y + 3.$$

Solution

a) $-(7xy^2 - 6xy - 4y + 3)$ Writing the opposite of P as $-P$

b) $-7xy^2 + 6xy + 4y - 3$ Writing the opposite of each term, or multiplying each term by -1

To subtract one polynomial from another, we add the opposite of the polynomial being subtracted.

EXAMPLE 12

Subtract: $(-9x^5 + 2x^2 + 4) - (2x^5 + 4x^3 - 3x^2)$.

Solution

$$(-9x^5 + 2x^2 + 4) - (2x^5 + 4x^3 - 3x^2)$$
$$= (-9x^5 + 2x^2 + 4) + (-2x^5 - 4x^3 + 3x^2) \quad \text{Adding the opposite of the polynomial being subtracted}$$

$$= -11x^5 - 4x^3 + 5x^2 + 4$$

After some practice, you will find that you can skip some steps, by mentally taking the opposite of each term and then collecting like terms. Eventually, all you will write is the answer.

We can also use columns for subtraction. We mentally change the signs of the polynomial being subtracted.

EXAMPLE 13

Subtract:

$$(4x^2y - 6x^3y^2 + x^2y^2) - (4x^2y + x^3y^2 + 3x^2y^3 - 8x^2y^2).$$

Solution

Write: (Subtract)

$$\begin{array}{r} 4x^2y - 6x^3y^2 \qquad\quad + \ x^2y^2 \\ -(4x^2y + \ x^3y^2 + 3x^2y^3 - 8x^2y^2) \\ \hline \end{array}$$

Think: (Add)

$$\begin{array}{r} 4x^2y - 6x^3y^2 \qquad\quad + \ x^2y^2 \\ -4x^2y - \ x^3y^2 - 3x^2y^3 + 8x^2y^2 \\ \hline - 7x^3y^2 - 3x^2y^3 + 9x^2y^2 \end{array}$$

Mentally, take the opposite of each term and add.

EXERCISE SET | 5.1

Determine the degree of each term and the degree of the polynomial.

1. $-11x^4 - x^3 + x^2 + 3x - 9$

2. $t^3 - 3t^2 + t + 1$

3. $y^3 + 2y^7 + x^2y^4 - 8$

4. $u^2 + 3v^5 - u^3v^4 - 7$

5. $a^5 + 4a^2b^4 + 6ab + 4a - 3$

6. $8p^6 + 2p^4t^4 - 7p^3t + 5p^2 - 14$

Arrange in descending order. Then find the leading term and the leading coefficient.

7. $23 - 4y^3 + 7y - 6y^2$

8. $5 - 8y + 6y^2 + 11y^3 - 18y^4$

9. $5x^2 + 3x^7 - x + 12$

10. $9 - 3x - 10x^4 + 7x^2$

11. $a + 5a^3 - a^7 - 19a^2 + 8a^5$

12. $a^3 - 7 + 11a^4 + a^9 - 5a^2$

Arrange in ascending powers of x.

13. $4x + 12 + 3x^4 - 5x^2$

14. $-5x^2 + 10x + 5$

15. $-9x^3y + 3xy^3 + x^2y^2 + 2x^4$

16. $5x^2y^2 - 9xy + 8x^3y^2 - 5x^4$

17. $4ax - 7ab + 4x^6 - 7ax^2$

18. $5xy^8 - 3ax^5 + 4ax^3 - 12a + 5x^5$

Find the specified function values.

19. Find $P(4)$ and $P(0)$: $P(x) = 4x^2 - 3x + 2$.

20. Find $Q(3)$ and $Q(-1)$: $Q(x) = -5x^3 + 7x^2 - 12$.

21. Find $P(-2)$ and $P\left(\frac{1}{3}\right)$: $P(y) = 8y^3 - 12y - 5$.

22. Find $Q(-3)$ and $Q(0)$:

$$Q(y) = 9y^3 + 8y^2 - 4y - 9.$$

Evaluate the polynomial for $x = 4$.

23. $-5x + 2$

24. $-3x + 1$

25. $2x^2 - 5x + 7$

26. $3x^2 + x + 7$

27. $x^3 - 5x^2 + x$

28. $7 - x + 3x^2$

Evaluate the polynomial function for $x = -1$.

29. $f(x) = x^2 - 2x + 1$

30. $g(x) = 5x - 6 + x^2$

31. $g(x) = -3x^3 + 7x^2 - 3x - 2$

32. $f(x) = -2x^3 - 5x^2 + 4x + 3$

Daily accidents. The daily number of accidents (the average number of accidents per day) involving drivers of age a is approximated by the polynomial function

$$P(a) = 0.4a^2 - 40a + 1039.$$

33. Find the number of daily accidents involving an 18-year-old driver.

34. Find the number of daily accidents involving a 20-year-old driver.

Falling distance. The distance s, in feet, traveled by a body falling freely from rest in t seconds is approximated by the function.

$$s(t) = 16t^2.$$

35. A stone is dropped from a cliff and takes 8 sec to hit the ground. How high is the cliff?

$S = 16t^2$

36. A brick is dropped from the top of a building and takes 3 sec to hit the ground. How high is the building?

Total revenue.   An electronics firm is marketing a new kind of stereo. The firm determines that when it sells x stereos, its total revenue is

$$R(x) = 280x - 0.4x^2 \text{ dollars.}$$

37. What is the total revenue from the sale of 75 stereos?

38. What is the total revenue from the sale of 100 stereos?

Total cost.   The electronics firm determines that the total cost, in dollars, of producing x stereos is given by

$$C(x) = 5000 + 0.6x^2.$$

39. What is the total cost of producing 75 stereos?

40. What is the total cost of producing 100 stereos?

Surface area of a right circular cylinder.   The surface area of a right circular cylinder is given by the polynomial

$$2\pi rh + 2\pi r^2,$$

where h = the height and r = the radius of the base.

41. ▦ A 12-oz beverage can has height 4.7 in. and radius 1.2 in. Find the surface area of the can. (Use 3.14 as an approximation for π. Give your answer to the nearest hundredth.)

42. ▦ A 16-oz beverage can has height 6.3 in. and radius 1.2 in. Find the surface area of the can. (Use 3.14 as an approximation for π.)

Collect like terms.

43. $6x^2 - 7x^2 + 3x^2$

44. $-2y^2 - 7y^2 + 5y^2$

45. $5a + 7 - 4 + 2a - 6a + 3$

46. $9x + 12 - 8 - 7x + 5x + 10$

47. $3a^2b + 4b^2 - 9a^2b - 6b^2$

48. $5x^2y^2 + 4x^3 - 8x^2y^2 - 12x^3$

49. $8x^2 - 3xy + 12y^2 + x^2 - y^2 + 5xy + 4y^2$

50. $a^2 - 2ab + b^2 + 9a^2 + 5ab - 4b^2 + a^2$

Add.

51. $(3x^2 + 5y^2 + 6) + (2x^2 - 3y^2 - 1)$

52. $(9y^2 + 8y - 4) + (12y^2 - 5y + 8)$

53. $(2a + 3b - c) + (4a - 2b + 2c)$

54. $(5x - 4y + 2z) + (9x + 12y - 8z)$

55. $(a^2 - 3b^2 + 4c^2) + (-5a^2 + 2b^2 - c^2)$

56. $(x^2 - 5y^2 - 9z^2) + (-6x^2 + 9y^2 - 2z^2)$

57. $(x^2 + 2x - 3xy - 7) + (-3x^2 - x + 2xy + 6)$

58. $(3a^2 - 2b + ab + 6) + (-a^2 + 5b - 5ab - 2)$

59. $(7x^2y - 3xy^2 + 4xy) + (-2x^2y - xy^2 + xy)$

60. $(7ab - 3ac + 5bc) + (13ab - 15ac - 8bc)$

61. $(2r^2 + 12r - 11) + (6r^2 - 2r + 4) + (r^2 - r - 2)$

62. $(5x^2 + 19x - 23) + (-7x^2 - 11x + 12) + (-x^2 - 9x + 8)$

63. $\left(\frac{2}{3}xy + \frac{5}{6}xy^2 + 5.1x^2y\right) + \left(-\frac{4}{5}xy + \frac{3}{4}xy^2 - 3.4x^2y\right)$

64. $\left(\frac{1}{8}xy - \frac{3}{5}x^3y^2 + 4.3y^3\right) + \left(-\frac{1}{3}xy - \frac{3}{4}x^3y^2 - 2.9y^3\right)$

Write two equivalent expressions for the opposite, or additive inverse, of the polynomial.

65. $5x^3 - 7x^2 + 3x - 6$

66. $-8y^4 - 18y^3 + 4y - 9$

67. $-12y^5 + 4ay^4 - 7by^2$

68. $7ax^3y^2 - 8by^4 - 7abx - 12ay$

Subtract.

69. $(8x - 4) - (-5x + 2)$

70. $(9y + 3) - (-4y - 2)$

71. $(-3x^2 + 2x + 9) - (x^2 + 5x - 4)$

72. $(-9y^2 + 4y + 8) - (4y^2 + 2y - 3)$

73. $(5a - 2b + c) - (3a + 2b - 2c)$

74. $(8x - 4y + z) - (4x + 6y - 3z)$

75. $(3x^2 - 2x - x^3) - (5x^2 - 8x - x^3)$

76. $(8y^2 - 3y - 4y^3) - (3y^2 - 9y - 7y^3)$

77. $(5a^2 + 4ab - 3b^2) - (9a^2 - 4ab + 2b^2)$

78. $(9y^2 - 14yz - 8z^2) - (12y^2 - 8yz + 4z^2)$

79. $(6ab - 4a^2b + 6ab^2) - (3ab^2 - 10ab - 12a^2b)$

80. $(10xy - 4x^2y^2 - 3y^3) - (-9x^2y^2 + 4y^3 - 7xy)$

81. $(0.09y^4 - 0.052y^3 + 0.93) - (0.03y^4 - 0.084y^3 + 0.94y^2)$

82. $(1.23x^4 - 3.122x^3 + 1.11x) - (0.79x^4 - 8.734x^3 + 0.04x^2 + 6.71x)$

83. $\left(\frac{5}{8}x^4 - \frac{1}{4}x^2 - \frac{1}{2}\right) - \left(-\frac{3}{8}x^4 + \frac{3}{4}x^2 + \frac{1}{2}\right)$

84. $\left(\frac{5}{6}y^4 - \frac{1}{2}y^2 - 7.8y + \frac{1}{3}\right) - \left(-\frac{3}{8}y^4 + \frac{3}{4}y^2 + 3.4y - \frac{1}{5}\right)$

Total profit. Total profit is defined as Total revenue minus Total cost. In Exercises 85 and 86, use the functions for revenue and cost to find the total profit, $P(x)$, from the sale of x stereos.

85. $R(x) = 280x - 0.4x^2$, $C(x) = 5000 + 0.6x^2$

86. $R(x) = 280x - 0.7x^2$, $C(x) = 8000 + 0.5x^2$

87. Use the result of Exercise 85 to find the profit from the sale of 70 stereos.

88. Use the result of Exercise 86 to find the profit from the sale of 100 stereos.

Skill Maintenance

89. Multiply: $3(y - 2)$.

90. Simplify: $3(x - 4) - 5(x + 16)$.

Synthesis

91. ◈ Write a problem in which revenue and cost functions are given and a profit function, $P(x)$, is required. Devise the problem so that $P(0) < 0$ and $P(100) > 0$.

92. ◈ Write a problem in which revenue and cost functions are given and a profit function, $P(x)$, is required. Devise the problem so that $P(10) < 0$ and $P(50) > 0$.

For the polynomial functions $P(x)$ and $Q(x)$ given below, find each of the following.

$$P(x) = 13x^5 - 22x^4 - 36x^3 + 40x^2 - 16x + 75,$$
$$Q(x) = 42x^5 - 37x^4 + 50x^3 - 28x^2 + 34x + 100$$

93. $2[P(x)] + Q(x)$

94. $3[P(x)] - Q(x)$

95. $2[Q(x)] - 3[P(x)]$

96. $4[P(x)] + 3[Q(x)]$

97. Find a polynomial function that gives the surface area of a box like this one, with an open top and dimensions as shown.

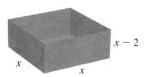

Perform the indicated operation. Assume that the exponents are natural numbers.

98. $(2x^{2a} + 4x^a + 3) + (6x^{2a} + 3x^a + 4)$

99. $(47x^{4a} + 3x^{3a} + 22x^{2a} + x^a + 1) + (37x^{3a} + 8x^{2a} + 3)$

100. $(3x^{6a} - 5x^{5a} + 4x^{3a} + 8) - (2x^{6a} + 4x^{4a} + 3x^{3a} + 2x^{2a})$

101. $(2x^{5b} + 4x^{4b} + 3x^{3b} + 8) - (x^{5b} + 2x^{3b} + 6x^{2b} + 9x^b + 8)$

5.2

Multiplication of Polynomials

Multiplying Monomials   •   Multiplying Monomials and Binomials   •   Multiplying Any Two Polynomials   •   The Product of Two Binomials: FOIL   •   Squares of Binomials   •   Products of Sums and Differences   •   Manipulating Function Notation

Just like numbers, polynomials can be multiplied. The product of two polynomials $P(x)$ and $Q(x)$ is a polynomial $R(x)$ that gives the same value as $P(x) \cdot Q(x)$ for any replacement of x.

Multiplying Monomials

To multiply monomials, we first multiply their coefficients. Then we multiply the variables using the rules for exponents and the commutative and associative laws that we studied in Chapter 1.

EXAMPLE 1

Multiply and simplify: **(a)** $(-8x^4y^7)(5x^3y^2)$; **(b)** $(3x^2yz^5)(-6x^5y^{10}z^2)$.

Solution

a) $(-8x^4y^7)(5x^3y^2) = -8 \cdot 5 \cdot x^4 \cdot x^3 \cdot y^7 \cdot y^2$ Using the associative and commutative laws

$= -40x^{4+3}y^{7+2}$ Multiplying coefficients; adding exponents

$= -40x^7y^9$

b) $(3x^2yz^5)(-6x^5y^{10}z^2) = 3 \cdot (-6) \cdot x^2 \cdot x^5 \cdot y \cdot y^{10} \cdot z^5 \cdot z^2$

$= -18x^7y^{11}z^7$ Multiplying coefficients; adding exponents

❑

You should try to work mentally, writing only the answer.

Multiplying Monomials and Binomials

The distributive law is the basis for multiplying polynomials other than monomials. We first multiply a monomial and a binomial.

EXAMPLE 2

Multiply: **(a)** $2x(3x - 5)$; **(b)** $3a^2b(a^2 - b^2)$.

Solution

a) $2x \cdot (3x - 5) = 2x \cdot 3x - 2x \cdot 5$ Using the distributive law

$= 6x^2 - 10x$ Multiplying monomials

b) $3a^2b \cdot (a^2 - b^2) = 3a^2b \cdot a^2 - 3a^2b \cdot b^2$ Using the distributive law

$$= 3a^4b - 3a^2b^3 \qquad \qquad \square$$

To multiply two binomials, we use the distributive law twice. First we consider one of the binomials as a single expression and multiply it by each term of the other binomial.

EXAMPLE 3

Multiply: $(y^3 - 5)(2y^3 + 4)$.

Solution

$(y^3 - 5)(2y^3 + 4) = (y^3 - 5)2y^3 + (y^3 - 5)4$ Using the distributive law to "distribute" the $y^3 - 5$

$= 2y^3(y^3 - 5) + 4(y^3 - 5)$ Using the commutative law for multiplication. This step is optional.

$= 2y^3 \cdot y^3 - 2y^3 \cdot 5 + 4 \cdot y^3 - 4 \cdot 5$ Using the distributive law

$= 2y^6 - 10y^3 + 4y^3 - 20$ Multiplying the monomials

$= 2y^6 - 6y^3 - 20$ Collecting like terms

$\qquad \qquad \square$

Multiplying Any Two Polynomials

Repeated use of the distributive law enables us to multiply any two polynomials.

EXAMPLE 4

Multiply: $(p + 2)(p^4 - 2p^3 + 3)$.

Solution   By the distributive law, we have

$(p + 2)(p^4 - 2p^3 + 3) = (p + 2)(p^4) - (p + 2)(2p^3) + (p + 2)(3)$

$= p^4(p + 2) - 2p^3(p + 2) + 3(p + 2)$

$= p^4 \cdot p + p^4 \cdot 2 - 2p^3 \cdot p - 2p^3 \cdot 2 + 3 \cdot p + 3 \cdot 2$

$= p^5 + 2p^4 - 2p^4 - 4p^3 + 3p + 6$

$= p^5 - 4p^3 + 3p + 6.$ Collecting like terms $\square$

To multiply any polynomials P and Q, multiply each term of P by each term of Q and collect like terms.

We can use columns for long multiplications. We multiply each term at the top by every term at the bottom, keeping like terms in columns and leaving spaces for missing terms. Then we add.

EXAMPLE 5

Multiply: $(5x^3 + x - 4)(-2x^2 + 3x + 6)$.

Solution

$$
\begin{array}{r}
5x^3 + x - 4 \\
-2x^2 + 3x + 6 \\
\hline
30x^3 + 6x - 24 \\
15x^4 + 3x^2 - 12x \\
-10x^5 - 2x^3 + 8x^2 \\
\hline
-10x^5 + 15x^4 + 28x^3 + 11x^2 - 6x - 24
\end{array}
$$

Multiplying by 6
Multiplying by $3x$
Multiplying by $-2x^2$

Adding ☐

EXAMPLE 6

Multiply $4x^4y - 7x^2y + 3y$ by $2y - 3x^2y$.

Solution

$$
\begin{array}{r}
4x^4y - 7x^2y + 3y \\
-3x^2y + 2y \\
\hline
8x^4y^2 - 14x^2y^2 + 6y^2 \\
-12x^6y^2 + 21x^4y^2 - 9x^2y^2 \\
\hline
-12x^6y^2 + 29x^4y^2 - 23x^2y^2 + 6y^2
\end{array}
$$

Writing descending powers of x

Multiplying by $2y$
Multiplying by $-3x^2y$

Adding ☐

The Product of Two Binomials: FOIL

We now consider what are called *special products*. These lead to faster ways to multiply in certain situations.

Let us find a faster special-product rule for the product of two binomials. Consider $(x + 7)(x + 4)$. We multiply each term of $(x + 7)$ by each term of $(x + 4)$:

$$(x + 7)(x + 4) = x \cdot x + 4 \cdot x + 7 \cdot x + 7 \cdot 4.$$

This multiplication illustrates a pattern that occurs any time two binomials are multiplied:

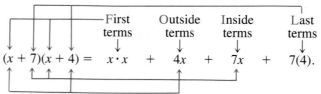

We use the mnemonic device FOIL to remember this method for multiplying.

The FOIL Method

To multiply two binomials $A + B$ and $C + D$, multiply the First terms AC, the Outside terms AD, the Inside terms BC, and then the Last terms BD. Then collect like terms, if possible.

$$(A + B)(C + D) = AC + AD + BC + BD$$

1. Multiply First terms: AC.
2. Multiply Outside terms: AD.
3. Multiply Inside terms: BC.
4. Multiply Last terms: BD.

↓

FOIL

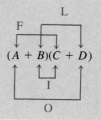

EXAMPLE 7

Multiply: **(a)** $(x + 5)(x - 8)$; **(b)** $(2x - 3)(y + 2)$; **(c)** $(2x + 3y)(x - 4y)$; **(d)** $(3xy + 2x)(x^2 + 2xy^2)$.

Solution

$$\quad\quad\quad\quad\quad\quad \text{F} \quad \text{O} \quad \text{I} \quad \text{L}$$

a) $(x + 5)(x - 8) = x^2 - 8x + 5x - 40$
$$= x^2 - 3x - 40 \quad \text{Collecting like terms}$$

We write the result in descending order because the original binomials are in descending order.

$$\quad\quad\quad\quad\quad\quad \text{F} \quad \text{O} \quad \text{I} \quad \text{L}$$

b) $(2x - 3)(y + 2) = 2xy + 4x - 3y - 6$

c) $(2x + 3y)(x - 4y) = 2x^2 - 8xy + 3xy - 12y^2 = 2x^2 - 5xy - 12y^2$

d) $(3xy + 2x)(x^2 + 2xy^2) = 3x^3y + 6x^2y^3 + 2x^3 + 4x^2y^2$ There are no like terms to combine. ❑

Squares of Binomials

We can use FOIL to develop a fast method for squaring a binomial. Note the following:

$$(A + B)^2 = (A + B)(A + B) \quad\quad\quad (A - B)^2 = (A - B)(A - B)$$
$$= A^2 + AB + AB + B^2 \quad\quad\quad\quad = A^2 - AB - AB + B^2$$
$$= A^2 + 2AB + B^2 \quad\quad\quad\quad\quad = A^2 - 2AB + B^2$$

$$(A + B)^2 = A^2 + 2AB + B^2;$$
$$(A - B)^2 = A^2 - 2AB + B^2$$

The square of a binomial is the square of the first term, plus twice the product of the two terms, plus the square of the last term.

EXAMPLE 8

Multiply: **(a)** $(y - 5)^2$; **(b)** $(2x + 3y)^2$; **(c)** $\left(\frac{1}{2}x - 3y^4\right)^2$.

Solution

$$(A - B)^2 = A^2 - 2 \cdot A \cdot B + B^2$$

a) $(y - 5)^2 = y^2 - 2 \cdot y \cdot 5 + 5^2$

It can be helpful to memorize the words of the rules and say them while you are calculating.

$$= y^2 - 10y + 25$$

b) $(2x + 3y)^2 = (2x)^2 + 2 \cdot 2x \cdot 3y + (3y)^2$
$$= 4x^2 + 12xy + 9y^2 \quad \text{Raising a product to a power}$$

c) $\left(\frac{1}{2}x - 3y^4\right)^2 = \left(\frac{1}{2}x\right)^2 - 2 \cdot \frac{1}{2}x \cdot 3y^4 + (3y^4)^2$
$$= \frac{1}{4}x^2 - 3xy^4 + 9y^8 \quad \text{Raising a product to a power; multiplying powers} \quad ❑$$

Products of Sums and Differences

Another pattern emerges when we are multiplying a sum and a difference. Note the following:

$$\begin{array}{cccc} F & O & I & L \\ \downarrow & \downarrow & \downarrow & \downarrow \end{array}$$
$$(A + B)(A - B) = A^2 - AB + AB - B^2$$
$$= A^2 - B^2.$$

$$(A + B)(A - B) = A^2 - B^2$$

The product of the sum and difference of the same two terms is the square of the first term minus the square of the second term.

EXAMPLE 9

Multiply.

a) $(y + 5)(y - 5)$

b) $(2xy^2 + 3x)(2xy^2 - 3x)$

c) $(0.2t - 1.4m)(0.2t + 1.4m)$

d) $\left(\frac{2}{3}n - m^3\right)\left(\frac{2}{3}n + m^3\right)$

Solution

$$\begin{array}{cccccc} (A & + & B)(A & - & B) & = A^2 - B^2 \\ \downarrow & \downarrow & \downarrow & \downarrow & \downarrow & \downarrow \end{array}$$

a) $(y + 5)(y - 5) = y^2 - 5^2$ Replacing A with y and B with 5

$\qquad\qquad\qquad = y^2 - 25$ Try to do problems like this mentally.

b) $(2xy^2 + 3x)(2xy^2 - 3x) = (2xy^2)^2 - (3x)^2$

$\qquad\qquad\qquad\qquad = 4x^2y^4 - 9x^2$ Raising a product to a power

c) $(0.2t - 1.4m)(0.2t + 1.4m) = (0.2t)^2 - (1.4m)^2$ Say the rule as you work.

$\qquad\qquad\qquad\qquad = 0.04t^2 - 1.96m^2$

d) $\left(\frac{2}{3}n - m^3\right)\left(\frac{2}{3}n + m^3\right) = \left(\frac{2}{3}n\right)^2 - (m^3)^2$

$\qquad\qquad\qquad\qquad = \frac{4}{9}n^2 - m^6$ ◻

EXAMPLE 10

Multiply: **(a)** $(5y + 4 + 3x)(5y + 4 - 3x)$; **(b)** $(3xy^2 + 4y)(-3xy^2 + 4y)$.

Solution

a) $(5y + 4 + 3x)(5y + 4 - 3x) = (5y + 4)^2 - (3x)^2$

$\qquad\qquad\qquad\qquad = 25y^2 + 40y + 16 - 9x^2$

Here we treated $5y + 4$ as the first expression A and $3x$ as B. This product could have been done by columns, but not as quickly.

b) $(3xy^2 + 4y)(-3xy^2 + 4y) = (4y + 3xy^2)(4y - 3xy^2)$ Rewriting

$\qquad\qquad\qquad\qquad = (4y)^2 - (3xy^2)^2$

$\qquad\qquad\qquad\qquad = 16y^2 - 9x^2y^4$ ◻

EXAMPLE 11

Multiply: $(a - 5b)(a + 5b)(a^2 - 25b^2)$.

Solution   We first note that $a - 5b$ and $a + 5b$ can be multiplied using the rule $(A - B)(A + B) = A^2 - B^2$.

$$(a - 5b)(a + 5b)(a^2 - 25b^2) = (a^2 - 25b^2)(a^2 - 25b^2)$$
$$= (a^2 - 25b^2)^2$$
$$= (a^2)^2 - 2(a^2)(25b^2) + (25b^2)^2 \qquad \text{Squaring a binomial}$$
$$= a^4 - 50a^2b^2 + 625b^4 \qquad\qquad \square$$

CAUTION! Keep in mind the following:

$$(A + B)^2 \neq A^2 + B^2 \quad \text{and} \quad (A - B)^2 \neq A^2 - B^2.$$

Manipulating Function Notation

Let's stop for a moment and look back at what we have done in this section. We have shown, for example, that

$$x^2 - 4 = (x - 2)(x + 2).$$

That is, $x^2 - 4$ and $(x - 2)(x + 2)$ are equivalent expressions. This means that they name the same number for any replacement of x. For example, if we replace x by 3, we get

$$x^2 - 4 = 3^2 - 4 = 9 - 4 = 5$$

and

$$(x - 2)(x + 2) = (3 - 2)(3 + 2) = 1 \cdot 5 = 5.$$

From the standpoint of functions, consider

$$f(x) = x^2 - 4 \quad \text{and} \quad g(x) = (x - 2)(x + 2).$$

For any given input x, the outputs $f(x)$ and $g(x)$ are identical. Thus the graphs of these functions are identical and we say that f and g represent the same function. Functions like these are graphed in detail in Chapter 8.

x	$f(x)$	$g(x)$
3	5	5
2	0	0
1	-3	-3
0	-4	-4
-1	-3	-3
-2	0	0
-3	5	5

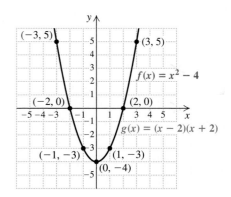

Our work with multiplying can be used when evaluating functions.

EXAMPLE 12

Given $f(x) = x^2 - 4x + 5$, find and simplify: **(a)** $f(a + 3)$; **(b)** $f(a + h) - f(a)$.

Solution

a) To find $f(a + 3)$, we replace each occurrence of x by $a + 3$. Then we simplify:

$$f(a + 3) = (a + 3)^2 - 4(a + 3) + 5$$
$$= a^2 + 6a + 9 - 4a - 12 + 5$$
$$= a^2 + 2a + 2.$$

b) $f(a + h) - f(a) = [(a + h)^2 - 4(a + h) + 5] - [a^2 - 4a + 5]$
$$= a^2 + 2ah + h^2 - 4a - 4h + 5 - a^2 + 4a - 5$$
$$= 2ah + h^2 - 4h$$

❑

EXERCISE SET | 5.2

Multiply.

1. $2y^2 \cdot 5y$

2. $-3x^2 \cdot 2xy$

3. $5x(-4x^2y)$

4. $-3ab^2(2a^2b^2)$

5. $(2x^3y^2)(-5x^2y^4)$

6. $(7a^2bc^4)(-8ab^3c^2)$

7. $2x(3 - x)$

8. $4a(a^2 - 5a)$

9. $3ab(a + b)$

10. $2xy(2x - 3y)$

11. $5cd(3c^2d - 5cd^2)$

12. $a^2(2a^2 - 5a^3)$

13. $(2x + 3)(3x - 4)$

14. $(2a - 3b)(4a - b)$

15. $(s + 3t)(s - 3t)$

16. $(y + 4)(y - 4)$

17. $(x - y)(x - y)$

18. $(a + 2b)(a + 2b)$

19. $(y + 8x)(2y - 7x)$

20. $(x + y)(x - 2y)$

21. $(a^2 - 2b^2)(a^2 - 3b^2)$

22. $(2m^2 - n^2)(3m^2 - 5n^2)$

23. $(x - 4)(x^2 + 4x + 16)$

24. $(y + 3)(y^2 - 3y + 9)$

25. $(x + y)(x^2 - xy + y^2)$

26. $(a - b)(a^2 + ab + b^2)$

27. $(a^2 + a - 1)(a^2 + 4a - 5)$

28. $(x^2 - 2x + 1)(x^2 + x + 2)$

29. $(4a^2b - 2ab + 3b^2)(ab - 2b + a)$

30. $(2x^2 + y^2 - 2xy)(x^2 - 2y^2 - xy)$

31. $\left(x - \frac{1}{2}\right)\left(x - \frac{1}{4}\right)$

32. $\left(b - \frac{1}{3}\right)\left(b - \frac{1}{3}\right)$

33. $(1.3x - 4y)(2.5x + 7y)$

34. $(40a - 0.24b)(0.3a + 10b)$

35. Let $P(x) = 3x^2 - 5$ and $Q(x) = 4x^2 - 7x + 2$. Find $P(x) \cdot Q(x)$.

36. Let $P(x) = x^2 - x + 1$ and $Q(x) = x^3 + x^2 + x + 2$. Find $P(x) \cdot Q(x)$.

Multiply.

37. $(a + 2)(a + 3)$

38. $(x + 5)(x + 8)$

39. $(y + 3)(y - 2)$

40. $(y - 4)(y + 7)$

41. $(x + 3)^2$

42. $(y - 7)^2$

43. $(x - 2y)^2$

44. $(2s + 3t)^2$

45. $\left(b - \frac{1}{3}\right)\left(b - \frac{1}{2}\right)$

46. $\left(x - \frac{3}{2}\right)\left(x - \frac{2}{3}\right)$

47. $(2x + 9)(x + 2)$

48. $(3b + 2)(2b - 5)$

49. $(10a - 0.12b)^2$

50. $(10p^2 + 2.3q)^2$

51. $(2x - 3y)(2x + y)$

52. $(2a - 3b)(2a - b)$

53. $\left(2a + \frac{1}{3}\right)^2$

54. $(3c - \frac{1}{2})^2$

55. $(2x^3 - 3y^2)^2$

56. $(3s^2 + 4t^3)^2$

57. $(a^2b^2 + 1)^2$

58. $(x^2y - xy^2)^2$

59. $(20a^5 - 0.16b)^2$

60. $(2p^5 + 3.1q)^2$

61. Let $P(x) = 3x - 4$. Find $P(x) \cdot P(x)$.

62. Let $Q(x) = 5x^2 - 11$. Find $Q(x) \cdot Q(x)$.

Multiply.

63. $(c + 2)(c - 2)$

64. $(x - 3)(x + 3)$

65. $(2a + 1)(2a - 1)$

66. $(3 - 2x)(3 + 2x)$

67. $(3m - 2n)(3m + 2n)$ **68.** $(3x + 5y)(3x - 5y)$

69. $(x^3 + yz)(x^3 - yz)$

70. $(2a^3 + 5ab)(2a^3 - 5ab)$

71. $(-mn + m^2)(mn + m^2)$

72. $(1.6 + pq)(-1.6 + pq)$

73. $\left(\frac{1}{2}p^4 - \frac{2}{3}q\right)\left(\frac{1}{2}p^4 + \frac{2}{3}q\right)$

74. $\left(\frac{3}{5}ab + 4c^5\right)\left(\frac{3}{5}ab - 4c^5\right)$

75. $(x + 1)(x - 1)(x^2 + 1)$

76. $(y - 2)(y + 2)(y^2 + 4)$

77. $(a - b)(a + b)(a^2 - b^2)$

78. $(2x - y)(2x + y)(4x^2 - y^2)$

79. $(a + b + 1)(a + b - 1)$

80. $(m + n + 2)(m + n - 2)$

81. $(2x + 3y + 4)(2x + 3y - 4)$

82. $(3a - 2b + c)(3a - 2b - c)$

83. Suppose that P dollars is invested in a savings account at interest rate i, compounded annually, for 2 years. The amount A in the account after 2 years is given by

$$A = P(1 + i)^2.$$

Find an equivalent expression for A.

84. Suppose that P dollars is invested in a savings account at interest rate i, compounded semiannually, for 1 year. The amount A in the account after 1 year is given by

$$A = P\left(1 + \frac{i}{2}\right)^2.$$

Find an equivalent expression for A.

85. Given $f(x) = 5x + x^2$, find and simplify.

 a) $f(t - 1)$

 b) $f(a + h) - f(a)$

86. Given $f(x) = 4 + 3x - x^2$, find and simplify.

 a) $f(p + 1)$

 b) $f(a + h) - f(a)$

Skill Maintenance

87. Takako worked a total of 17 days last month at her father's restaurant. She earned \$25 a day during the week and \$35 a day during the weekend. Last month Takako earned \$485. How many weekdays did she work?

88. The perimeter of a triangle is 174. The lengths of the three sides are consecutive even numbers. What are the lengths of the sides of the triangle?

89. In a factory, there are three machines A, B, and C. When all three are running, they produce 222 suitcases per day. If A and B work but C does not,

they produce 159 suitcases per day. If B and C work but A does not, they produce 147 suitcases. What is the daily production of each machine?

90. There are 50 dimes in a roll of dimes, 40 nickels in a roll of nickels, and 40 quarters in a roll of quarters. A student has 13 rolls of coins, which have a total value of \$89. There are three more rolls of dimes than nickels. How many of each type of roll does the student have?

Synthesis

91. ◈ Find two binomials whose product is $x^2 - 25$ and explain how you decided on those two binomials.

92. ◈ Find two binomials whose product is $x^2 - 6x + 9$ and explain how you decided on those two binomials.

Multiply. Assume that variables in exponents represent natural numbers.

93. $(6y)^2\left(-\frac{1}{3}x^2y^3\right)^3$

94. $[(a^{2n})^{2n}]^4$

95. $(-r^6s^2)^3\left(-\frac{r^2}{6}\right)^2(9s^4)^2$

96. $\left(-\frac{3}{8}xy^2\right)\left(-\frac{7}{9}x^3y\right)\left(-\frac{8}{7}x^2y\right)^2$

97. $(z^{n^2})^{n^3}(z^{4n^3})^{n^2}$

98. $(a^xb^{2y})\left(\frac{1}{2}a^{3x}b\right)^2$

99. $[(2x - 1)^2 - 1]^2$

100. $y^3z^n(y^{3n}z^3 - 4yz^{2n})$

101. $[(a + b)(a - b)][5 - (a + b)][5 + (a + b)]$

102. $[x + y + 1][x^2 - x(y + 1) + (y + 1)^2]$

103. $(r^2 + s^2)^2(r^2 + 2rs + s^2)(r^2 - 2rs + s^2)$

104. $(y - 1)^6(y + 1)^6$

105. $(a - b + c - d)(a + b + c + d)$

106. $\left(\frac{2}{3}x + \frac{1}{3}y + 1\right)\left(\frac{2}{3}x - \frac{1}{3}y - 1\right)$

107. $[2(y - 3) - 6(x + 4)][5(y - 3) - 4(x + 4)]$

108. $\left(x - \frac{1}{7}\right)\left(x^2 + \frac{1}{7}x + \frac{1}{49}\right)$

109. $(4x^2 + 2xy + y^2)(4x^2 - 2xy + y^2)$

110. $(x^2 - 7x + 12)(x^2 + 7x + 12)$

111. $(x^a + y^b)(x^a - y^b)(x^{2a} + y^{2b})$

112. $[1 + \frac{1}{5}(x + 1)^2]^2$

113. $[a - (b - 1)][(b - 1)^2 + a(b - 1) + a^2]$

114. $(x - 1)(x^2 + x + 1)(x^3 + 1)$

115. $\left[\left(\frac{1}{3}x^3 - \frac{2}{3}y^2\right)\left(\frac{1}{3}x^3 + \frac{2}{3}y^2\right)\right]^2$

116. $10(0.1x^4y - 0.01xy^4)^2$

117. $(x^{a - b})^{a + b}$

118. $(M^{x + y})^{x + y}$

5.3

Common Factors and Factoring by Grouping

Terms with Common Factors • Factoring by Grouping

Factoring is the reverse of multiplication. To **factor** an expression means to write an equivalent expression that is a product. Skill at factoring will assist us when working with polynomial functions and solving polynomial equations later in this chapter.

Terms with Common Factors

When factoring, we look for factors common to every term in an expression and then use the distributive law.

EXAMPLE 1

Factor out a common factor: $4y^2 - 8$.

Solution

$$4y^2 - 8 = 4 \cdot y^2 - 4 \cdot 2 \qquad \text{Noting that 4 is a common factor}$$
$$= 4(y^2 - 2) \qquad \text{Using the distributive law} \qquad \square$$

In some cases, there is more than one common factor. In Example 2(a) below, for instance, 5 is a common factor, x^3 is a common factor, and $5x^3$ is a common factor. If there is more than one common factor, we factor out the *largest common factor,* that is, the factor that has the largest coefficient and the highest degree. In Example 2(a), the largest common factor is $5x^3$.

EXAMPLE 2

Factor out a common factor.

a) $5x^4 - 20x^3$ **b)** $12x^2y - 20x^3y$ **c)** $10p^6q^2 - 4p^5q^3 + 2p^4q^4$

Solution

a) $5x^4 - 20x^3 = 5x^3(x - 4)$ Try to write your answer directly. Multiply mentally to check your answer.

b) $12x^2y - 20x^3y = 4x^2y(3 - 5x)$
Check: $4x^2y \cdot 3 = 12x^2y$ and $4x^2y(-5x) = -20x^3y$,
  so $4x^2y(3 - 5x) = 12x^2y - 20x^3y$.

c) $10p^6q^2 - 4p^5q^3 + 2p^4q^4 = 2p^4q^2(5p^2 - 2pq + q^2)$
The check is left to the student. $\square$

The polynomials in Examples 1 and 2 have been **factored completely** since they cannot be factored further. The factors are said to be **prime polynomials.**
  Consider the polynomial

$$-6x^2 + 3.$$

We can factor this as

$$-6x^2 + 3 = 3(-2x^2 + 1)$$

or as

$$-6x^2 + 3 = -3(2x^2 - 1).$$

In certain situations, the latter factorization will be more helpful since it allows the leading coefficient of the polynomial factor to be positive.

EXAMPLE 3

Factor out a common factor with a negative coefficient: **(a)** $-4x - 24$; **(b)** $-2x^3 + 6x^2 - 10x$.

Solution

a) $-4x - 24 = -4(x + 6)$

b) $-2x^3 + 6x^2 - 10x = -2x(x^2 - 3x + 5)$ ❑

EXAMPLE 4

Height of a thrown object.   Suppose that an object is thrown upward from ground level with an initial velocity of 64 ft/sec. Its height after t seconds is a function h given by

$$h(t) = -16t^2 + 64t.$$

Find an equivalent expression for $h(t)$ by factoring out a common factor.

Solution   We factor out $-16t$ as follows:

$$h(t) = -16t^2 + 64t = -16t(t - 4).$$

Note that we can obtain function values using either expression for $h(t)$, since factoring forms *equivalent expressions*. For example,

$$h(1) = -16 \cdot 1^2 + 64 \cdot 1 = 48$$

and $h(1) = -16 \cdot 1(1 - 4) = 48$.   Using the factorization   ❑

In Example 4, we could have evaluated $-16t^2 + 64t$ and $-16t(t - 4)$ using any value for t. The result should always be the same. Thus a quick (but not always foolproof) way to check your factoring is to evaluate the factorization and the original polynomial using the same replacement.

Factoring by Grouping

The largest common factor may be a *binomial* factor.

EXAMPLE 5

Factor: $(a - b)(x + 5) + (a - b)(x - y^2)$.

Solution   Here the largest common factor is the binomial $a - b$:

$$(a - b)(x + 5) + (a - b)(x - y^2) = (a - b)[(x + 5) + (x - y^2)]$$
$$= (a - b)[2x + 5 - y^2].$$

Sometimes in order to identify a common binomial factor, we first regroup into two groups of two terms each.

EXAMPLE 6

Factor: **(a)** $y^3 + 3y^2 + 4y + 12$; **(b)** $4x^3 - 15 + 20x^2 - 3x$.

Solution

a) $y^3 + 3y^2 + 4y + 12 = (y^3 + 3y^2) + (4y + 12)$ Grouping
$$= y^2(y + 3) + 4(y + 3)$$ Factoring out common factors
$$= (y + 3)(y^2 + 4)$$ Factoring out $y + 3$

b) When we try grouping $4x^3 - 15 + 20x^2 - 3x$ as

$$(4x^3 - 15) + (20x^2 - 3x),$$

we are unable to factor $4x^3 - 15$. When this happens, we can rearrange the polynomial and try a different grouping:

$$4x^3 - 15 + 20x^2 - 3x = 4x^3 + 20x^2 - 3x - 15$$ Using the commutative law to rearrange the terms
$$= 4x^2(x + 5) - 3(x + 5)$$
$$= (x + 5)(4x^2 - 3)$$

In Section 1.3, we saw that the expressions $b - a$ and $-(a - b)$ or $-1(a - b)$ are equivalent. Remembering this can help if we wish to reverse a subtraction.

EXAMPLE 7

Factor: $ax - bx + by - ay$.

Solution

$$ax - bx + by - ay = (ax - bx) + (by - ay)$$ Grouping
$$= x(a - b) + y(b - a)$$ Factoring out common factors
$$= x(a - b) + y(-1)(a - b)$$ Factoring out -1 to reverse the subtraction
$$= x(a - b) - y(a - b)$$ Simplifying
$$= (a - b)(x - y)$$ Factoring out $a - b$

Not all polynomials with four terms can be factored by grouping. An example is
$$x^3 + x^2 + 3x - 3.$$

Not only is there no common monomial factor, but no matter how we group the terms, there is no common binomial factor. For example,

$$x^3 + x^2 + 3x - 3 = x^2(x + 1) + 3(x - 1);$$ No common factor

$$x^3 + 3x + x^2 - 3 = x(x^2 + 3) + (x^2 - 3);$$ No common factor

$$x^3 - 3 + x^2 + 3x = (x^3 - 3) + x(x + 3).$$ No common factor

---

EXERCISE SET | 5.3

Factor.

1. $4a^2 + 2a$

2. $6y^2 + 3y$

3. $y^2 - 5y$

4. $x^2 + 9x$

5. $y^3 + 9y^2$

6. $x^3 + 8x^2$

7. $6x^2 - 3x^4$

8. $8y^2 + 4y^4$

9. $4x^2y - 12xy^2$

10. $5x^2y^3 + 15x^3y^2$

11. $3y^2 - 3y - 9$

12. $5x^2 - 5x + 15$

13. $4ab - 6ac + 12ad$

14. $8xy + 10xz - 14xw$

15. $10a^4 + 15a^2 - 25a - 30$

16. $12t^5 - 20t^4 + 8t^2 - 16$

Factor out a factor with a negative coefficient.

17. $-3x + 12$

18. $-5x - 40$

19. $-6y - 72$

20. $-8t + 72$

21. $-2x^2 + 4x - 12$

22. $-2x^2 + 12x + 40$

23. $-3y^2 + 24x$

24. $-7x^2 - 56y$

25. $-3y^3 + 12y^2 - 15y$

26. $-4m^4 - 32m^3 + 64m$

27. $-x^2 + 3x - 7$

28. $-p^3 - 4p^2 + 11$

29. $-a^4 + 2a^3 - 13a$

30. $-m^3 - m^2 + m - 2$

31. *Height of a thrown object.*   Suppose that an object is thrown upward from ground level with an initial velocity of 80 ft/sec. Its height after t seconds is a function h given by

$$h(t) = -16t^2 + 80t.$$

   a) Find an equivalent expression for $h(t)$ by factoring out a common factor with a negative coefficient.

   b) Check your factoring by evaluating both expressions for $h(t)$ at $t = 3$.

32. *Height of a thrown object.*   Suppose that an object is thrown upward from ground level with an

initial velocity of 96 ft/sec. Its height after t seconds is a function h given by

$$h(t) = -16t^2 + 96t.$$

   a) Find an equivalent expression for $h(t)$ by factoring out a common factor with a negative coefficient.

   b) Check your factoring by evaluating both expressions for $h(t)$ at $t = 2$.

33. *Counting spheres in a pile.*   The number N of spheres in a pile like the one shown here is a polynomial function given by

$$N(x) = \tfrac{1}{6}x^3 + \tfrac{1}{2}x^2 + \tfrac{1}{3}x,$$

where x is the number of layers and $N(x)$ is the number of spheres. Find an equivalent expression for the function by factoring out a common factor.

34. *Number of games in a league.*   If there are n teams in a league and each team plays the others once in a season, we can find the total number of games played by the polynomial function $f(n) = \tfrac{1}{2}(n^2 - n)$. Find an equivalent expression for the function by factoring out a common factor.

35. *Number of diagonals.*   The number of diagonals of a polygon having n sides is given by the poly-

nomial function
$$P(n) = \tfrac{1}{2}n^2 - \tfrac{3}{2}n.$$

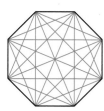

Find an equivalent expression for the function by factoring out a common factor.

36. *Surface area of a right circular cylinder.* The surface area of a right circular cylinder is given by the polynomial

$$2\pi rh + 2\pi r^2,$$

where h = the height and r = the radius of the base. Find an equivalent expression for the function by factoring out a common factor.

37. *Total revenue.* An electronics firm is marketing a new kind of stereo. The firm determines that when it sells x stereos, the total revenue R is given by the polynomial function

$$R(x) = 280x - 0.4x^2 \text{ dollars}.$$

Find an equivalent expression for the function by factoring out $0.4x$.

38. *Total cost.* An electronics firm determines that the total cost C of producing x stereos is given by the polynomial function

$$C(x) = 0.18x + 0.6x^2.$$

Find an equivalent expression for the function by factoring out $0.6x$.

Factor.

39. $a(b - 2) + c(b - 2)$

40. $a(x^2 - 3) - 2(x^2 - 3)$

41. $(x - 2)(x + 5) + (x - 2)(x + 8)$

42. $(m - 4)(m + 3) + (m - 4)(m - 3)$

43. $a^2(x - y) + a^2(y - x)$

44. $3x^2(x - 6) + 3x^2(6 - x)$

45. $ac + ad + bc + bd$ **46.** $xy + xz + wy + wz$

47. $b^3 - b^2 + 2b - 2$ **48.** $y^3 - y^2 + 3y - 3$

49. $a^3 - 3a^2 + 6 - 2a$ **50.** $t^3 + 6t^2 - 2t - 12$

51. $24x^3 - 36x^2 + 72x$

52. $12a^4 - 21a^3 - 6a$

53. $x^6 - x^5 - x^3 + x^4$

54. $y^4 - y^3 - y + y^2$

55. $2y^4 + 6y^2 + 5y^2 + 15$

56. $2xy - x^2y - 6 + 3x$

Skill Maintenance

57. A collection of nickels, dimes, and quarters is worth $18.10. The number of quarters is two more than twice the number of dimes. There are 120 coins in all. How many of each type of coin are there?

58. The first angle of a triangle is four times the second angle. The third angle is 75°. What are the measures of the first two angles?

Synthesis

59. ◈ Just after Example 4, we mentioned that checking by replacement is not foolproof. Write an example of an *incorrectly* factored polynomial and explain how evaluating both the polynomial and the factorization might not catch the mistake.

60. ◈ Is it always true that if all of a polynomial's coefficients and exponents are prime numbers, then the polynomial itself is prime? Why or why not?

Nested evaluation. An alternative procedure for evaluating a polynomial function is as follows. Given a polynomial function such as $P(x) = 3x^4 - 5x^3 + 4x^2 - 1$, successively factor out x, as shown:

$$P(x) = x(x(x(3x - 5) + 4) + 0) - 1.$$

Given a value for x, substitute it in the innermost parentheses and work your way out, at each step multiplying, then adding or subtracting.

Find nested form.

61. $P(x) = x^4 - 3x^3 - 5x^2 + 4x + 2$

62. $P(x) = x^4 + 5x^3 + 7x^2 - 2x - 4$

63. $P(x) = 5x^5 - 3x^3 + 4x^2 - 5x + 1$

64. $Q(x) = 4x^5 + 2x^4 - 5x^2 - 7$

Find nested form and then find the indicated function values.

65. $P(x) = 2x^4 - 3x^3 + 5x^2 + 6x - 4$;
$P(5), P(-2), P(10)$

66. $Q(x) = 2x^5 - 3x^4 + 5x^2 - 7x + 12$;
$Q(-4), Q(7), Q(11)$

Factor. Assume all exponents are natural numbers.

67. $4y^{4a} + 12y^{2a} + 10y^{2a} + 30$

68. $2x^{4p} + 6x^{2p} + 5x^{2p} + 15$

69. $4x^{a+b} + 7x^{a-b}$

70. $7y^{2a+b} - 5y^{a+b} + 3y^{a+2b}$

5.4

Factoring Trinomials

Factoring Trinomials of the Type $x^2 + bx + c$ •
Factoring Trinomials of the Type $ax^2 + bx + c$, $a \neq 1$

To begin our study of the factoring of trinomials, we first examine trinomials of the type $x^2 + bx + c$. We then study trinomials of the type $ax^2 + bx + c$, where $a \neq 1$.

Factoring Trinomials of the Type $x^2 + bx + c$

When trying to factor trinomials of the type $x^2 + bx + c$, we can use a trial-and-error procedure.

Constant Term Positive

Recall the FOIL method of multiplying two binomials:

$$\begin{array}{cccc} & \text{F} & \text{O} \quad \text{I} & \text{L} \\ (x+3)(x+5) & = x^2 & + 5x + 3x & + 15 \end{array}$$

$$= x^2 + \quad 8x \quad + 15.$$

The product is a trinomial in which the leading coefficient is 1. The constant term is positive. To factor $x^2 + 8x + 15$, we think of FOIL: The first term, x^2, is the product of the First terms of the binomial factors so the first term in each binomial will be x. The challenge is to find two numbers p and q such that

$$x^2 + 8x + 15 = (x + p)(x + q).$$

Note that the Outer and Inner products, qx and px, can be written as $(p + q)x$. The Last product, pq, will be a constant. Thus the numbers p and q must be selected so that their product is 15 and their sum is 8. In this case, we know from above that the numbers are 3 and 5. Thus the factorization is

$$(x + 3)(x + 5), \quad \text{or} \quad (x + 5)(x + 3)$$

by the commutative law of multiplication. In general,

$$(x + p)(x + q) = x^2 + (p + q)x + pq.$$

To factor $x^2 + (p + q)x + pq$, we use FOIL in reverse:

$$x^2 + (p + q)x + pq = (x + p)(x + q).$$

EXAMPLE 1

Factor: $x^2 + 9x + 8$.

Solution   We think of FOIL in reverse. The first term of each factor is x. We are looking for numbers p and q such that

$$x^2 + 9x + 8 = (x + p)(x + q) = x^2 + (p + q)x + pq.$$

Thus we look for factors of 8 whose sum is 9.

Pair of Factors	Sum of Factors
2, 4	6
1, 8	9 ←

————The numbers we need are 1 and 8.

The factorization is thus $(x + 1)(x + 8)$. We can check by multiplying to see if the product is the original trinomial.   ❑

When we are factoring trinomials with a leading coefficient of 1, it suffices to list all pairs of factors along with their sums, as we did above. At times, however, you may wish to simply form factors without calculating any sums. It is essential that you check any attempt made in this manner! For example, if we attempt the factorization

$$x^2 + 9x + 8 \overset{?}{=} (x + 2)(x + 4),$$

a check reveals that $(x + 2)(x + 4) = x^2 + 6x + 8 \neq x^2 + 9x + 8$. This type of trial-and-error procedure becomes easier to use with time. As you gain experience, you will find that many trials can be performed mentally.

When the constant term of a trinomial is positive, the constant terms in the binomial factors both have the same sign. This assures a positive product. The sign used is that of the trinomial's middle term.

EXAMPLE 2

Factor: $y^2 - 9y + 20$.

Solution   Since the constant term is positive and the coefficient of the middle term is negative, we look for a factorization of 20 in which both factors are negative. Their sum must be -9.

Pair of Factors	Sum of Factors
$-1, -20$	-21
$-2, -10$	-12
$-4, -5$	-9 ←

————The numbers we need are -4 and -5.

The factorization is $(y - 4)(y - 5)$.   ❑

Constant Term Negative

When the constant term of a trinomial is negative, we look for two factors whose product is negative. One factor must be positive and the other negative. Their sum must still be the coefficient of the middle term.

EXAMPLE 3

Factor: $x^3 - x^2 - 30x$.

Solution   *Always* look first for a common factor! This time there is one, x. We first factor it out:

$$x^3 - x^2 - 30x = x(x^2 - x - 30).$$

Now we consider $x^2 - x - 30$. We look for a factorization of -30 in which one factor is positive, the other factor is negative, and the sum of the factors is -1. Since the sum is to be negative, the negative factor must be further from 0 than the positive factor is. Thus we need only consider pairs of factors in which the negative term has the larger absolute value.

Pair of Factors	Sum of Factors
1, -30	-29
3, -10	-7
5, -6	-1 ← The numbers we want are 5 and -6.

The factorization of $x^2 - x - 30$ is

$$(x + 5)(x - 6).$$

Don't forget to include the factor that was factored out earlier! In this case, the factorization of the original trinomial is

$$x(x + 5)(x - 6). \qquad \square$$

EXAMPLE 4

Factor: $2x^2 + 34x - 220$.

Solution   *Always* look first for a common factor! This time we can factor out 2:

$$2x^2 + 34x - 220 = 2(x^2 + 17x - 110).$$

To factor $x^2 + 17x - 110$, we look for a factorization of -110 in which one factor is positive, the other factor is negative, and the sum of the factors is 17. Since the sum is to be positive, the positive factor must be further from 0 than the negative factor is. Thus we consider only pairs of factors in which the positive term has the larger absolute value.

Pair of Factors	Sum of Factors
-2, 55	53
-10, 11	1
-5, 22	17 ← The numbers we need are -5 and 22.

The factorization of $x^2 + 17x - 110$ is $(x - 5)(x + 22)$. The factorization of the original trinomial, $2x^2 + 34x - 220$, is $2(x - 5)(x + 22)$. $\qquad \square$

Some polynomials are not factorable using integers.

EXAMPLE 5

Factor: $x^2 - x - 7$.

Solution   There are no factors of -7 whose sum is -1. This trinomial is *not* factorable into binomials with integer coefficients. $\qquad \square$

To factor $x^2 + bx + c$:

1. First arrange in descending order. Use a trial-and-error procedure, looking for factors of c whose sum is b.
2. If c is positive, the signs of the factors are the same as the sign of b.
3. If c is negative, one factor is positive and the other is negative. If the sum of the two factors is the opposite of b, changing the signs of each factor will give the desired factors whose sum is b.
4. Check the result by multiplying the binomials.

The procedure considered here can also be applied to a trinomial with more than one variable.

EXAMPLE 6

Factor: $x^2 - 2xy - 48y^2$.

Solution   We look for numbers p and q such that

$$x^2 - 2xy - 48y^2 = (x + py)(x + qy).$$

Our thinking is much the same as if we were factoring $x^2 - 2x - 48$. We look for factors of -48 whose sum is -2. Those factors are 6 and -8. Then

$$x^2 - 2xy - 48y^2 = (x + 6y)(x - 8y).$$

We leave the check to the student.   ❑

Sometimes a trinomial like $x^6 + 2x^3 - 15$ can be factored if we first think of it as $(x^3)^2 + 2x^3 - 15$. To do this, make a substitution (perhaps just mentally) in which $u = x^3$. The trinomial then becomes

$$u^2 + 2u - 15.$$

We try factoring this trinomial and if a factorization is found, we replace all occurrences of u by x^3. Since

$$u^2 + 2u - 15 = (u - 3)(u + 5),$$

the original polynomial can be factored as

$$x^6 + 2x^3 - 15 = (x^3 - 3)(x^3 + 5).$$

Factoring Trinomials of the Type $ax^2 + bx + c, a \neq 1$

Now we look at trinomials in which the leading coefficient is not 1. We consider two methods. Use the one that works best for you or the one that your instructor chooses for you.

Method 1: The FOIL Method

We first consider the **FOIL method** for factoring trinomials of the type

$$ax^2 + bx + c, \quad \text{where } a \neq 1.$$

Consider the following multiplication.

$$\begin{array}{cccc} F & O & I & L \\ \downarrow & \downarrow & \downarrow & \downarrow \end{array}$$

$$(3x + 2)(4x + 5) = 12x^2 + 15x + 8x + 10$$

$$= 12x^2 + 23x + 10$$

To factor $12x^2 + 23x + 10$, we do the reverse of what we just did. We look for two binomials whose product is this trinomial. The product of the First terms must be $12x^2$. The product of the Outside terms plus the product of the Inside terms must be $23x$. The product of the Last terms must be 10. We know from the preceding discussion that the answer is

$$(3x + 2)(4x + 5).$$

Generally, however, finding such an answer is a trial-and-error process. It turns out that

$$(-3x - 2)(-4x - 5)$$

is also a correct answer, but we usually choose an answer in which the first coefficients are positive.

We use the following method:

To factor $ax^2 + bx + c$, $a \neq 1$, using the FOIL method:

1. Factor out the largest common factor, if one exists.

2. Find two First terms whose product is ax^2:

$$(\blacksquare x + \ )(\blacksquare x + \ ) = ax^2 + bx + c.$$
$$\underline{\qquad\qquad}\text{FOIL}$$

3. Find two Last terms whose product is c:

$$(\ x + \blacksquare)(\ x + \blacksquare) = ax^2 + bx + c.$$
$$\underline{\qquad\qquad}\text{FOIL}$$

4. Perform steps (2) and (3) until a combination is found for which the sum of the Outer and Inner products is bx:

$$(\blacksquare x + \blacksquare)(\blacksquare x + \blacksquare) = ax^2 + bx + c.$$
$$\text{I}$$
$$\text{O} \qquad\qquad \text{FOIL}$$

EXAMPLE 7

Factor: $3x^2 + 10x - 8$.

Solution

1. First we observe that there is no common factor (other than 1 or -1).

2. Next we factor the first term, $3x^2$. The only possibility for factors is $3x \cdot x$. The desired factorization is then of the form

$$(3x + \blacksquare)(x + \blacksquare),$$

where we must find the right numbers for the blanks.

3. We then factor the last term, -8. The possibilities are $(-8)(1)$, $8(-1)$, $2(-4)$, and $(-2)4$, as well as $(1)(-8)$, $(-1)8$, $(-4)2$, and $4(-2)$. We look for factors such that the sum of the products (the "outside" and "inside" parts of FOIL) is the middle term, $10x$. We try some possibilities for factorization and check by multiplying:

$$(3x - 8)(x + 1) = 3x^2 - 5x - 8.$$

This gives us a middle term that is negative. We avoid such possibilities and look for possibilities that give a positive middle term:

$$(3x + 8)(x - 1) = 3x^2 + 5x - 8.$$

Note that changing the signs in the binomials has the effect of changing the sign of the middle term. We try again:

$$(3x - 2)(x + 4) = 3x^2 + 10x - 8.$$

Thus the desired factorization is $(3x - 2)(x + 4)$. ❏

EXAMPLE 8

Factor: $6x^6 - 19x^5 + 10x^4$.

Solution

1. First we factor out the common factor x^4:

$$x^4(6x^2 - 19x + 10).$$

2. Next, consider that $6x^2 = 6x \cdot x$ and $6x^2 = 3x \cdot 2x$. Thus, $6x^2 - 19x + 10$ may factor into

$$(3x + \blacksquare)(2x + \blacksquare) \quad \text{or} \quad (6x + \blacksquare)(x + \blacksquare).$$

3. We factor the last term, 10. The possibilities are $10 \cdot 1$, $(-10)(-1)$, $5 \cdot 2$, and $(-5)(-2)$, as well as $1 \cdot 10$, $(-1)(-10)$, $2 \cdot 5$, and $(-2)(-5)$.

We see that there are 8 possibilities for each factorization in step (2). We look for factors such that the sum of the products (the "outer" and "inner" parts of FOIL) is the middle term, $-19x$. We try a possible factorization and check by multiplying:

$$(3x - 10)(2x - 1) = 6x^2 - 23x + 10.$$

We try again:

$$(3x - 5)(2x - 2) = 6x^2 - 16x + 10.$$

Actually this last attempt could have been rejected by simply noting that $2x - 2$ has a common factor, 2. Since we removed the *largest* common factor in step (1), no other common factors can exist. We try again, reversing the -5 and -2:

$$(3x - 2)(2x - 5) = 6x^2 - 19x + 10.$$

The factorization of $6x^2 - 19x + 10$ is $(3x - 2)(2x - 5)$. But do not forget the common factor! We must include it to get a factorization of the original trinomial:

$$6x^6 - 19x^5 + 10x^4 = x^4(3x - 2)(2x - 5).$$ ❏

Here is another tip that might speed up your factoring. Suppose in Example 8 that we considered the possibility

$$(3x + 2)(2x + 5) = 6x^2 + 19x + 10.$$

We might have tried this before noticing that using all plus signs would give us a plus sign for the middle term. If we change *both* signs, however, we get the correct answer before including the common factor:

$$(3x - 2)(2x - 5) = 6x^2 - 19x + 10.$$

Tips for Factoring with FOIL

1. If the largest common factor has been factored out of the original trinomial, then no binomial factor can have a common factor (other than 1 or -1).
2. If a and c are both positive, then the signs in the factors will be the same as the sign of b.
3. When a possible factoring produces the opposite of the middle term being sought, reverse the signs in both factors.
4. Be systematic about your trials. Keep track of those possibilities that you have tried and those that you have not.

Keep in mind that this method of factoring trinomials of the type $ax^2 + bx + c$ involves trial and error. As you practice, you will find that you can make better and better guesses.

Method 2: The Grouping Method

The second method for factoring trinomials of the type $ax^2 + bx + c$, $a \neq 1$, is known as the *grouping method*. It involves not only trial and error and FOIL but also factoring by grouping. We know that

$$\begin{aligned} x^2 + 7x + 10 &= x^2 + 2x + 5x + 10 \\ &= x(x + 2) + 5(x + 2) \\ &= (x + 2)(x + 5), \end{aligned}$$

but what if the leading coefficient is not 1? Consider $6x^2 + 23x + 20$. The method is similar to what we just did with $x^2 + 7x + 10$, but we need two more steps. We first multiply the leading coefficient, 6, and the constant, 20, and get 120. Then we look for a factorization of 120 in which the sum of the factors is the coefficient of the middle term: 23. Next we split the middle term into a sum or difference using these factors.

$$6x^2 + 23x + 20$$

(1) Multiply 6 and 20: $6 \cdot 20 = 120$.
(2) Factor 120: $120 = 8 \cdot 15$, and $8 + 15 = 23$.
(3) Split the middle term: $23x = 8x + 15x$.
(4) Factor by grouping.

We factor by grouping as follows:

$$\begin{aligned} 6x^2 + 23x + 20 &= 6x^2 + 8x + 15x + 20 \\ &= 2x(3x + 4) + 5(3x + 4) \\ &= (3x + 4)(2x + 5). \end{aligned}$$

Factoring by grouping

To factor $ax^2 + bx + c$, using the grouping method:

1. Make sure that any common factors have been factored out.
2. Multiply the leading coefficient a and the constant c.
3. Try to factor the product ac such that the sum of the factors is b. That is, find integers p and q such that $pq = ac$ and $p + q = b$.
4. Split the middle term. That is, write bx as $px + qx$.
5. Factor by grouping.

EXAMPLE 9

Factor: $3x^2 + 10x - 8$.

Solution

1. First we look for a common factor. There is none (other than 1 or -1).
2. We multiply the leading coefficient and the constant, 3 and -8:

$$3(-8) = -24.$$

3. We try to factor -24 so that the sum of the factors is 10:

$$-24 = 12(-2) \quad \text{and} \quad 12 + (-2) = 10.$$

4. We split $10x$ using the results of step (3):

$$10x = 12x - 2x.$$

5. We then factor by grouping:

$$3x^2 + 10x - 8 = 3x^2 + 12x - 2x - 8$$
$$= 3x(x + 4) - 2(x + 4)$$
$$= (x + 4)(3x - 2).$$

Factoring by grouping

EXERCISE SET | 5.4

Factor.

1. $x^2 + 9x + 20$
2. $x^2 + 8x + 15$
3. $t^2 - 8t + 15$
4. $y^2 - 12y + 27$
5. $x^2 - 27 - 6x$
6. $t^2 - 15 - 2t$
7. $2y^2 - 16y + 32$
8. $2a^2 - 20a + 50$
9. $p^3 + 3p^2 - 54p$
10. $m^3 + m^2 - 72m$
11. $14x + x^2 + 45$
12. $12y + y^2 + 32$
13. $y^2 + 2y - 63$
14. $p^2 + 3p - 40$
15. $t^2 - 11t + 28$
16. $y^2 - 14y + 45$
17. $3x + x^2 - 10$
18. $x + x^2 - 6$
19. $x^2 + 5x + 6$
20. $y^2 + 8y + 7$
21. $56 + x - x^2$
22. $32 + 4y - y^2$
23. $32y + 4y^2 - y^3$
24. $56x + x^2 - x^3$
25. $x^4 + 11x^2 - 80$
26. $y^4 + 5y^2 - 84$
27. $x^2 - 3x + 7$
28. $x^2 + 12x + 13$
29. $x^2 + 12xy + 27y^2$
30. $p^2 - 5pq - 24q^2$
31. $x^2 - 14xy + 49y^2$
32. $y^2 + 8yz + 16z^2$
33. $x^4 + 50x^2 + 49$
34. $p^4 + 80p^2 + 79$
35. $x^6 + 2x^3 - 63$
36. $x^8 - 7x^4 + 10$
37. $3x^2 - 16x - 12$
38. $6x^2 - 5x - 25$
39. $6x^3 - 15x - x^2$
40. $10y^3 - 12y - 7y^2$
41. $3a^2 - 10a + 8$
42. $12a^2 - 7a + 1$

43. $35y^2 + 34y + 8$

44. $9a^2 + 18a + 8$

45. $4t + 10t^2 - 6$

46. $8x + 30x^2 - 6$

47. $8x^2 - 16 - 28x$

48. $18x^2 - 24 - 6x$

49. $12x^3 - 31x^2 + 20x$

50. $15x^3 - 19x^2 - 10x$

51. $14x^4 - 19x^3 - 3x^2$

52. $70x^4 - 68x^3 + 16x^2$

53. $3a^2 - a - 4$

54. $6a^2 - 7a - 10$

55. $9x^2 + 15x + 4$

56. $6y^2 - y - 2$

57. $3 + 35z - 12z^2$

58. $8 - 6a - 9a^2$

59. $-8t^2 - 8t + 30$

60. $-36a^2 + 21a - 3$

61. $3x^3 - 5x^2 - 2x$

62. $18y^3 - 3y^2 - 10y$

63. $24x^2 - 2 - 47x$

64. $15y^2 - 10 - 47y$

65. $63x^3 + 111x^2 + 36x$

66. $50y^3 + 115y^2 + 60y$

67. $40x^4 + 16x^2 - 12$

68. $24y^4 + 2y^2 - 15$

69. $12a^2 - 17ab + 6b^2$

70. $20p^2 - 23pq + 6q^2$

71. $2x^2 + xy - 6y^2$

72. $8m^2 - 6mn - 9n^2$

73. $6x^2 - 29xy + 28y^2$

74. $10p^2 + 7pq - 12q^2$

75. $9x^2 - 30xy + 25y^2$

76. $4p^2 + 12pq + 9q^2$

77. $6x^6 + x^3 - 2$

78. $2p^8 + 11p^4 + 15$

Skill Maintenance

79. Suppose that an object is thrown upward with an initial velocity of 80 ft/sec from a height of 224 ft. Its height after t seconds is a function h given by

$$h(t) = -16t^2 + 80t + 224.$$

What is the height of the object after 0 sec, 1 sec, 3 sec, 4 sec, and 6 sec?

80. Suppose that an object is thrown upward with an initial velocity of 96 ft/sec from a height of 880 ft. Its height after t seconds is a function h given by

$$h(t) = -16t^2 + 96t + 880.$$

What is the height of the object after 0 sec, 1 sec, 3 sec, 8 sec, and 10 sec?

Synthesis

81. ◈ Explain how to conclude that $x^2 + 5x + 200$ is a prime polynomial without performing any trials.

82. ◈ Explain how to conclude that $x^2 - 59x + 6$ is a prime polynomial without performing any trials.

Factor. Assume that variables in exponents represent positive integers.

83. $p^2q^2 + 7pq + 12$

84. $2x^4y^6 - 3x^2y^3 - 20$

85. $x^2 - \frac{4}{25} + \frac{3}{5}x$

86. $y^2 - \frac{8}{49} + \frac{2}{7}y$

87. $y^2 + 0.4y - 0.05$

88. $t^2 + 0.6t - 0.27$

89. $7a^2b^2 + 6 + 13ab$

90. $9x^2y^2 - 4 + 5xy$

91. $216x + 78x^2 + 6x^3$

92. $x^{2a} + 5x^a - 24$

93. $4x^{2a} - 4x^a - 3$

94. $x^2 + ax + bx + ab$

95. $bdx^2 + adx + bcx + ac$

96. $a^2p^{2a} + a^2p^a - 2a^2$

97. $2ar^2 + 4asr + as^2 - asr$

98. $(x + 3)^2 - 2(x + 3) - 35$

99. $6(x - 7)^2 + 13(x - 7) - 5$

100. Find all integers m for which $x^2 + mx + 75$ can be factored.

101. Find all integers q for which $x^2 + qx - 32$ can be factored.

102. One of the factors of $x^2 - 345x - 7300$ is $x + 20$. Find the other factor.

103. To better understand factoring $ax^2 + bx + c$ by grouping, suppose that $ax^2 + bx + c = (mx + r)(nx + s)$. Show that if $P = ms$ and $Q = rn$, then $P + Q = b$ and $PQ = ac$.

5.5

Factoring Trinomial Squares and Differences of Squares

Trinomial Squares   •   Differences of Squares   •
More Factoring by Grouping

We now introduce a faster way to factor trinomials that are squares of binomials. We also develop a method for factoring differences of squares.

Trinomial Squares

Consider the trinomial

$$x^2 + 6x + 9.$$

To factor it, we can use the method considered in the preceding section. We look for factors of 9 whose sum is 6. We see that these factors are 3 and 3 and the factorization is

$$x^2 + 6x + 9 = (x + 3)(x + 3) = (x + 3)^2.$$

Note that the result is the square of a binomial. We also say that $x^2 + 6x + 9$ is a **trinomial square.** We can certainly use the trial-and-error procedure to factor trinomial squares, but we want to develop a faster procedure.

In order to do so, we must first be able to recognize when a trinomial is a square.

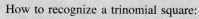

How to recognize a trinomial square:

i) Two of the terms must be squares, such as A^2 and B^2.
ii) There must be no minus sign before A^2 or B^2.
iii) Multiplying A and B (which are the square roots of these expressions) and doubling the result should give the remaining term, $2AB$, or its opposite, $-2AB$.

EXAMPLE 1

Determine whether the polynomial is a trinomial square: **(a)** $x^2 + 10x + 25$; **(b)** $4x + 16 + 3x^2$; **(c)** $100y^2 + 81 - 180y$.

Solution

a)   i) Two of the terms in $x^2 + 10x + 25$ are squares: x^2 and 25.
   ii) There is no minus sign before either x^2 or 25.
   iii) If we multiply the square roots, x and 5, and double the product, we get $10x$, the remaining term.

Thus the trinomial is a square.

b)   In $4x + 16 + 3x^2$, only one term, 16, is a square ($3x^2$ is not a square because 3 is not a perfect square integer and $4x$ is not a square because x is not a square).
Thus the trinomial is not a square.

c)   It can help to first write the polynomial in descending order:

$$100y^2 - 180y + 81.$$

   i) Two of the terms, $100y^2$ and 81, are squares.

   ii) There is no minus sign before either $100y^2$ or 81.

   iii) If we multiply the square roots, $10y$ and 9, and double the product, we get the opposite of the remaining term: $2(10y)(9) = 180y$, which is the opposite of $-180y$.

Thus, $100y^2 + 81 - 180y$ is a trinomial square.   ❑

To factor a trinomial square, we use the same equations that we used for squaring a binomial. The pattern learned in Section 5.2 is reversed.

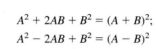

$$A^2 + 2AB + B^2 = (A + B)^2;$$
$$A^2 - 2AB + B^2 = (A - B)^2$$

EXAMPLE 2

Factor: **(a)** $x^2 - 10x + 25$;   **(b)** $16y^2 + 49 + 56y$;   **(c)** $-20xy + 4y^2 + 25x^2$.

Solution

a) $x^2 - 10x + 25 = (x - 5)^2$ We find the square terms and write the quantities that were squared with a minus sign between them.

       Note the sign!

b) $16y^2 + 49 + 56y = 16y^2 + 56y + 49$ Using a commutative law

$$= (4y + 7)^2$$

    We find the square terms and write the quantities that were squared with a plus sign between them.

c) $-20xy + 4y^2 + 25x^2 = 4y^2 - 20xy + 25x^2$ Writing in descending order of y

$$= (2y - 5x)^2$$

This square can also be expressed as

$$25x^2 - 20xy + 4y^2 = (5x - 2y)^2.$$

As always, we can check our factorization by multiplying:

$$(5x - 2y)^2 = (5x - 2y)(5x - 2y) = 25x^2 - 20xy + 4y^2. \qquad ❑$$

In factoring, we should always remember to *first* look for a factor common to all the terms.

EXAMPLE 3

Factor: **(a)** $2x^2 - 12xy + 18y^2$;   **(b)** $-4y^2 - 144y^8 + 48y^5$.

Solution

a)   We first look for a common factor. This time, there is a common factor, 2.

$$2x^2 - 12xy + 18y^2 = 2(x^2 - 6xy + 9y^2) \qquad \text{Factoring out the 2}$$
$$= 2(x - 3y)^2 \qquad \text{Factoring the trinomial square}$$

b) $-4y^2 - 144y^8 + 48y^5 = -4y^2(1 + 36y^6 - 12y^3)$ Factoring out the common factor

$= -4y^2(1 - 12y^3 + 36y^6)$ Changing order. Note that $(y^3)^2 = y^6$.

$= -4y^2(1 - 6y^3)^2$ Factoring the trinomial square   ❏

Differences of Squares

Any time that we recognize an expression like $x^2 - 9$ as a difference of two squares, we can reverse another pattern first seen in Section 5.2.

$$A^2 - B^2 = (A + B)(A - B)$$

To factor a difference of two squares, write the product of the sum and the difference of the two quantities being squared.

You will find it useful to memorize this rule.

EXAMPLE 4

Factor: **(a)** $x^2 - 9$; **(b)** $25y^6 - 49x^2$.

Solution

a) $x^2 - 9 = x^2 - 3^2 = (x + 3)(x - 3)$

$$A^2 \quad - \quad B^2 \;\; = (A \;\; + \;\; B)(A \;\; - \;\; B)$$

b) $25y^6 - 49x^2 = (5y^3)^2 - (7x)^2 = (5y^3 + 7x)(5y^3 - 7x)$ ❏

As always, the first step in factoring is to look for common factors.

EXAMPLE 5

Factor: **(a)** $5 - 5x^2y^6$; **(b)** $2x^4 - 8y^4$; **(c)** $16x^4y - 81y$.

Solution

a) $5 - 5x^2y^6 = 5(1 - x^2y^6)$ Factoring out the common factor

$= 5[1^2 - (xy^3)^2]$ Rewriting x^2y^6 as a quantity squared

$= 5(1 + xy^3)(1 - xy^3)$ Factoring the difference of squares

b) $2x^4 - 8y^4 = 2(x^4 - 4y^4)$ Factoring out the common factor

$= 2[(x^2)^2 - (2y^2)^2]$

$= 2(x^2 + 2y^2)(x^2 - 2y^2)$ Factoring the difference of squares

c) $16x^4y - 81y = y(16x^4 - 81)$ Factoring out the common factor

$= y[(4x^2)^2 - 9^2]$

$= y(4x^2 + 9)(4x^2 - 9)$ Factoring the difference of squares

$= y(4x^2 + 9)(2x + 3)(2x - 3)$ Factoring $4x^2 - 9$, which is *also* a difference of squares   ❏

In Example 5(c), it may be tempting to try to factor $(4x^2 + 9)$. Note that it is a sum of two expressions that are squares. If the largest common factor of a sum of squares has been removed, then the remaining sum cannot be factored using real numbers. Note also in Example 5(c) that $4x^2 - 9$ could be factored further. Whenever a factor itself can be factored, you should do so. That way you will be factoring *completely*.

More Factoring by Grouping

Sometimes when factoring a polynomial with four terms, we may be able to factor further.

EXAMPLE 6

Factor: $x^3 + 3x^2 - 4x - 12$.

Solution

$$x^3 + 3x^2 - 4x - 12 = x^2(x + 3) - 4(x + 3)$$
$$= (x + 3)(x^2 - 4)$$
$$= (x + 3)(x + 2)(x - 2) \qquad \square$$

A difference of squares can have more than two terms. For example, one of the squares may be a trinomial. We can factor by a different type of grouping.

EXAMPLE 7

Factor: $x^2 + 6x + 9 - y^2$.

Solution

$$x^2 + 6x + 9 - y^2 = (x^2 + 6x + 9) - y^2 \qquad \text{Grouping as a trinomial minus } y^2 \text{ to show a difference of squares}$$
$$= (x + 3)^2 - y^2$$
$$= (x + 3 + y)(x + 3 - y) \qquad \square$$

EXAMPLE 8

Factor: $a^2 - b^2 + 8b - 16$.

Solution   Grouping into two groups of two terms each does not yield a common binomial factor, so we look for a trinomial square. In this case, the trinomial square is being subtracted from a^2:

$$a^2 - b^2 + 8b - 16 = a^2 - (b^2 - 8b + 16) \qquad \text{Factoring out } -1 \text{ and rewriting as subtraction}$$
$$= a^2 - (b - 4)^2 \qquad \text{Factoring the trinomial square}$$
$$= (a + (b - 4))(a - (b - 4)) \qquad \text{Factoring a difference of squares}$$
$$= (a + b - 4)(a - b + 4) \qquad \text{Removing parentheses} \quad \square$$

EXERCISE SET | 5.5

Factor.

1. $y^2 - 6y + 9$

2. $x^2 - 8x + 16$

3. $x^2 + 14x + 49$

4. $x^2 + 16x + 64$

5. $x^2 + 1 + 2x$

6. $x^2 + 1 - 2x$

7. $2a^2 + 8a + 8$

8. $4a^2 - 16a + 16$

9. $y^2 + 36 - 12y$

10. $y^2 + 36 + 12y$

11. $-18y^2 + y^3 + 81y$ **12.** $24a^2 + a^3 + 144a$

13. $12a^2 + 36a + 27$ **14.** $20y^2 + 100y + 125$

15. $2x^2 - 40x + 200$ **16.** $32x^2 + 48x + 18$

17. $1 - 8d + 16d^2$ **18.** $64 + 25y^2 - 80y$

19. $y^4 + 8y^2 + 16$ **20.** $a^4 - 10a^2 + 25$

21. $0.25x^2 + 0.30x + 0.09$

22. $0.04x^2 - 0.28x + 0.49$

23. $p^2 - 2pq + q^2$ **24.** $m^2 + 2mn + n^2$

25. $a^2 + 4ab + 4b^2$ **26.** $49p^2 - 14pq + q^2$

27. $25a^2 - 30ab + 9b^2$ **28.** $49p^2 - 84pq + 36q^2$

29. $x^4 + 2x^2y^2 + y^4$ **30.** $p^8 + 2p^4q^4 + q^8$

31. $x^2 - 16$ **32.** $y^2 - 25$

33. $p^2 - 49$ **34.** $m^2 - 64$

35. $p^2q^2 - 25$ **36.** $a^2b^2 - 81$

37. $6x^2 - 6y^2$ **38.** $8x^2 - 8y^2$

39. $4xy^4 - 4xz^4$ **40.** $25ab^4 - 25az^4$

41. $4a^3 - 49a$ **42.** $9x^3 - 25x$

43. $3x^8 - 3y^8$ **44.** $9a^4 - a^2b^2$

45. $9a^4 - 25a^2b^4$ **46.** $16x^6 - 121x^2y^4$

47. $\frac{1}{25} - x^2$ **48.** $\frac{1}{16} - y^2$

49. $0.04x^2 - 0.09y^2$ **50.** $0.01x^2 - 0.04y^2$

51. $m^3 - 7m^2 - 4m + 28$

52. $x^3 + 8x^2 - x - 8$

53. $a^3 - ab^2 - 2a^2 + 2b^2$

54. $p^2q - 25q + 3p^2 - 75$

55. $(a + b)^2 - 100$ **56.** $(p - 7)^2 - 144$

57. $a^2 + 2ab + b^2 - 9$ **58.** $x^2 - 2xy + y^2 - 25$

59. $r^2 - 2r + 1 - 4s^2$

60. $c^2 + 4cd + 4d^2 - 9p^2$

61. $50a^2 - 2m^2 - 4mn - 2n^2$

62. $3x^2 - 12y^2 - 12y - 3$

63. $9 - a^2 + 2ab - b^2$ **64.** $16 - x^2 + 2xy - y^2$

Skill Maintenance _____

Solve.

65. $\begin{aligned} x - y + z &= 6, \\ 2x + y - z &= 0, \\ x + 2y + z &= 3 \end{aligned}$ **66.** $|5 - 7x| \geq 9$

67. $|5 - 7x| \leq 9$ **68.** $5 - 7x > -9 + 12x$

Synthesis _____

69. ◈ Explain how it might be possible for a certain sum of two squares to be factorable.

70. ◈ Without finding the entire factorization, determine the number of factors of $x^{256} - 1$. Explain how you arrived at your answer.

Factor. Assume that variables in exponents represent positive integers.

71. $-225x + x^3$

72. $4y^3 - 96y^2 + 576y$

73. $3xy^2 - 150xy + 1875x$

74. $7x^2 - 4375y^2$

75. $12x^2 - 72xy + 108y^2$

76. $-\frac{3}{4}p^2 + \frac{6}{5}p - \frac{12}{25}$

77. $-\frac{8}{27}r^2 - \frac{10}{9}rs - \frac{1}{6}s^2 + \frac{2}{3}rs$

78. $x^4y^4 - 8x^2y^2 + 16$

79. $-24ab + 16a^2 + 9b^2$

80. $\frac{1}{36}x^8 + \frac{2}{9}x^4 + \frac{4}{9}$

81. $0.09x^2 + 0.48x + 0.64$

82. $x^{2a} - y^2$

83. $x^{4a} - y^{2b}$

84. $4y^{4a} + 20y^{2a} + 20y^{2a} + 100$

85. $25y^{2a} - (x^{2b} - 2x^b + 1)$

86. $8(a - 3)^2 - 64(a - 3) + 128$

87. $3(x + 1)^2 + 12(x + 1) + 12$

88. $5c^{100} - 80d^{100}$

89. $9x^{2n} - 6x^n + 1$

90. $c^{2w + 1} + 2c^{w + 1} + c$

91. If $P(x) = x^2$, use factoring to simplify
$$P(a + h) - P(a).$$

92. The volume of a carpet that is rolled up can be estimated by the polynomial $\pi R^2h - \pi r^2h$.

a) Factor the polynomial.

b) Use both the original and the factored forms to find the volume of a roll for which $R = 50$ cm, $r = 10$ cm, and $h = 4$ m.

5.6

Factoring Sums or Differences of Cubes

Formulas for Factoring Sums or Differences of Cubes •
Using the Formulas

Although a sum of two squares cannot be factored using real-number coefficients (unless it has a common factor), a sum of two cubes can. In this section, we develop a method for factoring sums or differences of two cubes.

Consider the following products:

$$(A + B)(A^2 - AB + B^2) = A(A^2 - AB + B^2) + B(A^2 - AB + B^2)$$
$$= A^3 - A^2B + AB^2 + A^2B - AB^2 + B^3$$
$$= A^3 + B^3 \qquad \text{Combining like terms}$$

and

$$(A - B)(A^2 + AB + B^2) = A(A^2 + AB + B^2) - B(A^2 + AB + B^2)$$
$$= A^3 + A^2B + AB^2 - A^2B - AB^2 - B^3$$
$$= A^3 - B^3. \qquad \text{Combining like terms}$$

These equations (reversed) allow us to factor a sum or a difference of two cubes.

$$A^3 + B^3 = (A + B)(A^2 - AB + B^2)$$
$$A^3 - B^3 = (A - B)(A^2 + AB + B^2)$$

The table of cubes below can help in the following examples.

N	0.2	0.1	0	1	2	3	4	5	6	7	8	9	10
N^3	0.008	0.001	0	1	8	27	64	125	216	343	512	729	1000

EXAMPLE 1

Factor: $x^3 - 27$.

Solution We have

$$x^3 - 27 = x^3 - 3^3.$$

In one set of parentheses, we write the first quantity that was cubed, x, then a minus sign, and then the second quantity that was cubed, 3. This gives us the expression $x - 3$:

$$(x - 3)(\qquad).$$

To get the other factor, we think of $x - 3$ and do the following:

Square the first term: x^2.
Multiply the terms and then change the sign: $3x$.
Square the second term: $(-3)^2$, or 9.

$$(x - 3)(x^2 + 3x + 9).$$

Note that we cannot factor $x^2 + 3x + 9$. (It is not a trinomial square nor can it be factored by trial and error.) ❑

EXAMPLE 2

Factor: **(a)** $125x^3 + y^3$; **(b)** $128y^7 - 250x^6y$; **(c)** $a^6 - b^6$.

Solution

a) We have

$$125x^3 + y^3 = (5x)^3 + y^3.$$

In one set of parentheses, we write the first quantity that is cubed, $5x$, then a plus sign, and then the second quantity that is cubed, y:

$$(5x + y)(\qquad).$$

To get the other factor, we think of $5x + y$ and do the following:

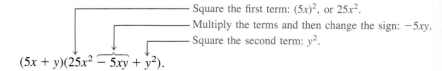
Square the first term: $(5x)^2$, or $25x^2$.
Multiply the terms and then change the sign: $-5xy$.
Square the second term: y^2.

$$(5x + y)(25x^2 - 5xy + y^2).$$

b) We have

$$128y^7 - 250x^6y = 2y(64y^6 - 125x^6) \qquad \text{Remember: } \textit{Always} \text{ look for a common factor.}$$

$$128y^7 - 250x^6y = 2y[(4y^2)^3 - (5x^2)^3] \qquad \text{Rewriting as quantities cubed}$$

$$= 2y(4y^2 - 5x^2)(16y^4 + 20x^2y^2 + 25x^4).$$

c) We have

$$a^6 - b^6 = (a^3)^2 - (b^3)^2.$$

We factor as follows:

$$(a^3 + b^3)(a^3 - b^3).$$

One factor is a sum of two cubes, and the other factor is a difference of two cubes. We factor them:

$$(a + b)(a^2 - ab + b^2)(a - b)(a^2 + ab + b^2). \qquad ❑$$

In Example 2(c), had we thought of factoring first as a difference of two cubes, we would have had

$$(a^2)^3 - (b^2)^3 = (a^2 - b^2)(a^4 + a^2b^2 + b^4)$$
$$= (a + b)(a - b)(a^4 + a^2b^2 + b^4).$$

In this case, we might have missed some factors; $a^4 + a^2b^2 + b^4$ can be factored as $(a^2 - ab + b^2)(a^2 + ab + b^2)$, but we probably would never have suspected that such a factorization exists.

EXAMPLE 3

Factor: $64a^6 - 729b^6$.

Solution   We have

$$64a^6 - 729b^6 = (8a^3 - 27b^3)(8a^3 + 27b^3) \qquad \text{Factoring a difference of squares}$$

$$= [(2a)^3 - (3b)^3][(2a)^3 + (3b)^3].$$

Each factor is a sum or a difference of cubes. We factor each:

$$= (2a - 3b)(4a^2 + 6ab + 9b^2)(2a + 3b)(4a^2 - 6ab + 9b^2). \qquad \square$$

Remember the following:

Sum of cubes:	$A^3 + B^3 = (A + B)(A^2 - AB + B^2)$;
Difference of cubes:	$A^3 - B^3 = (A - B)(A^2 + AB + B^2)$;
Difference of squares:	$A^2 - B^2 = (A + B)(A - B)$;
Sum of squares:	$A^2 + B^2$ cannot be factored using real numbers if the largest common factor has been removed.

EXERCISE SET | 5.6

Factor.

1. $x^3 + 8$

2. $c^3 + 27$

3. $y^3 - 64$

4. $z^3 - 1$

5. $w^3 + 1$

6. $x^3 + 125$

7. $8a^3 + 1$

8. $27x^3 + 1$

9. $y^3 - 8$

10. $p^3 - 27$

11. $8 - 27b^3$

12. $64 - 125x^3$

13. $64y^3 + 1$

14. $125x^3 + 1$

15. $8x^3 + 27$

16. $27y^3 + 64$

17. $a^3 - b^3$

18. $x^3 - y^3$

19. $a^3 + \frac{1}{8}$

20. $b^3 + \frac{1}{27}$

21. $2y^3 - 128$

22. $3z^3 - 3$

23. $24a^3 + 3$

24. $54x^3 + 2$

25. $rs^3 + 64r$

26. $ab^3 + 125a$

27. $5x^3 - 40z^3$

28. $2y^3 - 54z^3$

29. $x^3 + 0.001$

30. $y^3 + 0.125$

31. $64x^6 - 8t^6$

32. $125c^6 - 8d^6$

33. $2y^4 - 128y$

34. $3z^5 - 3z^2$

35. $z^6 - 1$

36. $t^6 + 1$

37. $t^6 + 64y^6$

38. $p^6 - q^6$

Skill Maintenance

39. The width of a rectangle is 7 ft less than its length. If the width is increased by 2 ft, the perimeter is then 66 ft. What is the area of the original rectangle?

Solve.

40. $|x| = 27$

41. $|5x - 6| \leqslant 39$

42. $|5x - 6| > 39$

Synthesis

43. ◈ Explain how the formula for factoring a *difference* of two cubes can be used to factor $x^3 + 8$.

44. ◈ How could you use factoring to convince someone that $x^3 + y^3 \neq (x + y)^3$?

Factor. Assume that variables in exponents represent natural numbers.

45. $x^{6a} + y^{3b}$

46. $a^3x^3 - b^3y^3$

47. $3x^{3a} + 24y^{3b}$

48. $\frac{8}{27}x^3 + \frac{1}{64}y^3$

49. $\frac{1}{24}x^3y^3 + \frac{1}{3}z^3$

50. $\frac{1}{16}x^{3a} + \frac{1}{2}y^{6a}z^{9b}$

51. $7x^3 + \frac{7}{8}$

52. $[(c - d)^3 - d^3]^2$

53. $(x + y)^3 - x^3$

54. $(1 - x)^3 + (x - 1)^6$

55. $(a + 2)^3 - (a - 2)^3$

56. $y^4 - 8y^3 - y + 8$

57. If $P(x) = x^3$, use factoring to simplify
$$P(a + h) - P(a).$$

58. If $Q(x) = x^6$, use factoring to simplify
$$Q(a + h) - Q(a).$$

59. Using one set of axes, graph each function.
  a) $f(x) = x^3$
  b) $g(x) = x^3 - 8$
  c) $h(x) = (x - 2)^3$

5.7

Factoring: A General Strategy

Mixed Factoring Problems

Factoring is an important algebraic skill. Once you know what kind of expression you have to factor, the factoring can be done without too much difficulty. Below is a general strategy for factoring.

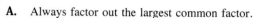

A. Always factor out the largest common factor.

B. Look at the number of terms.

Two terms: Try factoring as a difference of squares first. Next, try factoring as a sum or a difference of cubes. Do *not* try to factor a *sum* of squares.

Three terms: Try factoring as a trinomial square. Next, try trial and error, using the FOIL method or the grouping method.

Four or more terms: Try factoring by grouping and factoring out a common binomial factor. Next, try grouping into a difference of squares, one of which is a trinomial.

C. Always *factor completely*. If a factor with more than one term can itself be factored further, do so.

EXAMPLE 1

Factor: $10a^2x - 40b^2x$.

Solution

A. Look first for a common factor:

$$10x(a^2 - 4b^2).$$ Factoring out the largest common factor

B. The factor $a^2 - 4b^2$ has two terms. It is a difference of squares. We factor it, keeping the common factor:

$$10x(a + 2b)(a - 2b).$$

C. Have we factored completely? Yes, because no factor with more than one term can be factored further.   ❑

EXAMPLE 2

Factor: $x^6 - 64$.

Solution

A. Look for a common factor. There is none (other than 1 or -1).

B. There are two terms, a difference of squares: $(x^3)^2 - (8)^2$. We factor it:

$$(x^3 + 8)(x^3 - 8).$$

C. One factor is a sum of two cubes, and the other factor is a difference of two cubes. We factor them:

$$(x + 2)(x^2 - 2x + 4)(x - 2)(x^2 + 2x + 4).$$

The factorization is complete because no factor can be factored further.   ❑

EXAMPLE 3

Factor: $10x^6 + 40y^2$.

Solution

A. Factor out the largest common factor: $10(x^6 + 4y^2)$.
B. In the parentheses, there are two terms, a sum of squares, which cannot be factored.
C. We cannot factor further. ❏

EXAMPLE 4

Factor: $2x^2 + 50a^2 - 20ax$.

Solution

A. Factor out the largest common factor: $2(x^2 + 25a^2 - 10ax)$.
B. We rearrange the trinomial in descending powers of x: $2(x^2 - 10ax + 25a^2)$. The trinomial is a square that we can factor:

$$2(x - 5a)^2.$$

C. No factor with more than one term can be factored further. ❏

EXAMPLE 5

Factor: $6x^2 - 20x - 16$.

Solution

A. Factor out the largest common factor: $2(3x^2 - 10x - 8)$.
B. In the parentheses, there are three terms. The trinomial is not a square. We factor by trial: $2(x - 4)(3x + 2)$.
C. We cannot factor further. ❏

EXAMPLE 6

Factor: $3x + 12 + ax^2 + 4ax$.

Solution

A. There is no common factor (other than 1 or -1).
B. There are four terms. We try grouping to find a common binomial factor:

$3(x + 4) + ax(x + 4)$ Factoring two grouped binomials
  $= (x + 4)(3 + ax).$ Removing the common binomial factor

C. We cannot factor further. ❏

EXAMPLE 7

Factor: $y^2 - 9a^2 + 12y + 36$.

Solution

A. There is no common factor (other than 1 or -1).
B. There are four terms. We try grouping to remove a common binomial factor, but find none. Next, we try grouping as a difference of squares:

$(y^2 + 12y + 36) - 9a^2$ Grouping
  $= (y + 6)^2 - (3a)^2$ Factoring the trinomial square
  $= (y + 6 + 3a)(y + 6 - 3a).$ Factoring the difference of squares

C. No factor with more than one term can be factored further. ❏

EXAMPLE 8

Factor: $x^3 - xy^2 + x^2y - y^3$.

Solution

A. There is no common factor (other than 1 or -1).

B. There are four terms. We try grouping to remove a common binomial factor:

$$x(x^2 - y^2) + y(x^2 - y^2) \qquad \text{Factoring two grouped binomials}$$
$$= (x^2 - y^2)(x + y). \qquad \text{Removing the common binomial factor}$$

C. The factor $x^2 - y^2$ can be factored further:

$$(x + y)(x - y)(x + y). \qquad \text{Factoring a difference of squares}$$

No factor can be factored further, so we have factored completely. ❑

EXERCISE SET | 5.7

Factor completely.

1. $x^2 - 144$

2. $y^2 - 81$

3. $2x^2 + 11x + 12$

4. $8a^2 + 18a - 5$

5. $3x^4 - 12$

6. $2xy^2 - 50x$

7. $a^2 + 25 + 10a$

8. $p^2 + 64 + 16p$

9. $2x^2 - 10x - 132$

10. $3y^2 - 15y - 252$

11. $9x^2 - 25y^2$

12. $16a^2 - 81b^2$

13. $m^6 - 1$

14. $64t^6 - 1$

15. $x^2 + 6x - y^2 + 9$

16. $t^2 + 10t - p^2 + 25$

17. $250x^3 - 128y^3$

18. $27a^3 - 343b^3$

19. $8m^3 + m^6 - 20$

20. $-37x^2 + x^4 + 36$

21. $ac + cd - ab - bd$

22. $xw - yw + xz - yz$

23. $4c^2 - 4cd + d^2$

24. $70b^2 - 3ab - a^2$

25. $-7x^2 + 2x^3 + 4x - 14$

26. $9m^2 + 3m^3 + 8m + 24$

27. $2x^3 + 6x^2 - 8x - 24$

28. $3x^3 + 6x^2 - 27x - 54$

29. $16x^3 + 54y^3$

30. $250a^3 + 54b^3$

31. $36y^2 - 35 + 12y$

32. $2b - 28a^2b + 10ab$

33. $a^8 - b^8$

34. $2x^4 - 32$

35. $a^3b - 16ab^3$

36. $x^3y - 25xy^3$

37. $a(b - 2) + c(b - 2)$

38. $(x - 2)(x + 5) + (x - 2)(x + 8)$

39. $7a^4 - 14a^3 + 21a^2 - 7a$

40. $a^3 - ab^2 + a^2b - b^3$

41. $42ab + 27a^2b^2 + 8$

42. $-23xy + 20x^2y^2 + 6$

43. $8y^4 - 125y$

44. $64p^4 - p$

45. $a^2 - b^2 - 6b - 9$

46. $m^2 - n^2 - 8n - 16$

Skill Maintenance

47. There are 75 questions on a college entrance exam-ination. Two points are awarded for each correct answer, and one half point is deducted for each wrong answer. A score of 100 indicates how many correct and how many wrong answers, assuming that all questions are answered?

48. A pentagon with all five sides the same size has the same perimeter as an octagon in which all eight sides are the same size. One side of the pentagon is 2 less than 3 times the length of one side of the octagon. Find the perimeters.

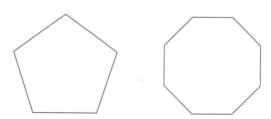

Synthesis

49. ◈ In your own words, outline a procedure that can be used to factor any polynomial.

50. ◈ Explain how one could construct a polynomi-al that is a trinomial square and contains a sum of two cubes and a difference of two cubes as factors.

Factor. Assume that variables in exponents represent natural numbers.

51. $20x^2 - 4x - 39$

52. $30y^4 - 97xy^2 + 60x^2$

53. $72a^2 + 284a - 160$

54. $3x^2y^2z + 25xyz^2 + 28z^3$

55. $7x^3 - \frac{7}{8}$

56. $-16 + 17(5 - y^2) - (5 - y^2)^2$

57. $[a^3 - (a + b)^3]$

58. $(x - p)^2 - p^2$

59. $x^4 - 50x^2 + 49$

60. $(y - 1)^4 - (y - 1)^2$

61. $27x^{6s} + 64y^{3t}$

62. $x^6 - 2x^5 + x^4 - x^2 + 2x - 1$

63. $4x^2 + 4xy + y^2 - r^2 + 6rs - 9s^2$

64. $(1 - x)^3 - (x - 1)^6$

65. $c^{2w + 1} - 2c^{w + 1} + c$

66. $24x^{2a} - 6$

67. $y^9 - y$

68. $1 - \dfrac{x^{27}}{1000}$

69. $3a^2 + 3b^2 - 3c^2 - 3d^2 + 6ab - 6cd$

70. $3(x + 1)^2 + 9(x + 1) - 12$

71. $(m - 1)^3 + (m + 1)^3$

72. $3(a - 2)^2 - 30(a - 2) + 75$

73. Suppose that $\left(x + \dfrac{2}{x}\right)^2 = 6$. Then find $x^3 + \dfrac{8}{x^3}$.

5.8

Polynomial Equations and Problem Solving

The Principle of Zero Products • **Polynomial Functions and Graphs** • **Problem Solving**

Whenever two polynomials are set equal to each other, we have a **polynomial equation.** Some examples of polynomial equations are

$$4x^3 + x^2 + 5x = 6x - 3, \qquad x^2 - x = 6, \quad \text{and} \quad 3y^4 + 2y^2 + 2 = 0.$$

The *degree of a polynomial equation* is the same as the highest degree of any term in the equation. Thus, from left to right, the degree of each equation listed above is 3, 2, and 4. A second-degree polynomial equation in one variable is often called a **quadratic equation.** Of the equations listed above, only $x^2 - x = 6$ is a quadratic equation.

Polynomial equations, and quadratic equations in particular, occur frequently in applications, so the ability to solve them is an important skill. One way of solving certain polynomial equations involves factoring.

The Principle of Zero Products

When we multiply two or more numbers, the product will be 0 if one of the factors is 0. Conversely, if a product is 0, then at least one of the factors must be 0. This property of real numbers gives us a new principle for solving equations.

The Principle of Zero Products

For any real numbers a and b:

If $ab = 0$, then $a = 0$ or $b = 0$ (or both).

If $a = 0$ or $b = 0$, then $ab = 0$.

In order to use this principle to solve a polynomial equation, we write the equation in *standard form,* with 0 on one side of the equation and the leading coefficient positive.

EXAMPLE 1

Solve: $x^2 - x = 6$.

Solution   To apply the principle of zero products, we must have 0 on one side of the equation, so we add -6 on both sides:

$$x^2 - x - 6 = 0. \qquad \text{Getting 0 on one side}$$

In order to express the polynomial as a product, we factor the polynomial:

$$(x - 3)(x + 2) = 0. \qquad \text{Factoring}$$

Since $(x - 3)(x + 2)$ is 0, the principle of zero products says that at least one factor is 0. Thus,

$$x - 3 = 0 \quad or \quad x + 2 = 0. \qquad \text{Using the principle of zero products}$$

We now have two linear equations. We solve them separately:

$$x = 3 \quad or \quad x = -2.$$

We check as follows:

Check:
$$
\begin{array}{c|c}
x^2 - x = 6 & x^2 - x = 6 \\
\hline
3^2 - 3 \; ? \; 6 & (-2)^2 - (-2) \; ? \; 6 \\
9 - 3 & 4 + 2 \\
6 \; | \; 6 \;\; \text{TRUE} & 6 \; | \; 6 \;\; \text{TRUE}
\end{array}
$$

Both 3 and -2 are solutions. The solution set is $\{3, -2\}$.   ❏

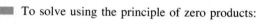

To solve using the principle of zero products:

1. Obtain a 0 on one side of the equation using the addition principle for equations.
2. Factor the nonzero side of the equation.
3. Set each factor equal to 0.
4. Solve the resulting equations.

EXAMPLE 2

Solve: $7y + 3y^2 = -2$.

Solution   Since there must be a 0 on one side of the equation, we add 2 to get 0 on one side and arrange in descending order. Then we factor and use the principle of zero products:

$$7y + 3y^2 = -2$$
$$3y^2 + 7y + 2 = 0$$
$$(3y + 1)(y + 2) = 0 \qquad \text{Factoring}$$
$$3y + 1 = 0 \quad or \quad y + 2 = 0 \qquad \text{Using the principle of zero products}$$
$$y = -\tfrac{1}{3} \quad or \quad y = -2.$$

The solutions are $-\tfrac{1}{3}$ and -2. The solution set is $\left\{-\tfrac{1}{3}, -2\right\}$.   ❏

CAUTION! When using the principle of zero products, you must make sure that there is a 0 on one side of the equation. If neither side of the equation is 0, the procedure will not work.

For example, consider $x^2 - x = 6$ in Example 1 as

$$x(x - 1) = 6.$$

Suppose we reasoned as follows, setting factors equal to 6:

$$x = 6 \quad or \quad x - 1 = 6$$
$$x = 7.$$

Neither 6 nor 7 check, as shown below:

$x(x - 1) = 6$		$x(x - 1) = 6$	
$6(6 - 1)$? 6		$7(7 - 1)$? 6	
$6(5)$		$7(6)$	
30	6 FALSE	42	6 FALSE

EXAMPLE 3

Solve the following equations.

a) $5b^2 = 10b$ **b)** $x^2 - 6x + 9 = 0$ **c)** $3x^3 - 9x^2 = 30x$

Solution

a) We have

$$5b^2 = 10b$$
$$5b^2 - 10b = 0 \qquad \text{Getting 0 on one side}$$
$$5b(b - 2) = 0 \qquad \text{Factoring}$$
$$5b = 0 \quad or \quad b - 2 = 0 \qquad \text{Using the principle of zero products}$$
$$b = 0 \quad or \quad b = 2.$$

The solutions are 0 and 2. The solution set is $\{0, 2\}$.

b) We have

$$x^2 - 6x + 9 = 0$$
$$(x - 3)(x - 3) = 0 \qquad \text{Factoring}$$
$$x - 3 = 0 \quad or \quad x - 3 = 0 \qquad \text{Using the principle of zero products}$$
$$x = 3 \quad or \quad x = 3.$$

There is only one solution, 3. The solution set is $\{3\}$.

c) We have

$$3x^3 - 9x^2 = 30x$$
$$3x^3 - 9x^2 - 30x = 0 \qquad \text{Getting 0 on one side}$$
$$3x(x^2 - 3x - 10) = 0 \qquad \text{Factoring out a common factor}$$
$$3x(x + 2)(x - 5) = 0 \qquad \text{Factoring the trinomial}$$
$$3x = 0 \quad or \quad x + 2 = 0 \quad or \quad x - 5 = 0 \qquad \text{Using the principle of zero products}$$
$$x = 0 \quad or \quad x = -2 \quad or \quad x = 5$$

The solutions are 0, -2, and 5. The solution set is $\{0, -2, 5\}$.   ❑

TECHNOLOGY
CONNECTION

A grapher allows us to solve any quadratic equation by graphing a quadratic function and zooming in on x-intercepts, if they exist. This is an important technique that will work whether the function can be factored or not. For example, let's solve the equation $x^2 - 1.83x - 6.64 = 0$ by graphing the function $f(x) = x^2 - 1.83x - 6.64$. If we use a $[-10, 10, -10, 10]$ window, the graph should resemble the following:

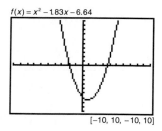

$f(x) = x^2 - 1.83x - 6.64$

$[-10, 10, -10, 10]$

There appears to be an x-intercept between -2 and -1, and another one between 3 and 4. Let's first examine the negative intercept using the window $[-2, -1, -1, 1]$, or the Zoom feature.

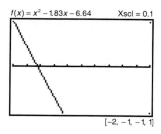

$f(x) = x^2 - 1.83x - 6.64$ Xscl = 0.1

$[-2, -1, -1, 1]$

When we use the Trace feature to check coordinates, it becomes apparent that the intercept is between -1.9 and -1.8. Changing the window dimensions to $[-1.9, -1.8, -0.1, 0.1]$, or zooming in again, we see that the intercept is very close to -1.82.

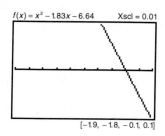

$f(x) = x^2 - 1.83x - 6.64$ Xscl = 0.01

$[-1.9, -1.8, -0.1, 0.1]$

We have determined that one zero of the function $f(x) = x^2 - 1.83x - 6.64$ is approximately -1.82. If we substitute this value into the function, we find $f(-1.82) = 0.003$, which is close enough to zero to show that this value approximates a zero, or root, of the function. Repeat the process of zooming in on the positive x-intercept to confirm that the other zero is about 3.65.

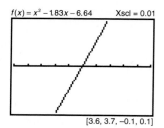

$f(x) = x^2 - 1.83x - 6.64$ Xscl = 0.01

$[3.6, 3.7, -0.1, 0.1]$

Use a grapher to find any solutions that exist accurate to two decimal places.

TC1. $x^2 + 1.14x - 3.21 = 0$

TC2. $x^2 - 14.60x + 53.29 = 0$

TC3. $-3.55x^2 + 15.73x - 22.44 = 0$

TC4. $-0.65x^2 - 8.80x - 26.62 = 0$

TC5. $x^3 - 11.22x^2 + 15.94x + 37.49 = 0$

TC6. $2.75x^3 - 18.98x^2 + 13.68x + 72.77 = 0$

Use a grapher to graph a function representing each side of the following equations. Then locate the x-values that are solutions.

TC7. $1.2x^2 + 3.25x = 4.3$

TC8. $-0.9x^2 + 4.13x = 1.87$

Polynomial Functions and Graphs

Let us return for a moment to the equation in Example 1, $x^2 - x = 6$. One way to look for a solution to this equation is to make a graph. We could either graph by hand or use computer graphics or a graphing calculator to draw the graph of the function $f(x) = x^2 - x$ and then check visually for an x-value that is paired with 6. See figure (a) below.

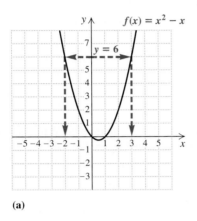

(a)

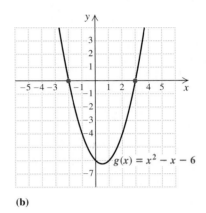

(b)

Equivalently, we could graph the function $g(x) = x^2 - x - 6$ and look for values of x for which $g(x) = 0$. See figure (b) above. Here you can visualize what we call the *roots*, or *zeros*, of a polynomial function.

It appears from the graph that $f(x) = 6$ and $g(x) = 0$ when $x \approx -2$ or $x \approx 3$. Although making a graph is not the fastest or most precise method of solving a polynomial equation, it gives us a visualization and may be useful with problems that are more difficult to solve algebraically.

Problem Solving

Some problems can be translated to quadratic equations, which we can now solve. The problem-solving process is the same as for other kinds of equations.

EXAMPLE 4

Fireworks are typically launched from a mortar with an upward velocity (initial speed) of about 64 ft/sec. The height in feet $h(t)$ of a "weeping willow" display, launched from a rooftop 80 ft above the ground, is given by

$$h(t) = -16t^2 + 64t + 80,$$

where t is the number of seconds from launch. After what amount of time will the cardboard shell from the fireworks reach the ground?

Solution

1. **FAMILIARIZE.** We draw a picture and label it, using the information provided (see the figure at the top of the following page). If we wanted to, we could evaluate $h(t)$ for a few values of t. Note that t cannot be negative, since it represents time from launch.

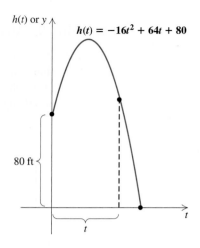

2. TRANSLATE. The relevant function has been provided. Since we are asked to determine how long it will take for the shell to hit the ground, we are interested in the value of t for which $h(t) = 0$:

$$-16t^2 + 64t + 80 = 0.$$

3. CARRY OUT. We solve by factoring:

$$-16t^2 + 64t + 80 = 0$$
$$\left.\begin{array}{l}-16(t^2 - 4t - 5) = 0 \\ -16(t - 5)(t + 1) = 0\end{array}\right\} \quad \text{Factoring}$$
$$t - 5 = 0 \quad or \quad t + 1 = 0$$
$$t = 5 \quad or \quad t = -1.$$

The solutions appear to be 5 and -1.

4. CHECK. Since t cannot be negative, we need check only 5. Thus,

$$h(5) = -16 \cdot 5^2 + 64 \cdot 5 + 80 = 0.$$

The number 5 checks.

5. STATE. The cardboard shell will hit the ground about 5 sec after the firework display has been launched. ❏

The following problem involves the **Pythagorean theorem,** which relates the lengths of the sides of a right triangle. A **right triangle** has a 90° angle. The side opposite the 90° angle is called the **hypotenuse.** The other sides are called **legs.**

The Pythagorean Theorem

The sum of the squares of the legs of a right triangle is equal to the square of the hypotenuse:

$$a^2 + b^2 = c^2.$$

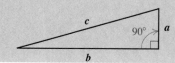

EXAMPLE 5

In order to build a deck at a right angle to their house, Lucinda and Felipe decide to plant a stake in the ground a precise distance from the back wall of their house. This stake will combine with two marks on the house to form a right triangle. From a course in geometry, Lucinda remembers that there are three consecutive integers that can work as sides of a right triangle. Find the measurements of that triangle.

Solution

1. FAMILIARIZE. We make a drawing and let

   x = the distance between the marks on the house.

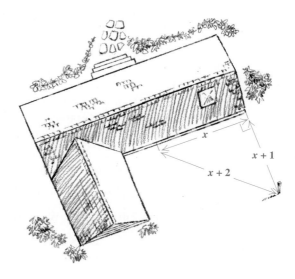

Since the lengths are consecutive integers and the hypotenuse is the longest length, we know that

   $x + 1$ = the length of the other leg

and

   $x + 2$ = the length of the hypotenuse.

2. TRANSLATE. Applying the Pythagorean theorem, we obtain the following translation:

$$a^2 + b^2 = c^2$$
$$x^2 + (x + 1)^2 = (x + 2)^2.$$

3. CARRY OUT. We solve the equation as follows:

$x^2 + (x^2 + 2x + 1) = x^2 + 4x + 4$	Squaring the binomials
$2x^2 + 2x + 1 = x^2 + 4x + 4$	Collecting like terms
$x^2 - 2x - 3 = 0$	Adding $-x^2$ and $-4x$ and -4 on both sides
$(x - 3)(x + 1) = 0$	Factoring
$x - 3 = 0 \quad or \quad x + 1 = 0$	Using the principle of zero products
$x = 3 \quad or \qquad x = -1.$	

4. **CHECK.** The integer -1 cannot be a length of a side because it is negative. When $x = 3$, $x + 1 = 4$, and $x + 2 = 5$; and $3^2 + 4^2 = 5^2$. So 3, 4, and 5 check.

5. **STATE.** Lucinda and Felipe should use a triangle with sides having a ratio of 3:4:5. Thus, if the marks on the house are 3 ft apart, they should locate the stake at the point in the yard that is precisely 4 ft from one mark and 5 ft from the other mark. ❑

EXAMPLE 6

Dovetail Woodworking determines that the revenue R, in thousands of dollars, from the sale of x sets of cabinets is given by $R(x) = 2x^2 + x$. If the cost C, in thousands of dollars, of producing x sets of cabinets is given by $C(x) = x^2 - 2x + 10$, how many sets must be produced and sold in order for the company to break even?

Solution

1. **FAMILIARIZE.** We recall that in order for a firm to break even, its revenue must equal its cost of production. Alternatively, we may recall that a firm breaks even when its profits are 0.

   We let $x =$ the number of sets of cabinets produced and sold.

2. **TRANSLATE.** Since revenue must *equal* cost in order for the firm to break even, we form the equation

   $$2x^2 + x = x^2 - 2x + 10.$$

3. **CARRY OUT.** We solve the equation:

   $$2x^2 + x = x^2 - 2x + 10$$
   $$x^2 + 3x - 10 = 0 \qquad \text{Adding } -x^2 \text{ and } 2x \text{ and } -10 \text{ on both sides}$$
   $$(x + 5)(x - 2) = 0 \qquad \text{Factoring}$$
   $$x + 5 = 0 \quad or \quad x - 2 = 0 \qquad \text{Using the principle of zero products}$$
   $$x = -5 \quad or \quad x = 2.$$

4. **CHECK.** We check the possible solutions in the original problem. The number -5 is not a solution since the number of sets cannot be negative. If 2 sets of cabinets are produced, the revenue is $R(2) = 2 \cdot 2^2 + 2 = 10$ thousand dollars, and the cost is $C(2) = 2^2 - 2 \cdot 2 + 10 = 10$ thousand dollars. We have a solution.

5. **STATE.** Dovetail Woodworking breaks even when 2 sets of cabinets are produced and sold. ❑

EXAMPLE 7

A valuable stamp is 4 cm wide and 5 cm long. The stamp is to be mounted on a sheet of paper that is $5\frac{1}{2}$ times the area of the stamp. Determine the dimensions of the paper that will ensure a uniform border around the stamp.

Solution

1. **FAMILIARIZE.** We make a drawing and label it with both known and unknown information (see the figure at the top of the following page). We let x represent the width of the border. Since the border extends uniformly around the entire stamp, the length of the sheet of paper is given by $5 + 2x$ and the width is given by $4 + 2x$.

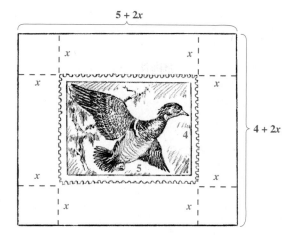

2. TRANSLATE. We rephrase the information given and translate as follows:

Area of sheet is $5\frac{1}{2}$ times area of stamp.

$$(5 + 2x)(4 + 2x) = 5\frac{1}{2} \cdot 5 \cdot 4$$

3. CARRY OUT. We solve the equation:

$(5 + 2x)(4 + 2x) = 5\frac{1}{2} \cdot 5 \cdot 4$	
$20 + 10x + 8x + 4x^2 = 110$	Multiplying
$4x^2 + 18x - 90 = 0$	Finding standard form
$2x^2 + 9x - 45 = 0$	Multiplying by $\frac{1}{2}$ on both sides
$(2x + 15)(x - 3) = 0$	Factoring
$2x + 15 = 0 \quad or \quad x - 3 = 0$	Principle of zero products
$x = -7\frac{1}{2} \quad or \quad x = 3.$	

4. CHECK. We now check in the original problem. We see that $-7\frac{1}{2}$ is not a solution because the width of the border cannot be negative.

Let's see whether 3 checks. If the border is 3 cm wide, the paper will have a length of $5 + 2 \cdot 3$, or 11 cm. It will have a width of $4 + 2 \cdot 3$, or 10 cm. The area of the sheet of paper is thus $11 \cdot 10$, or 110 cm^2. Since the area of the stamp is 20 cm^2 and 110 cm^2 is $5\frac{1}{2}$ times 20 cm^2, the number 3 checks.

5. STATE. The sheet of paper should be 11 cm long and 10 cm wide. ❑

EXERCISE SET | 5.8

Solve.

1. $x^2 + 3x = 28$

2. $y^2 - 4y = 45$

3. $y^2 + 16 = 8y$

4. $r^2 + 1 = 2r$

5. $x^2 - 12x + 36 = 0$

6. $y^2 + 16y + 64 = 0$

7. $9x + x^2 + 20 = 0$

8. $8y + y^2 + 15 = 0$

9. $x^2 + 8x = 0$

10. $t^2 + 9t = 0$

11. $x^2 - 9 = 0$

12. $p^2 - 16 = 0$

13. $z^2 = 36$

14. $y^2 = 81$

15. $x^2 + 14x + 45 = 0$

16. $y^2 + 12y + 32 = 0$

17. $y^2 + 2y = 63$

18. $a^2 + 3a = 40$

19. $p^2 - 11p = -28$

20. $x^2 - 14x = -45$

21. $32 + 4x - x^2 = 0$

22. $27 + 12t + t^2 = 0$

23. $3b^2 + 8b + 4 = 0$

24. $9y^2 + 15y + 4 = 0$

25. $8y^2 - 10y + 3 = 0$

26. $4x^2 + 11x + 6 = 0$

27. $6z - z^2 = 0$

28. $8y - y^2 = 0$

29. $12z^2 + z = 6$

30. $6x^2 - 7x = 10$

31. $5x^2 - 20 = 0$

32. $6y^2 - 54 = 0$

33. $2x^2 - 15x = -7$

34. $x^2 - 9x = -8$

35. $21r^2 + r - 10 = 0$

36. $12a^2 - 5a - 28 = 0$

37. $15y^2 = 3y$

38. $18x^2 = 9x$

39. $x^2 - \frac{1}{25} = 0$

40. $y^2 - \frac{1}{64} = 0$

41. $2x^3 - 2x^2 = 12x$

42. $50y + 5y^3 = 35y^2$

Problem Solving

43. The square of a number plus the number is 132. What is the number?

44. The square of a number plus the number is 156. What is the number?

45. A book is 5 cm longer than it is wide. Find the length and the width if the area is 84 cm².

46. An envelope is 4 cm longer than it is wide. The area is 96 cm². Find the length and the width.

47. If each of the sides of a square is lengthened by 4 cm, the area becomes 49 cm². Find the length of a side of the original square.

48. If each of the sides of a square is lengthened by 6 m, the area becomes 144 m². Find the length of a side of the original square.

49. A picture frame measures 12 cm by 20 cm, and 84 cm² of picture shows. Find the width of the frame.

50. A picture frame measures 14 cm by 20 cm, and 160 cm² of picture shows. Find the width of the frame.

51. The width of a rectangle is 5 m less than the length. The area is 24 m². Find the length and the width.

52. The width of a rectangle is 4 m less than the length. The area is 12 m². Find the length and the width.

53. A rectangular lawn measures 60 ft by 80 ft. Part of the lawn is torn up to install a sidewalk of uniform width around it. The area of the new lawn is one-sixth the area of the old. How wide is the sidewalk?

54. A rectangular garden is 30 ft by 40 ft. Part of the garden is removed in order to install a walkway of uniform width around it. The area of the new garden is one-half the area of the old garden. How wide is the walkway?

55. A wire is stretched from the ground to the top of an antenna tower, as shown. The wire is 20 ft long. The height of the tower is 4 ft greater than the distance d from the tower's base to the end of the wire. Find the distance d and the height of the tower.

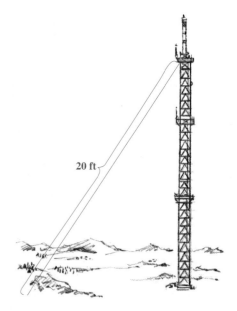

20 ft

56. The length of a rectangular parking lot is 50 ft longer than the width. Determine the dimensions of the parking lot if the parking lot measures 250 ft diagonally.

57. The base of a triangular sail is 9 m longer than the height. The area is 56 m². Find the height and the base of the sail.

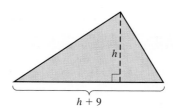

h

$h + 9$

58. The triangular entrance to a tent is 2 ft taller than it is wide. The area is 12 ft². Find the height and the base.

59. Three consecutive even integers are such that the square of the first plus the square of the third is 136. Find the three integers.

60. Three consecutive even integers are such that the square of the third is 76 more than the square of the second. Find the three integers.

61. The foot of an extension ladder is 9 ft from a wall. The height that the ladder reaches on the wall and the length of the ladder are consecutive integers. How long is the ladder?

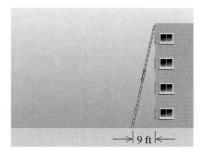

62. The foot of an extension ladder is 10 ft from a wall. The ladder is 2 ft longer than the height that it reaches on the wall. How far up the wall does the ladder reach?

63. Ignacio is planning a garden that is 25 m longer than it is wide. The garden will have an area of 7500 m². What will its dimensions be?

64. A flower bed is to be 3 m longer than it is wide. The flower bed will have an area of 108 m². What will its dimensions be?

65. The sum of the squares of two consecutive odd positive integers is 202. Find the integers.

66. The sum of the squares of two consecutive odd positive integers is 394. Find the integers.

67. Suppose that a flare is launched upward with an initial velocity of 80 ft/sec from a height of 224 ft. Its height after t seconds is a function h given by

$$h(t) = -16t^2 + 80t + 224.$$

After how long will the flare reach the ground?

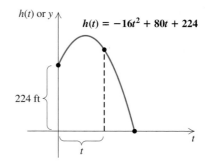

68. Suppose that a bottle rocket is launched upward with an initial velocity of 96 ft/sec from a height of 880 ft. Its height after t seconds is a function h given by

$$h(t) = -16t^2 + 96t + 880.$$

After how long will the rocket reach the ground?

69. The revenue R from the sale of x clocks is given by $R(x) = \frac{4}{9}x^2 + x$, where $R(x)$ is in hundreds of dollars. If the cost of producing x clocks is given by $C(x) = \frac{1}{3}x^2 + x + 1$, where $C(x)$ is in hundreds of dollars, how many clocks must be produced and sold in order for the firm to break even?

70. Suppose that the cost C of making x video cameras is given by $C(x) = \frac{1}{9}x^2 + 2x + 1$, where $C(x)$ is in thousands of dollars. If the revenue from the sale of x video cameras is given by $R(x) = \frac{5}{36}x^2 + 2x$, where $R(x)$ is in thousands of dollars, how many cameras must be sold in order for the firm to break even?

Skill Maintenance

71. At noon, two cars start from the same location going in opposite directions at different speeds. After 7 hr, they are 651 mi apart. If one car is traveling 15 mph slower than the other car, what are their respective speeds?

72. At the beginning of the month, an appliance store had 150 televisions in stock. During the month, they sold 45% of their color televisions and 60% of their black-and-white televisions. If they sold a total of 78 televisions, how many of each type did they sell?

Solve.

73. $2x - 14 + 9x > -8x + 16 + 10x$

74. $x + y = 0,$
$z - y = -2,$
$x - z = 6$

Synthesis ─────────────────────

Solve.

75. $(3x^2 - 7x - 20)(x - 5) = 0$

76. $(8x + 11)(12x^2 - 5x - 2) = 0$

77. $(x + 1)^3 = (x - 1)^3 + 26$

78. $(x - 2)^3 = x^3 - 2$

79. $3x^3 + 6x^2 - 27x - 54 = 0$

80. $2x^3 + 6x^2 = 8x + 24$

81. A square and an equilateral triangle have the same perimeter. The area of the square is 9 cm^2. What is the length of a side of the triangle?

82. The sum of two numbers is 17, and the sum of their squares is 205. Find the numbers.

83. A tugboat and a freighter leave the same port at the same time at right angles. The freighter travels 7 km/h slower than the tugboat. After 4 hr, they are 68 km apart. Find the speed of each boat.

84. A rectangular piece of tin is twice as long as it is wide. Squares 2 cm on a side are cut out of each corner, and the ends are turned up to make a box whose volume is 480 cm^3. What are the dimensions of the piece of tin?

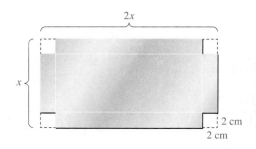

In Exercises 85–90, use a grapher to find any solutions that exist accurate to two decimal places.

85. $x^2 - 2.41x - 8.25 = 0$

86. $-x^2 - 2.65x - 3.25 = 0$

87. $-x^2 + 13.80x = 47.61$

88. $-x^2 + 3.63x + 34.34 = x^2$

89. $x^3 - 3.48x^2 + x = 3.48$

90. $-x^3 + 3.60x^2 + 8.05x - 8.23 = x^3$

SUMMARY AND REVIEW | 5

KEY TERMS

Polynomial, p. 218
Term, p. 218
Monomial, p. 218
Coefficient, p. 218
Degree, p. 218
Leading term, p. 218
Leading coefficient,
  p. 218
Linear, p. 219
Quadratic, p. 219
Cubic, p. 219
Binomial, p. 219
Trinomial, p. 219
Descending order,
  p. 219

Ascending order, p. 219
Similar, or like terms,
  p. 220
FOIL, p. 228
Square of a binomial,
  p. 229
Factor, p. 234
Factored completely,
  p. 234
Prime polynomial,
  p. 234
Factoring by grouping,
  p. 236
Trinomial square,
  p. 248

Difference of squares,
  p. 250
Sum or difference of
  cubes, p. 253
Polynomial equation,
  p. 259
Quadratic equation,
  p. 259
Standard form, p. 260
Root, or zero, p. 263
Pythagorean theorem,
  p. 264
Right triangle, p. 264
Hypotenuse, p. 264
Leg, p. 264

IMPORTANT PROPERTIES AND FORMULAS

Factoring Formulas

$$A^2 - B^2 = (A + B)(A - B);$$
$$A^2 + 2AB + B^2 = (A + B)^2;$$
$$A^2 - 2AB + B^2 = (A - B)^2;$$
$$A^3 + B^3 = (A + B)(A^2 - AB + B^2);$$
$$A^3 - B^3 = (A - B)(A^2 + AB + B^2)$$

To factor a polynomial:

A. Always factor out the largest common factor.

B. Look at the number of terms.

Two terms: Try factoring as a difference of squares first. Next, try factoring as a sum or a difference of cubes. Do *not* try to factor a *sum* of squares.

Three terms: Try factoring as a trinomial square. Next, try trial and error, using the FOIL method or the grouping method.

Four or more terms: Try factoring by grouping and factoring out a common binomial factor. Next, try grouping into a difference of squares, one of which is a trinomial.

C. Always *factor completely*. If a factor with more than one term can itself be factored further, do so.

The Principle of Zero Products

For any real numbers a and b:

If $ab = 0$, then $a = 0$ or $b = 0$ (or both).

If $a = 0$ or $b = 0$, then $ab = 0$.

REVIEW EXERCISES

This chapter's review and test include Skill Maintenance exercises from Sections 3.4, 3.5, 4.1, and 4.3.

1. Find $P(0)$ and $P(-1)$:

$$P(x) = x^3 - x^2 + 4x.$$

2. Evaluate the polynomial function for $x = -2$:

$$P(x) = 4 - 2x - x^2.$$

3. Given the polynomial

$$3x^6y - 7x^8y^3 + 2x^3 - 3x^2,$$

determine the degree of each term and the degree of the polynomial.

4. Given the polynomial

$$4x - 5x^3 + 2x^2 - 7,$$

determine the leading term and the leading coefficient.

5. Arrange in ascending powers of x:
$$3x^6y - 7x^8y^3 + 2x^3 - 3x^2.$$

6. Arrange in descending powers of y:
$$3x^6y - 7x^8y^3 + 2x^3 - 3x^2.$$

Collect like terms.

7. $4x^2y - 3xy^2 - 5x^2y + xy^2$

8. $3ab - 10 + 5ab^2 - 2ab + 7ab^2 + 14$

Add.

9. $(-6x^3 - 4x^2 + 3x + 1) + (5x^3 + 2x + 6x^2 + 1)$

10. $(3x^4 + 3x^3 - 8x + 9) + (-6x^4 + 4x + 7 + 3x)$

11. $(4x^3 - 2x^2 - 7x + 5) + (8x^2 - 3x^3 - 9 + 6x)$

12. $(-9xy^2 - xy + 6x^2y) + (-5x^2y - xy + 4xy^2) + (12x^2y - 3xy^2 + 6xy)$

Subtract.

13. $(3x - 5) - (-6x + 2)$

14. $(4a - b + 3c) - (6a - 7b - 4c)$

15. $(9p^2 - 4p + 4) - (-7p^2 + 4p + 4)$

16. $(8x^2 - 4xy + y^2) - (2x^2 + 3xy - 2y^2)$

Multiply.

17. $(3x^2y)(-6xy^3)$

18. $(x^4 - 2x^2 + 3)(x^4 + x^2 - 1)$

19. $(4ab + 3c)(2ab - c)$

20. $(2x + 5y)(2x - 5y)$

21. $(2x - 5y)^2$

22. $(5x^2 - 7x + 3)(4x^2 + 2x - 9)$

23. $(x^2 + 4y^3)^2$

24. $(x - 5)(x^2 + 5x + 25)$

25. $\left(x - \frac{1}{3}\right)\left(x - \frac{1}{6}\right)$

Factor.

26. $6x^2 + 5x$

27. $9y^4 - 3y^2$

28. $15x^4 - 18x^3 + 21x^2 - 9x$

29. $a^2 - 12a + 27$

30. $3m^2 + 14m + 8$

31. $25x^2 + 20x + 4$

32. $4y^2 - 16$

33. $a^2 - 81$

34. $ax + 2bx - ay - 2by$

35. $3y^3 + 6y^2 - 5y - 10$

36. $a^4 - 81$

37. $4x^4 + 4x^2 + 20$

38. $27x^3 - 8$

39. $0.064b^3 - 0.125c^3$

40. $y^5 - y$

41. $2z^8 - 16z^6$

42. $54x^6y - 2y$

43. $1 + a^3$

44. $36x^2 - 120x + 100$

45. $6t^2 + 17pt + 5p^2$

46. $x^3 + 2x^2 - 9x - 18$

47. $a^2 - 2ab + b^2 - 4t^2$

Solve.

48. $x^2 - 20x = -100$

49. $6b^2 - 13b + 6 = 0$

50. $8y^2 + 5 = 14y$

51. The area of a square is 5 more than 4 times the length of a side. What is the length of a side of the square?

52. The sum of the squares of three consecutive odd numbers is 83. Find the numbers.

53. A photograph is 3 in. longer than it is wide. When a 2-in. border is placed around the photograph, the total area of the photograph and the border is 108 in². Find the dimensions of the photograph.

54. Given $P(x) = x^2 + 3x$, find $P(a + h)$.

Skill Maintenance

Solve.

55. $3x + 2y + z = 3,$
$2x - y + 2z = 16,$
$x + y - z = -9$

56. $-19x + 10 + 15x > 2x - 4 - 12x$

57. $|10 - 3x| \le 14$

58. $|10 - 3x| \ge 14$

59. There are three machines A, B, and C in a factory. When all three work, they produce 287 screws per hour. When only A and C work, they produce 197 screws per hour. When only A and B work, they produce 202 screws per hour. How many screws per hour can each produce alone?

Synthesis

60. ◈ Explain how to find the roots of a polynomial function from its graph.

61. ◈ Explain in your own words why there must be a 0 on one side of an equation before you can use the principle of zero products.

Factor.

62. $128x^6 - 2y^6$

63. $(x - 1)^3 - (x + 1)^3$

Multiply.

64. $[a - (b - 1)][(b - 1)^2 + a(b - 1) + a^2]$

65. $(z^{n^2})^{n^3}(z^{4n^3})^{n^2}$

66. Solve: $64x^3 = x$.

CHAPTER TEST   5

1. Given $P(x) = 2x^3 + 3x^2 - x + 4$, find $P(0)$ and $P(-2)$.

2. Given $P(x) = x^2 - 5x$, find and simplify
$$P(a + h) - P(a).$$

Given the polynomial $3xy^3 - 4x^2y + 5x^5y^4 - 2x^4y$.

3. Determine the degree of the polynomial.

4. Arrange in descending powers of x.

5. Determine the leading term of the polynomial $8a - 2 + a^2 - 4a^3$.

6. Collect like terms:
$$5xy - 2xy^2 - 2xy + 5xy^2.$$

Add.

7. $(-6x^3 + 3x^2 - 4y) + (3x^3 - 2y - 7y^2)$

8. $(4a^3 - 2a^2 + 6a - 5) + (3a^3 - 3a + 2 - 4a^2)$

9. $(5m^3 - 4m^2n - 6mn^2 - 3n^3) + (9mn^2 - 4n^3 + 2m^3 + 6m^2n)$

Subtract.

10. $(9a - 4b) - (3a + 4b)$

11. $(4x^2 - 3x + 7) - (-3x^2 + 4x - 6)$

12. $(6y^2 - 2y - 5y^3) - (4y^2 - 7y - 6y^3)$

Multiply.

13. $(-4x^2y)(-16xy^2)$

14. $(6a - 5b)(2a + b)$

15. $(x - y)(x^2 - xy - y^2)$

16. $(3m^2 + 4m - 2)(-m^2 - 3m + 5)$

17. $(4y - 9)^2$

18. $(x - 2y)(x + 2y)$

Factor.

19. $9x^2 + 7x$

20. $24y^3 + 16y^2$

21. $y^3 + 5y^2 - 4y - 20$

22. $p^2 - 12p - 28$

23. $12m^2 + 20m + 3$

24. $9y^2 - 25$

25. $3r^3 - 3$

26. $9x^2 + 25 - 30x$

27. $(z + 1)^2 - b^2$

28. $x^8 - y^8$

29. $y^2 + 8y + 16 - 100t^2$

30. $20a^2 - 5b^2$

31. $24x^2 - 46x + 10$

32. $16a^7b + 54ab^7$

Solve.

33. $x^2 - 18 = 3x$

34. $5y^2 - 125 = 0$

35. $2x^2 + 21 = -17x$

36. A photograph is 3 cm longer than it is wide. Its area is 40 cm^2. Find its length and its width.

Skill Maintenance

Solve.

37. $|3x + 8| < 10$

38. $|3x + 8| > 10$

39. $-3x + 4 - 5x > 8 - 9x - 3$

40. $2x - y + z = 9,$
$x - y + z = 4,$
$x + 2y - z = 5$

41. There are 70 questions on a test. The questions are either multiple-choice, true–false, or fill-in. There are twice as many true–false as fill-in and five more multiple-choice than true–false. How many of each type of question are there on the test?

Synthesis

42. **a)** Multiply: $(x^2 + x + 1)(x^3 - x^2 + 1)$.
    b) Factor: $x^5 + x + 1$.

43. Factor: $6x^{2n} - 7x^n - 20$.

CHAPTER 6

Rational Expressions, Equations, and Functions

The formula

$$\frac{1}{R} = \frac{1}{r_1} + \frac{1}{r_2}$$

gives the resistance R of two resistors r_1 and r_2 connected in parallel. Solve for r_2.

This problem appears in Example 4 of Section 6.8.

Michael Holzhausen
ELECTRICIAN

"All electrician training programs are rich in math, and provide an important background for electrical work. Electricians regularly use math when laying out work, bending conduit, sizing wire, and more."

A rational expression is an expression that indicates division, as the fractional symbols in arithmetic do. In this chapter, you will learn to add, subtract, multiply, and divide rational expressions, as well as to use them in equations and functions. Then we will use rational expressions to solve problems that we could not have solved before.

In addition to material from this chapter, the review and test for Chapter 6 include material from Sections 2.2, 4.4, 5.7, and 5.8.

6.1

Rational Expressions: Multiplying and Dividing

Rational Functions • Multiplying • Simplifying Rational Expressions • Dividing and Simplifying

An expression that is a quotient of two polynomials is called a **rational expression.** Whereas a rational number can be expressed as a quotient of two integers, a rational expression is expressed as a quotient, or ratio, of two polynomials. The following are rational expressions:

$$\frac{7}{8}, \quad \frac{a}{b}, \quad \frac{8}{y+5}, \quad \frac{x^2 + 7xy - 4}{x^3 - y^3}, \quad \frac{1 + z^3}{1 - z^6}.$$

Rational Functions

Like polynomials, rational expressions can be used to define or describe functions. A function that can be described by a rational expression is known as a **rational function.**

EXAMPLE 1

The function

$$T(t) = \frac{t^2 + 5t}{2t + 5}$$

gives the time required for two machines, working together, to complete a job that the first machine could do alone in t hours and the other machine could do in $t + 5$ hours. How long will the two machines, working together, require for the job if the first machine alone would take **(a)** 1 hour? **(b)** 5 hours?

Solution

a) $T(1) = \dfrac{1^2 + 5 \cdot 1}{2 \cdot 1 + 5} = \dfrac{1 + 5}{2 + 5} = \dfrac{6}{7}$ hr

b) $T(5) = \dfrac{5^2 + 5 \cdot 5}{2 \cdot 5 + 5} = \dfrac{25 + 25}{10 + 5} = \dfrac{50}{15} = \dfrac{10}{3}$ hr ☐

In some rational expressions, certain substitutions are impossible, since division by 0 is undefined. Thus, as was the case in Section 2.6, the domain of a rational function may exclude certain numbers.

EXAMPLE 2

Find the domain of the function given by

$$g(x) = \frac{x^2 + 5x + 6}{2x - 3}.$$

Solution   To avoid division by 0, we must determine which x-value causes $2x - 3$ to be 0. We will then exclude that value from the domain of g. We set

$$2x - 3 = 0$$
$$2x = 3 \qquad \text{Adding 3 on both sides}$$
$$x = \tfrac{3}{2}. \qquad \text{Multiplying by } \tfrac{1}{2}$$

Thus the domain of $g = \left\{x \mid x \text{ is a real number and } x \neq \tfrac{3}{2}\right\}$. ❏

EXAMPLE 3

Find the domain of the function given by

$$f(x) = \frac{x^2 - 4}{x^2 - 3x - 28}.$$

TECHNOLOGY
CONNECTION

Determining the domain of a rational function is a relatively simple matter on a grapher. Consider the function that was used in Example 3:

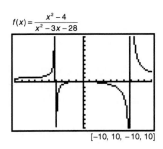

$f(x) = \dfrac{x^2 - 4}{x^2 - 3x - 28}$

[−10, 10, −10, 10]

Note the pair of vertical lines near $x = -4$ and $x = 7$. We know that these are the two values for which the function is not defined. This means there should be *nothing* at these two values. However, graphers usually try to connect the dots to produce a smoother looking graph. When those dots are off the top and bottom of the screen, the result is often the vertical lines that you see in this graph. Some graphers allow you to select whether or not you want the dots connected. If you select the *dot* mode instead of the *connected* mode, you would see the graph at the top of the next column. Note that the vertical lines no longer appear. By zooming in repeatedly or adjusting the size of the window, we see that the x-values −4.00 and 7.00 are not in the domain.

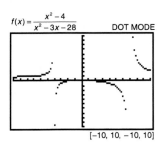

$f(x) = \dfrac{x^2 - 4}{x^2 - 3x - 28}$ DOT MODE

[−10, 10, −10, 10]

Use a grapher to determine the domain of each of the following rational functions. Round all values to two decimal places. (*Note:* Use parentheses carefully when writing rational expressions on a grapher. For example, $\dfrac{x + 1}{x - 1}$ must be expressed as $(x + 1)/(x - 1)$.)

TC1. $f(x) = \dfrac{x^2 + 7.55}{x^2 - 12.25}$

TC2. $f(x) = \dfrac{9.5x}{x^2 - 1.6x - 14.57}$

TC3. $f(x) = \dfrac{8.7}{x^2 - 4x + 3.94}$

TC4. $f(x) = \dfrac{4x^2 - 24.8}{5.7x - 12.35}$

Solution   We set

$$x^2 - 3x - 28 = 0$$

$$(x - 7)(x + 4) = 0 \qquad \text{Factoring the trinomial}$$

$$x - 7 = 0 \quad or \quad x + 4 = 0 \qquad \begin{array}{l}\text{Using the principle of} \\ \text{zero products}\end{array}$$

$$x = 7 \quad or \qquad x = -4.$$

Thus the domain of $f = \{x \mid x$ is a real number and $x \neq 7$ and $x \neq -4\}$.   ❑

Multiplying

Most of the calculations that we do with rational expressions are very much like the calculations that we do with fractional notation in arithmetic. By proceeding as in arithmetic, we can be certain that when one rational expression is obtained from another, the two expressions are equivalent.

To multiply two rational expressions, multiply numerators and multiply denominators:

$$\frac{A}{B} \cdot \frac{C}{D} = \frac{AC}{BD}, \quad \text{where } B \neq 0, D \neq 0.$$

For example,

$$\frac{x + 3}{y - 4} \cdot \frac{x^3}{y + 5} = \frac{(x + 3)x^3}{(y - 4)(y + 5)}. \qquad \begin{array}{l}\text{Multiplying the numerators and} \\ \text{multiplying the denominators}\end{array}$$

Note that we do not carry out the multiplications, because it is easier to simplify if we do not. In order to learn to simplify, we must first consider multiplying by 1. Recall that 1 is the identity for multiplication — multiplying any number a by 1 gives the same number a.

Any rational expression with the same numerator and denominator names the number 1:

$$\frac{y + 5}{y + 5}, \qquad \frac{4x^2 - 5}{4x^2 - 5}, \qquad \frac{-1}{-1}. \qquad \text{All name the number 1.}$$

We can multiply by 1 to get equivalent expressions. For example, let us multiply $(x + y)/5$ by 1:

$$\frac{x + y}{5} \cdot \frac{x - y}{x - y} = \frac{(x + y)(x - y)}{5(x - y)}. \qquad \text{Multiplying by } \frac{x - y}{x - y}, \text{ which is 1}$$

We know that

$$\frac{x + y}{5} \quad \text{and} \quad \frac{(x + y)(x - y)}{5(x - y)}$$

are equivalent. This means that they will name the same number for all replacements that do not make a denominator 0.

EXAMPLE 4

Multiply to obtain equivalent expressions.

a) $\dfrac{x^2 + 3}{x - 1} \cdot \dfrac{x + 1}{x + 1}$

b) $\dfrac{-1}{-1} \cdot \dfrac{x - 4}{x - y}$

Solution

a) $\dfrac{x^2 + 3}{x - 1} \cdot \dfrac{x + 1}{x + 1} = \dfrac{(x^2 + 3)(x + 1)}{(x - 1)(x + 1)}$ Multiplying by $\dfrac{x + 1}{x + 1}$, which is 1

b) $\dfrac{-1}{-1} \cdot \dfrac{x - 4}{x - y} = \dfrac{-1 \cdot (x - 4)}{-1 \cdot (x - y)}$ Multiplying by $\dfrac{-1}{-1}$, which is 1   ❑

Simplifying Rational Expressions

We can simplify rational expressions by reversing the procedure of multiplying by 1. To "remove" a factor of 1, we first factor the numerator and the denominator. Then we factor the rational expression, finding a factor that is equal to 1. Use the largest common factor of the numerator and the denominator.

EXAMPLE 5

Simplify by removing factors equal to 1.

a) $\dfrac{5x^2}{x}$

b) $\dfrac{4a + 8}{2}$

c) $\dfrac{2x^2 + 4x}{6x^2 + 2x}$

d) $\dfrac{x^2 - 4}{2x^2 - 3x - 2}$

e) $\dfrac{9x^2 + 6xy - 3y^2}{12x^2 - 12y^2}$

Solution

a) $\dfrac{5x^2}{x} = \dfrac{5x \cdot x}{1 \cdot x}$ Factoring the numerator and the denominator

$= \dfrac{5x}{1} \cdot \dfrac{x}{x}$ Factoring the rational expression

$= 5x \cdot 1$ $\dfrac{x}{x} = 1$ for $x \neq 0$

$= 5x$ Removing a factor of 1

In this example, we wrote a 1 in the denominator. This can always be done, if necessary.

b) $\dfrac{4a + 8}{2} = \dfrac{2(2a + 4)}{2 \cdot 1}$ Factoring the numerator and the denominator

$= \dfrac{2}{2} \cdot \dfrac{2a + 4}{1}$ Factoring the rational expression; $\dfrac{2}{2} = 1$

$= \dfrac{2a + 4}{1}$ Removing a factor of 1

$= 2a + 4$

c) $\dfrac{2x^2 + 4x}{6x^2 + 2x} = \dfrac{2x(x + 2)}{2x(3x + 1)}$ Factoring the numerator and the denominator

$= \dfrac{2x}{2x} \cdot \dfrac{x + 2}{3x + 1}$ Factoring the rational expression; $\dfrac{2x}{2x} = 1$ for $x \neq 0$

$= \dfrac{x + 2}{3x + 1}$ Removing a factor of 1

d) $\dfrac{x^2 - 4}{2x^2 - 3x - 2} = \dfrac{(x - 2)(x + 2)}{(2x + 1)(x - 2)}$ Factoring the numerator and the denominator

$= \dfrac{x - 2}{x - 2} \cdot \dfrac{x + 2}{2x + 1}$ Factoring the rational expression; $\dfrac{x - 2}{x - 2} = 1$ for $x \neq 2$

$= \dfrac{x + 2}{2x + 1}$ Removing a factor of 1

e) $\dfrac{9x^2 + 6xy - 3y^2}{12x^2 - 12y^2} = \dfrac{3(x + y)(3x - y)}{12(x + y)(x - y)}$ Factoring the numerator and the denominator

$= \dfrac{3(x + y)}{3(x + y)} \cdot \dfrac{3x - y}{4(x - y)}$ Factoring the rational expression

$= \dfrac{3x - y}{4(x - y)}$ Removing a factor of 1

For purposes of later work, we usually do not multiply out the numerator and the denominator. ❏

Canceling

Canceling is a shortcut that you may have used for removing a factor of 1 when working with fractional notation or rational expressions. With great concern, we mention it here as a possible way to speed up your work. Canceling can be done to remove factors of 1 in products. It *cannot* be done in sums or when adding expressions together. Our concern is that canceling be done with care and understanding. Example 5(e) might have been done faster as follows:

$\dfrac{9x^2 + 6xy - 3y^2}{12x^2 - 12y^2} = \dfrac{3(x + y)(3x - y)}{3 \cdot 4(x + y)(x - y)}$ When a factor of 1 is noted, it is "canceled" as shown.

$= \dfrac{3x - y}{4(x - y)}.$ Removing a factor of 1: $\dfrac{3(x + y)}{3(x + y)} = 1$

CAUTION! Canceling is often performed incorrectly in situations such as the following:

$\dfrac{x + 3}{x} = 3,$ $\dfrac{4a + 3}{2} = 2a + 3,$ $\dfrac{5}{5 + x} = \dfrac{1}{x}$

Wrong!   Wrong!   Wrong!

In each of these situations, the expressions canceled are *not* factors of 1. Factors are parts of products. For example, in $x \cdot 3$, x and 3 are factors, but in $x + 3$, x and 3 are *not* factors. **If you can't factor, you can't cancel!** If in doubt, don't cancel!

After multiplying, we ordinarily simplify, if possible. That is why we leave the numerator and the denominator in factored form. Even so, we might need to factor them further in order to simplify.

EXAMPLE 6

Multiply. Then simplify by removing a factor of 1.

a) $\dfrac{x+2}{x-3} \cdot \dfrac{x^2-4}{x^2+x-2}$

b) $\dfrac{1-a^3}{a^2} \cdot \dfrac{a^5}{a^2-1}$

Solution

a) $\dfrac{x+2}{x-3} \cdot \dfrac{x^2-4}{x^2+x-2} = \dfrac{(x+2)(x^2-4)}{(x-3)(x^2+x-2)}$ Multiplying the numerators and also the denominators

$= \dfrac{(x+2)(x-2)(x+2)}{(x-3)(x+2)(x-1)}$ Factoring the numerator and the denominator and looking for common factors

$= \dfrac{(x+2)(x+2)(x-2)}{(x-3)(x+2)(x-1)}$ Removing a factor of 1: $\dfrac{x+2}{x+2}=1$

$= \dfrac{(x+2)(x-2)}{(x-3)(x-1)}$ Simplifying

b) $\dfrac{1-a^3}{a^2} \cdot \dfrac{a^5}{a^2-1} = \dfrac{(1-a^3)a^5}{a^2(a^2-1)}$

$= \dfrac{(1-a)(1+a+a^2)a^5}{a^2(a-1)(a+1)}$ Factoring a difference of two cubes and a difference of two squares

$= \dfrac{-1(a-1)(1+a+a^2)a^5}{a^2(a-1)(a+1)}$ Factoring out -1 to reverse the subtraction

$= \dfrac{(a-1)a^2 \cdot a^3(-1)(1+a+a^2)}{(a-1)a^2(a+1)}$ Rewriting a^5 as $a^2 \cdot a^3$; removing a factor of 1: $\dfrac{(a-1)a^2}{(a-1)a^2}=1$

$= \dfrac{-a^3(1+a+a^2)}{a+1}$ Simplifying   ❑

Dividing and Simplifying

Two expressions are reciprocals of each other if their product is 1. As in arithmetic, to find the reciprocal of a rational expression, we interchange numerator and denominator.

The reciprocal of $\dfrac{x}{x^2+3}$ is $\dfrac{x^2+3}{x}$.

The reciprocal of $y-8$ is $\dfrac{1}{y-8}$.

For any rational expressions A/B and C/D, with B, C, $D \neq 0$,

$$\frac{A}{B} \div \frac{C}{D} = \frac{A}{B} \cdot \frac{D}{C}.$$

(To divide two rational expressions, multiply by the reciprocal of the divisor. We often say that we "*invert* and multiply.")

EXAMPLE 7

Divide. Simplify by removing a factor of 1 if possible.

a) $\dfrac{x-2}{x+1} \div \dfrac{x+5}{x-3}$ **b)** $\dfrac{a^2-1}{a-1} \div \dfrac{a^2-2a+1}{a+1}$

Solution

a) $\dfrac{x-2}{x+1} \div \dfrac{x+5}{x-3} = \dfrac{x-2}{x+1} \cdot \dfrac{x-3}{x+5}$ Multiplying by the reciprocal

$\qquad = \dfrac{(x-2)(x-3)}{(x+1)(x+5)}$ Multiplying the numerators and the denominators

b) $\dfrac{a^2-1}{a-1} \div \dfrac{a^2-2a+1}{a+1} = \dfrac{a^2-1}{a-1} \cdot \dfrac{a+1}{a^2-2a+1}$ Multiplying by the reciprocal

$\qquad = \dfrac{(a^2-1)(a+1)}{(a-1)(a^2-2a+1)}$ Multiplying the numerators and the denominators

$\qquad = \dfrac{(a+1)(a-1)(a+1)}{(a-1)(a-1)(a-1)}$ Factoring the numerator and the denominator

$\qquad = \dfrac{(a+1)(a-1)(a+1)}{(a-1)(a-1)(a-1)}$ Removing a factor of 1: $\dfrac{a-1}{a-1} = 1$

$\qquad = \dfrac{(a+1)(a+1)}{(a-1)(a-1)}$ Simplifying   ☐

EXAMPLE 8

Perform the indicated operations and simplify:

$$\left[\frac{c^3-d^3}{(c+d)^2} \div (c-d) \right] \cdot (c+d).$$

Solution We have

$\left[\dfrac{c^3-d^3}{(c+d)^2} \div (c-d) \right] \cdot (c+d)$

$= \dfrac{c^3-d^3}{(c+d)^2} \cdot \dfrac{1}{c-d} \cdot (c+d)$ Multiplying by the reciprocal

$= \dfrac{(c-d)(c^2+cd+d^2)(c+d)}{(c+d)(c+d)(c-d)}$ Factoring; rewriting $(c+d)^2$ as $(c+d)(c+d)$

$= \dfrac{(c-d)(c^2+cd+d^2)(c+d)}{(c+d)(c+d)(c-d)}$ $\dfrac{(c-d)(c+d)}{(c-d)(c+d)} = 1$

$= \dfrac{c^2+cd+d^2}{c+d}.$ ☐

Keep in mind that the procedures we learn in this chapter are by their nature rather long. It may help you to write out lots of steps as you do the problems. If you have difficulty, consider taking a clean sheet of paper and starting over.

EXERCISE SET | 6.1

For each rational function, find the function values indicated, provided the value exists.

1. $v(t) = \dfrac{4t^2 - 5t + 2}{t + 3}$; $v(0)$, $v(3)$, $v(7)$

2. $s(x) = \dfrac{5x^2 + 4x - 12}{6 - x}$; $s(4)$, $s(-1)$, $s(3)$

3. $r(y) = \dfrac{3y^3 - 2y}{y - 5}$; $r(0)$, $r(4)$, $r(5)$

4. $f(r) = \dfrac{\pi r^2 + 2\pi r}{r - 1}$; (use 3.14 for π)

$f(2)$, $f(5)$, $f(1)$

5. $g(x) = \dfrac{2x^3 - 9}{x^2 - 4x + 4}$; $g(0)$, $g(2)$, $g(-1)$

6. $r(t) = \dfrac{t^2 - 5t + 4}{t^2 - 9}$; $r(1)$, $r(2)$, $r(-4)$

7. $f(t) = \dfrac{9 - t^2}{5 - 6t + t^2}$; $f(-3)$, $f(0)$, $f(1)$

8. $g(y) = \dfrac{y^3 - 8}{y^2 - 8y + 16}$; $g(2)$, $g(4)$, $g(-2)$

For each rational function, find the domain.

9. $f(x) = \dfrac{7}{3 - x}$

10. $g(t) = \dfrac{3}{t - 5}$

11. $v(t) = \dfrac{t - 7}{t^2 - 4t}$

12. $r(x) = \dfrac{9 - x}{x^2 - 3x}$

13. $s(x) = \dfrac{5}{x^2 - 4}$

14. $m(y) = \dfrac{9}{y^2 - 25}$

15. $F(x) = \dfrac{x^2 - 4}{x^2 - 8x + 12}$

16. $f(x) = \dfrac{9 - x^2}{x^2 - 6x + 8}$

Multiply to obtain equivalent expressions. Do not simplify.

17. $\dfrac{3x}{3x} \cdot \dfrac{x + 1}{x + 3}$

18. $\dfrac{4 - y^2}{6 - y} \cdot \dfrac{-1}{-1}$

19. $\dfrac{t - 3}{t + 2} \cdot \dfrac{t + 3}{t + 3}$

20. $\dfrac{p - 4}{p - 5} \cdot \dfrac{p + 5}{p + 5}$

21. $\dfrac{x^2 - 3}{x - 6} \cdot \dfrac{x + 6}{x + 6}$

22. $\dfrac{t^2 - 9}{3 - t} \cdot \dfrac{t + 2}{t + 2}$

Simplify by removing a factor of 1.

23. $\dfrac{9y^2}{15y}$

24. $\dfrac{6x^3}{18x^2}$

25. $\dfrac{8t^3}{4t^7}$

26. $\dfrac{27y^7}{18y^9}$

27. $\dfrac{2a - 6}{2}$

28. $\dfrac{3a - 6}{3}$

29. $\dfrac{6x - 9}{12}$

30. $\dfrac{25a - 30}{15}$

31. $\dfrac{4y - 12}{4y + 12}$

32. $\dfrac{8x + 16}{8x - 16}$

33. $\dfrac{6x - 12}{5x - 10}$

34. $\dfrac{7x - 21}{3x - 9}$

35. $\dfrac{12 - 6x}{5x - 10}$

36. $\dfrac{21 - 7x}{3x - 9}$

37. $\dfrac{t^2 - 16}{t^2 - 8t + 16}$

38. $\dfrac{p^2 - 25}{p^2 + 10p + 25}$

39. $\dfrac{x^2 + 9x + 8}{x^2 - 3x - 4}$

40. $\dfrac{t^2 - 8t - 9}{t^2 + 5t + 4}$

41. $\dfrac{16 - t^2}{t^2 - 8t + 16}$

42. $\dfrac{25 - p^2}{p^2 + 10p + 25}$

Multiply and simplify.

43. $\dfrac{5x^2}{3t^5} \cdot \dfrac{9t^8}{25x}$

44. $\dfrac{7a^3}{10b^7} \cdot \dfrac{5b^3}{3a}$

45. $\dfrac{3x - 6}{5x} \cdot \dfrac{x^3}{5x - 10}$

46. $\dfrac{5t^3}{4t - 8} \cdot \dfrac{6t - 12}{10t}$

47. $\dfrac{y^2 - 16}{2y + 6} \cdot \dfrac{y + 3}{y - 4}$

48. $\dfrac{m^2 - n^2}{4m + 4n} \cdot \dfrac{m + n}{m - n}$

49. $\dfrac{x^2 - 16}{x^2} \cdot \dfrac{x^2 - 4x}{x^2 - x - 12}$

50. $\dfrac{y^2 + 10y + 25}{y^2 - 9} \cdot \dfrac{y^2 + 3y}{y + 5}$

51. $\dfrac{6-2t}{t^2+4t+4} \cdot \dfrac{t^3+2t^2}{t^8-9t^6}$

52. $\dfrac{x^2-6x+9}{12-4x} \cdot \dfrac{x^6-9x^4}{x^3-3x^2}$

53. $\dfrac{x^2-2x-35}{2x^3-3x^2} \cdot \dfrac{4x^3-9x}{7x-49}$

54. $\dfrac{y^2-10y+9}{y^2-1} \cdot \dfrac{y+4}{y^2-5y-36}$

55. $\dfrac{c^3+8}{c^5-4c^3} \cdot \dfrac{c^6-4c^5+4c^4}{c^2-2c+4}$

56. $\dfrac{x^3-27}{x^4-9x^2} \cdot \dfrac{x^5-6x^4+9x^3}{x^2+3x+9}$

57. $\dfrac{a^3-b^3}{3a^2+9ab+6b^2} \cdot \dfrac{a^2+2ab+b^2}{a^2-b^2}$

58. $\dfrac{x^3+y^3}{x^2+2xy-3y^2} \cdot \dfrac{x^2-y^2}{3x^2+6xy+3y^2}$

59. $\dfrac{4x^2-9y^2}{8x^3-27y^3} \cdot \dfrac{4x^2+6xy+9y^2}{4x^2+12xy+9y^2}$

60. $\dfrac{3x^2-3y^2}{27x^3-8y^3} \cdot \dfrac{6x^2+5xy-6y^2}{6x^2+12xy+6y^2}$

Divide and simplify.

61. $\dfrac{16a^7}{3b^5} \div \dfrac{8a^3}{6b}$

62. $\dfrac{9x^5}{8y^2} \div \dfrac{3x}{16y^9}$

63. $\dfrac{3y+15}{y^7} \div \dfrac{y+5}{y^2}$

64. $\dfrac{6x+12}{x^8} \div \dfrac{x+2}{x^3}$

65. $\dfrac{y^2-9}{y^2} \div \dfrac{y^5+3y^4}{y+2}$

66. $\dfrac{x^2-4}{x^3} \div \dfrac{x^5-2x^4}{x+4}$

67. $\dfrac{4a^2-1}{a^2-4} \div \dfrac{2a-1}{a-2}$

68. $\dfrac{25x^2-4}{x^2-9} \div \dfrac{5x-2}{x+3}$

69. $\dfrac{x^2-y^2}{4x+4y} \div \dfrac{3y-3x}{12x^2}$

70. $\dfrac{5y-5x}{15y^3} \div \dfrac{x^2-y^2}{3x+3y}$

71. $\dfrac{x^2-16}{x^2-10x+25} \div \dfrac{3x-12}{x^2-3x-10}$

72. $\dfrac{y^2-36}{y^2-8y+16} \div \dfrac{3y-18}{y^2-y-12}$

73. $\dfrac{y^3+3y}{y^2-9} \div \dfrac{y^2+5y-14}{y^2+4y-21}$

74. $\dfrac{a^3+4a}{a^2-16} \div \dfrac{a^2+8a+15}{a^2+a-20}$

75. $\dfrac{x^3-64}{x^3+64} \div \dfrac{x^2-16}{x^2-4x+16}$

76. $\dfrac{8y^3-27}{64y^3-1} \div \dfrac{4y^2-9}{16y^2+4y+1}$

77. $\dfrac{8a^3+b^3}{2a^2+3ab+b^2} \div \dfrac{8a^2-4ab+2b^2}{4a^2+4ab+b^2}$

78. $\dfrac{x^3+8y^3}{2x^2+5xy+2y^2} \div \dfrac{x^3-2x^2y+4xy^2}{8x^2-2y^2}$

Skill Maintenance

79. Solve by substitution:
$$3x+y=13,$$
$$x=y+1.$$

80. Evaluate:
$$\begin{vmatrix} 3 & -2 \\ 4 & 7 \end{vmatrix}.$$

81. Solve: $\frac{2}{3}(3x-4)=8$.

82. A concert committee needs to take in \$4000 from ticket sales in order to break even. If a total of 400 tickets is to be sold at full price and 200 tickets sold at half price, how should the tickets be priced?

Synthesis

83. ◇ A student *incorrectly* simplifies $\dfrac{x+2}{x}$ as
$$\dfrac{x+2}{x}=\dfrac{\cancel{x}+2}{\cancel{x}}=1+2=3.$$
The student insists this is correct because it checked when x was replaced by 1. Explain the student's misconception.

84. ◇ A student *incorrectly* argues that since
$$\dfrac{a^2-4}{a-2}=\dfrac{a^2}{a}+\dfrac{-4}{-2}=a+2,$$
it follows that
$$\dfrac{x^2+9}{x+1}=\dfrac{x^2}{x}+\dfrac{9}{1}=x+9.$$
Explain the student's misconception.

85. Graph the function given by
$$f(x)=\dfrac{x^2-9}{x-3}.$$
(*Hint:* Determine the domain of f and simplify.)

86. Let
$$g(x)=\dfrac{2x+3}{4x-1}.$$
Determine each of the following.
a) $g(x+h)$
b) $g(2x-2)\cdot g(x)$
c) $g\left(\frac{1}{2}x+1\right)\cdot g(x)$

Perform the indicated operations and simplify.

87. $\left[\dfrac{r^2 - 4s^2}{r + 2s} \div (r + 2s) \right] \cdot \dfrac{2s}{r - 2s}$

88. $\left[\dfrac{d^2 - d}{d^2 - 6d + 8} \cdot \dfrac{d - 2}{d^2 + 5d} \right] \div \dfrac{5d}{d^2 - 9d + 20}$

Simplify.

89. $\dfrac{x(x + 1) - 2(x + 3)}{(x + 1)(x + 2)(x + 3)}$

90. $\dfrac{2x - 5(x + 2) - (x - 2)}{x^2 - 4}$

91. $\dfrac{m^2 - t^2}{m^2 + t^2 + m + t + 2mt}$

92. $\dfrac{a^3 - 2a^2 + 2a - 4}{a^3 - 2a^2 - 3a + 6}$

93. $\dfrac{x^3 + x^2 - y^3 - y^2}{x^2 - 2xy + y^2}$

94. $\dfrac{u^6 + v^6 + 2u^3v^3}{u^3 - v^3 + u^2v - uv^2}$

95. $\dfrac{x^5 - x^3 + x^2 - 1 - (x^3 - 1)(x + 1)^2}{(x^2 - 1)^2}$

96. Let

$$f(x) = \frac{4}{x^2 - 1}$$

and

$$g(x) = \frac{4x^2 + 8x + 4}{x^3 - 1}.$$

Find each of the following.

a) $(f \cdot g)(x)$
b) $(f/g)(x)$
c) $(g/f)(x)$

Use a grapher to determine the domain of each of the following rational functions. Round all values to two decimal places.

97. $f(x) = \dfrac{17.5x^2 - 25.2}{2.1x^3 + 18.27x^2}$

98. $f(x) = \dfrac{5.6}{x^2 + 6.9x + 11.22}$

99. $f(x) = \dfrac{18.64x^2 - 36.75}{x^2 - 4.07x - 6.65}$

100. $f(x) = \dfrac{50.3x}{x^3 - 1.1x^2 - 18.25x + 33.92}$

6.2

Rational Expressions: Adding and Subtracting

When Denominators Are the Same   •
When Denominators Are Different

We add and subtract rational expressions as we do using fractional notation in arithmetic.

When Denominators Are the Same

To add or subtract when denominators are the same, add or subtract the numerators and keep the same denominator.

$$\frac{A}{C} + \frac{B}{C} = \frac{A + B}{C} \quad \text{and} \quad \frac{A}{C} - \frac{B}{C} = \frac{A - B}{C}, \quad \text{where } C \neq 0.$$

EXAMPLE 1

Add: $\dfrac{3 + x}{x} + \dfrac{4}{x}$.

Solution

$$\frac{3 + x}{x} + \frac{4}{x} = \frac{7 + x}{x}$$ This expression cannot be simplified further because x is not a factor of $7 + x$.

Example 1 shows that

$$\frac{3 + x}{x} + \frac{4}{x} \quad \text{and} \quad \frac{7 + x}{x}$$

are equivalent expressions. They name the same number for all replacements of x except 0. As functions,

$$\frac{3 + x}{x} + \frac{4}{x}$$

and

$$\frac{7 + x}{x}$$

are the same function. Their graphs are identical. The graph never crosses the vertical line $x = 0$ because 0 is not in the domain of the function.

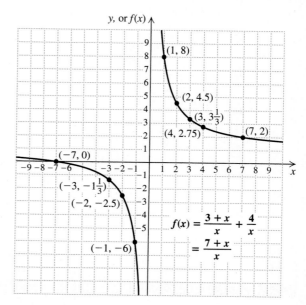

$$f(x) = \frac{3 + x}{x} + \frac{4}{x}$$
$$= \frac{7 + x}{x}$$

EXAMPLE 2

Add: $\dfrac{4x^2 - 5xy}{x^2 - y^2} + \dfrac{2xy - y^2}{x^2 - y^2}$.

Solution

$$\frac{4x^2 - 5xy}{x^2 - y^2} + \frac{2xy - y^2}{x^2 - y^2} = \frac{4x^2 - 3xy - y^2}{x^2 - y^2}$$ Adding the numerators and combining like terms. The denominator is unchanged.

$$= \frac{(x - y)(4x + y)}{(x - y)(x + y)}$$ Factoring the numerator and the denominator and looking for common factors

$$= \frac{(x - y)(4x + y)}{(x - y)(x + y)}$$ Removing a factor of 1: $\dfrac{x - y}{x - y} = 1$

$$= \frac{4x + y}{x + y}$$ Simplifying

Recall from Chapter 1 that a fraction bar is a grouping symbol. Thus, when a numerator containing a polynomial is subtracted, care must be taken to subtract, or change the sign of, each term in that polynomial.

EXAMPLE 3

Subtract: $\dfrac{4x + 5}{x + 3} - \dfrac{x - 2}{x + 3}$.

Solution

$$\dfrac{4x + 5}{x + 3} - \dfrac{x - 2}{x + 3} = \dfrac{4x + 5 - (x - 2)}{x + 3}$$

The parentheses are important to make sure that you subtract the entire quantity.

$$= \dfrac{4x + 5 - x + 2}{x + 3}$$

$$= \dfrac{3x + 7}{x + 3}$$ ❑

TECHNOLOGY
CONNECTION

To check Example 3 with a grapher, simply graph the equations

$$y_1 = \dfrac{4x + 5}{x + 3} - \dfrac{x - 2}{x + 3} \quad \text{and} \quad y_2 = \dfrac{3x + 7}{x + 3}$$

on the same set of axes. Since the equations are equivalent, one curve (it has two branches) should appear.

If denominators are opposites of each other, we multiply one of the rational expressions by $-1/-1$. This gives us a common denominator.

EXAMPLE 4

Add: $\dfrac{a}{2a} + \dfrac{a^3}{-2a}$.

Solution

$$\dfrac{a}{2a} + \dfrac{a^3}{-2a} = \dfrac{a}{2a} + \dfrac{-1}{-1} \cdot \dfrac{a^3}{-2a}$$

Multiplying by $\dfrac{-1}{-1}$

$$= \dfrac{a}{2a} + \dfrac{-a^3}{2a}$$

This is equal to 1 (not -1).

$$= \dfrac{a - a^3}{2a}$$

Adding the numerators

$$= \dfrac{a(1 - a^2)}{2a}$$

Factoring the numerator and looking for common factors

$$= \dfrac{\cancel{a}(1 - a^2)}{2\cancel{a}}$$

Removing a factor of 1: $\dfrac{a}{a} = 1$

$$= \dfrac{1 - a^2}{2}$$ ❑

EXAMPLE 5

Subtract: $\dfrac{5x}{x - 2y} - \dfrac{3y - 7}{2y - x}$.

Solution

$$\dfrac{5x}{x - 2y} - \dfrac{3y - 7}{2y - x} = \dfrac{5x}{x - 2y} - \dfrac{-1}{-1} \cdot \dfrac{3y - 7}{2y - x}$$

$$= \dfrac{5x}{x - 2y} - \dfrac{7 - 3y}{x - 2y}$$

Performing the multiplication.
Note: $-1(2y - x) = -2y + x$
$= x - 2y$.

$$= \dfrac{5x - (7 - 3y)}{x - 2y}$$

Subtracting the numerators

$$= \dfrac{5x - 7 + 3y}{x - 2y}$$ ❑

In Example 5, you may have noticed that the expression $3y - 7$ is multiplied by -1 and then subtracted. This resulted in $-7 + 3y$, which is equivalent to the original $3y - 7$. Thus, instead of multiplying the numerator by -1 and then subtracting, we could have simply *added* $3y - 7$ to $5x$, as in the following:

$$\frac{5x}{x - 2y} - \frac{3y - 7}{2y - x} = \frac{5x}{x - 2y} + (-1) \cdot \frac{3y - 7}{2y - x} \qquad \text{Rewriting subtraction as addition}$$

$$= \frac{5x}{x - 2y} + \frac{1}{-1} \cdot \frac{3y - 7}{2y - x} \qquad \text{Writing } -1 \text{ as } \frac{1}{-1}$$

$$= \frac{5x}{x - 2y} + \frac{3y - 7}{x - 2y} \qquad \text{The opposite of } 2y - x \text{ is } x - 2y.$$

$$= \frac{5x + 3y - 7}{x - 2y}.$$

When Denominators Are Different

In order to add rational expressions such as

$$\frac{7}{12xy^2} + \frac{8}{15x^3y}$$

or

$$\frac{x}{x^2 - y^2} + \frac{y}{x^2 - 4xy + 3y^2},$$

we must first find common denominators. As was the case with fractional notation, our work will be easier if we use the *least common multiple* (LCM) of the denominators involved.

> To find the least common multiple, or LCM, use each factor the greatest number of times that it occurs in any one prime factorization.

EXAMPLE 6

Find the least common multiple of each pair of polynomials.

a) $21x$ and $3x^2$

b) $x^2 + x - 12$ and $x^2 - 16$

Solution

a) We write the prime factorization of $21x$:

$$21x = 3 \cdot 7 \cdot x.$$

The factors 3, 7, and x must appear in the LCM if $21x$ is to be a factor of the LCM. Note that $3x^2$, the other polynomial, is not a factor of $3 \cdot 7 \cdot x$. This is because the prime factors of $3x^2$ — namely 3, x, and x — do not all appear in $3 \cdot 7 \cdot x$. By including a second factor of x, we obtain a product that contains both $21x$ and $3x^2$ as factors:

$$\text{LCM} = 3 \cdot 7 \cdot x \cdot x$$

$21x$ is a factor.

$3x^2$ is a factor.

Note that each factor is used the greatest number of times that it occurs as a factor of either $21x$ or $3x^2$.

b)   We factor both expressions:

$$x^2 + x - 12 = (x - 3)(x + 4),$$
$$x^2 - 16 = (x + 4)(x - 4).$$

The LCM must include all the factors of *both* polynomials. By multiplying the factors of $x^2 + x - 12$ by $x - 4$, we obtain a product that contains both $x^2 + x - 12$ and $x^2 - 16$ as factors:

$$\text{LCM} = (x - 3)(x + 4)(x - 4)$$

$x^2 + x - 12$ is a factor.

$x^2 - 16$ is a factor.

To add or subtract rational expressions with unlike denominators, we first find the *least common denominator,* or LCD, by finding the LCM of the denominators. Then we multiply by 1, as needed, to write both rational expressions using the LCD.

EXAMPLE 7

Add: $\dfrac{2}{21x} + \dfrac{5}{3x^2}.$

Solution   In Example 6(a), we found that the LCM is $3 \cdot 7 \cdot x \cdot x$, or $21x^2$. We now multiply each rational expression by 1. We want to use an expression for 1 that gives us the LCD in each expression. In this case, we use x/x and $7/7$:

$$\frac{2}{21x} \cdot \frac{x}{x} + \frac{5}{3x^2} \cdot \frac{7}{7} = \frac{2x}{21x^2} + \frac{35}{21x^2}$$

$$= \frac{2x + 35}{21x^2}. \qquad \text{This expression cannot be simplified.}$$

Note that multiplying by x/x in the first expression gave us the LCD, $21x^2$. Similarly, multiplying by $7/7$ in the second expression also gave us a denominator of $21x^2$.

EXAMPLE 8

Add: $\dfrac{3x^2 + 3xy}{x^2 - y^2} + \dfrac{2 - 3x}{x - y}.$

Solution   We first find the LCD:

$$\left. \begin{array}{l} x^2 - y^2 = (x + y)(x - y) \\ x - y = x - y \end{array} \right\} \qquad \text{The LCD is } (x + y)(x - y).$$

Next we multiply by 1 to get the LCD in the second expression. Then we add and simplify if possible.

$$\frac{3x^2 + 3xy}{(x + y)(x - y)} + \frac{2 - 3x}{x - y} \cdot \frac{x + y}{x + y}$$

Multiplying by 1 to get the LCD

$$= \frac{3x^2 + 3xy}{(x + y)(x - y)} + \frac{(2 - 3x)(x + y)}{(x - y)(x + y)}$$

$$= \frac{3x^2 + 3xy}{(x + y)(x - y)} + \frac{2x + 2y - 3x^2 - 3xy}{(x - y)(x + y)}$$

Multiplying in the numerator

$$= \frac{3x^2 + 3xy + 2x + 2y - 3x^2 - 3xy}{(x + y)(x - y)}$$

Adding the numerators

$$= \frac{2x + 2y}{(x + y)(x - y)} = \frac{2(x + y)}{(x + y)(x - y)}$$

Combining like terms and then factoring the numerator

$$= \frac{2\cancel{(x + y)}}{\cancel{(x + y)}(x - y)}$$

Removing a factor of 1: $\frac{x + y}{x + y} = 1$

$$= \frac{2}{x - y}$$

❑

EXAMPLE 9

Subtract: $\dfrac{2y + 1}{y^2 - 7y + 6} - \dfrac{y + 3}{y^2 - 5y - 6}$.

Solution

$$\frac{2y + 1}{y^2 - 7y + 6} - \frac{y + 3}{y^2 - 5y - 6}$$

$$= \frac{2y + 1}{(y - 6)(y - 1)} - \frac{y + 3}{(y - 6)(y + 1)}$$

The LCD is $(y - 6)(y - 1)(y + 1)$.

$$= \frac{2y + 1}{(y - 6)(y - 1)} \cdot \frac{y + 1}{y + 1} - \frac{y + 3}{(y - 6)(y + 1)} \cdot \frac{y - 1}{y - 1}$$

Multiplying by 1 to get the LCD in each expression

$$= \frac{(2y + 1)(y + 1) - (y + 3)(y - 1)}{(y - 6)(y - 1)(y + 1)}$$

$$= \frac{2y^2 + 3y + 1 - (y^2 + 2y - 3)}{(y - 6)(y - 1)(y + 1)}$$

The parentheses are important.

$$= \frac{2y^2 + 3y + 1 - y^2 - 2y + 3}{(y - 6)(y - 1)(y + 1)}$$

$$= \frac{y^2 + y + 4}{(y - 6)(y - 1)(y + 1)}$$

❑

There is no need to multiply out the denominator in the last step of Example 9. In fact, leaving denominators in factored form will ease our later work on solving equations.

EXAMPLE 10

Perform the indicated operations and simplify:

$$\frac{2x}{x^2 - 4} + \frac{5}{2 - x} - \frac{1}{2 + x}.$$

Solution

$$\frac{2x}{x^2 - 4} + \frac{5}{2 - x} - \frac{1}{2 + x}$$

$$= \frac{2x}{(x - 2)(x + 2)} + \frac{5}{2 - x} - \frac{1}{2 + x}$$

$$= \frac{2x}{(x - 2)(x + 2)} + \frac{-1}{-1} \cdot \frac{5}{(2 - x)} - \frac{1}{x + 2}$$ Multiplying by $\dfrac{-1}{-1}$ to reverse subtraction

$$= \frac{2x}{(x - 2)(x + 2)} + \frac{-5}{x - 2} - \frac{1}{x + 2}$$ The LCM is $(x - 2)(x + 2)$.

$$= \frac{2x}{(x - 2)(x + 2)} + \frac{-5}{x - 2} \cdot \frac{x + 2}{x + 2} - \frac{1}{x + 2} \cdot \frac{x - 2}{x - 2}$$ Multiplying by 1 to get the LCD in each expression

$$= \frac{2x - 5(x + 2) - (x - 2)}{(x - 2)(x + 2)} = \frac{2x - 5x - 10 - x + 2}{(x - 2)(x + 2)}$$

$$= \frac{-4x - 8}{(x - 2)(x + 2)} = \frac{-4(x + 2)}{(x - 2)(x + 2)}$$

$$= \frac{-4(x + 2)}{(x - 2)(x + 2)}$$ Removing a factor of 1: $\dfrac{x + 2}{x + 2} = 1$

$$= \frac{-4}{x - 2}, \quad \text{or} \quad -\frac{4}{x - 2}$$

Another correct answer is $4/(2 - x)$. It is found by multiplying $-4/(x - 2)$ by $-1/-1$.   ❑

EXERCISE SET | 6.2

Perform the indicated operations. Simplify when possible.

1. $\dfrac{3}{2a} + \dfrac{5}{2a}$

2. $\dfrac{4}{3y} + \dfrac{8}{3y}$

3. $\dfrac{3}{4a^2b} - \dfrac{7}{4a^2b}$

4. $\dfrac{5}{3m^2n^2} - \dfrac{4}{3m^2n^2}$

5. $\dfrac{a - 3b}{a + b} + \dfrac{a + 5b}{a + b}$

6. $\dfrac{x - 5y}{x + y} + \dfrac{x + 7y}{x + y}$

7. $\dfrac{4y + 2}{y - 2} - \dfrac{y - 3}{y - 2}$

8. $\dfrac{3t + 2}{t - 4} - \dfrac{t - 2}{t - 4}$

9. $\dfrac{3x - 4}{x^2 - 5x + 4} + \dfrac{3 - 2x}{x^2 - 5x + 4}$

10. $\dfrac{5x - 4}{x^2 - 6x - 7} + \dfrac{5 - 4x}{x^2 - 6x - 7}$

11. $\dfrac{3a - 8}{a^2 - 9} - \dfrac{2a - 5}{a^2 - 9}$

12. $\dfrac{4a - 7}{a^2 - 25} - \dfrac{3a - 2}{a^2 - 25}$

13. $\dfrac{a^2}{a - b} + \dfrac{b^2}{b - a}$

14. $\dfrac{s^2}{r - s} + \dfrac{r^2}{s - r}$

15. $\dfrac{3}{x} - \dfrac{8}{-x}$

16. $\dfrac{2}{a} - \dfrac{5}{-a}$

17. $\dfrac{2x - 9}{x^2 - 25} - \dfrac{4 - x}{25 - x^2}$

18. $\dfrac{y - 9}{y^2 - 16} - \dfrac{7 - y}{16 - y^2}$

19. $\dfrac{t^2 + 3}{t^4 - 16} + \dfrac{7}{16 - t^4}$

20. $\dfrac{y^2 - 5}{y^4 - 81} + \dfrac{4}{81 - y^4}$

21. $\dfrac{m - 3n}{m^3 - n^3} - \dfrac{2n}{n^3 - m^3}$

22. $\dfrac{r - 6s}{r^3 - s^3} - \dfrac{5s}{s^3 - r^3}$

23. $\dfrac{y - 2}{y + 4} + \dfrac{y + 3}{y - 5}$

24. $\dfrac{x - 2}{x + 3} + \dfrac{x + 2}{x - 4}$

25. $2 + \dfrac{x - 3}{x + 1}$

26. $3 + \dfrac{y + 2}{y - 5}$

27. $\dfrac{4xy}{x^2 - y^2} + \dfrac{x - y}{x + y}$

28. $\dfrac{5ab}{a^2 - b^2} + \dfrac{a + b}{a - b}$

29. $\dfrac{9x + 2}{3x^2 - 2x - 8} + \dfrac{7}{3x^2 + x - 4}$

30. $\dfrac{3y + 2}{2y^2 - y - 10} + \dfrac{8}{2y^2 - 7y + 5}$

31. $\dfrac{4}{x + 1} + \dfrac{x + 2}{x^2 - 1} + \dfrac{3}{x - 1}$

32. $\dfrac{-2}{y + 2} + \dfrac{5}{y - 2} + \dfrac{y + 3}{y^2 - 4}$

33. $\dfrac{x - 1}{3x + 15} - \dfrac{x + 3}{5x + 25}$

34. $\dfrac{y - 2}{4y + 8} - \dfrac{y + 6}{5y + 10}$

35. $\dfrac{5ab}{a^2 - b^2} - \dfrac{a - b}{a + b}$

36. $\dfrac{6xy}{x^2 - y^2} - \dfrac{x + y}{x - y}$

37. $\dfrac{x}{x^2 + 9x + 20} - \dfrac{4}{x^2 + 7x + 12}$

38. $\dfrac{x}{x^2 + 11x + 30} - \dfrac{5}{x^2 + 9x + 20}$

39. $\dfrac{3y}{y^2 - 7y + 10} - \dfrac{2y}{y^2 - 8y + 15}$

40. $\dfrac{5x}{x^2 - 6x + 8} - \dfrac{3x}{x^2 - x - 12}$

41. $\dfrac{y}{y^2 - y - 20} + \dfrac{2}{y + 4}$

42. $\dfrac{2t + 9}{t^2 - t - 6} + \dfrac{1}{t + 2}$

43. $\dfrac{3y + 2}{y^2 + 5y - 24} + \dfrac{7}{y^2 + 4y - 32}$

44. $\dfrac{3x + 2}{x^2 - 7x + 10} + \dfrac{2x}{x^2 - 8x + 15}$

45. $\dfrac{3x - 1}{x^2 + 2x - 3} - \dfrac{x + 4}{x^2 - 9}$

46. $\dfrac{3p - 2}{p^2 + 2p - 24} - \dfrac{p - 3}{p^2 - 16}$

47. $\dfrac{2}{a^2 - 5a + 4} + \dfrac{-2}{a^2 - 4}$

48. $\dfrac{3}{a^2 - 7a + 6} + \dfrac{-3}{a^2 - 9}$

49. $3 + \dfrac{t}{t + 2} - \dfrac{2}{t^2 - 4}$

50. $2 + \dfrac{t}{t - 3} - \dfrac{3}{t^2 - 9}$

51. $\dfrac{1}{x + 1} - \dfrac{x}{x - 2} + \dfrac{x^2 + 2}{x^2 - x - 2}$

52. $\dfrac{2}{y + 3} - \dfrac{y}{y - 1} + \dfrac{y^2 + 2}{y^2 + 2y - 3}$

53. $\dfrac{4x}{x^2 - 1} + \dfrac{3x}{1 - x} - \dfrac{4}{x - 1}$

54. $\dfrac{5y}{1 - 2y} - \dfrac{2y}{2y + 1} + \dfrac{3}{4y^2 - 1}$

55. $\dfrac{1}{t^2 + 5t + 6} - \dfrac{2}{t^2 + 3t + 2} + \dfrac{1}{t^2 - 3t - 4}$

56. $\dfrac{2}{x^2 - 5x + 6} - \dfrac{4}{x^2 - 2x - 3} + \dfrac{2}{x^2 + 4x + 3}$

Skill Maintenance _____

57. Simplify. Use only positive exponents in your answer.

$$\frac{15x^{-7}y^{12}z^4}{35x^{-2}y^6z^{-3}}$$

58. Find an equation for the line that passes through the point $(-2, 3)$ and is perpendicular to the line $f(x) = -\frac{4}{5}x + 7$.

59. There are 50 dimes in a roll of dimes, 40 nickels in a roll of nickels, and 40 quarters in a roll of quarters. Robert has a total of 12 rolls of coins with a total value of $70.00. If he has three more rolls of nickels than dimes, how many of each roll of coins does he have?

60. Anna wants to buy tapes to record her favorite music. She needs some 30-min tapes and some 60-min tapes. If she buys 12 tapes with a total recording time of 10 hr, how many tapes of each length did she buy?

Synthesis _____

61. ◈ Examine Example 8 and explain how the expressions can be added using $x - y$ as the LCD.

62. ◈ Many students make the mistake of always multiplying denominators when looking for a common denominator. Use Example 9 to explain why this approach can yield results that are more difficult to simplify.

Find the LCM.

63. $x^8 - x^4, \ x^5 - x^2, \ x^5 - x^3, \ x^5 + x^2$

64. $2a^3 + 2a^2b + 2ab^2, \ a^6 - b^6,$
$2b^2 + ab - 3a^2, \ 2a^2b + 4ab^2 + 2b^3$

65. The LCM of two expressions is $8a^4b^7$. One of the expressions is $2a^3b^7$. List all the possibilities for the other expression.

Planet orbits and LCMs. The earth, Jupiter, Saturn, and Uranus all revolve around the sun. The earth takes 1 year, Jupiter takes 12 years, Saturn takes 30 years, and Uranus takes 84 years.

66. How often will the earth, Jupiter, and Saturn line up with each other?

67. In how many years will these four planets align themselves exactly as they are this evening?

If

$$f(x) = \frac{x^3}{x^2 - 4} \quad \text{and} \quad g(x) = \frac{x^2}{x^2 + 3x - 10},$$

find each of the following.

68. $(f + g)(x)$

69. $(f - g)(x)$

70. $(f \cdot g)(x)$

71. $(f/g)(x)$

72. The domain of $f + g$

Perform the indicated operations and simplify.

73. $2x^{-2} + 3x^{-2}y^{-2} - 7xy^{-1}$

74. $5(x - 3)^{-1} + 4(x + 3)^{-1} - 2(x + 3)^{-2}$

75. $4(y - 1)(2y - 5)^{-1} + 5(2y + 3)(5 - 2y)^{-1} + (y - 4)(2y - 5)^{-1}$

76. $\dfrac{x + 4}{6x^2 - 20x} \cdot \left(\dfrac{x}{x^2 - x - 20} + \dfrac{2}{x + 4} \right)$

77. $\dfrac{x^2 - 7x + 12}{x^2 - x - 29/3} \cdot \left(\dfrac{3x + 2}{x^2 + 5x - 24} + \dfrac{7}{x^2 + 4x - 32} \right)$

78. $\dfrac{8t^5}{2t^2 - 10t + 12} \div \left(\dfrac{2t}{t^2 - 8t + 15} - \dfrac{3t}{t^2 - 7t + 10} \right)$

79. $\dfrac{9t^3}{3t^3 - 12t^2 + 9t} \div \left(\dfrac{t + 4}{t^2 - 9} - \dfrac{3t - 1}{t^2 + 2t - 3} \right)$

6.3

Complex Rational Expressions

Multiplying by the LCD • **Adding or Subtracting Within the Complex Rational Expression**

A **complex rational expression** is a rational expression that contains rational expressions within its numerator and/or its denominator. Here are some examples:

$$\frac{x + \dfrac{5}{x}}{4x}, \qquad \frac{\dfrac{x - y}{x + y}}{\dfrac{2x - y}{3x + y}}, \qquad \frac{\dfrac{2}{3}}{\dfrac{4}{5}}, \qquad \frac{\dfrac{3x}{5} - \dfrac{2}{x}}{\dfrac{4x}{3} + \dfrac{7}{6x}}.$$

The rational expressions within each complex rational expression are red.

We will consider two methods that can be used to simplify complex rational expressions.

Multiplying by the LCD

One method of simplifying a complex rational expression is to multiply the entire expression by 1. To write 1, we use the LCD of the expressions within the complex rational expression.

EXAMPLE 1

Simplify:

$$\frac{\dfrac{1}{a} + \dfrac{1}{b}}{\dfrac{1}{a^3} + \dfrac{1}{b^3}}.$$

Solution   Since the denominators within the complex rational expression are a, b, a^3, and b^3, the LCD is a^3b^3. We multiply by 1, using $(a^3b^3)/(a^3b^3)$:

$$\frac{\dfrac{1}{a} + \dfrac{1}{b}}{\dfrac{1}{a^3} + \dfrac{1}{b^3}} = \frac{\dfrac{1}{a} + \dfrac{1}{b}}{\dfrac{1}{a^3} + \dfrac{1}{b^3}} \cdot \frac{a^3b^3}{a^3b^3}$$ Multiplying by 1

$$= \frac{\left(\dfrac{1}{a} + \dfrac{1}{b}\right)a^3b^3}{\left(\dfrac{1}{a^3} + \dfrac{1}{b^3}\right)a^3b^3}$$ Multiplying the numerators and the denominators. Remember to use parentheses.

$$= \frac{\dfrac{1}{a} \cdot a^3b^3 + \dfrac{1}{b} \cdot a^3b^3}{\dfrac{1}{a^3} \cdot a^3b^3 + \dfrac{1}{b^3} \cdot a^3b^3}$$ Using the distributive law to carry out the multiplications

$$= \frac{\dfrac{\cancel{a}}{\cancel{a}} \cdot a^2b^3 + \dfrac{\cancel{b}}{\cancel{b}} \cdot a^3b^2}{\dfrac{\cancel{a^3}}{\cancel{a^3}} \cdot b^3 + \dfrac{\cancel{b^3}}{\cancel{b^3}} \cdot a^3}$$ Removing factors of 1; study this carefully

$$= \frac{a^2b^3 + a^3b^2}{b^3 + a^3}$$ Simplifying

$$= \frac{a^2b^2(b + a)}{(b + a)(b^2 - ab + a^2)}$$ Factoring and looking for a factor of 1

$$= \frac{a^2b^2(\cancel{b + a})}{(\cancel{b + a})(b^2 - ab + a^2)}$$ Removing a factor of 1

$$= \frac{a^2b^2}{b^2 - ab + a^2}.$$ Simplifying   ❑

To simplify a complex rational expression by using the LCD:

1. Find the LCD of all expressions *within* the complex rational expression.
2. Multiply the complex rational expression by 1, using the LCD to form the expression for 1.
3. Distribute and simplify. Neither the numerator nor the denominator of the complex rational expression should contain a rational expression.
4. Factor and simplify, if possible.

Note that we choose the LCD to form the number 1 in order to clear the numerator and the denominator of the complex rational expression of all rational expressions.

| EXAMPLE 2 | Simplify:

$$\frac{\dfrac{3}{2x-2} - \dfrac{1}{x+1}}{\dfrac{1}{x-1} + \dfrac{x}{x^2-1}}.$$

Solution Note that to find the LCD, we may have to factor first:

$$\frac{\dfrac{3}{2x-2} - \dfrac{1}{x+1}}{\dfrac{1}{x-1} + \dfrac{x}{x^2-1}} = \frac{\dfrac{3}{2(x-1)} - \dfrac{1}{x+1}}{\dfrac{1}{x-1} + \dfrac{x}{(x-1)(x+1)}}$$ The LCD is $2(x-1)(x+1)$.

$$= \frac{\dfrac{3}{2(x-1)} - \dfrac{1}{x+1}}{\dfrac{1}{x-1} + \dfrac{x}{(x-1)(x+1)}} \cdot \frac{2(x-1)(x+1)}{2(x-1)(x+1)}$$ Multiplying by 1

$$= \frac{\dfrac{3}{2(x-1)} \cdot 2(x-1)(x+1) - \dfrac{1}{x+1} \cdot 2(x-1)(x+1)}{\dfrac{1}{x-1} \cdot 2(x-1)(x+1) + \dfrac{x}{(x-1)(x+1)} \cdot 2(x-1)(x+1)}$$ Using the distributive law

$$= \frac{\dfrac{2(x-1)}{2(x-1)} \cdot 3(x+1) - \dfrac{x+1}{x+1} \cdot 2(x-1)}{\dfrac{x-1}{x-1} \cdot 2(x+1) + \dfrac{(x-1)(x+1)}{(x-1)(x+1)} \cdot 2x}$$ Removing factors of 1

$$= \frac{3(x+1) - 2(x-1)}{2(x+1) + 2x}$$ Simplifying

$$= \frac{3x+3 - 2x+2}{2x+2 + 2x}$$ Using the distributive law

$$= \frac{x+5}{4x+2}.$$ ❑

Adding or Subtracting Within the Complex Rational Expression

Another method for simplifying a complex rational expression involves first adding or subtracting, as necessary, to get one rational expression in the numerator and one rational expression in the denominator. The problem is thereby simplified to one involving the division of two rational expressions.

EXAMPLE 3

Simplify:

$$\frac{\dfrac{3}{x} - \dfrac{2}{x^2}}{\dfrac{3}{x-2} + \dfrac{1}{x^2}}.$$

Solution

$$\frac{\dfrac{3}{x} - \dfrac{2}{x^2}}{\dfrac{3}{x-2} + \dfrac{1}{x^2}} = \frac{\dfrac{3}{x}\cdot\dfrac{x}{x} - \dfrac{2}{x^2}}{\dfrac{3}{x-2}\cdot\dfrac{x^2}{x^2} + \dfrac{1}{x^2}\cdot\dfrac{x-2}{x-2}}$$

→ Multiplying $3/x$ by 1 to get the common denominator x^2

→ Multiplying $3/(x-2)$ and $1/x^2$ by 1 to get the common denominator $x^2(x-2)$

$$= \frac{\dfrac{3x}{x^2} - \dfrac{2}{x^2}}{\dfrac{3x^2}{(x-2)x^2} + \dfrac{x-2}{x^2(x-2)}}$$

There is now a common denominator in the numerator and a common denominator in the denominator of the complex rational expression.

$$= \frac{\dfrac{3x-2}{x^2}}{\dfrac{3x^2+x-2}{(x-2)x^2}}$$

Subtracting in the numerator and adding in the denominator. We now have one rational expression divided by another rational expression.

$$= \frac{3x-2}{x^2}\cdot\frac{(x-2)x^2}{3x^2+x-2}$$

To divide, multiply by the reciprocal of the divisor.

$$= \frac{(3x-2)(x-2)x^2}{x^2(3x-2)(x+1)}$$

Factoring

$$= \frac{(3x-2)(x-2)x^2}{x^2(3x-2)(x+1)}$$

Removing a factor of 1: $\dfrac{x^2(3x-2)}{x^2(3x-2)} = 1$

$$= \frac{x-2}{x+1}$$

□

To simplify a complex rational expression by first adding or subtracting:

1. Add or subtract, as necessary, to get a single rational expression in the numerator.
2. Add or subtract, as necessary, to get a single rational expression in the denominator.
3. Perform the indicated division (invert and multiply).
4. Simplify, if possible, by removing any factors of 1.

EXAMPLE 4

Simplify:

$$\frac{1 + \dfrac{1}{x}}{1 - \dfrac{1}{x^2}}.$$

Solution

$$\frac{1 + \dfrac{1}{x}}{1 - \dfrac{1}{x^2}} = \frac{\dfrac{x}{x} + \dfrac{1}{x}}{\dfrac{x^2}{x^2} - \dfrac{1}{x^2}} \quad \longleftarrow \text{Finding a common denominator} \atop \longleftarrow \text{Finding a common denominator}$$

$$= \frac{\dfrac{x + 1}{x}}{\dfrac{x^2 - 1}{x^2}} \qquad \begin{array}{l} \text{Adding in the numerator} \\[18pt] \text{Subtracting in the denominator} \end{array}$$

$$= \frac{x + 1}{x} \cdot \frac{x^2}{x^2 - 1} \qquad \text{Multiplying by the reciprocal}$$

$$= \frac{(x + 1) \cdot x^2}{x(x + 1)(x - 1)} \qquad \text{Factoring}$$

$$= \frac{\cancel{(x + 1)} x \cdot x}{\cancel{x} \cancel{(x + 1)} (x - 1)} \qquad \text{Removing a factor of 1: } \frac{(x + 1)x}{(x + 1)x} = 1$$

$$= \frac{x}{x - 1} \qquad \text{Simplifying} \qquad\qquad \Box$$

Example 4 shows that the expressions

$$\frac{1 + \dfrac{1}{x}}{1 - \dfrac{1}{x^2}} \quad \text{and} \quad \frac{x}{x - 1}$$

are equivalent. In terms of functions, if

$$f(x) = \frac{1 + \dfrac{1}{x}}{1 - \dfrac{1}{x^2}} \quad \text{and} \quad g(x) = \frac{x}{x - 1},$$

then f and g are equal except that the domain of g excludes the number 1, whereas the domain of f excludes the values -1 and 0, in addition to 1, since those values make a denominator 0 (see the graphs).

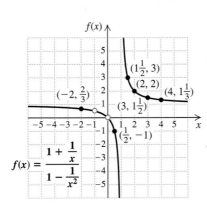

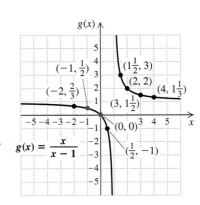

If negative exponents occur, we first find an equivalent expression using positive exponents and then proceed as above.

EXAMPLE 5

Simplify:

$$\frac{a^{-1} + b^{-1}}{a^{-3} + b^{-3}}.$$

Solution

$$\frac{a^{-1} + b^{-1}}{a^{-3} + b^{-3}} = \frac{\dfrac{1}{a} + \dfrac{1}{b}}{\dfrac{1}{a^3} + \dfrac{1}{b^3}}$$

Rewriting with positive exponents. The problem is now identical to Example 1.

$$= \frac{\dfrac{1}{a} \cdot \dfrac{b}{b} + \dfrac{1}{b} \cdot \dfrac{a}{a}}{\dfrac{1}{a^3} \cdot \dfrac{b^3}{b^3} + \dfrac{1}{b^3} \cdot \dfrac{a^3}{a^3}}$$

Finding a common denominator

Finding a common denominator

$$= \frac{\dfrac{b}{ab} + \dfrac{a}{ab}}{\dfrac{b^3}{a^3b^3} + \dfrac{a^3}{a^3b^3}}$$

$$= \frac{\dfrac{b + a}{ab}}{\dfrac{b^3 + a^3}{a^3b^3}}$$

Adding in the numerator

Adding in the denominator

$$= \frac{b + a}{ab} \cdot \frac{a^3b^3}{b^3 + a^3}$$

Multiplying by the reciprocal of the denominator

$$= \frac{(b + a) \cdot ab \cdot a^2b^2}{ab(b + a)(b^2 - ab + a^2)}$$

Factoring and looking for common factors

$$= \frac{\cancel{(b + a)} \cdot \cancel{ab} \cdot a^2b^2}{\cancel{ab}\cancel{(b + a)}(b^2 - ab + a^2)}$$

Removing a factor of 1: $\dfrac{(b + a)ab}{(b + a)ab} = 1$

$$= \frac{a^2b^2}{b^2 - ab + a^2}$$

It is difficult to say which method is best to use. To simplify expressions such as

$$\frac{\dfrac{3x + 1}{x - 5}}{\dfrac{2 - x}{x + 3}} \quad \text{or} \quad \frac{\dfrac{3}{x} - \dfrac{2}{x}}{\dfrac{1}{x + 1} + \dfrac{5}{x + 1}},$$

the second method is probably easier to use since it is little or no work to write the expression as a quotient of two rational expressions. We then invert and multiply.

On the other hand, expressions such as

$$\frac{\dfrac{3}{a^2b} - \dfrac{4}{bc^3}}{\dfrac{1}{b^3c} + \dfrac{2}{ac^4}} \quad \text{or} \quad \frac{\dfrac{5}{a^2 - b^2} + \dfrac{2}{a^2 + 2ab + b^2}}{\dfrac{1}{a - b} + \dfrac{4}{a + b}}$$

require fewer steps if we use the first method. Either method works for any complex rational expression.

EXERCISE SET | 6.3   ✱ Trouble

Simplify.

1. $\dfrac{\dfrac{1}{x} + 4}{\dfrac{1}{x} - 3}$

2. $\dfrac{\dfrac{1}{y} + 7}{\dfrac{1}{y} - 5}$

3. $\dfrac{x - x^{-1}}{x + x^{-1}}$

4. $\dfrac{y + y^{-1}}{y - y^{-1}}$

5. $\dfrac{\dfrac{3}{x} + \dfrac{4}{y}}{\dfrac{4}{x} - \dfrac{3}{y}}$

6. $\dfrac{\dfrac{2}{y} + \dfrac{5}{z}}{\dfrac{1}{y} - \dfrac{4}{z}}$

7. $\dfrac{\dfrac{x^2 - y^2}{xy}}{\dfrac{x - y}{y}}$

8. $\dfrac{\dfrac{a^2 - b^2}{ab}}{\dfrac{a - b}{b}}$

9. $\dfrac{a - \dfrac{3a}{b}}{b - \dfrac{b}{a}}$

10. $\dfrac{1 - \dfrac{2}{3x}}{x - \dfrac{4}{9x}}$

11. $\dfrac{a^{-1} + b^{-1}}{\dfrac{a^2 - b^2}{ab}}$

12. $\dfrac{x^{-1} + y^{-1}}{\dfrac{x^2 - y^2}{xy}}$

13. $\dfrac{\dfrac{1}{x + h} - \dfrac{1}{x}}{h}$

14. $\dfrac{\dfrac{1}{a - h} - \dfrac{1}{a}}{h}$

15. $\dfrac{\dfrac{y^2 - y - 6}{y^2 - 5y - 14}}{\dfrac{y^2 + 6y + 5}{y^2 - 6y - 7}}$

16. $\dfrac{\dfrac{x^2 - x - 12}{x^2 - 2x - 15}}{\dfrac{x^2 + 8x + 12}{x^2 - 5x - 14}}$

17. $\dfrac{\dfrac{1}{x - 2} + \dfrac{3}{x - 1}}{\dfrac{2}{x - 1} + \dfrac{5}{x - 2}}$

18. $\dfrac{\dfrac{2}{y - 3} + \dfrac{1}{y + 1}}{\dfrac{3}{y + 1} + \dfrac{4}{y - 3}}$

19. $\dfrac{a(a + 3)^{-1} - 2(a - 1)^{-1}}{a(a + 3)^{-1} - (a - 1)^{-1}}$

20. $\dfrac{a(a + 2)^{-1} - 3(a - 3)^{-1}}{a(a + 2)^{-1} - (a - 3)^{-1}}$

21. $\dfrac{\dfrac{x}{x^2 + 3x - 4} - \dfrac{1}{x^2 + 3x - 4}}{\dfrac{x}{x^2 + 6x + 8} + \dfrac{3}{x^2 + 6x + 8}}$

22. $\dfrac{\dfrac{x}{x^2 + 5x - 6} + \dfrac{6}{x^2 + 5x - 6}}{\dfrac{x}{x^2 - 5x + 4} - \dfrac{2}{x^2 - 5x + 4}}$

23. $\dfrac{\dfrac{y}{y^2 - 1} + \dfrac{3}{1 - y^2}}{\dfrac{y^2}{y^2 - 1} + \dfrac{9}{1 - y^2}}$

24. $\dfrac{\dfrac{y}{y^2 - 4} + \dfrac{5}{4 - y^2}}{\dfrac{y^2}{y^2 - 4} + \dfrac{25}{4 - y^2}}$

25. $\dfrac{\dfrac{2}{a^2 - 1} + \dfrac{1}{a + 1}}{\dfrac{3}{a^2 - 1} + \dfrac{2}{a - 1}}$

26. $\dfrac{\dfrac{3}{a^2 - 9} + \dfrac{2}{a + 3}}{\dfrac{4}{a^2 - 9} + \dfrac{1}{a + 3}}$

27. $\dfrac{\dfrac{5}{x^2 - 4} - \dfrac{3}{x - 2}}{\dfrac{4}{x^2 - 4} - \dfrac{2}{x + 2}}$

28. $\dfrac{\dfrac{4}{x^2 - 1} - \dfrac{3}{x + 1}}{\dfrac{5}{x^2 - 1} - \dfrac{2}{x - 1}}$

29. $\dfrac{\dfrac{y^2}{y^2-9}-\dfrac{y}{y+3}}{\dfrac{y}{y^2-9}-\dfrac{1}{y-3}}$

30. $\dfrac{\dfrac{y^2}{y^2-25}-\dfrac{y}{y-5}}{\dfrac{y}{y^2-25}-\dfrac{1}{y+5}}$

31. $\dfrac{\dfrac{a}{a+3}+\dfrac{4}{5a}}{\dfrac{a}{2a+6}+\dfrac{3}{a}}$

32. $\dfrac{\dfrac{a}{a+2}+\dfrac{5}{a}}{\dfrac{a}{2a+4}+\dfrac{1}{3a}}$

33. $\dfrac{\dfrac{x}{x+y}+\dfrac{x}{y}}{\dfrac{x}{3x+3y}+\dfrac{y}{x}}$

34. $\dfrac{\dfrac{x}{x+y}+\dfrac{y}{x}}{\dfrac{x}{5x+5y}+\dfrac{x}{y}}$

35. $\dfrac{\dfrac{1}{x^2-1}+\dfrac{1}{x^2+4x+3}}{\dfrac{1}{x^2-1}+\dfrac{1}{x^2-3x+2}}$

36. $\dfrac{\dfrac{1}{x^2-4}+\dfrac{1}{x^2+3x+2}}{\dfrac{1}{x^2-4}+\dfrac{1}{x^2-4x+4}}$

37. $\dfrac{\dfrac{y}{y^2-4}-\dfrac{2y}{y^2+y-6}}{\dfrac{2y}{y^2-4}-\dfrac{y}{y^2+5y+6}}$

38. $\dfrac{\dfrac{y}{y^2-1}-\dfrac{3y}{y^2+5y+4}}{\dfrac{3y}{y^2-1}-\dfrac{y}{y^2-4y+3}}$

39. $\dfrac{\dfrac{1}{a^2+7a+12}+\dfrac{1}{a^2+a-6}}{\dfrac{1}{a^2+2a-8}+\dfrac{1}{a^2+5a+4}}$

40. $\dfrac{\dfrac{1}{a^2-5a+6}+\dfrac{1}{a^2-4a+3}}{\dfrac{1}{a^2-3a+2}+\dfrac{1}{a^2+3a-10}}$

41. $\dfrac{\dfrac{2}{x^2-7x+12}-\dfrac{1}{x^2+7x+10}}{\dfrac{2}{x^2-x-6}-\dfrac{1}{x^2+x-20}}$

42. $\dfrac{\dfrac{3}{x^2+2x-3}-\dfrac{1}{x^2-3x-10}}{\dfrac{3}{x^2-6x+5}-\dfrac{1}{x^2+5x+6}}$

Skill Maintenance

43. Solve for x: $\dfrac{a}{x+y}=b$.

44. If $f(x)=x^2+2$, find $f(-3)$.

45. The length of one rectangle is 3 less than the length of a second rectangle. The width of the first rectangle is 4 less than the width of the second. If the perimeter of the second rectangle is 1 less than twice the perimeter of the first, what are their respective perimeters?

46. A waitress received $14 in tips on Monday, $11 in tips on Tuesday, and $18 in tips on Wednesday. How much will she have to earn in tips on Thursday if her average tips for the four days is to be $15?

Synthesis

47. ◈ Explain how the graphs of

$$f(x)=\frac{1}{x} \quad \text{and} \quad g(x)=\frac{\dfrac{1}{x+2}}{\dfrac{x}{x+2}}$$

differ. You need not draw a graph.

48. ◈ Use Method 1 to explain why we "invert and multiply" when dividing one rational expression by another.

Simplify.

49. $\dfrac{5x^{-1}-5y^{-1}+10x^{-1}y^{-1}}{6x^{-1}-6y^{-1}+12x^{-1}y^{-1}}$

50. $\left[\dfrac{\dfrac{x+3}{x-3}+1}{\dfrac{x+3}{x-3}-1}\right]^4$

51. $(a^2-ab+b^2)^{-1}(a^2b^{-1}+b^2a^{-1})\times$
$\quad (a^{-2}-b^{-2})(a^{-2}+2a^{-1}b^{-1}+b^{-2})^{-1}$

Find the reciprocal and simplify.

52. $x^2-\dfrac{1}{x}$

53. $1+\dfrac{1}{1+\dfrac{1}{1+\dfrac{1}{1+\dfrac{1}{x}}}}$

54. For $f(x)=\dfrac{1}{1-x}$, find $f(f(x))$ and $f(f(f(x)))$ and simplify.

Find and simplify

$$\frac{f(x+h)-f(x)}{h}$$

for each rational function f in Exercises 55–58.

55. $f(x)=\dfrac{3}{x^2}$

56. $f(x)=\dfrac{5}{x}$

57. $f(x)=\dfrac{1}{1-x}$

58. $f(x)=\dfrac{x}{1+x}$

59. If

$$F(x)=\frac{3+\dfrac{1}{x}}{2-\dfrac{8}{x^2}},$$

find the domain of F.

60. If

$$G(x)=\frac{x-\dfrac{1}{x^2-1}}{\dfrac{1}{9}-\dfrac{1}{x^2-16}},$$

find the domain of G.

61. ▨ Use a grapher to check the answers to Exercises 1, 10, 25, 35, and 59.

6.4

Rational Equations

Solving Rational Equations • Rational Equations and Graphs

Solving Rational Equations

In Sections 6.1–6.3, we learned how to *simplify* expressions. We now learn to *solve* a new type of equation. A **rational equation** is an equation that contains one or more rational expressions. Here are some examples:

$$\frac{2}{3}-\frac{5}{6}=\frac{1}{x},\qquad \frac{x-1}{x-5}=\frac{4}{x^2-25},\qquad x^3+\frac{6}{x}=5.$$

As you will see in Section 6.5, equations of this type occur frequently in applications. To solve rational equations, recall from Section 6.3 that one way to *clear fractions* is to multiply by the LCD.

To solve a rational equation, we multiply on both sides by the LCD. This is called *clearing fractions*.

EXAMPLE 1   Solve: $\dfrac{x+4}{3x}+\dfrac{x+8}{5x}=2$.

Solution   The LCD is $3\cdot5\cdot x$. We multiply both sides of the equation by the LCD to clear all fractions:

$$3 \cdot 5 \cdot x \left(\frac{x+4}{3x} + \frac{x+8}{5x} \right) = 3 \cdot 5 \cdot x \cdot 2 \qquad \text{Multiplying by the LCD}$$

$$3 \cdot 5 \cdot x \cdot \frac{x+4}{3x} + 3 \cdot 5 \cdot x \cdot \frac{x+8}{5x} = 3 \cdot 5 \cdot x \cdot 2 \qquad \text{Using the distributive law}$$

$$\frac{3 \cdot 5 \cdot x \cdot (x+4)}{3x} + \frac{3 \cdot 5 \cdot x \cdot (x+8)}{5x} = 30x \qquad \text{Locating factors of 1}$$

$$5(x+4) + 3(x+8) = 30x \qquad \text{Removing factors of 1:}$$
$$\frac{3x}{3x} = 1; \frac{5x}{5x} = 1$$

$$5x + 20 + 3x + 24 = 30x \qquad \text{Using the distributive law}$$

$$8x + 44 = 30x$$

$$44 = 22x$$

$$2 = x.$$

Check:

$$\frac{x+4}{3x} + \frac{x+8}{5x} = 2$$

$$\frac{2+4}{3 \cdot 2} + \frac{2+8}{5 \cdot 2} \; ? \; 2$$

$$\frac{6}{6} + \frac{10}{10}$$

$$2 \; \Big| \; 2 \quad \text{TRUE}$$

The number 2 is the solution.   ❑

Note that when we clear fractions, all denominators "disappear." Then we have an equation without rational expressions, which we know how to solve.

When solving rational equations, it is extremely important to check possible solutions in the original equation. They may not check, even if we have made no error.

EXAMPLE 2

Solve: $\dfrac{x-1}{x-5} = \dfrac{4}{x-5}$.

Solution   Since the LCD is $x - 5$, we multiply by $x - 5$ on both sides:

$$(x-5) \cdot \frac{x-1}{x-5} = (x-5) \cdot \frac{4}{x-5}$$

$$x - 1 = 4$$

$$x = 5.$$

Check:

$$\frac{x-1}{x-5} = \frac{4}{x-5}$$

$$\frac{5-1}{5-5} \; ? \; \frac{4}{5-5}$$

$$\frac{4}{0} \; \Big| \; \frac{4}{0} \qquad \text{Division by 0 is undefined.}$$

We know that 5 is *not* a solution of the original equation because it results in division by 0. The equation has no solution. ❏

To help see why 5 is not a solution to Example 2, consider the fact that the multiplication principle for equations requires that we multiply by a *nonzero* number on both sides if we are to form an equivalent equation. When both sides of an equation are multiplied by an expression containing variables, it is possible that certain replacements will make that expression equal to 0. Thus it is safe to say that, *if* a solution of

$$\frac{x-1}{x-5} = \frac{4}{x-5}$$

exists, then the solution is a solution of $x - 1 = 4$. We *cannot* conclude that every solution of $x - 1 = 4$ is a solution of the original equation.

EXAMPLE 3

Solve: $\dfrac{x^2}{x-3} = \dfrac{9}{x-3}$.

Solution   Since the LCD is $x - 3$, we multiply by $x - 3$ on both sides:

$$(x-3)\cdot\frac{x^2}{x-3} = (x-3)\cdot\frac{9}{x-3}$$

$$x^2 = 9 \qquad \text{Simplifying}$$

$$x^2 - 9 = 0 \qquad \text{Getting 0 on one side}$$

$$(x+3)(x-3) = 0 \qquad \text{Factoring}$$

$$x = 3 \quad or \quad x = -3. \qquad \text{Using the principle of zero products}$$

Check: For 3:

$$\frac{x^2}{x-3} = \frac{9}{x-3}$$

$$\frac{3^2}{3-3} \;?\; \frac{9}{3-3}$$

$$\frac{9}{0} \;\Big|\; \frac{9}{0} \qquad \text{UNDEFINED}$$

For −3:

$$\frac{x^2}{x-3} = \frac{9}{x-3}$$

$$\frac{(-3)^2}{-3-3} \;?\; \frac{9}{-3-3}$$

$$\frac{9}{-6} \;\Big|\; \frac{9}{-6} \qquad \text{TRUE}$$

The number −3 is a solution, but 3 is not. ❏

EXAMPLE 4

Solve: $x + \dfrac{6}{x} = 5$.

Solution   Since the LCD is x, we multiply on both sides by x:

$$x\left(x + \frac{6}{x}\right) = 5\cdot x \qquad \text{Multiplying on both sides by } x$$

$$x\cdot x + x\cdot\frac{6}{x} = 5x \qquad \text{Using the distributive law}$$

$$x^2 + 6 = 5x \qquad \text{Simplifying}$$

$$x^2 - 5x + 6 = 0 \qquad \text{Getting 0 on one side}$$

$$(x-3)(x-2) = 0 \qquad \text{Factoring}$$

$$x = 3 \quad or \quad x = 2. \qquad \text{Using the principle of zero products}$$

Check: For 3:

$$x + \dfrac{6}{x} = 5$$

$$\begin{array}{c|c} 3 + \dfrac{6}{3} \;?\; 5 & \\[2ex] 3 + 2 & \\[1ex] 5 & 5 \;\;\text{TRUE} \end{array}$$

For 2:

$$x + \dfrac{6}{x} = 5$$

$$\begin{array}{c|c} 2 + \dfrac{6}{2} \;?\; 5 & \\[2ex] 2 + 3 & \\[1ex] 5 & 5 \;\;\text{TRUE} \end{array}$$

The solutions are 2 and 3.   ❑

EXAMPLE 5

Solve: $\dfrac{2}{x-1} = \dfrac{3}{x+1}$.

Solution   We multiply on both sides by the LCD, which is $(x-1)(x+1)$:

$$(x-1)(x+1) \cdot \dfrac{2}{(x-1)} = (x-1)(x+1) \cdot \dfrac{3}{x+1} \qquad \text{Multiplying}$$

$$2(x+1) = 3(x-1) \qquad\qquad \text{Simplifying}$$

$$2x + 2 = 3x - 3$$

$$5 = x.$$

As the student should confirm, 5 checks in the original equation. The number 5 is the solution.   ❑

EXAMPLE 6

Solve: $\dfrac{2}{x+5} + \dfrac{1}{x-5} = \dfrac{16}{x^2-25}$.

Solution   The LCD is $(x+5)(x-5)$. We multiply by $(x+5)(x-5)$ and then use the distributive law:

$$(x+5)(x-5) \cdot \left(\dfrac{2}{x+5} + \dfrac{1}{x-5} \right) = (x+5)(x-5) \cdot \dfrac{16}{x^2-25}$$

$$(x+5)(x-5) \cdot \dfrac{2}{x+5} + (x+5)(x-5) \cdot \dfrac{1}{x-5} = (x+5)(x-5) \cdot \dfrac{16}{x^2-25}$$

$$2(x-5) + (x+5) = 16$$

$$2x - 10 + x + 5 = 16$$

$$3x - 5 = 16$$

$$3x = 21$$

$$x = 7.$$

The check is left to the student. The solution is 7.   ❑

Rational Equations and Graphs

Let us return briefly to Example 1:

$$\dfrac{x+4}{3x} + \dfrac{x+8}{5x} = 2.$$

One way to look for a solution to this equation is to make a graph. We could use a

TECHNOLOGY CONNECTION

To solve Example 6 using a grapher, graph the curves

$$y_1 = \dfrac{2}{x+5} + \dfrac{1}{x-5}$$

and

$$y_2 = \dfrac{16}{x^2-25}$$

on the same set of axes. By using the Zoom and/or the Range features, the point at which the curves intersect can be found in the window $[6, 8, -0.5, 1]$. Rounded to the nearest hundredth, the x-value at this point is 7.00, confirming our work in Example 6.

TC1.   Use a grapher to solve the equations in each of Examples 1–5.

computer, a calculator, or a hand-drawn sketch of

$$f(x) = \frac{x+4}{3x} + \frac{x+8}{5x}$$

and then check visually for an x-value that is paired with the number 2. (Note that no y-value is paired with 0, since 0 is not in the domain of f.)

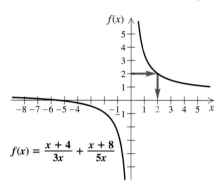

$$f(x) = \frac{x+4}{3x} + \frac{x+8}{5x}$$

It appears from the graph that $f(x) = 2$ when $x \approx 2$. Although making a graph is not the fastest or most precise method of solving a rational equation, it gives us a visualization and may be useful with problems that are more difficult to solve algebraically.

EXERCISE SET | 6.4

Solve.

1. $\dfrac{2}{5} + \dfrac{7}{8} = \dfrac{y}{20}$

2. $\dfrac{4}{5} + \dfrac{1}{3} = \dfrac{t}{9}$

3. $\dfrac{x}{3} - \dfrac{x}{4} = 12$

4. $\dfrac{y}{5} - \dfrac{y}{3} = 15$

5. $\dfrac{1}{3} - \dfrac{5}{6} = \dfrac{1}{x}$

6. $\dfrac{5}{8} - \dfrac{2}{5} = \dfrac{1}{y}$

7. $\dfrac{2}{3} - \dfrac{1}{5} = \dfrac{7}{3x}$

8. $\dfrac{1}{2} - \dfrac{2}{7} = \dfrac{3}{2x}$

9. $\dfrac{2}{6} + \dfrac{1}{2x} = \dfrac{1}{3}$

10. $\dfrac{12}{15} - \dfrac{1}{3x} = \dfrac{4}{5}$

11. $\dfrac{4}{z} + \dfrac{2}{z} = 3$

12. $\dfrac{4}{3y} - \dfrac{3}{y} = \dfrac{10}{3}$

13. $y + \dfrac{5}{y} = -6$

14. $x + \dfrac{4}{x} = -5$

15. $2x - \dfrac{6}{x} = 1$

16. $2x - \dfrac{15}{x} = 1$

17. $\dfrac{y-1}{y-3} = \dfrac{2}{y-3}$

18. $\dfrac{x-2}{x-4} = \dfrac{2}{x-4}$

19. $\dfrac{x+1}{x} = \dfrac{3}{2}$

20. $\dfrac{y+2}{y} = \dfrac{5}{3}$

21. $\dfrac{x-3}{x+2} = \dfrac{1}{5}$

22. $\dfrac{y-5}{y+1} = \dfrac{3}{5}$

23. $\dfrac{3}{y+1} = \dfrac{2}{y-3}$

24. $\dfrac{4}{x-1} = \dfrac{3}{x+2}$

25. $\dfrac{7}{5x-2} = \dfrac{5}{4x}$

26. $\dfrac{5}{y+4} = \dfrac{3}{y-2}$

27. $\dfrac{2}{x} - \dfrac{3}{x} + \dfrac{4}{x} = 5$

28. $\dfrac{4}{y} - \dfrac{6}{y} + \dfrac{8}{y} = 8$

29. $\dfrac{1}{2} - \dfrac{4}{9x} = \dfrac{4}{9} - \dfrac{1}{6x}$

30. $\dfrac{1}{3} - \dfrac{5}{4y} = \dfrac{3}{4} - \dfrac{1}{6y}$

31. $\dfrac{z}{z-1} = \dfrac{6}{z+1}$

32. $\dfrac{2y}{y+3} = \dfrac{-4}{y-7}$

33. $\dfrac{60}{x} - \dfrac{60}{x-5} = \dfrac{2}{x}$

34. $\dfrac{50}{y} - \dfrac{50}{y-2} = \dfrac{4}{y}$

35. $\dfrac{x}{x-2} + \dfrac{x}{x^2-4} = \dfrac{x+3}{x+2}$

36. $\dfrac{3}{y-2} + \dfrac{2y}{4-y^2} = \dfrac{5}{y+2}$

37. $\dfrac{a}{2a-6} - \dfrac{3}{a^2-6a+9} = \dfrac{a-2}{3a-9}$

38. $\dfrac{2}{x+4} + \dfrac{2x-1}{x^2+2x-8} = \dfrac{1}{x-2}$

39. $\dfrac{2x+3}{x-1} = \dfrac{10}{x^2-1} + \dfrac{2x-3}{x+1}$

40. $\dfrac{5}{y+1} + \dfrac{3y+5}{y^2+4y+3} = \dfrac{2}{y+3}$

Skill Maintenance _____

41. Factor completely: $81x^4 - y^4$.

42. Determine whether each of the following systems is consistent or inconsistent.
   a) $2x - 3y = 4,$
   $4x - 6y = 7$
   b) $x + 3y = 2,$
   $2x - 3y = 1$

43. There are 70 questions on a test. The questions are either multiple-choice, true–false, or fill-ins. There are twice as many true–false as fill-in and 5 fewer multiple-choice than true–false. How many of each type of question are there on the test?

44. Find two consecutive even numbers whose product is 288.

Synthesis _____

45. ◈ When checking a possible solution of a rational equation, is it sufficient to check that the "solution" does not make any denominator equal 0? Why or why not?

46. ◈ Explain how one can easily produce rational equations for which no solution exists. (*Hint:* Examine Example 2.)

For each pair of functions f and g, find all values of a for which $f(a) = g(a)$.

47. $f(x) = \dfrac{x - \frac{3}{2}}{x + \frac{2}{3}}, \quad g(x) = \dfrac{x + \frac{1}{2}}{x - \frac{2}{3}}$

48. $f(x) = \dfrac{2 + \frac{x}{2}}{2 - \frac{x}{4}}, \quad g(x) = \dfrac{2}{\frac{x}{4} - 2}$

Solve.

49. $\left(\dfrac{1}{1+x} + \dfrac{x}{1-x}\right) \div \left(\dfrac{x}{1+x} - \dfrac{1}{1-x}\right) = -1$

50. $\dfrac{x+3}{x+2} - \dfrac{x+4}{x+3} = \dfrac{x+5}{x+4} - \dfrac{x+6}{x+5}$

51. ▦ $\dfrac{2.315}{y} - \dfrac{12.6}{17.4} = \dfrac{6.71}{7} + 0.763$

52. ▦ $\dfrac{6.034}{x} - 43.17 = \dfrac{0.793}{x} + 18.15$

Equations that are true for any possible replacement of the variables are called *identities*. Determine whether each of the following equations is an identity.

53. $\dfrac{x^2+6x-16}{x-2} = x+8, \, x \neq 2$

54. $\dfrac{x^3+8}{x^2-4} = \dfrac{x^2-2x+4}{x-2}, \, x \neq -2, x \neq 2$

55. ◩ Use a grapher to solve Exercises 13, 25, 31, 39, 51, and 52.

6.5

Problem Solving Using Rational Equations

Problems Involving Work • **Problems Involving Motion**

Now that we are able to solve rational equations, we can use that skill to solve certain problems that we could not have handled before. The problem-solving steps are the same as before.

EXAMPLE 1

If a certain number is added to 5 times the reciprocal of 2 more than that number, the result is 4. Find the number.

Solution

1. FAMILIARIZE. We let y = the number in question.

2. TRANSLATE.

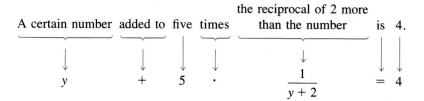

3. CARRY OUT. We solve as follows:

$$y + 5 \cdot \frac{1}{y + 2} = 4$$

$$(y + 2) \cdot y + (y + 2) \cdot 5 \cdot \frac{1}{y + 2} = (y + 2) \cdot 4 \qquad \text{Multiplying by the LCD on both sides}$$

$$y^2 + 2y + 5 = 4y + 8 \qquad \text{Simplifying}$$

$$y^2 - 2y - 3 = 0 \qquad \text{Collecting all like terms on one side}$$

$$(y - 3)(y + 1) = 0 \qquad \text{Factoring}$$

$$y - 3 = 0 \quad or \quad y + 1 = 0 \qquad \text{Principle of zero products}$$

$$y = 3 \quad or \qquad y = -1.$$

4. CHECK. The possible solutions are 3 and -1. We check both in the conditions of the problem.

Number:	3	-1
2 more than the number:	3 + 2, or 5	$-1 + 2$, or 1
Reciprocal of 2 more than the number:	$\frac{1}{5}$	1
5 times the reciprocal of 2 more than the number:	$5 \cdot \frac{1}{5}$, or 1	$5 \cdot 1$, or 5
Sum of number and 5 times the reciprocal of 2 more than the number:	3 + 1 = 4	$-1 + 5 = 4$

The numbers check.

5. STATE. There are two numbers satisfying the conditions of the problem: 3 and -1.

Problems Involving Work

EXAMPLE 2

Lon can mow a lawn in 4 hr. Penny can mow the same lawn in 5 hr. How long would it take both of them, working together, to mow the lawn?

Solution

1. FAMILIARIZE. We familiarize ourselves with the problem by considering two *incorrect* ways of translating the problem to mathematical language.

a) A common *incorrect* way to translate the problem is to add the two times:

$$4 \text{ hr} + 5 \text{ hr} = 9 \text{ hr}.$$

Now think about this. Lon can do the job alone in 4 hr. If Lon and Penny work together, whatever time it takes them should be *less* than 4 hr. Thus we reject 9 hr as a solution, but we do have a partial check on any answer we get. The answer should be less than 4 hr.

b) Another *incorrect* way to translate the problem is as follows. Suppose the two people split up the mowing job in such a way that Lon mows half the lawn and Penny mows the other half. Then

Lon mows $\frac{1}{2}$ the lawn in $\frac{1}{2}(4 \text{ hr})$, or 2 hr,

and

Penny mows $\frac{1}{2}$ the lawn in $\frac{1}{2}(5 \text{ hr})$, or $2\frac{1}{2}$ hr.

But time would be wasted since Lon would finish $\frac{1}{2}$ hr earlier than Penny. In effect, they have not worked together to get the job done as fast as possible. Note that if Lon helps Penny after completing his half, the entire job should be done in a time between 2 hr and $2\frac{1}{2}$ hr.

We proceed to a translation by considering how much of the job is finished in 1 hr, 2 hr, 3 hr, and so on. It takes Lon 4 hr to mow the entire lawn. Thus in 1 hr, he can mow $\frac{1}{4}$ of the lawn. It takes Penny 5 hr to mow the entire lawn, so in 1 hr, she can mow $\frac{1}{5}$ of the lawn. Working together, they can mow

$\frac{1}{4} + \frac{1}{5}$, or $\frac{9}{20}$ of the lawn in 1 hr.

In 2 hr, Lon can mow $2\left(\frac{1}{4}\right)$ of the lawn and Penny can mow $2\left(\frac{1}{5}\right)$ of the lawn. Working together, they can mow

$2\left(\frac{1}{4}\right) + 2\left(\frac{1}{5}\right)$, or $\frac{9}{10}$ of the lawn in 2 hr.

Continuing this reasoning, we can form a table like the following one.

Time	Fraction of the Lawn Mowed		
	By Lon	**By Penny**	**Together**
1 hr	$\frac{1}{4}$	$\frac{1}{5}$	$\frac{1}{4} + \frac{1}{5}$, or $\frac{9}{20}$
2 hr	$2\left(\frac{1}{4}\right)$	$2\left(\frac{1}{5}\right)$	$2\left(\frac{1}{4}\right) + 2\left(\frac{1}{5}\right)$, or $\frac{9}{10}$
3 hr	$3\left(\frac{1}{4}\right)$	$3\left(\frac{1}{5}\right)$	$3\left(\frac{1}{4}\right) + 3\left(\frac{1}{5}\right)$, or $1\frac{7}{20}$
t hr	$t\left(\frac{1}{4}\right)$	$t\left(\frac{1}{5}\right)$	$t\left(\frac{1}{4}\right) + t\left(\frac{1}{5}\right)$

From the table, we see that if they work for 3 hr, they will have mowed $1\frac{7}{20}$ lawns, which is more than needs to be done. We want to find the number of hours t required for Lon and Penny to mow exactly one lawn.

2. TRANSLATE.   From the table, we see that t must be some number for which

$$t\left(\frac{1}{4}\right) + t\left(\frac{1}{5}\right) = 1,$$

Portion of work done by Lon in t hr   Portion of work done by Penny in t hr

$$\left(\frac{t}{4}\right) + \left(\frac{t}{5}\right) = 1,$$

or

where 1 represents the idea that one entire job is completed in t hours.

3. CARRY OUT.   We solve the equation:

$$\frac{t}{4} + \frac{t}{5} = 1$$

$$20\left(\frac{t}{4} + \frac{t}{5}\right) = 20 \cdot 1 \qquad \text{Multiplying by the LCD}$$

$$20 \cdot \frac{t}{4} + 20 \cdot \frac{t}{5} = 20 \qquad \text{Using the distributive law}$$

$$5t + 4t = 20$$

$$9t = 20$$

$$t = \frac{20}{9}, \quad \text{or } 2\frac{2}{9} \text{ hr.}$$

4. CHECK.   In $\frac{20}{9}$ hr, Lon mows $\frac{20}{9} \cdot \frac{1}{4}$, or $\frac{5}{9}$, of the lawn and Penny mows $\frac{20}{9} \cdot \frac{1}{5}$, or $\frac{4}{9}$, of the lawn. Together, they mow $\frac{5}{9} + \frac{4}{9}$, or 1 lawn. The fact that our solution is between 2 and $2\frac{1}{2}$ hr (see step 1 above) is also a check.

5. STATE.   It will take $2\frac{2}{9}$ hr for Lon and Penny, together, to mow the lawn.

❏

EXAMPLE 3

It takes Red 9 hr longer to build a wall than it takes Mort. If they work together, they can build the wall in 20 hr. How long would it take each, working alone, to build the wall?

Solution

1. FAMILIARIZE.   Unlike Example 2, this problem does not provide us with the times required by the individuals to do the job alone. Let's have $t =$ the amount of time it would take Mort working alone and $t + 9 =$ the amount of time it would take Red working alone.

2. TRANSLATE. With the same reasoning that we used in Example 2, we see that Mort can build $1/t$ of a wall in 1 hr and Red can build $1/(t + 9)$ of a wall in 1 hr. In 20 hr, Mort builds $20(1/t)$ or $20/t$ of the wall and Red builds $20[1/(t + 9)]$ or $20/(t + 9)$ of the wall. Since Mort and Red complete 1 entire wall in 20 hr, we have

Mort's portion of the work $\qquad \dfrac{20}{t} + \dfrac{20}{t + 9} = 1.$ Red's portion of the work

3. CARRY OUT. We solve the equation:

$$\frac{20}{t} + \frac{20}{t + 9} = 1$$

$$t(t + 9)\left(\frac{20}{t} + \frac{20}{t + 9}\right) = t(t + 9)1 \qquad \text{Multiplying by the LCD}$$

$$(t + 9)20 + t \cdot 20 = t(t + 9) \qquad \text{Distributing and simplifying}$$

$$40t + 180 = t^2 + 9t$$

$$0 = t^2 - 31t - 180 \qquad \text{Getting 0 on one side}$$

$$0 = (t - 36)(t + 5) \qquad \text{Factoring}$$

$$t - 36 = 0 \quad or \quad t + 5 = 0 \qquad \text{Principle of zero products}$$

$$t = 36 \quad or \qquad t = -5.$$

4. CHECK. Since negative time has no meaning in the problem, -5 is not a solution of the original problem. The number 36 checks since, if Mort took 36 hr alone and Red took $36 + 9 = 45$ hr alone, in 20 hr they would have completed

$$\frac{20}{36} + \frac{20}{45} = \frac{5}{9} + \frac{4}{9} = 1 \text{ wall.}$$

5. STATE. It would take Mort 36 hr to build the wall alone, and Red 45 hr.

The equations used in Examples 2 and 3 can be generalized, as follows.

If

$a =$ the time needed for A to complete the work alone,

$b =$ the time needed for B to complete the work alone, and

$t =$ the time needed for A and B to complete the work together,

then

$$\frac{t}{a} + \frac{t}{b} = 1.$$

Problems Involving Motion

Problems dealing with distance, rate (or speed), and time are called **motion problems**. To translate them, we use either the basic motion formula, $d = rt$, or either of two formulas $r = d/t$ or $t = d/r$, which can be derived from $d = rt$.

EXAMPLE 4

A racer is bicycling 15 km/h faster than a person on a touring bike. In the time it takes the racer to travel 80 km, the person on the touring bike has gone 50 km. Find the speed of each biker.

Solution

1. FAMILIARIZE. Let's guess that the person on the touring bike is going 10 km/h. The racer would then be traveling 10 + 15, or 25 km/h. At 25 km/h, the racer will travel 80 km in $\frac{80}{25} = 3.2$ hr. Going 10 km/h, the touring bike will cover 50 km in $\frac{50}{10} = 5$ hr. Since 3.2 ≠ 5, our guess was wrong, but we can see that if r = the rate, in kilometers per hour, of the slower bike, then the rate of the racer = $r + 15$.

   Drawing a sketch and constructing a table can be helpful.

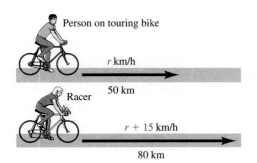

	Distance	Speed	Time
Touring Bike	50	r	t
Racing Bike	80	$r + 15$	t

2. TRANSLATE. By looking at how we checked our guess, we see that in the **Time** column of the table the t's can be replaced, using the formula Time = Distance/Rate:

	Distance	Speed	Time
Touring Bike	50	r	$50/r$
Racing Bike	80	$r + 15$	$80/(r + 15)$

   Since we are told that the times must be the same, we can write an equation:

   $$\frac{50}{r} = \frac{80}{r + 15}.$$

3. CARRY OUT. We solve the equation:

   $$\frac{50}{r} = \frac{80}{r + 15}$$

   $$r(r + 15)\frac{50}{r} = r(r + 15)\frac{80}{r + 15} \qquad \text{Multiplying by the LCD}$$

   $$50r + 750 = 80r \qquad \text{Simplifying}$$

   $$750 = 30r$$

   $$25 = r.$$

4. CHECK. If our answer checks, the touring bike is going 25 km/h and the racing bike is going 25 + 15 = 40 km/h.

   Traveling 80 km at 40 km/h, the racer is riding for $\frac{80}{40} = 2$ hr. Traveling 50 km at 25 km/h, the person on the touring bike is riding for $\frac{50}{25} = 2$ hr. Our answer checks since the two times are the same.

5. STATE.   The racer's speed is 40 km/h, and the person on the touring bike is traveling at a speed of 25 km/h.   ❑

In the following example, although the distance is the same in both directions, the key to the translation lies in an additional piece of given information.

EXAMPLE 5

Sandy's tugboat goes 10 mph in still water. It travels 24 mi upstream and back in a total time of 5 hr. What is the speed of the current?

Solution

1. FAMILIARIZE.   Let's guess that the speed of the current is 4 mph. The tugboat would then be moving $10 - 4 = 6$ mph upstream and $10 + 4 = 14$ mph downstream. The tugboat would require $\frac{24}{6} = 4$ hr to travel 24 mi upstream and $\frac{24}{14} = 1\frac{5}{7}$ hr to travel 24 mi downstream. Since the total time, $4 + 1\frac{5}{7} = 5\frac{5}{7}$ hr, is not the 5 hr mentioned in the problem, we know that our guess is wrong.

Suppose that the current's speed $= c$ mph. From our guess, we see that the tugboat would then travel $10 - c$ mph when going upstream and $10 + c$ mph when going downstream.

A sketch and table can help display the information.

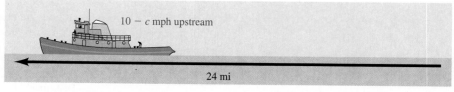

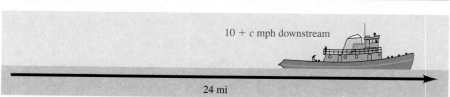

	Distance	Speed	Time
Upstream	24	$10 - c$	t_1
Downstream	24	$10 + c$	t_2

2. TRANSLATE.   From examining our guess, we see that the time traveled can be represented using the formula Time = Distance/Rate:

	Distance	Speed	Time
Upstream	24	$10 - c$	$24/(10 - c)$
Downstream	24	$10 + c$	$24/(10 + c)$

Since the total time upstream and back is 5 hr, we use the last column of the table to form an equation:

$$\frac{24}{10-c}+\frac{24}{10+c}=5.$$

3. CARRY OUT. We solve the equation:

$$\frac{24}{10-c}+\frac{24}{10+c}=5$$

$$(10-c)(10+c)\left[\frac{24}{10-c}+\frac{24}{10+c}\right]=(10-c)(10+c)5 \qquad \text{Multiplying by the LCD}$$

$$24(10+c)+24(10-c)=(100-c^2)5$$

$$480=500-5c^2 \qquad \text{Simplifying}$$

$$5c^2-20=0$$

$$5(c^2-4)=0$$

$$5(c-2)(c+2)=0$$

$$c=2 \quad or \quad c=-2.$$

4. CHECK. Since speed cannot be negative in this problem, -2 cannot be a solution. The student should confirm that 2 checks in the original problem.

5. STATE. The speed of the current is 2 mph.   ❑

EXERCISE SET | 6.5

Solve.

1. The reciprocal of 5 plus the reciprocal of 7 is the reciprocal of what number?

2. The reciprocal of 3 plus the reciprocal of 6 is the reciprocal of what number?

3. The sum of a number and 6 times its reciprocal is -5. Find the number.

4. The sum of a number and 21 times its reciprocal is -10. Find the number.

5. The reciprocal of the product of two consecutive integers is $\frac{1}{72}$. Find the two integers.

6. The reciprocal of the product of two consecutive integers is $\frac{1}{42}$. Find the two integers.

7. Sam, an experienced shipping clerk, can fill a certain order in 5 hr. Willy, a new clerk, needs 9 hr to complete the same job. Working together, how long will it take them to fill the order?

8. Paul can paint a room in 4 hr. Sally can paint the same room in 3 hr. Working together, how long will it take them to paint the room?

9. A swimming pool can be filled in 12 hr if water enters through a pipe alone or in 30 hr if water enters through a hose alone. If water is entering through both the pipe and the hose, how long will it take to fill the pool?

10. A tank can be filled in 18 hr by pipe A alone and in 22 hr by pipe B alone. How long will it take to fill the tank if both pipes are working?

11. Bill can clear a lot in 5.5 hr. His partner can complete the same job in 7.5 hr. How long will it take them to clear the lot working together?

12. Pronto Press can print an order of booklets in 4.5 hr. Red Dot Printers can do the same job in 5.5 hr. How long will it take if both presses are used?

13. Claudia can paint a neighbor's house 4 times as fast as Jan can. The year they worked together it took them 8 days. How long would it take each to paint the house alone?

14. Zsuzanna can deliver papers 3 times as fast as Stan can. If they work together, it takes them 1 hr. How long would it take each to deliver the papers alone?

15. Rosita can wax her car in 2 hr. When she works together with Helga, they can wax the car in 45 min. How long would it take Helga, working by herself, to wax the car?

16. Hannah can sand the living room floor in 3 hr. When she works together with Henri, the job takes 2 hr. How long would it take Henri, working by himself, to sand the floor?

17. Jake can cut and split a cord of firewood in 6 fewer hr than Skyler can. When they work together, it takes them 4 hr. How long would it take each of them to do the job alone?

18. Sara takes 3 hr longer to paint a floor than it takes Kate. When they work together, it takes them 2 hr. How long would each take to do the job alone?

19. Together, it takes John and Deb 2 hr 55 min to sort recyclables. Alone, John would require 2 more hr than Deb. How long would it take Deb to do the job alone? (*Hint:* Convert minutes to hours or hours to minutes.)

20. Together, Larry and Mo require 4 hr 48 min to pave a driveway. Alone, Larry would require 4 hr more than Mo. How long would it take Mo to do the job alone? (*Hint:* Convert minutes to hours.)

21. A new photocopier works twice as fast as an old one. When the machines work together, a university can produce all its staff manuals in 15 hr. Find

the time it would take each machine, working alone, to complete the same job.

22. Working together, Hans and Gina can paint a floor in 1.5 hr. Hans takes 4 hr longer than Gina does when working alone. How long would it take Gina alone to paint the floor?

23. The speed of a stream is 3 mph. A boat travels 4 mi upstream in the same time it takes to travel 10 mi downstream. What is the speed of the boat in still water?

24. The speed of a stream is 4 mph. A boat travels 6 mi upstream in the same time it takes to travel 12 mi downstream. What is the speed of the boat in still water?

25. The speed of a moving sidewalk at an airport is 7 ft/sec. Once on a moving sidewalk, a person can walk 80 ft forward in the same time it takes to walk 15 ft in the opposite direction. At what rate would the person walk on a nonmoving sidewalk?

26. The speed of a moving sidewalk at an airport is 6 ft/sec. Once on a moving sidewalk, a person can walk 70 ft forward in the same time it takes to walk 20 ft in the opposite direction. At what rate would the person walk on a nonmoving sidewalk?

27. Rosanna walks 2 mph slower than Simone. In the time it takes Simone to walk 8 mi, Rosanna walks 5 mi. Find the speed of each person.

28. A local bus travels 7 mph slower than the express. If the express travels 90 mi in the time it takes the local to travel 75 mi, find the speed of each bus.

29. The A train goes 12 mph slower than the B train. The A train travels 230 mi in the same time that the B train travels 290 mi. Find the speed of each train.

30. The speed of a passenger train is 14 mph faster

than the speed of a freight train. The passenger train travels 400 mi in the same time that the freight train travels 330 mi. Find the speed of each train.

31. A steamboat travels 10 km/h faster than a freighter. The steamboat travels 75 km in the same time that the freighter travels 50 km. Find the speed of each boat.

32. Jaime's moped travels 8 km/h faster than Mara's. Jaime travels 69 km in the same time that Mara travels 45 km. Find the speed of each person's moped.

33. Suzie has a boat that travels 15 km/h in still water. She rides 140 km downriver in the same time it takes to ride 35 km upstream. What is the speed of the river?

34. A paddleboat travels 2 km/h in still water. The boat is paddled 4 km downstream in a river in the same time it takes to go 1 km upstream. What is the speed of the river?

35. A barge moves 7 km/h in still water. It travels 45 km upriver and 45 km downriver in a total time of 14 hr. What is the speed of the current?

36. Janet bicycles 9 mph with no wind. She bikes 24 mi against the wind and 24 mi with the wind in a total time of 6 hr. Find the wind speed.

37. A plane travels 100 mph in still air. It travels 240 mi into the wind and 240 mi with the wind in a total time of 5 hr. Find the wind speed.

38. Al swims 55 m per minute in still water. He swims 150 m upstream and 150 m downstream in a total time of 5.5 min. What is the speed of the current?

39. A car traveled 120 mi at a certain speed. If the speed had been 10 mph faster, the trip could have been made in 2 hr less time. Find the speed.

40. A boat travels 45 mi upstream and 45 mi back. The time required for the round trip is 8 hr. The speed of the stream is 3 mph. Find the speed of the boat in still water.

Skill Maintenance

41. 8% of what number is 480?

42. Determine the domain of f if
$$f(x) = \frac{x-5}{x^2-4x-5}.$$

43. Solve: $|x-2| = 9$.

44. Combine similar terms:
$$4y - 5xy^2 + 6xy - 3xy^2 - 2y.$$

Synthesis

45. ◈ Write a work problem for a classmate to solve. Devise the problem so that the solution is "Jen takes 5 hr and Pablo takes 6 hr to complete the job alone."

46. ◈ Write a work problem for a classmate to solve. Devise the problem so that the solution is "Liane and Michele will take 4 hr to complete the job, working together."

47. A boat travels 96 km downstream in 4 hr. It travels 28 km upstream in 7 hr. Find the speed of the boat and the speed of the stream.

48. An airplane carries enough fuel for 6 hr of flight time, and its speed in still air is 240 mph. It leaves an airport against a wind of 40 mph and returns to the same airport with a wind of 40 mph. How far can it fly under those conditions without refueling?

49. A motor boat travels 3 times as fast as the current. A trip up the river and back takes 10 hr, and the total distance of the trip is 100 km. Find the speed of the current.

50. Melissa drives to work at 50 mph and arrives 1 min late. She drives to work at 60 mph and arrives 5 min early. How far does Melissa live from work?

51. A tank can be filled in 9 hr and drained in 11 hr. How long will it take to fill the tank if the drain is left open?

52. A tub can be filled in 10 min and drained in 8 min. How long will it take to empty a full tub if the water is left on?

53. At what time after 4:00 will the minute hand and the hour hand of a clock first be in the same position?

54. At what time after 10:30 will the hands of a clock first be perpendicular?

Average speed is defined as *total distance divided by total time.*

55. Lenore drove 200 km. For the first 100 km of the trip, she drove at a speed of 40 km/h. For the second half of the trip, she traveled at a speed of 60 km/h. What was the average speed for the entire trip? (It was *not* 50 km/h.)

56. For the first 50 mi of a 100-mi trip, Chip drove 40 mph. What speed would he have to travel for the last half of the trip so that the average speed for the entire trip would be 45 mph?

6.6

Division of Polynomials

Divisor a Monomial • Divisor a Polynomial

Recall that rational expressions indicate division. Division of polynomials, like division of real numbers, makes great use of our multiplication and subtraction skills.

Divisor a Monomial

We first consider division by a monomial. When we are dividing a monomial by a monomial, we can use our rules of exponents and subtract exponents when variables in the bases are the same. (We studied this in Section 1.5.) For example,

$$\frac{45x^{10}}{3x^4} = 15x^{10-4} = 15x^6, \qquad \frac{48a^2b^5}{-3ab^2} = \frac{48}{-3}a^{2-1}b^{5-2} = -16ab^3.$$

To divide a monomial into a polynomial, we break up the division into a sum of quotients of monomials. This uses the rule for addition using fractional notation in reverse. That is, since

$$\frac{A}{C} + \frac{B}{C} = \frac{A+B}{C}, \quad \text{we know that} \quad \frac{A+B}{C} = \frac{A}{C} + \frac{B}{C}.$$

EXAMPLE 1

Divide $12x^3 + 8x^2 + x + 4$ by $4x$.

Solution

$$\frac{12x^3 + 8x^2 + x + 4}{4x} \qquad \text{Writing a fractional expression}$$

$$= \frac{12x^3}{4x} + \frac{8x^2}{4x} + \frac{x}{4x} + \frac{4}{4x} \qquad \text{Writing as a sum of quotients}$$

$$= 3x^2 + 2x + \frac{1}{4} + \frac{1}{x} \qquad \text{Performing the four indicated divisions} \qquad ❏$$

EXAMPLE 2

Divide: $(8x^4y^5 - 3x^3y^4 + 5x^2y^3) \div x^2y^3$.

Solution

$$\frac{8x^4y^5 - 3x^3y^4 + 5x^2y^3}{x^2y^3} = \frac{8x^4y^5}{x^2y^3} - \frac{3x^3y^4}{x^2y^3} + \frac{5x^2y^3}{x^2y^3}$$

$$= 8x^2y^2 - 3xy + 5 \qquad ❏$$

You should try to write only the answer.

To divide a polynomial by a monomial, divide each term of the polynomial by the monomial.

Divisor a Polynomial

When the divisor has more than one term, we use a procedure very similar to long division in arithmetic.

EXAMPLE 3

Divide $2x^2 - 7x - 15$ by $x - 5$.

Solution   We have

$$
\begin{array}{r}
2x \\
x - 5 \overline{\smash{)}2x^2 - 7x - 15} \\
-(2x^2 - 10x) \\
\hline
3x
\end{array}
$$

Divide $2x^2$ by x: $2x^2/x = 2x$.

Multiply $x - 5$ by $2x$.

Change signs mentally and add:
$-7x + 10x = 3x$.

We now "bring down" the other term in the dividend, -15.

$$
\begin{array}{r}
2x + 3 \\
x - 5 \overline{\smash{)}2x^2 - 7x - 15} \\
2x^2 - 10x \\
\hline
3x - 15 \\
-(3x - 15) \\
\hline
0
\end{array}
$$

Divide $3x$ by x: $3x/x = 3$.

Multiply $x - 5$ by 3.

Subtract.

The quotient is $2x + 3$.

Check: $(x - 5)(2x + 3) = 2x^2 - 7x - 15$. The answer checks.   ❏

To understand why we perform long division as we do, note that Example 3 amounts to "filling in" a missing polynomial:

$$(x - 5)(\ \ ? \ \ ) = 2x^2 - 7x - 15.$$

We see that $2x$ must be in the unknown polynomial if we are to get the first term, $2x^2$, from the multiplication. To see what else is needed, note that

$$(x - 5)(2x) = 2x^2 - 10x \neq 2x^2 - 7x - 15.$$

The $2x^2 - 10x$ can be considered a (poor) approximation of $2x^2 - 7x - 15$. To see how far off our approximation is, we subtract:

$$
\begin{array}{r}
2x^2 - 7x - 15 \\
-(2x^2 - 10x) \\
\hline
3x - 15
\end{array}
$$

To get the needed terms, $3x - 15$, we include the term $+ 3$. We use 3 because $(x - 5) \cdot 3$ is $3x - 15$:

$$
\begin{aligned}
(x - 5)(2x + 3) &= 2x^2 - 10x + 3x - 15 \\
&= 2x^2 - 7x - 15.
\end{aligned}
$$

If we subtract now, we have a remainder of 0.

When a nonzero remainder occurs, how long should we keep dividing? We continue until the degree of the remainder is less than the degree of the divisor, as in the following example.

EXAMPLE 4

Divide $x^2 + 5x + 8$ by $x + 3$.

Solution   We have

$$
\begin{array}{r}
x \\
x + 3 \overline{)x^2 + 5x + 8} \\
x^2 + 3x \\
\hline
2x
\end{array}
$$

Divide the first term of the dividend by the first term of the divisor: $x^2/x = x$.

Multiply x above by $x + 3$.

Subtract.

The subtraction we have done is $(x^2 + 5x) - (x^2 + 3x)$. Remember: To subtract, add the opposite (change the sign of every term, then add).

We now "bring down" the next term of the dividend—in this case, 8—and repeat the process:

$$
\begin{array}{r}
x \; + 2 \\
x + 3 \overline{)x^2 + 5x + 8} \\
x^2 + 3x \\
\hline
2x + 8 \\
2x + 6 \\
\hline
2
\end{array}
$$

Divide the first term by the first term: $2x/x = 2$.

The 8 has been "brought down."

Multiply 2 by $x + 3$.

Subtract: $(2x + 8) - (2x + 6)$.

The quotient is $x + 2$, with remainder 2. Note that the degree of 2, the remainder, is less than the degree of $x + 3$, the divisor.

Check: $(x + 3)(x + 2) + 2 = x^2 + 5x + 6 + 2$ Add the remainder to the product.

$$= x^2 + 5x + 8$$

We can write our answer as $x + 2$, R2, or we can write our answer as

$$
\text{Quotient} + \cfrac{\text{Remainder}}{\text{Divisor}}
$$

$$
\underbrace{x + 2}_{} \;\; + \;\; \left(\cfrac{2}{x + 3} \right),
$$

which is how answers for the problems in this section will appear at the back of the book. Answers given in this form can be checked by multiplying:

$$(x + 3)\left[(x + 2) + \frac{2}{x + 3} \right] = (x + 3)(x + 2) + (x + 3)\frac{2}{x + 3}$$ Using the distributive law

$$= x^2 + 5x + 6 + 2$$

$$= x^2 + 5x + 8.$$ This was the dividend above.   ❑

You may have noticed that in each of our preceding examples all polynomials were written in descending order. When this is not the case, we rearrange terms before dividing.

Always remember:

1. Arrange polynomials in descending order.
2. If there are missing terms in the dividend, either write them with 0 coefficients or leave space for them.
3. Continue the long division process until the degree of the remainder is less than the degree of the divisor.

EXAMPLE 5

Divide: $(125y^3 - 8) \div (5y - 2)$.

Solution

$$
\begin{array}{r}
25y^2 + 10y + 4 \\
5y - 2\overline{)125y^3 + 0y^2 + 0y - 8} \\
\underline{125y^3 - 50y^2} \\
50y^2 + 0y \\
\underline{50y^2 - 20y} \\
20y - 8 \\
\underline{20y - 8} \\
0
\end{array}
$$

When there are missing terms, we can write them in as in this example, or leave space as in Example 6.

This subtraction is $125y^3 - (125y^3 - 50y^2)$. We get $50y^2$.

The answer is $25y^2 + 10y + 4$.

EXAMPLE 6

Divide: $(9x^2 + x^3 - 5) \div (x^2 - 1)$.

Solution

$(x^3 + 9x^2 - 5) \div (x^2 - 1)$ Rewriting in descending order

$$
\begin{array}{r}
x + 9 \\
x^2 - 1\overline{)x^3 + 9x^2 \quad\quad - 5} \\
\underline{x^3 \quad\quad - x} \\
9x^2 + x - 5 \\
\underline{9x^2 \quad\quad - 9} \\
x + 4
\end{array}
$$

Here we've left space for the missing term.

Subtracting: $x^3 + 9x^2 - (x^3 - x) = 9x^2 + x$

The degree of the remainder is less than the degree of the divisor, so we are finished.

The answer is $x + 9 + \dfrac{x + 4}{x^2 - 1}$.

EXERCISE SET | 6.6

Divide and check.

1. $\dfrac{30x^8 - 15x^6 + 40x^4}{5x^4}$

2. $\dfrac{24y^6 + 18y^5 - 36y^2}{6y^2}$

3. $\dfrac{-14a^3 + 28a^2 - 21a}{7a}$

4. $\dfrac{-32x^4 - 24x^3 - 12x^2}{4x}$

5. $(9y^4 - 18y^3 + 27y^2) \div 9y$

6. $(24a^3 + 28a^2 - 20a) \div 2a$

7. $(36x^6 - 18x^4 - 12x^2) \div (-6x)$

8. $(18y^7 - 27y^4 - 3y^2) \div (-3y^2)$

9. $(a^2b - a^3b^3 - a^5b^5) \div a^2b$

10. $(x^3y^2 - x^3y^3 - x^4y^2) \div x^2y^2$

11. $(6p^2q^2 - 9p^2q + 12pq^2) \div (-3pq)$

12. $(16y^4z^2 - 8y^6z^4 + 12y^8z^3) \div 4y^4z$

13. $(x^2 + 10x + 21) \div (x + 3)$

14. $(y^2 - 8y + 16) \div (y - 4)$

15. $(a^2 - 8a - 16) \div (a + 4)$

16. $(y^2 - 10y - 25) \div (y - 5)$

17. $(x^2 - 11x + 23) \div (x - 5)$

18. $(x^2 - 11x + 23) \div (x - 7)$

19. $(y^2 - 25) \div (y + 5)$

20. $(a^2 - 81) \div (a - 9)$
21. $(y^3 - 4y^2 + 3y - 6) \div (y - 2)$
22. $(x^3 - 5x^2 + 4x - 7) \div (x - 3)$
23. $(2x^3 + 3x^2 - x - 3) \div (x + 2)$
24. $(3x^3 - 5x^2 - 3x - 2) \div (x - 2)$
25. $(a^3 - a + 12) \div (a - 4)$
26. $(x^3 - x + 6) \div (x + 2)$
27. $(8x^3 + 27) \div (2x + 3)$
28. $(64y^3 - 8) \div (4y - 2)$
29. $(x^4 - x^2 - 42) \div (x^2 - 7)$
30. $(y^4 - y^2 - 54) \div (y^2 - 3)$
31. $(x^4 - x^2 - x + 2) \div (x - 1)$
32. $(y^4 - y^2 - y + 3) \div (y + 1)$
33. $(10y^3 + 6y^2 - 9y + 10) \div (5y - 2)$
34. $(6x^3 - 11x^2 + 11x - 2) \div (2x - 3)$
35. $(2x^4 - x^3 - 5x^2 + x - 6) \div (x^2 + 2)$
36. $(3x^4 + 2x^3 - 11x^2 - 2x + 5) \div (x^2 - 2)$
37. $(2x^5 + x^4 + 2x^3 + x) \div (x^2 + 1)$
38. $(2x^5 - 3x^3 + x^2 + 4) \div (x^2 - 1)$

Skill Maintenance

Solve.

39. $x^2 - 5x = 0$ **40.** $25y^2 = 64$

41. Find three consecutive positive integers such that the product of the first and second integers is 26 less than the product of the second and third integers.

42. If $f(x) = 2x^3$, find $f(-3a)$.

Synthesis

43. Explain how you could construct a second-degree polynomial that has a remainder of 3 when divided by $x + 1$.

44. Can the quotient of two sums always be rewritten as a sum of two quotients? Why or why not?

Divide.

45. $(x^4 - x^3y + x^2y^2 + 2x^2y - 2xy^2 + 2y^3) \div (x^2 - xy + y^2)$
46. $(4a^3b + 5a^2b^2 + a^4 + 2ab^3) \div (a^2 + 2b^2 + 3ab)$
47. $(x^4 - y^4) \div (x - y)$
48. $(a^7 + b^7) \div (a + b)$

Solve.

49. Find k so that when $x^3 - kx^2 + 3x + 7k$ is divided by $x + 2$, the remainder will be 0.

50. When $x^2 - 3x + 2k$ is divided by $x + 2$, the remainder is 7. Find k.

51. Let

$$f(x) = \frac{3x + 7}{x + 2}.$$

a) Use division to find an expression equivalent to $f(x)$. Then graph f.
b) On the same set of axes, sketch both $g(x) = 1/(x + 2)$ and $h(x) = 1/x$.
c) How do the graphs of f, g, and h compare?

6.7

Synthetic Division

Streamlining Long Division • **Using Synthetic Division**

To divide a polynomial by a binomial of the type $x - a$, we can streamline the usual procedure to develop a process called *synthetic division*.

Compare the following. In each stage, we attempt to write a bit less than in the previous stage, while retaining enough essentials to solve the problem. At the end, we will return to the usual polynomial notation.

Stage 1

When a polynomial is written in descending order, the coefficients provide the essential information:

$$
\begin{array}{r}
4x^2 + 5x + 11 \\
x - 2\overline{)4x^3 - 3x^2 + x + 7} \\
\underline{4x^3 - 8x^2} \\
5x^2 + x \\
\underline{5x^2 - 10x} \\
11x + 7 \\
\underline{11x - 22} \\
29
\end{array}
\qquad
\begin{array}{r}
4 + 5 + 11 \\
1 - 2\overline{)4 - 3 + 1 + 7} \\
\underline{4 - 8} \\
5 + 1 \\
\underline{5 - 10} \\
11 + 7 \\
\underline{11 - 22} \\
29
\end{array}
$$

Because the coefficient of x is 1 in the divisor, each time we multiply the divisor by a term in the answer, the leading coefficient of that product duplicates a coefficient in the answer. In the next stage, we don't bother to duplicate these numbers. We also show where -2 is used in the problem and stop writing 1 in the divisor.

Stage 2

$$
\begin{array}{r}
4x^2 + 5x + 11 \\
x - 2\overline{)4x^3 - 3x^2 + x + 7} \\
\underline{4x^3 - 8x^2} \\
5x^2 + x \\
\underline{5x^2 - 10x} \\
11x + 7 \\
\underline{11x - 22} \\
29
\end{array}
$$

$$
\begin{array}{r}
4 + 5 + 11 \\
-2\overline{)4 - 3 + 1 + 7} \\
(-2)4 \\
5 + 1 \\
(-2)5 \\
11 + 7 \\
(-2)11 \\
29
\end{array}
$$

Multiply: $4(-2) = -8$.
Subtract: $-3 - (-8) = 5$.
Multiply: $5(-2) = -10$.
Subtract: $1 - (-10) = 11$.
Multiply: $11(-2) = -22$.
Subtract: $7 - (-22) = 29$.

To simplify further, we now change the sign of the -2 in the divisor and, in exchange, *add* at each step in the long division.

Stage 3

$$
\begin{array}{r}
4x^2 + 5x + 11 \\
x - 2\overline{)4x^3 - 3x^2 + x + 7} \\
\underline{4x^3 - 8x^2} \\
5x^2 + x \\
\underline{5x^2 - 10x} \\
11x + 7 \\
\underline{11x - 22} \\
29
\end{array}
$$

$$
\begin{array}{r}
4 + 5 + 11 \\
2\overline{)4 - 3 + 1 + 7} \\
8 \\
5 + 1 \\
10 \\
11 + 7 \\
22 \\
29
\end{array}
$$

Replace the -2 with 2.
Multiply: $4 \cdot 2 = 8$.
Add: $-3 + 8 = 5$.
Multiply: $5 \cdot 2 = 10$.
Add: $1 + 10 = 11$.
Multiply: $11 \cdot 2 = 22$.
Add: $7 + 22 = 29$.

As you can see from the blue numbers, there is still some duplication that we can eliminate.

Stage 4

$$
\begin{array}{r}
4x^2 + 5x + 11 \\
x - 2\overline{)4x^3 - 3x^2 + x + 7} \\
\underline{4x^3 - 8x^2} \\
5x^2 + x \\
\underline{5x^2 - 10x} \\
11x + 7 \\
\underline{11x - 22} \\
29
\end{array}
$$

$$
\begin{array}{rrrr}
4 & 5 & 11 & \\
2\overline{)4} & -3 & 1 & 7 \\
& 8 & 10 & 22 \\
\hline
5 & 11 & 29 &
\end{array}
$$

Don't lose sight of how the products 8, 10, and 22 are found. Also, keep in mind that the 5 and 11 preceding the remainder 29 coincide with the 5 and 11 following the 4 on the top line. If we write a 4 to the left of 5 on the bottom line, we can dispense with the top line and read our answer from the bottom line. This final stage is commonly called **synthetic division.**

Stage 5

$$
\begin{array}{r}
4x^2 + 5x + 11 \\
x - 2\overline{)4x^3 - 3x^2 + x + 7} \\
\underline{4x^3 - 8x^2} \\
5x^2 + x \\
\underline{5x^2 - 10x} \\
11x + 7 \\
\underline{11x - 22} \\
29
\end{array}
$$

$$
\begin{array}{r|rrrr}
2 & 4 & -3 & 1 & 7 \\
& & 8 & 10 & 22 \\
\hline
& 4 & 5 & 11 & \boxed{29}
\end{array}
$$
← This is the remainder.

This is the zero-degree coefficient.

This is the first-degree coefficient.

This is the second-degree coefficient.

The quotient is $4x^2 + 5x + 11$. The remainder is 29.

> Remember that in order for this method to work, the divisor must be of the form $x - a$, that is, a variable minus a constant. The coefficient of the variable must be 1.

EXAMPLE 1

Use synthetic division to divide:

$$(x^3 + 6x^2 - x - 30) \div (x - 2).$$

Solution

A.
$$
\begin{array}{r|rrrr}
2 & 1 & 6 & -1 & -30 \\
& & & & \\
\hline
& 1 & & &
\end{array}
$$
Write the 2 of $x - 2$ and the coefficients of the dividend.

Bring down the first coefficient.

B.
$$
\begin{array}{r|rrrr}
2 & 1 & 6 & -1 & -30 \\
& & 2 & & \\
\hline
& 1 & 8 & &
\end{array}
$$
Multiply 1 by 2 to get 2.

Add 6 and 2.

C.
$$
\begin{array}{r|rrrr}
2 & 1 & 6 & -1 & -30 \\
& & 2 & 16 & \\
\hline
& 1 & 8 & 15 &
\end{array}
$$
Multiply 8 by 2.

Add −1 and 16.

D. $2\big|$ 1 6 −1 −30
          2 16   30   Multiply 15 by 2 and add.
       ‾‾‾‾‾‾‾‾‾‾‾‾‾‾‾
       1 8 15   0

The answer is $x^2 + 8x + 15$ with R0, or just $x^2 + 8x + 15$. ❑

EXAMPLE 2

Use synthetic division to divide.

a) $(2x^3 + 7x^2 - 5) \div (x + 3)$ **b)** $(x^3 + 4x^2 - x - 4) \div (x + 4)$
c) $(8x^5 - 6x^3 + x - 8) \div (x - 2)$

Solution

a) $(2x^3 + 7x^2 - 5) \div (x + 3)$

The dividend has no x-term, so we must write a 0 for its coefficient of x. Note that $x + 3 = x - (-3)$.

$-3\big|$ 2   7   0   −5
            −6 −3   9
         ‾‾‾‾‾‾‾‾‾‾‾‾‾‾‾
         2   1 −3 $\big|$ 4

The answer is $2x^2 + x - 3$, with R4, or $2x^2 + x - 3 + \dfrac{4}{x+3}$.

b) $(x^3 + 4x^2 - x - 4) \div (x + 4)$

$-4\big|$ 1   4 −1 −4
             −4   0   4
          ‾‾‾‾‾‾‾‾‾‾‾‾‾‾
          1   0 −1 $\big|$ 0

The answer is $x^2 - 1$.

c) $(8x^5 - 6x^3 + x - 8) \div (x - 2)$

$2\big|$ 8   0 −6   0   1   −8
            16 32 52 104 210
         ‾‾‾‾‾‾‾‾‾‾‾‾‾‾‾‾‾‾‾‾‾‾‾‾
         8 16 26 52 105 $\big|$ 202

The answer is $8x^4 + 16x^3 + 26x^2 + 52x + 105$ with R202, or

$$8x^4 + 16x^3 + 26x^2 + 52x + 105 + \frac{202}{x-2}.$$ ❑

EXERCISE SET | 6.7

Use synthetic division to divide.

1. $(x^3 - 2x^2 + 2x - 5) \div (x - 1)$
2. $(x^3 - 2x^2 + 2x - 5) \div (x + 1)$
3. $(a^2 + 11a - 19) \div (a + 4)$
4. $(a^2 + 11a - 19) \div (a - 4)$

5. $(x^3 - 7x^2 - 13x + 3) \div (x - 2)$
6. $(x^3 - 7x^2 - 13x + 3) \div (x + 2)$
7. $(3x^3 + 7x^2 - 4x + 3) \div (x + 3)$
8. $(3x^3 + 7x^2 - 4x + 3) \div (x - 3)$
9. $(y^3 - 3y + 10) \div (y - 2)$

10. $(x^3 - 2x^2 + 8) \div (x + 2)$

11. $(3x^4 - 25x^2 - 18) \div (x - 3)$

12. $(6y^4 + 15y^3 + 28y + 6) \div (y + 3)$

13. $(x^3 - 27) \div (x - 3)$

14. $(y^3 + 27) \div (y + 3)$

15. $(y^5 - 1) \div (y - 1)$

16. $(x^5 - 32) \div (x - 2)$

17. $(3x^4 + 8x^3 + 2x^2 - 7x - 4) \div (x + 2)$

18. $(2x^4 - x^3 - 5x^2 + x + 7) \div (x + 1)$

19. $(3x^3 + 7x^2 - x + 1) \div \left(x + \frac{1}{3}\right)$

20. $(8x^3 - 6x^2 + 7x - 1) \div \left(x - \frac{1}{2}\right)$

Skill Maintenance

Graph on a plane.

21. $2x - 3y < 6$

22. $5x + 3y \leq 15$

23. $y > 4$

24. $x \leq -2$

Synthesis

25. ◈   Describe how synthetic division can be used when the divisor is a linear polynomial $ax + b$, with $a > 1$.

26. ◈   For the polynomial function
$$P(x) = 8x^5 - 3x^4 + 7x - 4:$$

  a) By synthetic division, find the remainder when $P(x)$ is divided by $x - 2$.

  b) Using nested evaluation, find $P(2)$. (See Exercise Set 5.3.)

  c) Compare the answers and procedures from parts (a) and (b).

27. Let $f(x) = x^3 - 5x^2 + 5x - 4$.

  a) Use synthetic division to show that $x - 4$ is a factor of $x^3 - 5x^2 + 5x - 4$.

  b) Why does the result in part (a) indicate that $f(4) = 0$?

  c) Check that $f(4) = 0$ by substituting 4 into the given polynomial function.

6.8

Formulas and Applications

Solving for a Letter   •   Solving and Evaluating

Formulas occur frequently in applications of mathematics. Many formulas contain rational expressions, and to solve such formulas for a specified letter, we proceed as we do in solving rational equations.

EXAMPLE 1

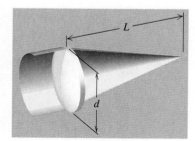

Optics.   The formula $f = L/d$ tells how to calculate a camera's "f-stop." In this formula, f is the f-stop, L is the focal length (approximately the distance from the lens to the film), and d is the diameter of the lens. Solve this formula for d.

Solution   We solve this rational equation using the method developed in Section 6.4:

$$f = \frac{L}{d}$$

$$d \cdot f = d \cdot \frac{L}{d} \qquad \text{Multiplying by the LCD to clear fractions}$$

$$df = L$$

$$df \cdot \frac{1}{f} = L \cdot \frac{1}{f} \qquad \text{Multiplying by } \frac{1}{f} \text{ on both sides}$$

$$d = \frac{L}{f}. \qquad \text{Simplifying}$$

The formula $d = L/f$ can now be used to determine the diameter of a lens if we know the focal length and the f-stop. ❏

EXAMPLE 2

Astronomy.   The formula

$$L = \frac{dR}{D - d},$$

where D is the diameter of the sun, d is the diameter of the earth, R is the earth's distance from the sun, and L is some fixed distance, is used in calculating when lunar eclipses occur. Solve for D.

Solution   We first clear fractions by multiplying by the LCD, which is $D - d$:

$$(D - d)L = (D - d)\frac{dR}{D - d}$$

$$(D - d)L = dR.$$

We do *not* multiply the factors on the left since we wish to get D all alone. Instead we multiply both sides by $1/L$ and then add d:

$$D - d = \frac{dR}{L} \qquad \text{Multiplying by } \frac{1}{L}$$

$$D = \frac{dR}{L} + d. \qquad \text{Adding } d$$

We now have D all alone on one side of the equation, and D does not appear on the other side, so we have solved the formula for D. ❏

Sometimes additional steps may be required to isolate a desired letter.

EXAMPLE 3

Solve the formula for p:

$$I = \frac{pT}{M + pn}.$$

Solution   We first clear fractions by multiplying by the LCD, $M + pn$:

$$I = \frac{pT}{M + pn}$$

$$I(M + pn) = \frac{pT}{M + pn}(M + pn)$$

$$IM + Ipn = pT. \qquad \text{Here, because } p \text{ appears on both sides,} \\ \text{we } did \text{ distribute on the left side.}$$

We must now get all terms containing p alone on one side:

$$IM = pT - Ipn \qquad \text{Adding } -Ipn$$

$$IM = p(T - In) \qquad \text{Factoring out } p$$

$$\frac{IM}{T - In} = p. \qquad \text{Multiplying by } \frac{1}{T - In}$$

We now have p alone on one side, and p does *not* appear on the other side, so we have solved the formula for p. ❏

EXAMPLE 4

Resistance. The formula

$$\frac{1}{R} = \frac{1}{r_1} + \frac{1}{r_2}$$

gives the resistance R of two resistors r_1 and r_2 connected in parallel.

a) Solve for r_2.

b) Find r_2 when $R = 3.75$ ohms and $r_1 = 6$ ohms.

Solution

a) To solve for r_2, we multiply by the LCD, Rr_1r_2:

$$\frac{1}{R} = \frac{1}{r_1} + \frac{1}{r_2}$$

$$Rr_1r_2 \cdot \frac{1}{R} = Rr_1r_2 \cdot \left[\frac{1}{r_1} + \frac{1}{r_2}\right] \qquad \text{Multiplying by the LCD}$$

$$Rr_1r_2 \cdot \frac{1}{R} = Rr_1r_2 \cdot \frac{1}{r_1} + Rr_1r_2 \cdot \frac{1}{r_2} \qquad \text{Using the distributive law}$$

$$\left. \begin{array}{l} \dfrac{\cancel{R}r_1r_2}{\cancel{R}} = \dfrac{R\cancel{r_1}r_2}{\cancel{r_1}} + \dfrac{Rr_1\cancel{r_2}}{\cancel{r_2}} \\[2mm] r_1r_2 = Rr_2 + Rr_1. \end{array} \right\} \qquad \begin{array}{l} \text{Simplifying by} \\ \text{removing factors of 1:} \\ \dfrac{R}{R} = 1; \dfrac{r_1}{r_1} = 1; \dfrac{r_2}{r_2} = 1 \end{array}$$

You might be tempted at this point to multiply by $1/r_1$ to get r_2 alone on the left, but note that there is an r_2 on the right. We must get all the terms involving r_2 on the *same side* of the equation:

$$r_1r_2 - Rr_2 = Rr_1 \qquad \text{Adding } -Rr_2$$

$$r_2(r_1 - R) = Rr_1. \qquad \text{Factoring out } r_2$$

Multiplying by $1/(r_1 - R)$ on both sides, we obtain the equation

$$r_2 = \frac{Rr_1}{r_1 - R}.$$

b) We use the equation for r_2, replacing R with 3.75 and r_1 with 6:

$$r_2 = \frac{3.75 \cdot 6}{6 - 3.75}$$

$$r_2 = 10.$$

The second resistor has a resistance of 10 ohms. ◻

To solve a formula containing rational expressions:

1. Clear fractions.
2. Multiply if necessary to remove parentheses.
3. Get all terms with the unknown alone on one side of the equation.
4. Factor out the unknown if it appears in more than one term.
5. Use the multiplication principle to get the unknown alone on one side.

EXERCISE SET | 6.8

Solve the formula for the specified letter.

1. $\dfrac{W_1}{W_2} = \dfrac{d_1}{d_2}$; d_1

2. $\dfrac{W_1}{W_2} = \dfrac{d_1}{d_2}$; W_1

3. $s = \dfrac{(v_1 + v_2)t}{2}$; t

4. $s = \dfrac{(v_1 + v_2)t}{2}$; v_1

5. $\dfrac{1}{R} = \dfrac{1}{r_1} + \dfrac{1}{r_2}$; r_1

6. $\dfrac{1}{R} = \dfrac{1}{r_1} + \dfrac{1}{r_2}$; R

7. $R = \dfrac{gs}{g + s}$; s

8. $R = \dfrac{gs}{g + s}$; g

9. $I = \dfrac{2V}{R + 2r}$; r

10. $I = \dfrac{2V}{R + 2r}$; R

11. $\dfrac{1}{p} + \dfrac{1}{q} = \dfrac{1}{f}$; q

12. $\dfrac{1}{p} + \dfrac{1}{q} = \dfrac{1}{f}$; p

13. $I = \dfrac{nE}{R + nr}$; r

14. $I = \dfrac{nE}{R + nr}$; n

15. $S = \dfrac{H}{m(t_1 - t_2)}$; H

16. $S = \dfrac{H}{m(t_1 - t_2)}$; t_1

17. $\dfrac{E}{e} = \dfrac{R + r}{r}$; e

18. $\dfrac{E}{e} = \dfrac{R + r}{r}$; r

19. $S = \dfrac{a - ar^n}{1 - r}$; a

20. $S = \dfrac{a}{1 - r}$; r

Problem Solving

21. The formula $A = 9R/I$ gives a pitcher's earned run average, where A is the earned run average, R is the number of earned runs, and I is the number of innings pitched. How many earned runs were given up if a pitcher's earned run average is 2.4 after 45 innings?

22. Two resistors are connected in parallel. Their resistances are, respectively, 8 ohms and 15 ohms. What is the resistance of the combination?

23. A resistor has a resistance of 50 ohms. What size resistor should be put with it, in parallel, in order to obtain a resistance of 5 ohms?

24. The area of a certain trapezoid is 25 cm². Its height is 5 cm and the length of one base is 4 cm. Find the length of the other base.

25. The formula

$$\frac{1}{t} = \frac{1}{u} + \frac{1}{v}$$

gives the total time t required for two workers to complete a job, if the workers' individual times are u and v. Solve for t.

26. The formula

$$A = \frac{2Tt + Qq}{2T + Q}$$

gives a student's average A after T tests and Q quizzes, where each test counts as 2 quizzes and t is the test average and q is the quiz average. Solve for Q.

27. The formula

$$v = \frac{d_2 - d_1}{t_2 - t_1}$$

gives an object's average speed v when it has gone d_1 miles in t_1 hours and d_2 miles in t_2 hours. Solve for t_2.

28. The formula

$$P = \frac{A}{1 + r}$$

can be used to determine the principal P that must be invested for one year at $(100 \cdot r)\%$ simple interest in order to have A dollars after a year. Solve for r.

29. At what yearly interest rate should $1600 be invested in order to have a total of $1712 after one year? (See Exercise 28.)

30. At what time will a car, averaging a speed of 60 mph, reach Philadelphia if it leaves New York at 2:00 A.M. and New York is 105 mi from Philadelphia? (See Exercise 27.)

31. The formula

$$I_t = \frac{I_f}{1 - T}$$

gives the *taxable interest rate* I_t equivalent to the *tax-free interest rate* I_f for a person in the $(100 \cdot T)\%$ tax bracket. Solve for T.

32. Solve the formula of Example 2 for d, the diameter of the earth.

33. The formula

$$\frac{V^2}{R^2} = \frac{2g}{R + h}$$

is used to find a satellite's *escape velocity V,* where R is a planet's radius, h is the satellite's height above the planet, and g is the planet's acceleration due to gravity. Solve for h.

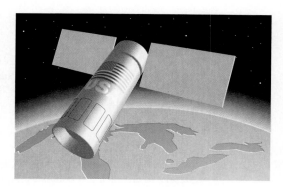

34. The formula

$$a = \frac{v_2 - v_1}{t_2 - t_1}$$

gives a particle's *average acceleration* when its velocity changes from v_1 at time t_1 to v_2 at time t_2. Solve for t_1.

35. A student with a test average of 79 and a quiz average of 90 has a grade calculated by using the formula found in Exercise 26. If a student's overall average is 84 and 5 quizzes were taken, how many tests were taken?

36. A line passing through two points has a slope of $-\frac{2}{5}$. If the coordinates of the points are $(x_1, 2)$ and $(2x_1, 8)$, find the coordinates of both points.

Skill Maintenance

37. Graph on a plane: $6x - y < 6$.

38. If $f(x) = x^3 - x$, find $f(2a)$.

39. Factor: $t^3 + 8b^3$.

40. Solve: $6x^2 = 11x + 35$.

Synthesis

41. [calculator] A satellite's escape velocity is 6.5 mi/sec, the radius of the earth is 3960 mi, and the acceleration due to gravity is 32.2 ft/sec^2. How far is the satellite from the center of the earth? (See Exercise 33.)

42. The *harmonic mean* of two numbers a and b is a number M such that the reciprocal of M is the average of the reciprocals of a and b. Find a formula for the harmonic mean.

43. Solve for x:

$$x^2\left(1 - \frac{2pq}{x}\right) = \frac{2p^2q^3 - pq^2x}{-q}.$$

44. The formula

$$a = \frac{\dfrac{d_4 - d_3}{t_4 - t_3} - \dfrac{d_2 - d_1}{t_2 - t_1}}{t_4 - t_2}$$

can be used to approximate average acceleration where the d's are distances and the t's are the corresponding times. Solve for t_1.

SUMMARY AND REVIEW | 6

IMPORTANT PROPERTIES AND FORMULAS

Addition: $\dfrac{A}{C} + \dfrac{B}{C} = \dfrac{A+B}{C}$ Subtraction: $\dfrac{A}{C} - \dfrac{B}{C} = \dfrac{A-B}{C}$

Multiplication: $\dfrac{A}{B} \cdot \dfrac{C}{D} = \dfrac{AC}{BD}$ Division: $\dfrac{A}{B} \div \dfrac{C}{D} = \dfrac{A}{B} \cdot \dfrac{D}{C}$

To find the least common multiple, LCM, use each factor the greatest number of times that it occurs in any one prime factorization.

Simplifying Complex Rational Expressions

I: By using the LCD

1. Find the LCD of all expressions *within* the complex rational expression.
2. Multiply the complex rational expression by 1, using the LCD to form the expression for 1.
3. Distribute and simplify. Neither the numerator nor the denominator of the complex rational expression should contain a rational expression.
4. Factor and simplify, if possible.

II: By first adding or subtracting

1. Add or subtract, as necessary, to get a single rational expression in the numerator.
2. Add or subtract, as necessary, to get a single rational expression in the denominator.
3. Perform the indicated division (invert and multiply).
4. Simplify, if possible, by removing any factors of 1.

To solve an equation involving rational expressions:

1. Multiply on both sides by the LCD.
2. Solve the resulting equation.
3. Check the answer in the original problem.

REVIEW EXERCISES

This chapter's review and test include Skill Maintenance exercises from Sections 2.2, 4.4, 5.7, and 5.8.

1. Find the domain of the function given by

$$f(x) = \frac{x^2 - 6x - 7}{x^2 - 5x - 6}.$$

2. If

$$g(t) = \frac{t^2 - 16}{t^2 - 4t + 3},$$

find the following function values.

   a) $g(-1)$ **b)** $g(4)$ **c)** $g(2)$

Find the LCD.

3. $\dfrac{5}{6x^3}, \ \dfrac{y}{16x^2}$

4. $\dfrac{1}{x^2 + x - 20}, \ \dfrac{7}{x^2 + 3x - 10}$

Perform the indicated operations and simplify.

5. $\dfrac{x^3}{x + 2} + \dfrac{8}{x + 2}$

6. $\dfrac{4x - 2}{x^2 - 5x + 4} - \dfrac{3x - 1}{x^2 - 5x + 4}$

7. $\dfrac{3a^2b^3}{5c^3d^2} \cdot \dfrac{15c^9d^4}{9a^7b}$

8. $\dfrac{1}{6m^2n^3p} + \dfrac{2}{9mn^4p^2}$

9. $\dfrac{y^2 - 64}{2y + 10} \cdot \dfrac{y + 5}{y + 8}$

10. $\dfrac{x^3 - 8}{x^2 - 25} \cdot \dfrac{x^2 + 10x + 25}{x^2 + 2x + 4}$

11. $\dfrac{9a^2 - 1}{a^2 - 9} \div \dfrac{3a + 1}{a + 3}$

12. $\dfrac{x^3 - 64}{x^2 - 16} \div \dfrac{x^2 + 5x + 6}{x^2 - 3x - 18}$

13. $\dfrac{x}{x^2 + 5x + 6} - \dfrac{2}{x^2 + 3x + 2}$

14. $\dfrac{9xy}{x^2 - y^2} + \dfrac{x + y}{x - y}$

15. $\dfrac{2x^2}{x - y} + \dfrac{2y^2}{x + y}$

16. $\dfrac{3}{y + 4} - \dfrac{y}{y - 1} + \dfrac{y^2 + 3}{y^2 + 3y - 4}$

Simplify.

17. $\dfrac{3 + \dfrac{3}{y}}{4 + \dfrac{4}{y}}$

18. $\dfrac{\dfrac{2}{a} + \dfrac{2}{b}}{\dfrac{4}{a^3} + \dfrac{4}{b^3}}$

19. $\dfrac{\dfrac{y^2 + 4y - 77}{y^2 - 10y + 25}}{\dfrac{y^2 - 5y - 14}{y^2 - 25}}$

20. $\dfrac{\dfrac{5}{x^2 - 9} - \dfrac{3}{x + 3}}{\dfrac{4}{x^2 + 6x + 9} + \dfrac{2}{x - 3}}$

Solve.

21. $\dfrac{6}{x} + \dfrac{4}{x} = 5$

22. $\dfrac{x}{7} + \dfrac{x}{4} = 1$

23. $\dfrac{5}{3x + 2} = \dfrac{3}{2x}$

24. $\dfrac{4x}{x + 1} + \dfrac{4}{x} + 9 = \dfrac{4}{x^2 + x}$

25. $\dfrac{90}{x^2 - 3x + 9} - \dfrac{5x}{x + 3} = \dfrac{405}{x^3 + 27}$

Solve.

26. Kim can paint a house in 12 hr. Kelly can paint the same house in 9 hr. How long would it take them working together to paint the house?

27. A river's current is 6 mph. A boat travels 50 mi downstream in the same time that it takes to travel 30 mi upstream. What is the speed of the boat in still water?

28. A car and a motorcycle leave a rest area at the same time, with the car traveling 8 mph faster than the motorcycle. The car then travels 105 mi in the time it takes the motorcycle to travel 93 mi. Find the speed of each vehicle.

Divide.

29. $(20r^2s^3 - 15r^2s^2 - 10r^3s^3) \div 5r^2s$

30. $(y^3 - 64) \div (y - 4)$

31. $(4x^3 + 3x^2 - 5x - 2) \div (x^2 + 1)$

Divide using synthetic division.

32. $(x^3 + 3x^2 + 2x - 6) \div (x - 3)$

33. $(4x^3 + 6x^2 - 5) \div (x + 3)$

34. Solve $R = \dfrac{gs}{g + s}$ for s.

35. Solve $S = \dfrac{H}{m(t_1 - t_2)}$ for m.

36. Solve $\dfrac{1}{ac} = \dfrac{2}{ab} - \dfrac{3}{bc}$ for c.

37. Solve $T = \dfrac{A}{v(t_2 - t_1)}$ for t_1.

Skill Maintenance _____

Graph on a plane.

38. $y - 2x \geqslant 4$ **39.** $x > -3$

40. Factor: $125x^3 - 8y^3$.

41. Factor: $6x^2 + 29x - 42$.

42. Solve: $42 = 29x + 6x^2$.

43. If $f(x) = 7x^2 - 6x$, find $f(-2)$.

Synthesis _____

44. ◈   Discuss at least three different uses of the LCD studied in this chapter.

45. ◈   Explain the difference between a rational expression and a rational equation.

Solve.

46. $\dfrac{5}{x - 13} - \dfrac{5}{x} = \dfrac{65}{x^2 - 13x}$

47. $\dfrac{\dfrac{x}{x^2 - 25} + \dfrac{2}{x - 5}}{\dfrac{3}{x - 5} - \dfrac{4}{x^2 - 10x + 25}} = 1$

48. One summer, Anna sold 4 sweepers for every 3 sweepers sold by her brother Franz. Together they sold 98 sweepers. How many did each sell?

CHAPTER TEST   6

Simplify.

1. $\dfrac{4y^2 - 4}{3y + 9} \cdot \dfrac{y + 3}{y + 1}$

2. $\dfrac{x^3 + 27}{x^2 - 16} \div \dfrac{x^2 + 8x + 15}{x^2 + x - 20}$

Find the LCD.

3. $\dfrac{1}{x^2 - 16}, \dfrac{7}{x^3 - 64}$

4. $\dfrac{3x}{x^2 + 8x - 33}, \dfrac{x + 1}{x^2 - 12x + 27}$

Perform the indicated operation and simplify if possible.

5. $\dfrac{25x}{x + 5} + \dfrac{x^3}{x + 5}$ **6.** $\dfrac{3a^2}{a - b} - \dfrac{3b^2 - 6ab}{b - a}$

7. $\dfrac{4ab}{a^2 - b^2} + \dfrac{a^2 + b^2}{a + b}$ **8.** $\dfrac{6}{x^3 - 64} - \dfrac{4}{x^2 - 16}$

9. $\dfrac{4}{y + 3} - \dfrac{y}{y - 2} + \dfrac{y^2 + 4}{y^2 + y - 6}$

Simplify.

10. $\dfrac{\dfrac{5}{x} - \dfrac{3}{y}}{\dfrac{2}{x} + \dfrac{3}{y}}$ **11.** $\dfrac{\dfrac{x^2 - 5x - 36}{x^2 - 36}}{\dfrac{x^2 + x - 12}{x^2 - 12x + 36}}$

12. $\dfrac{\dfrac{4}{x + 3} - \dfrac{2}{x^2 - 3x + 2}}{\dfrac{3}{x - 2} + \dfrac{1}{x^2 + 2x - 3}}$

Let $f(x) = \dfrac{x - 3}{x^2 - x - 12}$.

13. Calculate $f(2)$.   **14.** Calculate $f(3)$.

15. Determine the domain of f.

Solve.

16. $\dfrac{1}{x} + \dfrac{3}{x} = 4$

17. $\dfrac{6}{5a + 3} = \dfrac{4}{2a - 5}$

18. Tom can mow the yard in 3.5 hr. Geoff can mow the yard in 4.5 hr. How long will it take them, working together, to mow the yard?

Divide.

19. $(16ab^3c - 10ab^2c^2 + 12a^2b^2c) \div (4a^2b)$

20. $(y^2 - 20y + 64) \div (y - 6)$

21. $(6x^4 + 3x^2 + 5x + 4) \div (x^2 + 2)$

22. Divide using synthetic division:
$$(x^3 + 5x^2 + 4x - 7) \div (x - 4).$$

23. Solve the formula for the specified letter:
$$A = \frac{h(b_1 + b_2)}{2}, \quad \text{for } b_1.$$

24. The product of the reciprocals of two consecutive integers is $\frac{1}{30}$. Find both integers.

25. A biker can travel 12 mph with no wind. The same rider can bicycle 8 mi against the wind in the same time that it takes to bicycle 14 mi with the wind. What is the speed of the wind?

26. Two electrical resistors are connected in parallel. The resistance of one of them is 10 ohms and the resistance of the combination is 2 ohms. What is the resistance of the other resistor?

Skill Maintenance _____

27. If $f(x) = x^2 - 3$, find $f(a + 1)$.

28. Factor: $16t^2 - 24t - 72$.

29. Solve: $16t^2 = 24t + 72$.

30. Graph on a plane: $2x + 5y > 10$.

Synthesis _____

31. Solve: $\dfrac{6}{x - 15} - \dfrac{6}{x} = \dfrac{90}{x^2 - 15x}$.

32. Find the LCM: $1 - t^6, 1 + t^6$.

33. Find the x- and y-intercepts for the function given by
$$f(x) = \frac{5}{x + 4} - \frac{3}{x - 2}.$$

CUMULATIVE REVIEW | 1–6

1. Evaluate
$$\frac{2x - y^2}{x + y}$$
for $x = 3$ and $y = -4$.

2. Convert to scientific notation: 5,760,000,000.

3. Determine the slope and the y-intercept for the line given by $7x - 4y = 12$.

4. Find an equation for the line that passes through the points $(-1, 7)$ and $(2, -3)$.

5. Solve the system
$$5x - 2y = -23,$$
$$3x + 4y = 7.$$

6. Solve the system
$$-3x + 4y + z = -5,$$
$$x - 3y - z = 6,$$
$$2x + 3y + 5z = -8.$$

7. Luigi's House of Pizza sold 27 pizzas during one day. Small pizzas sold for $7.00 and large pizzas sold for $10.00. The total receipts that day from the pizzas were $234. How many of each size pizza were sold?

8. The sum of three numbers is 20. The first number is 3 less than twice the third number. The second number minus the third number is -7. What are the numbers?

9. Evaluate: $\begin{vmatrix} 5 & -3 \\ 4 & 6 \end{vmatrix}$.

Solve.

10. $8x = 1 + 16x^2$

11. $625 = 49y^2$

12. $20 > 2 - 6x$

13. $-0.5y \leqslant 25$

14. $\frac{1}{3}x - \frac{1}{5} \geqslant \frac{1}{5}x - \frac{1}{3}$

15. $-8 < x + 2 < 15$

16. $3x - 2 < -6 \text{ or } x + 3 > 9$

17. $|x| > 6.4$

18. $|3x - 6| = 2$

19. $|4x - 1| \leqslant 14$

20. $\frac{2}{n} - \frac{7}{n} = 3$

21. $\frac{6}{x - 5} = \frac{2}{2x}$

22. $\frac{3x}{x - 2} - \frac{6}{x + 2} = \frac{24}{x^2 - 4}$

23. $\frac{3x^2}{x + 2} + \frac{5x - 22}{x - 2} = \frac{-48}{x^2 - 4}$

24. Solve $5m - 3n = 4m + 12$ for n.

25. Solve $P = \dfrac{3a}{a + b}$ for a.

Graph on a plane.

26. $4x \geqslant 5y + 20$

27. $y < -2$

Perform the indicated operations and simplify.

28. $(2x^2 - 3x + 1) + (6x - 3x^3 + 7x^2 - 4)$

29. $(5x^3y^2)(-3xy^2)$

30. $(3a + b - 2c) - (-4b + 3c - 2a)$

31. $(5x^2 - 2x + 1)(3x^2 + x - 2)$

32. $(2x^2 - y)^2$

33. $(2x^2 - y)(2x^2 + y)$

34. $(-5m^3n^2 - 3mn^3) + (-4m^2n^2 + 4m^3n^2) - (2mn^3 - 3m^2n^2)$

35. $\dfrac{y^2 - 36}{2y + 8} \cdot \dfrac{y + 4}{y + 6}$

36. $\dfrac{x^4 - 1}{x^2 - x - 2} \div \dfrac{x^2 + 1}{x - 2}$

37. $\dfrac{5ab}{a^2 - b^2} + \dfrac{a + b}{a - b}$

38. $\dfrac{2}{m + 1} + \dfrac{3}{m - 5} - \dfrac{m^2 - 1}{m^2 - 4m - 5}$

39. $y - \dfrac{2}{3y}$

40. Simplify: $\dfrac{\frac{1}{x} - \frac{1}{y}}{x + y}$.

41. Divide: $(9x^3 + 5x^2 + 2) \div (x + 2)$.

Factor.

42. $4x^3 + 18x^2$

43. $x^2 + 8x - 84$

44. $16y^2 - 81$

45. $64x^3 + 8$

46. $t^2 - 16t + 64$

47. $x^6 - x^2$

48. $0.027b^3 - 0.008c^3$

49. $20x^2 + 7x - 3$

50. $3x^2 - 17x - 28$

51. $x^5 - x^3y + x^2y - y^2$

52. If $f(x) = x^2 - 4$ and $g(x) = x^2 - 7x + 10$, find the domain of f/g.

Solve.

53. Ed's tractor can plow a field in 3 hr. Nell's tractor can plow the field in 1.5 hr. Working together, how long should it take them to plow the field?

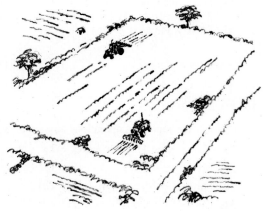

Hours required to mow the field when working together = t

54. The length of a rectangle is 3 ft longer than the width. The area is 54 ft^2. Find the perimeter of the rectangle.

55. The sum of the squares of three consecutive even integers is equal to 8 more than 3 times the square of the second number. Find the integers.

Synthesis _____

56. Multiply: $(x - 4)^3$.

57. Find all roots for $f(x) = x^4 - 34x^2 + 225$.

Solve.

58. $4 \leq |3 - x| \leq 6$

59. $\dfrac{18}{x - 9} + \dfrac{10}{x + 5} = \dfrac{28x}{x^2 - 4x - 45}$

60. $16x^3 = x$

CHAPTER 7

Exponents and Radicals

Police can estimate the speed at which a car was traveling by measuring its skid marks. The formula

$$r = 2\sqrt{5L}$$

can be used, where r is the speed in miles per hour and L is the length of the skid mark in feet. Estimate the speed of a car that left a skid mark 70 ft long.

This problem appears as Exercise 119 in Section 7.3.

Mark Michael
STATE POLICE OFFICER

"A police officer needs a strong math background. In my work, I use math for everything from daily activity reports to breathalizer tests. Algebra skills are essential for a successful trooper."

In this chapter, we study square roots, cube roots, fourth roots, and so on. We study these in connection with radical expressions that involve these roots, and solve problems involving such roots. We also define and use fractional exponents and complex numbers.

In addition to material from this chapter, the review and test for Chapter 7 include material from Sections 5.8, 6.1, and 6.4.

7.1

Radical Expressions

Square Roots • Square-Root Functions • Evaluating Expressions of the Form $\sqrt{a^2}$ • Cube Roots • Odd and Even kth Roots

In this section, we consider roots, such as square roots and cube roots. We look at the symbolism that is used for them and the ways in which we can manipulate symbols to get equivalent expressions. All of this will be important in problem solving.

Square Roots

When a number is raised to the second power, the number is squared. Often we need to know what number was squared in order to produce some value a. If such a number can be found, we call that number a *square root* of a.

Square Root

The number c is a *square root* of a if $c^2 = a$.

For example,

> 5 is a square root of 25 because $5 \cdot 5 = 25$;
>
> -5 is a square root of 25 because $(-5)(-5) = 25$;
>
> -4 does not have a real-number square root because there is no real number c such that $c^2 = -4$.

Later in this chapter, we will see that there is a number system, different from the real-number system, in which negative numbers do have square roots.

Every positive real number has two real-number square roots. The number 0 has just one square root, 0 itself. Negative numbers do not have real-number square roots.

EXAMPLE 1

Find the two square roots of 64.

Solution   The square roots are 8 and -8, because $8^2 = 64$ and $(-8)^2 = 64$.   ❑

Principal Square Root

The *principal square root* of a nonnegative number is its nonnegative square root. The symbol $\sqrt{a}$ represents the principal square root of a. To name the negative square root of a, we write $-\sqrt{a}$.

EXAMPLE 2

Simplify each of the following: **(a)** $\sqrt{25}$; **(b)** $\sqrt{\dfrac{25}{64}}$; **(c)** $-\sqrt{64}$; **(d)** $\sqrt{0.0049}$.

Solution

a) $\sqrt{25} = 5$ $\sqrt{}$ indicates the principal square root.

b) $\sqrt{\dfrac{25}{64}} = \dfrac{5}{8}$ Since $\dfrac{5}{8} \cdot \dfrac{5}{8} = \dfrac{25}{64}$

c) $-\sqrt{64} = -8$ Since $\sqrt{64} = 8$, $-\sqrt{64} = -8$.

d) $\sqrt{0.0049} = 0.07$ $(0.07)(0.07) = 0.0049$

The symbol $\sqrt{}$ is called a *radical sign*. An expression written with a radical sign is called a *radical expression*. The expression written under the radical sign is called the *radicand*.

The following are radical expressions:

$$\sqrt{5}, \qquad \sqrt{a}, \qquad -\sqrt{5x}, \qquad \sqrt{\dfrac{y^2 + 7}{\sqrt{x}}}.$$

An expression like $\sqrt{5}$ can be read as "radical 5," "the square root of 5," or simply "root 5."

Square-Root Functions

For every nonnegative real number x, there is a principal square root $\sqrt{x}$. Thus there is a *square-root function*,

$$f(x) = \sqrt{x}.$$

The domain is the set of all nonnegative real numbers, and the range is the set of all nonnegative real numbers.

x	$\sqrt{x}$	$(x, f(x))$
0	0	(0, 0)
1	1	(1, 1)
4	2	(4, 2)
9	3	(9, 3)

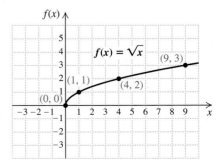

There is another square-root function,

$$g(x) = -\sqrt{x}.$$

The domain of this function is the set of all nonnegative real numbers, and the range is the set of all nonpositive real numbers.

x	$-\sqrt{x}$	$(x, g(x))$
0	0	$(0, 0)$
1	-1	$(1, -1)$
4	-2	$(4, -2)$
9	-3	$(9, -3)$

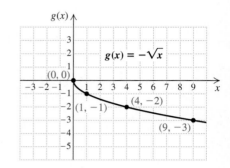

Years ago, the symbol $\sqrt{x}$ was used to represent *all* the square roots of x. That usage has almost completely disappeared, because today we pay a lot more attention to functions. Recall that functions must have exactly one output for each member of the domain.

EXAMPLE 3

For each function, find the indicated function value.

a) $f(x) = \sqrt{3x - 2}$, input 1   **b)** $g(z) = -\sqrt{6z + 4}$, input 2

Solution

a) $f(1) = \sqrt{3 \cdot 1 - 2}$ Substituting

$\qquad = \sqrt{1} = 1$ Simplifying and taking the square root

b) $g(2) = -\sqrt{6 \cdot 2 + 4}$ Substituting

$\qquad = -\sqrt{16} = -4$ Simplifying, taking the square root, and then taking the opposite   ❑

▦ **Function Values, Calculators, and Tables.**   All but the simplest calculators give values for square roots, but they are, for the most part, approximations. For example, if you enter 5 in your calculator and then press the $\boxed{\sqrt{}}$ key, you will obtain something like

2.23606798,

depending on how your calculator rounds. The exact value is not given by any repeating or terminating decimal. The same will be true of the square root of any whole number that is not a perfect square. We discussed such *irrational numbers* in Chapter 1.

Table 1 at the back of the book contains approximate values of square roots. That table can, of course, be used to find approximate function values for square-root functions. We will have more to say about the table later.

Evaluating Expressions of the Form $\sqrt{a^2}$

We will often have reason to evaluate radical expressions. Note that when 5 or -5 is substituted into the expression $\sqrt{x^2}$, the result is the same.

EXAMPLE 4

Evaluate the expression $\sqrt{x^2}$ for the following values: **(a)** 5; **(b)** 0; **(c)** -5.

Solution

a) $\sqrt{5^2} = \sqrt{25} = 5$
b) $\sqrt{0^2} = \sqrt{0} = 0$ Remember that $\sqrt{m}$ means the
c) $\sqrt{(-5)^2} = \sqrt{25} = 5$ nonnegative square root of m.

We have the following general rule that we can use in simplifying radical expressions.

> For any real number a,
> $$\sqrt{a^2} = |a|.$$
> (The principal square root of a^2 is the absolute value of a.)

When an expression contains perfect squares, like $25x^2$ or $(m - 3)^2$, in the radicand, we need to use absolute-value signs when simplifying unless we know that the quantities being squared do not represent negative numbers.

EXAMPLE 5

Simplify each expression. Assume that the variable can represent any real number.

a) $\sqrt{(x + 1)^2}$ **b)** $\sqrt{x^2 - 8x + 16}$ **c)** $\sqrt{(3b)^2}$

Solution

a) $\sqrt{(x + 1)^2} = |x + 1|$ Since $x + 1$ might be negative (for example, if $x = -3$), absolute-value notation is necessary.

b) $\sqrt{x^2 - 8x + 16} = \sqrt{(x - 4)^2} = |x - 4|$ Since $x - 4$ might be negative, absolute-value notation is necessary.

c) $\sqrt{(3b)^2} = |3b|$, or $3|b|$

> $|3b|$ can be simplified to $3|b|$ because the absolute value of any product is the product of the absolute values. That is, $|a \cdot b| = |a| \cdot |b|$. In this case, $|3b| = |3| \cdot |b|$, or $3|b|$.

EXAMPLE 6

Simplify each expression. Assume that no radicands were formed by raising negative quantities to even powers.

a) $\sqrt{y^2}$ **b)** $\sqrt{(5x + 2)^2}$
c) $\sqrt{x^2 - 2x + 1}$ **d)** $\sqrt{a^6}$

Solution

a) $\sqrt{y^2} = y$ We are assuming that y is nonnegative, so no absolute-value notation is necessary. When y *is* negative, $\sqrt{y^2} \neq y$.

b) $\sqrt{(5x + 2)^2} = 5x + 2$ Assuming that $5x + 2$ is nonnegative

c) $\sqrt{x^2 - 2x + 1} = \sqrt{(x - 1)^2} = x - 1$ Assuming that $x - 1$ is nonnegative

d) $\sqrt{a^6} = \sqrt{(a^3)^2} = a^3$ Assuming that a^3 is nonnegative

Cube Roots

We often need to know what number was cubed in order to produce a certain value. When such a number is found, we say that we have found a *cube root*.

Cube Root

The number c is the *cube root* of a if $c^3 = a$. In symbols, we write $\sqrt[3]{a}$ to denote the cube root of a.

For example,

> 2 is the cube root of 8 because $2^3 = 2 \cdot 2 \cdot 2 = 8$;
> -4 is the cube root of -64 because $(-4)^3 = (-4)(-4)(-4) = -64$.

In the real-number system, every number has exactly one cube root. The cube root of a positive number is positive, and the cube root of a negative number is negative. In simplifying expressions involving cube roots, we need not use absolute-value signs.

EXAMPLE 7

For each function, find the indicated function value.

a) $f(y) = \sqrt[3]{y}$, input 125   **b)** $g(x) = \sqrt[3]{x - 3}$, input -24

Solution

a) $f(125) = \sqrt[3]{125} = 5$ Since $5 \cdot 5 \cdot 5 = 125$

b) $g(-24) = \sqrt[3]{-24 - 3}$
$\qquad\qquad = \sqrt[3]{-27}$
$\qquad\qquad = -3$ Since $(-3)(-3)(-3) = -27$ ☐

EXAMPLE 8

Simplify: $\sqrt[3]{-8y^3}$.

Solution

$$\sqrt[3]{-8y^3} = -2y \qquad \text{Since } (-2y)(-2y)(-2y) = -8y^3 \qquad \square$$

Odd and Even kth Roots

The fifth root of a number a is the number c for which $c^5 = a$. There are also 6th roots, 7th roots, and so on. We write $\sqrt[k]{a}$ for the kth root. The number k is called the *index* (plural, indices). When the index is 2, we do not write it.

   If k is odd, we say that we are taking an odd root. Every number has just one root for an odd k. If a number is positive, its root is positive. If a number is negative, its root is negative.

EXAMPLE 9

Find each of the following: **(a)** $\sqrt[5]{32}$; **(b)** $\sqrt[5]{-32}$; **(c)** $-\sqrt[5]{32}$; **(d)** $-\sqrt[5]{-32}$;
(e) $\sqrt[7]{x^7}$; **(f)** $\sqrt[9]{(x - 1)^9}$.

Solution

a) $\sqrt[5]{32} = 2$ Since $2^5 = 32$

b) $\sqrt[5]{-32} = -2$ Since $(-2)^5 = -32$

c) $-\sqrt[5]{32} = -2$ Taking the opposite of $\sqrt[5]{32}$

d) $-\sqrt[5]{-32} = -(-2) = 2$ Taking the opposite of $\sqrt[5]{-32}$

e) $\sqrt[7]{x^7} = x$

f) $\sqrt[9]{(x-1)^9} = x - 1$

Absolute-value signs are never needed when we are finding odd roots.

When the index k in $\sqrt[k]{a}$ is an even number, we say that we are taking an *even* root. Every positive real number has two real kth roots when k is even. One of those roots is positive and one is negative. When k is even, the notation $\sqrt[k]{a}$ indicates the positive kth root. As in the case of square roots, negative numbers do not have real kth roots when k is even. Also, when we are finding even kth roots, absolute-value signs will sometimes be necessary.

EXAMPLE 10

Simplify each expression. Assume that variables can represent any real number.

(a) $\sqrt[4]{16}$; **(b)** $-\sqrt[4]{16}$; **(c)** $\sqrt[4]{-16}$; **(d)** $\sqrt[4]{81x^4}$; **(e)** $\sqrt[6]{(y+7)^6}$

Solution

a) $\sqrt[4]{16} = 2$ Since $2^4 = 16$

b) $-\sqrt[4]{16} = -2$ Taking the opposite of $\sqrt[4]{16}$

c) $\sqrt[4]{-16}$ cannot be simplified.   No real-number even root exists.

d) $\sqrt[4]{81x^4} = 3|x|$ Use absolute value since x could represent a negative number.

e) $\sqrt[6]{(y+7)^6} = |y+7|$ Use absolute value since $y + 7$ could be negative.

For any real number a:

a) $\sqrt[k]{a^k} = |a|$ when k is even. We use absolute value when k is even unless a is known to be nonnegative.

b) $\sqrt[k]{a^k} = a$ when k is odd. We do not use absolute value when k is odd.

Observe that for $f(x) = \sqrt{5x - 12}$, we have $f(1) = \sqrt{5 \cdot 1 - 12} = \sqrt{-7}$, which is not a real number. Thus 1 is not in the domain of f. To determine the domain of a function such as $f(x) = \sqrt{5x - 12}$, we find all x-values for which the radicand is nonnegative.

EXAMPLE 11

Determine the domain of $f(x) = \sqrt{5x - 12}$.

Solution   We need to find all x-values for which $5x - 12$ is nonnegative. To do so, we solve the inequality $5x - 12 \geqslant 0$:

$$5x - 12 \geqslant 0$$
$$5x \geqslant 12$$
$$x \geqslant \tfrac{12}{5}.$$

The domain of f is the set of all real numbers greater than or equal to $\frac{12}{5}$—that is,

Domain of $f = \left\{ x \mid x \geqslant \tfrac{12}{5} \right\}$.

EXAMPLE 12

Determine the domain of $g(x) = \sqrt[6]{7 - 3x}$.

Solution   Since the index is even, the radicand must be nonnegative. We solve the inequality:

$$7 - 3x \geqslant 0$$
$$-3x \geqslant -7$$
$$x \leqslant \tfrac{7}{3}. \qquad \text{Multiplying by } -\tfrac{1}{3} \text{ on both sides} \atop \text{and reversing the inequality}$$

Thus,

$$\text{Domain of } g = \left\{ x \mid x \leqslant \tfrac{7}{3} \right\}.$$

EXERCISE SET | 7.1

Find the square roots of the number.

1. 16   **2.** 225   **3.** 144

4. 9   **5.** 400   **6.** 81

7. 49   **8.** 900

Simplify.

9. $-\sqrt{\dfrac{49}{36}}$ **10.** $-\sqrt{\dfrac{361}{9}}$ **11.** $\sqrt{196}$

12. $\sqrt{441}$ **13.** $-\sqrt{\dfrac{16}{81}}$ **14.** $-\sqrt{\dfrac{81}{144}}$

15. $\sqrt{0.09}$ **16.** $\sqrt{0.36}$

17. $-\sqrt{0.0049}$ **18.** $\sqrt{0.0144}$

Identify the radicand in the expression.

19. $5\sqrt{p^2 + 4}$ **20.** $-7\sqrt{y^2 - 8}$

21. $x^2 y^2 \sqrt[3]{\dfrac{x}{y + 4}}$ **22.** $a^2 b^3 \sqrt[3]{\dfrac{a}{a^2 - b}}$

For each function, determine whether the specified inputs are members of the domain. If so, find the function values.

23. $f(y) = \sqrt{5y - 10}$;
   Inputs: 6, 2, 0, −3

24. $g(x) = \sqrt{x^2 - 25}$;
   Inputs: −5, 5, 0, 6, −6

25. $t(x) = -\sqrt{2x + 1}$;
   Inputs: 4, −4, 0, 12

26. $p(z) = \sqrt{2z^2 - 20}$;
   Inputs: 0, 5, −5, 10, −10

27. $f(t) = \sqrt{t^2 + 1}$;
   Inputs: 5, −5, 0, 10, −10

28. $g(x) = -\sqrt{(x + 1)^2}$;
   Inputs: 1, −1, 3, −3, 5, −5

29. $g(x) = \sqrt{x^3 + 9}$;
   Inputs: 2, −2, 3, −3

30. $f(t) = \sqrt{t^3 - 10}$;
   Inputs: 2, −2, 3, −3

Simplify. Assume that variables can represent any real number.

31. $\sqrt{16x^2}$ **32.** $\sqrt{25t^2}$

33. $\sqrt{(-7c)^2}$ **34.** $\sqrt{(-6b)^2}$

35. $\sqrt{(a + 1)^2}$ **36.** $\sqrt{(5 - b)^2}$

37. $\sqrt{x^2 - 4x + 4}$ **38.** $\sqrt{y^2 + 16y + 64}$

39. $\sqrt{4x^2 + 28x + 49}$ **40.** $\sqrt{9x^2 - 30x + 25}$

41. $\sqrt[4]{625}$ **42.** $-\sqrt[4]{256}$

43. $\sqrt[5]{-1}$ **44.** $-\sqrt[5]{-32}$

45. $\sqrt[5]{-\dfrac{32}{243}}$ **46.** $\sqrt[5]{-\dfrac{1}{32}}$

47. $\sqrt[6]{x^6}$ **48.** $\sqrt[8]{y^8}$

49. $\sqrt[4]{(5a)^4}$ **50.** $\sqrt[4]{(7b)^4}$

51. $\sqrt[10]{(-6)^{10}}$ **52.** $\sqrt[12]{(-10)^{12}}$

53. $\sqrt[414]{(a + b)^{414}}$ **54.** $\sqrt[1976]{(2a + b)^{1976}}$

55. $\sqrt[7]{y^7}$ **56.** $\sqrt[3]{(-6)^3}$

57. $\sqrt[5]{(x - 2)^5}$ **58.** $\sqrt[9]{(2xy)^9}$

Simplify. Assume that no radicands were formed by raising negative quantities to even powers.

59. $\sqrt{16x^2}$

60. $\sqrt{25t^2}$

61. $\sqrt{(-6b)^2}$

62. $\sqrt{(-7c)^2}$

63. $\sqrt{(a+1)^2}$

64. $\sqrt{(5+b)^2}$

65. $\sqrt{4x^2+8x+4}$

66. $\sqrt{9x^2+36x+36}$

67. $\sqrt{9t^2-12t+4}$

68. $\sqrt{25t^2-20t+4}$

69. $\sqrt[3]{27}$

70. $-\sqrt[3]{64}$

71. $\sqrt[4]{16x^4}$

72. $\sqrt[4]{(-3x)^4}$

73. $\sqrt[3]{-216}$

74. $-\sqrt[5]{-100{,}000}$

75. $-\sqrt[3]{-125y^3}$

76. $-\sqrt[3]{-64x^3}$

77. $\sqrt[5]{0.00032(x+1)^5}$

78. $\sqrt[3]{0.000008(y-2)^3}$

79. $\sqrt[6]{64x^{12}}$

80. $\sqrt[4]{81y^{12}}$

For each function, determine whether the specified inputs are members of the domain. If so, find the function values.

81. $f(x) = \sqrt[3]{x+1}$;
Inputs: 7, 26, -9, -65

82. $g(x) = -\sqrt[3]{2x-1}$;
Inputs: 0, -62, -13, 63

83. $g(t) = \sqrt[4]{t-3}$;
Inputs: 19, -13, 1, 84

84. $f(t) = \sqrt[4]{t+1}$;
Inputs: 0, 15, -82, 80

Determine the domain of the function.

85. $f(x) = \sqrt{x+7}$

86. $g(x) = \sqrt{x-10}$

87. $g(t) = \sqrt[4]{2t-9}$

88. $f(x) = \sqrt[4]{x+1}$

89. $g(x) = \sqrt[4]{5-x}$

90. $g(t) = \sqrt[3]{2t-5}$

91. $f(t) = \sqrt[5]{3t+7}$

92. $f(t) = \sqrt[6]{2t+5}$

93. $h(z) = -\sqrt[6]{5z+3}$

94. $d(x) = -\sqrt[4]{7x-5}$

95. $f(t) = 7 + 2\sqrt[8]{3t-5}$

96. $g(t) = 9 - 3\sqrt[8]{5t-4}$

Skill Maintenance

Simplify.

97. $(a^3b^2c^5)^3$

98. $(5a^7b^8)(2a^3b)$

Multiply.

99. $(x-3)(x+3)$

100. $(a+bx)(a-bx)$

Synthesis

101. ◈ If the domain of $f = \{x \mid x \le 4\}$, write an equation for $f(x)$ and explain how other equations for $f(x)$ could be formulated.

102. ◈ If the domain of $g = \{x \mid x \ge 6\}$, write an equation for $g(x)$ and explain how other equations for $g(x)$ could be formulated.

103. *Spaces in a parking lot.*   A parking lot has attendants to park the cars. The number N of stalls needed for waiting cars before attendants can get to them is given by the formula $N = 2.5\sqrt{A}$, where A is the number of arrivals in peak hours. Find the number of spaces needed for the given number of arrivals in peak hours: **(a)** 25; **(b)** 36; **(c)** 49; **(d)** 64.

Graph the function.

104. $y = \sqrt{x+3}$

105. $y = \sqrt{x} + 3$

106. $y = \sqrt{x-1}$

107. $y = \sqrt{x} - 2$

7.2

Rational Numbers as Exponents

Rational Exponents • **Negative Rational Exponents** • **Laws of Exponents** • **Simplifying Radical Expressions**

Rational Exponents

Expressions containing rational exponents, like $a^{1/2}$, $5^{-1/4}$, and $(2y)^{4/5}$, have not yet been defined. We will define such expressions so that the usual properties of exponents hold.

Consider $a^{1/2} \cdot a^{1/2}$. If we still want to multiply by adding exponents, it must follow that $a^{1/2} \cdot a^{1/2} = a^{1/2 + 1/2}$, or a^1. Thus we should define $a^{1/2}$ to be a square root of a. Similarly, $a^{1/3} \cdot a^{1/3} \cdot a^{1/3} = a^{1/3 + 1/3 + 1/3}$, or a^1, so $a^{1/3}$ should be defined to mean $\sqrt[3]{a}$.

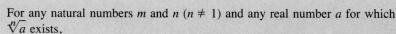

For any nonnegative number a and any index n, $a^{1/n}$ means $\sqrt[n]{a}$ (the nonnegative nth root of a). If a is negative, then n must be odd.

EXAMPLE 1

Rewrite without rational exponents: **(a)** $x^{1/2}$; **(b)** $(-27)^{1/3}$; **(c)** $(abc)^{1/5}$.

Solution

a) $x^{1/2} = \sqrt{x}$

b) $(-27)^{1/3} = \sqrt[3]{-27}$
$\qquad\qquad\; = -3$

c) $(abc)^{1/5} = \sqrt[5]{abc}$ ❑

EXAMPLE 2

Rewrite with rational exponents: **(a)** $\sqrt[5]{7xy}$; **(b)** $\sqrt[7]{x^3y/9}$.

Solution   Parentheses are required to indicate the base.

a) $\sqrt[5]{7xy} = (7xy)^{1/5}$

b) $\sqrt[7]{\dfrac{x^3y}{9}} = \left(\dfrac{x^3y}{9}\right)^{1/7}$ ❑

How should we define $a^{2/3}$? If the properties of exponents are to hold, we must have $a^{2/3} = (a^{1/3})^2$ and $a^{2/3} = (a^2)^{1/3}$. This would suggest that $a^{2/3} = (\sqrt[3]{a})^2$ and $a^{2/3} = \sqrt[3]{a^2}$. We make our definition accordingly.

For any natural numbers m and n ($n \neq 1$) and any real number a for which $\sqrt[n]{a}$ exists,
$$a^{m/n} \quad \text{means} \quad (\sqrt[n]{a})^m, \quad \text{or} \quad \sqrt[n]{a^m}.$$

EXAMPLE 3

Rewrite without rational exponents: **(a)** $27^{2/3}$; **(b)** $25^{3/2}$.

Solution

a) $27^{2/3} = \sqrt[3]{27^2}$, or $(\sqrt[3]{27})^2$ This is easier to compute using $(\sqrt[3]{27})^2$.
$\qquad\quad\; = 3^2$, or 9

b) $25^{3/2} = \sqrt[2]{25^3}$, or $(\sqrt[2]{25})^3$ We normally omit the index 2.
$\qquad\quad\; = 5^3$, or 125 Taking the square root and cubing ❑

EXAMPLE 4

Rewrite with rational exponents: **(a)** $\sqrt[3]{9^4}$; **(b)** $(\sqrt[4]{7xy})^5$.

Solution

a) $\sqrt[3]{9^4} = 9^{4/3}$

b) $\left(\sqrt[4]{7xy}\right)^5 = (7xy)^{5/4}$

The index is the denominator of the rational exponent.

Negative Rational Exponents

Negative rational exponents have a meaning similar to that of negative integer exponents.

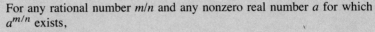

For any rational number m/n and any nonzero real number a for which $a^{m/n}$ exists,

$$a^{-m/n} \quad \text{means} \quad \frac{1}{a^{m/n}}.$$

EXAMPLE 5

Rewrite with positive exponents: **(a)** $9^{-1/2}$; **(b)** $(5xy)^{-4/5}$.

Solution

a) $9^{-1/2} = \dfrac{1}{9^{1/2}}$ $9^{-1/2}$ is the reciprocal of $9^{1/2}$.

Since $9^{1/2} = \sqrt{9} = 3$, the answer simplifies to $\dfrac{1}{3}$.

b) $(5xy)^{-4/5} = \dfrac{1}{(5xy)^{4/5}}$ $(5xy)^{-4/5}$ is the reciprocal of $(5xy)^{4/5}$.

CAUTION! Example 5 shows that a negative exponent does *not* indicate that the expression represents a negative quantity.

Laws of Exponents

The same laws hold for rational-number exponents as for integer exponents. We list them for review.

For any real numbers a and b and any rational exponents m and n for which a^m, a^n, and b^m are defined:

1. $a^m \cdot a^n = a^{m+n}$ In multiplying, add exponents if the bases are the same.

2. $\dfrac{a^m}{a^n} = a^{m-n}$ In dividing, subtract exponents if the bases are the same. (Assume $a \neq 0$.)

3. $(a^m)^n = a^{m \cdot n}$ To raise a power to a power, multiply the exponents.

4. $(ab)^m = a^m b^m$ To raise a product to a power, raise each factor to the power and multiply.

EXAMPLE 6

Use the laws of exponents to simplify: **(a)** $3^{1/5} \cdot 3^{3/5}$; **(b)** $7^{1/4}/7^{1/2}$; **(c)** $(7.2^{2/3})^{3/4}$; **(d)** $(a^{-1/3}b^{2/5})^{1/2}$.

Solution

a) $3^{1/5} \cdot 3^{3/5} = 3^{1/5 \,+\, 3/5} = 3^{4/5}$ Adding exponents

b) $\dfrac{7^{1/4}}{7^{1/2}} = 7^{1/4 \,-\, 1/2} = 7^{1/4 \,-\, 2/4} = 7^{-1/4}$ Subtracting exponents after finding a common denominator

c) $(7.2^{2/3})^{3/4} = 7.2^{2/3 \,\cdot\, 3/4} = 7.2^{6/12}$ Multiplying exponents

$\qquad\qquad\quad = 7.2^{1/2}$ Using arithmetic to simplify the exponent

d) $(a^{-1/3}b^{2/5})^{1/2} = a^{-1/3 \,\cdot\, 1/2} \cdot b^{2/5 \,\cdot\, 1/2}$ Raising a product to a power and multiplying exponents

$\qquad\qquad\qquad = a^{-1/6}b^{1/5}$ □

Simplifying Radical Expressions

Rational exponents can be used to simplify some radical expressions. The procedure is as follows.

1. Convert radical expressions to exponential expressions.
2. Use arithmetic and the laws of exponents to simplify.
3. Convert back to radical notation when appropriate.

EXAMPLE 7

Use rational exponents to simplify: **(a)** $\sqrt[6]{x^3}$; **(b)** $\sqrt[3]{8r^{12}}$; **(c)** $\sqrt[8]{a^2b^4}$.

Solution

a) $\sqrt[6]{x^3} = x^{3/6}$ Converting to an exponential expression

$\qquad\quad = x^{1/2}$ Using arithmetic to simplify the exponent

$\qquad\quad = \sqrt{x}$ Converting back to radical notation

b) $\sqrt[3]{8r^{12}} = (8r^{12})^{1/3}$ Converting to exponential notation

$\qquad\quad = 8^{1/3}(r^{12})^{1/3}$ Raising a product to a power

$\qquad\quad = 2r^4$ $\sqrt[3]{8} = 2;\ r^{12 \,\cdot\, 1/3} = r^4$

c) $\sqrt[8]{a^2b^4} = (a^2b^4)^{1/8}$ Converting to exponential notation

$\qquad\quad = a^{2/8} \cdot b^{4/8}$ Raising a product to a power and multiplying exponents

$\qquad\quad = a^{1/4} \cdot b^{1/2}$ Using arithmetic to simplify the exponents

$\qquad\quad = a^{1/4} \cdot b^{2/4}$ Converting the exponents to fractions that have the least common denominator of the fractions

$\qquad\quad = (a^1b^2)^{1/4}$ Multiplying exponents; raising a product to a power (in reverse)

$\qquad\quad = \sqrt[4]{ab^2}$ Converting back to radical notation   □

EXERCISE SET | 7.2

Note: Assume for all exercises that even roots are of nonnegative quantities and that all denominators are nonzero.

Rewrite without fractional exponents.

1. $x^{1/4}$ **2.** $y^{1/5}$ **3.** $(8)^{1/3}$

4. $(16)^{1/2}$ **5.** $81^{1/4}$ **6.** $64^{1/3}$

7. $9^{1/2}$ **8.** $25^{1/2}$ **9.** $(xyz)^{1/3}$

10. $(ab)^{1/4}$ **11.** $(a^2b^2)^{1/5}$ **12.** $(x^3y^3)^{1/4}$

13. $a^{2/3}$ **14.** $b^{3/2}$ **15.** $16^{3/4}$

16. $4^{7/2}$ **17.** $49^{3/2}$ **18.** $27^{4/3}$

19. $9^{5/2}$ **20.** $81^{3/2}$ **21.** $(81x)^{3/4}$

22. $(125a)^{2/3}$ **23.** $(25x^4)^{3/2}$ **24.** $(9y^6)^{3/2}$

25. $(8a^2b^4)^{2/3}$ **26.** $(27a^5b)^{2/3}$

Rewrite with fractional exponents.

27. $\sqrt[3]{20}$ **28.** $\sqrt[3]{19}$ **29.** $\sqrt{17}$

30. $\sqrt{6}$ **31.** $\sqrt{x^3}$ **32.** $\sqrt{a^5}$

33. $\sqrt[5]{m^2}$ **34.** $\sqrt[5]{n^4}$ **35.** $\sqrt[4]{cd}$

36. $\sqrt[5]{xy}$ **37.** $\sqrt[5]{xy^2z}$ **38.** $\sqrt[7]{x^3y^2z^2}$

39. $\left(\sqrt{3mn}\right)^3$ **40.** $\left(\sqrt[3]{7xy}\right)^4$ **41.** $\left(\sqrt[7]{8x^2y}\right)^5$

42. $\left(\sqrt[6]{2a^5b}\right)^7$

Rewrite with positive exponents.

43. $x^{-1/3}$ **44.** $y^{-1/4}$ **45.** $(2rs)^{-3/4}$

46. $(5xy)^{-5/6}$ **47.** $\left(\dfrac{1}{10}\right)^{-2/3}$ **48.** $\left(\dfrac{1}{8}\right)^{-3/4}$

49. $\dfrac{1}{x^{-2/3}}$ **50.** $\dfrac{1}{x^{-5/6}}$ **51.** $\dfrac{5}{\sqrt[4]{x}}$

52. $\dfrac{8}{\sqrt[3]{a}}$ **53.** $\dfrac{3}{5m^{-1/2}}$ **54.** $\dfrac{2}{7x^{-1/3}}$

Use the properties of exponents to simplify.

55. $5^{3/4} \cdot 5^{1/8}$ **56.** $11^{2/3} \cdot 11^{1/2}$ **57.** $\dfrac{7^{5/8}}{7^{3/8}}$

58. $\dfrac{9^{9/11}}{9^{7/11}}$ **59.** $\dfrac{8.3^{3/4}}{8.3^{2/5}}$ **60.** $\dfrac{3.9^{3/5}}{3.9^{1/4}}$

61. $(10^{3/5})^{2/5}$ **62.** $(5^{5/4})^{3/7}$ **63.** $a^{2/3} \cdot a^{5/4}$

64. $x^{3/4} \cdot x^{2/3}$ **65.** $(x^{2/3})^{3/7}$ **66.** $(a^{3/2})^{2/5}$

67. $(m^{2/3}n^{1/2})^{1/4}$ **68.** $(x^{1/3}y^{2/5})^{1/4}$

69. $(a^{-2/3}b^{-1/4})^{-6}$ **70.** $(m^{-1/5}n^{-5/6})^{-10}$

Use fractional exponents to simplify.

71. $\sqrt[6]{a^4}$ **72.** $\sqrt[6]{y^2}$ **73.** $\sqrt[3]{8y^6}$

74. $\sqrt{x^4y^6}$ **75.** $\sqrt[4]{32}$ **76.** $\sqrt[8]{81}$

77. $\sqrt[6]{4x^2}$ **78.** $\sqrt[4]{16x^4y^2}$

79. $\sqrt[5]{32c^{10}d^{15}}$ **80.** $\sqrt[4]{16x^{12}y^{16}}$

81. $\sqrt[6]{\dfrac{m^{12}n^{24}}{64}}$ **82.** $\sqrt[5]{\dfrac{x^{15}y^{20}}{32}}$ **83.** $\sqrt[8]{r^4s^2}$

84. $\sqrt[12]{64t^6s^6}$ **85.** $\sqrt[3]{27a^3b^9}$ **86.** $\sqrt[4]{81x^8y^8}$

Skill Maintenance

Solve.

87. $x^2 - 1 = 8$ **88.** $3x - 4 = 5x + 7$

89. $\dfrac{1}{x} + 2 = 5$

90. For homes under \$100,000, the real-estate transfer tax in Vermont is 0.5% of the selling price. Find the selling price of a home that had a transfer tax of \$467.50.

Synthesis

91. ◈ If $f(x) = (x + 5)^{1/2}(x + 7)^{-1/2}$, find the domain of f. Explain how you found your answer.

92. ◈ How does the graph of $f(x) = x^{1/2}$ compare with the graph of $g(x) = -x^{1/2}$?

Use rational exponents to simplify.

93. $\sqrt[5]{x^2y\sqrt{xy}}$ **94.** $\sqrt{x^5\sqrt[3]{x^4}}$

95. $\sqrt[4]{\sqrt[3]{8x^3y^6}}$ **96.** $\sqrt[12]{p^2 + 2pq + q^2}$

97. ▦ *Road pavement messages.* In a psychological study, it was determined that the proper length L of the letters of a word printed on pavement is given by

$$L = \dfrac{0.000169d^{2.27}}{h},$$

where d is the distance of a car from the lettering and h is the height of the eye above the surface of the road. All units are in meters. This formula says that if a person is h meters above the surface of the road and is to be able to recognize a message d meters away, that message will be the most recognizable if the length of the letters is L. Find L to the nearest tenth of a meter, given d and h.

a) $h = 1$ m, $d = 60$ m
b) $h = 0.9906$ m, $d = 75$ m
c) $h = 2.4$ m, $d = 80$ m
d) $h = 1.1$ m, $d = 100$ m

98. Graph the function $f(x) = x^{3/2}$, for $x \geqslant 0$, and compare it to the graphs of $y = x$ and $y = x^2$.

99. The function $r(t) = 10^{-12}2^{-t/5700}$ expresses the ratio of carbon isotopes to carbon atoms in a fossil that is t years old. What ratio of carbon isotopes to carbon atoms would a 1900-year-old bone have?

7.3

Multiplying and Simplifying with Radical Expressions

Multiplying Radical Expressions   •   Simplifying by Factoring   •
Multiplying and Simplifying   •   Approximating Square Roots

Multiplying Radical Expressions

Note that $\sqrt{4}\,\sqrt{25} = 2 \cdot 5 = 10$. Also $\sqrt{4 \cdot 25} = \sqrt{100} = 10$. Likewise,

$$\sqrt[3]{27}\,\sqrt[3]{8} = 3 \cdot 2 = 6 \quad \text{and} \quad \sqrt[3]{27 \cdot 8} = \sqrt[3]{216} = 6.$$

These examples suggest the following.

The Product Rule for Radicals

For any real numbers $\sqrt[k]{a}$ and $\sqrt[k]{b}$,

$$\sqrt[k]{a} \cdot \sqrt[k]{b} = \sqrt[k]{a \cdot b}.$$

(To multiply, multiply the radicands.)

EXAMPLE 1

Multiply: **(a)** $\sqrt{3} \cdot \sqrt{5}$; **(b)** $\sqrt{x+3}\,\sqrt{x-3}$; **(c)** $\sqrt[3]{4} \cdot \sqrt[3]{5}$; **(d)** $\sqrt[4]{\dfrac{y}{5}} \cdot \sqrt[4]{\dfrac{7}{x}}$.

Solution

a) $\sqrt{3} \cdot \sqrt{5} = \sqrt{3 \cdot 5} = \sqrt{15}$

b) $\sqrt{x+3}\,\sqrt{x-3} = \sqrt{(x+3)(x-3)} = \sqrt{x^2 - 9}$ ←

CAUTION!
$\sqrt{x^2 - 9} \neq \sqrt{x^2} - \sqrt{9}$.

c) $\sqrt[3]{4}\,\sqrt[3]{5} = \sqrt[3]{4\cdot5} = \sqrt[3]{20}$

d) $\sqrt[4]{\dfrac{y}{5}}\cdot\sqrt[4]{\dfrac{7}{x}} = \sqrt[4]{\dfrac{y}{5}\cdot\dfrac{7}{x}} = \sqrt[4]{\dfrac{7y}{5x}}$ ❑

It is important to remember that the product rule for radicals can be used only when radicals have the same index. When indices differ, rational exponents can be useful. Study the steps in the following example carefully.

EXAMPLE 2

Multiply: **(a)** $\sqrt[3]{5}\cdot\sqrt{2}$; **(b)** $\sqrt{x-2}\cdot\sqrt[4]{3y}$.

Solution

a) $\sqrt[3]{5}\cdot\sqrt{2} = 5^{1/3}\cdot2^{1/2}$ Converting to exponential notation

$\qquad = 5^{2/6}\cdot2^{3/6}$ Rewriting so that exponents have a common denominator

$\qquad = (5^2\cdot2^3)^{1/6}$ Using the laws of exponents

$\qquad = \sqrt[6]{5^2\cdot2^3}$ Converting back to radical notation

$\qquad = \sqrt[6]{200}$ Multiplying under the radical

b) $\sqrt{x-2}\cdot\sqrt[4]{3y} = (x-2)^{1/2}(3y)^{1/4}$ Converting to exponential notation

$\qquad = (x-2)^{2/4}(3y)^{1/4}$ Writing exponents with a common denominator

$\qquad = [(x-2)^2(3y)]^{1/4}$ Using the laws of exponents

$\qquad = \sqrt[4]{(x^2-4x+4)\cdot3y}$ Converting back to radical notation

$\qquad = \sqrt[4]{3x^2y-12xy+12y}$ Multiplying under the radical   ❑

Simplifying by Factoring

Reading the product rule from right to left, we have

$$\sqrt[k]{ab} = \sqrt[k]{a}\cdot\sqrt[k]{b}.$$

This shows a way to factor and thus simplify radical expressions. Consider $\sqrt{20}$. The number 20 has the factor 4, which is a perfect square. Therefore,

$\sqrt{20} = \sqrt{4\cdot5}$ Factoring the radicand (4 is a perfect square)

$\qquad = \sqrt{4}\cdot\sqrt{5}$ Factoring into two radicals

$\qquad = 2\sqrt{5}.$ Taking the square root of 4

To simplify a radical expression by factoring, look for the largest factors of the radicand that are perfect kth powers (where k is the index). Then take the kth root of the resulting factors. A radical expression, with index k, is *simplified* when its radicand has no factors that are perfect kth powers.

EXAMPLE 3

Simplify by factoring: **(a)** $\sqrt{200}$; **(b)** $\sqrt[3]{32}$; **(c)** $\sqrt[4]{48}$.

Solution

a) $\sqrt{200} = \sqrt{100 \cdot 2} = \sqrt{100} \cdot \sqrt{2} = 10\sqrt{2}$

    This is the largest perfect square factor of 200.

b) $\sqrt[3]{32} = \sqrt[3]{8 \cdot 4} = \sqrt[3]{8} \cdot \sqrt[3]{4} = 2\sqrt[3]{4}$

    This is the largest perfect cube (third power) factor of 32.

c) $\sqrt[4]{48} = \sqrt[4]{16 \cdot 3} = \sqrt[4]{16} \cdot \sqrt[4]{3} = 2\sqrt[4]{3}$

    This is the largest fourth-power factor of 48.   ❑

In many situations, we can assume that no radicands were formed by raising negative quantities to even powers. We will make this assumption and henceforth refrain from using absolute-value notation when taking even roots.

EXAMPLE 4

Simplify by factoring: **(a)** $\sqrt{5x^2}$; **(b)** $\sqrt{2x^2 - 4x + 2}$; **(c)** $\sqrt{216x^5y^3}$.

Solution

a) $\sqrt{5x^2} = \sqrt{x^2 \cdot 5}$ Factoring the radicand

    $= \sqrt{x^2} \cdot \sqrt{5}$ Factoring into two radicals

    $= x \cdot \sqrt{5}$ Taking the square root of x^2. We assume $x \geq 0$.

b) $\sqrt{2x^2 - 4x + 2} = \sqrt{2(x - 1)^2}$ Factoring the radicand

    $= \sqrt{(x - 1)^2} \cdot \sqrt{2}$ Factoring into two radicals

    $= (x - 1) \cdot \sqrt{2}$ Taking the square root of $(x - 1)^2$. We assume $x - 1 \geq 0$.

c) $\sqrt{216x^5y^3} = \sqrt{36 \cdot 6 \cdot x^4 \cdot x \cdot y^2 \cdot y}$

    $= \sqrt{36 \cdot x^4 \cdot y^2 \cdot 6 \cdot x \cdot y}$ Factoring the radicand

    $= \sqrt{36}\sqrt{x^4}\sqrt{y^2}\sqrt{6xy}$ Factoring into several radicals

    $= 6x^2y\sqrt{6xy}$ Taking square roots. Note that $36 = 6^2$ and $x^4 = (x^2)^2$. Assume $x, y \geq 0$.

Note: Had we not seen that $216 = 36 \cdot 6$, where 36 is the largest square factor of 216, we could have found the prime factorization

$$2 \cdot 2 \cdot 2 \cdot 3 \cdot 3 \cdot 3.$$

Each pair of identical factors makes a square, so

$$\sqrt{2 \cdot 2 \cdot 2 \cdot 3 \cdot 3 \cdot 3} = \sqrt{2^2 \cdot 3^2 \cdot 2 \cdot 3}$$
$$= 2 \cdot 3\sqrt{2 \cdot 3} = 6\sqrt{6}.$$

    ❑

Recall that when a quantity raised to a power is itself raised to a power, the result can be found by multiplying powers. Using this reasoning in reverse, we can simplify by finding factors in the radicand that have powers that are multiples of the index.

EXAMPLE 5

Simplify: $\sqrt{x^7y^{11}z^9}$.

Solution

$$\sqrt{x^7y^{11}z^9} = \sqrt{x^6 \cdot x \cdot y^{10} \cdot y \cdot z^8 \cdot z}$$

Factoring the radicand. Because we're taking the second (square) root, we look for powers that are multiples of 2.

$$= \sqrt{x^6}\sqrt{y^{10}}\sqrt{z^8}\sqrt{xyz}$$

Factoring into several radicals

$$= x^3y^5z^4\sqrt{xyz}$$

Note that $x^6 = (x^3)^2$, $y^{10} = (y^5)^2$, and $z^8 = (z^4)^2$. Assume x, y, $z \geq 0$.

Check:

$$(x^3y^5z^4\sqrt{xyz})^2 = (x^3)^2(y^5)^2(z^4)^2(\sqrt{xyz})^2$$
$$= x^6 \cdot y^{10} \cdot z^8 \cdot xyz$$
$$= x^7y^{11}z^9$$

Our check shows that $x^3y^5z^4\sqrt{xyz}$ is the square root of $x^7y^{11}z^9$. ❏

EXAMPLE 6

Simplify: $\sqrt[3]{16a^7b^{11}}$.

Solution

$$\sqrt[3]{16a^7b^{11}} = \sqrt[3]{8 \cdot 2 \cdot a^6 \cdot a \cdot b^9 \cdot b^2}$$

Factoring the radicand. We look for the largest powers that are multiples of 3. Note that $8 = 2^3$.

$$= \sqrt[3]{8} \cdot \sqrt[3]{a^6} \cdot \sqrt[3]{b^9} \cdot \sqrt[3]{2ab^2}$$

Factoring into radicals

$$= 2a^2b^3\sqrt[3]{2ab^2}$$

Taking cube roots

Check:

$$(2a^2b^3\sqrt[3]{2ab^2})^3 = 2^3(a^2)^3(b^3)^3(\sqrt[3]{2ab^2})^3$$
$$= 8 \cdot a^6 \cdot b^9 \cdot 2ab^2$$
$$= 16a^7b^{11}$$

We see that $2a^2b^3\sqrt[3]{2ab^2}$ is the cube root of $16a^7b^{11}$. ❏

Problems like Examples 5 and 6 can also be solved using rational exponents. Let's redo Example 6.

EXAMPLE 6A

Simplify: $\sqrt[3]{16a^7b^{11}}$.

Solution

$$\sqrt[3]{16a^7b^{11}} = (2^4a^7b^{11})^{1/3}$$

Converting to exponential notation

$$= 2^{4/3}a^{7/3}b^{11/3}$$

Multiplying exponents

$$= 2^{1+1/3} \cdot a^{2+1/3} \cdot b^{3+2/3}$$

Rewriting powers as mixed numbers

$$= 2 \cdot 2^{1/3} \cdot a^2 \cdot a^{1/3} \cdot b^3 \cdot b^{2/3}$$

Factoring

$$= 2a^2b^3(2ab^2)^{1/3}$$

Using the laws of exponents

$$= 2a^2b^3\sqrt[3]{2ab^2}$$

❏

Multiplying and Simplifying

Sometimes after multiplying, we can simplify by factoring.

EXAMPLE 7

Multiply and simplify: **(a)** $\sqrt{15}\sqrt{6}$; **(b)** $3\sqrt[3]{25} \cdot 2\sqrt[3]{5}$; **(c)** $\sqrt[4]{8x^3y^5}\sqrt[4]{4x^2y^3}$.

Solution

a) $\sqrt{15}\,\sqrt{6} = \sqrt{15\cdot 6} = \sqrt{90} = \sqrt{9\cdot 10} = 3\sqrt{10}$

b) $3\sqrt[3]{25}\cdot 2\sqrt[3]{5} = 6\cdot\sqrt[3]{25\cdot 5}$ ⎫

           $= 6\cdot\sqrt[3]{125}$ ⎬   Multiplying radicands

           $= 6\cdot 5$, or 30   Finding the cube root of 125

c) $\sqrt[4]{8x^3y^5}\,\sqrt[4]{4x^2y^3} = \sqrt[4]{8x^3y^5\cdot 4x^2y^3}$

                     $= \sqrt[4]{32x^5y^8}$ Multiplying radicands

                     $= \sqrt[4]{16x^4y^8\cdot 2x}$ Factoring the radicand

                     $= \sqrt[4]{16}\,\sqrt[4]{x^4}\,\sqrt[4]{y^8}\,\sqrt[4]{2x}$ Factoring into radicals

                     $= 2xy^2\,\sqrt[4]{2x}$ Finding the fourth root

The checks are left for the student.   ❑

When radicals have different indices, we multiply as in Example 2 and then simplify.

EXAMPLE 8

Multiply and simplify: $\sqrt{x^3}\,\sqrt[3]{x}$.

Solution

      $\sqrt{x^3}\,\sqrt[3]{x} = x^{3/2}\cdot x^{1/3}$ Converting to exponential notation

               $= x^{11/6}$ Adding exponents: $\frac{3}{2}+\frac{1}{3}=\frac{9}{6}+\frac{2}{6}$

               $= x^{1\,+\,5/6}$ Writing 11/6 as a mixed number

               $= x\cdot x^{5/6}$ Factoring

               $= x\sqrt[6]{x^5}$ Converting back to radical notation   ❑

Approximating Square Roots

We often need to use rational numbers to approximate square roots that are irratio-nal.* Such approximations can be found using a table such as Table 1 on p. 603. They can also be found on a calculator with a square root key. For example, if we were to approximate $\sqrt{37}$ using a calculator, we might get

     $\sqrt{37} \approx 6.082762530.$ Using a calculator with a 10-digit readout

Different calculators give different numbers of digits in their readouts. This may cause some variance in their answers. We might round to the third decimal place. Then

     $\sqrt{37} \approx 6.083.$ This can also be found in Table 1.

Now consider $\sqrt{275}$. To approximate such a root, we can use a calculator, or we can factor and use Table 1. Different procedures can lead to variance in approximations.

---

*Rational and irrational numbers are discussed in Section 1.1.

EXAMPLE 9

Use a calculator or Table 1 to approximate $\sqrt{275}$. Round to three decimal places.

Solution   Using a calculator gives us

$$\sqrt{275} \approx 16.58312395 \approx 16.583.$$

Using factoring and Table 1, we get

$$\sqrt{275} = \sqrt{25 \cdot 11} \qquad \text{Factoring the radicand}$$

$$= \sqrt{25} \cdot \sqrt{11} \qquad \text{Factoring into two radicals}$$

$$= 5\sqrt{11} \qquad \text{Simplifying}$$

$$\approx 5 \times 3.317 \qquad \text{Using Table 1}$$

$$= 16.585.$$

Note the variance in the answers. Because calculators are so much a part of our everyday life, the answers at the back of the book are found using a calculator. If you use a table and get an answer slightly different from the one given at the back, keep in mind that your work may not be wrong.   ❑

EXAMPLE 10

Approximate to the nearest thousandth: $\dfrac{16 - \sqrt{640}}{4}$.

Solution

Method 1.   Using a calculator, we get

$$\frac{16 - \sqrt{640}}{4} \approx \frac{16 - 25.29822128}{4}$$

$$= \frac{-9.298221280}{4}$$

$$= -2.324555320$$

$$\approx -2.325.$$

Method 2.   Using factoring and Table 1 gives us

$$\frac{16 - \sqrt{640}}{4} = \frac{16 - \sqrt{64 \cdot 10}}{4} \qquad \text{Factoring the radicand}$$

$$= \frac{16 - \sqrt{64} \cdot \sqrt{10}}{4} \qquad \text{Factoring into two radicals}$$

$$= \frac{16 - 8\sqrt{10}}{4} \longleftarrow \boxed{\text{A common error here is to divide 4 into 16 but } not \text{ into 8, and get } 4 - 8\sqrt{10}. \text{ Always remember to } remove~a~factor~of~1.}$$

$$= \frac{4(4 - 2\sqrt{10})}{4} = 4 - 2\sqrt{10} \qquad \text{Removing a factor of 1: } \tfrac{4}{4} = 1$$

$$\approx 4 - 2(3.162) \qquad \text{Using Table 1}$$

$$= 4 - 6.324$$

$$= -2.324.$$

Note again the variance in answers.   ❑

EXAMPLE 11

⊞ Use a calculator to approximate $\sqrt{0.000000005768}$.

Solution   If the number will not fit into a calculator, we factor:

$$\sqrt{0.000000005768} = \sqrt{57.68 \times 10^{-10}}$$ Factoring the radicand. (Note that the exponent, -10, is divisible by 2.)

$$= \sqrt{57.68} \times \sqrt{10^{-10}}$$ Factoring the expression

$$\approx 7.595 \times 10^{-5}$$ Approximating $\sqrt{57.68}$ with a calculator and finding $\sqrt{10^{-10}}$

$$= 0.00007595.$$ ❑

EXERCISE SET | 7.3

Note: Assume that no radicands were formed by raising negative quantities to even powers.

Multiply.

1. $\sqrt{3}\,\sqrt{2}$
2. $\sqrt{5}\,\sqrt{7}$
3. $\sqrt[3]{2}\,\sqrt[3]{5}$
4. $\sqrt[3]{7}\,\sqrt[3]{2}$
5. $\sqrt[4]{8}\,\sqrt[4]{9}$
6. $\sqrt[4]{6}\,\sqrt[4]{3}$
7. $\sqrt{3a}\,\sqrt{10b}$
8. $\sqrt{2x}\,\sqrt{13y}$
9. $\sqrt[5]{9t^2}\,\sqrt[5]{2t}$
10. $\sqrt[5]{8y^3}\,\sqrt[5]{10y}$
11. $\sqrt{x-a}\,\sqrt{x+a}$
12. $\sqrt{y-b}\,\sqrt{y+b}$
13. $\sqrt[3]{0.3x}\,\sqrt[3]{0.2x}$
14. $\sqrt[3]{0.7y}\,\sqrt[3]{0.3y}$
15. $\sqrt[4]{x-1}\,\sqrt[4]{x^2+x+1}$
16. $\sqrt[5]{x-2}\,\sqrt[5]{(x-2)^2}$
17. $\sqrt{\dfrac{6}{x}}\,\sqrt{\dfrac{y}{5}}$
18. $\sqrt{\dfrac{7}{t}}\,\sqrt{\dfrac{s}{11}}$
19. $\sqrt[7]{\dfrac{x-3}{4}}\,\sqrt[7]{\dfrac{5}{x+2}}$
20. $\sqrt[6]{\dfrac{a}{b-2}}\,\sqrt[6]{\dfrac{3}{b+2}}$

Use rational exponents to write a single radical expression.

21. $\sqrt[3]{7}\cdot\sqrt{2}$
22. $\sqrt[3]{7}\cdot\sqrt[4]{5}$
23. $\sqrt{x}\,\sqrt[3]{2x}$
24. $\sqrt[3]{y}\,\sqrt[5]{3y}$
25. $\sqrt{x}\,\sqrt[3]{x-2}$
26. $\sqrt[4]{3x}\,\sqrt{y+4}$
27. $\sqrt[5]{yx^2}\,\sqrt{xy}$
28. $\sqrt{ab}\,\sqrt[3]{2a^2b^2}$
29. $\sqrt[4]{x(y+1)^2}\,\sqrt[3]{x^2(y+1)}$
30. $\sqrt[5]{2a^2b}\,\sqrt{4ab}$
31. $\sqrt[3]{a^2bc^3}\,\sqrt[3]{ab^2c}$
32. $\sqrt[3]{x^2(y-1)}\,\sqrt[4]{x(y-1)^2}$

Simplify by factoring.

33. $\sqrt{27}$
34. $\sqrt{28}$
35. $\sqrt{45}$
36. $\sqrt{12}$
37. $\sqrt{8}$
38. $\sqrt{18}$
39. $\sqrt{24}$
40. $\sqrt{20}$
41. $\sqrt{180x^4}$
42. $\sqrt{175y^6}$
43. $\sqrt[3]{800}$
44. $\sqrt[3]{270}$
45. $\sqrt[3]{-16x^6}$
46. $\sqrt[3]{-32a^6}$
47. $\sqrt[3]{54x^8}$
48. $\sqrt[3]{40y^3}$
49. $\sqrt[3]{80x^8}$
50. $\sqrt[3]{108m^5}$
51. $\sqrt[4]{32}$
52. $\sqrt[4]{80}$
53. $\sqrt[4]{810}$
54. $\sqrt[4]{160}$
55. $\sqrt[4]{96a^8}$
56. $\sqrt[4]{240x^8}$
57. $\sqrt[4]{162c^4d^6}$
58. $\sqrt[4]{243x^8y^{10}}$
59. $\sqrt[3]{(x+y)^4}$
60. $\sqrt[3]{(a-b)^5}$
61. $\sqrt[3]{8000(m+n)^8}$
62. $\sqrt[3]{-1000(x+y)^{10}}$
63. $\sqrt[5]{-a^6b^{11}c^{17}}$
64. $\sqrt[5]{x^{13}y^8z^{22}}$

Multiply and simplify by factoring.

65. $\sqrt{3}\,\sqrt{6}$
66. $\sqrt{5}\,\sqrt{10}$
67. $\sqrt{15}\,\sqrt{12}$
68. $\sqrt{2}\,\sqrt{32}$
69. $\sqrt{6}\,\sqrt{8}$
70. $\sqrt{18}\,\sqrt{14}$
71. $\sqrt[3]{3}\,\sqrt[3]{18}$
72. $\sqrt{45}\,\sqrt{60}$
73. $\sqrt{5b^3}\,\sqrt{10c^4}$
74. $\sqrt[3]{-6a}\,\sqrt[3]{20a^4}$
75. $\sqrt[3]{10x^5}\,\sqrt[3]{-75x^2}$
76. $\sqrt{2x^3y}\,\sqrt{12xy}$
77. $\sqrt[3]{y^4}\,\sqrt[3]{16y^5}$
78. $\sqrt[3]{5^2t^4}\,\sqrt[3]{5^4t^6}$
79. $\sqrt[3]{(b+3)^4}\,\sqrt[3]{(b+3)^2}$

80. $\sqrt[3]{(x+y)^3}\ \sqrt[3]{(x+y)^5}$

81. $\sqrt{12a^3b}\ \sqrt{8a^4b^2}$

82. $\sqrt{18a^2b^5}\ \sqrt{30a^3b^4}$

83. $\sqrt[5]{a^2(b+c)^4}\ \sqrt[5]{a^4(b+c)^7}$

84. $\sqrt[5]{x^3(y-z)^7}\ \sqrt[5]{x^6(y-z)^9}$

Multiply and simplify. Write the answer in radical notation.

85. $\sqrt[5]{a^3b}\ \sqrt{ab}$

86. $\sqrt{xy^3}\ \sqrt[3]{x^2y}$

87. $\sqrt[3]{4xy^2}\ \sqrt{2x^3y^3}$

88. $\sqrt{3a^4b}\ \sqrt[4]{9ab^3}$

89. $\sqrt[4]{x^3y^5}\ \sqrt{xy}$

90. $\sqrt[5]{a^3b}\ \sqrt{ab}$

91. $\sqrt{a^4b^3c^4}\ \sqrt[3]{ab^2c}$

92. $\sqrt[3]{xy^2z}\ \sqrt{x^3yz^2}$

93. $\sqrt[3]{x^2yz^2}\ \sqrt[4]{xy^3z^3}$

94. $\sqrt[4]{a^2bc^3}\ \sqrt[3]{a^2b^2c^2}$

95. $\sqrt[3]{4a^2(b-5)^2}\ \sqrt{8a(b-5)^3}$

96. $\sqrt{27x^3(y-2)}\ \sqrt[3]{81x(y-2)^5}$

Approximate to the nearest thousandth using a calculator or Table 1.

97. $\sqrt{180}$

98. $\sqrt{124}$

99. $\dfrac{8+\sqrt{480}}{4}$

100. $\dfrac{12-\sqrt{450}}{3}$

101. $\dfrac{16-\sqrt{48}}{20}$

102. $\dfrac{25-\sqrt{250}}{10}$

103. $\dfrac{24+\sqrt{128}}{8}$

104. $\dfrac{96-\sqrt{90}}{12}$

Use a calculator to approximate each of the following.

105. $\sqrt{24{,}500{,}000{,}000}$

106. $\sqrt{16{,}500{,}000{,}000}$

107. $\sqrt{468{,}200{,}000{,}000}$

108. $\sqrt{99{,}400{,}000{,}000}$

109. $\sqrt{0.0000000395}$

110. $\sqrt{0.0000001543}$

111. $\sqrt{0.0000005001}$

112. $\sqrt{0.000010101}$

Skill Maintenance

113. During a one-hour television show, there were 12 commercials. Some of the commercials were 30 sec long and the others were 60 sec long. If the number of 30-sec commercials was 6 less than the total number of minutes of commercial time during the show, how many 60-sec commercials were used during the hour?

114. Multiply: $(2x-3)(2x+3)$.

Factor.

115. $4x^2-49$

116. $2x^2-26x+72$

Synthesis

117. ◈ Why are mixed numbers important when multiplying expressions like $\sqrt{ab}\ \sqrt[3]{a^2b^2}$?

118. ◈ Is the statement $\sqrt{500}=22.36067977$ true? Why or why not?

119. ▦ *Speed of a skidding car.* Police can estimate the speed at which a car was traveling by measuring its skid marks. The formula

$$r=2\sqrt{5L}$$

can be used, where r is the speed in miles per hour and L is the length of a skid mark in feet. Estimate (to the nearest tenth mile per hour) the speed of a car that left skid marks **(a)** 20 ft long; **(b)** 70 ft long; **(c)** 90 ft long.

120. ▦ *Wind chill temperature.* In cold weather we feel colder if there is wind than if there is not. When the temperature is T degrees Celsius and the wind speed is v meters per second, the *wind chill temperature*, T_w, is the temperature that it feels like. Here is a formula for finding wind chill temperature:

$$T_w=33-\frac{(10.45+10\sqrt{v}-v)(33-T)}{22}.$$

Find the wind chill temperature to the nearest tenth of a degree for the given actual temperatures and wind speeds.

a) $T=7°C,\ v=8$ m/sec
b) $T=0°C,\ v=12$ m/sec
c) $T=-5°C,\ v=14$ m/sec
d) $T=-23°C,\ v=15$ m/sec

121. Solve $\sqrt[3]{5x^{k+1}}\ \sqrt[3]{25x^k}=5x^7$ for k.

122. Solve $\sqrt[5]{4a^{3k+2}}\ \sqrt[5]{8a^{6-k}}=2a^4$ for k.

123. What assumption do we make about x if $\sqrt{(2x+3)^2}=2x+3$?

124. Graph the function given by

$$f(x)=\sqrt{(x-2)^2}.$$

What is the domain of f?

7.4

Dividing and Simplifying Radical Expressions

Dividing Radical Expressions • Roots of Quotients •
Powers and Roots Combined

Dividing Radical Expressions

Note that

$$\frac{\sqrt[3]{27}}{\sqrt[3]{8}} = \frac{3}{2} \quad \text{and} \quad \sqrt[3]{\frac{27}{8}} = \frac{3}{2}.$$

This example suggests the following.

The Quotient Rule for Radicals

For any real numbers $\sqrt[k]{a}$ and $\sqrt[k]{b}$, $b \neq 0$,

$$\frac{\sqrt[k]{a}}{\sqrt[k]{b}} = \sqrt[k]{\frac{a}{b}}.$$

(To divide, we divide the radicands. After doing this, we can sometimes simplify by taking roots.)

To help understand the quotient rule for radicals, note that

$$\frac{\sqrt[k]{a}}{\sqrt[k]{b}} = \frac{a^{1/k}}{b^{1/k}} = \left(\frac{a}{b}\right)^{1/k} = \sqrt[k]{\frac{a}{b}}.$$

EXAMPLE 1

Divide and simplify by taking roots, if possible.

a) $\dfrac{\sqrt{80}}{\sqrt{5}}$ b) $\dfrac{3\sqrt{2}}{5\sqrt{3}}$ c) $\dfrac{5\sqrt[3]{32}}{\sqrt[3]{2}}$

d) $\dfrac{\sqrt{72xy}}{2\sqrt{2}}$ e) $\dfrac{\sqrt[4]{33a^9b^3}}{\sqrt[4]{2b^{-1}}}$

Solution

a) $\dfrac{\sqrt{80}}{\sqrt{5}} = \sqrt{\dfrac{80}{5}} = \sqrt{16} = 4$ ⟵ We divide the radicands.

b) $\dfrac{3\sqrt{2}}{5\sqrt{3}} = \dfrac{3}{5} \cdot \dfrac{\sqrt{2}}{\sqrt{3}} = \dfrac{3}{5} \cdot \sqrt{\dfrac{2}{3}}$ ⟵

c) $\dfrac{5\sqrt[3]{32}}{\sqrt[3]{2}} = 5\sqrt[3]{\dfrac{32}{2}} = 5\sqrt[3]{16} = 5\sqrt[3]{8 \cdot 2} = 5\sqrt[3]{8}\sqrt[3]{2} = 5 \cdot 2\sqrt[3]{2} = 10\sqrt[3]{2}$

d) $\dfrac{\sqrt{72xy}}{2\sqrt{2}} = \dfrac{1}{2}\,\dfrac{\sqrt{72xy}}{\sqrt{2}} = \dfrac{1}{2}\,\sqrt{\dfrac{72xy}{2}} = \dfrac{1}{2}\,\sqrt{36xy} = \dfrac{1}{2}\,\sqrt{36}\,\sqrt{xy}$

$\qquad\qquad = \dfrac{1}{2}\cdot 6\sqrt{xy} = 3\sqrt{xy}$

e) $\dfrac{\sqrt[4]{33a^9b^3}}{\sqrt[4]{2b^{-1}}} = \sqrt[4]{\dfrac{33a^9b^3}{2b^{-1}}} = \sqrt[4]{\dfrac{33a^9b^4}{2}}$

$\qquad\qquad = \sqrt[4]{a^8b^4}\,\sqrt[4]{\dfrac{33a}{2}} = a^2b\,\sqrt[4]{\dfrac{33a}{2}}$

Note that 8 is the largest power less than 9 that is a multiple of the index 4. Assume that $a \geqslant 0$, $b > 0$. ❏

Remember that when we divide radical expressions by dividing the radicands, both radicals must have the same index. When the indices differ, rational exponents are useful.

EXAMPLE 2

Use rational exponents to write as a single radical expression.

a) $\dfrac{\sqrt[3]{a^2b^4}}{\sqrt{ab}}$ $\qquad$ **b)** $\dfrac{\sqrt[4]{x^3y^2}}{\sqrt[3]{x^2y}}$ $\qquad$ **c)** $\dfrac{\sqrt[4]{(x+y)^3}}{\sqrt{x+y}}$

Solution

a) $\dfrac{\sqrt[3]{a^2b^4}}{\sqrt{ab}} = \dfrac{(a^2b^4)^{1/3}}{(ab)^{1/2}}$ $\qquad$ Converting to exponential notation

$\qquad = \dfrac{a^{2/3}b^{4/3}}{a^{1/2}b^{1/2}}$ $\qquad$ Using the product and power rules

$\qquad = a^{2/3-1/2}b^{4/3-1/2}$ $\qquad$ Subtracting exponents

$\left.\begin{aligned} &= a^{1/6}b^{5/6} \\ &= (ab^5)^{1/6} \\ &= \sqrt[6]{ab^5} \end{aligned}\right\}$ $\qquad$ Converting back to radical notation

b) $\dfrac{\sqrt[4]{x^3y^2}}{\sqrt[3]{x^2y}} = \dfrac{(x^3y^2)^{1/4}}{(x^2y)^{1/3}}$ $\qquad$ Converting to exponential notation

$\qquad = \dfrac{x^{3/4}y^{2/4}}{x^{2/3}y^{1/3}}$ $\qquad$ Using the product and power rules

$\qquad = x^{3/4-2/3}y^{2/4-1/3}$ $\qquad$ Subtracting exponents

$\left.\begin{aligned} &= x^{1/12}y^{2/12} \\ &= (xy^2)^{1/12} \\ &= \sqrt[12]{xy^2} \end{aligned}\right\}$ $\qquad$ Converting back to radical notation

c) $\dfrac{\sqrt[4]{(x+y)^3}}{\sqrt{x+y}} = \dfrac{(x+y)^{3/4}}{(x+y)^{1/2}}$ $\qquad$ Converting to exponential notation

$\qquad = (x+y)^{3/4-1/2}$ $\qquad$ Subtracting exponents

$\left.\begin{aligned} &= (x+y)^{1/4} \\ &= \sqrt[4]{x+y} \end{aligned}\right\}$ $\qquad$ Converting back to radical notation

❏

Roots of Quotients

We can reverse the quotient rule to simplify a quotient. We simplify the root of a quotient by taking the roots of the numerator and of the denominator separately.

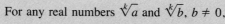

For any real numbers $\sqrt[k]{a}$ and $\sqrt[k]{b}$, $b \neq 0$,

$$\sqrt[k]{\frac{a}{b}} = \frac{\sqrt[k]{a}}{\sqrt[k]{b}}.$$

EXAMPLE 3

Simplify by taking the roots of the numerator and the denominator.

a) $\sqrt[3]{\dfrac{27}{125}}$ b) $\sqrt{\dfrac{25}{y^2}}$ c) $\sqrt{\dfrac{16x^3}{y^8}}$ d) $\sqrt[3]{\dfrac{27y^{14}}{343x^3}}$

Solution

a) $\sqrt[3]{\dfrac{27}{125}} = \dfrac{\sqrt[3]{27}}{\sqrt[3]{125}} = \dfrac{3}{5}$ Taking the cube roots of the numerator and the denominator

b) $\sqrt{\dfrac{25}{y^2}} = \dfrac{\sqrt{25}}{\sqrt{y^2}} = \dfrac{5}{y}$ Taking the square roots of the numerator and the denominator. Assume $y > 0$.

c) $\sqrt{\dfrac{16x^3}{y^8}} = \dfrac{\sqrt{16x^3}}{\sqrt{y^8}} = \dfrac{\sqrt{16x^2 \cdot x}}{\sqrt{y^8}} = \dfrac{4x\sqrt{x}}{y^4}$ Assume $x \geq 0$, $y \neq 0$.

d) $\sqrt[3]{\dfrac{27y^{14}}{343x^3}} = \dfrac{\sqrt[3]{27y^{14}}}{\sqrt[3]{343x^3}} = \dfrac{\sqrt[3]{27y^{12}y^2}}{\sqrt[3]{343x^3}} = \dfrac{\sqrt[3]{27y^{12}}\sqrt[3]{y^2}}{\sqrt[3]{343x^3}} = \dfrac{3y^4\sqrt[3]{y^2}}{7x}$ Assume $x \neq 0$.

Powers and Roots Combined

We saw in Section 7.2 that $a^{m/n}$ means $(\sqrt[n]{a})^m$, or $\sqrt[n]{a^m}$. Thus, if $\sqrt[n]{a}$ and a^m both exist, $(\sqrt[n]{a})^m$ and $\sqrt[n]{a^m}$ are the same number. This can be illustrated by noting that

$$(\sqrt[3]{8})^2 = 2^2 = 4 \quad \text{and} \quad \sqrt[3]{8^2} = \sqrt[3]{64} = 4.$$

The Power–Root Rule

For any index k for which $\sqrt[k]{a}$ exists, and any integer m for which a^m exists,

$$(\sqrt[k]{a})^m = \sqrt[k]{a^m}.$$

(We can raise to a power and then take a root, or we can take a root and then raise to a power.)

Often, one way of calculating is easier than the other.

EXAMPLE 4

Simplify each expression. Then use the power–root rule to calculate a second way. **(a)** $\sqrt[3]{27^2}$; **(b)** $\sqrt[3]{2^6}$; **(c)** $(\sqrt{5x})^3$; **(d)** $(\sqrt[3]{16x^3y^2})^2$

Solution

a1) $\sqrt[3]{27^2} = \sqrt[3]{729} = 9$ Finding 27^2 and then taking the cube root

a2) $(\sqrt[3]{27})^2 = (3)^2 = 9$ Taking the cube root and then squaring

b1) $\sqrt[3]{2^6} = \sqrt[3]{64} = 4$ Finding 2^6 and then taking the cube root

b2) $(\sqrt[3]{2})^6 = \sqrt[3]{2}\ \sqrt[3]{2}\ \sqrt[3]{2} \cdot \sqrt[3]{2}\ \sqrt[3]{2}\ \sqrt[3]{2} = 2 \cdot 2 = 4$

Note that in Example 4(b) we could have easily used fractional exponents: $\sqrt[3]{2^6} = (2^6)^{1/3} = 2^2$ and $(\sqrt[3]{2})^6 = (2^{1/3})^6 = 2^2$.

$\left.\begin{array}{l}\textbf{c1)}\quad (\sqrt{5x})^3 = \sqrt{5x}\ \sqrt{5x}\ \sqrt{5x} = 5x\sqrt{5x} \\[2mm] \textbf{c2)}\quad \sqrt{(5x)^3} = \sqrt{5^3x^3} = \sqrt{5^2x^2}\ \sqrt{5x} = 5x\sqrt{5x}\end{array}\right\}$ Assume $x \geq 0$.

d1) $(\sqrt[3]{16x^3y^2})^2 = (\sqrt[3]{8x^3 \cdot 2y^2})^2$

$\qquad\qquad\qquad = (2x\sqrt[3]{2y^2})^2$ Simplifying the cube root

$\qquad\qquad\qquad = 2x\sqrt[3]{2y^2} \cdot 2x\sqrt[3]{2y^2}$

$\qquad\qquad\qquad = 4x^2\sqrt[3]{4y^4}$ Multiplying and combining radicals

$\left.\begin{array}{l}\qquad\qquad\qquad = 4x^2\sqrt[3]{y^3 \cdot 4y} \\[2mm] \qquad\qquad\qquad = 4x^2y\ \sqrt[3]{4y}\end{array}\right\}$ Simplifying the radical

d2) $\sqrt[3]{(16x^3y^2)^2} = \sqrt[3]{256x^6y^4}$

$\qquad\qquad\qquad = \sqrt[3]{64x^6y^3 \cdot 4y}$ Factoring the radicand

$\qquad\qquad\qquad = \sqrt[3]{64x^6y^3}\sqrt[3]{4y}$

$\qquad\qquad\qquad = 4x^2y\ \sqrt[3]{4y}$ ❑

EXERCISE SET | 7.4

Divide. Then simplify by taking roots, if possible. Assume that all radicands represent positive numbers.

1. $\dfrac{\sqrt{21a}}{\sqrt{3a}}$

2. $\dfrac{\sqrt{28y}}{\sqrt{4y}}$

3. $\dfrac{\sqrt[3]{54}}{\sqrt[3]{2}}$

4. $\dfrac{\sqrt[3]{40}}{\sqrt[3]{5}}$

5. $\dfrac{\sqrt{40xy^3}}{\sqrt{8x}}$

6. $\dfrac{\sqrt{56ab^3}}{\sqrt{7a}}$

7. $\dfrac{\sqrt[3]{96a^4b^2}}{\sqrt[3]{12a^2b}}$

8. $\dfrac{\sqrt[3]{189x^5y^7}}{\sqrt[3]{7x^2y^2}}$

9. $\dfrac{\sqrt{144xy}}{2\sqrt{2}}$

10. $\dfrac{\sqrt{75ab}}{3\sqrt{3}}$

11. $\dfrac{\sqrt[4]{48x^9y^{13}}}{\sqrt[4]{3xy^5}}$

12. $\dfrac{\sqrt[5]{64a^{11}b^{28}}}{\sqrt[5]{2ab^2}}$

13. $\dfrac{\sqrt{x^3 - y^3}}{\sqrt{x - y}}$

14. $\dfrac{\sqrt{r^3 + s^3}}{\sqrt{r + s}}$

Hint: Factor and then simplify.

15. $\dfrac{\sqrt[3]{a^2}}{\sqrt[4]{a}}$

16. $\dfrac{\sqrt[3]{x^2}}{\sqrt[5]{x}}$

17. $\dfrac{\sqrt[4]{x^2y^3}}{\sqrt[3]{xy^2}}$

18. $\dfrac{\sqrt[5]{a^3b^4}}{\sqrt[3]{ab^2}}$

19. $\dfrac{\sqrt{ab^3c}}{\sqrt[5]{a^2b^3c}}$

20. $\dfrac{\sqrt[5]{x^3y^4z^9}}{\sqrt{xyz^3}}$

21. $\dfrac{\sqrt[4]{(3x-1)^3}}{\sqrt[5]{(3x-1)^3}}$

22. $\dfrac{\sqrt[4]{(5-3x)^3}}{\sqrt[3]{(5-3x)^2}}$

Simplify by taking the roots of the numerator and the denominator. Assume that all radicands represent positive numbers.

23. $\sqrt{\dfrac{16}{25}}$

24. $\sqrt{\dfrac{100}{81}}$

25. $\sqrt[3]{\dfrac{64}{27}}$

26. $\sqrt[3]{\dfrac{343}{512}}$

27. $\sqrt{\dfrac{49}{y^2}}$

28. $\sqrt{\dfrac{121}{x^2}}$

29. $\sqrt{\dfrac{25y^3}{x^4}}$

30. $\sqrt{\dfrac{36a^5}{b^6}}$

31. $\sqrt[3]{\dfrac{8x^5}{27y^3}}$

32. $\sqrt[3]{\dfrac{64x^7}{216y^6}}$

33. $\sqrt[4]{\dfrac{16a^4}{81}}$

34. $\sqrt[4]{\dfrac{81x^4}{y^8}}$

35. $\sqrt[4]{\dfrac{a^5b^8}{c^{10}}}$

36. $\sqrt[4]{\dfrac{x^9y^{12}}{z^6}}$

37. $\sqrt[5]{\dfrac{32x^6}{y^{11}}}$

38. $\sqrt[5]{\dfrac{243a^9}{b^{13}}}$

39. $\sqrt[6]{\dfrac{x^6y^8}{z^{15}}}$

40. $\sqrt[6]{\dfrac{a^9b^{12}}{c^{13}}}$

Calculate as shown. Then use the power–root rule to calculate another way. Assume that all radicands represent nonnegative numbers.

41. $\sqrt{(6a)^3}$

42. $\sqrt{(7y)^3}$

43. $\left(\sqrt{16b^2}\right)^3$

44. $\left(\sqrt{25r^2}\right)^3$

45. $\sqrt{(18a^2b)^3}$

46. $\sqrt{(12x^2y)^3}$

47. $\left(\sqrt[3]{3c^2d}\right)^4$

48. $\left(\sqrt[3]{2x^2y}\right)^4$

49. $\sqrt[3]{(5x^2y)^2}$

50. $\sqrt[3]{(6ab^2)^2}$

51. $\sqrt[4]{(x^2y)^3}$

52. $\sqrt[4]{(2a^3)^3}$

53. $\left(\sqrt[3]{8a^4b}\right)^2$

54. $\left(\sqrt[3]{27xy^5}\right)^2$

55. $\left(\sqrt[4]{16x^2y^3}\right)^2$

56. $\left(\sqrt[4]{16xy^5}\right)^3$

Skill Maintenance

Solve.

57. $\dfrac{12x}{x-4} - \dfrac{3x^2}{x+4} = \dfrac{384}{x^2-16}$

58. $\dfrac{2}{3} + \dfrac{1}{t} = \dfrac{4}{5}$

59. The width of a rectangle is one fourth the length. The area is twice the perimeter. Find the dimensions of the rectangle.

Synthesis

60. ◈ Explain why no assumptions need be made regarding the numbers that x and y represent in Example 4(d).

61. ◈ Explain why $\sqrt[3]{x^6} = x^2$ for any value x, whereas $\sqrt[2]{x^6} = x^3$ only when $x \geq 0$.

62. 🖩 *Pendulums.* The *period* of a pendulum is the time it takes to complete one cycle, swinging to and fro. If a pendulum consists of a weight on a string, the period T is given by the formula

$$T = 2\pi\sqrt{\dfrac{L}{980}},$$

where T is in seconds and L is the length of the pendulum in centimeters. Find to the nearest hundredth of a second the period of a pendulum of length **(a)** 65 cm; **(b)** 98 cm; **(c)** 120 cm. Use 3.14 for π.

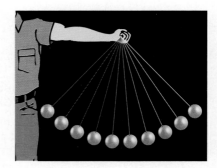

Divide and simplify.

63. $\dfrac{7\sqrt{a^2b}\,\sqrt{25xy}}{5\sqrt{a^{-4}b^{-1}}\,\sqrt{49x^{-1}y^{-3}}}$

64. $\dfrac{\left(\sqrt[3]{81mn^2}\right)^2}{\left(\sqrt[3]{mn}\right)^2}$

65. $\dfrac{\sqrt{44x^2y^9z}\,\sqrt{22y^9z^6}}{\left(\sqrt{11xy^8z^2}\right)^2}$

7.5	

Addition, Subtraction, and More Multiplication

Addition and Subtraction Involving Radicals •
More Multiplication with Radicals

Addition and Subtraction Involving Radicals

Any two real numbers can be added. For instance, the sum of 7 and $\sqrt{3}$ can be expressed as

$$7 + \sqrt{3}.$$

We cannot simplify this name for the sum. **Like radicals,** however, are radicals having the same index and radicand, and they can be simplified using the distributive law.

EXAMPLE 1

Simplify by collecting like radical terms.

a) $6\sqrt{7} + 4\sqrt{7}$

b) $8\sqrt[3]{2} - 7x\sqrt[3]{2} + 5\sqrt[3]{2}$

c) $6\sqrt[5]{4x} + 4\sqrt[5]{4x} - \sqrt[3]{4x}$

Solution

a) $6\sqrt{7} + 4\sqrt{7} = (6 + 4)\sqrt{7}$ Using the distributive law (factoring out $\sqrt{7}$)

$\qquad\qquad\qquad = 10\sqrt{7}$

b) $8\sqrt[3]{2} - 7x\sqrt[3]{2} + 5\sqrt[3]{2} = (8 - 7x + 5)\sqrt[3]{2}$ Factoring out $\sqrt[3]{2}$

$\qquad\qquad\qquad\qquad\qquad = (13 - 7x)\sqrt[3]{2}$ These parentheses are important!

c) $6\sqrt[5]{4x} + 4\sqrt[5]{4x} - \sqrt[3]{4x} = (6 + 4)\sqrt[5]{4x} - \sqrt[3]{4x}$ Try to do this step mentally.

$\qquad\qquad\qquad\qquad\qquad = 10\sqrt[5]{4x} - \sqrt[3]{4x}$ ❑

> Note that these expressions have the *same* radicand but are *not* like radicals because they have *different* indices!

One way to think of problems like Example 1(a) is as follows: When 4 square roots of 7 are added to 6 square roots of 7, the result is 10 square roots of 7.

Sometimes we need to factor one or more of the radicals in order to have like radical terms.

EXAMPLE 2

Simplify by collecting like radical terms, if possible.

a) $3\sqrt{8} - 5\sqrt{2}$

b) $5\sqrt{2} - 4\sqrt{3}$

c) $5\sqrt[3]{16y^4} + 7\sqrt[3]{2y}$

Solution

a) $3\sqrt{8} - 5\sqrt{2} = 3\sqrt{4 \cdot 2} - 5\sqrt{2}$ Factoring 8

$= 3\sqrt{4} \cdot \sqrt{2} - 5\sqrt{2}$ Factoring $\sqrt{4 \cdot 2}$ into two radicals

$= 3 \cdot 2\sqrt{2} - 5\sqrt{2}$ Taking the square root of 4

$= 6\sqrt{2} - 5\sqrt{2}$

$= (6 - 5)\sqrt{2}$ Factoring out $\sqrt{2}$ to collect like radical terms

$= \sqrt{2}$

b) $5\sqrt{2} - 4\sqrt{3}$ cannot be simplified.

c) $5\sqrt[3]{16y^4} + 7\sqrt[3]{2y} = 5\sqrt[3]{8y^3 \cdot 2y} + 7\sqrt[3]{2y}$ ⎫
$= 5\sqrt[3]{8y^3} \cdot \sqrt[3]{2y} + 7\sqrt[3]{2y}$ ⎬ Factoring the first radical
$= 5 \cdot 2y \cdot \sqrt[3]{2y} + 7\sqrt[3]{2y}$ ⎭ Taking the cube root
$= 10y\sqrt[3]{2y} + 7\sqrt[3]{2y}$
$= (10y + 7)\sqrt[3]{2y}$ Factoring to collect like radical terms   ❏

More Multiplication with Radicals

To multiply expressions in which some factors contain more than one term, we use the procedures for multiplying polynomials.

EXAMPLE 3   Multiply.

a) $\sqrt{3}(x - \sqrt{5})$ **b)** $\sqrt[3]{y}(\sqrt[3]{y^2} + \sqrt[3]{2})$ **c)** $(4\sqrt{3} + \sqrt{2})(\sqrt{3} - 5\sqrt{2})$

Solution

a) $\sqrt{3}(x - \sqrt{5}) = \sqrt{3} \cdot x - \sqrt{3} \cdot \sqrt{5}$ Using the distributive law

$= x\sqrt{3} - \sqrt{15}$ Multiplying radicals

b) $\sqrt[3]{y}(\sqrt[3]{y^2} + \sqrt[3]{2}) = \sqrt[3]{y} \cdot \sqrt[3]{y^2} + \sqrt[3]{y} \cdot \sqrt[3]{2}$ Using the distributive law

$= \sqrt[3]{y^3} + \sqrt[3]{2y}$ Multiplying radicals

$= y + \sqrt[3]{2y}$ Simplifying $\sqrt[3]{y^3}$

$\qquad\qquad\qquad\qquad\qquad\qquad\qquad$ F $\qquad$ O $\qquad$ I $\qquad$ L

c) $(4\sqrt{3} + \sqrt{2})(\sqrt{3} - 5\sqrt{2}) = 4(\sqrt{3})^2 - 20\sqrt{3} \cdot \sqrt{2} + \sqrt{2} \cdot \sqrt{3} - 5(\sqrt{2})^2$

$= 4 \cdot 3 - 20\sqrt{6} + \sqrt{6} - 5 \cdot 2$ Multiplying radicals

$= 12 - 20\sqrt{6} + \sqrt{6} - 10$

$= 2 - 19\sqrt{6}$ Collecting like terms   ❏

EXAMPLE 4   Multiply. Assume that all radicands are nonnegative.

a) $(\sqrt{r} + \sqrt{3})(\sqrt{s} + \sqrt{3})$ **b)** $(\sqrt{3} + x)^2$

c) $(2 - \sqrt{7})(2 + \sqrt{7})$ **d)** $(\sqrt{a} + \sqrt{b})(\sqrt{a} - \sqrt{b})$

Solution

a) $(\sqrt{r} + \sqrt{3})(\sqrt{s} + \sqrt{3}) = \sqrt{r}\sqrt{s} + \sqrt{r}\sqrt{3} + \sqrt{3}\sqrt{s} + \sqrt{3}\sqrt{3}$ Using FOIL

$= \sqrt{rs} + \sqrt{3r} + \sqrt{3s} + 3$ Multiplying radicals

b) $(\sqrt{3} + x)^2 = (\sqrt{3})^2 + 2x\sqrt{3} + x^2$ Squaring a binomial

$= 3 + 2x\sqrt{3} + x^2$

c) $(2 - \sqrt{7})(2 + \sqrt{7}) = 2^2 - (\sqrt{7})^2$ This is in the same form as a difference of two squares.

$= 4 - 7$

$= -3$

d) $(\sqrt{a} + \sqrt{b})(\sqrt{a} - \sqrt{b}) = (\sqrt{a})^2 - (\sqrt{b})^2$

$= a - b$ ❑

Pairs of expressions in the form $\sqrt{a} + \sqrt{b}$ and $\sqrt{a} - \sqrt{b}$ or $c - \sqrt{b}$ and $c + \sqrt{b}$ are called **conjugates.** Their product is always an expression that has no radicals. Conjugates play an important role in the work we will do in Section 7.6.

As in our earlier work, when different indices appear, rational exponents are useful.

EXAMPLE 5

Multiply: $\sqrt[3]{x^2}(\sqrt{x} + \sqrt[4]{xy^3})$. Assume $x, y \geqslant 0$.

Solution

$\sqrt[3]{x^2}(\sqrt{x} + \sqrt[4]{xy^3}) = x^{2/3}(x^{1/2} + (xy^3)^{1/4})$ Converting to exponential notation

$= x^{2/3}(x^{1/2} + x^{1/4}y^{3/4})$ Using the laws of exponents

$= x^{2/3} \cdot x^{1/2} + x^{2/3} \cdot x^{1/4}y^{3/4}$ Using the distributive law

$= x^{2/3 + 1/2} + x^{2/3 + 1/4}y^{3/4}$ Adding exponents

$= x^{7/6} + x^{11/12}y^{3/4}$

$= x^{1 + 1/6} + x^{11/12}y^{9/12}$ Writing a mixed number; finding a common denominator

$= x^1 x^{1/6} + (x^{11}y^9)^{1/12}$ Using the laws of exponents

$= x\sqrt[6]{x} + \sqrt[12]{x^{11}y^9}$ Converting back to radical notation   ❑

EXERCISE SET | 7.5

Add or subtract. Simplify by collecting like radical terms, if possible. Assume that all variables and radicands represent nonnegative numbers.

1. $6\sqrt{3} + 2\sqrt{3}$

2. $8\sqrt{5} + 9\sqrt{5}$

3. $9\sqrt[3]{5} - 6\sqrt[3]{5}$

4. $14\sqrt[5]{2} - 6\sqrt[5]{2}$

5. $4\sqrt[3]{y} + 9\sqrt[3]{y}$

6. $6\sqrt[4]{t} - 3\sqrt[4]{t}$

7. $8\sqrt{2} - 6\sqrt{2} + 5\sqrt{2}$

8. $2\sqrt{6} + 8\sqrt{6} - 3\sqrt{6}$

9. $4\sqrt[3]{3} - \sqrt{5} + 2\sqrt[3]{3} + \sqrt{5}$

10. $5\sqrt{7} - 8\sqrt[4]{11} + \sqrt{7} + 9\sqrt[4]{11}$

11. $8\sqrt{27} - 3\sqrt{3}$

12. $9\sqrt{50} - 4\sqrt{2}$

13. $8\sqrt{45} + 7\sqrt{20}$

14. $9\sqrt{12} + 16\sqrt{27}$

15. $18\sqrt{72} + 2\sqrt{98}$

16. $12\sqrt{45} - 8\sqrt{80}$

17. $3\sqrt[3]{16} + \sqrt[3]{54}$

18. $\sqrt[3]{27} - 5\sqrt[3]{8}$

19. $\sqrt{5a} + 2\sqrt{45a^3}$

20. $4\sqrt{3x^3} - \sqrt{12x}$

21. $\sqrt[3]{24x} - \sqrt[3]{3x^4}$

22. $\sqrt[3]{54x} - \sqrt[3]{2x^4}$

23. $\sqrt{8y - 8} + \sqrt{2y - 2}$

24. $\sqrt{12t + 12} + \sqrt{3t + 3}$

25. $\sqrt{x^3 - x^2} + \sqrt{9x - 9}$

26. $\sqrt{4x - 4} - \sqrt{x^3 - x^2}$

27. $5\sqrt[3]{32} - \sqrt[3]{108} + 2\sqrt[3]{256}$

28. $3\sqrt[3]{8x} - 4\sqrt[3]{27x} + 2\sqrt[3]{64x}$

29. $\sqrt{x^3 + x^2} + \sqrt{4x^3 + 4x^2} - \sqrt{9x^3 + 9x^2}$

30. $\sqrt{5x^2 + 4} - 5\sqrt{45x^2 + 36} + 3\sqrt{20x^2 + 16}$

31. $\sqrt[4]{x^5 - x^4} + 3\sqrt[4]{x^9 - x^8}$

32. $\sqrt[4]{16a^4 + 16a^5} - 2\sqrt[4]{a^8 + a^9}$

Multiply. Assume that all variables represent nonnegative real numbers.

33. $\sqrt{6}(2 - 3\sqrt{6})$

34. $\sqrt{3}(4 + \sqrt{3})$

35. $\sqrt{2}(\sqrt{3} - \sqrt{5})$

36. $\sqrt{5}(\sqrt{5} - \sqrt{2})$

37. $\sqrt{3}(2\sqrt{5} - 3\sqrt{4})$

38. $\sqrt{2}(3\sqrt{10} - 2\sqrt{2})$

39. $\sqrt[3]{2}(\sqrt[3]{4} - 2\sqrt[3]{32})$

40. $\sqrt[3]{3}(\sqrt[3]{9} - 4\sqrt[3]{21})$

41. $\sqrt[3]{a}(\sqrt[3]{2a^2} + \sqrt[3]{16a^2})$

42. $\sqrt[3]{x}(\sqrt[3]{3x^2} - \sqrt[3]{81x^2})$

43. $\sqrt[4]{x}(\sqrt[4]{x^7} + \sqrt[4]{3x^2})$

44. $\sqrt[4]{a}(\sqrt[4]{2a} - \sqrt[4]{a^{11}})$

45. $(5 - \sqrt{7})(5 + \sqrt{7})$

46. $(3 + \sqrt{5})(3 - \sqrt{5})$

47. $(\sqrt{5} + \sqrt{8})(\sqrt{5} - \sqrt{8})$

48. $(\sqrt{3} - \sqrt{5})(\sqrt{3} + \sqrt{5})$

49. $(3 - 2\sqrt{7})(3 + 2\sqrt{7})$

50. $(4 - 3\sqrt{2})(4 + 3\sqrt{2})$

51. $(\sqrt{a} + \sqrt{2})(\sqrt{a} + \sqrt{3})$

52. $(2 - \sqrt{x})(1 - \sqrt{x})$

53. $(3 - \sqrt[3]{5})(2 + \sqrt[3]{5})$

54. $(2 + \sqrt[3]{6})(4 - \sqrt[3]{6})$

55. $(2\sqrt{7} - 4\sqrt{2})(3\sqrt{7} + 6\sqrt{2})$

56. $(4\sqrt{5} + 3\sqrt{3})(3\sqrt{5} - 4\sqrt{3})$

57. $(2\sqrt[3]{3} + \sqrt[3]{2})(\sqrt[3]{3} - 2\sqrt[3]{2})$

58. $(3\sqrt[4]{7} + \sqrt[4]{6})(2\sqrt[4]{9} - 3\sqrt[4]{6})$

59. $(1 + \sqrt{3})^2$

60. $(\sqrt{5} + 1)^2$

61. $(a + \sqrt{b})^2$

62. $(x - \sqrt{y})^2$

63. $(2x - \sqrt[4]{y})^2$

64. $(3a + \sqrt[4]{b})^2$

65. $(\sqrt{m} + \sqrt{n})^2$

66. $(\sqrt{r} - \sqrt{s})^2$

67. $\sqrt{a}(\sqrt[3]{a} + \sqrt[4]{ab})$

68. $\sqrt{x}(\sqrt[3]{xy} + \sqrt[4]{x^2y})$

69. $\sqrt[3]{x^2y}(\sqrt{xy} - \sqrt[5]{xy^3})$

70. $\sqrt[4]{a^2b}(\sqrt[3]{a^2b} - \sqrt[5]{a^2b^2})$

71. $(m + \sqrt[3]{n^2})(2m + \sqrt[4]{n})$

72. $(r - \sqrt[4]{s^3})(3r - \sqrt[5]{s})$

73. $(a + \sqrt[4]{b})(a - \sqrt[3]{c})$

74. $(x - \sqrt[3]{y})(x + \sqrt[4]{z})$

75. $(\sqrt{x} - \sqrt[3]{yz})(\sqrt[4]{x} + \sqrt[5]{xz})$

76. $(\sqrt{a} + \sqrt[4]{3b})(\sqrt[3]{2a} - \sqrt[5]{bc})$

Skill Maintenance

77. Solve:
$$\frac{5}{x - 1} + \frac{9}{x^2 + x + 1} = \frac{15}{x^3 - 1}.$$

78. Divide and simplify:
$$\frac{2x^2 - x - 6}{x^2 + 4x + 3} \div \frac{2x^2 + x - 3}{x^2 - 1}.$$

Synthesis

Perform the indicated operations and simplify. Assume that all variables represent nonnegative real numbers unless otherwise indicated.

79. $\sqrt{432} - \sqrt{6125} + \sqrt{845} - \sqrt{4800}$

80. $\sqrt{1250x^3y} - \sqrt{1800xy^3} - \sqrt{162x^3y^3}$

81. $\frac{1}{2}\sqrt{36a^5bc^4} - \frac{1}{2}\sqrt[3]{64a^4bc^6} + \frac{1}{6}\sqrt{144a^3bc^2}$

82. $7x\sqrt{(x + y)^3} - 5xy\sqrt{x + y} - 2y\sqrt{(x + y)^3}$
   (Assume $x + y \geqslant 0$.)

83. $\sqrt{9 + 3\sqrt{5}}\sqrt{9 - 3\sqrt{5}}$

84. $(\sqrt{x + 2} - \sqrt{x - 2})^2$

85. $(\sqrt{3} + \sqrt{5} - \sqrt{6})^2$

86. $\sqrt[3]{y}(1 - \sqrt[3]{y})(1 + \sqrt[3]{y})$

87. $(\sqrt[3]{9} - 2)(\sqrt[3]{9} + 4)$

88. $\left[\sqrt{3 + \sqrt{2 + \sqrt{1}}}\right]^4$

7.6

Rationalizing Numerators and Denominators

Rationalizing Denominators   •   Rationalizing Numerators   •
Rationalizing When There Are Two Terms

Rationalizing Denominators

Sometimes in mathematics it is useful to find an equivalent expression without a radical in the denominator. This provides a standard notation for expressing results. The procedure for finding such an expression is called **rationalizing a denominator.** We carry this out by multiplying by 1 in either of two ways.

One way is to multiply by 1 *under* the radical to make the denominator of the radicand a perfect power.

EXAMPLE 1

Rationalize the denominator: **(a)** $\sqrt{\dfrac{7}{3}}$;   **(b)** $\sqrt[3]{\dfrac{5}{16}}$.

Solution

a)   We multiply by 1 under the radical, using $\frac{3}{3}$. We do this so that the denominator of the radicand will be a perfect square:

$$\sqrt{\frac{7}{3}} = \sqrt{\frac{7}{3} \cdot \frac{3}{3}}$$

$$= \sqrt{\frac{21}{9}}$$

$$= \frac{\sqrt{21}}{\sqrt{9}}$$

$$= \frac{\sqrt{21}}{3}.$$

b)   Note that $16 = 4^2$. Thus, to make the denominator a perfect cube, we multiply under the radical by $\frac{4}{4}$:

$$\sqrt[3]{\frac{5}{16}} = \sqrt[3]{\frac{5}{4 \cdot 4} \cdot \frac{4}{4}}$$

$$= \sqrt[3]{\frac{20}{4^3}}$$

$$= \frac{\sqrt[3]{20}}{\sqrt[3]{4^3}}$$

$$= \frac{\sqrt[3]{20}}{4}.$$

Another way to rationalize a denominator is to multiply by 1 *outside* the radical in order to eliminate the need for a radical in the denominator.

EXAMPLE 2

Rationalize the denominator.

a) $\sqrt{\dfrac{4}{5b}}$ b) $\dfrac{\sqrt[3]{a}}{\sqrt[3]{9x}}$ c) $\dfrac{3x}{\sqrt[5]{2x^2y^3}}$

Solution

a) We rewrite the expression as a quotient of two radicals. Then we simplify and multiply by 1:

$$\sqrt{\frac{4}{5b}} = \frac{\sqrt{4}}{\sqrt{5b}} = \frac{2}{\sqrt{5b}} \qquad \text{We assume } b > 0.$$

$$= \frac{2}{\sqrt{5b}} \cdot \frac{\sqrt{5b}}{\sqrt{5b}} \qquad \text{Multiplying by 1}$$

$$= \frac{2\sqrt{5b}}{(\sqrt{5b})^2} \qquad \text{Try to do this step mentally.}$$

$$= \frac{2\sqrt{5b}}{5b}.$$

b) To rationalize the denominator $\sqrt[3]{9x}$, we observe that $9x$ is $3 \cdot 3 \cdot x$. To make $9x$ a cube, we need another factor of 3 and two more factors of x. Thus, we multiply by 1, using $\sqrt[3]{3x^2}/\sqrt[3]{3x^2}$:

$$\frac{\sqrt[3]{a}}{\sqrt[3]{9x}} = \frac{\sqrt[3]{a}}{\sqrt[3]{9x}} \cdot \frac{\sqrt[3]{3x^2}}{\sqrt[3]{3x^2}} \qquad \text{Multiplying by 1}$$

$$= \frac{\sqrt[3]{3ax^2}}{\sqrt[3]{27x^3}} \longleftarrow \text{This radicand is a perfect cube.}$$

$$= \frac{\sqrt[3]{3ax^2}}{3x}.$$

c) For the radicand $2x^2y^3$ to be a perfect fifth power, it needs four more factors of 2, three more factors of x, and two more factors of y. Thus we multiply by 1 using $\sqrt[5]{2^4x^3y^2}/\sqrt[5]{2^4x^3y^2}$, or $\sqrt[5]{16x^3y^2}/\sqrt[5]{16x^3y^2}$:

$$\frac{3x}{\sqrt[5]{2x^2y^3}} = \frac{3x}{\sqrt[5]{2x^2y^3}} \cdot \frac{\sqrt[5]{16x^3y^2}}{\sqrt[5]{16x^3y^2}} \qquad \text{Multiplying by 1}$$

$$= \frac{3x\sqrt[5]{16x^3y^2}}{\sqrt[5]{32x^5y^5}} \longleftarrow \text{This radicand is a perfect fifth power.}$$

$$= \frac{3x\sqrt[5]{16x^3y^2}}{2xy} = \frac{3\sqrt[5]{16x^3y^2}}{2y} \qquad \text{Always simplify if possible.} \quad \square$$

Rationalizing Numerators

Sometimes in calculus it is necessary to rationalize a numerator. To do so, we multiply by 1 to make the radicand in the *numerator* a perfect power.

EXAMPLE 3

Rationalize the numerator.

a) $\sqrt{\dfrac{7}{5}}$

b) $\dfrac{\sqrt[3]{4a^2}}{\sqrt[3]{5b}}$

Solution

a) $\sqrt{\dfrac{7}{5}} = \dfrac{\sqrt{7}}{\sqrt{5}}$

$= \dfrac{\sqrt{7}}{\sqrt{5}} \cdot \dfrac{\sqrt{7}}{\sqrt{7}}$ Multiplying by 1

$= \dfrac{\sqrt{49}}{\sqrt{35}}$ The radicand in the numerator is a perfect square.

$= \dfrac{7}{\sqrt{35}}$

b) $\dfrac{\sqrt[3]{4a^2}}{\sqrt[3]{5b}} = \dfrac{\sqrt[3]{4a^2}}{\sqrt[3]{5b}} \cdot \dfrac{\sqrt[3]{2a}}{\sqrt[3]{2a}}$ Multiplying by 1

$= \dfrac{\sqrt[3]{8a^3}}{\sqrt[3]{10ba}}$ ⟵ This radicand is a perfect cube.

$= \dfrac{2a}{\sqrt[3]{10ab}}$ ❏

Rationalizing When There Are Two Terms

Recall from Section 7.5 that when a pair of conjugates are multiplied, the product has no radicals in it. Thus when the denominator to be rationalized has two terms, we use the conjugate of the denominator to write a symbol for 1.

EXAMPLE 4

For each expression, write the symbol for 1 that can be used to rationalize the denominator.

a) $\dfrac{3}{x + \sqrt{7}}$

b) $\dfrac{\sqrt{7} + 4}{3 - 2\sqrt{5}}$

Solution

Expression   *Symbol for 1*

a) $\dfrac{3}{x + \sqrt{7}}$ $\dfrac{x - \sqrt{7}}{x - \sqrt{7}}$

Change the operation sign to obtain the conjugate. Use the conjugate for the numerator and denominator of the symbol for 1.

b) $\dfrac{\sqrt{7} + 4}{3 - 2\sqrt{5}}$ $\dfrac{3 + 2\sqrt{5}}{3 + 2\sqrt{5}}$ ❏

EXAMPLE 5

Rationalize the denominator: $\dfrac{4 + \sqrt{2}}{\sqrt{5} - \sqrt{2}}$.

Solution

$$\frac{4 + \sqrt{2}}{\sqrt{5} - \sqrt{2}} = \frac{4 + \sqrt{2}}{\sqrt{5} - \sqrt{2}} \cdot \frac{\sqrt{5} + \sqrt{2}}{\sqrt{5} + \sqrt{2}}$$

Multiplying by 1, using the conjugate of $\sqrt{5} - \sqrt{2}$, which is $\sqrt{5} + \sqrt{2}$

$$= \frac{(4 + \sqrt{2})(\sqrt{5} + \sqrt{2})}{(\sqrt{5} - \sqrt{2})(\sqrt{5} + \sqrt{2})}$$

Multiplying numerators and denominators

$$= \frac{4\sqrt{5} + 4\sqrt{2} + \sqrt{2}\sqrt{5} + (\sqrt{2})^2}{(\sqrt{5})^2 - (\sqrt{2})^2}$$

Using FOIL

$$= \frac{4\sqrt{5} + 4\sqrt{2} + \sqrt{10} + 2}{5 - 2}$$

Squaring in the denominator and the numerator

$$= \frac{4\sqrt{5} + 4\sqrt{2} + \sqrt{10} + 2}{3}$$ ❑

EXAMPLE 6

Rationalize the denominator: $\dfrac{4}{\sqrt{3} + x}$.

Solution

$$\frac{4}{\sqrt{3} + x} = \frac{4}{\sqrt{3} + x} \cdot \frac{\sqrt{3} - x}{\sqrt{3} - x}$$ Multiplying by 1

$$= \frac{4(\sqrt{3} - x)}{(\sqrt{3} + x)(\sqrt{3} - x)}$$

$$= \frac{4(\sqrt{3} - x)}{(\sqrt{3})^2 - x^2}$$

$$= \frac{4\sqrt{3} - 4x}{3 - x^2}$$ ❑

To rationalize a numerator with more than one term, we use the conjugate of the numerator.

EXAMPLE 7

Rationalize the numerator: $\dfrac{4 + \sqrt{2}}{\sqrt{5} - \sqrt{2}}$.

Solution

$$\frac{4 + \sqrt{2}}{\sqrt{5} - \sqrt{2}} = \frac{4 + \sqrt{2}}{\sqrt{5} - \sqrt{2}} \cdot \frac{4 - \sqrt{2}}{4 - \sqrt{2}}$$ Multiplying by 1

$$= \frac{16 - (\sqrt{2})^2}{4\sqrt{5} - \sqrt{5}\sqrt{2} - 4\sqrt{2} + (\sqrt{2})^2}$$

$$= \frac{14}{4\sqrt{5} - \sqrt{10} - 4\sqrt{2} + 2}$$ ❑

EXERCISE SET | 7.6

Rationalize the denominator.

1. $\sqrt{\dfrac{6}{5}}$ **2.** $\sqrt{\dfrac{11}{6}}$ **3.** $\sqrt{\dfrac{10}{7}}$

4. $\sqrt{\dfrac{22}{3}}$ **5.** $\dfrac{6\sqrt{5}}{5\sqrt{3}}$ **6.** $\dfrac{2\sqrt{3}}{5\sqrt{2}}$

7. $\sqrt[3]{\dfrac{16}{9}}$ **8.** $\sqrt[3]{\dfrac{2}{9}}$ **9.** $\dfrac{\sqrt[3]{3a}}{\sqrt[3]{5c}}$

10. $\dfrac{\sqrt[3]{7x}}{\sqrt[3]{3y}}$ **11.** $\dfrac{\sqrt[3]{5y^4}}{\sqrt[3]{6x^4}}$ **12.** $\dfrac{\sqrt[3]{3a^4}}{\sqrt[3]{7b^2}}$

13. $\dfrac{1}{\sqrt[3]{xy}}$ **14.** $\dfrac{1}{\sqrt[3]{ab}}$ **15.** $\sqrt{\dfrac{7a}{18}}$

16. $\sqrt{\dfrac{3x}{10}}$ **17.** $\sqrt{\dfrac{9}{20x^2y}}$ **18.** $\sqrt{\dfrac{5}{32ab^2}}$

19. $\sqrt[3]{\dfrac{9}{100x^2y^5}}$ **20.** $\sqrt[3]{\dfrac{7}{36a^4b}}$

Rationalize the numerator.

21. $\dfrac{\sqrt{7}}{\sqrt{3x}}$ **22.** $\dfrac{\sqrt{6}}{\sqrt{5x}}$ **23.** $\sqrt{\dfrac{14}{21}}$

24. $\sqrt{\dfrac{12}{15}}$ **25.** $\dfrac{4\sqrt{13}}{3\sqrt{7}}$ **26.** $\dfrac{5\sqrt{21}}{2\sqrt{5}}$

27. $\dfrac{\sqrt[3]{7}}{\sqrt[3]{2}}$ **28.** $\dfrac{\sqrt[3]{5}}{\sqrt[3]{4}}$ **29.** $\sqrt{\dfrac{7x}{3y}}$

30. $\sqrt{\dfrac{6a}{5b}}$ **31.** $\dfrac{\sqrt[3]{5y^4}}{\sqrt[3]{6x^5}}$ **32.** $\dfrac{\sqrt[3]{3a^5}}{\sqrt[3]{7b^2}}$

33. $\dfrac{\sqrt{ab}}{3}$ **34.** $\dfrac{\sqrt{xy}}{5}$

35. $\sqrt{\dfrac{x^3y}{2}}$ **36.** $\sqrt{\dfrac{ab^5}{3}}$

37. $\dfrac{\sqrt[3]{a^2b}}{\sqrt[3]{5}}$ **38.** $\dfrac{\sqrt[3]{xy^2}}{\sqrt[3]{7}}$

39. $\sqrt[3]{\dfrac{x^4y^2}{3}}$ **40.** $\sqrt[3]{\dfrac{a^5b}{2}}$

Rationalize the denominator. Assume that all variables represent nonnegative numbers and that no denominators are 0.

41. $\dfrac{5}{8 - \sqrt{6}}$ **42.** $\dfrac{7}{9 + \sqrt{10}}$

43. $\dfrac{-4\sqrt{7}}{\sqrt{5} - \sqrt{3}}$ **44.** $\dfrac{-3\sqrt{2}}{\sqrt{3} - \sqrt{5}}$

45. $\dfrac{\sqrt{5} - 2\sqrt{6}}{\sqrt{3} - 4\sqrt{5}}$ **46.** $\dfrac{\sqrt{6} - 3\sqrt{5}}{\sqrt{3} - 2\sqrt{7}}$

47. $\dfrac{\sqrt{x} - \sqrt{y}}{\sqrt{x} + \sqrt{y}}$ **48.** $\dfrac{\sqrt{a} + \sqrt{b}}{\sqrt{a} - \sqrt{b}}$

49. $\dfrac{5\sqrt{3} - 3\sqrt{2}}{3\sqrt{2} - 2\sqrt{3}}$ **50.** $\dfrac{7\sqrt{2} + 4\sqrt{3}}{4\sqrt{3} - 3\sqrt{2}}$

51. $\dfrac{3\sqrt{x} + \sqrt{y}}{2\sqrt{x} + 3\sqrt{y}}$ **52.** $\dfrac{2\sqrt{a} - \sqrt{b}}{3\sqrt{a} + 2\sqrt{b}}$

Rationalize the numerator. Assume that all variables represent nonnegative numbers and that no denominators are 0.

53. $\dfrac{\sqrt{3} + 5}{8}$ **54.** $\dfrac{3 - \sqrt{2}}{5}$

55. $\dfrac{\sqrt{3} - 5}{\sqrt{2} + 5}$ **56.** $\dfrac{\sqrt{6} - 3}{\sqrt{3} + 7}$

57. $\dfrac{\sqrt{x} - \sqrt{y}}{\sqrt{x} + \sqrt{y}}$ **58.** $\dfrac{\sqrt{x} + \sqrt{y}}{\sqrt{x} - \sqrt{y}}$

59. $\dfrac{4\sqrt{6} - 5\sqrt{3}}{2\sqrt{3} + 7\sqrt{6}}$ **60.** $\dfrac{8\sqrt{2} + 5\sqrt{3}}{5\sqrt{3} - 7\sqrt{2}}$

61. $\dfrac{\sqrt{3} + 2\sqrt{x}}{\sqrt{3} - \sqrt{x}}$ **62.** $\dfrac{\sqrt{5} - 3\sqrt{x}}{\sqrt{5} + \sqrt{x}}$

63. $\dfrac{a + b\sqrt{c}}{a + \sqrt{c}}$ **64.** $\dfrac{a\sqrt{b} - c}{\sqrt{b} - c}$

Skill Maintenance ⎯⎯⎯⎯⎯⎯⎯⎯⎯⎯⎯⎯

65. Solve:
$$\dfrac{1}{2} - \dfrac{1}{3} = \dfrac{1}{t}.$$

66. Divide and simplify:

$$\frac{1}{x^3 - y^3} \div \frac{1}{(x - y)(x^2 + xy + y^2)}.$$

Synthesis

67. ◈ Explain why it is easier to approximate

$$\frac{\sqrt{2}}{2} \quad \text{than} \quad \frac{1}{\sqrt{2}}$$

if no calculator is available and $\sqrt{2} \approx$ 1.414213562.

68. ◈ Use what we know about factoring a difference of two cubes to present a method for rationalizing any denominator of the form $\sqrt[3]{a} - \sqrt[3]{b}$.

For Exercises 69–77, assume that all radicands are positive and that no denominators are 0.

Rationalize the denominator.

69. $\dfrac{a - \sqrt{a + b}}{\sqrt{a + b} - b}$

70. $\dfrac{3\sqrt{y} + 4\sqrt{yz}}{5\sqrt{y} - 2\sqrt{z} + y}$

71. $\dfrac{b + \sqrt{b}}{1 + b + \sqrt{b}}$

Rationalize the numerator.

72. $\dfrac{\sqrt{y + 18} - \sqrt{y}}{18}$

73. $\dfrac{\sqrt{x + 6} - 5}{\sqrt{x + 6} + 5}$

Simplify.

74. $\sqrt{a^2 - 3} - \dfrac{a^2}{\sqrt{a^2 - 3}}$

75. $5\sqrt{\dfrac{x}{y}} + 4\sqrt{\dfrac{y}{x}} - \dfrac{3}{\sqrt{xy}}$

76. $\dfrac{\dfrac{1}{\sqrt{w}} - \sqrt{w}}{\dfrac{\sqrt{w} + 1}{\sqrt{w}}}$

77. $\dfrac{1}{4 + \sqrt{3}} + \dfrac{1}{\sqrt{3}} + \dfrac{1}{\sqrt{3} - 4}$

7.7

Solving Radical Equations

The Principle of Powers • Equations with Two Radical Terms

The Principle of Powers

A **radical equation** has variables in one or more radicands. These are radical equations:

$$\sqrt[3]{2x} + 1 = 5, \qquad \sqrt{x} + \sqrt{4x - 2} = 7.$$

To solve such equations, we need a new principle. Suppose an equation $a = b$ is true. If we square both sides, we get another true equation: $a^2 = b^2$. This can be generalized.

The Principle of Powers

If $a = b$, then $a^n = b^n$ for any natural number n.

Note that the principle of powers is an "if–then" statement. The statement obtained by interchanging the two parts of the sentence — "if $a^n = b^n$ for some natural number n, then $a = b$" — *is not always true*. For example, $3^2 = (-3)^2$ is true, but $3 = -3$ is *not* true. This means that we must always check our solution in the original problem when we use the principle of powers, because $a = b$ and $a^n = b^n$ are not always equivalent equations.

EXAMPLE 1

Solve: $\sqrt{x} - 3 = 4$.

Solution

$$\sqrt{x} - 3 = 4$$
$$\sqrt{x} = 7 \qquad \text{Adding to isolate the radical}$$
$$\left(\sqrt{x}\right)^2 = 7^2 \qquad \text{Using the principle of powers}$$
$$x = 49$$

Check:
$$\begin{array}{c|c} \sqrt{x} - 3 = 4 \\ \hline \sqrt{49} - 3 \ ? \ 4 \\ 7 - 3 \\ 4 & 4 \ \text{TRUE} \end{array}$$

The solution is 49. □

EXAMPLE 2

Solve: $\sqrt{x} - 3 = -5$.

Solution

$$\sqrt{x} - 3 = -5$$
$$\sqrt{x} = -2 \qquad \text{Adding to isolate the radical}$$

> The equation $\sqrt{x} = -2$ has no solution because the principal square root of a number is never negative. We continue as in Example 1 for comparison.

$$\left(\sqrt{x}\right)^2 = (-2)^2 \qquad \text{Using the principle of powers (squaring)}$$
$$x = 4$$

Check:
$$\begin{array}{c|c} \sqrt{x} - 3 = -5 \\ \hline \sqrt{4} - 3 \ ? \ -5 \\ 2 - 3 \\ -1 & -5 \ \text{FALSE} \end{array}$$

The number 4 does not check. Thus the equation $\sqrt{x} - 3 = -5$ has no real-number solution. □

The principle of powers does not always give equivalent equations. For this reason, a check is a must!

Note in Example 2 that the equation $x = 4$ has solution 4, but that $\sqrt{x} - 3 = -5$ has *no* solution. Thus the equations $x = 4$ and $\sqrt{x} - 3 = -5$ are *not* equivalent.

EXAMPLE 3

TECHNOLOGY
CONNECTION

To solve Example 3 with a grapher, graph the curves $y_1 = x$ and $y_2 = (x + 7)^{1/2} + 5$ on the same set of axes.

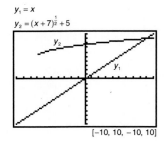

$y_1 = x$
$y_2 = (x + 7)^{\frac{1}{2}} + 5$

$[-10, 10, -10, 10]$

By using the Trace and Zoom features, try to determine the point of intersection. If you lack a Zoom feature, adjust the window with the Range feature. The intersection should occur when $x = 9$.

TC1. Use a grapher to solve Examples 1, 2, 4, 5, and 6. Compare your answers with those found using the algebraic methods shown.

Solve: $x = \sqrt{x + 7} + 5$.

Solution

$$x = \sqrt{x + 7} + 5$$

$$x - 5 = \sqrt{x + 7}$$
Adding -5 on both sides to isolate the radical term

$$(x - 5)^2 = \left(\sqrt{x + 7}\right)^2$$
Using the principle of powers; squaring both sides

$$x^2 - 10x + 25 = x + 7$$

$$x^2 - 11x + 18 = 0$$
Adding $-x - 7$ on both sides to write the quadratic equation in standard form

$$(x - 9)(x - 2) = 0$$
Factoring

$$x = 9 \quad or \quad x = 2$$
Using the principle of zero products

The possible solutions are 9 and 2. Let us check.

Check:

For 9:

$$\frac{x = \sqrt{x + 7} + 5}{9 \; ? \; \sqrt{9 + 7} + 5}$$
$$9 \; | \; 9 \qquad\qquad \text{TRUE}$$

For 2:

$$\frac{x = \sqrt{x + 7} + 5}{2 \; ? \; \sqrt{2 + 7} + 5}$$
$$2 \; | \; 8 \qquad\qquad \text{FALSE}$$

Since 9 checks but 2 does not, the solution is 9. ❑

Suppose in Example 3 that we had used the principle of powers *before* we added -5 to each side. We then would have had the expression $\left(\sqrt{x + 7} + 5\right)^2$ or $x + 7 + 10\sqrt{x + 7} + 25$ on the right side, and the radical would still have been in the problem.

EXAMPLE 4

Solve: $(2x + 1)^{1/3} + 5 = 0$.

Solution   We rewrite using radical notation and solve as above:

$$(2x + 1)^{1/3} + 5 = 0$$

$$\sqrt[3]{2x + 1} + 5 = 0$$
Converting to radical notation. This is sometimes performed mentally.

$$\sqrt[3]{2x + 1} = -5$$
Adding -5; this isolates the radical term.

$$\left(\sqrt[3]{2x + 1}\right)^3 = (-5)^3$$
Using the principle of powers; cubing both sides

$$2x + 1 = -125$$

$$2x = -126$$
Adding -1

$$x = -63$$

Check:
$$(2x + 1)^{1/3} + 5 = 0$$

$$\frac{}{\quad(2(-63) + 1)^{1/3} + 5 \;?\; 0\quad}$$
$$(-125)^{1/3} + 5$$
$$-5 + 5$$
$$0 \;\bigg|\; 0 \quad \text{TRUE}$$

The solution is -63.   ❑

Equations with Two Radical Terms

A general strategy for solving equations with two radical terms is as follows.

1. Isolate one of the radical terms.
2. Use the principle of powers.
3. If a radical remains, perform steps (1) and (2) again.
4. Check possible solutions.

EXAMPLE 5

Solve: $\sqrt{x - 3} + \sqrt{x + 5} = 4$.

Solution

$$\sqrt{x - 3} + \sqrt{x + 5} = 4$$

$$\sqrt{x - 3} = 4 - \sqrt{x + 5} \qquad \text{Adding } -\sqrt{x + 5}\text{; this isolates one of the radical terms.}$$

$$(\sqrt{x - 3})^2 = (4 - \sqrt{x + 5})^2 \qquad \text{Using the principle of powers (squaring both sides)}$$

This is like squaring a binomial. We square 4, then find twice the product of 4 and $-\sqrt{x + 5}$, and then the square of $\sqrt{x + 5}$.

$$x - 3 = 16 - 8\sqrt{x + 5} + (x + 5)$$
$$-3 = 21 - 8\sqrt{x + 5} \qquad \text{Adding } -x \text{ and collecting like terms}$$
$$-24 = -8\sqrt{x + 5} \qquad \text{Isolating the remaining radical term}$$
$$3 = \sqrt{x + 5} \qquad \text{Dividing by } -8$$
$$3^2 = (\sqrt{x + 5})^2 \qquad \text{Squaring}$$
$$9 = x + 5$$
$$4 = x$$

The number 4 checks and is the solution.   ❑

CAUTION! A common error in solving equations like $\sqrt{x - 3} + \sqrt{x + 5} = 4$ is to obtain $(x - 3) + (x + 5)$ when squaring the left side. This is wrong because the square of a sum is *not* the sum of the squares. For example, $(\sqrt{9} + \sqrt{16})^2 = 7^2$, or 49, whereas $(\sqrt{9})^2 + (\sqrt{16})^2 = 9 + 16$, or 25.

EXAMPLE 6

Solve: $\sqrt{2x-5} = 1 + \sqrt{x-3}$.

Solution

$$\sqrt{2x-5} = 1 + \sqrt{x-3}$$

$$(\sqrt{2x-5})^2 = (1 + \sqrt{x-3})^2$$ One radical is already isolated; we square both sides.

$$2x - 5 = 1 + 2\sqrt{x-3} + (\sqrt{x-3})^2$$

$$2x - 5 = 1 + 2\sqrt{x-3} + (x-3)$$

$$x - 3 = 2\sqrt{x-3}$$ Isolating the remaining radical term

$$(x-3)^2 = (2\sqrt{x-3})^2$$ Squaring both sides

$$x^2 - 6x + 9 = 4(x-3)$$

$$x^2 - 6x + 9 = 4x - 12$$

$$x^2 - 10x + 21 = 0$$

$$(x-7)(x-3) = 0$$ Factoring

$$x = 7 \quad or \quad x = 3$$ Using the principle of zero products

The numbers 7 and 3 check and are the solutions. ❏

EXERCISE SET | 7.7

Solve.

1. $\sqrt{2x-3} = 1$

2. $\sqrt{x+3} = 6$

3. $\sqrt{3x+1} = 7$

4. $\sqrt{2x-1} = 7$

5. $\sqrt{y+1} - 5 = 8$

6. $\sqrt{x-2} - 7 = -4$

7. $\sqrt{y-3} + 4 = 2$

8. $\sqrt{y+4} + 6 = 7$

9. $\sqrt[3]{x+5} = 2$

10. $\sqrt[3]{x-2} = 3$

11. $\sqrt[4]{y-3} = 2$

12. $\sqrt[4]{x+3} = 3$

13. $\sqrt{3y+1} = 9$

14. $\sqrt{2y+1} = 13$

15. $3\sqrt{x} = 6$

16. $8\sqrt{y} = 2$

17. $2y^{1/2} - 7 = 9$

18. $3x^{1/2} + 12 = 9$

19. $\sqrt[3]{x} = -3$

20. $\sqrt[3]{y} = -4$

21. $\sqrt{y+3} - 20 = 0$

22. $\sqrt{x+4} - 11 = 0$

23. $(x+2)^{1/2} = -4$

24. $(y-3)^{1/2} = -2$

25. $\sqrt{2x+3} - 5 = -2$

26. $\sqrt{3x+1} - 4 = -1$

27. $8 = x^{-1/2}$

28. $3 = y^{-1/2}$

29. $\sqrt[3]{6x+9} + 8 = 5$

30. $\sqrt[3]{3y+6} + 2 = 3$

31. $\sqrt{3y+1} = \sqrt{2y+6}$

32. $\sqrt{5x-3} = \sqrt{2x+3}$

33. $2\sqrt{1-x} = \sqrt{5}$

34. $2\sqrt{2y-3} = \sqrt{4y}$

35. $2\sqrt{t-1} = \sqrt{3t-1}$

36. $\sqrt{y+10} = 3\sqrt{2y+3}$

37. $\sqrt{y-5} + \sqrt{y} = 5$

38. $\sqrt{x-9} + \sqrt{x} = 1$

39. $3 + \sqrt{z-6} = \sqrt{z+9}$

40. $\sqrt{4x-3} = 2 + \sqrt{2x-5}$

41. $\sqrt{20-x} + 8 = \sqrt{9-x} + 11$

42. $4 + \sqrt{10-x} = 6 + \sqrt{4-x}$

43. $\sqrt{x+2} + \sqrt{3x+4} = 2$

44. $\sqrt{6x+7} - \sqrt{3x+3} = 1$

45. $\sqrt{4y+1} - \sqrt{y-2} = 3$

46. $\sqrt{y+15} - \sqrt{2y+7} = 1$

47. $\sqrt{3x-5}+\sqrt{2x+3}+1=0$

48. $\sqrt{2m-3}=\sqrt{m+7}-2$

49. $2\sqrt{3x+6}-\sqrt{4x+9}=5$

50. $2\sqrt{x-3}+\sqrt{3x-5}=8$

51. $3\sqrt{t+1}-\sqrt{2t-5}=7$

52. $3\sqrt{7x+1}-\sqrt{12x+21}=9$

Skill Maintenance _____

53. Solve:

$$\frac{3}{2x}+\frac{1}{x}=\frac{2x+3.5}{3x}.$$

54. The base of a triangle is 2 in. longer than the height. The area is $31\frac{1}{2}$ in^2. Find the height and the base.

Synthesis _____

55. ◈ The principle of powers is an "if–then" statement that becomes false when the sentence parts are interchanged. Give an example of another such if–then statement.

56. ◈ Explain a method that could be used to solve

$$\sqrt{x}+\sqrt{2x-1}-\sqrt{3x+2}+\sqrt{2+x}=0.$$

Sighting to the horizon. The function $V(h)=1.2\sqrt{h}$ can be used to approximate the distance V, in miles, that a person can see to the horizon from a height h, in feet.

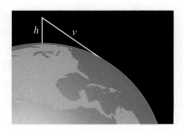

57. ▧ How far can you see to the horizon through an airplane window at a height of 30,000 ft?

58. ▧ How high above sea level must a sailor climb in order to see 10.2 mi out to sea?

Solve.

59. $\dfrac{x+\sqrt{x+1}}{x-\sqrt{x+1}}=\dfrac{5}{11}$

60. $\left(\dfrac{z}{4}\right)^{1/3}-10=2$

61. $(z^2+17)^{1/4}=3$

62. $\sqrt{\sqrt{\sqrt{y}+49}}=7$

63. $\sqrt[3]{x^2+x+15}-3=0$

64. $x^2-5x-\sqrt{x^2-5x-2}=4$

65. $\sqrt{8-b}=b\sqrt{8-b}$

66. $\sqrt{x-2}-\sqrt{x+2}+2=0$

67. $6\sqrt{y}+6y^{-1/2}=37$

68. $\sqrt{a^2+30a}=a+\sqrt{5a}$

7.8

Geometric Applications

Using the Pythagorean Theorem • Two Special Triangles

There are many kinds of problems that involve powers and roots. Many also involve right triangles and the Pythagorean theorem, which we studied in Section 5.8.

EXAMPLE 1

A baseball diamond is actually a square 90 ft on a side. Suppose a catcher fields a bunt along the third-base line 10 ft from home plate. How far would the catcher have to throw the ball to first base? Give an exact answer and an approximation to three decimal places.

Solution   We first make a drawing and let d = the distance, in feet, to first base. We see that a right triangle is formed in which the length of the leg from home to first base is 90 ft. The length of the leg from home to where the catcher fields the ball is 10 ft. We substitute these values into the Pythagorean equation to find d:

$$d^2 = 90^2 + 10^2$$
$$d^2 = 8100 + 100$$
$$d^2 = 8200$$
$$d = \sqrt{8200}.$$

Exact answer: $d = \sqrt{8200}$ ft

Approximation: $d \approx 90.554$ ft

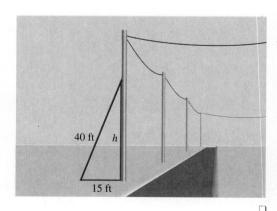

If you use Table 1 to find an approximation, you will need to simplify before finding an approximation in the table:

$$d = \sqrt{8200}$$
$$= \sqrt{100 \cdot 82}$$
$$= 10\sqrt{82}$$
$$\approx 10(9.055) = 90.550 \text{ ft.}$$

Note that we get a variance in the third decimal place. ❑

E X A M P L E   2

The base of a 40-ft–long guy wire is located 15 ft from the telephone pole that it is anchoring. How high up the pole does the guy wire reach? Give an exact answer and an approximation to three decimal places.

Solution   We make a drawing and let h represent the height on the pole that the guy wire reaches. We have a right triangle in which the length of one leg is 15 ft and the length of the hypotenuse is 40 ft. Substituting into the Pythagorean equation, we find h:

$$h^2 + 15^2 = 40^2$$
$$h^2 + 225 = 1600$$
$$h^2 = 1375$$
$$h = \sqrt{1375}.$$

Exact answer:

$$h = \sqrt{1375} \text{ ft}$$

Approximation:

$$h \approx 37.081 \text{ ft}$$

Using a calculator ❑

Two Special Triangles

When both legs of a right triangle are the same size, we say that the triangle is an *isosceles right triangle*. If one leg of an isosceles right triangle has length a, we can

find the length of the hypotenuse as follows:

$$c^2 = a^2 + b^2$$
$$c^2 = a^2 + a^2 \qquad \text{Because the triangle is isosceles, both legs are the same size.}$$
$$c^2 = 2a^2 \qquad \text{Collecting like terms}$$
$$c = \sqrt{2a^2}$$
$$c = \sqrt{a^2 \cdot 2} = a\sqrt{2}.$$

EXAMPLE 3

One leg of an isosceles right triangle measures 7 cm. Find the length of the hypotenuse. Give an exact answer and an approximation to three decimal places.

Solution   We substitute:

$$c = a\sqrt{2} \qquad \text{This equation should be memorized.}$$
$$c = 7\sqrt{2}.$$

Exact answer: $c = 7\sqrt{2}$ cm

Approximation: $c \approx 9.899$ cm

When the hypotenuse of an isosceles right triangle is known, the lengths of the legs can be found.

EXAMPLE 4

The hypotenuse of an isosceles right triangle is 5 ft long. Find the length of a leg. Give an exact answer and an approximation to three decimal places.

Solution   We replace c with 5 and solve for a:

$$5 = a\sqrt{2} \qquad \text{Substituting}$$
$$\frac{5}{\sqrt{2}} = a \qquad \text{Multiplying by } 1/\sqrt{2}$$
$$\frac{5\sqrt{2}}{2} = a. \qquad \text{Rationalizing the denominator}$$

Exact answer: $a = \dfrac{5\sqrt{2}}{2}$ ft

Approximation: $a \approx 3.536$ ft

A second special triangle is known as a 30–60–90 right triangle, so named because of the measures of its angles. Note that in an equilateral triangle, all sides have the same length and all angles are 60°. An altitude, drawn dashed in the figure, bisects, or splits, one angle and one side. Two 30–60–90 right triangles are thus formed.

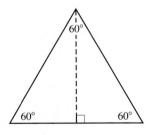

As a result of how the altitude is drawn, if a represents the length of the shorter leg in a 30–60–90 right triangle, then $2a$ represents the length of the hypotenuse. We have

$$a^2 + b^2 = (2a)^2 \quad \text{Using the Pythagorean equation}$$

$$a^2 + b^2 = 4a^2$$

$$b^2 = 3a^2 \quad \text{Adding } -a^2 \text{ on both sides}$$

$$b = \sqrt{3a^2}$$

$$b = \sqrt{a^2 \cdot 3} = a\sqrt{3}.$$

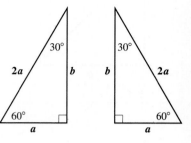

EXAMPLE 5

The shorter leg of a 30–60–90 right triangle measures 8 in. Find the lengths of the other sides. Give exact answers and, where appropriate, approximate to three decimal places.

Solution   The hypotenuse is twice as long as the shorter leg, so we have

$$c = 2a \quad \text{This equation should be memorized.}$$

$$c = 2 \cdot 8 = 16 \text{ in.}$$

The length of the longer leg is the length of the shorter leg times $\sqrt{3}$. This gives us

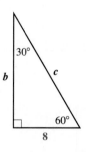

$$b = a\sqrt{3} \quad \text{This should also be memorized.}$$

$$b = 8\sqrt{3} \text{ in.}$$

Exact answer: $c = 16$ in., $b = 8\sqrt{3}$ in.

Approximation: $b \approx 13.856$ in.   □

EXAMPLE 6

The length of the longer leg of a 30–60–90 right triangle is 14 cm. Find the length of the hypotenuse. Give an exact answer and an approximation to three decimal places.

Solution   The length of the hypotenuse is twice the length of the shorter leg. We first find a, the length of the shorter leg, by using the length of the longer leg:

$$14 = a\sqrt{3} \quad \text{Substituting 14 for } b \text{ in } b = a\sqrt{3}$$

$$\frac{14}{\sqrt{3}} = a \quad \text{Multiplying by } 1/\sqrt{3}$$

$$\frac{14\sqrt{3}}{3} = a. \quad \text{Rationalizing the denominator}$$

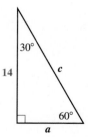

Since the hypotenuse is twice as long as the shorter leg, we have

$$c = 2a$$

$$c = 2 \cdot \frac{14\sqrt{3}}{3} \quad \text{Substituting}$$

$$c = \frac{28\sqrt{3}}{3} \text{ cm}$$

Exact answer: $c = \dfrac{28\sqrt{3}}{3}$ cm

Approximation: $c \approx 16.166$ cm   ❑

Lengths Within Isosceles and 30–60–90 Right Triangles

The length of the hypotenuse in an isosceles right triangle is the length of a leg times $\sqrt{2}$.

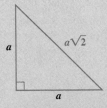

The length of the longer leg in a 30–60–90 right triangle is the length of the shorter leg times $\sqrt{3}$. The hypotenuse is twice as long as the shorter leg.

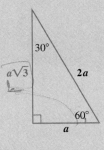

EXERCISE SET | 7.8

In a right triangle, find the length of the side not given. Give an exact answer and, where appropriate, an approximation to three decimal places.

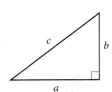

1. $a = 3, b = 5$ 　　　**2.** $a = 8, b = 10$

3. $a = 12, b = 12$ 　　**4.** $a = 10, b = 10$

5. $b = 12, c = 13$ 　　**6.** $a = 5, c = 12$
7. $c = 6, a = \sqrt{5}$ 　**8.** $c = 8, a = 4\sqrt{3}$
9. $b = 1, c = \sqrt{13}$ 　**10.** $a = 1, c = \sqrt{20}$
11. $a = 1, c = \sqrt{n}$ 　**12.** $c = 2, a = \sqrt{n}$

In the following problems, give an exact answer and, where appropriate, an approximation to three decimal places.

13. *Guy wire.* How long is a guy wire reaching from the top of a 15-ft pole to a point on the ground 10 ft from the pole?

14. *Softball diamond.* A slow-pitch softball diamond

is actually a square 65 ft on a side. How far is it from home to second base?

15. Suppose the catcher in Example 1 makes a throw to second base. How far is that throw?

16. *Speaker placement.*   A stereo receiver is in a corner of a 12-ft by 14-ft room. Speaker wire will run under a rug, diagonally, to a speaker in the far corner. Allowing for 4 ft of slack on both ends, how long a piece of wire should be purchased?

17. *Television sets.*   What does it mean to refer to a 20-in. TV set or a 25-in. TV set? Such units refer to the diagonal of the screen. A 20-in. TV set also has a width of 16 in. What is its height?

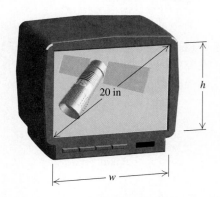

18. *Television sets.*   A 25-in. TV set has a screen with a height of 15 in. What is its width?

19. *Vegetable garden.*   Benito and Dominique are planting a 30-ft by 40-ft vegetable garden and are laying it out using string. They would like to know the length of a diagonal to make sure that right angles are formed. Find the length of a diagonal.

20. *Distance over water.*   To determine the width of a pond, a surveyor locates two stakes at either end of the pond and uses instrumentation to place a third stake so that the distance across the pond is the length of a hypotenuse. If the third stake is 90 m from one stake and 70 m from the other, how wide is the pond?

For each triangle, find the missing length(s). Give an exact answer and, where appropriate, an approximation to three decimal places.

21.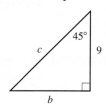

22.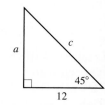

23.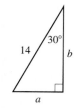

24.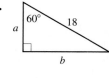

25.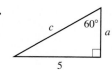

26.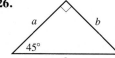

27.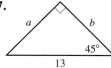

28.

29.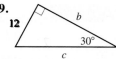

30.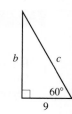

31.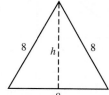

32.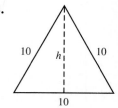

33.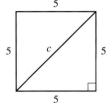

34.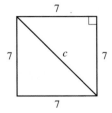

35.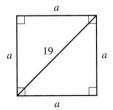

36.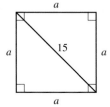

In the following problems, give an exact answer and, where appropriate, an approximation to three decimal places.

37. Triangle *ABC* has sides of lengths 25 ft, 25 ft, and 30 ft. Triangle *PQR* has sides of lengths 25 ft, 25 ft, and 40 ft. Which triangle has the greater area and by how much?

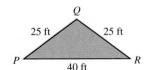

38. *Bridge expansion.* During the summer heat, a 2-mi bridge expands 2 ft in length. Assuming the bulge occurs straight up the middle, how high is the bulge? (The answer may surprise you. In reality, bridges are built with expansion spaces to avoid such buckling.)

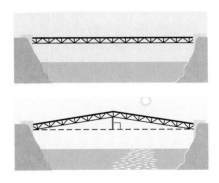

39. *Camping tent.* The entrance to a pup tent is the shape of an equilateral triangle. If the base of the tent is 4 ft wide, how tall is the tent?

40. Each side of a regular octagon has length *s*. Find a formula for the distance *d* between the parallel sides of the octagon.

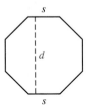

41. The diagonal of a square has length $8\sqrt{2}$ ft. Find the length of a side of the square.

42. The length and the width of a rectangle are given by consecutive integers. The area of the rectangle is 90 cm². Find the length of a diagonal of the rectangle.

43. Find all points on the *y*-axis of a Cartesian coordinate system that are 5 units from the point (3, 0).

44. Find all points on the *x*-axis of a Cartesian coordinate system that are 5 units from the point (0, 4).

Skill Maintenance _____

Solve.

45. $x^2 - 11x + 24 = 0$ **46.** $2x^2 + 11x - 21 = 0$

Synthesis _____

47. ◈ Write a problem for a classmate to solve in which the solution is: "The height of the tepee is $5\sqrt{3}$ ft."

48. ◈ Write a problem for a classmate to solve in which the solution is: "The height of the window is $15\sqrt{3}$ ft."

49. A cube measures 5 cm on each side. How long is the diagonal that connects two opposite corners of the cube? Give an exact answer.

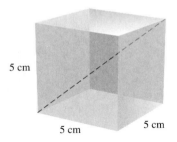

50. Kit's house is 24 ft wide and 32 ft long. The peak of the roof is 6 ft higher than the sides. Kit plans to reshingle the roof. If each packet of shingles covers 100 square feet, how many packets should Kit buy?

6 ft

32 ft

24 ft

7.9

The Complex Numbers

Imaginary and Complex Numbers • **Addition and Subtraction** • **Multiplication** • **Powers of** *i* • **Conjugates and Division** • **Solutions of Equations**

Imaginary and Complex Numbers

Negative numbers do not have square roots in the real-number system. However, a larger number system that contains the real-number system is designed so that negative numbers *do* have square roots. That system is called the **complex-number system,** and it makes use of a number that is a square root of -1. We call this new number *i*.

> We define the number $i = \sqrt{-1}$. That is, $i = \sqrt{-1}$ and $i^2 = -1$.

To express roots of negative numbers in terms of *i*, we can use the fact that in the complex numbers, $\sqrt{-p} = \sqrt{-1}\sqrt{p}$ when *p* is a positive real number.

EXAMPLE 1

Express each number in terms of *i*: **(a)** $\sqrt{-7}$; **(b)** $\sqrt{-16}$; **(c)** $-\sqrt{-13}$; **(d)** $-\sqrt{-64}$; **(e)** $\sqrt{-48}$.

Solution

a) $\sqrt{-7} = \sqrt{-1 \cdot 7} = \sqrt{-1} \cdot \sqrt{7} = i\sqrt{7}$, or $\sqrt{7}i$ *i* is *not* under the radical.

b) $\sqrt{-16} = \sqrt{-1 \cdot 16} = \sqrt{-1} \cdot \sqrt{16} = i \cdot 4 = 4i$

c) $-\sqrt{-13} = -\sqrt{-1 \cdot 13} = -\sqrt{-1} \cdot \sqrt{13} = -i\sqrt{13}$, or $-\sqrt{13}i$

d) $-\sqrt{-64} = -\sqrt{-1 \cdot 64} = -\sqrt{-1} \cdot \sqrt{64} = -i \cdot 8 = -8i$

e) $\sqrt{-48} = \sqrt{-1 \cdot 48} = \sqrt{-1} \cdot \sqrt{48} = i\sqrt{48} = i \cdot 4\sqrt{3} = 4\sqrt{3}i$, or $4i\sqrt{3}$ ❑

**Imaginary
Number**

> An *imaginary number* is a number that can be written $a + bi$, where a and b are real numbers, $b \neq 0$.

Don't let the name "imaginary" fool you. Imaginary numbers appear in such fields as engineering and the physical sciences. The following are examples of imaginary numbers:

$$\left.\begin{array}{l} 5 + 4i, \\ \sqrt{5} - \pi i, \end{array}\right\} \qquad \text{Here } a \neq 0, \, b \neq 0.$$

$$17i. \qquad\qquad \text{Here } a = 0, \, b \neq 0.$$

When a and b are real numbers and b is allowed to be 0, the number $a + bi$ is said to be **complex.**

Complex Number

> A *complex number* is any number that can be written $a + bi$, where a and b are any real numbers. (Note that a and b both can be 0.)

The following are examples of complex numbers:

$7 + 3i$ (here $a \neq 0$, $b \neq 0$); $4i$ (here $a = 0$, $b \neq 0$);

8 (here $a \neq 0$, $b = 0$); 0 (here $a = 0$, $b = 0$).

Complex numbers like $17i$ or $4i$, in which $a = 0$ and $b \neq 0$, are imaginary numbers with no real part. Such numbers are called *pure imaginary* numbers.

Note that when $b = 0$, $a + 0i = a$, so every real number is a complex number. The relationships among various real and complex numbers are shown below.

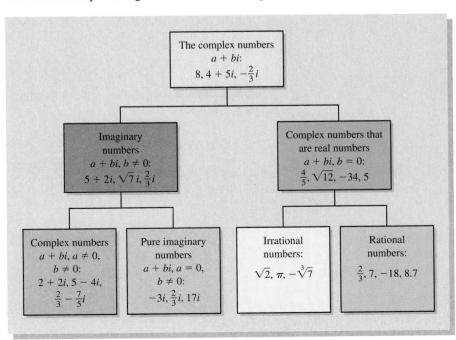

It is important to keep in mind some comparisons between numbers that have real-number roots and those that have complex-number roots that are not real. For example, $\sqrt{-48}$ is a complex number that is not a real number because we are taking the square root of a negative number. *But,* $\sqrt[3]{-48}$ is a real number because each real number has a cube root that is a real number.

Addition and Subtraction

The complex numbers obey the commutative, associative, and distributive laws. Thus we can add and subtract them as we do binomials.

EXAMPLE 2

Add or subtract and simplify.

a) $(8 + 6i) + (3 + 2i)$
b) $(4 + 5i) - (6 - 3i)$

Solution

a) $(8 + 6i) + (3 + 2i) = (8 + 3) + (6i + 2i)$
Collecting the real parts and the imaginary parts

$$= 11 + (6 + 2)i = 11 + 8i$$

b) $(4 + 5i) - (6 - 3i) = (4 - 6) + [5i - (-3i)]$
Note that the 6 and the $-3i$ are both being subtracted.

$$= -2 + 8i$$ ❑

Multiplication

For complex numbers, the property $\sqrt{a}\,\sqrt{b} = \sqrt{ab}$ does *not* hold in general, but it does hold when $a = -1$ and b is a nonnegative number. To multiply square roots of negative real numbers, we first express them in terms of i. For example,

$$\sqrt{-2}\cdot\sqrt{-5} = \sqrt{-1}\cdot\sqrt{2}\cdot\sqrt{-1}\cdot\sqrt{5} = i\sqrt{2}\cdot i\sqrt{5}$$
$$= i^2\sqrt{10} = -1\sqrt{10} = -\sqrt{10}\quad\text{is correct!}$$

But

$$\sqrt{-2}\cdot\sqrt{-5} = \sqrt{(-2)(-5)} = \sqrt{10}\quad\text{is wrong!}$$

Keeping this and the fact that $i^2 = -1$ in mind, we multiply in much the same way that we do with real numbers.

EXAMPLE 3

Multiply and simplify.

a) $\sqrt{-16}\cdot\sqrt{-25}$ **b)** $\sqrt{-5}\cdot\sqrt{-7}$ **c)** $-3i\cdot 8i$
d) $-4i(3 - 5i)$ **e)** $(1 + 2i)(1 + 3i)$

Solution

a) $\sqrt{-16}\cdot\sqrt{-25} = \sqrt{-1}\cdot\sqrt{16}\cdot\sqrt{-1}\cdot\sqrt{25}$

$$= i\cdot 4\cdot i\cdot 5$$
$$= i^2\cdot 20$$
$$= -1\cdot 20\qquad i^2 = -1$$
$$= -20$$

b) $\sqrt{-5} \cdot \sqrt{-7} = \sqrt{-1} \cdot \sqrt{5} \cdot \sqrt{-1} \cdot \sqrt{7}$

$= i \cdot \sqrt{5} \cdot i \cdot \sqrt{7}$

$= i^2 \cdot \sqrt{35}$

$= -1 \cdot \sqrt{35} \qquad i^2 = -1$

$= -\sqrt{35}$

c) $-3i \cdot 8i = -24 \cdot i^2$

$= -24 \cdot (-1) \qquad i^2 = -1$

$= 24$

d) $-4i(3 - 5i) = -4i \cdot 3 + (-4i)(-5i)$ Using the distributive law

$= -12i + 20i^2$

$= -12i - 20 \qquad\qquad\qquad i^2 = -1$

$= -20 - 12i \qquad\qquad\qquad$ Writing in the form $a + bi$

e) $(1 + 2i)(1 + 3i) = 1 + 3i + 2i + 6i^2$ Multiplying each term of one number by every term of the other (FOIL)

$= 1 + 3i + 2i - 6 \qquad\qquad i^2 = -1$

$= -5 + 5i \qquad\qquad\qquad\quad$ Collecting like terms ❑

Powers of i

We now want to simplify certain expressions involving higher powers of i. To do so, we recall that -1 raised to an *even* power is 1, and -1 raised to an *odd* power is -1. Simplifying powers of i can then be done by using the fact that $i^2 = -1$ and expressing the given power of i in terms of i^2. Consider the following:

i, or $\sqrt{-1}$,

$i^2 = -1$,

$i^3 = i^2 \cdot i = (-1)i = -i$,

$i^4 = (i^2)^2 = (-1)^2 = 1$,

$i^5 = i^4 \cdot i = (i^2)^2 \cdot i = (-1)^2 \cdot i = i$,

$i^6 = (i^2)^3 = (-1)^3 = -1$.

Note that the powers of i cycle themselves through the values i, -1, $-i$, and 1.

EXAMPLE 4

Simplify: **(a)** i^{37}; **(b)** i^{58}; **(c)** i^{75}; **(d)** i^{80}.

Solution

a) $i^{37} = i^{36} \cdot i = (i^2)^{18} \cdot i = (-1)^{18} \cdot i = 1 \cdot i = i$

b) $i^{58} = (i^2)^{29} = (-1)^{29} = -1$

c) $i^{75} = i^{74} \cdot i = (i^2)^{37} \cdot i = (-1)^{37} \cdot i = -1 \cdot i = -i$

d) $i^{80} = (i^2)^{40} = (-1)^{40} = 1$ ❑

Now let us simplify other expressions.

EXAMPLE 5

Write the expression in the form $a + bi$: **(a)** $8 - i^2$; **(b)** $17 + 6i^3$; **(c)** $i^{22} - 67i^2$; **(d)** $i^{23} + i^{48}$.

Solution

a) $8 - i^2 = 8 - (-1) = 8 + 1 = 9$

b) $17 + 6i^3 = 17 + 6 \cdot i^2 \cdot i = 17 + 6(-1)i = 17 - 6i$

c) $i^{22} - 67i^2 = (i^2)^{11} - 67(-1) = (-1)^{11} + 67 = -1 + 67 = 66$

d) $i^{23} + i^{48} = (i^{22}) \cdot i + (i^2)^{24}$

$$= (i^2)^{11} \cdot i + (-1)^{24} = (-1)^{11} \cdot i + (-1)^{24}$$

$$= -i + 1 = 1 - i$$

Conjugates and Division

Conjugates of complex numbers are defined as follows.

Conjugate of a Complex Number

The *conjugate* of a complex number $a + bi$ is $a - bi$, and the *conjugate* of $a - bi$ is $a + bi$.

EXAMPLE 6

Find the conjugate: **(a)** $5 + 7i$; **(b)** $14 - 3i$; **(c)** $-3 - 9i$; **(d)** $4i$.

Solution

a) $5 + 7i$ The conjugate is $5 - 7i$.

b) $14 - 3i$ The conjugate is $14 + 3i$.

c) $-3 - 9i$ The conjugate is $-3 + 9i$.

d) $4i$ The conjugate is $-4i$. Note that $4i = 0 + 4i$.

When a complex number is multiplied by its conjugate, we get a real number.

EXAMPLE 7

Multiply: **(a)** $(5 + 7i)(5 - 7i)$; **(b)** $(2 - 3i)(2 + 3i)$.

Solution

a) $(5 + 7i)(5 - 7i) = 5^2 - (7i)^2$ Using $(A + B)(A - B) = A^2 - B^2$

$$= 25 - 49i^2$$

$$= 25 - 49(-1) \qquad i^2 = -1$$

$$= 25 + 49$$

$$= 74$$

b) $(2 - 3i)(2 + 3i) = 2^2 - (3i)^2$

$$= 4 - 9i^2$$

$$= 4 - 9(-1) \qquad i^2 = -1$$

$$= 4 + 9$$

$$= 13$$

Conjugates are used when dividing complex numbers.

EXAMPLE 8

Divide and simplify to the form $a + bi$: **(a)** $\dfrac{-5 + 9i}{1 - 2i}$; **(b)** $\dfrac{7 + 3i}{5i}$.

Solution

a) Rewriting i as $\sqrt{-1}$, we can multiply by 1 using the conjugate, as in Section 7.6:

$$\frac{-5+9i}{1-2i} = \frac{-5+9\sqrt{-1}}{1-2\sqrt{-1}} = \frac{-5+9\sqrt{-1}}{1-2\sqrt{-1}} \cdot \frac{1+2\sqrt{-1}}{1+2\sqrt{-1}}. \qquad \text{Multiplying by 1}$$

By writing 1 as $(1 + 2i)/(1 + 2i)$, we can use the same method without converting to radical notation:

$$\frac{-5+9i}{1-2i} = \frac{-5+9i}{1-2i} \cdot \frac{1+2i}{1+2i} \qquad \begin{array}{l}\text{Multiplying by 1 using the conjugate of}\\ \text{the denominator in the symbol for 1}\end{array}$$

$$= \frac{(-5+9i)(1+2i)}{(1-2i)(1+2i)}$$

$$= \frac{-5-10i+9i+18i^2}{1^2-4i^2}$$

$$= \frac{-5-i-18}{1-4(-1)} \qquad i^2 = -1$$

$$= \frac{-23-i}{5}$$

$$= -\frac{23}{5} - \frac{1}{5}i \qquad \text{Writing in the form } a + bi$$

b) $\dfrac{7+3i}{5i} = \dfrac{7+3i}{5i} \cdot \dfrac{-5i}{-5i}$ $\qquad \begin{array}{l}\text{Multiplying by 1 using the conjugate of}\\ 0 + 5i \text{ in the symbol for 1}\end{array}$

$$= \frac{-35i-15i^2}{-25i^2} \qquad \text{Multiplying}$$

$$= \frac{-35i-15(-1)}{-25(-1)} \qquad i^2 = -1$$

$$= \frac{15-35i}{25}$$

$$= \frac{15}{25} - \frac{35}{25}i$$

$$= \frac{3}{5} - \frac{7}{5}i$$

Solutions of Equations

The equation $x^2 + 1 = 0$ has no real-number solution, but it has two nonreal complex solutions.

EXAMPLE 9

Determine whether i is a solution of the equation $x^2 + 1 = 0$.

Solution   We substitute i for x in the equation.

$$
\begin{array}{c}
x^2 + 1 = 0 \\
\hline
i^2 + 1 \; ? \; 0 \\
-1 + 1 \quad \Big| \\
0 \;\Big|\; 0 \;\text{TRUE}
\end{array}
$$

The number i is a solution.   ❑

Any polynomial equation in one variable has complex-number solutions. Sometimes it is not easy to find the solutions, but they always exist.

EXAMPLE 10

Determine whether $1 + i$ is a solution of the equation $x^2 - 2x + 2 = 0$.

Solution   We substitute $1 + i$ for x in the equation.

$$
\begin{array}{c}
x^2 - 2x + 2 = 0 \\
\hline
(1 + i)^2 - 2(1 + i) + 2 \; ? \; 0 \\
1 + 2i + i^2 - 2 - 2i + 2 \quad \Big| \\
1 + 2i - 1 - 2 - 2i + 2 \quad \Big| \\
(1 - 1 - 2 + 2) + (2 - 2)i \quad \Big| \\
0 + 0i \quad \Big| \\
0 \;\Big|\; 0 \;\text{TRUE}
\end{array}
$$

The number $1 + i$ is a solution.   ❑

EXERCISE SET | 7.9

Express in terms of i.

1. $\sqrt{-15}$ **2.** $\sqrt{-17}$ **3.** $\sqrt{-16}$

4. $\sqrt{-25}$ **5.** $-\sqrt{-12}$ **6.** $-\sqrt{-20}$

7. $\sqrt{-3}$ **8.** $\sqrt{-4}$ **9.** $\sqrt{-81}$

10. $\sqrt{-27}$ **11.** $\sqrt{-98}$ **12.** $-\sqrt{-18}$

13. $-\sqrt{-49}$ **14.** $-\sqrt{-125}$

15. $4 - \sqrt{-60}$ **16.** $6 - \sqrt{-84}$

17. $\sqrt{-4} + \sqrt{-12}$

18. $-\sqrt{-76} + \sqrt{-125}$

Add or subtract and simplify.

19. $(3 + 2i) + (5 - i)$

20. $(-2 + 3i) + (7 + 8i)$

21. $(4 - 3i) + (5 - 2i)$

22. $(-2 - 5i) + (1 - 3i)$

23. $(9 - i) + (-2 + 5i)$

24. $(6 + 4i) + (2 - 3i)$

25. $(3 - i) - (5 + 2i)$

26. $(-2 + 8i) - (7 + 3i)$

27. $(4 - 2i) - (5 - 3i)$

28. $(-2 - 3i) - (1 - 5i)$

29. $(9 + 5i) - (-2 - i)$

30. $(6 - 3i) - (2 + 4i)$

Multiply. Write the answer in the form $a + bi$.

31. $\sqrt{-25}\,\sqrt{-36}$ **32.** $\sqrt{-81}\,\sqrt{-49}$

33. $\sqrt{-6}\,\sqrt{-5}$

34. $\sqrt{-7}\,\sqrt{-10}$

35. $\sqrt{-50}\,\sqrt{-3}$

36. $\sqrt{-72}\,\sqrt{-3}$

37. $\sqrt{-48}\,\sqrt{-6}$

38. $\sqrt{-15}\,\sqrt{-75}$

39. $5i \cdot 8i$

40. $6i \cdot 9i$

41. $5i \cdot (-7i)$

42. $7i \cdot (-4i)$

43. $5i(3 - 2i)$

44. $4i(5 - 6i)$

45. $-3i(7 - 4i)$

46. $-7i(9 - 3i)$

47. $(3 + 2i)(1 + i)$

48. $(4 + 3i)(2 + 5i)$

49. $(2 + 3i)(6 - 2i)$

50. $(5 + 6i)(2 - i)$

51. $(6 - 5i)(3 + 4i)$

52. $(5 - 6i)(2 + 5i)$

53. $(7 - 2i)(2 - 6i)$

54. $(-4 + 5i)(3 - 4i)$

55. $(5 - 3i)(4 - 5i)$

56. $(7 - 3i)(4 - 7i)$

57. $(-2 + 3i)(-2 + 5i)$

58. $(-3 + 6i)(-3 + 4i)$

59. $(-5 - 4i)(3 + 7i)$

60. $(2 + 9i)(-3 - 5i)$

61. $(3 - 2i)^2$

62. $(5 - 2i)^2$

63. $(2 + 3i)^2$

64. $(4 + 2i)^2$

65. $(-2 + 3i)^2$

66. $(-5 - 2i)^2$

Simplify.

67. i^7

68. i^{11}

69. i^{24}

70. i^{35}

71. i^{42}

72. i^{64}

73. i^9

74. $(-i)^{71}$

75. $(-i)^6$

76. $(-i)^4$

77. $(5i)^3$

78. $(-3i)^5$

Simplify to the form $a + bi$.

79. $7 + i^4$

80. $-18 + i^3$

81. $i^4 - 26i$

82. $i^5 + 37i$

83. $i^2 + i^4$

84. $5i^5 + 4i^3$

85. $i^5 + i^7$

86. $i^{84} - i^{100}$

87. $1 + i + i^2 + i^3 + i^4$

88. $i - i^2 + i^3 - i^4 + i^5$

89. $5 - \sqrt{-64}$

90. $\sqrt{-12} + 36i$

Divide and simplify to the form $a + bi$.

91. $\dfrac{5}{3 - i}$

92. $\dfrac{3}{5 + i}$

93. $\dfrac{2i}{7 + 3i}$

94. $\dfrac{4i}{2 - 5i}$

95. $\dfrac{7}{6i}$

96. $\dfrac{3}{10i}$

97. $\dfrac{8 - 3i}{7i}$

98. $\dfrac{3 + 8i}{5i}$

99. $\dfrac{3 + 2i}{2 + i}$

100. $\dfrac{4 + 5i}{5 - i}$

101. $\dfrac{5 - 2i}{2 + 5i}$

102. $\dfrac{3 - 2i}{4 + 3i}$

103. $\dfrac{3 - 5i}{3 - 2i}$

104. $\dfrac{2 - 7i}{5 - 4i}$

Determine whether the complex number is a solution of the equation.

105. $1 + 2i$;
$x^2 - 2x + 5 = 0$

106. $1 - 2i$;
$x^2 - 2x + 5 = 0$

107. $1 - i$;
$x^2 + 2x + 2 = 0$

108. $2 + i$;
$x^2 - 4x - 5 = 0$

Skill Maintenance

Solve.

109. $\dfrac{196}{x^2 - 7x + 49} - \dfrac{2x}{x + 7} = \dfrac{2058}{x^3 + 343}$

110. $\dfrac{5}{t} - \dfrac{3}{2} = \dfrac{4}{7}$

111. $28 = 3x^2 - 17x$

Synthesis

112. A function g is given by
$$g(z) = \frac{z^4 - z^2}{z - 1}.$$
Find $g(2i)$; $g(i + 1)$; $g(2i - 1)$.

113. Evaluate
$$\frac{1}{w - w^2}$$
when
$$w = \frac{1 - i}{10}.$$

Simplify.

114. $\dfrac{i^5 + i^6 + i^7 + i^8}{(1 - i)^4}$

115. $(1 - i)^3(1 + i)^3$

116. $\dfrac{5 - \sqrt{5}i}{\sqrt{5}i}$

117. $\dfrac{6}{1 + \dfrac{3}{i}}$

118. $\left(\dfrac{1}{2} - \dfrac{1}{3}i\right)^2 - \left(\dfrac{1}{2} + \dfrac{1}{3}i\right)^2$

119. $\dfrac{i - i^{38}}{1 + i}$

Example: $\sqrt{8} + 3\sqrt{2} = \sqrt{4} \cdot \sqrt{2} + 3\sqrt{2} = 2\sqrt{2} + 3\sqrt{2} = 5\sqrt{2}$

7

4. *Rationalizing denominators.* Radical expressions are usually considered simpler if there are no radicals in the denominator.

Example: $\dfrac{1}{\sqrt{2}} = \dfrac{1}{\sqrt{2}} \cdot \dfrac{\sqrt{2}}{\sqrt{2}} = \dfrac{\sqrt{2}}{2}$

The Principle of Powers

If $a = b$, then $a^n = b^n$ for any natural number n.

A general strategy for solving equations with two radical terms is as follows.

1. Isolate one of the radical terms.
2. Use the principle of powers.
3. If a radical remains, perform steps (1) and (2) again.
4. Check possible solutions.

Special Triangles

The length of the hypotenuse in an isosceles right triangle is the length of a leg times $\sqrt{2}$.

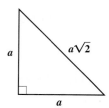

The length of the longer leg in a 30–60–90 right triangle is the length of the shorter leg times $\sqrt{3}$. The hypotenuse is twice as long as the shorter leg.

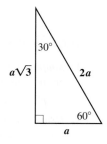

A complex number is any number that can be written $a + bi$, where a and b are any real numbers and $i = \sqrt{-1}$.

REVIEW EXERCISES

This chapter's review and test include Skill Maintenance exercises from Sections 5.8, 6.1, and 6.4.

Simplify.

1. $\sqrt{\dfrac{36}{81}}$

2. $\sqrt{0.0049}$

Let $f(x) = \sqrt{2x - 7}$. Find the following.

3. $f(16)$

4. The domain of f

Simplify. Assume that letters can represent *any* real number.

5. $\sqrt{81a^2}$

6. $\sqrt{(c+8)^2}$

7. $\sqrt{x^2 - 6x + 9}$

8. $\sqrt{4x^2 + 4x + 1}$

9. $\sqrt[5]{-32}$

10. $\sqrt[3]{-\dfrac{1}{27}}$

11. $\sqrt[10]{x^{10}}$

12. $-\sqrt[13]{(-3)^{13}}$

13. Rewrite with rational exponents: $\left(\sqrt[5]{8x^6y^2}\right)^4$.

14. Rewrite without rational exponents: $(5a)^{3/4}$.

Use rational exponents to simplify.

15. $x^{1/3} \cdot y^{1/4}$

16. $\sqrt[3]{\dfrac{x^9 y^{12}}{-8}}$

Approximate to the nearest thousandth using a calculator or Table 1.

17. $\sqrt{245}$

18. $\sqrt{112}$

19. $\dfrac{14 - \sqrt{245}}{7}$

Multiply and simplify by factoring. Assume that all variables are nonnegative.

20. $\sqrt{3x^2}\ \sqrt{6y^3}$

21. $\sqrt[3]{a^5b}\ \sqrt[3]{27b}$

22. $\sqrt{3x^3y^4}\ \sqrt[3]{9x^2y}$

Divide. Then simplify by taking roots, if possible.

23. $\dfrac{\sqrt[3]{60xy^3}}{\sqrt[3]{10x}}$

24. $\dfrac{\sqrt{75x}}{2\sqrt{3}}$ (Assume $x \geq 0$.)

Simplify.

25. $\left(\sqrt{8xy^2}\right)^2$ (Assume $x, y \geq 0$.)

26. $\left(\sqrt[3]{4a^2b}\right)^2$

Perform the indicated operation. Simplify by collecting like radical terms, if possible.

27. $12\sqrt[3]{135} - 3\sqrt[3]{40}$

28. $\sqrt{50} + 2\sqrt{18} + \sqrt{32}$

29. $\left(\sqrt[3]{27} - \sqrt[3]{2}\right)\left(\sqrt[3]{27} + \sqrt[3]{2}\right)$

30. $\left(\sqrt{5} - 3\sqrt{8}\right)\left(\sqrt{5} + 2\sqrt{8}\right)$

31. $\left(1 - \sqrt{7}\right)^2$

32. Rationalize the denominator. Assume that a and b represent positive numbers.

$$\dfrac{5\sqrt{12a}}{\sqrt{a} + \sqrt{b}}$$

33. Rationalize the numerator of the expression in Exercise 32.

Solve.

34. $\sqrt{3x - 3} = 1 + \sqrt{x}$

35. $\sqrt[4]{x + 3} = 2$

Solve. Give an exact answer and, where appropriate, an approximation to three decimal places.

36. The diagonal of a square has length $9\sqrt{2}$ cm. Find the length of a side of the square.

37. A bookcase is 1 foot taller than it is wide. A diagonal brace, 1 foot longer than the height of the bookcase, is needed for support. What is the length of the brace?

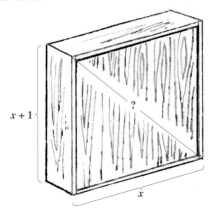

38. Express in terms of i and simplify: $-\sqrt{-8}$.

39. Add: $(-4 + 3i) + (2 - 12i)$.

40. Subtract: $(4 - 7i) - (3 - 8i)$.

Multiply.

41. $(2 + 5i)(2 - 5i)$ **42.** i^{13}

43. $(6 - 3i)(2 - i)$

Divide and simplify to the form $a + bi$.

44. $\dfrac{-3 + 2i}{5i}$ **45.** $\dfrac{6 - 3i}{2 + i}$

Skill Maintenance

46. Find three consecutive positive integers such that the product of the first and second integers is 26 less than the product of the second and third integers.

47. Solve:
$$\frac{7}{x + 2} + \frac{5}{x^2 - 2x + 4} = \frac{84}{x^3 + 8}.$$

48. Solve:
$$2x^2 + 3x - 27 = 0.$$

49. Multiply and simplify:
$$\frac{x^2 + 3x}{x^2 - y^2} \cdot \frac{x^2 - xy + 2x - 2y}{x^2 - 9}.$$

Synthesis

50. ◈ Explain why $\sqrt[k]{x^k} = |x|$ when k is even, but $\sqrt[k]{x^k} = x$ when k is odd.

51. ◈ What is the difference between real numbers and complex numbers?

52. Solve:
$$\sqrt{11x + \sqrt{6 + x}} = 6.$$

53. Simplify:
$$\frac{2}{1 - 3i} - \frac{3}{4 + 2i}.$$

CHAPTER TEST   7

1. Simplify: $\sqrt{\dfrac{100}{49}}$.

2. Determine the domain of f if
$$f(t) = \sqrt{2t + 10}.$$

In Questions 3–6, assume that letters can represent *any* real number. Simplify.

3. $\sqrt{36y^2}$ **4.** $\sqrt{x^2 + 10x + 25}$

5. $\sqrt[3]{-8}$ **6.** $\sqrt[10]{(-4)^{10}}$

7. Rewrite with fractional exponents: $\left(\sqrt{5xy^2}\right)^5$.

8. Use rational exponents to simplify: $\sqrt[3]{16a^5b^{10}}$.

9. Approximate using a calculator:
$$\sqrt{0.000001204}.$$

10. Multiply and simplify by factoring:
$$\sqrt[3]{x^4}\sqrt[3]{8x^5}.$$

11. Simplify by taking the roots of the numerator and the denominator. (Assume $x, y > 0$.)
$$\sqrt{\frac{25x^3}{36y^4}}$$

12. Simplify: $\left(\sqrt[3]{16a^2b}\right)^2$.

13. Simplify and add:
$$3\sqrt{128} + 2\sqrt{18} + 2\sqrt{32}.$$

14. Multiply and simplify:
$$\left(\sqrt{20} + 2\sqrt{5}\right)\left(\sqrt{20} - 3\sqrt{5}\right).$$

15. Multiply and simplify:
$$\left(2x - \sqrt[3]{y^2}\right)\left(x + \sqrt{y^3}\right).$$

16. Rationalize the denominator:
$$\frac{1 + \sqrt{2}}{3 - 5\sqrt{2}}.$$

17. Solve: $\sqrt{y - 6} = \sqrt{y + 9} - 3$.

18. The shorter leg of a 30–60–90 right triangle measures 10 cm. Find the lengths of the other sides. Give exact answers and, where appropriate, approximations to three decimal places.

19. Express in terms of i and simplify: $\sqrt{-18}$.

20. Subtract: $(5 + 8i) - (-2 + 3i)$.

21. Multiply: $\sqrt{-100}\,\sqrt{-9}$.

22. Multiply. Write the answer in the form $a + bi$: $(1 - i)^2$.

23. Divide and simplify to the form $a + bi$:
$$\frac{-7 + 14i}{6 - 8i}.$$

24. Simplify to the form $a + bi$: $5i^{25} - i^{12}$.

Skill Maintenance _____

25. Solve: $6x^2 = 13x + 5$.

26. Divide and simplify:
$$\frac{x^3 - 27}{x^2 - 16} \div \frac{x^2 + 3x + 9}{x + 4}.$$

27. Solve:
$$\frac{11x}{x + 3} + \frac{33}{x} + 12 = \frac{99}{x^2 + 3x}.$$

28. Find two consecutive even integers whose product is 288.

Synthesis _____

29. Solve:
$$\sqrt{2x - 2} + \sqrt{7x + 4} = \sqrt{13x + 10}.$$

30. Simplify:
$$\frac{1 - 4i}{4i(1 + 4i)^{-1}}.$$

Quadratic Equations and Functions

A formula relating an athlete's vertical leap V, measured in inches, to hang time T, measured in seconds, is

$$V = 48T^2.$$ ← This is a quadratic equation.

In Section 8.9, we show how to solve this equation for T:

$$T = \frac{\sqrt{3V}}{12}.$$

Lisa Sonntag, M.S.
EXERCISE PHYSIOLOGIST, NATIONALLY RANKED RACEWALKER

"Math comes up in my daily routine. It is important for interpreting research results and in problem solving. I use math when designing exercise prescriptions and work simulations. Math has also paid off in coaching and in my own racewalking."

n translating problem situations to mathematical language, we often obtain a function or equation containing a second-degree polynomial in one variable. Such functions or equations are said to be *quadratic*. In this chapter, we will study a variety of equations, inequalities, and applications for which we will need to solve quadratic equations or graph quadratic functions.

In addition to the material in this chapter, the review and test for Chapter 8 include material from Sections 3.3, 6.2, 7.3, and 7.7.

8.1

Quadratic Equations

The Principle of Square Roots • **Completing the Square** • **Problem Solving**

In Section 5.8, we solved quadratic equations like $3x^2 = 2 - x$ by factoring. Let's review that procedure.

EXAMPLE 1

Solve: $3x^2 = 2 - x$.

Solution

$$3x^2 = 2 - x$$
$$3x^2 + x - 2 = 0 \qquad \text{Adding } -2 + x \text{ on both sides to obtain standard form}$$
$$(3x - 2)(x + 1) = 0$$
$$3x - 2 = 0 \quad or \quad x + 1 = 0 \qquad \text{Using the principle of zero products}$$
$$3x = 2 \quad or \qquad x = -1$$
$$x = \tfrac{2}{3} \quad or \qquad x = -1$$

Check: For -1:

$$\begin{array}{c|c} \multicolumn{2}{c}{3x^2 = 2 - x} \\ \hline 3(-1)^2 \;?\; 2 - (-1) & \\ 3 \cdot 1 & 2 + 1 \\ 3 & 3 \qquad \text{TRUE} \end{array}$$

For $\tfrac{2}{3}$:

$$\begin{array}{c|c} \multicolumn{2}{c}{3x^2 = 2 - x} \\ \hline 3(\tfrac{2}{3})^2 \;?\; 2 - \tfrac{2}{3} & \\ 3 \cdot \tfrac{4}{9} & \tfrac{6}{3} - \tfrac{2}{3} \\ \tfrac{4}{3} & \tfrac{4}{3} \qquad \text{TRUE} \end{array}$$

The solutions are -1 and $\tfrac{2}{3}$.

EXAMPLE 2

Solve: $x^2 = 25$.

Solution

$$x^2 = 25$$
$$x^2 - 25 = 0 \qquad \text{Writing in standard form}$$
$$(x - 5)(x + 5) = 0 \qquad \text{Factoring}$$
$$x - 5 = 0 \quad or \quad x + 5 = 0 \qquad \text{Using the principle of zero products}$$
$$x = 5 \quad or \qquad x = -5$$

The solutions are 5 and -5. We leave the checks to the student.

The Principle of Square Roots

Consider the equation $x^2 = 25$ again. We know from Chapter 7 that the number 25 has two real-number square roots, namely, 5 and -5. Note that these are the solutions of the equation in Example 2. Thus square roots can provide a quick method for solving equations of the type $x^2 = k$.

The Principle of Square Roots

For any real number k, if $x^2 = k$, then $x = \sqrt{k}$ or $x = -\sqrt{k}$.

EXAMPLE 3

Solve: $3x^2 = 6$.

Solution

$$3x^2 = 6$$
$$x^2 = 2 \qquad \text{Multiplying by } \tfrac{1}{3}$$
$$x = \sqrt{2} \quad or \quad x = -\sqrt{2} \qquad \text{Using the principle of square roots}$$

We often use the symbol $\pm\sqrt{2}$ to represent both of the numbers $\sqrt{2}$ and $-\sqrt{2}$. We check as follows.

Check: For $\sqrt{2}$:   For $-\sqrt{2}$:

$$
\begin{array}{c|c}
3x^2 = 6 \\
\hline
3(\sqrt{2})^2 \ ? \ 6 \\
3 \cdot 2 \\
6 \ \bigм| \ 6 \ \text{TRUE}
\end{array}
\qquad
\begin{array}{c|c}
3x^2 = 6 \\
\hline
3(-\sqrt{2})^2 \ ? \ 6 \\
3 \cdot 2 \\
6 \ \big| \ 6 \ \text{TRUE}
\end{array}
$$

The solutions are $\sqrt{2}$ and $-\sqrt{2}$, or $\pm\sqrt{2}$. ❑

Sometimes we rationalize denominators to simplify answers.

EXAMPLE 4

Solve: $-5x^2 + 2 = 0$.

Solution

$$-5x^2 + 2 = 0$$
$$x^2 = \frac{2}{5} \qquad \text{Isolating } x^2$$

$$x = \sqrt{\frac{2}{5}} \quad or \quad x = -\sqrt{\frac{2}{5}} \qquad \text{Using the principle of square roots}$$

$$x = \sqrt{\frac{2}{5} \cdot \frac{5}{5}} \quad or \quad x = -\sqrt{\frac{2}{5} \cdot \frac{5}{5}} \qquad \text{Rationalizing the denominators}$$

$$x = \frac{\sqrt{10}}{5} \quad or \quad x = -\frac{\sqrt{10}}{5}$$

The solutions are $\dfrac{\sqrt{10}}{5}$ and $-\dfrac{\sqrt{10}}{5}$. This can also be written as $\pm\dfrac{\sqrt{10}}{5}$. We leave the checks to the student. ❑

Sometimes we get solutions that are imaginary numbers.

EXAMPLE 5

Solve: $4x^2 + 9 = 0$.

Solution

$$4x^2 + 9 = 0$$
$$x^2 = -\tfrac{9}{4} \qquad \text{Isolating } x^2$$
$$x = \sqrt{-\tfrac{9}{4}} \quad or \quad x = -\sqrt{-\tfrac{9}{4}} \qquad \text{Using the principle of square roots}$$
$$x = \tfrac{3}{2}i \quad\quad or \quad x = -\tfrac{3}{2}i \qquad \text{Simplifying}$$

Check: We check both solutions at once, since there is an x^2-term but no x-term in the equation.

$$\begin{array}{c|c} 4x^2 + 9 = 0 & \\ \hline 4\left(\pm\tfrac{3}{2}i\right)^2 + 9 \ ? \ 0 & \\ 4 \cdot \tfrac{9}{4} \cdot i^2 + 9 & \\ -9 + 9 & \\ 0 & 0 \quad \text{TRUE} \end{array}$$

The solutions are $\tfrac{3}{2}i$ and $-\tfrac{3}{2}i$, or $\pm\tfrac{3}{2}i$. ❏

Equations like $(x - 2)^2 = 7$ can also be solved using the principle of square roots.

EXAMPLE 6

Solve: $(x - 2)^2 = 7$.

Solution

$$(x - 2)^2 = 7$$
$$x - 2 = \sqrt{7} \quad or \quad x - 2 = -\sqrt{7} \qquad \text{Using the principle of square roots}$$
$$x = 2 + \sqrt{7} \quad or \quad x = 2 - \sqrt{7}$$

The solutions are $2 + \sqrt{7}$ and $2 - \sqrt{7}$, or $2 \pm \sqrt{7}$. We leave the checks to the student. ❏

In Example 6, one side of the equation is the square of a binomial and the other side is a constant. When an equation can be written in this form, we can proceed as we did in Example 6.

EXAMPLE 7

Solve: $x^2 + 6x + 9 = 2$.

Solution

$$x^2 + 6x + 9 = 2 \qquad \text{The left side is the square of a binomial.}$$
$$(x + 3)^2 = 2$$
$$x + 3 = \sqrt{2} \quad or \quad x + 3 = -\sqrt{2} \qquad \text{Using the principle of square roots}$$
$$x = -3 + \sqrt{2} \quad or \quad x = -3 - \sqrt{2}$$

The solutions are $-3 + \sqrt{2}$ and $-3 - \sqrt{2}$, or $-3 \pm \sqrt{2}$. ❏

Completing the Square

By using a method called *completing the square,* we can solve *any* quadratic equation.

EXAMPLE 8

Solve: $x^2 + 6x + 4 = 0$.

Solution

$$x^2 + 6x + 4 = 0$$

$$x^2 + 6x \quad = -4 \qquad \text{Adding } -4 \text{ on both sides}$$

$$x^2 + 6x + 9 = -4 + 9 \qquad \text{Adding 9 on both sides. We explain this shortly.}$$

$$(x + 3)^2 = 5 \qquad \text{Factoring the trinomial square}$$

$$x + 3 = \pm \sqrt{5} \qquad \text{Using the principle of square roots. Remember that } \pm\sqrt{5} \text{ represents two numbers.}$$

$$x = -3 \pm \sqrt{5} \qquad \text{Adding } -3 \text{ on both sides}$$

Check: For $-3 + \sqrt{5}$:

$$x^2 + 6x + 4 = 0$$
$$\overline{\left(-3 + \sqrt{5}\right)^2 + 6\left(-3 + \sqrt{5}\right) + 4 \;?\; 0}$$
$$9 - 6\sqrt{5} + 5 - 18 + 6\sqrt{5} + 4$$
$$9 + 5 - 18 + 4 - 6\sqrt{5} + 6\sqrt{5}$$
$$0 \;\Big|\; 0 \quad \text{TRUE}$$

For $-3 - \sqrt{5}$:

$$x^2 + 6x + 4 = 0$$
$$\overline{\left(-3 - \sqrt{5}\right)^2 + 6\left(-3 - \sqrt{5}\right) + 4 \;?\; 0}$$
$$9 + 6\sqrt{5} + 5 - 18 - 6\sqrt{5} + 4$$
$$9 + 5 - 18 + 4 + 6\sqrt{5} - 6\sqrt{5}$$
$$0 \;\Big|\; 0 \quad \text{TRUE}$$

The solutions are $-3 + \sqrt{5}$ and $-3 - \sqrt{5}$, or $-3 \pm \sqrt{5}$.   ❑

Let's examine how the above procedure works. The decision to add 9 on both sides in Example 8 was not made arbitrarily. We chose 9 because it made the left side a trinomial square. The 9 was obtained by taking half of the coefficient of x and squaring it — that is,

$$\left(\tfrac{1}{2} \cdot 6\right)^2 = 3^2, \quad \text{or 9}.$$

To help see why this procedure works, examine the following drawings.

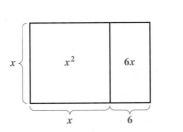 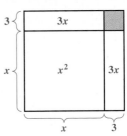

Note that both figures represent the same area, $x^2 + 6x$. However, only the figure on the right can be converted into a square with the addition of a constant term. The constant term, 9, can be interpreted as the area of the "missing" piece of the diagram on the right. It *completes* the square.

EXAMPLE 9

Complete the square.

a) $x^2 + 14x$ **b)** $x^2 - 5x$ **c)** $x^2 + \frac{3}{4}x$

Solution

a) We take half of the coefficient of x and square it.

$$x^2 + 14x$$

⟶ Half of 14 is 7, and $7^2 = 49$.
We add 49.

Thus, $x^2 + 14x + 49$ is a trinomial square. It is equivalent to $(x + 7)^2$. We must add 49 to $x^2 + 14x$ in order to get a trinomial square.

b) We take half of the coefficient of x and square it:

$$x^2 - 5x$$

⟶ $\frac{1}{2} \cdot (-5) = -\frac{5}{2}$, and $\left(-\frac{5}{2}\right)^2 = \frac{25}{4}$.

Thus, $x^2 - 5x + \frac{25}{4}$ is a trinomial square. It is equivalent to $\left(x - \frac{5}{2}\right)^2$. Note that for purposes of factoring, it is best *not* to write $\frac{25}{4}$ as a mixed number or a decimal.

c) We take half of the coefficient of x and square it:

$$x^2 + \frac{3}{4}x$$

⟶ $\frac{1}{2} \cdot \frac{3}{4} = \frac{3}{8}$, and $\left(\frac{3}{8}\right)^2 = \frac{9}{64}$.

Thus, $x^2 + \frac{3}{4}x + \frac{9}{64}$ is a trinomial square. It is equivalent to $\left(x + \frac{3}{8}\right)^2$. ❏

We can now use the method of completing the square to solve quadratic equations similar to the one in Example 8.

EXAMPLE 10

Solve: $x^2 - 8x - 7 = 0$.

Solution

$$x^2 - 8x - 7 = 0$$
$$x^2 - 8x = 7 \qquad \text{Adding 7 on both sides}$$
$$x^2 - 8x + 16 = 7 + 16 \qquad \text{Adding 16 on both sides to complete the square: } \tfrac{1}{2}(-8) = -4, \text{ and } (-4)^2 = 16$$
$$(x - 4)^2 = 23 \qquad \text{Factoring}$$
$$x - 4 = \pm\sqrt{23} \qquad \text{Using the principle of square roots}$$
$$x = 4 \pm \sqrt{23} \qquad \text{Adding 4 on both sides}$$

The solutions are $4 - \sqrt{23}$ and $4 + \sqrt{23}$, or $4 \pm \sqrt{23}$. The checks are left to the student. ❏

EXAMPLE 11

Solve: $x^2 + 5x - 3 = 0$.

Solution

$$x^2 + 5x - 3 = 0$$

$$x^2 + 5x = 3 \qquad \text{Adding 3 on both sides}$$

$$x^2 + 5x + \frac{25}{4} = 3 + \frac{25}{4} \qquad \text{Completing the square: } \frac{1}{2} \cdot 5 = \frac{5}{2}, \text{ and } \left(\frac{5}{2}\right)^2 = \frac{25}{4}$$

$$\left(x + \frac{5}{2}\right)^2 = \frac{37}{4} \qquad \text{Factoring}$$

$$x + \frac{5}{2} = \pm \frac{\sqrt{37}}{2}$$

$$x = \frac{-5 \pm \sqrt{37}}{2} \qquad \text{Adding } -\frac{5}{2} \text{ on both sides}$$

The checking of possible solutions such as these is quite cumbersome. When we use the method of completing the square or the quadratic formula of Section 8.2, we will never obtain any numbers that are not solutions of the original quadratic equation unless we have made a mistake. Thus it is often easier to check our work than to check by substituting into the original equation. The solutions are $(-5 - \sqrt{37})/2$ and $(-5 + \sqrt{37})/2$, or $(-5 \pm \sqrt{37})/2$. ❑

Before we can use the method of completing the square, the coefficient of x^2 must be 1. When it is not 1, we divide both sides of the equation by the x^2-coefficient.

| EXAMPLE 12

Solve: $3x^2 + 7x - 2 = 0$.

TECHNOLOGY
CONNECTION

As we saw in Section 5.8, a grapher can be used to find approximate solutions of any quadratic equation that has real-number solutions. To review this, check Example 10 by graphing $y = x^2 - 8x - 7$:

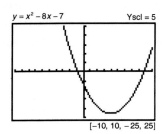

$y = x^2 - 8x - 7$ Yscl = 5

$[-10, 10, -25, 25]$

Use the Trace and Zoom features to find the x-intercepts. If you lack a Zoom feature, adjust the window size to magnify the region near each intercept.

TC1. Use a grapher to confirm the solutions in Examples 8 and 11.

TC2. Can a grapher be used to find *exact* solutions of Example 12? Why or why not?

Solution

$$3x^2 + 7x - 2 = 0$$

$$3x^2 + 7x = 2 \qquad \text{Adding 2 on both sides}$$

$$x^2 + \frac{7}{3}x = \frac{2}{3} \qquad \text{Dividing by 3 on both sides}$$

$$x^2 + \frac{7}{3}x + \frac{49}{36} = \frac{2}{3} + \frac{49}{36} \qquad \text{Completing the square: } \frac{1}{2} \cdot \frac{7}{3} = \frac{7}{6}, \text{ and } \left(\frac{7}{6}\right)^2 = \frac{49}{36}$$

$$\left(x + \frac{7}{6}\right)^2 = \frac{73}{36} \qquad \text{Factoring and simplifying}$$

$$x + \frac{7}{6} = \pm \frac{\sqrt{73}}{6} \qquad \text{Using the principle of square roots and the quotient rule for radicals}$$

$$x = \frac{-7 \pm \sqrt{73}}{6} \qquad \text{Adding } -\frac{7}{6} \text{ on both sides}$$

The solutions are

$$\frac{-7 - \sqrt{73}}{6} \quad \text{and} \quad \frac{-7 + \sqrt{73}}{6},$$

or

$$\frac{-7 \pm \sqrt{73}}{6}.$$
❑

We summarize the procedure used in Example 12. This can be used to solve *any* quadratic equation.

To solve a quadratic equation in x by completing the square:

1. Isolate the terms with variables on one side of the equation, and arrange them in descending order.
2. Divide by the coefficient of x^2 on both sides if that coefficient is not 1.
3. Complete the square by taking half of the coefficient of x and adding its square on both sides.
4. Express one side as the square of a binomial. Find a common denominator on the other side and simplify.
5. Use the principle of square roots.
6. Solve for x by adding appropriately on both sides.

Problem Solving

If you put money in a savings account, the bank will pay you interest. At the end of a year, the bank will start paying you interest on both the original amount and the interest already earned. This is called **compounding interest annually.**

The Compound-
Interest Formula

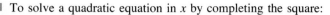

If an amount of money P is invested at interest rate r, compounded annually, then in t years, it will grow to the amount A given by

$$A = P(1 + r)^t.$$

We can use quadratic equations to solve certain interest problems.

EXAMPLE 13

$4000 was invested at interest rate r, compounded annually. In 2 years, it grew to $4410. What was the interest rate?

Solution

1. FAMILIARIZE. We are already familiar with the compound-interest formula. If we were not, we would need to consult an outside source.

2. TRANSLATE. The translation consists of substituting into the formula:

$$A = P(1 + r)^t$$
$$4410 = 4000(1 + r)^2.$$

3. CARRY OUT. We solve for r:

$$4410 = 4000(1 + r)^2$$
$$\tfrac{4410}{4000} = (1 + r)^2 \quad \text{Multiplying by } \tfrac{1}{4000} \text{ on both sides}$$
$$\tfrac{441}{400} = (1 + r)^2 \quad \text{Simplifying}$$
$$\pm\sqrt{\tfrac{441}{400}} = 1 + r \quad \text{Using the principle of square roots}$$
$$\pm\tfrac{21}{20} = 1 + r \quad \text{Simplifying}$$
$$-\tfrac{20}{20} \pm \tfrac{21}{20} = r$$
$$\tfrac{1}{20} = r \quad or \quad -\tfrac{41}{20} = r.$$

4. CHECK. Since the interest rate cannot be negative, we need only check $\frac{1}{20}$, or 5%. If $4000 were invested at 5% interest, compounded annually, then in 2 years it would grow to $4000(1.05)^2$, or $4410. The number 5% checks.

5. STATE. The interest rate was 5%. ❏

EXAMPLE 14

The formula $s = 16t^2$ is used to approximate the distance s, in feet, that an object falls freely from rest in t seconds. The RCA Building in New York City is 850 ft tall. How long will it take an object to fall from the top?

Solution

1. FAMILIARIZE. We make a sketch to help visualize the problem.

2. TRANSLATE. We substitute into the formula:

$$s = 16t^2$$
$$850 = 16t^2.$$

$s = 16t^2$

3. CARRY OUT. We solve for t:

$$850 = 16t^2$$
$$\frac{850}{16} = t^2$$
$$53.125 = t^2$$
$$\sqrt{53.125} = t \qquad \text{Using the principle of square roots; rejecting the negative square root since } t \text{ is not negative in this problem}$$
$$7.3 \approx t. \qquad \text{Using a calculator to approximate the square root and rounding to the nearest tenth}$$

4. CHECK. Since $16(7.3)^2 = 852.64 \approx 850$, our answer checks.

5. STATE. It takes about 7.3 sec for an object to fall freely from the top of the RCA Building. ❏

EXERCISE SET | 8.1

Solve.

1. $4x^2 = 20$
2. $3x^2 = 21$
3. $25x^2 + 4 = 0$
4. $9x^2 + 16 = 0$
5. $2x^2 - 3 = 0$
6. $3x^2 - 7 = 0$
7. $(x + 2)^2 = 49$
8. $(x - 1)^2 = 6$
9. $(x - 3)^2 = 21$
10. $(x + 3)^2 = 6$
11. $(x - 13)^2 = 8$
12. $(x - 13)^2 = 64$
13. $(x - 7)^2 = -4$
14. $(x + 1)^2 = -9$
15. $(x - 9)^2 = 34$
16. $(t - 2)^2 = 25$
17. $\left(x - \frac{3}{2}\right)^2 = \frac{7}{2}$
18. $\left(y + \frac{3}{4}\right)^2 = \frac{17}{16}$
19. $x^2 + 6x + 9 = 64$

20. $x^2 + 10x + 25 = 100$
21. $y^2 - 14y + 49 = 4$
22. $p^2 - 8p + 16 = 1$

Complete the square. Then write the trinomial square in factored form.

23. $x^2 + 10x$
24. $x^2 + 16x$
25. $x^2 - 8x$
26. $x^2 - 6x$
27. $x^2 - 24x$
28. $x^2 - 18x$
29. $x^2 + 9x$
30. $x^2 + 3x$
31. $x^2 - 7x$
32. $x^2 - 11x$
33. $x^2 + \frac{2}{3}x$
34. $x^2 + \frac{2}{5}x$
35. $x^2 - \frac{5}{6}x$
36. $x^2 - \frac{5}{3}x$
37. $x^2 + \frac{9}{5}x$
38. $x^2 + \frac{9}{4}x$

Solve by completing the square. Show your work.

39. $x^2 + 8x = -7$ **40.** $x^2 + 6x = 7$

41. $x^2 - 10x = 22$ **42.** $x^2 - 8x = -9$

43. $x^2 + 6x + 5 = 0$ **44.** $x^2 + 10x + 9 = 0$

45. $x^2 - 10x + 21 = 0$ **46.** $x^2 - 10x + 24 = 0$

47. $x^2 + 5x + 6 = 0$ **48.** $x^2 + 7x + 12 = 0$

49. $x^2 + 4x + 1 = 0$ **50.** $x^2 + 6x + 7 = 0$

51. $x^2 - 10x + 23 = 0$ **52.** $x^2 - 6x + 4 = 0$

53. $x^2 + 6x + 13 = 0$ **54.** $x^2 + 8x + 25 = 0$

55. $2x^2 - 5x - 3 = 0$ **56.** $3x^2 + 5x - 2 = 0$

57. $4x^2 + 8x + 3 = 0$ **58.** $9x^2 + 18x + 8 = 0$

59. $6x^2 - x - 2 = 0$ **60.** $6x^2 - x - 15 = 0$

61. $2x^2 + 4x + 1 = 0$ **62.** $2x^2 + 5x + 2 = 0$

63. $3x^2 - 5x - 3 = 0$ **64.** $4x^2 - 6x - 1 = 0$

Use $A = P(1 + r)^t$ for Exercises 65–70. What is the interest rate?

65. $2560 grows to $2890 in 2 years

66. $1000 grows to $1210 in 2 years

67. $1000 grows to $1440 in 2 years

68. $2560 grows to $3610 in 2 years

69. $6250 grows to $6760 in 2 years

70. $6250 grows to $7290 in 2 years

The formula $s = 16t^2$ is used to approximate the distance s, in feet, that an object falls freely from rest in t seconds. Use the formula for Exercises 71–74.

71. The height of the World Trade Center in New York is 1377 ft (excluding television towers and antennas). How long would it take an object to fall freely from the top?

72. The John Hancock Building in Chicago is 1107 ft tall. How long would it take an object to fall freely from the top?

73. The Sears Tower in Chicago is 1451 ft tall. How long would it take an object to fall freely from the top?

74. The Gateway Arch in St. Louis is 640 ft high. How long would it take an object to fall freely from the top?

Skill Maintenance

75. Graph: $y = 2x + 1$.

76. Simplify: $\sqrt{88}$.

77. Approximate to the nearest tenth: $14 - \sqrt{88}$.

78. Rationalize the denominator: $\sqrt{\frac{2}{5}}$.

Synthesis

79. ◈ Explain in your own words a sequence of steps that can be used to solve any quadratic equation in the quickest way.

80. ◈ Write an interest-rate problem for a classmate to solve. Devise the problem so that the solution is "The loan was made at 7% interest."

Find b such that the trinomial is a square.

81. $x^2 + bx + 64$ **82.** $x^2 + bx + 75$

Solve.

83. $x(2x^2 + 9x - 56)(3x + 10) = 0$

84. $\left(x - \frac{1}{3}\right)\left(x - \frac{1}{3}\right) + \left(x - \frac{1}{3}\right)\left(x + \frac{2}{9}\right) = 0$

85. Boats A and B leave the same point at the same time at right angles. B travels 7 km/h slower than A. After 4 hr, they are 68 km apart. Find the speed of each boat.

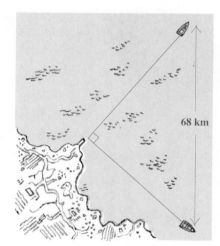

86. Find three consecutive integers such that the square of the first plus the product of the other two is 67.

87. ⌵ Use a grapher to check the solutions of Exercises 5, 45, 59, and 63.

88. ⌵ Problems like Exercises 17, 21, and 41 can be solved on a grapher without first rewriting in standard form. Simply graph two curves y_1 and y_2, where y_1 represents the left side of the equation and y_2 the right side. Then use a grapher to magnify the region near the points of intersection and to identify the x-coordinate at those points. Use a grapher to solve Exercises 17, 21, and 41 in this manner.

8.2

The Quadratic Formula

Solving Using the Quadratic Formula   •   Approximating Solutions

There are at least two reasons for learning to complete the square. One is to enhance your ability to graph certain equations that will be encountered later in this and other chapters. The other is to prove a general formula that can be used to solve quadratic equations.

Solving Using the Quadratic Formula

Each time you solve by completing the square, the procedure is the same. In mathematics, when we do the same kind of procedure many times, we look for a formula to speed up our work.

Consider any quadratic equation in standard form:

$$ax^2 + bx + c = 0, \quad a > 0.$$

Let's solve by completing the square. As we carry out the steps, compare them with Example 12 in the preceding section.

$$ax^2 + bx \quad\quad = -c \qquad \text{Adding } -c \text{ on both sides}$$

$$x^2 + \frac{b}{a}x \quad\quad = -\frac{c}{a} \qquad \text{Dividing by } a \text{ on both sides}$$

Half of $\dfrac{b}{a}$ is $\dfrac{b}{2a}$ and $\left(\dfrac{b}{2a}\right)^2$ is $\dfrac{b^2}{4a^2}$. We add $\dfrac{b^2}{4a^2}$ on both sides:

$$x^2 + \frac{b}{a}x + \frac{b^2}{4a^2} = -\frac{c}{a} + \frac{b^2}{4a^2} \qquad \text{Adding } \dfrac{b^2}{4a^2} \text{ to complete the square}$$

$$\left(x + \frac{b}{2a}\right)^2 = -\frac{4ac}{4a^2} + \frac{b^2}{4a^2}$$

$$\left(x + \frac{b}{2a}\right)^2 = \frac{b^2 - 4ac}{4a^2}$$

Factoring and simplifying; $-\dfrac{c}{a} = -\dfrac{4a}{4a} \cdot \dfrac{c}{a}$

$$x + \frac{b}{2a} = \pm\frac{\sqrt{b^2 - 4ac}}{2a}$$

Using the principle of square roots and the quotient rule for radicals; since $a > 0$, $\sqrt{4a^2} = 2a$

$$x = \frac{-b \pm \sqrt{b^2 - 4ac}}{2a}. \qquad \text{Adding } -\dfrac{b}{2a} \text{ on both sides}$$

A similar derivation yields the same result when a is negative. It is important that you remember the quadratic formula and know how to use it.

The Quadratic Formula

The solutions of $ax^2 + bx + c = 0$, $a \neq 0$, are given by

$$x = \frac{-b \pm \sqrt{b^2 - 4ac}}{2a}.$$

EXAMPLE 1

Solve $5x^2 + 8x = -3$ using the quadratic formula.

Solution   We first find standard form and determine a, b, and c:

$$5x^2 + 8x + 3 = 0;$$

$$a = 5, \quad b = 8, \quad c = 3.$$

We then use the quadratic formula:

$$x = \frac{-b \pm \sqrt{b^2 - 4ac}}{2a}$$

$$x = \frac{-8 \pm \sqrt{8^2 - 4 \cdot 5 \cdot 3}}{2 \cdot 5} \qquad \text{Substituting}$$

$$x = \frac{-8 \pm \sqrt{64 - 60}}{10}$$

> Be sure to write the fraction bar all the way across.

$$x = \frac{-8 \pm \sqrt{4}}{10}$$

$$x = \frac{-8 \pm 2}{10}$$

$$x = \frac{-8 + 2}{10} \quad or \quad x = \frac{-8 - 2}{10}$$

$$x = \frac{-6}{10} \quad or \quad x = \frac{-10}{10}$$

$$x = -\frac{3}{5} \quad or \quad x = -1.$$

The solutions are $-\frac{3}{5}$ and -1.   ❑

We could have solved the equation in Example 1 more easily by factoring as follows:

$$5x^2 + 8x + 3 = 0$$

$$(5x + 3)(x + 1) = 0$$

$$5x + 3 = 0 \quad or \quad x + 1 = 0$$

$$5x = -3 \quad or \qquad x = -1$$

$$x = -\tfrac{3}{5} \quad or \qquad x = -1.$$

To solve a quadratic equation:

1. Check for the form $ax^2 = p$ or $(x + k)^2 = d$. If it is in either of these forms, use the principle of square roots as in Section 8.1.
2. If it is not in the form of step (1), write it in standard form $ax^2 + bx + c = 0$.
3. Try factoring and using the principle of zero products.
4. If it is not possible to factor or factoring seems difficult, use the quadratic formula.

The solutions of a quadratic equation can always be found using the quadratic formula. They cannot always be found by factoring.

EXAMPLE 2

Solve: $5x^2 - 8x = 3$.

Solution   We first find standard form and determine a, b, and c:

$$5x^2 - 8x - 3 = 0;$$
$$a = 5, \quad b = -8, \quad c = -3$$

We then substitute into the quadratic formula:

$$x = \frac{-(-8) \pm \sqrt{(-8)^2 - 4 \cdot 5 \cdot (-3)}}{2 \cdot 5}$$

$$x = \frac{8 \pm \sqrt{64 + 60}}{10} = \frac{8 \pm \sqrt{124}}{10} = \frac{8 \pm \sqrt{4 \cdot 31}}{10}$$

$$x = \frac{8 \pm 2\sqrt{31}}{10} = \frac{2(4 \pm \sqrt{31})}{2 \cdot 5} = \frac{4 \pm \sqrt{31}}{5}. \qquad \text{Removing a factor of 1: } \frac{2}{2} = 1$$

> **CAUTION!** To avoid a common error in simplifying, remember to *factor the numerator and the denominator* and then remove a factor of 1.

The solutions are

$$\frac{4 + \sqrt{31}}{5} \quad \text{and} \quad \frac{4 - \sqrt{31}}{5}.$$

Some quadratic equations have solutions that are imaginary numbers.

EXAMPLE 3

Solve: $x^2 + x + 1 = 0$.

Solution   We have $a = 1$, $b = 1$, and $c = 1$. Substituting into the quadratic formula gives us

$$x = \frac{-1 \pm \sqrt{1^2 - 4 \cdot 1 \cdot 1}}{2 \cdot 1}$$

$$= \frac{-1 \pm \sqrt{1 - 4}}{2}$$

$$= \frac{-1 \pm \sqrt{-3}}{2}$$

$$= \frac{-1 \pm i\sqrt{3}}{2}.$$

The solutions are

$$\frac{-1 + i\sqrt{3}}{2} \quad \text{and} \quad \frac{-1 - i\sqrt{3}}{2}.$$

EXAMPLE 4

Solve: $2 + \dfrac{7}{x} = \dfrac{4}{x^2}$.

Solution   We first find standard form:

$$2x^2 + 7x = 4 \qquad \text{Multiplying by } x^2, \text{ the LCD}$$
$$2x^2 + 7x - 4 = 0. \qquad \text{Subtracting 4}$$

We saw in Sections 5.8 and 8.1 how graphers can solve quadratic equations. To determine whether quadratic equations are solved more quickly on a grapher or by using the quadratic formula, solve Examples 2 and 4 both ways. Which method is faster? Which method is more precise? Why?

We have $a = 2$, $b = 7$, and $c = -4$. Substituting then gives us

$$x = \frac{-7 \pm \sqrt{7^2 - 4 \cdot 2 \cdot (-4)}}{2 \cdot 2} = \frac{-7 \pm \sqrt{49 + 32}}{4}$$

$$x = \frac{-7 \pm \sqrt{81}}{4} = \frac{-7 \pm 9}{4}$$

$$x = \frac{-7 + 9}{4} = \frac{1}{2} \quad or \quad x = \frac{-7 - 9}{4} = -4.$$

The quadratic formula always gives correct results when we start with the standard form. In such cases, we need check only to detect errors in our work. In Example 4, since we started with a rational equation, checking is more essential. We cleared fractions before obtaining standard form, and this step could introduce numbers that do not check in the original equation. At least we need to show that neither result makes a denominator 0. Since neither does, the solutions are $\frac{1}{2}$ and -4. ❑

Approximating Solutions

Many solutions of quadratic equations are irrational numbers, because they involve square roots. In such cases, we may wish to find rational-number approximations to the solutions, using a calculator or table.

EXAMPLE 5

Approximate, to the nearest tenth, the solutions of the equation in Example 2.

Solution   Using a calculator, we find that $\sqrt{31} \approx 5.568$. Thus,

$$\frac{4 + \sqrt{31}}{5} \approx \frac{4 + 5.568}{5}$$

$$= \frac{9.568}{5} \qquad \text{Adding}$$

$$\approx 1.914 \qquad \text{Dividing}$$

$$\approx 1.9; \qquad \text{Rounding to the nearest tenth}$$

$$\frac{4 - \sqrt{31}}{5} \approx \frac{4 - 5.568}{5}$$

$$= \frac{-1.568}{5} \qquad \text{Subtracting}$$

$$\approx -0.314 \qquad \text{Dividing}$$

$$\approx -0.3. \qquad \text{Rounding to the nearest tenth} \qquad ❑$$

EXERCISE SET | 8.2

Solve.

1. $x^2 + 6x + 4 = 0$

2. $x^2 - 6x - 4 = 0$

3. $3p^2 = -8p - 5$

4. $3u^2 = 18u - 6$

5. $x^2 - x + 1 = 0$

6. $x^2 + x + 2 = 0$

7. $x^2 + 13 = 4x$

8. $x^2 + 13 = 6x$

9. $r^2 + 3r = 8$

10. $h^2 + 4 = 6h$

11. $1 + \dfrac{2}{x} + \dfrac{5}{x^2} = 0$

12. $1 + \dfrac{5}{x^2} = \dfrac{2}{x}$

13. $3x + x(x - 2) = 0$

14. $4x + x(x - 3) = 0$

15. $14x^2 + 9x = 0$

16. $19x^2 + 8x = 0$

17. $25x^2 - 20x + 4 = 0$

18. $36x^2 + 84x + 49 = 0$

19. $4x(x - 2) - 5x(x - 1) = 2$

20. $3x(x + 1) - 7x(x + 2) = 6$

21. $14(x - 4) - (x + 2) = (x + 2)(x - 4)$

22. $11(x - 2) + (x - 5) = (x + 2)(x - 6)$

23. $5x^2 = 13x + 17$

24. $25x = 3x^2 + 28$

25. $x^2 + 5 = 2x$

26. $x^2 + 5 = 4x$

27. $x + \dfrac{1}{x} = \dfrac{13}{6}$

28. $\dfrac{3}{x} + \dfrac{x}{3} = \dfrac{5}{2}$

29. $\dfrac{1}{x} + \dfrac{1}{x + 3} = \dfrac{1}{2}$

30. $\dfrac{1}{x} + \dfrac{1}{x + 4} = \dfrac{1}{7}$

31. $(2t - 3)^2 + 17t = 15$

32. $2y^2 - (y + 2)(y - 3) = 12$

33. $(x - 2)^2 + (x + 1)^2 = 0$

34. $(x + 3)^2 + (x - 1)^2 = 0$

35. $x^3 - 8 = 0$

(*Hint:* Factor the difference of cubes. Then use the quadratic formula.)

36. $x^3 + 1 = 0$

Use a calculator or Table 1 to approximate solutions to the nearest tenth.

37. $x^2 + 4x - 7 = 0$

38. $x^2 + 6x + 4 = 0$

39. $x^2 - 6x + 4 = 0$

40. $x^2 - 4x + 1 = 0$

41. $2x^2 - 3x - 7 = 0$

42. $3x^2 - 3x - 2 = 0$

43. $5x^2 = 3 + 8x$

44. $2y^2 + 2y - 3 = 0$

45. Twin Cities Roasters has Kenyan coffee worth $4.50 a pound and Peruvian coffee worth $7.50 a pound. How much of each kind should be mixed to obtain a 50-lb mixture that is worth $5.70 a pound?

Synthesis

46. ◈   Given the solutions of a quadratic equation, is it possible to reconstruct the original equation? Why or why not?

47. Let $g(x) = 2x^2 - 3x - 20$.

a) Find the x-intercepts of the graph of the function.

b) Find all x-values for which $g(x) = -12$.

48. Let $f(x) = x^2 + x - 8$.

a) Find the x-intercepts of the graph of the function.

b) Find all x-values for which $f(x) = 9$.

Solve.

49. ▦ $x^2 - 0.75x - 0.5 = 0$

50. ▦ $z^2 + 0.84z - 0.4 = 0$

51. $x^2 + x - \sqrt{2} = 0$

52. $x^2 - x - \sqrt{3} = 0$

53. $x^2 + \sqrt{5}x - \sqrt{3} = 0$

54. $\sqrt{2}x^2 + 5x + \sqrt{2} = 0$

55. $\left(1 + \sqrt{3}\right)x^2 - \left(3 + 2\sqrt{3}\right)x + 3 = 0$

56. $ix^2 - 2x + 1 = 0$

57. One solution of $kx^2 + 3x - k = 0$ is -2. Find the other.

58. ◺   Use a grapher to solve Exercises 7, 17, 27, 47, and 48.

8.3

Rational Equations and Problem Solving

Rational Equations That Are Quadratic   •   Solving Problems

When fractions are cleared in a rational equation, we sometimes get a quadratic equation. When we do, we proceed as in solving any quadratic equation, but it is especially important to check all possible solutions (see Section 6.4).

Rational Equations That Are Quadratic

Recall that to solve a rational equation, we multiply on both sides by the LCD.

EXAMPLE 1

Solve: $\dfrac{14}{x + 2} - \dfrac{1}{x - 4} = 1$.

Solution

$$(x + 2)(x - 4) \cdot \left[\dfrac{14}{x + 2} - \dfrac{1}{x - 4} \right] = (x + 2)(x - 4) \cdot 1 \qquad \text{The LCD is } (x + 2)(x - 4).$$

$$(x + 2)(x - 4) \dfrac{14}{x + 2} - (x + 2)(x - 4) \dfrac{1}{x - 4} = (x + 2)(x - 4)$$

$$14(x - 4) - (x + 2) = (x + 2)(x - 4) \qquad \text{Removing factors of 1: } (x + 2)/(x + 2) \text{ and } (x - 4)/(x - 4)$$

$$14x - 56 - x - 2 = x^2 - 2x - 8 \qquad \text{Multiplying}$$

$$13x - 58 = x^2 - 2x - 8 \qquad \text{Collecting like terms}$$

$$0 = x^2 - 15x + 50 \qquad \text{Writing standard form}$$

$$0 = (x - 5)(x - 10) \qquad \text{Factoring}$$

$$x = 5 \quad or \quad x = 10. \qquad \text{The principle of zero products}$$

Since we might have introduced some numbers that are not solutions of the original equation when we cleared fractions, we must check.

Check: For 5:

$$\dfrac{14}{x + 2} - \dfrac{1}{x - 4} = 1$$

$$\dfrac{14}{5 + 2} - \dfrac{1}{5 - 4} \ \overset{?}{\vert} \ 1$$

$$\dfrac{14}{7} - 1$$

$$2 - 1$$

$$1 \ \vert \ 1 \quad \text{TRUE}$$

For 10:

$$\dfrac{14}{x + 2} - \dfrac{1}{x - 4} = 1$$

$$\dfrac{14}{10 + 2} - \dfrac{1}{10 - 4} \ \overset{?}{\vert} \ 1$$

$$\dfrac{14}{12} - \dfrac{1}{6}$$

$$\dfrac{14}{12} - \dfrac{2}{12}$$

$$1 \ \vert \ 1 \quad \text{TRUE}$$

The solutions are 5 and 10.   □

If clearing the fractions of a rational equation produces a quadratic equation, solve that resulting equation in the usual way. Remember that possible solutions to rational equations should be checked in the original equation.

Solving Problems

As we saw in Section 6.5, some problems translate to rational equations. Sometimes the rational equation is actually quadratic.

EXAMPLE 2

Josif and Sally work together to type a short story, and it takes them 4 hr. It would take Sally 6 hr more than Josif to type the story alone. How long would each need to type the story if they worked alone?

Solution

1. FAMILIARIZE. Let's make a guess. Suppose that it takes Josif 5 hr. Then it would take Sally 6 more hr, or 11 hr. If this is a correct answer, then Josif will type $\frac{4}{5}$ of the story in 4 hr and Sally will type $\frac{4}{11}$ of the story in 4 hr:

$$\frac{4}{5} + \frac{4}{11} = \frac{44}{55} + \frac{20}{55}$$
$$= \frac{64}{55}.$$

We do not have the correct answer. If we did, they would complete $\frac{55}{55}$ of the story in 4 hr.

Suppose that Josif takes x hr. Then Sally would take 6 hr more, or $(x + 6)$ hr. Josif would type $1/x$ of the story per hour and Sally would type $1/(x + 6)$ of the story per hour. In 4 hr, Josif would type $4(1/x)$ of the story and Sally would type $4[1/(x + 6)]$ of the story.

2. TRANSLATE. Since we are told that working together Josif and Sally complete one entire story in 4 hr, we have an equation:

Fraction of the story typed by Josif in 4 hr	plus	fraction of the story typed by Sally in 4 hr	equals	one completed story
↓	↓	↓	↓	↓
$4\left(\dfrac{1}{x}\right)$	$+$	$4\left(\dfrac{1}{x+6}\right)$	$=$	$1.$

3. CARRY OUT. We solve the equation:

$$x(x + 6)\left[4\left(\frac{1}{x}\right) + 4\left(\frac{1}{x + 6}\right)\right] = x(x + 6) \cdot 1 \qquad \text{Multiplying by the LCD}$$

$$x(x + 6) \cdot 4\left(\frac{1}{x}\right) + x(x + 6) \cdot 4\left(\frac{1}{x + 6}\right) = x(x + 6) \cdot 1$$

$$4(x + 6) + 4x = x(x + 6) \qquad \text{Removing factors of 1: } x/x \text{ and } (x + 6)/(x + 6)$$

$$4x + 24 + 4x = x^2 + 6x$$

$$8x + 24 = x^2 + 6x$$

$$0 = x^2 - 2x - 24 \qquad \text{Standard form}$$

$$0 = (x - 6)(x + 4) \qquad \text{Factoring}$$

$$x = 6 \quad or \quad x = -4. \qquad \text{Principle of zero products}$$

4. CHECK. Since negative time has no meaning in this problem, -4 is not a solution. Let's see if 6 checks. This is the time for Josif to type the story alone. Then Sally would take 12 hr alone. Thus in 4 hr Josif would type $\frac{4}{6}$ of the story, Sally would type $\frac{4}{12}$ of it, and together they would complete $\frac{4}{6} + \frac{4}{12}$, or $\frac{2}{3} + \frac{1}{3}$, of the story. This is all of it, so the numbers check.

5. STATE. The answer is that it would take Josif 6 hr alone and Sally 12 hr alone to type the story. ❑

EXAMPLE 3

Makita's motorcycle traveled 300 mi at a certain speed. Had she gone 10 mph faster, the trip would have taken 1 hr less. Find the speed of the motorcycle.

Solution

1. FAMILIARIZE. We make a drawing, labeling it with the known and unknown information. We can also organize the information in a table. We let r and t represent the rate, in miles per hour, and time, in hours, respectively.

Distance	Speed	Time
300	r	t
300	$r + 10$	$t - 1$

300 miles

Time t Speed r

300 miles

Time $t - 1$ Speed $r + 10$

Recall that the definition of speed, $r = d/t$, relates the three quantities.

2. TRANSLATE. From the first line of the table, we obtain

$$r = \frac{300}{t}.$$

From the second line, we get

$$r + 10 = \frac{300}{t - 1}.$$

3. CARRY OUT. We now have a system of equations. We substitute for r from the first equation into the second and solve the resulting equation:

$$\frac{300}{t} + 10 = \frac{300}{t - 1} \qquad \text{Substituting } 300/t \text{ for } r$$

$$t(t - 1) \cdot \left[\frac{300}{t} + 10\right] = t(t - 1) \cdot \frac{300}{t - 1} \qquad \text{Multiplying by the LCD}$$

$$t(t - 1) \cdot \frac{300}{t} + t(t - 1) \cdot 10 = t(t - 1) \cdot \frac{300}{t - 1} \qquad \begin{array}{l}\text{Removing factors of 1: } t/t \\ \text{and } (t - 1)/(t - 1)\end{array}$$

$$300(t - 1) + 10(t^2 - t) = 300t$$

$$10t^2 - 10t - 300 = 0 \qquad \text{Standard form}$$

$$t^2 - t - 30 = 0 \qquad \text{Multiplying by } \frac{1}{10}$$

$$(t - 6)(t + 5) = 0 \qquad \text{Factoring}$$

$$t = 6 \quad or \quad t = -5. \qquad \text{Principle of zero products}$$

4. CHECK. Note that we have solved for t, not r as required. Since negative time has no meaning in this problem, we disregard the -5 and use 6 hr to find r:

$$r = \frac{300}{6} = 50 \text{ mph.}$$

> **CAUTION!** Always make sure that you find the quantity asked for in the problem.

To see if 50 mph checks, we increase the speed 10 mph to 60 mph and see how long the trip would have taken at that speed:

$$t = \frac{d}{r} = \frac{300}{60} = 5 \text{ hr.}$$

This is 1 hr less than the trip actually took, so we have an answer to the problem.

5. STATE. Makita's motorcycle traveled at a speed of 50 mph. ❏

EXERCISE SET | 8.3

Solve.

1. $\dfrac{1}{x} = \dfrac{x-2}{24}$

2. $\dfrac{x+3}{14} = \dfrac{2}{x}$

3. $\dfrac{1}{2x-1} - \dfrac{1}{2x+1} = \dfrac{1}{4}$

4. $\dfrac{1}{4-x} - \dfrac{1}{2+x} = \dfrac{1}{4}$

5. $\dfrac{50}{x} - \dfrac{50}{x-5} = -\dfrac{1}{2}$

6. $3x + \dfrac{10}{x-1} = \dfrac{16}{x-1}$

7. $\dfrac{x+2}{x} = \dfrac{x-1}{2}$

8. $\dfrac{x}{3} - \dfrac{6}{x} = 1$

9. $x - 6 = \dfrac{1}{x+6}$

10. $x + 7 = \dfrac{1}{x-7}$

11. $\dfrac{2}{x} = \dfrac{x+3}{5}$

12. $\dfrac{x+3}{x} = \dfrac{x-4}{3}$

13. $x + 5 = \dfrac{3}{x-5}$

14. $x - 8 = \dfrac{1}{x+8}$

15. $\dfrac{40}{x} - \dfrac{20}{x-3} = \dfrac{8}{7}$

16. $\dfrac{11}{x} + \dfrac{14}{x+2} = 9$

17. $\dfrac{5}{x+2} + \dfrac{3x}{x+6} = 2$

18. $\dfrac{5}{x+4} + \dfrac{14}{x+7} = 4$

19. $\dfrac{3}{3x+1} + \dfrac{6x}{11x-1} = 1$

20. $\dfrac{7}{x+2} - \dfrac{4}{2x+1} = 1$

21. $\dfrac{16}{5(x-2)(x+2)} + \dfrac{9}{5(x+2)(x+3)}$

$$= \dfrac{-x}{(x-2)(x+3)}$$

22. $\dfrac{19}{(x-3)(x+3)} - \dfrac{10}{(x+3)(x-2)} = \dfrac{x}{(x-3)(x-2)}$

23. $\dfrac{6}{x^2-4x-5} - \dfrac{6}{x^2-2x-3} = \dfrac{x}{x^2-8x+15}$

24. $\dfrac{4}{x^2-3x+2} - \dfrac{4}{x^2+2x-3} = \dfrac{x}{x^2+x-6}$

25. During the first part of a canoe trip, Tim covered 60 km at a certain speed. He then traveled 24 km at a speed that was 4 km/h slower. If the total time for the trip was 8 hr, what was the speed on each part of the trip?

26. During the first part of a trip, Meira's Honda traveled 120 mi at a certain speed. Meira then drove another 100 mi at a speed that was 10 mph slower. If Meira's total trip time was 4 hr, what was her speed on each part of the trip?

27. Sandi's Subaru travels 280 mi at a certain speed. If the car had gone 5 mph faster, the trip would have taken 1 hr less. Find Sandi's speed.

28. Petra's Plymouth travels 200 mi at a certain speed. If the car had gone 10 mph faster, the trip would have taken 1 hr less. Find Petra's speed.

29. A turbo-jet flies 50 mph faster than a super-prop plane. If a turbo-jet goes 2000 mi in 3 hr less time than it takes the super-prop to go 2800 mi, find the speed of each plane.

30. A Cessna flies 600 mi at a certain speed. A Beechcraft flies 1000 mi at a speed that is 50 mph faster, but takes 1 hr longer. Find the speed of each plane.

31. On a sales trip, Gail drives the 600 mi to Richmond at a certain speed. The return trip is made at a speed that is 10 mph slower. Total time for the round trip was 22 hr. How fast did Gail travel on each part of the trip?

32. Naoki travels the 40 mi to Hillsboro at a certain speed. The return trip is made at a speed that is 6 mph slower. Total time for the round trip was 14 hr. How fast did Naoki travel on each part of the trip?

33. A stream flows at 2 mph. A boat travels 24 mi upstream and returns in a total time of 5 hr. What is the speed of the boat in still water?

34. A river flows at 5 mph. A boat travels 60 mi upriver and returns in a total time of 9 hr. What is the speed of the boat in still water?

35. Two pipes are connected to the same tank. When working together, they can fill the tank in 2 hr. The larger pipe, working alone, can fill the tank in 3 hr less time than the smaller one. How long would the smaller one take, working alone, to fill the tank?

36. Two hoses are connected to a swimming pool. When working together, they can fill the pool in 4 hr. The larger hose, working alone, can fill the pool in 6 hr less time than the smaller one. How long would the smaller one take, working alone, to fill the pool?

37. ▦ Ellen paddles 1 mi upstream and 1 mi back in a total time of 1 hr. The speed of the river is 2 mph. Find Ellen's speed in still water.

38. ▦ Dan rows 10 km upstream and 10 km back in a total time of 3 hr. The speed of the river is 5 km/h. Find Dan's speed in still water.

Skill Maintenance

39. Solve: $\sqrt{3x+1} = \sqrt{2x-1} + 1$.

40. Add: $\dfrac{1}{x-1} + \dfrac{1}{x^2-3x+2}$.

41. Multiply and simplify: $\sqrt[3]{18y^3}\ \sqrt[3]{4x^2}$.

Synthesis

42. ◈ Write a problem for a classmate to solve. Devise the problem so that (a) the solution is found after solving a rational equation, and (b) the solution is "The express train travels 90 mph and the local travels 60 mph."

43. ◈ Under what circumstances would a negative value for t, time, have meaning?

Solve.

44. $\dfrac{12}{x^2-9} = 1 + \dfrac{3}{x-3}$

45. $\dfrac{x^2}{x-2} - \dfrac{x+4}{2} + \dfrac{2-4x}{x-2} + 1 = 0$

46. $\dfrac{4}{2x+i} - \dfrac{1}{x-i} = \dfrac{2}{x+i}$

47. Find a when the reciprocal of $a-1$ is $a+1$.

48. A discount store bought a quantity of beach towels for $250 and sold all but 15 at a profit of $3.50 per towel. With the total amount received, the manager could buy 4 more than twice as many as were bought before. Find the cost per towel.

49. ◪ Use a grapher to solve Exercises 1, 9, 19, and 39.

8.4

The Discriminant and Solutions to Quadratic Equations

The Discriminant • **Writing Equations from Solutions**

The Discriminant

From the quadratic formula, we know that the solutions x_1 and x_2 of a quadratic equation are given by

$$x_1 = \frac{-b + \sqrt{b^2 - 4ac}}{2a} \quad \text{and} \quad x_2 = \frac{-b - \sqrt{b^2 - 4ac}}{2a}.$$

The expression $b^2 - 4ac$ shows the nature of the solutions. This expression is called the **discriminant.** If it is 0, then it doesn't matter whether we choose the plus or minus sign in the formula; hence there is just one real solution. If the discriminant is positive, there will be two real solutions. If it is negative, we will be taking the square root of a negative number; hence there will be two imaginary-number solutions, and they will be complex conjugates. We summarize:

Discriminant $b^2 - 4ac$	Nature of Solutions
0	Only one solution; it is a real number
Positive	Two different real-number solutions
Negative	Two different imaginary-number solutions (complex conjugates)

The discriminant also gives information about solving. When the discriminant is a perfect square, the equation can be solved by factoring.

EXAMPLE 1

For each equation, determine the nature of the solutions: **(a)** $9x^2 - 12x + 4 = 0$; **(b)** $x^2 + 5x + 8 = 0$; **(c)** $x^2 + 5x + 6 = 0$.

Solution

a) We have

$$a = 9, \quad b = -12, \quad c = 4.$$

We substitute and compute the discriminant:

$$b^2 - 4ac = (-12)^2 - 4 \cdot 9 \cdot 4$$
$$= 144 - 144$$
$$= 0.$$

There is just one solution, and it is a real number. Since 0 is a perfect square, the equation can be solved by factoring.

b) We have

$$a = 1, \quad b = 5, \quad c = 8.$$

We substitute and compute the discriminant:

$$b^2 - 4ac = 5^2 - 4 \cdot 1 \cdot 8$$
$$= 25 - 32$$
$$= -7.$$

Since the discriminant is negative, there are two imaginary-number solutions. The equation cannot be solved by factoring because -7 is not a perfect square.

c) We have

$$a = 1, \quad b = 5, \quad c = 6;$$
$$b^2 - 4ac = 5^2 - 4 \cdot 1 \cdot 6 = 1.$$

Since the discriminant is positive, there are two real-number solutions. The equation can be solved by factoring since the discriminant is a perfect square. ❏

Writing Equations from Solutions

We know by the principle of zero products that $(x - 2)(x + 3) = 0$ has solutions 2 and -3. If we know the solutions of an equation, we can write an equation, using this principle in reverse.

EXAMPLE 2

Find a quadratic equation whose solutions are given.

a) 3 and $-\frac{2}{5}$ **b)** $2i$ and $-2i$ **c)** $\sqrt{3}$ and $-2\sqrt{3}$

Solution

a) We have

$$x = 3 \quad or \quad x = -\tfrac{2}{5}$$
$$x - 3 = 0 \quad or \quad x + \tfrac{2}{5} = 0 \qquad \text{Getting 0's on one side}$$
$$\left(x - 3\right)\left(x + \tfrac{2}{5}\right) = 0 \qquad \text{Using the principle of zero products (multiplying)}$$
$$x^2 + \tfrac{2}{5}x - 3x - 3 \cdot \tfrac{2}{5} = 0 \qquad \text{Using FOIL}$$
$$x^2 - \tfrac{13}{5}x - \tfrac{6}{5} = 0, \qquad \text{Collecting like terms}$$

or

$$5x^2 - 13x - 6 = 0. \qquad \text{Multiplying by 5 on both sides to clear fractions}$$

b) We have

$$x = 2i \quad or \quad x = -2i$$
$$x - 2i = 0 \quad or \quad x + 2i = 0 \qquad \text{Getting 0's on one side}$$
$$(x - 2i)(x + 2i) = 0 \qquad \text{Using the principle of zero products (multiplying)}$$
$$x^2 + 2ix - 2ix - (2i)^2 = 0 \qquad \text{Using FOIL}$$
$$x^2 - (2i)^2 = 0$$
$$x^2 - 4i^2 = 0$$
$$x^2 + 4 = 0.$$

c) We have

$$x = \sqrt{3} \quad or \quad x = -2\sqrt{3}$$
$$x - \sqrt{3} = 0 \quad or \quad x + 2\sqrt{3} = 0 \qquad \text{Getting 0's on one side}$$
$$(x - \sqrt{3})(x + 2\sqrt{3}) = 0 \qquad \text{Using the principle of zero products}$$
$$x^2 + 2\sqrt{3}x - \sqrt{3}x - 2(\sqrt{3})^2 = 0 \qquad \text{Using FOIL}$$
$$x^2 + \sqrt{3}x - 6 = 0. \qquad \text{Collecting like terms}$$

Note that in Example 2(a) we multiplied on both sides by the LCD, 5. We normally perform this step and thus clear the equation of fractions. Had we preferred, we could have multiplied $x + \frac{2}{5} = 0$ by 5 on both sides, thus clearing fractions *before* using FOIL.

EXERCISE SET | 8.4

Determine the nature of the solutions of the equation.

1. $x^2 - 6x + 9 = 0$
2. $x^2 + 10x + 25 = 0$
3. $x^2 + 7 = 0$
4. $x^2 + 2 = 0$
5. $x^2 - 2 = 0$
6. $x^2 - 5 = 0$
7. $4x^2 - 12x + 9 = 0$
8. $4x^2 + 8x - 5 = 0$
9. $x^2 - 2x + 4 = 0$
10. $x^2 + 3x + 4 = 0$
11. $a^2 - 10a + 21 = 0$
12. $t^2 - 8t + 16 = 0$
13. $6x^2 + 5x - 4 = 0$
14. $10x^2 - x - 2 = 0$
15. $9t^2 - 3t = 0$
16. $4m^2 + 7m = 0$
17. $x^2 + 5x = 7$
18. $x^2 + 4x = -6$
19. $y^2 = \frac{1}{2}y - \frac{3}{5}$
20. $y^2 + \frac{9}{4} = 4y$
21. $4x^2 - 4\sqrt{3}x + 3 = 0$
22. $6y^2 - 2\sqrt{3}y - 1 = 0$

Write a quadratic equation having the given numbers as solutions.

23. $-11, 9$
24. $-4, 4$
25. 7, only solution [*Hint:* It must be a "double" solution.]
26. -5, only solution
27. $-3, -5$
28. $-2, -7$
29. $4, \frac{2}{3}$
30. $5, \frac{3}{4}$
31. $\frac{1}{2}, \frac{1}{3}$
32. $-\frac{1}{4}, -\frac{1}{2}$
33. $-\frac{2}{5}, \frac{6}{5}$
34. $\frac{2}{7}, -\frac{3}{7}$
35. $\sqrt{2}, 3\sqrt{2}$
36. $-\sqrt{3}, 2\sqrt{3}$
37. $-\sqrt{5}, -2\sqrt{5}$
38. $-\sqrt{6}, -3\sqrt{6}$
39. $3i, -3i$
40. $4i, -4i$
41. $5 - 2i, 5 + 2i$
42. $2 - 7i, 2 + 7i$
43. $\frac{1 + 3i}{2}, \frac{1 - 3i}{2}$
44. $\frac{2 - i}{3}, \frac{2 + i}{3}$

Skill Maintenance

45. During a one-hour television show, there were 12 commercials. Some of the commercials were 30 sec long and the others were 60 sec long. The amount of time for 30-sec commercials was 6 min less than the total number of minutes of commercial time during the show. How many 30-sec commercials were used? How many 60-sec commercials were used?

Synthesis

46. ◈ Describe a procedure that could be used to write an equation having the first seven natural numbers as solutions.

47. ◈ Explain why each of the following statements is true.

a) The sum of the solutions of
$$ax^2 + bx + c = 0$$
is $-b/a$.

b) The product of the solutions of

$$ax^2 + bx + c = 0$$

is c/a.

48. Find k for which

$$kx^2 - 4x + (2k - 1) = 0$$

and the product of the solutions is 3.

49. The graph of an equation of the form

$$y = ax^2 + bx + c$$

is a curve similar to the one shown below. Determine a, b, and c from the information given.

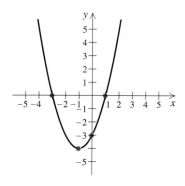

For each equation under the given condition, **(a)** find k, and **(b)** find the other solution.

50. $kx^2 - 17x + 33 = 0$; one solution is 3

51. $kx^2 - 2x + k = 0$; one solution is -3

52. $x^2 - kx + 2 = 0$; one solution is $1 + i$

53. $x^2 - (6 + 3i)x + k = 0$; one solution is 3

54. Find h and k, where $3x^2 - hx + 4k = 0$, the sum of the solutions is -12, and the product of the solutions is 20.

55. Find a quadratic equation for which the sum of the solutions is $\sqrt{3}$ and the product is 8.

56. Let $f(x) = x^2 - 2x - 8$.

  a) Find all x such that $f(x) = 0$.
  b) Find all x such that $f(x) = 17$.

57. Suppose that $f(x) = ax^2 + bx + c$, with $f(-3) = 0$, $f(\frac{1}{2}) = 0$, and $f(0) = -12$. Find a, b, and c.

58. The sum of the squares of the solutions of

$$x^2 + 2kx - 5 = 0$$

is 26. Find the absolute value of k.

59. When the solutions of each of the equations $x^2 + kx + 8 = 0$ and $x^2 - kx + 8 = 0$ are listed in increasing order, each solution of the second equation is 6 more than the corresponding solution of the first equation. Find k.

8.5

Equations Reducible to Quadratic

Recognizing Equations in Quadratic Form •
Using Substitution to Solve

Certain equations that are not really quadratic can be thought of in such a way that they can be solved as quadratic. For example, consider this fourth-degree equation:

$$x^4 \quad - \quad 9x^2 \quad + 8 = 0$$
$$\downarrow \qquad \downarrow \qquad \downarrow \quad \downarrow$$
$$(x^2)^2 - 9(x^2) + 8 = 0 \qquad \text{Thinking of } x^4 \text{ as } (x^2)^2$$
$$\downarrow \qquad \downarrow \qquad \downarrow \quad \downarrow$$
$$u^2 \quad - \quad 9u \quad + 8 = 0 \qquad \text{To make this clearer, write } u \text{ instead of } x^2.$$

The equation $u^2 - 9u + 8 = 0$ can be solved by factoring or by the quadratic formula. Then, remembering that $u = x^2$, we can solve for x. Equations that can be solved like this are said to be *reducible to quadratic*, or *in quadratic form*.

EXAMPLE 1

Solve: $x^4 - 9x^2 + 8 = 0$.

Solution   Let $u = x^2$. Then we solve the equation by substituting u for x^2:

$$u^2 - 9u + 8 = 0$$
$$(u - 8)(u - 1) = 0 \qquad \text{Factoring}$$
$$u - 8 = 0 \quad or \quad u - 1 = 0 \qquad \text{Principle of zero products}$$
$$u = 8 \quad or \qquad u = 1.$$

CAUTION! A common error is to solve for u and then forget to solve for x. Remember that you must find values for the *original* variable!

We substitute x^2 for u and solve these equations:

$$x^2 = 8 \qquad or \quad x^2 = 1$$
$$x = \pm\sqrt{8} \quad or \quad x = \pm 1$$
$$x = \pm 2\sqrt{2} \quad or \quad x = \pm 1.$$

To check, note that when $x = 2\sqrt{2}$, $x^2 = 8$ and $x^4 = 64$. Also, when $x = -2\sqrt{2}$, $x^2 = 8$ and $x^4 = 64$. Similarly, when $x = 1$, $x^2 = 1$ and $x^4 = 1$, and when $x = -1$, $x^2 = 1$ and $x^4 = 1$. Thus instead of making four checks, we need make only two.

Check:

For $\pm 2\sqrt{2}$:

$$\begin{array}{c|c} x^4 - 9x^2 + 8 = 0 \\ \hline (\pm 2\sqrt{2})^4 - 9(\pm 2\sqrt{2})^2 + 8 \;?\; 0 \\ 64 - 9 \cdot 8 + 8 \\ 0 \;|\; 0 \text{ TRUE} \end{array}$$

For ± 1:

$$\begin{array}{c|c} x^4 - 9x^2 + 8 = 0 \\ \hline (\pm 1)^4 - 9(\pm 1)^2 + 8 \;?\; 0 \\ 1 - 9 + 8 \\ 0 \;|\; 0 \text{ TRUE} \end{array}$$

The solutions are 1, -1, $2\sqrt{2}$, and $-2\sqrt{2}$. ❑

Solutions of equations like $x^4 - 9x^2 + 8 = 0$ will always check unless a mistake has been made in one of the steps. Sometimes, however, rational equations, radical equations, or equations containing fractional exponents are reducible to quadratic. As we saw in Chapters 6 and 7, answers to these equations should be checked in the original equation.

EXAMPLE 2

Solve: $x - 3\sqrt{x} - 4 = 0$.

Solution   Let $u = \sqrt{x}$. Then we solve the equation found by substituting u for $\sqrt{x}$ and u^2 for x:

$$u^2 - 3u - 4 = 0$$
$$(u - 4)(u + 1) = 0$$
$$u = 4 \quad or \quad u = -1.$$

Now we substitute $\sqrt{x}$ for u and solve these equations:

$$\sqrt{x} = 4 \quad or \quad \sqrt{x} = -1.$$

Squaring gives us $x = 16$ or $x = 1$.

Check: For 16:

$$x - 3\sqrt{x} - 4 = 0$$
$$\overline{16 - 3\sqrt{16} - 4 \ ? \ 0}$$
$$16 - 3 \cdot 4 - 4 \ \Big|$$
$$0 \ \Big| \ 0 \ \text{TRUE}$$

For 1:

$$x - 3\sqrt{x} - 4 = 0$$
$$\overline{1 - 3\sqrt{1} - 4 \ ? \ 0}$$
$$1 - 3 \cdot 1 - 4 \ \Big|$$
$$-6 \ \Big| \ 0 \ \text{FALSE}$$

The number 16 checks, but 1 does not. Had we noticed that $\sqrt{x} = -1$ has no solution (since principal roots are never negative), we could have solved only the equation $\sqrt{x} = 4$. The solution is 16. ❏

EXAMPLE 3

Solve: $(x^2 - 1)^2 - (x^2 - 1) - 2 = 0$.

Solution   Let $u = x^2 - 1$. Then we solve the equation found by substituting u for $x^2 - 1$:

$$u^2 - u - 2 = 0$$
$$(u - 2)(u + 1) = 0$$
$$u = 2 \qquad or \qquad u = -1.$$

Now we substitute $x^2 - 1$ for u and solve these equations:

$$x^2 - 1 = 2 \qquad or \quad x^2 - 1 = -1$$
$$x^2 = 3 \qquad or \qquad x^2 = 0$$
$$x = \pm\sqrt{3} \quad or \qquad x = 0.$$

The solutions are $-\sqrt{3}$, $\sqrt{3}$, and 0. ❏

Sometimes great care must be taken in deciding what substitution to make.

EXAMPLE 4

Solve: $y^{-2} - y^{-1} - 2 = 0$.

Solution   We rewrite the equation using positive exponents:

$$\frac{1}{y^2} - \frac{1}{y} - 2 = 0.$$

Note that if we let $u = 1/y$, then $u^2 = 1/y^2$. The equation can then be written as a quadratic:

$$u^2 - u - 2 = 0$$
$$(u - 2)(u + 1) = 0$$
$$u = 2 \quad or \quad u = -1.$$

Now we substitute $1/y$ for u and solve these equations:

$$\frac{1}{y} = 2 \quad or \quad \frac{1}{y} = -1.$$

Solving gives us

$$y = \frac{1}{2} \quad or \quad y = \frac{1}{(-1)} = -1.$$

The numbers $\frac{1}{2}$ and -1 both check. They are the solutions. ❏

EXAMPLE 5

Solve: $t^{2/5} - t^{1/5} - 2 = 0$.

Solution Note that $t^{2/5}$ can be rewritten as $(t^{1/5})^2$. The equation can thus be written as $(t^{1/5})^2 - t^{1/5} - 2 = 0$. We let $u = t^{1/5}$ and solve the resulting equation:

$$u^2 - u - 2 = 0$$
$$(u - 2)(u + 1) = 0$$
$$u = 2 \quad or \quad u = -1.$$

Now we substitute $t^{1/5}$ for u and solve:

$$t^{1/5} = 2 \quad or \quad t^{1/5} = -1$$
$$t = 32 \quad or \quad t = -1. \qquad \text{Principle of powers; raising to the 5th power}$$

For 32:		For -1:	
$t^{2/5} - t^{1/5} - 2 = 0$		$t^{2/5} - t^{1/5} - 2 = 0$	
$32^{2/5} - 32^{1/5} - 2 \ ?\ 0$		$(-1)^{2/5} - (-1)^{1/5} - 2 \ ?\ 0$	
$(32^{1/5})^2 - 32^{1/5} - 2$		$[(-1)^{1/5}]^2 - (-1)^{1/5} - 2$	
$2^2 - 2 - 2$		$(-1)^2 - (-1) - 2$	
0	0 TRUE	0	0 TRUE

Both numbers check. The solutions are 32 and -1. ❑

EXERCISE SET | 8.5

Solve.

1. $x^4 - 10x^2 + 25 = 0$
2. $x^4 - 3x^2 + 2 = 0$
3. $x^4 - 12x^2 + 27 = 0$
4. $x^4 - 9x^2 + 20 = 0$
5. $9x^4 - 14x^2 + 5 = 0$
6. $4x^4 - 19x^2 + 12 = 0$
7. $x - 10\sqrt{x} + 9 = 0$
8. $2x - 9\sqrt{x} + 4 = 0$
9. $3x + 10\sqrt{x} - 8 = 0$
10. $5x + 13\sqrt{x} - 6 = 0$
11. $(x^2 - 9)^2 + 3(x^2 - 9) + 2 = 0$
12. $(x^2 - 4)^2 + 5(x^2 - 4) + 6 = 0$
13. $(x^2 - 6x)^2 - 2(x^2 - 6x) - 35 = 0$
14. $(x^2 - 3x)^2 - 10(x^2 - 3x) + 24 = 0$
15. $(3 + \sqrt{x})^2 - 3(3 + \sqrt{x}) - 10 = 0$

16. $(1 + \sqrt{x})^2 + (1 + \sqrt{x}) - 6 = 0$
17. $x^{-2} - x^{-1} - 6 = 0$
18. $4x^{-2} - x^{-1} - 5 = 0$
19. $2x^{-2} + x^{-1} - 1 = 0$
20. $m^{-2} + 9m^{-1} - 10 = 0$
21. $t^{2/3} + t^{1/3} - 6 = 0$ (*Hint:* Let $u = t^{1/3}$.)
22. $w^{2/3} - 2w^{1/3} - 8 = 0$
23. $z^{1/2} - z^{1/4} - 2 = 0$
24. $m^{1/3} - m^{1/6} - 6 = 0$
25. $x^{2/5} + x^{1/5} - 6 = 0$
26. $x^{1/2} - x^{1/4} - 6 = 0$
27. $t^{1/3} + 2t^{1/6} = 3$
28. $m^{1/2} + 6 = 5m^{1/4}$
29. $\left(\dfrac{x + 3}{x - 3}\right)^2 - \left(\dfrac{x + 3}{x - 3}\right) - 6 = 0$
30. $\left(\dfrac{x - 4}{x + 1}\right)^2 - 2\left(\dfrac{x - 4}{x + 1}\right) - 35 = 0$

31. $9\left(\dfrac{x+2}{x+3}\right)^2 - 6\left(\dfrac{x+2}{x+3}\right) + 1 = 0$

32. $16\left(\dfrac{x-1}{x-8}\right)^2 + 8\left(\dfrac{x-1}{x-8}\right) + 1 = 0$

33. $\left(\dfrac{y^2-1}{y}\right)^2 - 4\left(\dfrac{y^2-1}{y}\right) - 12 = 0$

34. $\left(\dfrac{x^2-2}{x}\right)^2 - 7\left(\dfrac{x^2-2}{x}\right) - 18 = 0$

Skill Maintenance

35. Multiply and simplify: $\sqrt{3x^2}\,\sqrt{3x^3}$.

36. Solution A is 18% alcohol and solution B is 45% alcohol. How much of each should be mixed together to get 12 L of a solution that is 36% alcohol?

37. Subtract: $\dfrac{x+1}{x-1} - \dfrac{x+1}{x^2+x+1}$.

Synthesis

38. ◈ Explain how an equation of the form $ax + b\sqrt{x} + c = 0$ could have no solution.

39. Find all x-intercepts of the graph of the function f given by $f(x) = x^4 - 8x^2 + 7$.

Solve. Check possible solutions by substituting into the original equation.

40. ▥ $6.75x - 35\sqrt{x} - 5.36 = 0$

41. ▥ $\pi x^4 - \pi^2 x^2 - \sqrt{99.3} = 0$

42. $\dfrac{x}{x-1} - 6\sqrt{\dfrac{x}{x-1}} - 40 = 0$

43. $\left(\sqrt{\dfrac{x}{x-3}}\right)^2 - 24 = 10\sqrt{\dfrac{x}{x-3}}$

44. $\sqrt{x-3} - \sqrt[4]{x-3} = 12$

45. $a^3 - 26a^{3/2} - 27 = 0$

46. $x^6 - 28x^3 + 27 = 0$

47. $x^6 + 7x^3 - 8 = 0$

48. ⤳ Use a grapher to check your answers to Exercises 1, 3, 41, 46, and 47.

49. ⤳ Use a grapher to solve $x^4 - x^3 - 13x^2 + x + 12 = 0$.

8.6

Variation and Problem Solving

Direct Variation • **Inverse Variation** • **Combined Variation**

We extend our study of formulas and functions by examining three situations that frequently arise in problem solving: direct variation, inverse variation, and combined variation. These situations sometimes require the solution of a quadratic equation.

Direct Variation

Let's say that a worker earns $18 per hour. In 1 hr $18 is earned. In 2 hr $36 is earned. In 3 hr $54 is earned, and so on. This gives rise to a set of ordered pairs of numbers:

(1, 18), (2, 36), (3, 54), (4, 72),   and so on.

The ratio of earnings to time is $\frac{18}{1}$ in every case.

Whenever a situation gives rise to pairs of numbers in which the ratio is constant, we say that there is **direct variation.** Here the earnings *vary directly* as the time:

$$E = 18t \quad \text{or, using function notation,} \quad E(t) = 18t.$$

Direct Variation

Whenever a situation gives rise to a linear function $f(x) = kx$, or $y = kx$, where k is a nonzero constant, we say that there is *direct variation,* that y *varies directly as* x, or that y is *proportional to* x. The number k is called the *variation constant,* or *constant of proportionality.*

The graph of $y = kx$, $k > 0$, always goes through the origin and rises from left to right. Note that as x increases, y increases. The constant k is also the slope of the line.

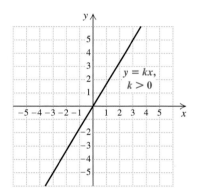

EXAMPLE 1

Find the variation constant and an equation of variation in which y varies directly as x, and $y = 32$ when $x = 2$.

Solution   We know that $(2, 32)$ is a solution of $y = kx$. Therefore,

$$32 = k \cdot 2 \qquad \text{Substituting}$$

$$\frac{32}{2} = k, \quad \text{or } k = 16. \qquad \text{Solving for } k$$

The variation constant is 16. The equation of variation is $y = 16x$. The notation $y(x) = 16x$ or $f(x) = 16x$ is also used. ❑

EXAMPLE 2

Water from melting snow.   The number of centimeters W of water produced from melting snow varies directly as the number of centimeters S of snow. Meteorologists have found that under certain conditions, 150 cm of snow will melt to 16.8 cm of water. To how many centimeters of water will 200 cm of snow melt under these conditions?

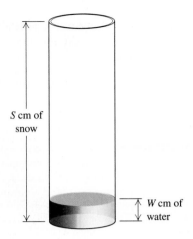

Solution

1. FAMILIARIZE. Because of the phrase "W . . . varies directly as . . . S," we decide to express the amount of water as a function of the amount of snow. Thus, $W(S) = kS$, where k is the variation constant. From the information provided, we know that $W(150) = 16.8$. That is, 150 cm of snow melts to 16.8 cm of water.

2. TRANSLATE. We find the variation constant using the data and then find the equation of variation:

$$W(S) = kS$$
$$W(150) = k \cdot 150 \qquad \text{Replacing } S \text{ with } 150$$
$$16.8 = k \cdot 150 \qquad \text{Substituting}$$
$$\frac{16.8}{150} = k \qquad \text{Solving for } k$$
$$0.112 = k. \qquad \text{This is the variation constant.}$$

The equation of variation is $W(S) = 0.112S$. This is the translation.

3. CARRY OUT. To find how much water 200 cm of snow will melt to, we compute $W(200)$:

$$W(S) = 0.112S$$
$$W(200) = 0.112(200) \qquad \text{Replacing } S \text{ with } 200$$
$$= 22.4.$$

4. CHECK. To check, we could reexamine all our calculations. Note that our answer seems reasonable since 200/22.4 and 150/16.8 are equal.

5. STATE. 200 cm of snow will melt into 22.4 cm of water. ❑

Inverse Variation

To see what we mean by inverse variation, consider the following situation.

A bus is traveling a distance of 20 mi. At a speed of 20 mph, the trip will take 1 hr. At 40 mph, it will take $\frac{1}{2}$ hr. At 60 mph, it will take $\frac{1}{3}$ hr, and so on. This gives rise to a set of pairs of numbers, all having the same product:

$(20, 1)$, $(40, \frac{1}{2})$, $(60, \frac{1}{3})$, $(80, \frac{1}{4})$,   and so on.

Whenever a situation gives rise to pairs of numbers whose product is constant, we say that there is **inverse variation.** The time t required for the bus to travel 20 mi at rate r is given by

$$t = \frac{20}{r} \quad \text{or, using function notation,} \quad t(r) = \frac{20}{r}.$$

Inverse Variation

Whenever a situation gives rise to a function $f(x) = k/x$, or $y = k/x$, where k is a nonzero constant, we say that there is *inverse variation,* that *y varies inversely as x,* or that *y is inversely proportional to x.* The number k is called the *variation constant,* or *constant of proportionality.*

Although we will not study such graphs until Chapter 10, it is helpful to look at the graph of $y = k/x$, for $k > 0$ and $x > 0$. The graph is like the one shown below. Note that as x increases, y decreases.

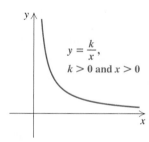

$$y = \frac{k}{x},$$
$$k > 0 \text{ and } x > 0$$

EXAMPLE 3

Find the variation constant and an equation of variation in which y varies inversely as x, and $y = 32$ when $x = 0.2$.

Solution   We know that $(0.2, 32)$ is a solution of

$$y = \frac{k}{x}.$$

Therefore,

$$32 = \frac{k}{0.2} \qquad \text{Substituting}$$
$$(0.2)32 = k$$
$$6.4 = k. \qquad \text{Solving for } k$$

The variation constant is 6.4. The equation of variation is

$$y = \frac{6.4}{x}.$$

There are many problems that translate to an equation of inverse variation.

EXAMPLE 4

The time t required to do a certain job varies inversely as the number of people P who work on the job (assuming that all do the same amount of work). It takes 4 hr for 12 people to build a woodshed. How long would it take 3 people to do the same job?

Solution

1. **FAMILIARIZE.**   Because of the phrase "t . . . varies inversely as . . . P," we decide to express the amount of time required, in hours, as a function of the number of people working. Thus we have $t(P) = k/P$. From the information provided, we know that $t(12) = 4$. That is, it takes 4 hr for 12 people to do the job.

2. **TRANSLATE.**   We find the variation constant using the data and then find the equation of variation:

$$t(P) = \frac{k}{P} \qquad \text{Using function notation}$$

$$t(12) = \frac{k}{12} \qquad \text{Replacing } P \text{ with } 12$$

$$4 = \frac{k}{12} \qquad \text{Substituting 4 for } t(12)$$

$$48 = k. \qquad \text{Solving for } k, \text{ the variation constant}$$

The equation of variation is $t(P) = 48/P$. This is the translation.

3. **CARRY OUT.**   To find how long it would take 3 people to do the job, we compute $t(3)$:

$$t(P) = \frac{48}{P}$$

$$t(3) = \frac{48}{3} \qquad \text{Replacing } P \text{ with } 3$$

$$t = 16. \qquad t = 16 \text{ when } P = 3$$

4. **CHECK.**   We could now recheck each step. Note that, as expected, as the number of people working goes *down,* the time required for the job goes *up.*

5. **STATE.**   It will take 3 people 16 hr to build a woodshed.   ❏

Combined Variation

Often one variable varies directly or inversely with more than one other variable. For example, in the formula for the volume of a right circular cylinder, $V = \pi r^2 h$, we say that V varies *jointly* as h and the square of r.

Joint Variation

y varies *jointly* as x and z if there is some nonzero constant k such that $y = kxz$.

EXAMPLE 5

Find an equation of variation in which y varies jointly as x and z, and $y = 42$ when $x = 2$ and $z = 3$.

Solution   We have

$$y = kxz, \quad \text{so } 42 = k \cdot 2 \cdot 3 \quad \text{and} \quad k = 7.$$

Thus, $y = 7xz$.   ❑

EXAMPLE 6

Find an equation of variation in which y varies jointly as x and z and inversely as the square of w, and $y = 105$ when $x = 3$, $z = 20$, and $w = 2$.

Solution   The equation of variation is of the form

$$y = k \cdot \frac{xz}{w^2},$$

so

$$105 = k \cdot \frac{3 \cdot 20}{2^2}, \quad \text{or} \quad 105 = k \cdot 15, \quad \text{and} \quad k = 7.$$

Thus, $y = 7 \cdot \dfrac{xz}{w^2}$.   ❑

EXAMPLE 7

The volume of a tree trunk.   The volume of wood V in a tree trunk varies jointly as the height h and the square of the girth g (girth is distance around). If the volume is 35 ft^3 when the height is 20 ft and the girth is 5 ft, what is the girth when the volume is 85.75 ft^3 and the height is 25 ft?

Solution

1. **FAMILIARIZE.**   We'll make a table, including the data from the problem and the data we need to find.

	Volume of Wood	Height of Tree	Girth of Tree
Smaller Tree	35 ft^3	20 ft	5 ft
Larger Tree	85.75 ft^3	25 ft	g

Let h, g, and V represent the height, girth, and volume of a tree, respectively. We wish to determine g when V is 85.75 ft^3 and h is 25 ft.

We know from the statement of the problem that in this situation the volume varies jointly as the height and the square of the girth.

2. TRANSLATE. First we find k using the first set of data. Then we solve for g using the second set of data:

$$V = khg^2$$
$$35 = k \cdot 20 \cdot 5^2$$
$$0.07 = k. \qquad \text{This is the variation constant.}$$

The equation of variation is $V = 0.07hg^2$.

3. CARRY OUT. The translation is $V = 0.07hg^2$. We substitute and solve for g:

$$85.75 = 0.07 \cdot 25 \cdot g^2$$

$$\left. \begin{array}{r} 85.75 = 1.75g^2 \\ 49 = g^2 \\ \pm 7 = g. \end{array} \right\} \quad \text{Solving the quadratic equation}$$

We could have first solved the formula for g and then substituted; either approach is valid.

4. CHECK. We should now recheck all our calculations and perhaps make an estimate to see whether our answer is reasonable. We leave this for the student to do. Since a tree's girth must be positive, we accept only 7 as a solution. This seems to be a reasonable figure.

5. STATE. The answer is that the girth of the tree is 7 ft. ❑

EXERCISE SET | 8.6

Find the variation constant and an equation of variation in which y varies directly as x and the following conditions exist.

1. $y = 24$ when $x = 3$ **2.** $y = 5$ when $x = 12$

3. $y = 3.6$ when $x = 1$ **4.** $y = 2$ when $x = 5$

5. $y = 30$ when $x = 8$ **6.** $y = 1$ when $x = \frac{1}{3}$

7. $y = 0.8$ when $x = 0.5$

8. $y = 0.6$ when $x = 0.4$

Solve.

9. *Ohm's law.* The electric current I, in amperes, in a circuit varies directly as the voltage V. When 12 volts are applied, the current is 4 amperes. What is the current when 18 volts are applied?

10. *Hooke's law.* Hooke's law states that the distance d that a spring is stretched by a hanging object varies directly as the mass m of the object. If the distance is 40 cm when the mass is 3 kg, what is the distance when the mass is 5 kg?

11. *Weekly allowance.* According to Fidelity Investments *Investment Vision Magazine,* the average weekly allowance A of children varies directly as their grade level, G. It is known that the average allowance of a 9th-grade student is $9.66 per week. What then is the average weekly allowance of a 4th-grade student?

12. *Use of aluminum cans.* The number N of aluminum cans used each year varies directly as the number of people using the cans. If 250 people use 60,000 cans in one year, how many cans are used each year in Dallas, which has a population of 850,000?

13. *Mass of water in body.* The number of kilograms W of water in a human body varies directly as the mass of the body. A 96-kg person contains 64 kg of water. How many kilograms of water are in a 75-kg person?

14. *Weight on Mars.*   The weight M of an object on Mars varies directly as its weight E on earth. A person who weighs 95 lb on earth weighs 38 lb on Mars. How much would a 100-lb person weigh on Mars?

15. *Relative aperture.*   The relative aperture, or *f*-stop, of a 23.5-mm lens is directly proportional to the focal length F of the lens. If a 150-mm focal length has an *f*-stop of 6.3, find the *f*-stop of a 23.5-mm lens with a focal length of 80 mm.

16. *Amount of air pollution.*   The amount of pollution A entering the atmosphere varies directly as the number of people N living in an area. If 60,000 people result in 42,600 tons of pollutants entering the atmosphere, how many tons enter the atmosphere in a city with a population of 750,000?

Find the variation constant and an equation of variation in which y varies inversely as x, and the following conditions exist.

17. $y = 6$ when $x = 10$ **18.** $y = 16$ when $x = 4$

19. $y = 4$ when $x = 3$ **20.** $y = 4$ when $x = 9$

21. $y = 12$ when $x = 3$ **22.** $y = 9$ when $x = 5$

23. $y = 27$ when $x = \frac{1}{3}$ **24.** $y = 81$ when $x = \frac{1}{9}$

Solve.

25. *Current and resistance.*   The current I in an electrical conductor varies inversely as the resistance R of the conductor. If the current is $\frac{1}{2}$ ampere when the resistance is 240 ohms, what is the current when the resistance is 540 ohms?

26. *Pumping rate.*   The time t required to empty a tank varies inversely as the rate r of pumping. If a pump can empty a tank in 45 min at the rate of 600 kL/min, how long will it take the pump to empty the same tank at the rate of 1000 kL/min?

27. *Volume and pressure.*   The volume V of a gas varies inversely as the pressure P upon it. The volume of a gas is 200 cm^3 under a pressure of 32 kg/cm^2. What will be its volume under a pressure of 40 kg/cm^2?

28. *Work rate.*   The time T required to do a job varies inversely as the number of people P working. It takes 5 hr for 7 bricklayers to complete a certain job. How long will it take 10 bricklayers to complete the job?

29. *Rate of travel.*   The time t required to drive a fixed distance varies inversely as the speed r. It takes 5 hr at a speed of 80 km/h to drive a fixed

distance. How long will it take to drive the fixed distance at a speed of 60 km/h?

30. *Wavelength and frequency.*   The wavelength W of a radio wave varies inversely as its frequency F. A wave with a frequency of 1200 kilohertz has a length of 300 meters. What is the length of a wave with a frequency of 800 kilohertz?

Find an equation of variation in which:

31. y varies directly as the square of x, and $y = 0.15$ when $x = 0.1$.

32. y varies directly as the square of x, and $y = 6$ when $x = 3$.

33. y varies inversely as the square of x, and $y = 0.15$ when $x = 0.1$.

34. y varies inversely as the square of x, and $y = 6$ when $x = 3$.

35. y varies jointly as x and z, and $y = 56$ when $x = 7$ and $z = 8$.

36. y varies directly as x and inversely as z, and $y = 4$ when $x = 12$ and $z = 15$.

37. y varies jointly as x and the square of z, and $y = 105$ when $x = 14$ and $z = 5$.

38. y varies jointly as x and z and inversely as w, and $y = \frac{3}{2}$ when $x = 2$, $z = 3$, and $w = 4$.

39. y varies jointly as w and the square of x and inversely as z, and $y = 49$ when $w = 3$, $x = 7$, and $z = 12$.

40. y varies directly as x and inversely as w and the square of z, and $y = 4.5$ when $x = 15$, $w = 5$, and $z = 2$.

41. y varies jointly as x and z and inversely as the product of w and p, and $y = \frac{3}{28}$ when $x = 3$, $z = 10$, $w = 7$, and $p = 8$.

42. y varies jointly as x and z and inversely as the square of w, and $y = \frac{12}{5}$ when $x = 16$, $z = 3$, and $w = 5$.

Solve.

43. *Stopping distance of a car.*   The stopping distance d of a car after the brakes have been applied varies directly as the square of the speed r. If a car traveling 60 mph can stop in 200 ft, how fast can a car go and still stop in 72 ft?

44. *Volume of a gas.*   The volume V of a given mass of a gas varies directly as the temperature T and inversely as the pressure P. If $V = 231$ cm^3 when $T = 42°$ and $P = 20$ kg/cm^2, what is the volume when $T = 30°$ and $P = 15$ kg/cm^2?

45. *Intensity of a signal.* The intensity I of a television signal varies inversely as the square of the distance d from the transmitter. If the intensity is 25 watts per square meter (W/m²) at a distance of 2 km, how far from the transmitter are you when the intensity is 2.56 W/m²?

46. *Distance of a fall.* The distance d that an object falls varies directly as the square of the amount of time t that it is falling. If an object falls 64 ft in 2 sec, how long will it take to fall 400 ft?

47. *Weight of an astronaut.* The weight W of an object varies inversely as the square of the distance d from the center of the earth. At sea level (6400 km from the center of the earth), an astronaut weighs 100 lb. How far *above the earth* must the astronaut be in order to weigh 64 lb?

48. *Intensity of light.* The intensity I of light from a light bulb varies inversely as the square of the distance d from the bulb. Suppose I is 90 W/m² when the distance is 5 m. How much *further* would it be to a point where the intensity is 40 W/m²?

49. *Electrical resistance.* At a fixed temperature, the resistance R of a wire varies directly as the length l and inversely as the square of its diameter d. If the resistance is 0.1 ohm when the diameter is 1 mm and the length is 50 cm, what is the diameter when the resistance is 1 ohm and the length is 2000 cm?

50. *Volume of a can.* The volume V of a can varies jointly as its height h and the square of its radius r. If a 12-fluid-ounce soda comes in a can that is 12 cm high with a 3.2-cm radius, what is the radius of a 9-fluid-ounce can that is 4 cm high?

51. *Drag force.* The drag force F on a boat varies jointly as the wetted surface area A and the square of the boat's velocity. If a boat going 6.5 mph experiences a drag force of 86 N (Newtons) when the wetted surface area is 41.2 ft², how fast must a boat with 28.5 ft² of wetted surface area go in order to experience a drag force of 94 N?

52. *Atmospheric drag.* Wind resistance, or atmospheric drag, tends to slow down moving objects. Atmospheric drag varies jointly as an object's surface area A and velocity v. If a car traveling at a speed of 40 mph with a surface area of 37.8 ft² experiences a drag of 222 N, how fast must a car with 51 ft² of surface area travel in order to experience a drag force of 430 N?

Skill Maintenance

53. Give an equation for a line with slope $-\frac{2}{3}$ and y-intercept $(0, -5)$.

54. Simplify:
$$\frac{\dfrac{1}{ab} - \dfrac{2}{bc}}{\dfrac{3}{ab} + \dfrac{4}{bc}}.$$

55. If $f(x) = x^3 - 2x^2$, find $f(3)$.

56. Multiply: $(3x - 2y)^2$.

Synthesis

57. If y varies directly as x^2, explain why doubling x would not cause y to be doubled as well.

58. If y varies directly as x^2, explain why tripling x would not cause y to be tripled as well.

59. Suppose that the number of customer complaints is inversely proportional to the number of employees hired. Will a firm reduce the number of complaints more by expanding from 5 to 10 employees, or from 20 to 25? Explain. Consider using a graph to help justify your answer.

60. *Volume and cost.* A peanut butter jar in the shape of a right circular cylinder is 4 in. high and 3 in. in diameter and sells for $1.20. Assuming the same ratio of volume to cost, how much should a jar 6 in. high and 6 in. in diameter cost?

61. If y varies inversely as the cube of x and x is multiplied by 0.5, what is the effect on y?

62. *The gravity model.* It has been determined that the average number of telephone calls in a day N, between two cities, is directly proportional to the populations P_1 and P_2 of the cities and inversely proportional to the square of the distance between the cities. This model is called the *gravity model* because the equation of variation resembles the equation that applies to Newton's law of gravity.

a) In 1986, the population of Indianapolis was 744,624 and the population of Cincinnati was 452,524. The average number of daily phone calls between the two cities was 11,153. Find the value k and write the equation of variation given that the cities are 174 km apart.

b) In 1986, the average number of daily phone calls between Indianapolis and New York was 4270, and the population of New York was 7,895,563. Estimate the distance between Indianapolis and New York.

63. *Tension of a stringed instrument.* The tension T on a string in a musical instrument varies jointly as the string's mass per unit length m, the square of its length l, and the square of its fundamental frequency f. A 2-m-long string of mass 5 gm/m with a fundamental frequency of 80 has a tension of 100 N. How long should the same string be if its tension is going to be changed to 72 N?

64. *Golf distance finder.* A device used in golf to estimate the distance d to a hole measures the size s that the 7-ft pin *appears* to be in a viewfinder. The viewfinder uses the principle, diagrammed here, that s gets bigger when d gets smaller. If $s = 0.56$ in. when $d = 50$ yd, find an equation of variation that expresses d as a function of s. What is d when $s = 0.40$ in.?

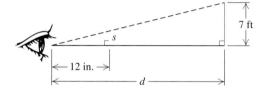

Describe, in words, the variation given by the equation.

65. $Q = \dfrac{kp^2}{q^3}$
66. $W = \dfrac{km_1M_1}{d^2}$

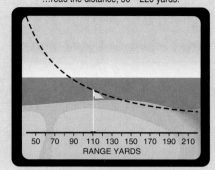

HOW IT WORKS:

Just sight the flagstick through the viewfinder... fit flag between top dashed line and the solid line below... ...read the distance, 50 – 220 yards.

RANGE YARDS
50 70 90 110 130 150 170 190 210

Nothing to focus.
•
Gives you exact distance that your ball lies from the flagstick.
•
Choose proper club on every approach shot.
•
Figure new pin placement instantly.
•
Train your naked eye for formal and tournament play.
•
Eliminate the need to remember every stake, tree, and bush on the course.

8.7

Quadratic Functions and Their Graphs

Graphs of $f(x) = ax^2$ • Graphs of $f(x) = a(x - h)^2$ •
Graphs of $f(x) = a(x - h)^2 + k$

The following bar graph shows the fall and rise of the rate of unemployment during a recent year. A curve drawn along the graph would approximate the graph of a *quadratic function*. We now consider such graphs.

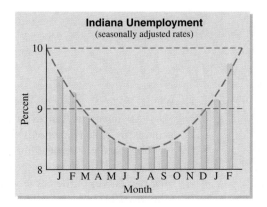

TECHNOLOGY
CONNECTION

To examine the effect of a when graphing $f(x) = ax^2$, draw the function $y_1 = x^2$ in a $[-5, 5, -10, 10]$ window. Without erasing this curve, graph the function $y_2 = 3x^2$. How do the graphs compare? Now include the graph of $y_3 = \frac{1}{3}x^2$. Find a rule that describes the effect of multiplying x^2 by a, when $a > 1$ and when $0 < a < 1$.

Clear the display and graph $y_1 = x^2$ again. Now include the graph of $y_2 = -x^2$. Notice how it differs from $f(x) = x^2$. Next graph $y_3 = \frac{2}{3}x^2$ and $y_4 = -\frac{2}{3}x^2$ and compare them. Find a rule that describes the effect of multiplying x^2 by a, when $a < -1$ and when $-1 < a < 0$.

Graphs of $f(x) = ax^2$

In Chapter 2, the notion of a function was introduced. We studied equations such as

$$f(x) = 3x + 2 \quad \text{or} \quad y = 3x + 2,$$

whose graphs are straight lines. We now consider equations (or functions) in which the right-hand side is a quadratic polynomial:

$$f(x) = ax^2 + bx + c, \quad a \neq 0.$$

A function of this type is referred to as a **quadratic function.**

EXAMPLE 1

Graph: $f(x) = x^2$.

Solution   We choose numbers for x, some positive and some negative, and for each number we compute $f(x)$.

We plot the ordered pairs and connect them with a smooth curve.

x	$f(x) = x^2$	$(x, f(x))$
-3	9	$(-3, 9)$
-2	4	$(-2, 4)$
-1	1	$(-1, 1)$
0	0	$(0, 0)$
1	1	$(1, 1)$
2	4	$(2, 4)$
3	9	$(3, 9)$

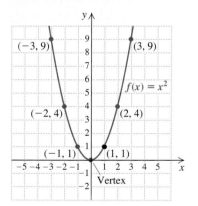

All quadratic functions have graphs similar to the one in Example 1. Such curves are called *parabolas*. They are smooth, cup-shaped curves that are symmetric with respect to a vertical line known as the parabola's *line of symmetry*. In the graph of $f(x) = x^2$, the y-axis is the line of symmetry. If the paper were folded on this line, the two halves of the curve would match. The point $(0, 0)$ is known as the *vertex* of this parabola.

By plotting points, we can see how the graphs of $g(x) = \frac{1}{2}x^2$ and $h(x) = 2x^2$ compare with the graph of $f(x) = x^2$.

x	$h(x) = 2x^2$
-3	18
-2	8
-1	2
0	0
1	2
2	8
3	18

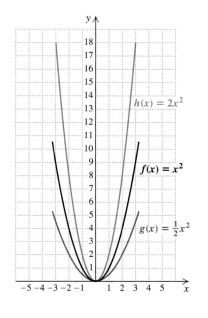

x	$g(x) = \frac{1}{2}x^2$
-3	$\frac{9}{2}$
-2	2
-1	$\frac{1}{2}$
0	0
1	$\frac{1}{2}$
2	2
3	$\frac{9}{2}$

Note that the graph of $g(x) = \frac{1}{2}x^2$ is a flatter parabola than the graph of $f(x) = x^2$, and the graph of $h(x) = 2x^2$ is narrower. The vertex and the line of symmetry, however, have not changed.

When we consider the graph of $k(x) = -\frac{1}{2}x^2$, we see that the parabola opens downward and is the same shape as the graph of $g(x) = \frac{1}{2}x^2$.

x	$k(x) = -\frac{1}{2}x^2$
-3	$-\frac{9}{2}$
-2	-2
-1	$-\frac{1}{2}$
0	0
1	$-\frac{1}{2}$
2	-2
3	$-\frac{9}{2}$

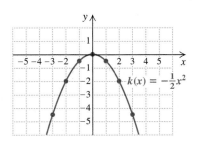

The graph of $g(x) = ax^2$ is a parabola with the vertical axis as its line of symmetry and its vertex at the origin.

If a is positive, the parabola opens upward; if a is negative, the parabola opens downward.

If $|a|$ is greater than 1, the parabola is narrower than $f(x) = x^2$.

If $|a|$ is between 0 and 1, the parabola is flatter than $f(x) = x^2$.

To investigate the effect of h on the graph of $f(x) = a(x - h)^2$, start by graphing $y_1 = 7.3x^2$ in the window $[-5, 5, -5, 5]$. On the same set of axes, graph the function $y_2 = 7.3(x - 1)^2$ and compare the graphs. Next graph $y_3 = 7.3(x - 2.7)^2$ and compare it with the first two graphs. Now replace y_2 and y_3 with

$$y_2 = 7.3(x + 1)^2 \quad \text{and} \quad y_3 = 7.3(x + 2.7)^2.$$

Find a rule that describes the effect of h in the function

$$f(x) = a(x - h)^2.$$

Graphs of $f(x) = a(x - h)^2$

Why not now consider graphs of

$$f(x) = ax^2 + bx + c,$$

where b and c are not both 0? In effect, we will do that, but in a disguised form. It turns out to be convenient to consider functions $f(x) = a(x - h)^2$, that is, where we start with ax^2 but then replace x by $x - h$, where h is some constant.*

EXAMPLE 2

Graph: $f(x) = (x - 3)^2$.

Solution   We choose some values for x and compute $f(x)$. Then we plot the points and draw the curve. Comparing these values of x and $f(x)$ with those found in Example 1, we see that when an input here is 3 more than an input for Example 1, the outputs match.

x	$f(x) = (x - 3)^2$
-1	16
0	9
1	4
2	1
3	0	← Vertex
4	1
5	4
6	9

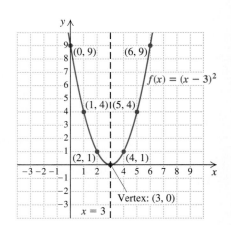

We note that $f(x) = 16$ when $x = -1$ and $f(x)$ gets larger and larger as x gets more and more negative. Thus we go back to positive values to fill out the table. Note that the line $x = 3$ is now the line of symmetry and the point $(3, 0)$ is the vertex. Had we observed that $x = 3$ is the line of symmetry at the outset, then we could have computed some values on one side, such as $(4, 1)$, $(5, 4)$, and $(6, 9)$, and then used symmetry to get their mirror images $(2, 1)$, $(1, 4)$, and $(0, 9)$ without further computation.   ❑

The graph of $f(x) = (x - 3)^2$ looks just like the graph of $f(x) = x^2$, except that it is moved, or translated, 3 units to the right.

---

*The letters h and k are often used to name functions, in which case the notation $h(x)$ and $k(x)$ is used. When h and k appear in expressions like $f(x) = a(x - h)^2 + k$, assume that they represent constants.

The graph of $g(x) = a(x - h)^2$ has the same shape as the graph of $y = ax^2$.
If h is positive, the graph of $y = ax^2$ is shifted h units to the right.
If h is negative, the graph of $y = ax^2$ is shifted $|h|$ units to the left.
The vertex is $(h, 0)$ and the line of symmetry is $x = h$.

EXAMPLE 3

Graph: $g(x) = -2(x + 3)^2$.

Solution   We express the equation in the equivalent form $g(x) = -2[x - (-3)]^2$. Then we know that the graph looks like that of $y = 2x^2$ translated 3 units to the left, and it will also open downward since $-2 < 0$. The vertex is $(-3, 0)$, and the line of symmetry is $x = -3$. Plotting points as needed, we obtain the graph shown here.

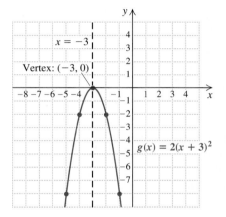

Graphs of $f(x) = a(x - h)^2 + k$

Given a graph of $f(x) = a(x - h)^2$, what happens to it if we add a constant k? Suppose we add 2. This increases each function value $f(x)$ by 2, so the curve is moved up. If k should be -3, the curve is moved down. The vertex of the parabola will be at the point (h, k), and the line of symmetry will be $x = h$.

Note that if a parabola opens upward ($a > 0$), the function value, or y-value, at the vertex is a least, or *minimum*, value. That is, it is less than the y-value at any other point on the graph. If the parabola opens downward ($a < 0$), the function value at the vertex will be a greatest, or *maximum*, value.

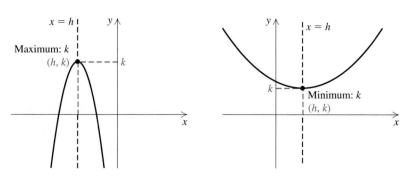

The graph of $f(x) = a(x - h)^2 + k$ looks like that of $y = a(x - h)^2$ translated up or down.

If k is positive, the graph of $y = a(x - h)^2$ is shifted k units up.

If k is negative, the graph of $y = a(x - h)^2$ is shifted $|k|$ units down.

The vertex is (h, k), and the line of symmetry is $x = h$.

For $a > 0$, k is the minimum function value. For $a < 0$, k is the maximum function value.

EXAMPLE 4

Graph $g(x) = (x - 3)^2 - 5$, and find the minimum function value.

Solution   We know that the graph looks like that of $f(x) = (x - 3)^2$ (see Example 2) but moved 5 units down. You can confirm this by plotting some points. For instance, $g(4) = (4 - 3)^2 - 5 = -4$, whereas $f(4) = (4 - 3)^2 = 1$.

The vertex is now $(3, -5)$, and the minimum function value is -5.

TECHNOLOGY CONNECTION

To consider the effect of k on the graph of $f(x) = a(x - h)^2 + k$, graph $y_1 = 7.3(x - 1)^2$ in the window $[-5, 5, -5, 5]$. On the same set of axes, graph the function $y_2 = 7.3(x - 1)^2 + 2$ and compare the graphs. Next graph $y_3 = 7.3(x - 1)^2 - 4$ and compare it with the first two graphs. Try other values of k, including fractions and decimals. Find a rule that describes the effect of adding a value k to the function $f(x) = a(x - h)^2$.

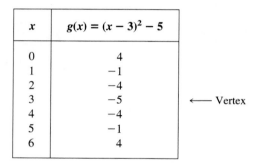

x	$g(x) = (x - 3)^2 - 5$
0	4
1	-1
2	-4
3	-5 ⟵ Vertex
4	-4
5	-1
6	4

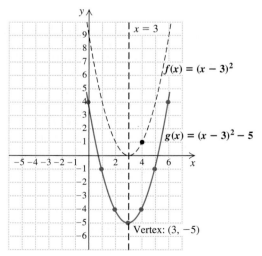

EXAMPLE 5

Graph $h(x) = \frac{1}{2}(x - 3)^2 + 5$, and find the minimum function value.

Solution   The graph looks just like that of $f(x) = \frac{1}{2}x^2$ but moved 3 units to the right and 5 units up. The vertex is $(3, 5)$, and the line of symmetry is $x = 3$. We draw $f(x) = \frac{1}{2}x^2$ and then move the curve over and up. We plot a few points as a check. The minimum function value is 5.

x	$h(x) = \frac{1}{2}(x - 3)^2 + 5$
0	$9\frac{1}{2}$
1	7
3	5
5	7
6	$9\frac{1}{2}$

⟵ Vertex

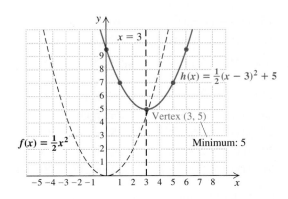

EXAMPLE 6

Graph $y = -2(x + 3)^2 + 5$. Find the vertex, the line of symmetry, and the maximum or minimum value.

Solution   We first express the equation in the equivalent form

$$y = -2[x - (-3)]^2 + 5.$$

The graph looks like that of $y = -2x^2$ translated 3 units to the left and 5 units up. The vertex is $(-3, 5)$, and the line of symmetry is $x = -3$. Since $-2 < 0$, we know that 5, the second coordinate of the vertex, is the maximum y-value.

We compute a few points as needed. The graph is shown here.

x	$y = -2(x + 3)^2 + 5$
-4	3
-3	5
-2	3

⟵ Vertex

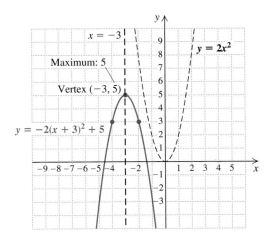

EXERCISE SET | 8.7

Graph.

1. $f(x) = x^2$

3. $f(x) = -4x^2$

5. $g(x) = \frac{1}{4}x^2$

7. $h(x) = -\frac{1}{3}x^2$

2. $f(x) = -x^2$

4. $f(x) = -3x^2$

6. $g(x) = \frac{1}{3}x^2$

8. $h(x) = -\frac{1}{4}x^2$

9. $f(x) = \frac{3}{2}x^2$

10. $f(x) = \frac{5}{2}x^2$

For each of the following, graph the function, label the vertex, and draw the line of symmetry.

11. $g(x) = (x + 1)^2$

13. $f(x) = (x - 4)^2$

12. $g(x) = (x + 4)^2$

14. $f(x) = (x - 1)^2$

15. $h(x) = (x - 3)^2$

16. $h(x) = (x - 7)^2$

17. $f(x) = -(x + 4)^2$

18. $f(x) = -(x - 2)^2$

19. $g(x) = -(x - 1)^2$

20. $g(x) = -(x + 5)^2$

21. $f(x) = 2(x - 1)^2$

22. $f(x) = 2(x + 4)^2$

23. $h(x) = -\frac{1}{2}(x - 3)^2$

24. $h(x) = -\frac{3}{2}(x - 2)^2$

25. $f(x) = \frac{1}{2}(x + 1)^2$

26. $f(x) = \frac{1}{3}(x + 2)^2$

27. $g(x) = -3(x - 2)^2$

28. $g(x) = -4(x - 7)^2$

29. $f(x) = -2(x + 9)^2$

30. $f(x) = 2(x + 7)^2$

31. $h(x) = -3\left(x - \frac{1}{2}\right)^2$

32. $h(x) = -2\left(x + \frac{1}{2}\right)^2$

For each of the following, graph the function and find the vertex, the line of symmetry, and the maximum value or the minimum value.

33. $f(x) = (x - 3)^2 + 1$

34. $f(x) = (x + 2)^2 - 3$

35. $f(x) = (x + 1)^2 - 2$

36. $f(x) = (x - 1)^2 + 2$

37. $g(x) = (x + 4)^2 + 1$

38. $g(x) = -(x - 2)^2 - 4$

39. $f(x) = \frac{1}{2}(x - 5)^2 + 2$

40. $f(x) = \frac{1}{2}(x + 1)^2 - 2$

41. $h(x) = -2(x - 1)^2 - 3$

42. $h(x) = -2(x + 1)^2 + 4$

43. $f(x) = -3(x + 4)^2 + 1$

44. $f(x) = -2(x - 5)^2 - 3$

45. $g(x) = -\frac{3}{2}(x - 1)^2 + 2$

46. $g(x) = \frac{3}{2}(x + 2)^2 - 1$

Without graphing, find the vertex, the line of symmetry, and the maximum value or the minimum value.

47. $f(x) = 8(x - 9)^2 + 5$

48. $f(x) = 10(x + 5)^2 - 8$

49. $h(x) = -\frac{2}{7}(x + 6)^2 + 11$

50. $h(x) = -\frac{3}{11}(x - 7)^2 - 9$

51. $f(x) = 5\left(x + \frac{1}{4}\right)^2 - 13$

52. $f(x) = 6\left(x - \frac{1}{4}\right)^2 + 19$

53. $f(x) = -7(x - 10)^2 - 20$

54. $f(x) = -9(x + 12)^2 + 23$

55. $f(x) = \sqrt{2}(x + 4.58)^2 + 65\pi$

56. $f(x) = 4\pi(x - 38.2)^2 - \sqrt{34}$

57. Solve the system

$$500 = 4a + 2b + c,$$
$$300 = a + b + c,$$
$$0 = c.$$

58. Solve: $6x^2 - 13x + 2 = 0$.

59. ◈ Explain, without plotting points, why the graph of $y = (x + 2)^2$ looks like the graph of $y = x^2$ translated 2 units to the left.

60. ◈ Explain, without plotting points, why the graph of $y = x^2 - 4$ looks like the graph of $y = x^2$ translated 4 units down.

For each of the following, write the equation of the parabola that has the shape of $f(x) = 2x^2$ or $g(x) = -2x^2$ and has a maximum or minimum value at the specified point.

61. Maximum: (0, 4)

62. Minimum: (2, 0)

63. Minimum: (6, 0)

64. Maximum: (0, 3)

65. Maximum: (3, 8)

66. Minimum: (-2, 3)

67. Minimum: (-3, 6)

Write an equation of the parabola that satisfies the following conditions.

68. The parabola has a minimum value at the same point as $f(x) = 3(x - 4)^2$, but for all x, the function values are twice the values obtained from $f(x) = 3(x - 4)^2$.

69. The parabola is the same shape as

$$f(x) = -\frac{1}{2}(x - 2)^2 + 4$$

and has a maximum value at the same point as

$$g(x) = -2(x - 1)^2 - 6.$$

Functions other than parabolas can be translated. For a function $f(x)$, if we replace x by $x - h$, where h is a constant, the graph will be moved horizontally. If we add a constant k to a function $f(x)$, the graph will be moved vertically.

Use the graph of the function $y = g(x)$ below for Exercises 70–75.

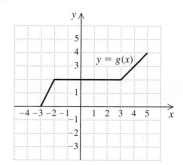

Draw a graph of each of the following.

70. $y = g(x + 2)$

71. $y = g(x - 3)$

72. $y - 3 = g(x)$ **73.** $y = g(x) + 4$

74. $y + 2 = g(x - 4)$ **75.** $y = g(x - 2) + 3$

76. 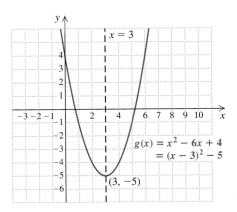 Use a grapher to check your graphs for Exercises 7, 25, 39, and 45.

77. Use the Trace feature of a grapher to confirm the maximum and minimum values given as answers to Exercises 35, 43, 51, and 53. Be sure to adjust the window appropriately.

8.8

More About Graphing Quadratic Functions

Completing the Square • Finding x-Intercepts

Completing the Square

By *completing the square* (see Section 8.1), we can always rewrite the polynomial $ax^2 + bx + c$ in the form $a(x - h)^2 + k$. Thus the procedures discussed in Section 8.7 enable us to graph any quadratic function.

EXAMPLE 1

Graph: $g(x) = x^2 - 6x + 4$.

Solution   We have

$$g(x) = x^2 - 6x + 4$$
$$= (x^2 - 6x) + 4.$$

To complete the square inside the parentheses, we take half the x-coefficient, $\frac{1}{2} \cdot (-6) = -3$, and square it to get $(-3)^2 = 9$. Then we add $9 - 9$ inside the parentheses:

$$g(x) = (x^2 - 6x + 9 - 9) + 4 \qquad \text{The effect is of adding 0.}$$
$$= (x^2 - 6x + 9) + (-9 + 4) \qquad \text{Using the associative law of addition to regroup}$$
$$= (x - 3)^2 - 5. \qquad \text{Factoring and simplifying}$$

This equation was graphed in Example 4 of Section 8.7. The vertex is $(3, -5)$, and the line of symmetry is $x = 3$.

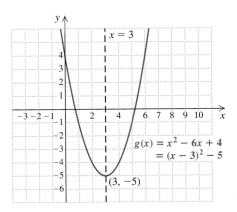

When the leading coefficient is not 1, we factor out that number from the first two terms. Then we complete the square.

EXAMPLE 2

Graph: $f(x) = 3x^2 + 12x + 13$.

Solution   Since the coefficient of x^2 is not 1, we need to factor out that number — in this case, 3 — from the first two terms. Remember that we want the form $f(x) = a(x - h)^2 + k$:

$$f(x) = 3x^2 + 12x + 13$$
$$= 3(x^2 + 4x) + 13.$$

Now we complete the square as before. We take half of the x-coefficient, $\frac{1}{2} \cdot 4 = 2$, and square it: $2^2 = 4$. Then we add $4 - 4$ inside the parentheses:

$$f(x) = 3(x^2 + 4x + 4 - 4) + 13.$$

This time we must distribute the 3 in order to rearrange terms:

$$f(x) = 3(x^2 + 4x + 4) + 3(-4) + 13 \qquad \text{Distributing to obtain a trinomial square}$$
$$= 3(x + 2)^2 + 1. \qquad \text{Factoring and simplifying}$$

The vertex is $(-2, 1)$, and the line of symmetry is $x = -2$. The coefficient of x^2 is 3, so the graph is narrow and opens upward. We choose a few x-values on either side of the vertex, compute y-values, and then graph the parabola.

x	$f(x) = 3(x + 2)^2 + 1$	
-2	1	← Vertex
-3	4	
-1	4	

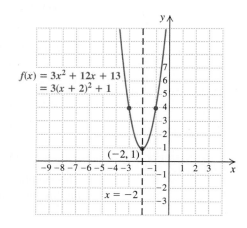

EXAMPLE 3

Graph: $f(x) = -2x^2 + 10x - 7$.

Solution   We first find the vertex by completing the square. We factor out -2 from the first two terms of the expression. This makes the coefficient of x^2 inside the parentheses 1:

$$f(x) = -2x^2 + 10x - 7$$
$$= -2(x^2 - 5x) - 7.$$

Now we complete the square as before. We take half of the x-coefficient and square it to get $\frac{25}{4}$. Then we add $\frac{25}{4} - \frac{25}{4}$ inside the parentheses:

$$f(x) = -2\left(x^2 - 5x + \tfrac{25}{4} - \tfrac{25}{4}\right) - 7$$
$$= -2\left(x^2 - 5x + \tfrac{25}{4}\right) - 2\left(-\tfrac{25}{4}\right) - 7 \qquad \text{Multiplying by } -2, \text{ using the distributive law, and rearranging terms}$$
$$= -2\left(x - \tfrac{5}{2}\right)^2 + \tfrac{11}{2}. \qquad \text{Factoring and simplifying}$$

The vertex is $\left(\frac{5}{2}, \frac{11}{2}\right)$, and the line of symmetry is $x = \frac{5}{2}$. The coefficient of x^2, -2, is negative, so the graph opens downward. We plot a few points on either side of the vertex, including the y-intercept, $f(0)$, and graph the parabola.

x	$f(x)$	
$\frac{5}{2}$	$\frac{11}{2}$	← Vertex
0	-7	← y-intercept
1	1	
4	1	

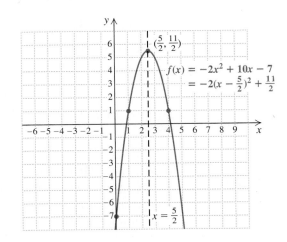

We can find a formula for computing the vertex. We do so by completing the square, as in Examples 1–3:

$$f(x) = ax^2 + bx + c$$
$$= a\left(x^2 + \frac{b}{a}x\right) + c. \qquad \text{Factoring } a \text{ out of the first two terms. Check by multiplying.}$$

Half of the x-coefficient, $\frac{b}{a}$, is $\frac{b}{2a}$. We square it to get $\frac{b^2}{4a^2}$ and add $\frac{b^2}{4a^2} - \frac{b^2}{4a^2}$ inside the parentheses. Then we multiply a back through, as follows, and factor:

$$f(x) = a\left(x^2 + \frac{b}{a}x + \frac{b^2}{4a^2} - \frac{b^2}{4a^2}\right) + c$$
$$= a\left(x^2 + \frac{b}{a}x + \frac{b^2}{4a^2}\right) + a\left(-\frac{b^2}{4a^2}\right) + c \qquad \text{Using the distributive law}$$
$$= a\left(x + \frac{b}{2a}\right)^2 + \frac{-b^2}{4a} + \frac{4ac}{4a} \qquad \text{Factoring and finding a common denominator}$$
$$= a\left[x - \left(-\frac{b}{2a}\right)\right]^2 + \frac{4ac - b^2}{4a}.$$

Thus we have the following.

For a parabola given by a quadratic function $f(x) = ax^2 + bx + c$, the vertex is

$$\left(-\frac{b}{2a}, \frac{4ac - b^2}{4a}\right).$$

The x-coordinate of the vertex is $-b/(2a)$. The line of symmetry is $x = -b/(2a)$. The second coordinate of the vertex can be found by substituting into the formula above, but is usually found most easily by evaluating the function at $-b/(2a)$.

Let us look back at Example 3 to see how we can find the vertex directly. From the above formula,

the x-coordinate of the vertex is $-\dfrac{b}{2a} = -\dfrac{10}{2(-2)} = \dfrac{5}{2}.$

Substituting $\frac{5}{2}$ into $f(x) = -2x^2 + 10x - 7$, we find the second coordinate of the vertex:

$$f\left(\tfrac{5}{2}\right) = -2\left(\tfrac{5}{2}\right)^2 + 10\left(\tfrac{5}{2}\right) - 7 = -2\left(\tfrac{25}{4}\right) + 25 - 7 = \tfrac{11}{2}.$$

The vertex is $\left(\tfrac{5}{2}, \tfrac{11}{2}\right)$. The line of symmetry is $x = \tfrac{5}{2}$.

We have actually developed two methods for finding the vertex. One is by completing the square and the other is by using a formula. You should consult with your instructor about which method to use.

Finding x-Intercepts

The points at which a graph crosses the x-axis are called **x-intercepts.** These are, of course, the points at which $y = 0$.

To find the x-intercepts of a quadratic function $f(x) = ax^2 + bx + c$, we solve the equation

$$0 = ax^2 + bx + c.$$

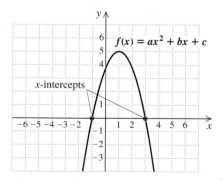

EXAMPLE 4

Find the x-intercepts of the graph of $f(x) = x^2 - 2x - 2$.

Solution   We solve the equation

$$0 = x^2 - 2x - 2.$$

The equation is hard to factor, so we use the quadratic formula and get $x = 1 \pm \sqrt{3}$. Thus the x-intercepts are $\left(1 - \sqrt{3}, 0\right)$ and $\left(1 + \sqrt{3}, 0\right)$.

If graphing, we would approximate, to get $(-0.7, 0)$ and $(2.7, 0)$.   ❏

The discriminant (see Section 8.4), $b^2 - 4ac$, tells us how many real-number solutions the equation $0 = ax^2 + bx + c$ has, so it also indicates how many x-intercepts there are. Compare the following graphs.

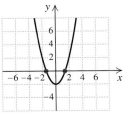

$y = ax^2 + bx + c$
$b^2 - 4ac > 0$
Two real solutions
Two x-intercepts

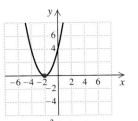

$y = ax^2 + bx + c$
$b^2 - 4ac = 0$
One real solution
One x-intercept

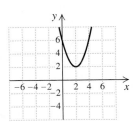

$y = ax^2 + bx + c$
$b^2 - 4ac < 0$
No real solutions
No x-intercepts

EXERCISE SET | 8.8

For each quadratic function, **(a)** find the vertex and the line of symmetry and **(b)** graph the function.

1. $f(x) = x^2 - 2x - 3$
2. $f(x) = x^2 + 2x - 5$
3. $g(x) = x^2 + 6x + 13$
4. $g(x) = x^2 - 4x + 5$
5. $f(x) = x^2 + 4x - 1$
6. $f(x) = x^2 - 10x + 21$
7. $h(x) = 2x^2 + 16x + 25$
8. $h(x) = 2x^2 - 16x + 23$
9. $f(x) = -x^2 + 4x + 6$
10. $f(x) = -x^2 - 4x + 3$
11. $g(x) = x^2 + 3x - 10$
12. $g(x) = x^2 + 5x + 4$
13. $f(x) = 3x^2 - 24x + 50$
14. $f(x) = 4x^2 + 8x - 3$
15. $h(x) = x^2 - 9x$
16. $h(x) = x^2 + x$
17. $f(x) = -2x^2 - 4x - 6$
18. $f(x) = -3x^2 + 6x + 2$
19. $g(x) = 2x^2 - 10x + 14$
20. $g(x) = 2x^2 + 6x + 8$

21. $f(x) = -3x^2 - 3x + 1$
22. $f(x) = -2x^2 + 2x + 1$
23. $h(x) = \frac{1}{2}x^2 + 4x + \frac{19}{3}$
24. $h(x) = \frac{1}{2}x^2 - 3x + 2$

Find the x-intercepts. If no x-intercepts exist, state so.

25. $f(x) = x^2 - 4x + 1$ **26.** $f(x) = x^2 + 6x + 10$
27. $g(x) = -x^2 + 2x + 3$
28. $g(x) = x^2 - 2x - 5$ **29.** $f(x) = x^2 - 3x - 4$
30. $f(x) = x^2 - 8x + 5$
31. $h(x) = -x^2 + 3x - 2$
32. $h(x) = 2x^2 - 4x + 6$ **33.** $f(x) = 2x^2 + 4x - 1$
34. $f(x) = x^2 - x + 2$ **35.** $g(x) = x^2 - x + 1$
36. $g(x) = 4x^2 + 12x + 9$

Skill Maintenance

Solve.

37. $\sqrt{4x - 4} = \sqrt{x + 4} + 1$
38. $\sqrt{5x - 4} + \sqrt{13 - x} = 7$

Synthesis

39. ◈ Suppose that the function $f(x) = ax^2 + bx + c$ has x_1 and x_2 as x-intercepts. Explain why the function $g(x) = -ax^2 - bx - c$ will also have x_1 and x_2 as x-intercepts.

40. ◈ Compare the graphs of $f(x) = a(x - h)^2 + k$ and $g(x) = -a(x - h)^2 + k$. What requirements, if any, must be placed on a, h, and k if both graphs are to have the same x-intercepts?

For each quadratic function, find **(a)** the maximum or minimum value and **(b)** the x-intercepts.

41. ▦ $f(x) = 2.31x^2 - 3.135x - 5.89$

42. ▦ $f(x) = -18.8x^2 + 7.92x + 6.18$

43. Graph the function

$$f(x) = x^2 - x - 6.$$

Then use the graph to approximate solutions to the following equations.

a) $x^2 - x - 6 = 2$
b) $x^2 - x - 6 = -3$

44. Graph the function

$$f(x) = \frac{x^2}{8} + \frac{x}{4} - \frac{3}{8}.$$

Then use the graph to approximate solutions to the following equations.

a) $\dfrac{x^2}{8} + \dfrac{x}{4} - \dfrac{3}{8} = 0$

b) $\dfrac{x^2}{8} + \dfrac{x}{4} - \dfrac{3}{8} = 1$

c) $\dfrac{x^2}{8} + \dfrac{x}{4} - \dfrac{3}{8} = 2$

Find an equivalent equation of the type $f(x) = a(x - h)^2 + k$.

45. $f(x) = mx^2 - nx + p$

46. $f(x) = 3x^2 + mx + m^2$

Graph.

47. $f(x) = |x^2 - 1|$

48. $f(x) = |3 - 2x - x^2|$

49. ◺ Use a grapher to check your solutions to Exercises 9, 23, 33, 41, 42, 43, and 44.

8.9

Problem Solving and Quadratic Functions

Maximum and Minimum Problems to Data • **Solving Formulas** • **Fitting Quadratic Functions**

Let's look now at some of the many situations in which quadratic functions are used for problem solving.

Maximum and Minimum Problems

For a quadratic function, the value $f(x)$ at the vertex will be either greater than any other $f(x)$ or less than any other $f(x)$. If the graph opens upward, f will achieve a minimum. If the graph opens downward, f will achieve a maximum. In certain problems, we want to find a maximum or a minimum. If the problem situation translates to a quadratic function, we can often solve by finding x or $f(x)$ at the vertex.

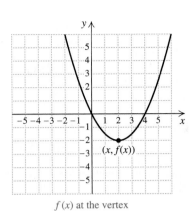

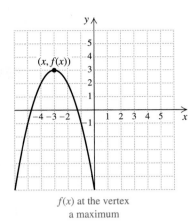

$f(x)$ at the vertex
a minimum

$f(x)$ at the vertex
a maximum

EXAMPLE 1

What are the dimensions of the largest rectangular pen that a farmer can enclose with 64 m of fence?

Solution

1. FAMILIARIZE. We make a drawing and label it.

Perimeter: $2w + 2l = 64$ m

Area: $A = l \cdot w$

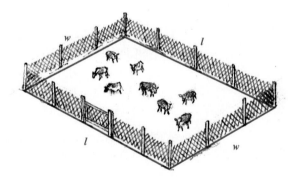

To get a better feel for the problem, we can look at some possible dimensions for a rectangular pen that can be enclosed with 64 m of fence.

l	w	A
22	10	220
20	12	240
18	14	252

2. TRANSLATE. We have two equations, one of which expresses area as a function of length and width:

$$2w + 2l = 64$$

$$A = l \cdot w.$$

3. CARRY OUT. We need to express A as a function of l or w but not both. To do so, we solve for l in the first equation to obtain $l = 32 - w$. Substituting for l in the second equation, we get a quadratic function, $A(w)$, or just A:

$$A = (32 - w)w \qquad \text{Substituting for } l$$
$$A = -w^2 + 32w. \qquad \text{This is a parabola opening down, so a maximum exists.}$$

Completing the square, we get

$$A = -(w^2 - 32w + 256 - 256) = -(w - 16)^2 + 256.$$

The maximum function value of 256 occurs when $w = 16$ and $l = 32 - 16$, or 16.

4. CHECK. Note that 256 is greater than any of the values for A found in the *Familiarize* step. To be more certain, we could check values other than those used in the *Familiarize* step. For example, if $w = 15$, then $l = 32 - 15 = 17$, and $A = 15 \cdot 17 = 255$. The same area results if $w = 17$ and $l = 15$. Since 256 is greater than 255, it looks as though we have a maximum.

5. STATE. For the pen to have maximal area, it should be 16 m by 16 m. ❏

EXAMPLE 2

What is the minimum product of two numbers whose difference is 5? What are the numbers?

Solution

1. FAMILIARIZE. We try some pairs of numbers that differ by 5 and compute their products:

$$1 \cdot 6 = 6,$$
$$0 \cdot 5 = 0,$$
$$(-1) \cdot 4 = -4.$$

We suspect that one of the two numbers will be negative and the other positive. Let x represent the larger number and $x - 5$ the other number.

2. TRANSLATE. We represent the product of the two numbers by

$$p = x(x - 5), \quad \text{or} \quad p(x) = x^2 - 5x.$$

3. CARRY OUT. The function $p(x) = x^2 - 5x$ represents a parabola opening up. Completing the square, we get

$$p(x) = x^2 - 5x + \frac{25}{4} - \frac{25}{4} = \left(x - \frac{5}{2}\right)^2 - \frac{25}{4}.$$

The minimum function value of $-\frac{25}{4}$ occurs when one number is $\frac{5}{2}$ and the other number is $\frac{5}{2} - 5 = -\frac{5}{2}$.

4. CHECK. Note that if $x = 2$, then $x - 5 = -3$ and $2(-3) = -6$. Also note that if $x = 3$, then $x - 5 = -2$ and $3(-2) = -6$. Thus, since $-\frac{25}{4} < -6$, it appears that we have a minimum.

5. STATE. The numbers $-\frac{5}{2}$ and $\frac{5}{2}$ differ by 5 and yield a minimum product of $-\frac{25}{4}$. ❏

Fitting Quadratic Functions to Data

Whenever we know that a quadratic function fits a certain situation, we can find that function if we know three inputs and their outputs. Each such ordered pair is called a *data point*.

The following example provides a good illustration.

EXAMPLE 3

The instruction booklet for a video cassette recorder (VCR) includes a table relating the time a tape has run to the counter reading.

Counter Reading	Time Tape Has Run (in Hours)
000	0
400	1
700	3
800	4

Find a quadratic function that fits the data. Then predict how long a tape has run when the counter reading is 500.

Solution

1. FAMILIARIZE. The statement of the problem leads us to look for a function of the form

$$T(n) = an^2 + bn + c,$$

where $T(n)$ represents the time, in hours, that the tape has run at counter reading n hundred. We need to determine values of a, b, and c that will fit the given data.

2. TRANSLATE. We substitute some values of n and T:

$$0 = a \cdot 0^2 + b \cdot 0 + c,$$
$$1 = a \cdot 4^2 + b \cdot 4 + c,$$
$$3 = a \cdot 7^2 + b \cdot 7 + c.$$

After simplifying, we see that we need to solve the system

$$0 = c,$$
$$1 = 16a + 4b + c,$$
$$3 = 49a + 7b + c.$$

3. CARRY OUT. Since $c = 0$, it suffices to solve the system

$$1 = 16a + 4b, \qquad \textbf{(1)}$$
$$3 = 49a + 7b. \qquad \textbf{(2)}$$

Multiplying the first equation by -7 and the second equation by 4, we obtain

$$-7 = -112a - 28b,$$
$$12 = 196a + 28b.$$

Thus,

$$5 = 84a \qquad \text{Adding equations}$$
$$\tfrac{5}{84} = a.$$

We substitute into one of the original equations to solve for b:

$$1 = 16 \cdot \tfrac{5}{84} + 4b \qquad \text{Substituting in Equation (1)}$$
$$1 = \tfrac{80}{84} + 4b$$
$$\tfrac{4}{84} = 4b$$
$$\tfrac{1}{84} = b.$$

Thus the function $T(n) = \tfrac{5}{84}n^2 + \tfrac{1}{84}n$ fits the given data. To find how long a tape has been running when the counter reading is 500, we find $T(5)$:

$$T(5) = \tfrac{5}{84} \cdot 5^2 + \tfrac{1}{84} \cdot 5$$
$$= \tfrac{125}{84} + \tfrac{5}{84}$$
$$= \tfrac{130}{84} \approx 1.55 \text{ hr.}$$

4. **CHECK.** Besides rechecking our calculations, we can check to see if the function we obtained will produce other values found in the table. As you should confirm,

$$T(8) \approx 3.9 \text{ hr.}$$

This last value should not bother us since it is fair to assume that the manufacturer of the VCR may have rounded off when making the table or that our model of the situation overlooked some details.

5. **STATE.** The answer is that when the counter reads 500, the tape has been running for just over $1\tfrac{1}{2}$ hours. ❏

Solving Formulas

Recall that to solve a formula for a certain letter, we use the principles for solving equations to get that letter alone on one side. When square roots appear, we can usually eliminate the radical signs by squaring both sides.

EXAMPLE 4

Period of a pendulum. The time T required for a pendulum of length l to swing back and forth (complete one period) is given by the formula $T = 2\pi\sqrt{l/g}$, where g is the gravitational constant. Solve for l.

Solution   We have

$$T = 2\pi \sqrt{\frac{l}{g}}$$

$$T^2 = \left(2\pi \sqrt{\frac{l}{g}} \right)^2 \qquad \text{Principle of powers (squaring)}$$

$$T^2 = 2^2\pi^2\frac{l}{g}$$

$$gT^2 = 4\pi^2 l \qquad \text{Clearing fractions}$$

$$\frac{gT^2}{4\pi^2} = l. \qquad \text{Multiplying by } \frac{1}{4\pi^2}$$

We now have l alone on one side and l does not appear on the other side, so the formula is solved for l.   ❑

In most formulas, the letters represent nonnegative numbers, so we do not need to use absolute-value signs when taking square roots.

EXAMPLE 5

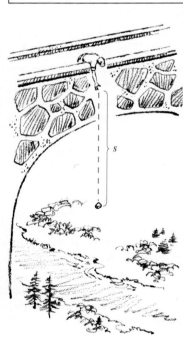

Hang time.*   A formula relating an athlete's vertical leap V, in inches, to hang time T, in seconds, is $V = 48T^2$. Solve for T.

Solution

$$48T^2 = V$$

$$T^2 = \frac{V}{48}$$ Multiplying by $\frac{1}{48}$ to get T^2 alone

$$T = \sqrt{\frac{V}{48}}$$ Taking the square root

$$T = \sqrt{\frac{V}{16 \cdot 3} \cdot \frac{3}{3}}$$ Factoring and multiplying by 1 to rationalize the denominator

$$T = \sqrt{\frac{3V}{144}}$$

$$T = \frac{\sqrt{3V}}{12}$$ ❑

EXAMPLE 6

Downward speed.   An object tossed downward with an initial speed (velocity) of v_0 will travel a distance of s meters, where $s = 4.9t^2 + v_0t$ and t is measured in seconds. Solve for t.

Solution   Since t is squared in one term and raised to the first power in the other term, the equation is quadratic in t. We find standard form and use the quadratic formula:

$$4.9t^2 + v_0t - s = 0$$ Writing standard form

$$a = 4.9, \quad b = v_0, \quad c = -s$$

$$t = \frac{-v_0 \pm \sqrt{v_0^2 - 4(4.9)(-s)}}{2(4.9)}.$$

Since the negative square root would yield a negative value for t, we use only the positive root:

$$t = \frac{-v_0 + \sqrt{v_0^2 + 19.6s}}{9.8}.$$ ❑

---

*This formula is taken from an article by Peter Brancazio, ''The Mechanics of a Slam Dunk,'' *Popular Mechanics,* November 1991. Courtesy of Professor Peter Brancazio, Brooklyn College.

The following list of steps should help you when solving formulas for a given letter. Try to remember that when solving a formula, you do the same things you would do to solve any equation.

To solve a formula for a letter, say, b:

1. Clear the fractions and use the principle of powers, as needed, until b does not appear in any radicand or denominator. (In some cases, you may clear the fractions first, and in some cases you may use the principle of powers first. Perform these steps until radicals containing b are gone and b is not in any denominator.)
2. Collect all terms with b^2 in them. Also collect all terms with b in them.
3. If b^2 does not appear, you can finish by using just the addition and multiplication principles. Get all terms containing b on one side of the equation; then factor out b. Dividing on both sides will then get b alone.
4. If b^2 appears but b does not appear to the first power, solve the equation for b^2. Then take square roots on both sides.
5. If there are terms containing both b and b^2, put the equation in standard form and use the quadratic formula.

EXERCISE SET | 8.9

Solve.

1. A rancher is fencing off a rectangular field with a perimeter of 76 ft. What dimensions will yield the maximum area? What is the maximum area?

2. A carpenter is building a rectangular room with a perimeter of 68 ft. What dimensions will yield the maximum area? What is the maximum area?

3. What is the maximum product of two numbers whose sum is 16? What numbers yield this product?

4. What is the maximum product of two numbers whose sum is 28? What numbers yield this product?

5. What is the maximum product of two numbers whose sum is 22? What numbers yield this product?

6. What is the maximum product of two numbers whose sum is 45? What numbers yield this product?

7. What is the minimum product of two numbers whose difference is 4? What are the numbers?

8. What is the minimum product of two numbers whose difference is 10? What are the numbers?

9. What is the minimum product of two numbers whose difference is 9? What are the numbers?

10. What is the minimum product of two numbers whose difference is 7? What are the numbers?

11. What is the maximum product of two numbers whose sum is -7? What numbers yield this product?

12. What is the maximum product of two numbers whose sum is -9? What numbers yield this product?

13. A farmer decides to enclose a rectangular garden, using the side of a barn as one side of the rectangle. What is the maximum area that the farmer can enclose with 40 ft of fence? What should the dimensions of the garden be to yield this area?

14. A stone mason has enough stones to enclose a rectangular patio with 60 ft of perimeter, assuming that the attached house forms one side of the rectangle. What is the maximum area that the mason can

enclose? What should the dimensions of the patio be to yield this area?

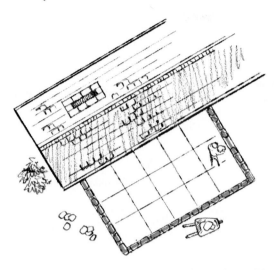

15. A rectangular compost container is to be formed in a corner of a fenced yard, with 8 ft of chicken wire completing the other two sides of the rectangle. If the chicken wire is 3 ft high, what dimensions of the base will maximize the container's volume?

16. A plastics manufacturer plans to produce a one-compartment vertical file by bending the long side of an 8 in. by 14 in. sheet of plastic along two lines to form a U shape. How tall should the file be to maximize the volume that the file can hold?

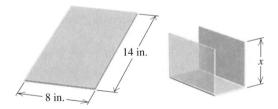

Find the quadratic function that fits the set of data points.

17. $(1, 4)$, $(-1, -2)$, $(2, 13)$

18. $(1, 4)$, $(-1, 6)$, $(-2, 16)$

19. $(2, 0)$, $(4, 3)$, $(12, -5)$

20. $(-3, -30)$, $(3, 0)$, $(6, 6)$

21. A business earns $38 in the first week, $66 in the second week, and $86 in the third week. The manager graphs the points $(1, 38)$, $(2, 66)$, and $(3, 86)$ and uses a quadratic function to describe the situation.

a) Find a quadratic function that fits the data.

b) Using the function, predict the earnings for the fourth week.

22. A business earns $1000 in its first month, $2000 in the second month, and $8000 in the third month. The manager plots the points $(1, 1000)$, $(2, 2000)$, and $(3, 8000)$ and uses a quadratic function to describe the situation.

a) Find a quadratic function that fits the data.

b) Using the function, predict the earnings for the fourth month.

23. a) Find a quadratic function that fits the following data.

Travel Speed (in Kilometers per Hour)	Number of Daytime Accidents (for Every 200 Million Kilometers)
60	100
80	130
100	200

b) Use the function to calculate the number of daytime accidents that occur at 50 km/h.

24. a) Find a quadratic function that fits the following data.

Travel Speed (in Kilometers per Hour)	Number of Nighttime Accidents (for Every 200 Million Kilometers)
60	400
80	250
100	250

b) Use the function to calculate the number of nighttime accidents that occur at 50 km/h.

25. 🖩   Pizza Unlimited has the following prices for pizzas.

Diameter	Price
8 in.	$ 6.00
12 in.	$ 8.50
16 in.	$11.50

Is price a quadratic function of diameter? It probably should be, because the price should be proportional to the area, and the area is a quadratic function of the diameter. (The area of a circular region is given by $A = \pi r^2$ or $(\pi/4) \cdot d^2$.)

a) Express price as a quadratic function of diameter using the data points $(8, 6)$, $(12, 8.50)$, and $(16, 11.50)$.

b) Use the function to find the price of a 14-in. pizza.

Recall that total profit P is the difference between total revenue R and total cost C. Given the following total-revenue and total-cost functions, find the total profit, the maximum value of the total profit, and the value of x at which it occurs.

26. $R(x) = 1000x - x^2$,
$C(x) = 3000 + 20x$

27. $R(x) = 200x - x^2$,
$C(x) = 5000 + 8x$

28. $R(x) = 300x - x^2$,
$C(x) = 50 + 80x$

Solve the formula for the indicated letter. Assume that all variables represent nonnegative numbers.

29. $A = 6s^2$, for s
(Surface area of a cube)

30. $A = 4\pi r^2$, for r
(Surface area of a sphere)

31. $F = \dfrac{Gm_1m_2}{r^2}$, for r

(Law of gravity)

32. $N = \dfrac{kQ_1Q_2}{s^2}$, for s

(Number of phone calls between two cities)

33. $E = mc^2$, for c
(Energy–mass relationship)

34. $A = \pi r^2$, for r
(Area of a circle)

35. $a^2 + b^2 = c^2$, for b
(Pythagorean formula in two dimensions)

36. $a^2 + b^2 + c^2 = d^2$, for c
(Pythagorean formula in three dimensions)

37. $N = \dfrac{k^2 - 3k}{2}$, for k

(Number of diagonals of a polygon)

38. $s = v_0t + \dfrac{gt^2}{2}$, for t

(A motion formula)

39. $A = 2\pi r^2 + 2\pi rh$, for r
(Surface area of a right cylindrical solid)

40. $A = \pi r^2 + \pi rs$, for r
(Surface area of a cone)

41. $N = \frac{1}{2}(n^2 - n)$, for n
(Number of games in a league of n teams)

42. $A = A_0(1 - r)^2$, for r
(A business formula)

43. $A = 2w^2 + 4lw$, for w
(Surface area of a rectangular solid)

44. $A = 4\pi r^2 + 2\pi rh$, for r
(Surface area of a capsule)

45. $T = 2\pi \sqrt{\dfrac{l}{g}}$, for g

(A pendulum formula)

46. $W = \sqrt{\dfrac{1}{LC}}$, for L

(An electricity formula)

47. $A = P_1(1 + r)^2 + P_2(1 + r)$, for r
(An investment formula)

48. $A = P_1\left(1 + \dfrac{r}{2}\right)^2 + P_2\left(1 + \dfrac{r}{2}\right)$, for r

(An investment formula)

49. $m = \dfrac{m_0}{\sqrt{1 - \dfrac{v^2}{c^2}}}$, for v

(A relativity formula)

50. Solve the formula given in Exercise 49 for c.

Solve. Examine Exercises 29–50 and Examples 4–6 for the appropriate formula.

51. a) An object is dropped 75 m from an airplane. How long does it take the object to reach the ground?

b) An object is thrown downward 75 m from the plane at an initial velocity of 30 m/sec. How long does it take the object to reach the ground?

c) How far will an object fall in 2 sec, if thrown downward at an initial velocity of 30 m/sec?

52. a) An object is dropped 500 m from an airplane. How long does it take the object to reach the ground?

b) An object is thrown downward 500 m from the plane at an initial velocity of 30 m/sec. How long does it take the object to reach the ground?

c) How far will an object fall in 5 sec, if thrown downward at an initial velocity of 30 m/sec?

53. ▦ Michael Jordan of the Chicago Bulls has a vertical leap of about 38 in. What is his hang time?

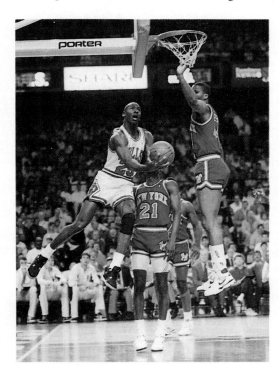

54. The surface area of a right cylindrical solid of height 3 m is 8π m². Find the radius of the solid.

55. In a softball league, each team plays each of the other teams once. If the league plays a total of 91 games, how many teams are in the league?

56. In a volleyball league, each team plays each of the other teams once. If a total of 66 games is played, how many teams are in the league?

57. Jesse is tied to one end of a 40-m elasticized (bungee) cord. The other end of the cord is tied to the middle of a train trestle. If Jesse jumps off the bridge, for how long will he be falling before the cord begins to stretch?

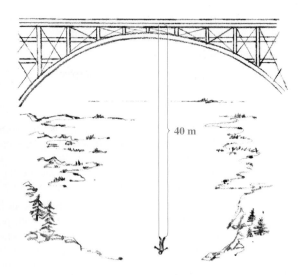

40 m

58. Sheila is tied to a bungee cord (see Exercise 57) and falls for 2.5 sec before her cord begins to stretch. How long is the bungee cord?

59. ▦ An object thrown downward from a 100-m cliff travels 51.6 m in 3 sec. What was the initial velocity of the object?

60. ▦ An object thrown downward from a 200-m cliff travels 91.2 m in 4 sec. What was the initial velocity of the object?

61. ▦ A firm invests $3000 in a savings account for two years. At the beginning of the second year, an additional $1700 is invested. If a total of $5253.70 is in the account at the end of the second year, what is the annual interest rate? (*Hint:* See Exercise 47.)

62. ▦ A business invests $10,000 in a savings account for two years. At the beginning of the second year, an additional $3500 is invested. If a total of $15,569.75 is in the account at the end of the second year, what is the annual interest rate? (*Hint:* See Exercise 47.)

Skill Maintenance _____

63. Express in terms of i: $\sqrt{-20}$.

64. Subtract:

$$\frac{x}{x^2 + 17x + 72} - \frac{8}{x^2 + 15x + 56}.$$

Synthesis _____

65. The sum of the base and the height of a triangle is 38 cm. Find the dimensions for which the area is a maximum, and find the maximum area.

66. The perimeter of a rectangle is 44 ft. Find the least possible length of a diagonal.

67. Solve for n:
$$mn^4 - r^2pm^3 - r^2n^2 + p = 0.$$

68. Solve for t:
$$rt^2 - rt - st^2 + s^2r - st = 0.$$

69. When a theater owner charges $2 for admission, she averages 100 people attending. For each 10¢ increase in admission price, the average number attending decreases by 1. What should the owner charge in order to make the most money?

70. An orange grower finds that she gets an average yield of 40 bushels (bu) per tree when she plants 20 trees on an acre of ground. Each time she adds a tree to an acre, the yield per tree decreases by 1 bu, due to congestion. How many trees per acre should she plant for maximum yield?

71. The height above the ground of a launched object is a quadratic function of the time that it is in the air. Suppose that a flare is launched from a cliff 64 ft above sea level. If 3 sec after being launched the flare is again level with the cliff, and if 2 sec after that it lands in the sea, what is the maximum height that the flare will reach?

72. The cables supporting a straight-line suspension bridge are parabolic in shape. Suppose that a suspension bridge is being designed to cross a river that is 160 ft wide and that the vertical cables are 30 ft above road level at the bridge's midpoint and are 80 ft above road level at a point 50 ft from the bridge's midpoint. How long are the longest vertical cables?

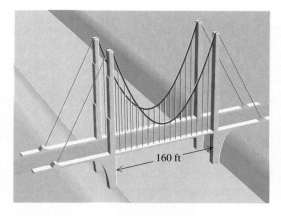

160 ft

73. Find a formula that expresses the diameter of a right cylindrical solid as a function of its surface area.

74. Find a formula that expresses the length of a cube's three-dimensional diagonal as a function of the cube's volume.

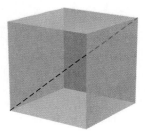

75. Find a formula that expresses the length of a cube's three-dimensional diagonal as a function of the cube's surface area.

8.10

Polynomial and Rational Inequalities

Quadratic and Other Polynomial Inequalities • Rational Inequalities

Quadratic and Other Polynomial Inequalities

Inequalities such as the following are called *polynomial inequalities*:
$$x^3 - 4x^2 + x - 6 > 0, \quad x^2 + 3x - 10 < 0, \quad 5x^2 - 3x + 2 \geq 0, \quad 3x - 1 \leq 0.$$

Second-degree polynomial inequalities in one variable are called *quadratic inequal-*

ities. We now consider three ways to solve polynomial inequalities. The first two provide understanding, and the last method is the fastest.

The first method for solving polynomial inequalities is to consider the graph of the related function in the plane.

EXAMPLE 1

Solve: $x^2 + 3x - 10 > 0$.

Solution   Consider the function $f(x) = x^2 + 3x - 10$ and its graph. Its graph opens upward since the leading coefficient is positive.

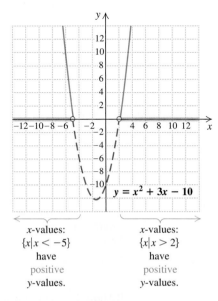

x-values: $\{x \mid x < -5\}$ have positive y-values.

x-values: $\{x \mid x > 2\}$ have positive y-values.

$$y = x^2 + 3x - 10$$

Values of y will be positive to the left and right of the x-intercepts, as shown. We find the intercepts by setting the polynomial equal to 0 and solving:

$$x^2 + 3x - 10 = 0$$
$$(x + 5)(x - 2) = 0$$
$$x + 5 = 0 \quad or \quad x - 2 = 0$$
$$x = -5 \quad or \quad x = 2.$$

Thus the solution set of the inequality is

$$\{x \mid x < -5 \ or \ x > 2\}, \quad or \quad (-\infty, -5) \cup (2, \infty).$$ ❑

We can solve any inequality by considering a graph of the related function and finding intercepts as in Example 1. In some cases, we may need to use the quadratic formula to find the intercepts.

EXAMPLE 2

Solve: $x^2 - 2x \leqslant 2$.

Solution   We first find standard form with 0 on one side:

$$x^2 - 2x - 2 \leqslant 0.$$

Consider $f(x) = x^2 - 2x - 2$. Its graph opens upward. Function values will be non-positive between and including its x-intercepts, as shown. We find the intercepts by solving $f(x) = 0$:

$$x = \frac{-b \pm \sqrt{b^2 - 4ac}}{2a} = \frac{-(-2) \pm \sqrt{(-2)^2 - 4 \cdot 1(-2)}}{2 \cdot 1}$$

$$= \frac{2 \pm \sqrt{12}}{2} = \frac{2 \pm 2\sqrt{3}}{2} = 1 \pm \sqrt{3}.$$

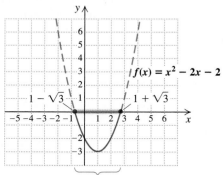

Inputs in this interval have negative or 0 outputs.

The x-intercepts are $1 + \sqrt{3}$ and $1 - \sqrt{3}$.
The solution set of the inequality is

$$\{x|\ 1 - \sqrt{3} \leqslant x \leqslant 1 + \sqrt{3}\}, \quad \text{or} \quad [1 - \sqrt{3}, 1 + \sqrt{3}].$$

It should be pointed out that we need not actually draw graphs as in the preceding examples. Visualizing the graph will usually suffice.

A second way to solve inequalities works for any polynomial that we can factor into a product of first-degree polynomials.

EXAMPLE 3

Solve: $x^2 + 2x - 3 > 0$.

Solution   We factor the inequality, obtaining $(x + 3)(x - 1) > 0$. The solutions of $(x + 3)(x - 1) = 0$ are -3 and 1. They are not solutions of the inequality, but they divide the real-number line in a natural way, as pictured here. The product $(x + 3)(x - 1)$ is positive or negative, for values other than -3 and 1, depending on the signs of the factors $x + 3$ and $x - 1$. We can determine this efficiently with a diagram as follows.

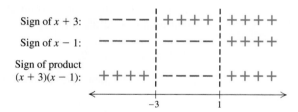

To set up the diagram, solve $x + 3 > 0$. We get $x > -3$. Thus, $x + 3$ is positive for all numbers to the right of -3. We indicate this with $+$ signs. Accordingly, $x + 3$ is negative for all numbers to the left of -3. We indicate that with the $-$ signs.

Similarly, we solve $x - 1 > 0$ and get $x > 1$. Thus, $x - 1$ is positive for all numbers to the right of 1 and negative for all numbers to the left of 1. We indicate this with the $+$ and $-$ signs.

In order for the product $(x + 3)(x - 1)$ to be positive, both factors must be positive or both must be negative. In the diagram, we see that this situation occurs when $x < -3$ and when $x > 1$. The solution set of the inequality is

$$\{x \mid x < -3 \ or \ x > 1\}, \quad or \quad (-\infty, -3) \cup (1, \infty). \qquad \square$$

EXAMPLE 4

Solve: $4x(x + 1)(x - 1) < 0$.

Solution   The solutions of $4x(x + 1)(x - 1) = 0$ are -1, 0, and 1. The product $4x(x + 1)(x - 1)$ is positive or negative, depending on the signs of the factors $4x$, $x + 1$, and $x - 1$. We determine this efficiently using a diagram.

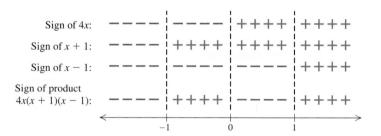

The product of three numbers is negative when it has an odd number of negative factors. We see from the diagram that the solution set is

$$\{x \mid x < -1 \ or \ 0 < x < 1\}, \quad or \quad (-\infty, -1) \cup (0, 1). \qquad \square$$

We now consider our final method for solving polynomial inequalities. In Example 4, we see that the intercepts divide the number line into intervals. If a particular function has a positive output for one number in an interval, it will be positive for *all* the numbers in the interval. Thus we can merely make a test substitution in each interval to solve the inequality.

EXAMPLE 5

Solve: $x^2 + 3x - 10 < 0$.

Solution   We set the polynomial equal to 0 and solve. The solutions of $x^2 + 3x - 10 = 0$ or $(x + 5)(x - 2) = 0$ are -5 and 2. We locate them on a number line as follows. Note that the numbers divide the number line into three intervals: A, B, and C.

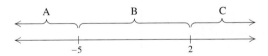

We pick a test number in interval A, say -7, and substitute -7 for x in the function $f(x) = x^2 + 3x - 10$:

$$f(-7) = (-7)^2 + 3(-7) - 10$$
$$= 49 - 21 - 10 = 18. \qquad f(-7) \ is \ positive.$$

Since $f(-7) > 0$, the function will be positive for any number in interval A.

Next we try a test number in interval B, say 1, and find $f(1)$:

$$f(1) = 1^2 + 3(1) - 10 = -6. \qquad f(1) \text{ is negative.}$$

Since $f(1) < 0$, the function will be negative for any number in interval B.

Next we try a test number in interval C, say 4, and find $f(4)$:

$$f(4) = (4)^2 + 3(4) - 10$$
$$= 16 + 12 - 10 = 18. \qquad f(4) \text{ is positive.}$$

Since $f(4) > 0$, the function will be positive for any number in interval C. We are looking for numbers x for which $x^2 + 3x - 10 < 0$. Thus any number x in interval B is a solution. If the inequality had been $\leq$ or $\geq$, we would also have included the intercepts -5 and 2. The solution set is $\{x | -5 < x < 2\}$. ❑

Let us review the last method.

To solve a polynomial inequality:

1. Get 0 on one side, set the polynomial on the other side equal to 0, and solve to find the intercepts.
2. Use the numbers found in step (1) to divide the number line into intervals.
3. Substitute a number from each interval into the related function. If the output is positive, then the function will be positive for all numbers in the interval. If the output is negative, then the function will be negative for all numbers in the interval.
4. Select the intervals for which the inequality is satisfied and write set-builder notation or interval notation for the solution set. Include the intercepts in the solution set if the inequality sign is $\leq$ or $\geq$.

TECHNOLOGY
CONNECTION

To solve the polynomial inequality $2.3x^2 \leq 9.11 - 2.94x$, we first rewrite the inequality in the form $2.3x^2 + 2.94x - 9.11 \leq 0$ and draw the graph of the function $f(x) = 2.3x^2 + 2.94x - 9.11$.

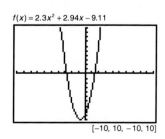
$f(x) = 2.3x^2 + 2.94x - 9.11$
[−10, 10, −10, 10]

To find the values of x for which $f(x) \leq 0$, we focus on the region in which the graph lies *on or below* the x-axis. From this graph, it appears that this region begins somewhere between -3 and -2, and continues to somewhere between

1 and 2. By zooming in on the region between -3 and -2, we can determine the left-hand endpoint. To two decimal places, it is -2.73. Similarly, we can determine that the right-hand endpoint is approximately 1.45. The solution set is approximately $\{x | -2.73 \leq x \leq 1.45\}$.

Had the inequality been $2.3x^2 > 9.11 - 2.94x$, we would look for portions of the graph that lie *above* the x-axis. An approximate solution set of such an inequality would be $\{x | x < -2.73 \text{ or } x > 1.45\}$.

Use a grapher to solve the following inequalities to the nearest hundredth.

TC1. $4.32x^2 - 3.54x - 5.34 \leq 0$

TC2. $7.34x^2 - 16.55x - 3.89 \geq 0$

TC3. $10.85x^2 + 4.28x + 4.44 > 7.91x^2 + 7.43x + 13.03$

TC4. $5.79x^3 - 5.68x^2 + 10.68x - 3.45$
$$> 2.11x^3 + 16.90x - 15.14$$

EXAMPLE 6

Solve: $7x(x + 3)(x - 2) \geqslant 0$.

Solution The solutions of $f(x) = 7x(x + 3)(x - 2) = 0$ are -3, 0, and 2. They divide the real-number line into four intervals as shown below.

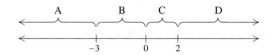

We try test numbers in each interval:

A: Test -5, $f(-5) = 7(-5)(-5 + 3)(-5 - 2) = -490$; ← Negative
B: Test -2, $f(-2) = 7(-2)(-2 + 3)(-2 - 2) = 56$; ← Positive
C: Test 1, $f(1) = 7(1)(1 + 3)(1 - 2) = -28$; ← Negative
D: Test 3, $f(3) = 7(3)(3 + 3)(3 - 2) = 126$. ← Positive

Since the inequality symbol is $\geqslant$, we must include the intercepts. The solution set of the inequality is

$$\{x|\ -3 \leqslant x \leqslant 0 \ or \ 2 \leqslant x\}, \quad or \quad [-3, 0] \cup [2, \infty).$$

Rational Inequalities

We adapt the preceding method when an inequality involves rational expressions. We call these **rational inequalities.**

EXAMPLE 7

Solve: $\dfrac{x - 3}{x + 4} \geqslant 2$.

Solution We write the related equation by changing the $\geqslant$ symbol to $=$:

$$\frac{x - 3}{x + 4} = 2.$$

Then we solve this related equation. We multiply on both sides of the equation by the LCD, which is $x + 4$:

$$(x + 4) \cdot \frac{x - 3}{x + 4} = (x + 4) \cdot 2$$
$$x - 3 = 2x + 8$$
$$-11 = x. \qquad \text{Solving for } x.$$

In the case of rational inequalities, we also need to determine those values that make the denominator 0. We set the denominator equal to 0 and solve:

$$x + 4 = 0$$
$$x = -4.$$

Now we use the numbers -11 and -4 to divide the number line into intervals:

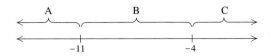

We try test numbers in each interval to see if each satisfies the original inequality

$$\frac{x-3}{x+4} \geqslant 2.$$

A: Test -15, $\dfrac{-15-3}{-15+4} = \dfrac{-18}{-11}$

$= \dfrac{18}{11} \not\geqslant 2$ -15 *is not* a solution, so interval A is not part of the solution set.

B: Test -8, $\dfrac{-8-3}{-8+4} = \dfrac{-11}{-4}$

$= \dfrac{11}{4} \geqslant 2$ -8 *is* a solution, so interval B is part of the solution set.

C: Test 1, $\dfrac{1-3}{1+4} = \dfrac{-2}{5}$

$= -\dfrac{2}{5} \not\geqslant 2$ 1 *is not* a solution, so interval C is not part of the solution set.

The solution set includes the interval B. The number -11 is also included since the inequality symbol is $\geqslant$ and -11 is a solution of the related equation. The number -4 is *not* included since $(x-3)/(x+4)$ is undefined for $x = -4$. Thus the solution set of the original inequality is

$$\{x|\ -11 \leqslant x < -4\}, \quad \text{or} \quad [-11, -4). \qquad \square$$

There is an interesting visual interpretation of Example 7. If we graph the function $f(x) = (x-3)/(x+4)$, we see that the solutions of the inequality $(x-3)/(x+4) \geqslant 2$ can be found by inspection. We simply sketch the line $y = 2$ and locate all x-values for which $f(x) \geqslant 2$.

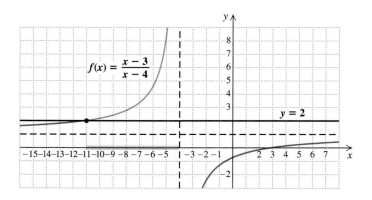

Because graphing rational functions can be very time-consuming, we normally just use test values.

To solve a rational inequality:

1. Change the inequality symbol to an equals sign and solve the related equation.
2. Find any replacements for which the rational expression is undefined.
3. Use the numbers found in steps (1) and (2) to divide the number line into intervals.
4. Substitute a number from each interval into the inequality. If the number is a solution, then the interval to which it belongs is part of the solution set.
5. Select the intervals for which the inequality is satisfied and write set-builder notation or interval notation for the solution set. If the inequality symbol is $\leq$ or $\geq$, then the solutions to step (1) should also be included in the solution set.

EXERCISE SET | 8.10

Solve.

1. $(x - 5)(x + 3) > 0$

2. $(x - 4)(x + 1) > 0$

3. $(x + 1)(x - 2) \leq 0$

4. $(x - 5)(x + 3) \leq 0$

5. $x^2 - x - 2 < 0$

6. $x^2 + x - 2 < 0$

7. $9 - x^2 \leq 0$

8. $4 - x^2 \geq 0$

9. $x^2 - 2x + 1 \geq 0$

10. $x^2 + 6x + 9 < 0$

11. $x^2 + 8 < 6x$

12. $x^2 - 12 > 4x$

13. $3x(x + 2)(x - 2) < 0$

14. $5x(x + 1)(x - 1) > 0$

15. $(x + 3)(x - 2)(x + 1) > 0$

16. $(x - 1)(x + 2)(x - 4) < 0$

17. $(x + 3)(x + 2)(x - 1) < 0$

18. $(x - 2)(x - 3)(x + 1) < 0$

19. $\dfrac{1}{x - 4} < 0$

20. $\dfrac{1}{x + 5} > 0$

21. $\dfrac{x + 1}{x - 3} > 0$

22. $\dfrac{x - 2}{x + 5} < 0$

23. $\dfrac{3x + 2}{x - 3} \leq 0$

24. $\dfrac{5 - 2x}{4x + 3} \leq 0$

25. $\dfrac{x - 1}{x - 2} > 3$

26. $\dfrac{x + 1}{2x - 3} < 1$

27. $\dfrac{(x - 2)(x + 1)}{x - 5} < 0$

28. $\dfrac{(x + 4)(x - 1)}{x + 3} > 0$

29. $\dfrac{x}{x - 2} \geq 0$

30. $\dfrac{x + 3}{x} \leq 0$

31. $\dfrac{x - 5}{x} < 1$

32. $\dfrac{x}{x - 1} > 2$

33. $\dfrac{x - 1}{(x - 3)(x + 4)} < 0$

34. $\dfrac{x + 2}{(x - 2)(x + 7)} > 0$

35. $2 < \dfrac{1}{x}$

36. $\dfrac{1}{x} \leq 3$

Skill Maintenance

37. Multiply and simplify: $\sqrt[5]{a^2 b} \ \sqrt[3]{ab^2}$.

38. The perimeter of an equilateral triangle is the same as that of a square. If the sides of the triangle are 2 units longer than the sides of the square, how long is each side of the square?

Synthesis

39. ◈ Explain how any quadratic inequality can be solved by examining a parabola that opens upward.

Solve.

40. $x^2 + 2x > 4$

41. $x^4 + 2x^2 \geq 0$

42. $x^4 + 3x^2 \leq 0$

43. $\left| \dfrac{x + 2}{x - 1} \right| < 3$

44. *Total profit.*   A company determines that its total-profit function is given by

$$P(x) = -3x^2 + 630x - 6000.$$

a) A company makes a profit for those nonnegative values of x for which $P(x) > 0$. Find the values of x for which the company makes a profit.

b) A company loses money for those nonnegative values of x for which $P(x) < 0$. Find the values of x for which the company loses money.

45. *Height of a thrown object.*   The function

$$S(t) = -16t^2 + 32t + 1920$$

gives the height S, in feet, of an object thrown from a cliff that is 1920 ft high. Here t is the time, in seconds, that the object is in the air.

a) For what times is the height greater than 1920 ft?

b) For what times is the height less than 640 ft?

46. *Number of handshakes.*   There are n people in a room. The number N of possible handshakes by the people is given by the function

$$N(n) = \frac{n(n-1)}{2}.$$

For what number of people n is

$$66 \leqslant N \leqslant 300?$$

47. *Number of diagonals.*   A polygon with n sides has D diagonals, where D is given by the function

$$D(n) = \frac{n(n-3)}{2}.$$

Find the number of sides n if

$$27 \leqslant D \leqslant 230.$$

Use a grapher to draw the function and find solutions of $f(x) = 0$. Then solve the inequalities $f(x) < 0$ and $f(x) > 0$.

48. $f(x) = x^3 - 2x^2 - 5x + 6$

49. $f(x) = \frac{1}{3}x^3 - x + \frac{2}{3}$

50. $f(x) = x + \frac{1}{x}$

51. $f(x) = x - \sqrt{x}, \; x \geqslant 0$

52. $f(x) = x^4 - 4x^3 - x^2 + 16x - 12$

53. $f(x) = \frac{x^3 + x^2 - 2x}{x^2 + x - 6}$

54. Use a grapher to solve Exercises 11, 25, and 35 by drawing two curves, one for each side of the inequality.

SUMMARY AND REVIEW | 8

KEY TERMS

Quadratic equation, p. 398
Principle of square roots, p. 399
Completing the square, p. 401
Compounding interest annually, p. 404
The quadratic formula, p. 407
Discriminant, p. 417
Reducible to quadratic, p. 420
Direct variation, p. 425

Variation constant, p. 425
Constant of proportionality, p. 425
Inverse variation, p. 426
Joint variation, p. 428
Quadratic function, p. 433
Parabola, p. 434
Line of symmetry, p. 434

Vertex, p. 434
Minimum value, p. 437
Maximum value, p. 437
x-intercept, p. 444
Polynomial inequality, p. 456
Quadratic inequality, p. 456
Rational inequality, p. 461

IMPORTANT PROPERTIES AND FORMULAS

The Principle of Square Roots

For any real number k, if $x^2 = k$, then $x = \sqrt{k}$ or $x = -\sqrt{k}$.

To solve a quadratic equation in x by completing the square:

1. Isolate the terms with variables on one side of the equation, and arrange them in descending order.
2. Divide by the coefficient of x^2 on both sides if that coefficient is not 1.
3. Complete the square by taking half of the coefficient of x and adding its square on both sides.
4. Express one side as the square of a binomial. Find a common denominator on the other side and simplify.
5. Use the principle of square roots.
6. Solve for x by adding appropriately on both sides.

The Quadratic Formula

The solutions of $ax^2 + bx + c = 0$, $a \neq 0$, are given by

$$x = \frac{-b \pm \sqrt{b^2 - 4ac}}{2a}.$$

To solve a quadratic equation:

1. Check for the form $ax^2 = p$ or $(x + k)^2 = d$. If it is in either of these forms, use the principle of square roots.
2. If it is not in the form of step (1), write it in standard form $ax^2 + bx + c = 0$.
3. Try factoring and using the principle of zero products.
4. If it is not possible to factor or factoring seems difficult, use the quadratic formula.

The solutions of a quadratic equation can always be found using the quadratic formula. They cannot always be found by factoring.

Discriminant $b^2 - 4ac$	Nature of Solutions
0	Only one solution; it is a real number
Positive	Two different real-number solutions
Negative	Two different imaginary-number solutions (complex conjugates)

Variation

y varies directly as x if there is some nonzero constant k such that $y = kx$.

y varies inversely as x if there is some nonzero constant k such that $y = k/x$.

y varies jointly as x and z if there is some nonzero constant k such that $y = kxz$.

The graph of $g(x) = ax^2$ is a parabola with the vertical axis as its line of symmetry and its vertex at the origin.

If a is positive, the parabola opens upward; if a is negative, the parabola opens downward.

If $|a|$ is greater than 1, the parabola is narrower than $f(x) = x^2$.

If $|a|$ is between 0 and 1, the parabola is flatter than $f(x) = x^2$.

The graph of $g(x) = a(x - h)^2$ has the same shape as the graph of $y = ax^2$.

If h is positive, the graph of $y = ax^2$ is shifted h units to the right.

If h is negative, the graph of $y = ax^2$ is shifted $|h|$ units to the left.

The vertex is $(h, 0)$, and the line of symmetry is $x = h$.

The graph of $f(x) = a(x - h)^2 + k$ looks like that of $y = a(x - h)^2$ translated up or down.

If k is positive, the graph of $y = a(x - h)^2$ is shifted k units up.

If k is negative, the graph of $y = a(x - h)^2$ is shifted $|k|$ units down.

The vertex is (h, k), and the line of symmetry is $x = h$.

For $a > 0$, k is the minimum function value. For $a < 0$, k is the maximum function value.

For a parabola given by a quadratic function $f(x) = ax^2 + bx + c$, the vertex of the parabola is

$$\left(-\frac{b}{2a}, \frac{4ac - b^2}{4a}\right).$$

The x-coordinate of the vertex is $-b/(2a)$. The line of symmetry is $x = -b/(2a)$. The second coordinate of the vertex can be found by substituting into the formula above, but is usually found most easily by evaluating the function at $-b/(2a)$.

To solve a formula for a letter, say, b:

1. Clear the fractions and use the principle of powers, as needed, until b does not appear in any radicand or denominator. (In some cases you may clear the fractions first, and in some cases you may use the principle of powers first. Perform these steps until radicals containing b are gone and b is not in any denominator.)
2. Collect all terms with b^2 in them. Also collect all terms with b in them.
3. If b^2 does not appear, you can finish by using just the addition and multiplication principles. Get all terms containing b on one side of the equation; then factor out b. Dividing on both sides will then get b alone.
4. If b^2 appears but b does not appear to the first pow-er, solve the equation for b^2. Then take square roots on both sides.
5. If there are terms containing both b and b^2, put the equation in standard form and use the quadratic formula.

To solve a polynomial inequality:

1. Get 0 on one side, set the polynomial on the other side equal to 0, and solve to find the intercepts.
2. Use the numbers found in step (1) to divide the number line into intervals.
3. Substitute a number from each interval into the related function. If the output is positive, then the function will be positive for all numbers in the interval. If the output is negative, then the function will be negative for all numbers in the interval.
4. Select the intervals for which the inequality is satisfied and write set-builder notation or interval notation for the solution set. Include the intercepts in the solution set if the inequality sign is $\leq$ or $\geq$.

To solve a rational inequality:

1. Change the inequality symbol to an equals sign and solve the related equation.
2. Find any replacements for which the rational expression is undefined.
3. Use the numbers found in steps (1) and (2) to divide the number line into intervals.
4. Substitute a number from each interval into the inequality. If the number is a solution, then the interval to which it belongs is part of the solution set.
5. Select the intervals for which the inequality is satisfied and write set-builder notation or interval notation for the solution set. If the inequality symbol is $\leq$ or $\geq$, then the solutions to step (1) should also be included in the solution set.

REVIEW EXERCISES

This chapter's review and test include Skill Maintenance exercises from Sections 3.3, 6.2, 7.3, and 7.7.

Solve.

1. $2x^2 - 7 = 0$ **2.** $14x^2 + 5x = 0$

3. $x^2 - 12x + 36 = 9$ **4.** $4x^2 + 3x + 1 = 0$

5. $x^2 - 7x + 13 = 0$

6. $4x(x - 1) + 15 = x(3x + 4)$

7. $x^2 + 4x + 1 = 0$.   Approximate the solutions to the nearest tenth.

Complete the square. Then write the trinomial square in factored form.

8. $x^2 - 12x$ **9.** $x^2 + \frac{3}{5}x$

Solve by completing the square. Show your work.

10. $x^2 - 2x - 8 = 0$ **11.** $x^2 - 6x + 1 = 0$

12. $2500 grows to $3025 in 2 years. Use the formula $A = P(1 + r)^t$ to find the interest rate.

13. The Peachtree Center Plaza in Atlanta, Georgia, is 723 ft tall. Use the formula $s = 16t^2$ to approximate how long it would take an object to fall from the top.

Solve.

14. $\dfrac{x}{x - 2} + \dfrac{4}{x - 6} = 0$ **15.** $\dfrac{x}{5} = \dfrac{x + 3}{x + 7}$

16. $\dfrac{x}{4} - \dfrac{4}{x} = 2$ **17.** $15 + \dfrac{6}{x - 2} = \dfrac{8}{x + 2}$

18. During the first part of a trip, Nina's Nissan traveled 50 mi at a certain speed. Nina drove 80 mi on the second part of the trip at a speed that was 10 mph slower. The total time for the trip was 3 hr. What was the speed on each part of the trip?

19. Working together, Jean and Stacy can cut and split a cord of wood in 4 hr. Working alone, Jean takes 6 hr more than Stacy. How long would it take Stacy to do this job alone?

Determine the nature of the solutions of the equation.

20. $x^2 + 3x - 6 = 0$ **21.** $x^2 + 2x + 5 = 0$

22. Write a quadratic equation having the solutions $\frac{1}{5}$, $-\frac{3}{5}$.

23. Write a quadratic equation having -4 as its only solution.

Solve.

24. $x^4 - 13x^2 + 36 = 0$

25. $15x^{-2} - 2x^{-1} - 1 = 0$

26. $(x^2 - 4)^2 - (x^2 - 4) - 6 = 0$

27. The power P expended by heat in an electric circuit of fixed resistance varies directly as the square of the current C in the circuit. A circuit expends 180 watts when a current of 6 amperes is flowing. What is the heat expended when the current is 10 amperes?

28. A warning dye is used by people in lifeboats to aid searching airplanes. The radius r of the circle formed by the dye varies directly as the square root of the volume V. It is found that 4 L of dye will spread to a circle of radius 5 m. How much dye is needed to form a circle with a 20-m radius?

29. **a)** Graph: $f(x) = -3(x + 2)^2 + 4$.
   b) Label the vertex.
   c) Draw the line of symmetry.
   d) Find the maximum or the minimum value.

30. For the function $f(x) = 2x^2 - 12x + 23$:
   a) find the vertex and the line of symmetry;
   b) graph the function.

31. Find the x-intercepts: $f(x) = x^2 - 9x + 14$.

32. An object is dropped 120.1 m from an airplane. How long does it take to reach the ground? Round to the nearest second.

33. Solve $N = 3\pi\sqrt{1/p}$ for p.

34. Solve $2A + T = 3T^2$ for T.

35. What is the minimum product of two numbers whose difference is 22? What numbers yield this product?

36. Find the quadratic function that fits the data points $(0, -2)$, $(1, 3)$, and $(3, 7)$.

Solve.

37. $x^2 < 6x + 7$ **38.** $\dfrac{x - 5}{x + 3} \leq 0$

Skill Maintenance _____

39. Metal alloy A is 75% silver. Metal alloy B is 25% silver. How much of each should be mixed in order to produce 300 kg of an alloy that is 60% silver?

40. Solve: $\sqrt{5x-1} + \sqrt{2x} = 5$.

41. Add: $\dfrac{x}{x^2 - 3x + 2} + \dfrac{2}{x^2 - 5x + 6}$.

42. Multiply and simplify: $\sqrt[3]{9t^6}\ \sqrt[3]{3s^4t^9}$.

Synthesis _____

43. ◈   Discuss two ways in which completing the square was used in this chapter.

44. ◈   Explain why, when solving a polynomial inequality, if a function has a positive output for one number in an interval, it will be positive for *all* the numbers in the interval.

45. Solve:
$$\frac{26}{x+13} - \frac{14}{x+7} = \frac{12x}{x^2 + 20x + 91}.$$

46. Find h and k if, for $3x^2 - hx + 4k = 0$, the sum of the solutions is 20 and the product is 80.

47. The average of two positive integers is 171. One of the numbers is the square root of the other. Find the integers.

CHAPTER TEST | 8

Solve.

1. $3x^2 - 4 = 0$

2. $4x(x-2) - 3x(x+1) = -18$

3. $x^2 + x + 1 = 0$

4. $x^2 + 4x = 2$. Approximate the solutions to the nearest tenth, using a calculator or Table 1.

5. $\dfrac{1}{4-x} + \dfrac{1}{2+x} = \dfrac{3}{4}$

6. $x^4 - 5x^2 + 5 = 0$

Complete the square. Then write the trinomial square in factored form.

7. $x^2 + 14x$ **8.** $x^2 - \frac{2}{7}x$

Solve by completing the square. Show your work.

9. $x^2 + 3x - 18 = 0$

10. $x^2 + 10x + 15 = 0$

Solve.

11. A river flows at a rate of 4 km/h. A boat travels 60 km upriver and returns in a total time of 8 hr. What is the speed of the boat in still water?

12. Two pipes can fill a tank in $1\frac{1}{2}$ hr. One pipe requires 4 hr longer running alone to fill the tank than the

other. How long would it take for the faster pipe, working alone, to fill the tank?

13. Determine the nature of the solutions of the equation $x^2 + 5x + 17 = 0$.

14. Write a quadratic equation having solutions $\sqrt{3}$ and $3\sqrt{3}$.

15. The surface area of a balloon varies directly as the square of its radius. The area is 3.4 in^2 when the radius is 5 in. What is the area when the radius is 7 in.?

16. **a)** Graph: $f(x) = 4(x-3)^2 + 5$.
  b) Label the vertex.
  c) Draw the line of symmetry.
  d) Find the maximum or the minimum function value.

17. For the function $f(x) = 2x^2 + 4x - 6$:
  a) find the vertex and the line of symmetry;
  b) graph the function.

18. Find the x-intercepts of $f(x) = x^2 - x - 6$.

19. Solve $V = \frac{1}{3}\pi(R^2 + r^2)$ for r.

20. What is the minimum product of two numbers having a difference of 8?

21. Find the quadratic function that fits the data points $(0, 0)$, $(3, 0)$, and $(5, 2)$.

22. Solve: $(x+2)(x-1)(x-2) > 0$.

Skill Maintenance _____

23. Solve: $\sqrt{x + 3} = x - 3$.

24. Multiply and simplify: $\sqrt[4]{2a^2b^3}\,\sqrt[4]{a^4b}$.

25. Subtract and simplify:

$$\frac{x}{x^2 + 15x + 56} - \frac{7}{x^2 + 13x + 42}.$$

26. The perimeter of a hexagon with all six sides the same length is the same as the perimeter of a square. One side of the hexagon is 3 less than a side of the square. Find the perimeter of each polygon.

Synthesis _____

27. One solution of $kx^2 + 3x - k = 0$ is -2. Find the other solution.

28. Solve:

$$\frac{88}{x - 11} - \frac{56}{x - 7} = \frac{32x}{x^2 - 18x + 77}.$$

Exponential and Logarithmic Functions

Because he objected to smoking, and because his first baseball card was issued in cigarette packs, the great shortstop Honus Wagner halted production of his card before many were produced. One of these cards was sold in 1986 for $110,000 and again in 1991 for $451,000.

In Section 9.7, we will use exponential functions to predict the card's value in future years.

Joshua Evans
LELAND'S AUCTION HOUSE

"Sports memorabilia ranging from cards to uniforms to seats in old stadiums often increase in value exponentially. By monitoring the past prices paid for certain items, we can use mathematics to estimate future prices for those same items."

The functions that we consider in this chapter are interesting not only from a purely intellectual point of view, but also for their rich applications to many fields. We will look at such applications as compound interest and population growth, to name just two.

The basis of the theory concerns exponents. We define some functions having variable exponents (*exponential functions*); the rest follows from those functions and their properties.

In addition to material from this chapter, the review and test for Chapter 9 include material from Sections 6.3, 7.9, 8.5, and 8.9.

9.1

Exponential Functions

Graphing Exponential Functions • **Equations with *x* and *y* Interchanged** • **Applications of Exponential Functions**

The following graph shows the number of cases of AIDS reported and the year in which that number was reported.

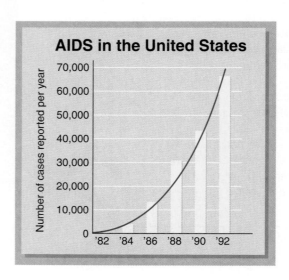

A curve drawn along the graph would approximate the graph of an *exponential function*. We now consider such graphs. We will also study graphs and properties of *logarithmic functions*, which are closely related to the exponential functions. These new functions will appear in a variety of applications.

Graphing Exponential Functions

In Chapter 7, we gave meaning to exponential expressions with rational number exponents, such as

$$5^{1/4}, \qquad 3^{-3/4}, \qquad 7^{2.34}, \qquad 5^{1.73}.$$

For example, $5^{1.73}$, or $5^{173/100}$, represents the 100th root of 5 raised to the 173rd power. We now give meaning to expressions with irrational exponents, such as

$$5^{\sqrt{3}}, \qquad 7^{\pi}, \qquad 9^{-\sqrt{2}}.$$

Consider $5^{\sqrt{3}}$. Let us think of rational numbers r close to $\sqrt{3}$ and look at 5^r. As r gets closer to $\sqrt{3}$, 5^r gets closer to some real number.

r closes in on $\sqrt{3}$.	5^r closes in on some real number p.
$1 < r < 2$	$5 = 5^1 < p < 5^2 = 25$
$1.7 < r < 1.8$	$15.426 = 5^{1.7} < p < 5^{1.8} = 18.119$
$1.73 < r < 1.74$	$16.189 = 5^{1.73} < p < 5^{1.74} = 16.452$
$1.732 < r < 1.733$	$16.241 = 5^{1.732} < p < 5^{1.733} = 16.267$

As r closes in on $\sqrt{3}$, 5^r closes in on some real number p. We define $5^{\sqrt{3}}$ to be the number p. To seven decimal places,

$$5^{\sqrt{3}} \approx 16.2424508.$$

Any positive irrational exponent can be defined in a similar way. Negative irrational exponents are then defined in the same way as negative integer exponents. Thus the expression a^x has meaning for *any* real number x. The general laws of exponents still hold, but we will not prove that here. We now define exponential functions.

Exponential Function

The function $f(x) = a^x$, where a is a positive constant, $a \neq 1$, is called the *exponential function*, base a.

We require the base a to be positive to avoid the imaginary numbers that would result from taking even roots of negative numbers. The restriction $a \neq 1$ is made to exclude the constant function $f(x) = 1^x$, or $f(x) = 1$.

The following are examples of exponential functions:

$$f(x) = 2^x, \qquad f(x) = \left(\tfrac{1}{2}\right)^x, \qquad f(x) = (0.4)^x.$$

Note that, in contrast to polynomial functions like $f(x) = x^2$ and $f(x) = x^3$, the variable in an exponential function is in the *exponent*. Let us consider graphs of exponential functions.

EXAMPLE 1

Graph the exponential function $y = f(x) = 2^x$.

Solution We compute some function values, thinking of y as $f(x)$, and list the results in a table. It is a good idea to start by letting $x = 0$.

$f(0) = 2^0 = 1;$ $\qquad f(-1) = 2^{-1} = \dfrac{1}{2^1} = \dfrac{1}{2};$

$f(1) = 2^1 = 2;$ $\qquad f(-2) = 2^{-2} = \dfrac{1}{2^2} = \dfrac{1}{4};$

$f(2) = 2^2 = 4;$

$f(3) = 2^3 = 8;$ $\qquad f(-3) = 2^{-3} = \dfrac{1}{2^3} = \dfrac{1}{8}.$

x	y, or $f(x)$
0	1
1	2
2	4
3	8
−1	$\frac{1}{2}$
−2	$\frac{1}{4}$
−3	$\frac{1}{8}$

Next, we plot these points and connect them with a smooth curve.

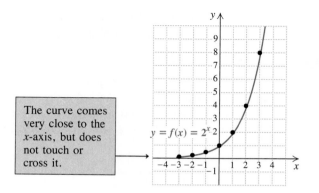

Be sure to plot enough points to determine how steeply the curve rises.

The curve comes very close to the x-axis, but does not touch or cross it.

$y = f(x) = 2^x$

Note that as x increases, the function values increase indefinitely. As x decreases, the function values decrease, getting very close to 0. The x-axis, or the line $y = 0$, is a horizontal *asymptote*, meaning that the curve gets closer and closer to this line the further we move to the left.   ❏

EXAMPLE 2

Graph the exponential function $y = f(x) = \left(\frac{1}{2}\right)^x$.

Solution   We compute some function values, thinking of y as $f(x)$, and list the results in a table. Before we do this, note that

$$y = f(x) = \left(\tfrac{1}{2}\right)^x = (2^{-1})^x = 2^{-x}.$$

Then we have

$f(0) = 2^{-0} = 1;$

$f(1) = 2^{-1} = \dfrac{1}{2^1} = \dfrac{1}{2};$

$f(2) = 2^{-2} = \dfrac{1}{2^2} = \dfrac{1}{4};$

$f(3) = 2^{-3} = \dfrac{1}{2^3} = \dfrac{1}{8};$

$f(-1) = 2^{-(-1)} = 2^1 = 2;$

$f(-2) = 2^{-(-2)} = 2^2 = 4;$

$f(-3) = 2^{-(-3)} = 2^3 = 8.$

x	y, or $f(x)$
0	1
1	$\frac{1}{2}$
2	$\frac{1}{4}$
3	$\frac{1}{8}$
−1	2
−2	4
−3	8

We plot these points and draw the curve. Note that this graph is a reflection across the y-axis of the graph in Example 1. The line $y = 0$ is again an asymptote.

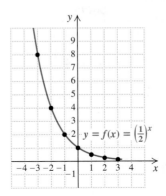

The preceding examples illustrate exponential functions with various bases. Let us list some of their characteristics.

A. When $a > 1$, the function $f(x) = a^x$ increases from left to right. The greater the value of a, the steeper the curve. (See the figure on the left, below.)

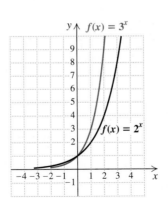
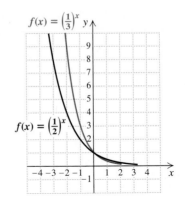

B. When $0 < a < 1$, the function $f(x) = a^x$ decreases from left to right. For smaller values of a, the curve becomes steeper. (See the figure on the right, above.)

C. All exponential functions of the form $f(x) = a^x$ go through the point $(0, 1)$. That is, the y-intercept is $(0, 1)$.

EXAMPLE 3

Graph: $y = f(x) = 2^{x-2}$.

Solution   We construct a table of values. Then we plot the points and connect them with a smooth curve. Be sure to note that $x - 2$ is the *exponent*.

$$f(0) = 2^{0-2} = 2^{-2} = \frac{1}{2^2} = \frac{1}{4} \qquad f(4) = 2^{4-2} = 2^2 = 4$$

$$f(1) = 2^{1-2} = 2^{-1} = \frac{1}{2^1} = \frac{1}{2} \qquad f(-1) = 2^{-1-2} = 2^{-3} = \frac{1}{2^3} = \frac{1}{8}$$

$$f(2) = 2^{2-2} = 2^0 = 1$$

$$f(3) = 2^{3-2} = 2^1 = 2 \qquad f(-2) = 2^{-2-2} = 2^{-4} = \frac{1}{2^4} = \frac{1}{16}$$

TECHNOLOGY CONNECTION

Graphers are especially helpful when graphing exponential functions because bases may not always be whole numbers and function values can quickly become very large. For example, consider the function $y = 5000(1.075)^x$. Because y-values will always be positive and will become quite large, an appropriate window for this function might be $[-10, 10, 0, 15000]$, where the *scale* of the y-axis is 1000.

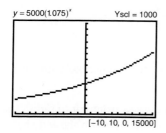

$y = 5000(1.075)^x$ Yscl = 1000

$[-10, 10, 0, 15000]$

Use a grapher to draw the graph of each function. Select an appropriate viewing box and scale.

TC1. $y = 9.34^x$

TC2. $y = 9.34^{-x}$

TC3. $y = 3.45^{x+5}$

TC4. $y = 8.11^{4.2-x}$

TC5. $y = 20,000(1.1225)^x$

TC6. $y = 100,000(0.25)^{0.5x}$

x	y, or $f(x)$
0	$\frac{1}{4}$
1	$\frac{1}{2}$
2	1
3	2
4	4
-1	$\frac{1}{8}$
-2	$\frac{1}{16}$

The graph looks just like the graph of $y = 2^x$, but it is translated 2 units to the right. The y-intercept of $y = 2^x$ is $(0, 1)$. The y-intercept of $y = 2^{x-2}$ is $(0, \frac{1}{4})$. The line $y = 0$ is again the asymptote.

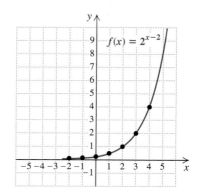

Equations with x and y Interchanged

It will be helpful in later work if we are able to graph an equation in which the x and y in $y = a^x$ have been interchanged.

EXAMPLE 4

Graph: $x = 2^y$.

Solution Note that x is alone on one side of the equation. We can find ordered pairs that are solutions by choosing values for y and then computing values for x.

For $y = 0$, $\quad x = 2^0 = 1$.

For $y = 1$, $\quad x = 2^1 = 2$.

For $y = 2$, $\quad x = 2^2 = 4$.

For $y = 3$, $\quad x = 2^3 = 8$.

For $y = -1$, $x = 2^{-1} = \frac{1}{2^1} = \frac{1}{2}$.

For $y = -2$, $x = 2^{-2} = \frac{1}{2^2} = \frac{1}{4}$.

For $y = -3$, $x = 2^{-3} = \frac{1}{2^3} = \frac{1}{8}$.

x	y
1	0
2	1
4	2
8	3
$\frac{1}{2}$	-1
$\frac{1}{4}$	-2
$\frac{1}{8}$	-3

(1) Choose values for y.
(2) Compute values for x.

We plot the points and connect them with a smooth curve.

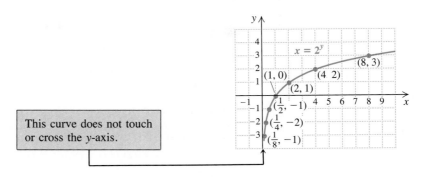

This curve does not touch or cross the y-axis.

Note too that this curve looks just like the graph of $y = 2^x$, except that it is reflected across the line $y = x$, as shown here.

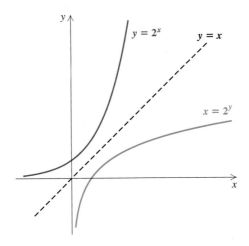

Applications of Exponential Functions

EXAMPLE 5

Interest compounded annually. The amount of money A that a principal P will be worth after t years at interest rate i, compounded annually, is given by the formula

$$A = P(1 + i)^t.$$

Suppose that $100,000 is invested at 8% interest, compounded annually.

a) Find a function for the amount in the account after t years.
b) Find the amount of money in the account at $t = 0$, $t = 4$, $t = 8$, and $t = 10$.
c) Graph the function.

Solution

a) If $P = \$100,000$ and $i = 8\% = 0.08$, we can substitute these values and form the following function:

$$A(t) = \$100,000(1 + 0.08)^t$$
$$= \$100,000(1.08)^t.$$

b) To find the function values, a calculator with a power key is helpful.

$$A(0) = \$100,000(1.08)^0$$
$$= \$100,000(1)$$
$$= \$100,000$$

$$A(4) = \$100,000(1.08)^4$$
$$= \$100,000(1.36048896)$$
$$\approx \$136,048.90$$

$$A(8) = \$100,000(1.08)^8$$
$$\approx \$100,000(1.85093021)$$
$$\approx \$185,093.02$$

$$A(10) = \$100,000(1.08)^{10}$$
$$\approx \$100,000(2.158924997)$$
$$\approx \$215,892.50$$

c) We use the function values computed in part (b), and others if we wish, and draw the graph as follows. Note that the axes are scaled differently because of the large numbers.

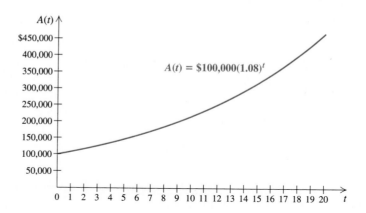

Graph.

1. $y = f(x) = 2^x$

2. $y = f(x) = 3^x$

3. $y = 5^x$

4. $y = 6^x$

5. $y = 2^{x+1}$

6. $y = 2^{x-1}$

7. $y = 3^{x-2}$

8. $y = 3^{x+2}$

9. $y = 2^x - 3$

10. $y = 2^x + 1$

11. $y = 5^{x+3}$

12. $y = 6^{x-4}$

13. $y = \left(\frac{1}{2}\right)^x$

14. $y = \left(\frac{1}{3}\right)^x$

15. $y = \left(\frac{1}{5}\right)^x$

16. $y = \left(\frac{1}{4}\right)^x$

17. $y = 2^{2x-1}$

18. $y = 3^{4-x}$

19. $y = 2^{x-1} - 3$

20. $y = 2^{x+3} - 4$

21. $x = 3^y$

22. $x = 6^y$

23. $x = \left(\frac{1}{2}\right)^y$

24. $x = \left(\frac{1}{3}\right)^y$

25. $x = 5^y$

26. $x = 4^y$

27. $x = \left(\frac{2}{3}\right)^y$

28. $x = \left(\frac{4}{3}\right)^y$

Graph both equations using the same set of axes.

29. $y = 2^x$, $x = 2^y$ **30.** $y = 3^x$, $x = 3^y$

31. $y = \left(\frac{1}{2}\right)^x$, $x = \left(\frac{1}{2}\right)^y$ **32.** $y = \left(\frac{1}{4}\right)^x$, $x = \left(\frac{1}{4}\right)^y$

Solve.

33. ▦ *Cases of AIDS.* The total number of Americans who have contracted AIDS is approximated by the exponential function

$$N(t) = 100,000(1.4)^t,$$

where $t = 0$ corresponds to 1989.

a) According to the function, how many Americans had been infected as of 1993?
b) Predict the total number of Americans who will have been infected by 1998.
c) Graph the function.

34. ▦ *Growth of bacteria.* The bacteria *Escherichi coli* are commonly found in the bladder of human beings. Suppose that 3000 of the bacteria are present at time $t = 0$. Then t minutes later, the number of bacteria present will be

$$N(t) = 3000(2)^{t/20}.$$

a) How many bacteria will be present after 10 min? 20 min? 30 min? 40 min? 60 min?
b) Graph the function.

35. ▦ *Recycling aluminum cans.* It is estimated that $\frac{2}{3}$ of all aluminum cans distributed will be recycled each year. A beverage company distributes 250,000 cans. The number still in use after time t, in years, is given by the exponential function

$$N(t) = 250,000\left(\frac{2}{3}\right)^t.$$

a) How many cans are still in use after 0 years? 1 year? 4 years? 10 years?
b) Graph the function.

36. ▦ *Salvage value.* A photocopier is purchased for $5200. Its value each year is about 80% of the value of the preceding year. Its value, in dollars, after t years is given by the exponential function

$$V(t) = 5200(0.8)^t.$$

a) Find the value of the machine after 0 years, 1 year, 2 years, 5 years, and 10 years.
b) Graph the function.

37. ▦ *Compact discs.* The number of compact discs purchased each year is increasing exponentially. The number N, in millions, purchased is given by the exponential function

$$N(t) = 7.5(6)^{0.5t},$$

where t is the number of years after 1985.

a) Find the number of compact discs sold in 1985, 1986, 1988, 1990, 1995, and 2000.
b) Graph the function.

38. ▦ *Turkey consumption.* The amount of turkey consumed by each person in this country is increasing exponentially. Assuming $t = 0$ corresponds to 1937, the amount of turkey, in pounds per person, consumed t years after 1937 is given by the exponential function

$$N(t) = 2.3(3)^{0.033t}.$$

a) How much turkey was consumed, per person, in 1940? in 1950? in 1980? in 1988?
b) How much will be consumed, per person, in 2000?
c) Graph the function.

Skill Maintenance

39. Multiply and simplify: $x^{-5} \cdot x^3$.

40. Simplify: $(x^{-3})^4$.

41. Divide and simplify: $\dfrac{x^{-3}}{x^4}$.

42. Simplify: 5^0.

Synthesis

43. ◈ Suppose that $1000 is invested for 5 years at 7% interest, compounded annually. In what year will the most interest be earned? Why?

44. ◈ Without using a calculator, explain why 2^π must be greater than 8 but less than 16.

Determine which of the two numbers is larger.

45. $\pi^{1.3}$ or $\pi^{2.4}$ **46.** $\sqrt{8^3}$ or $8^{\sqrt{3}}$

Graph each of the following. You will find a calculator with a power key most helpful.

47. $f(x) = (2.3)^x$ **48.** $f(x) = (3.8)^x$

49. $g(x) = (0.125)^x$ **50.** $g(x) = (0.9)^x$

Graph.

51. $y = 2^x + 2^{-x}$ **52.** $y = \left|\left(\frac{1}{2}\right)^x - 1\right|$

53. $y = 3^x + 3^{-x}$ **54.** $y = 2^{-(x-1)^2}$

55. $y = |2^{x^2} - 1|$ **56.** $y = |2^x - 2|$

Graph both equations using the same set of axes.

57. $y = 3^{-(x-1)}$, $x = 3^{-(y-1)}$

58. $y = 1^x$, $x = 1^y$

59. ▦ *Typing speed.* Jim is studying typing. After he has studied for t hours, Jim's speed, in words per minute, is given by the exponential function

$$S(t) = 200[1 - (0.99)^t].$$

a) How fast can Jim type after he has studied for 10 hours? 20 hours? 40 hours? 85 hours?
b) Graph the function.

9.2

Composite and Inverse Functions

Composite Functions • Inverses and One-to-One Functions •
Finding Formulas for Inverses • Graphing Functions and Their
Inverses • Inverse Functions and Composition

Composite Functions

In the real world, functions frequently occur in which some quantity depends on a variable that, in turn, depends on another variable. For instance, the number of employees hired by a firm may depend on the firm's profits, which may in turn depend on the number of items the firm produces. Functions like this are called **composite functions.**

For example, the function g that gives a correspondence between women's shoe sizes in the United States and those in Italy is given by $g(x) = 2x + 24$, where x is the U.S. size and $g(x)$ is the Italian size. Thus a U.S. size 4 corresponds to a shoe size of $g(4) = 2 \cdot 4 + 24$, or 32, in Italy.

There is also a function that gives a correspondence between women's shoe sizes in Italy and those in Britain. The function is given by $f(x) = \frac{1}{2}x - 14$, where x is the Italian size and $f(x)$ is the corresponding British size. Thus an Italian size 32 corresponds to a British size $f(32) = \frac{1}{2} \cdot 32 - 14$, or 2.

It seems reasonable to conclude that a shoe size of 4 in the United States corresponds to a size of 2 in Britain and that some function h describes this correspondence. Can we find a formula for h? If we look at the following tables, we might guess that such a formula is $h(x) = x - 2$, and that is indeed correct. But, for more complicated formulas, we would need to use algebra.

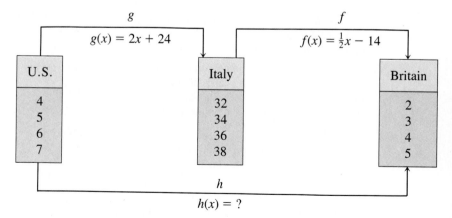

Size x shoes in the United States correspond to size $g(x)$ shoes in Italy, where

$$g(x) = 2x + 24.$$

Size n shoes in Italy correspond to size $f(n)$ shoes in Britain. Thus size $g(x)$ shoes in Italy correspond to size $f(g(x))$ shoes in Britain. Since the x in the expression

$f(g(x))$ represents a U.S. shoe size, we can find the British shoe size that corresponds to a U.S. size x:

$$f(g(x)) = f(2x + 24) = \tfrac{1}{2} \cdot (2x + 24) - 14 \qquad \text{Using } g(x) \text{ as an input}$$
$$= x + 12 - 14 = x - 2.$$

This gives a formula for h: $h(x) = x - 2$. Thus a shoe size of 4 in the United States corresponds to a shoe size of $h(4) = 4 - 2$, or 2, in Britain. The function h is called the *composition* of f and g and is denoted $f \circ g$.

Composition

The *composite function* $f \circ g$, the *composition* of f and g, is defined as

$$f \circ g(x) = f(g(x)).$$

We can visualize the composition of functions as follows.

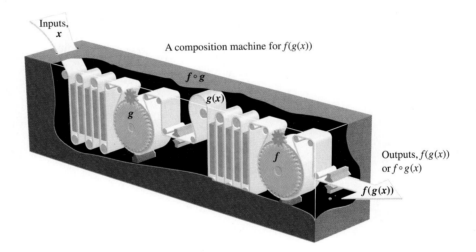

Inputs, x

A composition machine for $f(g(x))$

$f \circ g$

$g(x)$

g

f

Outputs, $f(g(x))$ or $f \circ g(x)$

$f(g(x))$

EXAMPLE 1

Given $f(x) = 3x$ and $g(x) = 1 + x^2$:

a) Find $f \circ g(5)$ and $g \circ f(5)$.
b) Find $f \circ g(x)$ and $g \circ f(x)$.

Solution   Consider each function separately:

$$f(x) = 3x \qquad \text{This function multiplies each input by 3.}$$

and

$$g(x) = 1 + x^2. \qquad \text{This function adds 1 to the square of each input.}$$

a) To find $f \circ g(5)$, we first find $g(5)$ by substituting in the formula for g: Square 5 and add 1, to get 26. We then use 26 as an input for f:

$$f \circ g(5) = f(g(5)) = f(1 + 5^2)$$
$$= f(26) = 3 \cdot 26 = 78.$$

To find $g \circ f(5)$, we first find $f(5)$ by substituting into the formula for f: Multiply 5 by 3, to get 15. We then use 15 as an input for g:

$$g \circ f(5) = g(f(5)) = g(3 \cdot 5)$$
$$= g(15) = 1 + 15^2 = 1 + 225 = 226.$$

b) We find $f \circ g(x)$ by substituting $g(x)$ for x in the equation for $f(x)$:

$$f \circ g(x) = f(g(x)) = f(1 + x^2) \qquad \text{Substituting } 1 + x^2 \text{ for } g(x)$$
$$= 3(1 + x^2) = 3 + 3x^2. \qquad \textit{These} \text{ parentheses indicate multiplication.}$$

To find $g \circ f(x)$, we substitute $f(x)$ for x in the equation for $g(x)$:

$$g \circ f(x) = g(f(x)) = g(3x) \qquad \text{Substituting } 3x \text{ for } f(x)$$
$$= 1 + (3x)^2$$
$$= 1 + 9x^2. \qquad\qquad \square$$

Note in Example 1 that $f \circ g(5) \neq g \circ f(5)$ and that, in general, $f \circ g(x) \neq g \circ f(x)$.

EXAMPLE 2

Given $f(x) = \sqrt{x}$ and $g(x) = x - 1$, find $f \circ g(x)$ and $g \circ f(x)$.

Solution

$$f \circ g(x) = f(g(x)) = f(x - 1) = \sqrt{x - 1}$$
$$g \circ f(x) = g(f(x)) = g(\sqrt{x}) = \sqrt{x} - 1 \qquad\qquad \square$$

In calculus, one needs to recognize how a function can be expressed as a composition.

EXAMPLE 3

If $h(x) = (7x + 3)^2$, find $f(x)$ and $g(x)$ so that $h(x) = f \circ g(x)$.

Solution   To find $h(x)$, we can think of two steps: forming $7x + 3$ and then squaring. This suggests that $g(x) = 7x + 3$ and $f(x) = x^2$. We check by forming the composition:

$$h(x) = f \circ g(x) = f(g(x)) = f(7x + 3) = (7x + 3)^2.$$

This is the most "obvious" answer to the question. There can be other less obvious answers. For example, if

$$f(x) = (x - 1)^2 \quad \text{and} \quad g(x) = 7x + 4,$$

then

$$h(x) = f \circ g(x) = f(g(x)) = f(7x + 4) = (7x + 4 - 1)^2 = (7x + 3)^2. \qquad \square$$

Inverses and One-to-One Functions

Let us consider the following two functions. We think of them as relations, or correspondences.

COST OF A 60-SECOND SUPER BOWL COMMERCIAL, BY YEAR	
Domain (Set of Inputs)	**Range** (Set of Outputs)
1967 ⟶	$80,000
1977 ⟶	$324,000
1981 ⟶	$550,000
1983 ⟶	$800,000
1988 ⟶	$1,350,000

U.S. SENATORS AND THEIR STATES

Domain (Set of Inputs)	**Range** (Set of Outputs)
Wellstone ⟶ Durenberger ⟶	Minnesota
Mack ⟶ Graham ⟶	Florida
Bradley ⟶ Lautenberg ⟶	New Jersey

Suppose we reverse the arrows. We obtain what is called the **inverse relation.** Are these inverse relations functions?

COST OF A 60-SECOND SUPER BOWL COMMERCIAL, BY YEAR	
Range (Set of Outputs)	**Domain** (Set of Inputs)
1967 ⟵	$80,000
1977 ⟵	$324,000
1981 ⟵	$550,000
1983 ⟵	$800,000
1988 ⟵	$1,350,000

U.S. SENATORS AND THEIR STATES

Range (Set of Outputs)	**Domain** (Set of Inputs)
Wellstone ⟵ Durenberger ⟵	Minnesota
Mack ⟵ Graham ⟵	Florida
Bradley ⟵ Lautenberg ⟵	New Jersey

We see that the inverse of the first correspondence is a function, but that the inverse of the second correspondence is not a function.

Recall that for each input, a function provides exactly one output. However, nothing in our definition of function prevents having the same output for two or more different inputs. Thus it is possible for different inputs to correspond to the same output in the range. Only when this possibility is *excluded* is the inverse also a function.

In the Super Bowl function, different inputs have different outputs. It is an example of a **one-to-one function**. In the U.S. Senator function, the inputs *Mack* and *Graham* both have the output *Florida*. Thus, the U.S. Senator function is not one-to-one.

One-to-One Function

A function f is *one-to-one* if different inputs have different outputs. That is, if for any $a \neq b$, we have $f(a) \neq f(b)$, the function f is one-to-one. If a function is one-to-one, then its inverse correspondence is also a function.

How can we tell graphically whether a function is one-to-one and thus has an inverse that is a function?

EXAMPLE 4

Shown here is the graph of a function. Determine whether the function is one-to-one and thus has an inverse that is a function.

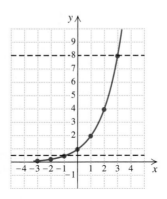

Solution   A function is one-to-one if different inputs have different outputs. In other words, no two x-values will have the same y-value. For this function, we cannot find two x-values that have the same y-value. Note also that no horizontal line can be drawn that will cross the graph more than once. The function is one-to-one so its inverse is a function.   ❑

The graph of any function must pass the vertical-line test. For a function to have an inverse that is a function, it must pass the *horizontal-line test* as well.

The
Horizontal-Line Test

A function is one-to-one and has an inverse that is a function if there is no horizontal line that crosses the graph more than once.

EXAMPLE 5

Determine whether the function $f(x) = x^2$ is one-to-one and has an inverse that is also a function.

Solution   The graph of $f(x) = x^2$ is shown here. Many horizontal lines cross the graph more than once, in particular the line $y = 4$.

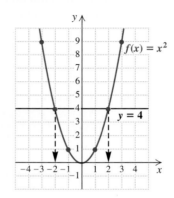

Note that where the line crosses, the first coordinates are -2 and 2. Although these are different inputs, they have the same output. That is, $-2 \neq 2$, but

$$f(-2) = (-2)^2 = 4 = 2^2 = f(2).$$

Thus the function is not one-to-one and no inverse function exists. ❑

Finding Formulas for Inverses

If the inverse of a function f is also a function, it can be named f^{-1} (read "f-inverse").

The -1 in f^{-1} is *not* an exponent!

Suppose a function is described by a formula. If it has an inverse that is a function, how do we find a formula for the inverse? For any equation in two variables, if we interchange the variables, we obtain an equation of the inverse correspondence. If it is a function, we proceed as follows to find a formula for f^{-1}.

If a function is one-to-one, a formula for its inverse can be found as follows:

1. Replace $f(x)$ by y.
2. Interchange x and y. (This gives the inverse function.)
3. Solve for y.
4. Replace y by $f^{-1}(x)$.

EXAMPLE 6

Determine if the function is one-to-one and if it is, find a formula for $f^{-1}(x)$:
(a) $f(x) = x + 2$; **(b)** $f(x) = 2x - 3$.

Solution

a) The graph of $f(x) = x + 2$ is shown below. It passes the horizontal-line test, so it is one-to-one. Thus its inverse is a function.

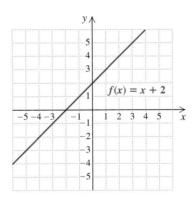

1. Replace $f(x)$ by y: $\qquad y = x + 2.$

2. Interchange x and y: $\qquad x = y + 2.$ This gives the inverse function.

3. Solve for y: $\qquad x - 2 = y.$

4. Replace y by $f^{-1}(x)$: $f^{-1}(x) = x - 2.$ We also "reversed" the equation.

In words, f adds 2 to all inputs. To "undo" f, f^{-1} subtracts 2 from all inputs.

b)   The function $f(x) = 2x - 3$ is also linear. Any linear function that is not constant will pass the horizontal-line test. Thus, f is one-to-one.

  1.   Replace $f(x)$ by y: $y = 2x - 3$.
  2.   Interchange x and y: $x = 2y - 3$.
  3.   Solve for y: $x + 3 = 2y$
$$\frac{x + 3}{2} = y.$$

  4.   Replace y by $f^{-1}(x)$: $f^{-1}(x) = \dfrac{x + 3}{2}.$ ❏

Let us consider inverses of functions in terms of a function machine. Suppose that a function f programmed into a machine has an inverse that is also a function. Suppose that the function machine has a reverse switch. When the switch is thrown, the machine performs the inverse function f^{-1}. Inputs then enter at the opposite end, and the entire process is reversed.

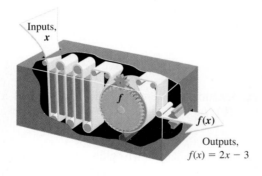

Inputs, x

$f(x)$

Outputs, $f(x) = 2x - 3$

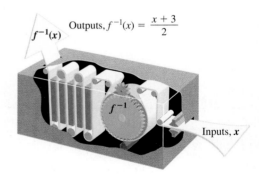

$f^{-1}(x)$

Outputs, $f^{-1}(x) = \dfrac{x + 3}{2}$

Inputs, x

Consider $f(x) = 2x - 3$ and $f^{-1}(x) = \dfrac{x + 3}{2}$ from Example 6(b). For the input 5, we have

$$f(5) = 2 \cdot 5 - 3 = 10 - 3 = 7.$$

The output is 7. Now we use 7 for the input in the inverse:

$$f^{-1}(7) = \frac{7 + 3}{2}$$

$$= \frac{10}{2} = 5.$$

The function f takes 5 to 7. The inverse function f^{-1} takes the number 7 back to 5.

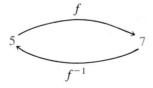

Graphing Functions and Their Inverses

How do the graphs of a function and its inverse compare?

EXAMPLE 7

Graph $f(x) = 2x - 3$ and $f^{-1}(x) = (x + 3)/2$ on the same axes. Then compare.

Solution The graph of each function follows. Note that the graph of f^{-1} can be drawn by reflecting the graph of f across the line $y = x$. That is, if we graph $f(x) = 2x - 3$ in wet ink and fold the paper along the line $y = x$, the graph of $f^{-1}(x) = (x + 3)/2$ will be formed from the impression made by f.

When we interchange y and x in finding a formula for the inverse, we are, in effect, flipping the graph of $f(x) = 2x - 3$ over the line $y = x$. For example, when the coordinates of the y-intercept of the graph of f, $(0, -3)$, are reversed, we get the x-intercept of the graph of f^{-1}, $(-3, 0)$.

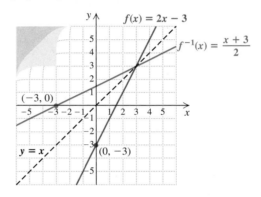

A grapher can provide a partial check of whether or not two functions are inverses of each other. First we plot the functions believed to be inverses of each other and the line $y = x$, all on the same set of axes. Then, using the fact that a function and its inverse are reflections of each other across the line $y = x$, we can decide if the two functions may be inverses of each other.

Before we do this, however, there is a problem that must be addressed. Most graphers have rectangular, rather than square, displays. This means that a "square" window, like the standard $[-10, 10, -10, 10]$, distorts a graph

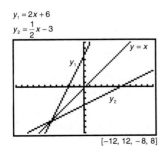

$y_1 = 2x + 6$
$y_2 = \frac{1}{2}x - 3$

$[-12, 12, -8, 8]$

because one unit in the x-direction is longer than one unit in the y-direction. To correct this, many graphers offer an option to "Square" axes, which guarantees that the unit length is the same in both the x- and y-directions.

To determine whether $y_1 = 2x + 6$ and $y_2 = \frac{1}{2}x - 3$ might be inverses of each other, we have drawn both functions, along with the line $y = x$, on a "squared" set of axes. It appears that y_1 and y_2 are inverses of each other.

Use a grapher to help match each function in Column A with its inverse from Column B.

Column A

TC1. $y = 5x^3 + 10$

TC2. $y = (5x + 10)^3$

TC3. $y = 5(x + 10)^3$

TC4. $y = (5x)^3 + 10$

Column B

A. $y = \dfrac{\sqrt[3]{x} - 10}{5}$

B. $y = \sqrt[3]{\dfrac{x}{5} - 10}$

C. $y = \sqrt[3]{\dfrac{x - 10}{5}}$

D. $y = \dfrac{\sqrt[3]{x - 10}}{5}$

The graph of f^{-1} is a reflection of the graph of f across the line $y = x$.

EXAMPLE 8

Consider $g(x) = x^3 + 2$.

a) Determine whether the function is one-to-one.
b) If it is one-to-one, find a formula for its inverse.
c) Graph the inverse, if it exists.

Solution

a) The graph of $g(x) = x^3 + 2$ follows. It passes the horizontal-line test and thus has an inverse.

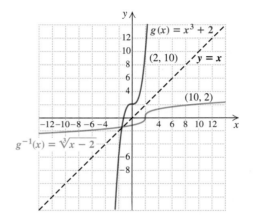

b) **1.** Replace $g(x)$ by y: $y = x^3 + 2$.
  2. Interchange x and y: $x = y^3 + 2$.
  3. Solve for y: $x - 2 = y^3$
                    $\sqrt[3]{x - 2} = y$.   Since a number has only one cube root, we can solve for y.
  4. Replace y by $g^{-1}(x)$: $g^{-1}(x) = \sqrt[3]{x - 2}$.

c) To find the graph, we reflect the graph of $g(x) = x^3 + 2$ across the line $y = x$, as we did in Example 7. It can also be found by substituting into $g^{-1}(x) = \sqrt[3]{x - 2}$ and plotting points. The graphs of g and g^{-1} are shown above using one set of axes.   ❑

Inverse Functions and Composition

Suppose that we used some input x for the function f and found its output, $f(x)$. The function f^{-1} would then take that output back to x. Similarly, if we began with an input x for the function f^{-1} and found its output, $f^{-1}(x)$, the original function f would then take that output back to x. This is summarized as follows.

If a function f is one-to-one, then f^{-1} is the unique function such that
$$f^{-1} \circ f(x) = x \quad \text{and} \quad f \circ f^{-1}(x) = x.$$

EXAMPLE 9

Let $f(x) = 2x + 1$. Show that $f^{-1}(x) = (x - 1)/2$.

Solution   We find $f^{-1} \circ f(x)$ and $f \circ f^{-1}(x)$ and check to see that each is x.

$$f^{-1} \circ f(x) = f^{-1}(f(x)) = f^{-1}(2x + 1) = \frac{(2x + 1) - 1}{2} = \frac{2x}{2} = x$$

$$f \circ f^{-1}(x) = f(f^{-1}(x)) = f\left(\frac{x - 1}{2}\right) = 2 \cdot \frac{x - 1}{2} + 1 = x - 1 + 1 = x$$

EXERCISE SET | 9.2

Find $f \circ g(x)$ and $g \circ f(x)$.

1. $f(x) = 3x^2 + 2$, $g(x) = 2x - 1$

2. $f(x) = 4x + 3$, $g(x) = 2x^2 - 5$

3. $f(x) = 4x^2 - 1$, $g(x) = 2/x$

4. $f(x) = 3/x$, $g(x) = 2x^2 + 3$

5. $f(x) = x^2 + 1$, $g(x) = x^2 - 1$

6. $f(x) = 1/x^2$, $g(x) = x + 2$

Find $f(x)$ and $g(x)$ such that $h(x) = f \circ g(x)$. Answers may vary.

7. $h(x) = (5 - 3x)^2$

8. $h(x) = 4(3x - 1)^2 + 9$

9. $h(x) = (3x^2 - 7)^5$ **10.** $h(x) = \sqrt{5x + 2}$

11. $h(x) = \dfrac{1}{x - 1}$ **12.** $h(x) = \dfrac{3}{x} + 4$

13. $h(x) = \dfrac{1}{\sqrt{7x + 2}}$ **14.** $h(x) = \sqrt{x - 7} - 3$

15. $h(x) = \dfrac{x^3 + 1}{x^3 - 1}$ **16.** $h(x) = \left(\sqrt{x} + 5\right)^4$

Determine whether the function is one-to-one.

17. $f(x) = 3x - 4$ **18.** $f(x) = 5 - 2x$

19. $f(x) = x^2 - 3$ **20.** $f(x) = 1 - x^2$

21. $g(x) = 3^x$ **22.** $g(x) = \left(\frac{1}{2}\right)^x$

23. $g(x) = |x|$ **24.** $h(x) = |x| - 1$

Given the function, **(a)** determine if it is one-to-one; **(b)** if it is one-to-one, find a formula for the inverse.

25. $f(x) = x + 4$ **26.** $f(x) = x + 7$

27. $f(x) = 5 - x$ **28.** $f(x) = 9 - x$

29. $g(x) = x - 5$ **30.** $g(x) = x - 8$

31. $f(x) = 3x$ **32.** $f(x) = 4x$

33. $g(x) = 3x + 2$ **34.** $g(x) = 4x + 7$

35. $h(x) = 7$ **36.** $h(x) = -3$

37. $f(x) = \dfrac{1}{x}$ **38.** $f(x) = \dfrac{3}{x}$

39. $f(x) = \dfrac{2x + 1}{3}$ **40.** $f(x) = \dfrac{3x + 2}{5}$

41. $f(x) = x^3 - 1$ **42.** $f(x) = x^3 + 5$

43. $g(x) = (x - 2)^3$ **44.** $g(x) = (x + 7)^3$

45. $f(x) = \sqrt{x}$ **46.** $f(x) = \sqrt{x - 1}$

47. $f(x) = 2x^2 + 3$, $x \geq 0$

48. $f(x) = 3x^2 - 2$, $x \geq 0$

Graph the function and its inverse using the same set of axes.

49. $f(x) = \frac{1}{2}x - 3$ **50.** $g(x) = x + 4$

51. $f(x) = x^3$ **52.** $f(x) = x^3 - 1$

53. $y = 2^x$ **54.** $y = 3^x$

55. $y = \left(\frac{1}{2}\right)^x$ **56.** $y = \left(\frac{2}{3}\right)^x$

57. $f(x) = 3 - x^2,\ x \geq 0$

58. $f(x) = x^2 - 1,\ x \leq 0$

59. Let $f(x) = \frac{4}{5}x$. Show that
$$f^{-1}(x) = \frac{5}{4}x.$$

60. Let $f(x) = (x + 7)/3$. Show that
$$f^{-1}(x) = 3x - 7.$$

61. Let $f(x) = (1 - x)/x$. Show that
$$f^{-1}(x) = \frac{1}{x + 1}.$$

62. Let $f(x) = x^3 - 5$. Show that
$$f^{-1}(x) = \sqrt[3]{x + 5}.$$

63. *Women's dress sizes in the United States and France.* A size-6 dress in the United States is size 38 in France. A function that will convert dress sizes in the United States to those in France is
$$f(x) = x + 32.$$

a) Find the dress sizes in France that correspond to sizes 8, 10, 14, and 18 in the United States.

b) Determine whether this function has an inverse that is a function. If so, find a formula for the inverse.

c) Use the inverse function to find dress sizes in the United States that correspond to sizes 40, 42, 46, and 50 in France.

64. *Women's dress sizes in the United States and Italy.* A size-6 dress in the United States is size 36 in Italy. A function that will convert dress sizes in the United States to those in Italy is
$$f(x) = 2(x + 12).$$

a) Find the dress sizes in Italy that correspond to sizes 8, 10, 14, and 18 in the United States.

b) Determine whether this function has an inverse that is a function. If so, find a formula for the inverse.

c) Use the inverse function to find dress sizes in the United States that correspond to sizes 40, 44, 52, and 60 in Italy.

Skill Maintenance _____

65. Find an equation of variation if y varies directly as x and $y = 7.2$ when $x = 0.8$.

66. Find an equation of variation if y varies inversely as x and $y = 3.5$ when $x = 6.1$.

Synthesis _____

67. ◈ Does the constant function $f(x) = 4$ have an inverse that is a function? If so, find a formula. If not, explain why.

68. ◈ An organization determines that the cost per person of chartering a bus is given by the function
$$C(x) = \frac{100 + 5x}{x},$$
where $x = $ the number of people in the group and $C(x)$ is in dollars. Determine $C^{-1}(x)$ and explain how this inverse function could be used.

69. Use the information in Exercises 63 and 64 to find a function for the French dress size that corresponds to a size x dress in Italy.

▧ In Exercises 70–73, use a grapher to help determine whether or not the given functions are inverses of each other.

70. $f(x) = 0.75x^2 + 2;\ g(x) = \sqrt{\dfrac{4(x - 2)}{3}}$

71. $f(x) = 1.4x^3 + 3.2;\ g(x) = \sqrt[3]{\dfrac{x - 3.2}{1.4}}$

72. $f(x) = \sqrt{2.5x + 9.25};$
$g(x) = 0.4x^2 - 3.7,\ x \geq 0$

73. $f(x) = 0.8x^{1/2} + 5.23;$
$g(x) = 1.25(x^2 - 5.23),\ x \geq 0$

9.3

Logarithmic Functions

Graphs of Logarithmic Functions • Converting Exponential and Logarithmic Equations • Solving Certain Logarithmic Equations

In this section we consider a type of function called a *logarithm function*, or *logarithmic function*. Such functions have many applications to problem solving.

Graphs of Logarithmic Functions

Consider the exponential function $f(x) = 2^x$. Does this function have an inverse that is a function? We see from the graph that this function is one-to-one so f^{-1} *is* a function.

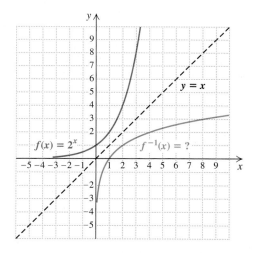

To find a formula for f^{-1}, we use the method of Section 9.2:

$$f(x) = 2^x.$$

1. Replace $f(x)$ by y: $y = 2^x.$

2. Interchange x and y: $x = 2^y.$

3. Solve for y: $y =$ the power to which we raise 2 to get $x.$

4. Replace y by $f^{-1}(x)$: $f^{-1}(x) =$ the power to which we raise to get $x.$

We now define a new symbol to replace the words "the power to which we raise 2 to get x":

> $\log_2 x$, read "the logarithm, base 2, of x", means "the power to which we raise 2 to get x."

Thus if $f(x) = 2^x$, then $f^{-1}(x) = \log_2 x$. Note that $f^{-1}(8) = \log_2 8 = 3$, because 3 is *the power to which we raise* 2 *to get* 8.

Although expressions like $\log_2 13$ cannot be simplified, we must remember that $\log_2 13$ represents *the power to which we raise 2 to get* 13.

For any exponential function $f(x) = a^x$, the inverse is called a **logarithmic function, base a.** The graph of the inverse can, of course, be drawn by reflecting the graph of $f(x) = a^x$ across the line $y = x$. It will be helpful to remember that the inverse of $f(x) = a^x$ is given by $f^{-1}(x) = \log_a x$. Normally, we use a number a that is greater than 1 for the logarithm base.

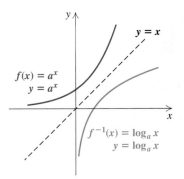

We define $y = \log_a x$ as that number y such that $a^y = x$, where $x > 0$ and a is a positive constant other than 1.

It is helpful in dealing with logarithmic functions to remember that the logarithm of a number is an *exponent*. It is the exponent y in a^y. You might also think to yourself, "The logarithm, base a, of a number x is the power to which a must be raised in order to get x."

A logarithm is an exponent.

The following is a comparison of exponential and logarithmic functions.

Exponential Function	Logarithmic Function
$y = a^x$	$x = a^y$
$f(x) = a^x$	$f(x) = \log_a x$
$a > 0, a \neq 1$	$a > 0, a \neq 1$
The input x can be any real number.	The output y can be any real number.
$y > 0$	$x > 0$
Outputs are positive.	Inputs are positive.

EXAMPLE 1

Graph: $y = f(x) = \log_5 x$.

Solution   The equation $y = \log_5 x$ is equivalent to $5^y = x$. We can find ordered pairs that are solutions by choosing values for y and computing the x-values.

For $y = 0$, $x = 5^0 = 1$.

For $y = 1$, $x = 5^1 = 5$.

For $y = 2$, $x = 5^2 = 25$.

For $y = -1$, $x = 5^{-1} = \frac{1}{5}$.

For $y = -2$, $x = 5^{-2} = \frac{1}{25}$.

x, or 5^y	y
1	0
5	1
25	2
$\frac{1}{5}$	-1
$\frac{1}{25}$	-2

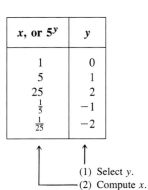

(1) Select y.

(2) Compute x.

We plot the set of ordered pairs and connect the points with a smooth curve. The graph of $y = 5^x$ has been shown only for reference.

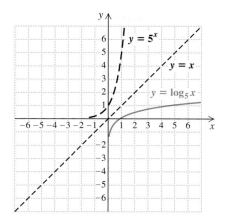

Converting Exponential and Logarithmic Equations

We use the definition of logarithm to convert from *exponential equations* to *logarithmic equations*:

$$y = \log_a x \quad \textbf{is equivalent to} \quad a^y = x.$$

> CAUTION! **Be sure to memorize this relationship!** It is probably the most important definition in the chapter. Many times this definition will serve as a justification for a step that we are considering.

EXAMPLE 2

Convert to logarithmic equations: **(a)** $8 = 2^x$; **(b)** $y^{-1} = 4$; **(c)** $a^b = c$.

Solution

a) $8 = 2^x$ is equivalent to $x = \log_2 8$ The exponent is the logarithm.
                                                        The base remains the same.

b) $y^{-1} = 4$ is equivalent to $-1 = \log_y 4$
c) $a^b = c$ is equivalent to $b = \log_a c$

We also use the definition of logarithm to convert from logarithmic equations to exponential equations.

EXAMPLE 3

Convert to exponential equations: **(a)** $y = \log_3 5$; **(b)** $-2 = \log_a 7$; **(c)** $a = \log_b d$.

Solution

a) $y = \log_3 5$ is equivalent to $3^y = 5$ The logarithm is the exponent.
                                                        The base remains the same.

b) $-2 = \log_a 7$ is equivalent to $a^{-2} = 7$
c) $a = \log_b d$ is equivalent to $b^a = d$

Solving Certain Logarithmic Equations

Some logarithmic equations can be solved by first converting to exponential equations.

EXAMPLE 4

Solve: **(a)** $\log_2 x = -3$; **(b)** $\log_x 16 = 2$.

Solution

a) $\log_2 x = -3$

$$2^{-3} = x \qquad \text{Converting to an exponential equation}$$
$$\tfrac{1}{8} = x \qquad \text{Computing } 2^{-3}$$

Check: For $\tfrac{1}{8}$ to be the solution, $\log_2 \tfrac{1}{8}$ should equal -3. Since $2^{-3} = \tfrac{1}{8}$, we know that $\tfrac{1}{8}$ checks and is the solution.

b) $\log_x 16 = 2$

$$x^2 = 16 \qquad\qquad \text{Converting to an exponential equation}$$
$$x = 4 \quad or \quad x = -4 \qquad \text{Principle of square roots}$$

Check: $\log_4 16 = 2$ because $4^2 = 16$. Thus, 4 is a solution. Because all logarithm bases must be positive, $\log_{-4} 16$ is not defined. Therefore, -4 is not a solution. Logarithm bases must be positive because logarithms are defined using exponential functions that are defined only for positive bases. ❑

Solving for x in $\log_b a = x$ amounts to finding the logarithm, base b, of the number a. To think of finding logarithms as solving equations may help.

EXAMPLE 5

Simplify: **(a)** $\log_{10} 1000$; **(b)** $\log_{10} 0.01$; **(c)** $\log_5 1$.

Solution

a) Method 1. Let $\log_{10} 1000 = x$. Then

$$10^x = 1000 \qquad \text{Converting to an exponential equation}$$
$$10^x = 10^3 \qquad \text{Writing 1000 as a power of 10}$$
$$x = 3. \qquad \text{The exponents must be the same.}$$

Therefore, $\log_{10} 1000 = 3$.

Method 2. Think of the meaning of $\log_{10} 1000$. It is the exponent to which we raise 10 to get 1000. That exponent is 3. Therefore, $\log_{10} 1000 = 3$.

b) Method 1. Let $\log_{10} 0.01 = x$. Then

$$10^x = 0.01 \qquad \text{Converting to an exponential equation}$$
$$10^x = \tfrac{1}{100}$$
$$10^x = 10^{-2} \qquad \text{Writing } \tfrac{1}{100} \text{ as a power of 10}$$
$$x = -2. \qquad \text{Equating exponents}$$

Therefore, $\log_{10} 0.01 = -2$.

Method 2. $\log_{10} 0.01$ is the exponent to which we raise 10 to get 0.01. If we note that $0.01 = 1/100 = 1/10^2$, it follows that the exponent is -2. Therefore, $\log_{10} 0.01 = -2$.

c) **Method 1.** Let $\log_5 1 = x$. Then

$$5^x = 1 \qquad \text{Converting to an exponential equation}$$
$$5^x = 5^0 \qquad \text{Writing 1 as a power of 5}$$
$$x = 0.$$

Therefore, $\log_5 1 = 0$.

Method 2. $\log_5 1$ is the exponent to which we raise 5 to get 1. That exponent is 0. Therefore, $\log_5 1 = 0$. ❑

Example 5(c) illustrates an important property of logarithms.

 The logarithm, base a, of 1 is always 0: $\log_a 1 = 0$.

This follows from the fact that $a^0 = 1$ is equivalent to the logarithmic equation $\log_a 1 = 0$.

Another property results from the fact that $a^1 = a$. This is equivalent to the equation $\log_a a = 1$.

 The logarithm, base a, of a is always 1: $\log_a a = 1$.

Thus, $\log_{10} 10 = 1$, $\log_8 8 = 1$, and so on.

EXERCISE SET | 9.3

Graph.

1. $y = \log_2 x$

2. $y = \log_{10} x$

3. $y = \log_6 x$

4. $y = \log_3 x$

5. $f(x) = \log_4 x$

6. $f(x) = \log_5 x$

7. $f(x) = \log_{1/2} x$

8. $f(x) = \log_{2.5} x$

Graph both functions using the same set of axes.

9. $f(x) = 3^x$, $f^{-1}(x) = \log_3 x$

10. $f(x) = 4^x$, $f^{-1}(x) = \log_4 x$

Convert to logarithmic equations.

11. $10^3 = 1000$

12. $10^2 = 100$

13. $5^{-3} = \frac{1}{125}$

14. $4^{-5} = \frac{1}{1024}$

15. $8^{1/3} = 2$

16. $16^{1/4} = 2$

17. $10^{0.3010} = 2$

18. $10^{0.4771} = 3$

19. $e^2 = t$

20. $p^k = 3$

21. $Q^t = x$

22. $p^m = V$

23. $e^2 = 7.3891$

24. $e^3 = 20.0855$

25. $e^{-2} = 0.1353$

26. $e^{-4} = 0.0183$

Convert to exponential equations.

27. $t = \log_3 8$

28. $h = \log_7 10$

29. $\log_5 25 = 2$

30. $\log_6 6 = 1$

31. $\log_{10} 0.1 = -1$

32. $\log_{10} 0.01 = -2$

33. $\log_{10} 7 = 0.845$

34. $\log_{10} 3 = 0.4771$

35. $\log_e 20 = 2.9957$

36. $\log_e 10 = 2.3026$

37. $\log_t Q = k$

38. $\log_m P = a$

39. $\log_e 0.25 = -1.3863$

40. $\log_e 0.989 = -0.0111$

41. $\log_r T = -x$

42. $\log_c M = -w$

Solve.

43. $\log_3 x = 2$

44. $\log_4 x = 3$

45. $\log_x 125 = 3$

46. $\log_x 64 = 3$

47. $\log_2 16 = x$

48. $\log_5 25 = x$

49. $\log_3 27 = x$

50. $\log_4 16 = x$

51. $\log_x 13 = 1$

52. $\log_x 23 = 1$

53. $\log_6 x = 0$

54. $\log_9 x = 1$

55. $\log_2 x = -1$

56. $\log_3 x = -2$

57. $\log_8 x = \frac{1}{3}$

58. $\log_{32} x = \frac{1}{5}$

Find each of the following.

59. $\log_{10} 100$

60. $\log_{10} 100{,}000$

61. $\log_{10} 0.1$

62. $\log_{10} 0.001$

63. $\log_{10} 1$

64. $\log_{10} 10$

65. $\log_5 625$

66. $\log_2 64$

67. $\log_5 \frac{1}{25}$

68. $\log_2 \frac{1}{16}$

69. $\log_3 1$

70. $\log_4 4$

71. $\log_e e$

72. $\log_e 1$

73. $\log_{27} 9$

74. $\log_8 2$

75. $\log_e e^3$

76. $\log_e e^{-4}$

77. $\log_{10} 10^t$

78. $\log_3 3^p$

Simplify.

79. $\dfrac{\dfrac{3}{x} - \dfrac{2}{xy}}{\dfrac{2}{x^2} + \dfrac{1}{xy}}$

80. $\dfrac{\dfrac{4 + x}{x^2 + 2x + 1}}{\dfrac{3}{x + 1} - \dfrac{2}{x + 2}}$

Rename.

81. 8^{-4}

82. $x^{4/5}$

83. $t^{-2/3}$

84. 5^1

85. ◈ Explain why 1 is excluded from being a logarithmic base.

86. ◈ Explain why the number $\log_2 13$ is between 3 and 4.

87. Graph both equations using the same set of axes: $y = \left(\frac{3}{2}\right)^x$, $y = \log_{3/2} x$.

Graph.

88. $y = \log_2 (x - 1)$

89. $y = \log_3 |x + 1|$

Solve.

90. $|\log_3 x| = 3$

91. $\log_{125} x = \frac{2}{3}$

92. $\log_4 (3x - 2) = 2$

93. $\log_8 (2x + 1) = -1$

94. $\log_{10} (x^2 + 21x) = 2$

Simplify.

95. $\log_{1/4} \frac{1}{64}$

96. $\log_{81} 3 \cdot \log_3 81$

97. $\log_{10} (\log_4 (\log_3 81))$

98. $\log_2 (\log_2 (\log_4 256))$

99. $\log_{1/5} 25$

9.4

Properties of Logarithmic Functions

Logarithms of Products • **Logarithms of Powers** •
Logarithms of Quotients • **Using the Properties Together**

Logarithmic functions are important in many applications and in more advanced mathematics. We now establish some basic properties that are useful in manipulating expressions involving logarithms. As the proofs of these properties reveal, the properties of logarithms are related to the properties of exponents.

Logarithms of Products

The first property we discuss is reminiscent of the property $a^m \cdot a^n = a^{m+n}$.

The Product Rule for Logarithms

For any positive numbers M, N, and a $(a \neq 1)$,

$$\log_a MN = \log_a M + \log_a N.$$

(The logarithm of a product is the sum of the logarithms of the factors.)

EXAMPLE 1

Express as a sum of logarithms: $\log_2 (4 \cdot 16)$.

Solution   We have

$$\log_2 (4 \cdot 16) = \log_2 4 + \log_2 16. \qquad \text{Using the product rule}$$

As a check, note that

$$\log_2 (4 \cdot 16) = \log_2 64 = 6$$

and that

$$\log_2 4 + \log_2 16 = 2 + 4 = 6. \qquad \qquad \qquad ❏$$

EXAMPLE 2

Express as a single logarithm: $\log_{10} 0.01 + \log_{10} 1000$.

Solution   We have

$$\log_{10} 0.01 + \log_{10} 1000 = \log_{10} (0.01 \times 1000) \qquad \text{Using the product rule}$$
$$= \log_{10} 10. \qquad \qquad \qquad ❏$$

A Proof of the Product Rule.   Let $\log_a M = x$ and $\log_a N = y$. Converting to exponential equations, we have $a^x = M$ and $a^y = N$.

Now we multiply the latter two equations, to obtain

$$MN = a^x \cdot a^y = a^{x+y}.$$

Converting back to a logarithmic equation, we get

$$\log_a MN = x + y.$$

Recalling what x and y represent, we get

$$\log_a MN = \log_a M + \log_a N. \qquad \qquad \qquad ❏$$

Logarithms of Powers

The second basic property is related to the property $(a^m)^n = a^{mn}$.

The Power Rule for Logarithms

For any positive numbers M and a $(a \neq 1)$, and any real number p,

$$\log_a M^p = p \cdot \log_a M.$$

(The logarithm of a power of M is the exponent times the logarithm of M.)

EXAMPLE 3

Express as a product: **(a)** $\log_a 9^{-5}$; **(b)** $\log_a \sqrt[4]{5}$.

Solution

a) $\log_a 9^{-5} = -5 \log_a 9$ Using the power rule

b) $\log_a \sqrt[4]{5} = \log_a 5^{1/4}$ Writing exponential notation

$\qquad\qquad = \frac{1}{4} \log_a 5$ Using the power rule   ❑

A Proof of the Power Rule. Let $x = \log_a M$. We then convert to an exponential equation, to get $a^x = M$. Raising both sides to the pth power, we obtain

$$(a^x)^p = M^p, \quad \text{or} \quad a^{xp} = M^p.$$

Converting back to a logarithmic equation gives us

$$\log_a M^p = xp.$$

But $x = \log_a M$, so substituting, we have

$$\log_a M^p = (\log_a M)p = p \cdot \log_a M.$$ ❑

Logarithms of Quotients

The third property that we study is similar to the property $a^m/a^n = a^{m-n}$.

**The Quotient Rule
for Logarithms**

For any positive numbers M, N, and a ($a \neq 1$),

$$\log_a \frac{M}{N} = \log_a M - \log_a N.$$

(The logarithm of a quotient is the logarithm of the dividend minus the logarithm of the divisor.)

EXAMPLE 4

Express as a difference of logarithms: $\log_t (6/U)$.

Solution

$$\log_t \frac{6}{U} = \log_t 6 - \log_t U \qquad \text{Using the quotient rule}$$ ❑

EXAMPLE 5

Express as a single logarithm: $\log_b 17 - \log_b 27$.

Solution

$$\log_b 17 - \log_b 27 = \log_b \frac{17}{27} \qquad \text{Using the quotient rule ``in reverse''}$$ ❑

A Proof of the Quotient Rule. Our proof uses both the product and power rules:

$$\log_a \frac{M}{N} = \log_a MN^{-1}$$

$$= \log_a M + \log_a N^{-1} \qquad \text{Using the product rule}$$

$$= \log_a M + (-1) \log_a N \qquad \text{Using the power rule}$$

$$= \log_a M - \log_a N.$$ ❑

Using the Properties Together

EXAMPLE 6

Express in terms of logarithms of x, y, z, m, and n.

a) $\log_a \dfrac{x^2 y^3}{z^4}$

b) $\log_a \sqrt[4]{\dfrac{xy}{z^3}}$

c) $\log_b \dfrac{xy}{m^3 n^4}$

Solution

a) $\log_a \dfrac{x^2 y^3}{z^4} = \log_a (x^2 y^3) - \log_a z^4$ Using the quotient rule

$= \log_a x^2 + \log_a y^3 - \log_a z^4$ Using the product rule

$= 2 \log_a x + 3 \log_a y - 4 \log_a z$ Using the power rule

b) $\log_a \sqrt[4]{\dfrac{xy}{z^3}} = \log_a \left(\dfrac{xy}{z^3} \right)^{1/4}$ Writing exponential notation

$= \dfrac{1}{4} \cdot \log_a \dfrac{xy}{z^3}$ Using the power rule

$= \dfrac{1}{4} (\log_a xy - \log_a z^3)$ Using the quotient rule. Parentheses are important.

$= \dfrac{1}{4} (\log_a x + \log_a y - 3 \log_a z)$ Using the product and power rules

c) $\log_b \dfrac{xy}{m^3 n^4} = \log_b xy - \log_b m^3 n^4$ Using the quotient rule

$= (\log_b x + \log_b y) - (\log_b m^3 + \log_b n^4)$ Using the product rule

$= \log_b x + \log_b y - \log_b m^3 - \log_b n^4$ Removing parentheses

$= \log_b x + \log_b y - 3 \log_b m - 4 \log_b n$ Using the power rule

❏

EXAMPLE 7

Express as a single logarithm.

a) $\dfrac{1}{2} \log_a x - 7 \log_a y + \log_a z$

b) $\log_a \dfrac{b}{\sqrt{x}} + \log_a \sqrt{bx}$

Solution

a) $\dfrac{1}{2} \log_a x - 7 \log_a y + \log_a z$

$= \log_a x^{1/2} - \log_a y^7 + \log_a z$ Using the power rule

$= (\log_a \sqrt{x} - \log_a y^7) + \log_a z$ Using parentheses to emphasize the order of operations; $x^{1/2} = \sqrt{x}$

$= \log_a \dfrac{\sqrt{x}}{y^7} + \log_a z$ Using the quotient rule

$= \log_a \dfrac{z\sqrt{x}}{y^7}$ Using the product rule

b) $\log_a \dfrac{b}{\sqrt{x}} + \log_a \sqrt{bx} = \log_a \dfrac{b}{\sqrt{x}} \sqrt{bx}$ Using the product rule

$\qquad\qquad\qquad\qquad\qquad = \log_a b\sqrt{b}$ Removing a factor of 1: $\sqrt{x}/\sqrt{x} = 1$

$\qquad\qquad\qquad\qquad\qquad = \log_a b^{3/2}, \quad \text{or} \quad \dfrac{3}{2}\log_a b$ Since $b\sqrt{b} = b^1 \cdot b^{1/2}$ ❑

EXAMPLE 8

Given $\log_a 2 = 0.301$ and $\log_a 3 = 0.477$, find each of the following: **(a)** $\log_a 6$; **(b)** $\log_a \frac{2}{3}$; **(c)** $\log_a 81$; **(d)** $\log_a \frac{1}{3}$; **(e)** $\log_a 2a$; **(f)** $\log_a 5$.

Solution

a) $\log_a 6 = \log_a (2 \cdot 3) = \log_a 2 + \log_a 3$ Using the product rule

$\qquad\qquad\qquad = 0.301 + 0.477 = 0.778$

b) $\log_a \frac{2}{3} = \log_a 2 - \log_a 3$ Using the quotient rule

$\qquad\qquad = 0.301 - 0.477 = -0.176$

c) $\log_a 81 = \log_a 3^4 = 4\log_a 3$ Using the power rule

$\qquad\qquad\qquad = 4(0.477) = 1.908$

d) $\log_a \frac{1}{3} = \log_a 1 - \log_a 3$ Using the quotient rule

$\qquad\qquad = 0 - 0.477 = -0.477$

e) $\log_a 2a = \log_a 2 + \log_a a$ Using the product rule

$\qquad\qquad = 0.301 + 1 = 1.301$

f) $\log_a 5$ *cannot be found using these properties.*
$(\log_a 5 \neq \log_a 2 + \log_a 3)$ ❑

A final property follows from the product rule: Since $\log_a a^k = k\log_a a$, and $\log_a a = 1$, we have $\log_a a^k = k$.

The Logarithm of the Base to a Power

For any base a,

$\qquad \log_a a^k = k$.

(The logarithm, base a, of a to a power is the power.)

This property also follows from the definition of logarithm: k is the power to which you raise a in order to get a^k.

EXAMPLE 9

Simplify: **(a)** $\log_3 3^7$; **(b)** $\log_{10} 10^{5.6}$; **(c)** $\log_e e^{-t}$.

Solution

a) $\log_3 3^7 = 7$ 7 is the power to which you raise 3 in order to get 3^7.

b) $\log_{10} 10^{5.6} = 5.6$

c) $\log_e e^{-t} = -t$ ❑

> CAUTION! Keep in mind that, in general,
> $$\log_a (M + N) \neq \log_a M + \log_a N,$$
> $$\log_a (M - N) \neq \log_a M - \log_a N,$$
> $$\log_a MN \neq (\log_a M)(\log_a N), \text{ and}$$
> $$\log_a (M/N) \neq (\log_a M) \div (\log_a N).$$

EXERCISE SET | 9.4

Express as a sum of logarithms.

1. $\log_2 (32 \cdot 8)$ **2.** $\log_3 (27 \cdot 81)$

3. $\log_4 (64 \cdot 16)$ **4.** $\log_5 (25 \cdot 125)$

5. $\log_c Bx$ **6.** $\log_t 5Y$

Express as a single logarithm.

7. $\log_a 6 + \log_a 70$ **8.** $\log_b 65 + \log_b 2$

9. $\log_c K + \log_c y$ **10.** $\log_t H + \log_t M$

Express as a product.

11. $\log_a x^3$ **12.** $\log_b t^5$

13. $\log_c y^6$ **14.** $\log_{10} y^7$

15. $\log_b C^{-3}$ **16.** $\log_c M^{-5}$

Express as a difference of logarithms.

17. $\log_a \dfrac{67}{5}$ **18.** $\log_t \dfrac{T}{7}$

19. $\log_b \dfrac{3}{4}$ **20.** $\log_a \dfrac{y}{x}$

Express as a single logarithm.

21. $\log_a 15 - \log_a 7$ **22.** $\log_b 42 - \log_b 7$

Express in terms of logarithms of w, x, y, and z.

23. $\log_a x^2 y^3 z$ **24.** $\log_a xy^4 z^3$

25. $\log_b \dfrac{xy^2}{z^3}$ **26.** $\log_b \dfrac{x^2 y^5}{w^4 z^7}$

27. $\log_c \sqrt[3]{\dfrac{x^4}{y^3 z^2}}$ **28.** $\log_a \sqrt{\dfrac{x^6}{y^5 z^8}}$

29. $\log_a \sqrt[4]{\dfrac{x^8 y^{12}}{a^3 z^5}}$ **30.** $\log_a \sqrt[3]{\dfrac{x^6 y^3}{a^2 z^7}}$

Express as a single logarithm and simplify, if possible.

31. $\frac{2}{3} \log_a x - \frac{1}{2} \log_a y$

32. $\frac{1}{2} \log_a x + 3 \log_a y - 2 \log_a x$

33. $\log_a 2x + 3(\log_a x - \log_a y)$

34. $\log_a x^2 - 2 \log_a \sqrt{x}$

35. $\log_a \dfrac{a}{\sqrt{x}} - \log_a \sqrt{ax}$

36. $\log_a (x^2 - 4) - \log_a (x - 2)$

Given $\log_b 3 = 1.099$ and $\log_b 5 = 1.609$, find each of the following.

37. $\log_b 15$ **38.** $\log_b \frac{3}{5}$ **39.** $\log_b \frac{5}{3}$

40. $\log_b \frac{1}{3}$ **41.** $\log_b \frac{1}{5}$ **42.** $\log_b \sqrt{b}$

43. $\log_b \sqrt{b^3}$ **44.** $\log_b 3b$ **45.** $\log_b 5b$

46. $\log_b 9$ **47.** $\log_b 25$ **48.** $\log_b 75$

Simplify.

49. $\log_t t^9$ **50.** $\log_p p^4$

51. $\log_e e^m$ **52.** $\log_Q Q^{-2}$

Solve for x.

53. $\log_3 3^4 = x$ **54.** $\log_5 5^7 = x$

55. $\log_e e^x = -7$ **56.** $\log_a a^x = 2.7$

Skill Maintenance

Compute and simplify. Express answers in the form $a + bi$, where $i^2 = -1$.

57. i^{29} **58.** $(2 + i)(2 - i)$

59. $\dfrac{2 + i}{2 - i}$

60. $(7 - 8i) - (-16 + 10i)$

Synthesis

61. ◈ Explain why $a^{\log_a 5} = 5$.

62. ◈ The product, power, and quotient rules enable us to simplify expressions like $\log_a (rs/pv)$. Explain why such expressions can always be simplified without ever using the quotient rule.

Express as a single logarithm and simplify, if possible.

63. $\log_a (x^8 - y^8) - \log_a (x^2 + y^2)$

64. $\log_a (x + y) + \log_a (x^2 - xy + y^2)$

Express as a sum or difference of logarithms.

65. $\log_a \sqrt{1 - s^2}$

66. $\log_a \dfrac{c - d}{\sqrt{c^2 - d^2}}$

67. If $\log_a x = 2$, $\log_a y = 3$, and $\log_a z = 4$, what is

$$\log_a \frac{\sqrt[3]{x^2 z}}{\sqrt[3]{y^2 z^{-2}}}?$$

68. If $\log_a x = 2$, what is $\log_a (1/x)$?

69. If $\log_a x = 2$, what is $\log_{1/a} x$?

Determine whether each of the following is true. Assume a, x, P, and $Q > 0$.

70. $\dfrac{\log_a P}{\log_a Q} = \log_a \dfrac{P}{Q}$

71. $\dfrac{\log_a P}{\log_a Q} = \log_a P - \log_a Q$

72. $\log_a 3x = \log_a 3 + \log_a x$

73. $\log_a 3x = 3 \log_a x$

74. $\log_a (P + Q) = \log_a P + \log_a Q$

75. $\log_a x^2 = 2 \log_a x$

76. Prove that for $a > 0$ and $x \geq \sqrt{3}$,

$$\log_a \frac{x + \sqrt{x^2 - 3}}{3} = -\log_a \left(x - \sqrt{x^2 - 3} \right).$$

9.5

Common and Natural Logarithms

Common Logarithms on a Calculator • **The Base e and Natural Logarithms on a Calculator** • **Changing Logarithmic Bases** • **Graphs of Exponential and Logarithmic Functions, Base e**

Any positive number different from 1 can be used as the base of a logarithmic function. However, some numbers are easier to use than others, and there are logarithm bases that fit into certain applications more naturally than others. Base-10 logarithms, called **common logarithms,** are useful because they are the same base as our "commonly" used decimal system for naming numbers.

Before calculators became so widely available, common logarithms were extensively used in calculations. In fact, that is why logarithms were invented. Another logarithm base widely used today is an irrational number named e. We will consider e and natural logarithms later in this section. We first consider common logarithms.

Common Logarithms on a Calculator

Before the invention of calculators, tables were developed to list common logarithms. Today we find common logarithms using calculators.

The abbreviation **log,** with no base written, is understood to mean logarithm base 10, or a common logarithm. Thus

$$\log 17 \quad \text{means} \quad \log_{10} 17.$$

On scientific calculators, the key for common logarithms is usually marked $\boxed{\text{LOG}}$. To find the common logarithm of a number, we enter that number and press the $\boxed{\text{LOG}}$ key. Table 2 at the back of this text can also be used.

EXAMPLE 1

Use a calculator to find each number: **(a)** log 53,128; **(b)** log 0.000128.

Solution

a) We enter 53,128 and then press the $\boxed{\text{log}}$ key. We find that

$$\log 53{,}128 \approx 4.7253. \qquad \text{Rounded to four decimal places}$$

b) We enter 0.000128 and then press the $\boxed{\text{log}}$ key. We find that

$$\log 0.000128 \approx -3.8928. \qquad \text{Rounded to four decimal places} \qquad \square$$

The inverse of a logarithmic function is, of course, an exponential function. The inverse of finding a logarithm is called finding an *antilogarithm* or an *inverse logarithm*. To find an antilogarithm, we evaluate a power of the base:

$$\text{for} \quad f(x) = \log x, \quad f^{-1}(x) = \text{antilog } x = 10^x.$$

Generally, there is no key on a calculator marked "antilog." It is up to you to know that to find the inverse, or antilogarithm, you must use the $\boxed{10^x}$ key, if there is one. If there is no such key, then you must raise 10 to the x power using a key marked $\boxed{x^y}$. Often the $\boxed{\text{log}}$ key serves as the $\boxed{10^x}$ key after a "shift" or "inverse" key is pushed.

EXAMPLE 2

Use a calculator to find each number.

a) antilog 2.1792   **b)** antilog (-4.678834)

Solution

a) We enter 2.1792 and then press the $\boxed{10^x}$ key. We find that

$$\text{antilog } 2.1792 = 10^{2.1792} \approx 151.078.$$

b) $\text{antilog } (-4.678834) = 10^{-4.678834} \approx 0.00002095 \qquad \square$

The Base *e* and Natural Logarithms on a Calculator

When interest is computed n times a year, the compound interest formula is

$$A = P\left(1 + \frac{i}{n}\right)^{nt},$$

where A is the amount that an initial investment P will be worth after t years at interest rate i. Suppose that \$1 is an initial investment at 100% interest for 1 year (no bank would pay this). The preceding formula becomes a function A defined in terms of the number of compounding periods n:

$$A(n) = \left(1 + \frac{1}{n}\right)^n.$$

Let us find some function values. We round to six decimal places, using a calculator with a power key $\boxed{x^y}$.

n	$A(n) = \left(1 + \frac{1}{n}\right)^{n}$
1 (compounded annually)	$2.00
2 (compounded semiannually)	$2.25
3	$2.370370
4 (compounded quarterly)	$2.441406
5	$2.488320
100	$2.704814
365 (compounded daily)	$2.714567
8760 (compounded hourly)	$2.718127

The numbers in this table get closer and closer to a very important number in mathematics, called e. Being irrational, its decimal representation does not terminate or repeat.

$$e \approx 2.7182818284 \ldots$$

Logarithms base e are called **natural logarithms,** or **Napierian logarithms,** in honor of John Napier (1550–1617), who first ''discovered'' logarithms.

The abbreviation ''ln'' is generally used with natural logarithms. Thus,

$\ln 53$ means $\log_e 53$.

The calculator key $\boxed{\text{ln}}$ is used for natural logarithms.

EXAMPLE 3

Find ln 4568.

Solution   We enter 4568 and then press the $\boxed{\text{ln}}$ key. We find that

$\ln 4568 \approx 8.4268.$ Rounded to four decimal places   ❏

To find the antilogarithm, base e, we use the $\boxed{e^x}$ key, if there is one. If not, we can use a power key $\boxed{x^y}$ and an approximation for e, say, 2.71828. Often the $\boxed{\text{ln}}$ key serves as the $\boxed{e^x}$ key after a ''shift'' or ''inverse'' key is pushed.

EXAMPLE 4

Find antilog$_e$ (-5.6734).

Solution   The problem gives the exponent. We enter -5.6734 and then press the $\boxed{e^x}$ key. We find that

$$\text{antilog}_e\, (-5.6734) = e^{-5.6734}$$
$$\approx 0.003436. \qquad ❏$$

Changing Logarithmic Bases

Most calculators give the values of both common logarithms and natural logarithms. To find a logarithm with some other base, we can use the following conversion formula.

The
Change-of-Base
Formula

For any logarithmic bases a and b, and any positive number M,

$$\log_b M = \frac{\log_a M}{\log_a b}.$$

Proof. Let $x = \log_b M$. Then, writing an equivalent exponential equation, we have $b^x = M$. Next we take the logarithm base a on both sides. This gives us

$$\log_a b^x = \log_a M.$$

By the power rule for logarithms,

$$x \log_a b = \log_a M,$$

and solving for x, we obtain

$$x = \frac{\log_a M}{\log_a b}.$$

But $x = \log_b M$, so we have

$$\log_b M = \frac{\log_a M}{\log_a b},$$

which is the change-of-base formula. ❏

EXAMPLE 5

Find $\log_5 8$ using common logarithms.

Solution   Let $a = 10$, $b = 5$, and $M = 8$. Then we substitute into the change-of-base formula:

$$\log_5 8 = \frac{\log_{10} 8}{\log_{10} 5} \qquad \text{Substituting into } \log_b M = \frac{\log_a M}{\log_a b}$$

$$\approx \frac{0.9031}{0.6990} \qquad \text{Using the } \boxed{\text{LOG}} \text{ key twice}$$

$$\approx 1.2920. \qquad \text{When using a calculator, you need not round before dividing.}$$

To check, we use a calculator with an $\boxed{x^y}$ key to verify that

$$5^{1.2920} \approx 8.$$ ❏

We can also use base e for a conversion.

EXAMPLE 6

Find $\log_4 31$ using natural logarithms.

Solution   Substituting e for a, 4 for b, and 31 for M, we have

$$\log_4 31 = \frac{\log_e 31}{\log_e 4} \qquad \text{Substituting into } \log_b M = \frac{\log_a M}{\log_a b}$$

$$= \frac{\ln 31}{\ln 4}$$

or

$$\log_4 31 \approx \frac{3.4340}{1.3863} \qquad \text{Using the } \boxed{\text{LN}} \text{ key twice}$$

$$\approx 2.4771.$$ ❑

Graphs of Exponential and Logarithmic Functions, Base e

E X A M P L E 7

Graph $f(x) = e^x$ and $g(x) = e^{-x}$.

Solution We use a calculator with an $\boxed{e^x}$ key to find approximate values of e^x and e^{-x}. Using these values, we can draw the graphs of the functions.

x	e^x	e^{-x}
0	1	1
1	2.7	0.4
2	7.4	0.1
−1	0.4	2.7
−2	0.1	7.4

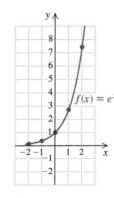
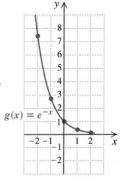

❑

E X A M P L E 8

Graph: $f(x) = e^{-0.5x}$.

Solution We find some solutions with a calculator, plot them, and then draw the graph. For example, $f(2) = e^{-0.5(2)} = e^{-1} \approx 0.4$.

x	$e^{-0.5x}$
0	1
1	0.6
2	0.4
3	0.2
−1	1.6
−2	2.7
−3	4.5

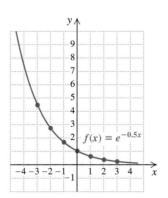

❑

E X A M P L E 9

Graph: **(a)** $g(x) = \ln x$; **(b)** $f(x) = \ln(x + 3)$.

Solution

a) We find some solutions with a calculator and then draw the graph. As expected, the graph is a reflection across the line $y = x$ of the graph of $y = e^x$.

x	ln x
1	0
4	1.4
7	1.9
0.5	−0.7

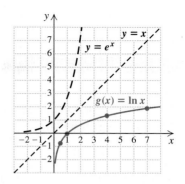

b) We find some solutions with a calculator, plot them, and then draw the graph.

x	ln (x + 3)
0	1.1
1	1.4
2	1.6
3	1.8
4	1.9
−1	0.7
−2	0
−2.5	−0.7

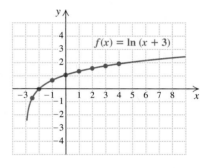

Note that the graph of $y = \ln (x + 3)$ is the graph of $y = \ln x$ translated 3 units to the left.

TECHNOLOGY CONNECTION

Graphs that involve e^x and/or ln x are easily drawn on a grapher because there are usually keys for these functions. Since these functions are inverses of each other, most graphers use the same key for both.

To graph a function like $y_1 = 3.4 \ln x - 0.25e^x$, check the domain and set the window accordingly. Since ln x is not defined for $x \le 0$, our window need not include any negative values of x. The figure at right shows the function drawn in the window [0, 5, −10, 10].

$y_1 = 3.4 \ln x - 0.25 e^x$

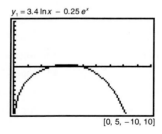

[0, 5, −10, 10]

Use a grapher to draw each function.

TC1. $f(x) = 2.3e^x - 3.8$

TC2. $f(x) = \dfrac{3e^x - 6.3}{e^x + 2.1}$

TC3. $f(x) = xe^{-2.3x} + 3x^2$

TC4. $f(x) = 7.4e^x \ln x$

TC5. $f(x) = 5.3 \ln (x - 2.1)$

TC6. $f(x) = 2x^3 \ln x$

EXERCISE SET | 9.5

Use a calculator to find each of the following logarithms and antilogarithms.

1. log 2

2. log 5

3. log 6.34

4. log 5.02

5. log 45

6. log 74

7. log 437

8. log 295

9. log 13,400

10. log 93,100

11. log 0.052

12. log 0.387

13. antilog 3

14. antilog 5

15. antilog 2.7

16. antilog 14.8

17. antilog 0.477133

18. antilog 0.06532

19. antilog (-0.5465)

20. antilog (-0.3404)

21. $10^{-2.9523}$

22. $10^{4.8982}$

23. ln 2

24. ln 3

25. ln 62

26. ln 30

27. ln 4365

28. ln 901.2

29. ln 0.0062

30. ln 0.00073

31. antilog$_e$ 3.6052

32. antilog$_e$ 4.9312

33. antilog$_e$ (-6.0751)

34. antilog$_e$ (-2.3001)

35. antilog$_e$ 0.00567

36. antilog$_e$ 0.01111

37. antilog$_e$ 34

38. antilog$_e$ 56

39. $e^{2.0325}$

40. $e^{-1.3783}$

Find each of the following logarithms using the change-of-base formula.

41. $\log_6 100$

42. $\log_3 18$

43. $\log_2 10$

44. $\log_7 50$

45. $\log_{200} 30$

46. $\log_{100} 30$

47. $\log_{0.5} 5$

48. $\log_{0.1} 3$

49. $\log_2 0.2$

50. $\log_2 0.08$

51. $\log_\pi 58$

52. $\log_\pi 200$

Graph.

53. $f(x) = e^x$

54. $f(x) = e^{3x}$

55. $f(x) = e^{-3x}$

56. $f(x) = e^{-x}$

57. $f(x) = e^{x-1}$

58. $f(x) = e^{x+2}$

59. $f(x) = e^{-x} - 3$

60. $f(x) = e^x + 3$

61. $f(x) = 5e^{0.2x}$

62. $f(x) = 8e^{0.6x}$

63. $f(x) = 20e^{-0.5x}$

64. $f(x) = 10e^{-0.4x}$

65. $f(x) = \ln (x + 4)$

66. $f(x) = \ln (x + 1)$

67. $f(x) = 3 \ln x$

68. $f(x) = 2 \ln x$

69. $f(x) = \ln (x - 2)$

70. $f(x) = \ln (x - 3)$

71. $f(x) = \ln x - 3$

72. $f(x) = \ln x + 2$

Skill Maintenance

Solve for x.

73. $4x^2 - 25 = 0$

74. $5x^2 - 7x = 0$

Solve.

75. $x^{1/2} - 6x^{1/4} + 8 = 0$

76. $2y - 7\sqrt{y} + 3 = 0$

Synthesis

77. ◈ Explain how the graph of $f(x) = e^x$ could be used to graph the function given by $g(x) = 1 + \ln x$.

78. ◈ Explain how the graph of $f(x) = \ln x$ could be used to graph the function given by $g(x) = e^{x-1}$.

79. Find a formula for converting common logarithms to natural logarithms.

80. Find a formula for converting natural logarithms to common logarithms.

Solve for x.

81. $\log 374x = 4.2931$

82. $\log 95x^2 = 3.0177$

83. $\log 692 + \log x = \log 3450$

84. $\dfrac{4.31}{\ln x} = \dfrac{28}{3.01}$

9.6

Solving Exponential and Logarithmic Equations

Solving Exponential Equations • Solving Logarithmic Equations

Solving Exponential Equations

Equations with variables in exponents, such as $5^x = 12$ and $2^{7x} = 64$, are called **exponential equations.** Sometimes, as in the case of $2^{7x} = 64$, we can write each side as a power of the same number:

$$2^{7x} = 2^6.$$

Since the base is the same, 2, the exponents are the same. We can equate exponents and then solve:

$$7x = 6$$
$$x = \tfrac{6}{7}.$$

We use the following property.

For any $a > 0$, $a \neq 1$,

$$a^x = a^y \quad \text{is equivalent to} \quad x = y.$$

This follows from the fact that $f(x) = a^x$ is a one-to-one function, so that if two outputs a^x and a^y are equal, their inputs are equal.

EXAMPLE 1

Solve: $2^{3x-5} = 16$.

Solution   Note that $16 = 2^4$. Thus we can write each side as a power of the same number:

$$2^{3x-5} = 2^4.$$

Since the base is the same, 2, the exponents must be the same. Thus,

$$3x - 5 = 4 \qquad \text{Equating exponents}$$
$$3x = 9$$
$$x = 3.$$

Check:
$$\begin{array}{c|c} 2^{3x-5} = 16 \\ \hline 2^{3 \cdot 3 - 5} \; ? \; 16 \\ 2^{9-5} \\ 2^4 \\ 16 \;\big|\; 16 \quad \text{TRUE} \end{array}$$

The solution is 3. ❑

When it does not seem possible to write each side as a power of the same base, we can take the common or natural logarithm on each side and then use the power rule for logarithms.

EXAMPLE 2

Solve: $5^x = 12$.

Solution

$$5^x = 12$$
$$\log 5^x = \log 12 \qquad \text{Taking the common logarithm on both sides}$$
$$x \log 5 = \log 12 \qquad \text{Using the power rule}$$
$$x = \frac{\log 12}{\log 5} \quad \leftarrow \boxed{\text{CAUTION! This is not } \log 12 - \log 5!}$$

This is an exact answer. We cannot simplify further, but we can approximate using a calculator or a table:

$$x = \frac{\log 12}{\log 5} \approx \frac{1.0792}{0.6990} \approx 1.544.$$

You can check this by finding $5^{1.544}$ using the $\boxed{x^y}$ key on a calculator.   ❏

If we prefer, we can take the logarithm with e as the base. This will often ease our work.

EXAMPLE 3

Solve: $e^{0.06t} = 1500$.

Solution   We take the natural logarithm on both sides:

$$\ln e^{0.06t} = \ln 1500 \qquad \text{Taking the natural logarithm on both sides}$$
$$0.06t = \ln 1500 \qquad \text{Finding the logarithm of the base to a power: } \log_a a^k = k$$
$$0.06t \approx 7.3132 \qquad \text{Using a calculator}$$
$$t \approx 121.89.$$ ❏

Solving Logarithmic Equations

Equations containing logarithmic expressions are called **logarithmic equations.** We solved some such equations in Section 9.3. We did so by converting to an equivalent exponential equation.

EXAMPLE 4

Solve: $\log_4 (8x - 6) = 3$.

Solution   We write an equivalent exponential equation:

$$8x - 6 = 4^3$$
$$8x = 70 \qquad 4^3 = 64 \text{ and } 64 + 6 = 70$$
$$x = \frac{70}{8}, \quad \text{or } \frac{35}{4}.$$

The check is left for the student. The solution is $\frac{35}{4}$.   ❏

To solve logarithmic equations, first try to obtain a single logarithmic expression on one side and then write an equivalent exponential equation.

EXAMPLE 5

Solve: $\log x + \log (x - 3) = 1$.

Solution   We have common logarithms here. It will help to write in the base, 10.

$\log_{10} x + \log_{10} (x - 3) = 1$

$\log_{10} [x(x - 3)] = 1$ Using the product rule for logarithms to obtain a single logarithm

$x(x - 3) = 10^1$ Writing an equivalent exponential equation

$x^2 - 3x = 10$

$x^2 - 3x - 10 = 0$

$(x + 2)(x - 5) = 0$ Factoring

$x + 2 = 0$ *or* $x - 5 = 0$ Principle of zero products

$x = -2$ *or* $x = 5$

Check: For -2:

$$\frac{\log x + \log (x - 3) = 1}{\log (-2) + \log (-2 - 3) \ ? \ 1}$$

The number -2 *does not check* because negative numbers do not have logarithms.

For 5:

$$\frac{\log x + \log (x - 3) = 1}{\begin{array}{c} \log 5 + \log (5 - 3) \ ? \ 1 \\ \log 5 + \log 2 \\ \log 10 \\ 1 \ \big| \ 1 \quad \text{TRUE} \end{array}}$$

The solution is 5.   ❑

TECHNOLOGY

CONNECTION

Exponential and logarithmic equations can be solved using a grapher by graphing the function on each side of the equals sign, making sure that the window is large enough to show all points of intersection. Then zoom in on any such points and use the Trace feature to determine their x-coordinates.

For example, to solve $e^{0.5x} - 7 = 2x + 6$, we graph the curves $y_1 = e^{0.5x} - 7$ and $y_2 = 2x + 6$ as shown in the figure at right. We then adjust the window and use the Trace feature to pinpoint any intersections. The x-coordinates at the intersections are approximately -6.48 and 6.52.

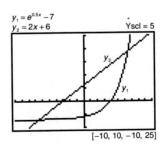

$y_1 = e^{0.5x} - 7$
$y_2 = 2x + 6$
$\text{Yscl} = 5$
$[-10, 10, -10, 25]$

Use a grapher to find solutions, accurate to the nearest hundredth, for each of the following equations.

TC1. $e^{7x} = 14$

TC2. $8e^{0.5x} = 3$

TC3. $xe^{3x - 1} = 5$

TC4. $4 \ln (x + 3.4) = 2.5$

TC5. $\ln 3x = 3x - 8$

TC6. $\ln x^2 = -x^2$

EXAMPLE 6

Solve: $\log_2 (x + 7) - \log_2 (x - 7) = 3$.

Solution

$$\log_2 (x + 7) - \log_2 (x - 7) = 3$$

$$\log_2 \frac{x + 7}{x - 7} = 3 \qquad \text{Using the quotient rule for logarithms to obtain a single logarithm}$$

$$\frac{x + 7}{x - 7} = 2^3 \qquad \text{Writing an equivalent exponential expression}$$

$$\frac{x + 7}{x - 7} = 8$$

$$x + 7 = 8(x - 7) \qquad \text{Multiplying by the LCD, } x - 7$$

$$x + 7 = 8x - 56 \qquad \text{Using the distributive law}$$

$$63 = 7x$$

$$\frac{63}{7} = x$$

$$9 = x$$

Check:
$$\begin{array}{c|c} \log_2 (x + 7) - \log_2 (x - 7) = 3 \\ \hline \log_2 (9 + 7) - \log_2 (9 - 7) \ ? \ 3 \\ \log_2 16 - \log_2 2 \\ 4 - 1 \\ 3 \ \bigg| \ 3 \quad \text{TRUE} \end{array}$$

The solution is 9.   ❑

EXERCISE SET | 9.6

Solve.

1. $2^x = 8$

2. $3^x = 81$

3. $4^x = 256$

4. $5^x = 125$

5. $2^{2x} = 32$

6. $4^{3x} = 64$

7. $3^{5x} = 27$

8. $5^{7x} = 625$

9. ▣ $2^x = 9$

10. ▣ $2^x = 30$

11. ▣ $2^x = 10$

12. ▣ $2^x = 33$

13. $5^{4x - 7} = 125$

14. $4^{3x + 5} = 16$

15. $3^{x^2} \cdot 3^{4x} = \frac{1}{27}$

16. $3^{5x} \cdot 3^{2x^2} = 27$

17. ▣ $4^x = 7$

18. ▣ $8^x = 10$

19. ▣ $e^t = 100$

20. ▣ $e^t = 1000$

21. ▣ $e^{-t} = 0.1$

22. ▣ $e^{-t} = 0.01$

23. ▣ $e^{-0.02t} = 0.06$

24. ▣ $e^{0.07t} = 2$

25. ▣ $2^x = 3^{x - 1}$

26. ▣ $3^{x + 2} = 5^{x - 1}$

27. ▣ $(2.8)^x = 41$

28. ▣ $(3.4)^x = 80$

29. ▣ $20 - (1.7)^x = 0$

30. ▣ $125 - (4.5)^y = 0$

31. $\log_3 x = 3$

32. $\log_5 x = 4$

33. $\log_2 x = -3$

34. $\log_4 x = \frac{1}{2}$

35. $\log x = 1$

36. $\log x = 3$

37. $\log x = -2$

38. $\log x = -3$

39. $\ln x = 2$

40. $\ln x = 1$

41. $\ln x = -1$

42. $\ln x = -3$

43. $\log_5 (2x - 7) = 3$

44. $\log_2 (7 - 6x) = 5$

45. $\log x + \log (x - 9) = 1$

46. $\log x + \log (x + 9) = 1$

47. $\log x - \log (x + 3) = -1$

48. $\log (x + 9) - \log x = 1$

49. $\log_2 (x + 1) + \log_2 (x - 1) = 3$

50. $\log_4 (x + 3) - \log_4 (x - 5) = 2$

51. $\log_4 (x + 6) - \log_4 x = 2$

52. $\log_2 x + \log_2 (x - 2) = 3$

53. $\log_4 (x + 3) + \log_4 (x - 3) = 2$

54. $\log_5 (x + 4) + \log_5 (x - 4) = 2$

Skill Maintenance _____

Simplify.

55. $(125x^7 y^{-2} z^6)^{-2/3}$ **56.** i^{79}

Solve.

57. $E = mc^2$, for c (Assume $E, m, c > 0$.)

58. $x^4 + 400 = 104x^2$

Synthesis _____

59. ◈ Explain how Exercises 35–38 could be solved using the graph of $f(x) = \log x$.

60. ◈ In Example 2, we took the common logarithm on both sides. What would have happened had we taken the natural logarithm on both sides?

Solve.

61. $8^x = 16^{3x + 9}$

62. $27^x = 81^{2x - 3}$

63. $\log_6 (\log_2 x) = 0$

64. $\log_x (\log_3 27) = 3$

65. $\log_5 \sqrt{x^2 - 9} = 1$

66. $x \log \frac{1}{8} = \log 8$

67. $\log (\log x) = 5$

68. $2^{x^2 + 4x} = \frac{1}{8}$

69. $\log x^2 = (\log x)^2$

70. $\log_5 |x| = 4$

71. $\log x^{\log x} = 25$

72. $\log \sqrt{2x} = \sqrt{\log 2x}$

73. $(81^{x - 2})(27^{x + 1}) = 9^{2x - 3}$

74. $3^{2x} - 8 \cdot 3^x + 15 = 0$

75. $3^{2x} - 3^{2x - 1} = 18$

76. Given that $2^y = 16^{x - 3}$ and $3^{y + 2} = 27^x$, find the value of $x + y$.

77. If $x = (\log_{125} 5)^{\log_5 125}$, what is the value of $\log_3 x$?

78. ◪ Use a grapher to check the solutions of Exercises 11, 19, 25, 35, and 47.

9.7

Applications of Exponential and Logarithmic Functions

Exponential Growth • **Exponential Decay**

We now consider applications of exponential and logarithmic functions. A calculator with logarithmic and power keys would be most helpful.

EXAMPLE 1 _____

Chemistry: pH of substances. In chemistry the pH of a substance is defined as follows:

$$pH = -\log [H^+],$$

where $[H^+]$ is the hydrogen ion concentration in moles per liter.

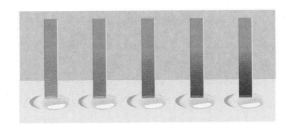

a) The hydrogen ion concentration of pineapple juice is 1.6×10^{-4} moles per liter. Find the pH.

b) The pH of a common hair rinse is 2.9. Find the hydrogen ion concentration.

Solution

a) To find the pH of pineapple juice, we substitute 1.6×10^{-4} for $[H^+]$ in the formula for pH:

$$pH = -\log \, [H^+] = -\log \, [1.6 \times 10^{-4}]$$
$$= -[\log 1.6 + \log 10^{-4}]$$
$$= -[\log 1.6 + (-4)] \qquad \text{log 1.6 is in Table 2.}$$
$$\approx -[0.2041 + (-4)]$$
$$\approx -(-3.7959) \approx 3.8 \qquad \text{This can be found directly using a calculator.}$$

The pH of pineapple juice is about 3.8.

b) To find the hydrogen ion concentration of the hair rinse, we substitute 2.9 for pH in the formula and solve for $[H^+]$:

$$2.9 = -\log \, [H^+]$$
$$-2.9 = \log \, [H^+] \qquad \text{Multiplying by } -1 \text{ on both sides}$$
$$10^{-2.9} = [H^+] \qquad \text{Writing an equivalent exponential equation}$$
$$0.0013 \approx [H^+]$$
$$[H^+] \approx 1.3 \times 10^{-3} \text{ moles per liter.} \qquad \text{Writing scientific notation: a number between 1 and 10, times a power of 10} \qquad \square$$

EXAMPLE 2

Earthquake magnitude.   The magnitude R (measured on the Richter scale) of an earthquake of intensity I is defined as

$$R = \log \frac{I}{I_0},$$

where I_0 is a minimum intensity used for comparison. We can regard I_0 as the intensity of the weakest earthquake that can be recorded on a seismograph. When one earthquake is 10 times as intense as another, its magnitude on the Richter scale is 1 higher. If an earthquake is 100 times as intense as another, its magnitude on the Richter scale is 2 higher, and so on. Thus an earthquake whose magnitude is 7 on the Richter scale is 10 times as intense as an earthquake whose magnitude is 6. The San Francisco (Loma Prieta) earthquake of 1989 had an intensity of $10^{7.2}I_0$. What was its magnitude on the Richter scale?

Solution   We substitute into the formula:

$$R = \log \frac{10^{7.2}I_0}{I_0} = \log \, 10^{7.2} = 7.2.$$

The magnitude on the Richter scale was 7.2. $\square$

EXAMPLE 3

Interest compounded annually.   Suppose that $30,000 is invested at 8% interest, compounded annually. In t years, it will grow to the amount A given by the function

$$A(t) = 30{,}000(1.08)^t.$$

(See Example 5 in Section 9.1.)

a) How long will it take until there is $150,000 in the account?

b) Let T = the amount of time it takes for the $30,000 to double itself. Find the *doubling time, T.*

Solution

a) We set $A(t) = 150,000$ and solve for t:

$$150,000 = 30,000(1.08)^t$$

$$\frac{150,000}{30,000} = 1.08^t \qquad \text{Dividing by 30,000 on both sides}$$

$$5 = 1.08^t$$

$$\log 5 = \log 1.08^t \qquad \text{Taking the common logarithm on both sides}$$

$$\log 5 = t \log 1.08 \qquad \text{Using the power rule for logarithms}$$

$$\frac{\log 5}{\log 1.08} = t.$$

We simplify further by using a calculator or Table 2 and approximating:

$$t = \frac{\log 5}{\log 1.08} \approx \frac{0.69897}{0.03342} \approx 20.9.$$

It will take about 20.9 years for the $30,000 to grow to $150,000.

Calculator Note: When doing a calculation like this on your calculator, it is best not to stop and round the approximate values of the logarithms. Just find and divide. Answers will be found that way in the exercises. You may notice some variation in the last one or two decimal places if you round as you go.

b) To find the doubling time T, we set $A(t) = 60,000$ and $t = T$ and solve for T:

$$60,000 = 30,000(1.08)^T$$

$$2 = (1.08)^T \qquad \text{Dividing by 30,000 on both sides}$$

$$\log 2 = \log (1.08)^T \qquad \text{Taking the common logarithm on both sides}$$

$$\log 2 = T \log 1.08 \qquad \text{Using the power rule for logarithms}$$

$$T = \frac{\log 2}{\log 1.08} \approx \frac{0.30103}{0.03342} \approx 9.0.$$

The doubling time is about 9 years. ◻

Exponential Growth

An equation of the form

$$P(t) = P_0 e^{kt}$$

can model the growth of many things, ranging from growing populations to investments that are increasing in value. In this equation, P_0 is the population at time 0, $P(t)$ is the population at time t, and k is a positive constant that depends on the situation. The constant k is often called the **exponential growth rate.** You should regard the exponential growth rate as a population's rate of growth at any *instant* in time. Since the population is continually growing, the percent of total growth after one year will exceed the exponential growth rate.

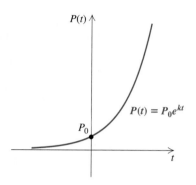

EXAMPLE 4

Growth of the United States.   In 1992, the population of the United States was 249 million and the exponential growth rate was 0.9% per year.

a) Find the exponential growth function.
b) What would you expect the population to be in the year 2000?

Solution

a) In 1992, at $t = 0$, the population was 249 million. We substitute 249 for P_0 and 0.9%, or 0.009, for k to obtain the exponential growth function:

$$P(t) = 249e^{0.009t}.$$

b) In 2000, we have $t = 8$. That is, 8 years have passed. To find the population in 2000, we substitute 8 for t:

$$P(8) = 249e^{0.009(8)} \qquad \text{Substituting 8 for } t$$
$$= 249e^{0.072}$$
$$\approx 249(1.0747) \qquad \text{Finding } e^{0.072} \text{ using a calculator}$$
$$\approx 267.6 \text{ million.}$$

The population of the United States in 2000 will be about 267.6 million.

❑

EXAMPLE 5

Spread of AIDS.   The number of people $N(t)$ infected with a contagious disease at time t usually increases exponentially. Through August 1989, 100,000 cases of AIDS had been reported in the United States. By the end of December 1991, the number had grown to 200,000.

a) Find the exponential growth rate and the exponential growth function.
b) Predict the year in which the 500,000th case will occur.

Solution

a) We use the equation $N(t) = N_0 e^{rt}$, where t is the number of months since August 1989. In August 1989, at $t = 0$, a total of 100,000 cases had been reported. We substitute 100,000 for N_0:

$$N(t) = 100,000e^{rt}.$$

To find the exponential growth rate r, observe that the 200,000th case was reported in December 1991, or 28 months after August 1989. Substituting, we can solve for r:

$$N(28) = 100,000e^{r \cdot 28}$$
$$200,000 = 100,000e^{28r}$$

$$2 = e^{28r} \qquad \text{Dividing by 100,000 on both sides}$$
$$\ln 2 = \ln e^{28r} \qquad \text{Taking the natural logarithm on both sides}$$
$$\ln 2 = 28r \qquad \text{Finding the logarithm of the base to a power}$$
$$\frac{\ln 2}{28} = r \qquad \text{Dividing by 28 on both sides}$$
$$0.025 \approx r. \qquad \text{Using a calculator and rounding}$$

The exponential function is $N(t) = 100,000e^{0.025t}$, where t is measured in months since August 1989.

b) To predict when the 500,000th case will occur, we replace $N(t)$ with 500,000 and solve for t:

$$500,000 = 100,000e^{0.025t}$$

$$5 = e^{0.025t} \qquad \text{Dividing by 100,000 on both sides}$$
$$\ln 5 = \ln e^{0.025t} \qquad \text{Taking the natural logarithm on both sides}$$
$$\ln 5 = 0.025t$$
$$\frac{\ln 5}{0.025} = t$$
$$64 \approx t. \qquad \text{Using a calculator}$$

Since 64 months is 5 years and 4 months, we predict that the 500,000th case will occur 5 years and 4 months from August 1989, or in December 1994. ❑

EXAMPLE 6

Interest compounded continuously.   Suppose an amount of money P_0 is invested in a savings account at interest rate k, compounded continuously. That is, suppose that interest is computed every "instant" and added to the amount in the account. The balance $P(t)$, after t years, is given by the exponential function

$$P(t) = P_0e^{kt}.$$

a) Suppose that $30,000 is invested and grows to $44,754.75 in 5 years. Find the exponential growth function.

b) After what amount of time will the $30,000 double itself?

Solution

a) At $t = 0$, $P(0) = 30,000$. Thus the exponential growth function is

$$P(t) = 30,000e^{kt}, \qquad \text{where } k \text{ must still be determined.}$$

We know that at $t = 5$, $P(5) = 44{,}754.75$. We substitute and solve for k:

$$44{,}754.75 = 30{,}000e^{k(5)} = 30{,}000e^{5k}$$

$$\frac{44{,}754.75}{30{,}000} = e^{5k} \qquad \text{Dividing on both sides by 30,000}$$

$$1.491825 = e^{5k}$$

$$\ln 1.491825 = \ln e^{5k} \qquad \text{Taking the natural logarithm on both sides}$$

$$0.4 \approx 5k \qquad \text{Finding } \ln 1.491825 \text{ on a calculator and simplifying } \ln e^{5k}$$

$$\frac{0.4}{5} = 0.08 \approx k.$$

The interest rate is about 0.08, or 8%, compounded continuously. Note that since interest is being compounded continuously, the interest earned each year is more than 8%. The exponential growth function is

$$P(t) = 30{,}000e^{0.08t}.$$

b) To find the doubling time T, we replace $P(T)$ with 60,000 and solve for T:

$$60{,}000 = 30{,}000e^{0.08T}$$

$$2 = e^{0.08T}$$

$$\ln 2 = \ln e^{0.08T} \qquad \text{Taking the natural logarithm on both sides}$$

$$\ln 2 = 0.08T$$

$$\frac{\ln 2}{0.08} = T \qquad \text{Dividing}$$

$$\frac{0.693147}{0.08} \approx T$$

$$8.7 \approx T.$$

Thus the original investment of \$30,000 will double in about 8.7 years.   ❏

Comparing Examples 3(b) and 6(b), we see that for any specified interest rate, continuous compounding gives the highest yield and the shortest doubling time.

Exponential Decay

The function

$$P(t) = P_0e^{-kt}, \qquad k > 0,$$

can model the decline, or decay, of a population or quantity. An example is the decay of a radioactive substance. Here P_0 is the amount of the substance at time $t = 0$, $P(t)$ is the amount of the substance remaining at time t, and k is a positive constant that depends on the situation. The constant k is called the **decay rate.** The **half-life** of a substance is the amount of time necessary for half of the substance to decay.

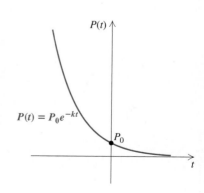

EXAMPLE 7

Carbon dating. The radioactive element carbon-14 has a half-life of 5750 years. The percentage of carbon-14 present in the remains of animal bones can be used to determine age. How old is an animal bone that has lost 40% of its carbon-14?

Solution   We first find k. To do so, we use the concept of half-life. When $t = 5750$ (the half-life), $P(t)$ will be half of P_0. Then

$$0.5P_0 = P_0 e^{-k(5750)}$$

$$0.5 = e^{-5750k} \qquad \text{Dividing by } P_0 \text{ on both sides}$$

$$\ln 0.5 = \ln e^{-5750k} \qquad \text{Taking the natural logarithm on both sides}$$

$$\ln 0.5 = -5750k$$

$$\frac{\ln 0.5}{-5750} = k$$

$$0.00012 \approx k.$$

Now we have the function

$$P(t) = P_0 e^{-0.00012t}.$$

(*Note:* This equation can be used for any subsequent carbon-dating problem.) If an animal bone has lost 40% of its carbon-14 from an initial amount P_0, then $60\%(P_0)$ is the amount present. To find the age t of the bone, we solve this equation for t:

$$0.6P_0 = P_0 e^{-0.00012t} \qquad \text{We want to find } t \text{ for which } P(t) = 0.6P_0.$$

$$0.6 = e^{-0.00012t}$$

$$\ln 0.6 = \ln e^{-0.00012t}$$

$$-0.5108 \approx -0.00012t$$

$$t \approx \frac{-0.5108}{-0.00012}$$

$$t \approx 4257.$$

The animal bone is about 4257 years old.   ❏

EXERCISE SET | 9.7 |

 Solve.

1. *Compact discs.* The number of compact discs purchased each year is increasing exponentially. The number N, in millions, purchased is given by

$$N(t) = 7.5(6)^{0.5t},$$

where $t = 0$ corresponds to 1985, $t = 1$ corresponds to 1986, and so on, t being the number of years after 1985.

a) After what amount of time will one billion compact discs be sold in a year?

b) What is the doubling time on the sale of compact discs?

2. *Spread of a rumor.* The number of people who have heard a rumor increases exponentially. If all who hear a rumor repeat it to two people a day, and if 20 people start the rumor, the number of people N who have heard the rumor after t days is given by

$$N(t) = 20(3)^t.$$

a) After what amount of time will 1000 people have heard the rumor?

b) What is the doubling time on the number of people who have heard the rumor?

3. *Interest compounded annually.* Suppose that $50,000 is invested at 6% interest, compounded annually. After t years, it grows to the amount A given by the function

$$A(t) = 50,000(1.06)^t.$$

a) After what amount of time will there be $450,000 in the account?
b) Find the doubling time.

4. *Turkey consumption.* The amount of turkey consumed by each person in one year in the United States is increasing exponentially. Assume that $t = 0$ corresponds to 1937. The amount of turkey, in pounds per person, consumed t years after 1937 is given by the function

$$N(t) = 2.3(3)^{0.033t}.$$

a) After what amount of time will the consumption rate be 20 lb of turkey per year?
b) What is the doubling time on the consumption of turkey?

5. *Recycling aluminum cans.* Approximately two thirds of all aluminum cans distributed will be recycled each year. A beverage company distributes 250,000 cans. The number still in use after t years is given by the function

$$N(t) = 250,000\left(\tfrac{2}{3}\right)^t.$$

a) After how many years will 60,000 cans still be in use?
b) After what amount of time will only 1000 cans still be in use?

6. *Salvage value.* An office machine is purchased for $5200. Its value each year is about 80% of the preceding year. Its value in dollars after t years is given by the exponential function

$$V(t) = 5200(0.8)^t.$$

a) After what amount of time will the salvage value be $1200?
b) After what amount of time will the salvage value be half the original value?

7. *Interest compounded continuously.* Suppose that P_0 is invested in a savings account where interest is compounded continuously at 9% per year. That is, the balance $P(t)$ after time t, in years, is an exponential function of the form

$$P(t) = P_0 e^{kt}.$$

a) Express $P(t)$ in terms of P_0 and 0.09.
b) Suppose that $1000 is invested. What is the balance after 1 year? after 2 years?
c) When will an investment of $1000 double itself?

8. *Interest compounded continuously.* Suppose that P_0 is invested in a savings account where interest is compounded continuously at 10% per year. That is, the balance $P(t)$ after time t, in years, is an exponential function of the form

$$P(t) = P_0 e^{kt}.$$

a) Express $P(t)$ in terms of P_0 and 0.10.
b) Suppose that $20,000 is invested. What is the balance after 1 year? after 2 years?
c) When will an investment of $20,000 double itself?

9. *Population growth.* The exponential growth rate of the population of Europe west of Russia is 1% per year. What is the doubling time?

10. *Population growth.* The exponential growth rate of the population of Central America is 3.5% per year (one of the highest in the world). What is the doubling time?

11. *World population growth.* The population of the world was 5.2 billion in 1990. The exponential growth rate was 1.6% per year.

a) Find the exponential growth function.
b) Estimate the population of the world in 2000.
c) When will the world population be 8.0 billion?

12. *Growth of bacteria.* The bacteria *Escherichi coli* are commonly found in the bladder of human beings. Suppose that 3000 of the bacteria are present at time $t = 0$. Then t minutes later, the number of bacteria present will be

$$N(t) = 3000(2)^{t/20}.$$

a) After what amount of time will there be 60,000 bacteria?
b) If 100,000,000 bacteria accumulate, a bladder infection can occur. What amount of time would have to pass in order for a possible infection to occur?
c) What is the doubling time?

13. *Forgetting.* Students in an English class took a final exam. They took equivalent forms of the exam in monthly intervals thereafter. The average score $S(t)$, in percent, after t months was found to be given by

$$S(t) = 68 - 20 \log (t + 1), \quad t \geqslant 0.$$

a) What was the average score when they initially took the test, $t = 0$?
b) What was the average score after 4 months? after 24 months?
c) Graph the function.
d) After what time t was the average score 50?

14. *Advertising.* A model for advertising response is given by

$$N(a) = 2000 + 500 \log a, \quad a \geq 1,$$

where $N(a)$ = the number of units sold and a = the amount spent on advertising, in thousands of dollars.

a) How many units were sold after spending $1000 ($a = 1$) on advertising?

b) How many units were sold after spending $8000?

c) Graph the function.

d) How much would have to be spent in order to sell 5000 units?

Consider the pH formula of Example 1 for Exercises 15–22.

Find the pH, given the hydrogen ion concentration.

15. A common brand of mouthwash:

$$[H^+] = 6.3 \times 10^{-7} \text{ moles per liter}$$

16. A common brand of insect repellent:

$$[H^+] = 4.0 \times 10^{-8} \text{ moles per liter}$$

17. Eggs:

$$[H^+] = 1.6 \times 10^{-8} \text{ moles per liter}$$

18. Tomatoes:

$$[H^+] = 6.3 \times 10^{-5} \text{ moles per liter}$$

Find the hydrogen ion concentration of each substance, given the pH.

19. Tap water: pH = 7

20. Rainwater: pH = 5.4

21. Orange juice: pH = 3.2

22. Wine: pH = 4.8

23. The San Francisco earthquake of 1906 had an intensity of $10^{8.25}$ times I_0. What was its magnitude on the Richter scale?

24. In 1986, there was an earthquake near Cleveland, Ohio. It had an intensity of 10^5 times I_0. What was its magnitude on the Richter scale?

25. *Oil demand.* The exponential growth rate of the demand for oil in the United States is 10% per year. When will the demand be double that of 1993?

26. *Coal demand.* The exponential growth rate of the demand for coal in the world is 4% per year. When will the demand be double that of 1993?

27. *Heart transplants.* In 1967, Dr. Christian Barnard of South Africa stunned the world by performing the first heart transplant. There was 1 transplant in 1967. In 1987, there were 1418 such transplants.

a) Find an exponential growth function that fits the data.

b) Use the function to predict the number of heart transplants in 1998.

28. *The cost of a first-class postage stamp.* The cost of a first-class postage stamp became 3¢ in 1932 and the exponential growth rate was 3.8% per year. The exponential growth function was

$$P(t) = 3e^{0.038t}.$$

a) The cost of first-class postage increased to 29¢ in 1991. Use the given function to see what the predicted cost was for 1991 and compare with the actual cost.

b) What will the cost of a first-class postage stamp be in 2000?

c) When will the cost of a first-class postage stamp be $1.00?

29. An ivory tusk has lost 20% of its carbon-14. How old is the tusk?

30. A piece of wood has lost 10% of its carbon-14. How old is the piece of wood?

31. The decay rate of iodine-131 is 9.6% per day. What is its half-life?

32. The decay rate of krypton-85 is 6.3% per year. What is its half-life?

33. The half-life of polonium is 3 minutes. What is its decay rate?

34. The half-life of lead is 22 years. What is its decay rate?

35. *Value of a sports card.* Because he objected to smoking, and because his first baseball card was issued in cigarette packs, the great shortstop Honus Wagner halted production of his card before many were produced. One of these cards was sold in 1986 for $110,000 and again in 1991 for $451,000. For the following questions, assume that the card's value increases exponentially.

a) Find an exponential function $V(t)$, if $V_0 = 110,000$.

b) Predict the card's value in 1998.

c) What is the doubling time for the value of the card?

d) In what year will the value of the card be $9,000,000?

36. *Value of a Van Gogh painting.* The Van Gogh painting *Irises,* shown on the following page, sold for $84,000 in 1947 and was sold again for $53,900,000 in 1987. Assume that the growth in the value V of the painting is exponential.

a) Find the exponential growth rate k and determine the exponential growth function, assuming $V_0 = 84,000$.

b) Estimate the value of the painting in 1997.

c) What is the doubling time for the value of the painting?

d) How long after 1947 will the value of the painting be $1 billion?

Van Gogh's *Irises*, a 28-by-32-inch oil on canvas.

Skill Maintenance

Simplify.

37. $\dfrac{\dfrac{x-5}{x+3}}{\dfrac{x}{x-3}+\dfrac{2}{x+3}}$

38. $\dfrac{\dfrac{3}{a}+\dfrac{5}{b}}{\dfrac{2}{a^2}-\dfrac{4}{b^2}}$

Synthesis

39. ◈ *Atmospheric pressure.* Atmospheric pressure P at altitude a is given by

$$P = P_0 e^{-0.00005a},$$

where $P_0 =$ the pressure at sea level ≈ 14.7 lb/in^2 (pounds per square inch). Explain how a barometer, or some other device for measuring atmospheric pressure, can be used to find the height of a skyscraper.

40. ◈ Examine the restriction on t in Exercise 13.

a) What upper limit might be placed on t?

b) In practice, would this upper limit ever be enforced? Why or why not?

41. ◈ Write a problem for a classmate to solve in which information is provided and the classmate is asked to find an exponential growth function. Make the problem as realistic as possible.

42. ▦ *Supply and demand.* The supply and demand for the sale of stereos by a sound company are given by

$$S(x) = e^x \quad \text{and} \quad D(x) = 162{,}755e^{-x},$$

where $S(x) =$ the price at which the company is willing to supply x stereos and $D(x) =$ the demand price for a quantity of x stereos. Find the equilibrium point. (For reference, see Section 3.8.)

SUMMARY AND REVIEW | 9

KEY TERMS

Exponential function, p. 473
Asymptote, p. 474
Composite functions, p. 480
Inverse relation, p. 483
One-to-one function, p. 483
Horizontal-line test, p. 484

Logarithmic function, p. 491
Exponential equation, p. 493
Logarithmic equation, p. 493
Common logarithm, p. 502
Antilogarithm, p. 503

Natural logarithm, p. 503
Exponential growth rate, p. 515
Decay rate, p. 518
Half-life, p. 518

IMPORTANT PROPERTIES AND FORMULAS

Exponential function:	$f(x) = a^x$
Interest compounded annually:	$A = P(1 + i)^t$
Composition of f and g:	$f \circ g(x) = f(g(x))$

To Find a Formula for the Inverse of a Function

If a function is one-to-one, a formula for its inverse can be found as follows:

1. Replace $f(x)$ by y.
2. Interchange x and y.
3. Solve for y.
4. Replace y by $f^{-1}(x)$.

Definition of logarithm:	$y = \log_a x$ is that number y such that $a^y = x$, where $x > 0$ and a is a positive constant other than 1

Properties of logarithms:

$$\log_a MN = \log_a M + \log_a N, \qquad \log_a \frac{M}{N} = \log_a M - \log_a N,$$

$$\log_a M^p = p \cdot \log_a M, \qquad \log_a 1 = 0,$$
$$\log_a a = 1, \qquad \log_a a^k = k,$$
$$\log M = \log_{10} M, \qquad e \approx 2.7182818284. \ldots \, ,$$

$$\ln M = \log_e M, \qquad \log_b M = \frac{\log_a M}{\log_a b}$$

Exponential growth:	$P(t) = P_0 e^{kt}$
Exponential decay:	$P(t) = P_0 e^{-kt}$
Interest compounded continuously:	$P(t) = P_0 e^{kt}$, where P_0 is the principal invested for t years at interest rate k
Carbon dating:	$P(t) = P_0 e^{-0.00012t}$

REVIEW EXERCISES

This chapter's review and test include Skill Maintenance exercises from Sections 6.3, 7.9, 8.5, and 8.9.

Graph.

1. $f(x) = 3^{x-2}$

2. $x = 3^{y-2}$

3. $y = \log_3 x$

4. $f(x) = e^{x+1}$

5. Find $f \circ g(x)$ and $g \circ f(x)$ if $f(x) = x^2$ and $g(x) = 3x - 5$.

6. Determine whether $f(x) = 4 - x^2$ is one-to-one.

Find a formula for the inverse.

7. $f(x) = x + 2$

8. $g(x) = \dfrac{2x - 3}{7}$

9. $f(x) = \dfrac{2}{x + 5}$

10. $g(x) = 8x^3$

Convert to an exponential equation.

11. $\log_4 16 = x$ **12.** $\log_{10} 2 = 0.3010$

13. $\log_{1/2} 8 = -3$ **14.** $\log_{16} 8 = \frac{3}{4}$

Convert to a logarithmic equation.

15. $10^4 = 10{,}000$ **16.** $25^{1/2} = 5$

17. $7^{-2} = \frac{1}{49}$ **18.** $(2.718)^3 = 20.1$

Express in terms of logarithms of x, y, and z.

19. $\log_a x^4 y^2 z^3$ **20.** $\log_a \dfrac{xy}{z^2}$

21. $\log \sqrt[4]{\dfrac{z^2}{x^3 y}}$ **22.** $\log_q \left(\dfrac{x^2 y^{1/3}}{z^4}\right)$

Express as a single logarithm.

23. $\log_a 8 + \log_a 15$ **24.** $\log_a 72 - \log_a 12$

25. $\frac{1}{2} \log a - \log b - 2 \log c$

26. $\frac{1}{3}[\log_a x - 2 \log_a y]$

Simplify.

27. $\log_m m$ **28.** $\log_m 1$

29. $\log_m m^{17}$ **30.** $\log_m m^{-7}$

Given $\log_a 2 = 1.8301$ and $\log_a 7 = 5.0999$, find each of the following.

31. $\log_a 14$ **32.** $\log_a \frac{2}{7}$ **33.** $\log_a 28$

34. $\log_a 3.5$ **35.** $\log_a \sqrt{7}$ **36.** $\log_a \frac{1}{4}$

Find each of the following using a calculator.

37. $\log 0.00627$ **38.** $\log 72{,}800{,}000$

39. antilog 4.4742 **40.** antilog (-1.4425)

41. antilog 2.3294 **42.** $\log 0.004937$

43. $\log 394{,}900$ **44.** antilog (-6.7889)

45. $\ln 23{,}912.2$ **46.** $\ln 0.06774$

47. antilog$_e$ (-10.56) **48.** antilog$_e$ 45

Find each of the following logarithms using the change-of-base formula and a calculator or Table 2.

49. $\log_5 2$ **50.** $\log_{12} 70$

Solve.

51. $\log_3 x = -2$ **52.** $\log_x 32 = 5$

53. $\log x = -4$ **54.** $\ln x = 2$

55. $4^{2x-5} = 16$ **56.** $4^x = 8.3$

57. $\log_4 16 = x$

58. $\log(x^2 - 9) - \log(x - 3) = 1$

59. $\log_4 x + \log_4 (x - 6) = 2$

60. $\log x + \log (x - 15) = 2$

61. $\log_3 (x - 4) = 3 - \log_3 (x + 4)$

62. ■   *Forgetting.* In a business class, students were tested at the end of the course on a final exam. They were tested again after 6 months. The forgetting formula was determined to be

$$S(t) = 62 - 18 \log (t + 1),$$

where t is the time, in months, after taking the first test.

  a) Determine the average score when they first took the test (when $t = 0$).
  b) What was the average score after 6 months?
  c) After what time was the average score 34?

63. ■   *Cost of a prime-rib dinner.* The average cost C of a prime-rib dinner was \$4.65 in 1962. In 1986, it was \$15.81. Assume that the cost increases exponentially.

  a) Find k and write the exponential growth function.
  b) Predict the cost of a prime-rib dinner in 2002.
  c) When will the average cost of a prime-rib dinner be \$30?
  d) What is the doubling time?

64. ■   The population of Riverton doubled in 16 years. What was the exponential growth rate?

65. ■   How long will it take \$7600 to double itself if it is invested at 8.4%, compounded continuously?

66. ■   How old is a skeleton that has lost 34% of its carbon-14?

67. ■   What is the pH of a substance whose hydrogen ion concentration is 2.3×10^{-7} moles per liter?

68. An earthquake has an intensity of $10^{8.3}$ times I_0. What is its amplitude on the Richter scale?

Skill Maintenance _____

69. Solve $aT^2 + bT = Q$ for T.

70. Solve: $x^4 + 80 = 21x^2$.

71. Divide: $\dfrac{4 - 5i}{1 + 3i}$.

72. Simplify:

$$\dfrac{\dfrac{1}{ab} - \dfrac{2}{bc}}{\dfrac{2}{ac} + \dfrac{3}{ab}}.$$

Synthesis

73. ◈ Explain why negative numbers do not have logarithms.

74. ◈ Explain why taking the natural or common logarithm on each side of an equation produces an equivalent equation.

Solve.

75. $\ln(\ln x) = 3$

76. $2^{x^2 + 4x} = \frac{1}{8}$

77. $5^{x+y} = 25,$
$2^{2x-y} = 64$

CHAPTER TEST   9

Graph.

1. $f(x) = 2^{x+3}$

2. $f(x) = \log_7 x$

3. Find $f \circ g(x)$ and $g \circ f(x)$ if $f(x) = x + x^2$ and $g(x) = 5x - 2$.

4. Determine whether $f(x) = 2 - |x|$ is one-to-one.

Find a formula for the inverse.

5. $f(x) = 4x - 3$

6. $g(x) = \dfrac{x-2}{4}$

7. $f(x) = \dfrac{x+1}{x-2}$

Convert to a logarithmic equation.

8. $4^{-3} = x$

9. $256^{1/2} = 16$

Convert to an exponential equation.

10. $\log_4 16 = 2$

11. $m = \log_7 49$

12. Express in terms of logarithms of a, b, and c:
$$\log \frac{a^3 b^{1/2}}{c^2}.$$

13. Express as a single logarithm:
$$\frac{1}{3}\log_a x - 3\log_a y + 2\log_a z.$$

Simplify.

14. $\log_t t^{23}$

15. $\log_p p$

16. $\log_c 1$

Given $\log_a 2 = 0.301$, $\log_a 6 = 0.778$, and $\log_a 7 = 0.845$, find each of the following.

17. $\log_a \frac{2}{7}$

18. $\log_a \sqrt{24}$

19. $\log_a 21$

Find each of the following using a calculator.

20. $\log 0.0123$

21. antilog 5.6484

22. antilog (-7.2614)

23. $\log 12{,}340$

24. $\ln 0.01234$

25. antilog$_e$ (5.6774)

26. Find $\log_{18} 31$ using the change-of-base formula and a calculator.

Solve.

27. $\log_x 25 = 2$

28. $\log_4 x = \frac{1}{2}$

29. $\log x = 4$

30. $5^{4-3x} = 125$

31. ▦ $7^x = 1.2$

32. ▦ $\ln x = \frac{1}{4}$

33. $\log(x^2 - 1) - \log(x - 1) = 1$

34. ▦ *Walking speed.* The average walking speed R of people living in a city of population P, in thousands, is given by
$$R = 0.37 \ln P + 0.05,$$
where R is in feet per second.

a) The population of Akron, Ohio, is 660,000. Find the average walking speed.
b) A city has an average walking speed of 2.6 ft/sec. Find the population.

35. ▦ *Population of Canada.* The population of Canada was 24 million in 1981, and the exponential growth rate was 1.2% per year.

a) Write an exponential function describing the population of Canada.
b) What will the population be in 1998? in 2010?
c) When will the population be 30 million?
d) What is the doubling time?

36. ▦ The population of Clay County doubled in 20 years. What was the exponential growth rate?

37. 🖩 How long will it take an investment to double itself if it is invested at 7.6%, compounded continuously?

38. 🖩 How old is an animal bone that has lost 43% of its carbon-14?

39. An earthquake has an intensity of $10^{8.34}$ times I_0. What is its magnitude on the Richter scale?

40. The hydrogen ion concentration of water is 1.0×10^{-7} moles per liter. What is the pH?

Skill Maintenance _____

41. Solve: $y - 9\sqrt{y} + 8 = 0$.

42. Solve $S = at^2 - bt$ for t.

43. Multiply: $(2 + 5i)(2 - 5i)$.

44. Simplify:

$$\frac{\dfrac{1}{x^2 - 4}}{\dfrac{1}{x + 2} + \dfrac{1}{x - 2}}.$$

Synthesis _____

45. Solve: $\log_5 |2x - 7| = 4$.

46. If $\log_a x = 2$, $\log_a y = 3$, and $\log_a z = 4$, find

$$\log_a \frac{\sqrt[3]{x^2 z}}{\sqrt[3]{y^2 z^{-1}}}.$$

CUMULATIVE REVIEW | 1–9

1. Evaluate $\dfrac{x^0 + y}{-z}$ when $x = 6$, $y = 9$, and $z = -5$.

Simplify.

2. $\left| -\dfrac{5}{2} + \left(-\dfrac{7}{2} \right) \right|$

3. $(-2x^2y^{-3})^{-4}$

4. $(-5x^4y^{-3}z^2)(-4x^2y^2)$

5. $\dfrac{3x^4y^6z^{-2}}{-9x^4y^2z^3}$

6. $2x - 3 - 2[5 - 3(2 - x)]$

7. $3^3 + 2^2 - (32 \div 4 - 16 \div 8)$

Solve.

8. $8(2x - 3) = 6 - 4(2 - 3x)$

9. $(5x - 2)(4x + 20) = 0$

10. $4x - 3y = 15,$
    $3x + 5y = 4$

11. $x + y - 3z = -1,$
    $2x - y + z = 4,$
    $-x - y + z = 1$

12. $5 = x^2 + 6x$

13. $x(x - 3) = 10$

14. $\dfrac{7}{x^2 - 5x} - \dfrac{2}{x - 5} = \dfrac{4}{x}$

15. $\dfrac{8}{x + 1} + \dfrac{11}{x^2 - x + 1} = \dfrac{24}{x^3 + 1}$

16. $\sqrt{x - 1} = \sqrt{x + 4} - 1$

17. $\sqrt[3]{2x} = 1$

18. $3x^2 + 75 = 0$

19. $x - 8\sqrt{x} + 15 = 0$

20. $x^4 - 13x^2 + 36 = 0$

21. $\log_8 x = 1$

22. $\log_x 49 = 2$

23. $9^x = 27$

24. $\log x - \log (x - 8) = 1$

25. $x^2 + 4x > 5$

26. $|2x - 3| \geq 9$

Solve.

27. $D = \dfrac{ab}{b + a}$, for a

28. $\dfrac{1}{p} + \dfrac{1}{q} = \dfrac{1}{f}$, for q

29. $M = \dfrac{2}{3}(A + B)$, for B

Evaluate.

30. $\begin{vmatrix} 6 & -5 \\ 4 & -3 \end{vmatrix}$

31. $\begin{vmatrix} 7 & -6 & 0 \\ -2 & 1 & 2 \\ -1 & 1 & -1 \end{vmatrix}$

Solve.

32. Twenty-four plus five times a number is eight times the number. Find the number.

33. The perimeter of a rectangular garden is 112 m. The length is 16 m more than the width. Find the length and the width.

34. In triangle ABC, the measure of angle B is three times the measure of angle A. The measure of angle C is 105° greater than the measure of angle A. Find the angle measures.

35. Phil can build a shed from a lumber kit in 10 hr. Jenny can build the same shed in 12 hr. How long would it take Phil and Jenny, working together, to build the shed?

36. Swim Clean is 30% muriatic acid. Pure Swim is 80% muriatic acid. How many liters of each should be mixed together in order to get 100 L of a solution that is 50% muriatic acid?

37. A boat can move at a speed of 5 km/h in still water. The boat travels 42 km downstream in the same time that it takes to travel 12 km upstream. What is the speed of the stream?

38. What is the minimum product of two numbers whose difference is 14? What are the numbers that yield this product?

39. The speed of a passenger train is 13 mph faster than that of a freight train. The passenger train travels 160 mi in the same time it takes the freight train to travel 108 mi. Find the speed of each train.

Forgetting. Students in a biology class took a final exam. A forgetting formula for the average exam grade was determined to be

$$S(t) = 78 - 15 \log (t + 1),$$

where t is the number of months after the final was taken.

40. The average score when the students first took the test is when $t = 0$. Find the students' average score on the final exam.

41. What would the average score be on a retest after 4 months?

Population growth. The population of Europe west of Russia was 430 million in 1961, and the exponential growth rate was 1% per year.

42. Write an exponential function describing the growth of the population of Europe.

43. 🖩 Predict what the population will be in 1997; in 2001.

44. y varies directly as the square of x and inversely as z, and $y = 2$ when $x = 5$ and $z = 100$. What is y when $x = 3$ and $z = 4$?

Perform the indicated operations and simplify.

45. $(5p^2q^3 - 4p^3q + 6pq - p^2 + 3)$
$+ (2p^2q^3 + 2p^3q + p^2 - 5pq - 9)$

46. $(11x^2 - 6x - 3) - (3x^2 + 5x - 2)$

47. $(3x^2 - 2y)^2$

48. $(5a + 3b)(2a - 3b)$

49. $\dfrac{x^2 + 8x + 16}{2x + 6} \div \dfrac{x^2 + 3x - 4}{x^2 - 9}$

50. $\dfrac{1 + \dfrac{3}{x}}{x - 1 - \dfrac{12}{x}}$

51. $\dfrac{a^2 - a - 6}{a^3 - 27} \cdot \dfrac{a^2 + 3a + 9}{6}$

52. $\dfrac{3}{x + 6} - \dfrac{2}{x^2 - 36} + \dfrac{4}{x - 6}$

Factor.

53. $xy - 2xz + xw$

54. $1 - 125x^3$

55. $6x^2 + 8xy - 8y^2$

56. $x^4 - 4x^3 + 7x - 28$

57. $a^2 - 10a + 25 - 81b^2$

58. $2m^2 + 12mn + 18n^2$

59. $x^4 - 16y^4$

60. For the function described by
$h(x) = -3x^2 + 4x + 8,$
find $h(-2)$.

61. Divide: $(x^4 - 5x^3 + 2x^2 - 6) \div (x - 3)$.

62. Multiply $(5.2 \times 10^4)(3.5 \times 10^{-6})$. Write scientific notation for the answer.

63. Divide: $\dfrac{3.4 \times 10^5}{6.8 \times 10^{-9}}$.

Write scientific notation for the answer.

For the radical expressions that follow, assume that all variables represent positive numbers.

64. Divide and simplify: $\dfrac{\sqrt[3]{40xy^8}}{\sqrt[3]{5xy}}$.

65. Multiply and simplify: $\sqrt{7xy^3} \cdot \sqrt{28x^2y}$.

66. Rewrite without fractional exponents: $(27a^6b)^{4/3}$.

67. Rationalize the denominator: $\dfrac{3 - \sqrt{y}}{2 - \sqrt{y}}$.

68. Divide and simplify: $\dfrac{\sqrt[5]{x + 5}}{\sqrt{x + 5}}$.

69. Multiply these complex numbers:
$(1 + i\sqrt{3})(6 - 2i\sqrt{3})$.

70. Divide these complex numbers: $\dfrac{3 - 2i}{4 - 3i}$.

71. Find the inverse of f if $f(x) = 7 - 2x$.

72. Find an equation of the line containing the points $(0, -3)$ and $(-1, 2)$.

73. Find an equation of the line containing the point $(-3, 5)$ and perpendicular to the line whose equation is $2x + y = 6$.

Graph.

74. $5x = 15 + 3y$

75. $y = 2x^2 - 4x - 1$

76. $y = \log_3 x$

77. $y = 3^x$

78. $-2x - 3y \le 6$

79. Graph: $f(x) = 2(x + 3)^2 + 1$.
  a) Label the vertex.
  b) Draw the line of symmetry.
  c) Find the maximum or minimum value.

80. Express in terms of logarithms of a, b, and c:
$\log\left(\dfrac{a^2c^3}{b}\right)$.

81. Express as a single logarithm:
$3 \log x - \dfrac{1}{2} \log y - 2 \log z$.

82. Convert to an exponential equation: $\log_a 5 = x$.

83. Convert to a logarithmic equation: $x^3 = t$.

Find each of the following using a calculator.

84. $\log 0.05566$

85. antilog 5.4453

86. $\ln 12.78$

87. antilog$_e$ (-3.6762)

88. 🖩 Solve: $3^{5x} = 7$.

Synthesis _____

89. Solve: $\dfrac{5}{3x - 3} + \dfrac{10}{3x + 6} = \dfrac{5x}{x^2 + x - 2}$.

90. Solve: $\log \sqrt{3x} = \sqrt{\log 3x}$.

91. A train travels 280 mi at a certain speed. If the speed had been increased 5 mph, the trip could have been made in 1 hr less time. Find the actual speed.

Conic Sections

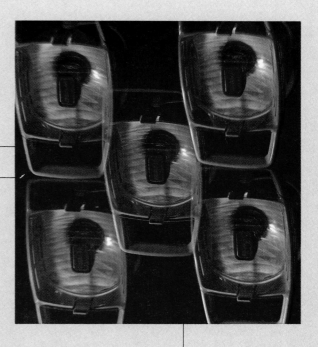

AN APPLICATION

The light source in a dental lamp shines against a reflector that is shaped like a portion of an *ellipse* in which the light source is one *focus* of the ellipse. Reflected light enters a patient's mouth at the other focus of the ellipse. If the ellipse from which the reflector was formed is 2 ft wide and 6 ft long, how far should the light source be from a patient's mouth?

This problem appears as Exercise 57 in Section 10.2.

Dr. Roschelle Major-Banks
DENTIST

"A good math background is critical for those working in dentistry. Dental assistants need math for their training, orthodontists use physics extensively in shaping braces, and dental technologists work with precise measurements when making dental appliances."

T he ellipse described in the chapter opening is a curve known as a *conic section*, meaning that the curve is formed as a cross section of a cone. In this chapter, we will study equations whose graphs are conic sections. We have already studied two conic sections, *lines* and *parabolas*, in some detail in Chapters 2 and 8. There are many applications involving conics, and we will consider some of them in this chapter.

In addition to material from this chapter, the review and test for Chapter 10 include material from Sections 6.5, 7.3, 7.6, and 8.2.

10.1

Conic Sections: Parabolas and Circles

Parabolas • **The Distance Formula** • **Midpoints of Segments** • **Circles**

In this section and the next, we study curves formed by cross sections of cones. These curves are graphs of second-degree equations in two variables. Some are shown below:

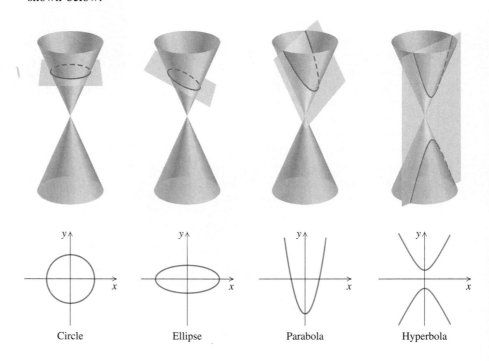

Circle   Ellipse   Parabola   Hyperbola

Parabolas

When a cone is cut as shown in the third figure above, the conic section formed is a **parabola.** Parabolas have many applications in electricity, mechanics, and optics. The cross section of a satellite dish is a parabola. Cables that support bridges that are straight lines are shaped like parabolas. (Free-hanging cables have a different shape, called a "catenary.")

Equation of a Parabola

Parabolas have equations as follows:

$y = ax^2 + bx + c$ (Line of symmetry parallel to the y-axis);

$x = ay^2 + by + c$ (Line of symmetry parallel to the x-axis).

We found in Chapter 8 that the graphs of quadratic functions of the form $f(x) = ax^2 + bx + c$ are parabolas. Our goal here is to review those graphs and extend those ideas to equations of the type $x = ay^2 + by + c$. These parabolas have lines of symmetry parallel to the x-axis and open to the right or to the left.

EXAMPLE 1

Graph: $y = x^2 - 4x + 8$.

Solution   We first find the vertex. We can do so in either of two ways. The first way is by completing the square:

$$y = (x^2 - 4x) + 8$$
$$= (x^2 - 4x + 4 - 4) + 8 \qquad \tfrac{1}{2}(-4) = -2; \; (-2)^2 = 4$$
$$= (x^2 - 4x + 4) + (-4 + 8) \qquad \text{Regrouping}$$
$$= (x - 2)^2 + 4. \qquad \text{Factoring and simplifying}$$

The vertex is $(2, 4)$.

A second way to find the vertex is to recall from Section 8.8 that *the x-coordinate of the vertex of the parabola given by $y = ax^2 + bx + c$ is $-b/(2a)$*. Thus, instead of completing the square as above, we could have evaluated the formula

$$x = -\frac{b}{2a} = -\frac{-4}{2(1)} = 2,$$

and found the y-coordinate of the vertex by substituting 2 for x:

$$y = x^2 - 4x + 8 = 2^2 - 4(2) + 8 = 4.$$

Either way we know that the vertex is $(2, 4)$. We choose some x-values on both sides of the vertex and compute the corresponding y-values. Then we plot the points and graph the parabola. Since the coefficient of x^2, 1, is positive, we know that the graph opens up. It is easy to find the y-intercept by finding y when $x = 0$.

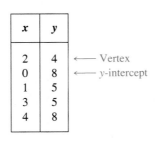

x	y	
2	4	← Vertex
0	8	← y-intercept
1	5	
3	5	
4	8	

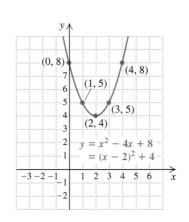

To graph an equation in the form $y = ax^2 + bx + c$:

1. Find the vertex (h, k) either by completing the square to find an equivalent equation $y = a(x - h)^2 + k$, or by using $-b/(2a)$ to find the x-coordinate and substituting to find the y-coordinate.
2. Choose other values for x on both sides of the vertex, and compute the corresponding y-values.
3. The graph opens upward for $a > 0$ and downward for $a < 0$.

EXAMPLE 2

Graph: $x = y^2 - 4y + 8$.

Solution   This equation is like that in Example 1 except that x and y are interchanged. The vertex is $(4, 2)$ instead of $(2, 4)$. To find ordered pairs, we first choose values for y on each side of the vertex. Then we compute values for x. A table is shown, together with the graph. Note that in this table the x- and y-values of the table in Example 1 are interchanged.

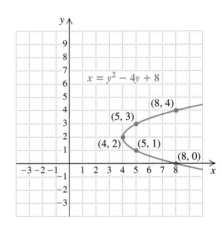

x	y	
4	2	← Vertex
8	0	← x-intercept
5	1	
5	3	
8	4	

(1) Choose these values for y.

(2) Compute these values for x.

To graph an equation in the form $x = ay^2 + by + c$:

1. Find the vertex (h, k) either by completing the square to find an equivalent equation $x = a(y - k)^2 + h$ or by using $-b/(2a)$ to find the y-coordinate and substituting to find the x-coordinate.
2. Choose other values for y that are above and below the vertex, and compute the corresponding x-values.
3. The graph opens to the right if $a > 0$ and to the left if $a < 0$.

EXAMPLE 3

Graph: $x = -2y^2 + 10y - 7$.

Solution   Completing the square, we have

$$x = -2y^2 + 10y - 7 = -2(y^2 - 5y) - 7$$
$$= -2\left(y^2 - 5y + \tfrac{25}{4}\right) - 7 - (-2)\tfrac{25}{4} \qquad \tfrac{1}{2}(-5) = \tfrac{-5}{2}; \left(\tfrac{-5}{2}\right)^2 = \tfrac{25}{4}; \text{ we add and}$$
$$\text{subtract } (-2)\tfrac{25}{4}$$
$$= -2\left(y - \tfrac{5}{2}\right)^2 + \tfrac{11}{2}. \qquad \text{Factoring and simplifying}$$

The vertex is $\left(\tfrac{11}{2}, \tfrac{5}{2}\right)$.

For practice, we also find the vertex by first computing the second coordinate, $y = -b/(2a)$, and then substituting to find the first coordinate:

$$y = -\frac{b}{2a} = -\frac{10}{2(-2)} = \frac{5}{2}$$

$$x = -2y^2 + 10y - 7 = -2\left(\tfrac{5}{2}\right)^2 + 10\left(\tfrac{5}{2}\right) - 7 = \tfrac{11}{2}.$$

To find ordered pairs, we first choose values for y and then compute values for x. A table is given, together with the graph. The graph opens to the left because the coefficient of y^2, -2, is negative.

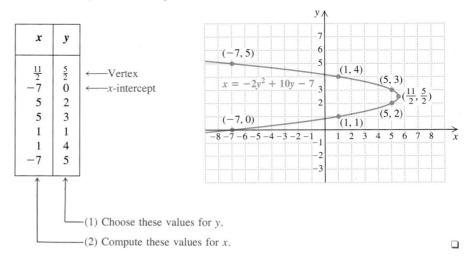

x	y	
$\frac{11}{2}$	$\frac{5}{2}$	←——Vertex
-7	0	←——x-intercept
5	2	
5	3	
1	1	
1	4	
-7	5	

(1) Choose these values for y.

(2) Compute these values for x.

The Distance Formula

Suppose that two points are on a horizontal line, and thus have the same second coordinate. We can find the distance between them by subtracting their first coordinates. This difference may be negative, depending on the order in which we subtract. So to make sure we get a positive number, we take the absolute value of this difference. The distance between two points on a horizontal line (x_1, y_1) and (x_2, y_1) is thus $|x_2 - x_1|$. Similarly, the distance between two points on a vertical line (x_2, y_1) and (x_2, y_2) is $|y_2 - y_1|$.

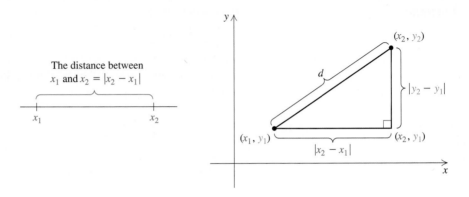

Now consider any two points (x_1, y_1) and (x_2, y_2). If $x_1 \neq x_2$ and $y_1 \neq y_2$, these points are vertices of a right triangle, as shown. The other vertex is then

(x_2, y_1). The lengths of the legs are $|x_2 - x_1|$ and $|y_2 - y_1|$. We find d, the length of the hypotenuse, by using the Pythagorean theorem:

$$d^2 = |x_2 - x_1|^2 + |y_2 - y_1|^2.$$

Since the square of a number is the same as the square of its opposite, we don't really need the absolute-value signs. Thus,

$$d^2 = (x_2 - x_1)^2 + (y_2 - y_1)^2.$$

Taking the principal square root, we obtain the distance between two points.

The Distance Formula

The distance d between any two points (x_1, y_1) and (x_2, y_2) is given by

$$d = \sqrt{(x_2 - x_1)^2 + (y_2 - y_1)^2}.$$

EXAMPLE 4

Find the distance between $(5, -1)$ and $(-4, 6)$. Find an exact answer and an approximation to three decimal places.

Solution   We substitute into the distance formula:

$$d = \sqrt{(-4 - 5)^2 + [6 - (-1)]^2} \qquad \text{Substituting}$$
$$= \sqrt{(-9)^2 + 7^2}$$
$$= \sqrt{130} \approx 11.402. \qquad \text{Using a calculator} \qquad \square$$

Midpoints of Segments

The distance formula can be used to verify or derive a formula for finding the coordinates of the *midpoint* of a segment when the coordinates of the endpoints are known. We state the formula and leave its proof to the exercises. Note that although the distance formula involves both subtraction and addition, the midpoint formula uses only addition.

The Midpoint Formula

If the endpoints of a segment are (x_1, y_1) and (x_2, y_2), then the coordinates of the midpoint are

$$\left(\frac{x_1 + x_2}{2}, \frac{y_1 + y_2}{2} \right).$$

(Average the coordinates of the endpoints to find the coordinates of the midpoint.)

EXAMPLE 5

Find the midpoint of the segment with endpoints $(-2, 3)$ and $(4, -6)$.

Solution   Using the midpoint formula, we obtain

$$\left(\frac{-2 + 4}{2}, \frac{3 + (-6)}{2} \right), \quad \text{or} \quad \left(\frac{2}{2}, \frac{-3}{2} \right), \quad \text{or} \quad \left(1, -\frac{3}{2} \right). \qquad \square$$

Circles

The distance formula can also be used to develop an equation for another conic section, shown in the figure at the beginning of this section, the circle. A **circle** is a set of points in a plane that are a fixed distance r, called the **radius,** from a fixed point (h, k), called the **center.** If a point (x, y) is on the circle, then by the definition of a circle and the distance formula, it must follow that

$$r = \sqrt{(x - h)^2 + (y - k)^2}.$$

Squaring both sides gives an equation of the circle in standard form: $(x - h)^2 + (y - k)^2 = r^2$. When $h = 0$ and $k = 0$, the circle is centered at the origin. Otherwise, we can think of that circle being translated $|h|$ units horizontally and $|k|$ units vertically.

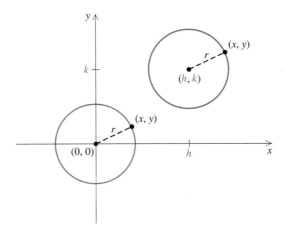

EXAMPLE 6

Find an equation of the circle having center $(4, -5)$ and radius 6.

Solution   Using the standard form, we obtain

$$(x - 4)^2 + [y - (-5)]^2 = 6^2, \qquad \text{Using } (x - h)^2 + (y - k)^2 = r^2$$

or

$$(x - 4)^2 + (y + 5)^2 = 36.$$

EXAMPLE 7

Find the center and the radius and then graph the circle.

a) $(x - 2)^2 + (y + 3)^2 = 4^2$
b) $x^2 + y^2 + 8x - 2y + 15 = 0$

Solution

a) We write standard form:

$$(x - 2)^2 + [y - (-3)]^2 = 4^2.$$

The center is $(2, -3)$ and the radius is 4. Now the graph is easy to draw using a compass.

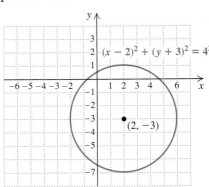

b) To write the equation $x^2 + y^2 + 8x - 2y + 15 = 0$ in standard form, we complete the square twice, once with $x^2 + 8x$ and once with $y^2 - 2y$:

$$x^2 + y^2 + 8x - 2y + 15 = 0$$
$$x^2 + 8x + y^2 - 2y + 15 = 0 \qquad \text{Using the commutative law}$$
$$x^2 + 8x \qquad + y^2 - 2y \qquad = -15 \qquad \text{Adding } -15 \text{ on both sides}$$
$$x^2 + 8x + 16 + y^2 - 2y + 1 = -15 + 16 + 1 \qquad \text{Adding } \left(\tfrac{8}{2}\right)^2, \text{ or 16, and } \left(-\tfrac{2}{2}\right)^2, \text{ or 1, on both sides}$$

$$(x + 4)^2 + (y - 1)^2 = 2 \qquad \text{Factoring}$$
$$[x - (-4)]^2 + (y - 1)^2 = \left(\sqrt{2}\right)^2. \qquad \text{Writing standard form}$$

The center is $(-4, 1)$ and the radius is $\sqrt{2}$.

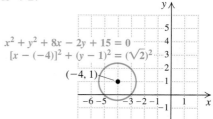

TECHNOLOGY CONNECTION

Graphing an equation of a circle on a grapher involves two steps:

1. Solve the equation for y. The result will include a $\pm$ sign in front of a radical.

2. Graph two functions, one including the $+$ sign and the other including the $-$ sign, on the same set of axes. (Because a grapher can graph only functions and a circle is not a function, we must divide it into two functions and graph both parts.)

For example, to graph the circle $x^2 + y^2 - 6x + 2y - 6 = 0$, we rewrite it as a quadratic equation in y, that is, $y^2 + 2y + (x^2 - 6x - 6) = 0$. The quadratic formula then gives us

$$y = \frac{-2 \pm \sqrt{4 - 4(x^2 - 6x - 6)}}{2},$$

which simplifies to

$$y = -1 \pm \sqrt{-x^2 + 6x + 7}$$

or

$$y_1 = -1 + \sqrt{-x^2 + 6x + 7} \quad \text{and}$$
$$y_2 = -1 - \sqrt{-x^2 + 6x + 7}.$$

When both functions are graphed (in a "squared" window,

to eliminate distortion), the result is the graph shown here:

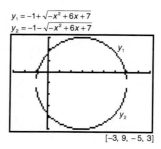

Most graphers have trouble connecting the two pieces because the segments become nearly vertical in this region.

Use a grapher to draw the graph of each of the following circles.

TC1. $x^2 + y^2 - 16 = 0$

TC2. $4x^2 + 4y^2 = 100$

TC3. $x^2 + y^2 + 14x - 16y + 54 = 0$

TC4. $x^2 + y^2 - 10x - 11 = 0$

EXERCISE SET | 10.1 |

Graph.

1. $y = x^2$
2. $x = y^2$
3. $x = y^2 + 4y + 1$
4. $y = x^2 - 2x + 3$
5. $y = -x^2 + 4x - 5$
6. $x = 4 - 3y - y^2$
7. $x = y^2 + 1$
8. $x = 2y^2$
9. $x = -1 \cdot y^2$
10. $x = y^2 - 1$
11. $x = -y^2 + 2y$
12. $x = y^2 + y - 6$
13. $x = 8 - y - y^2$
14. $y = x^2 + 2x + 1$
15. $y = x^2 - 2x + 1$
16. $y = -\frac{1}{2}x^2$
17. $x = -y^2 + 2y + 3$
18. $x = -y^2 - 2y + 3$
19. $x = -2y^2 - 4y + 1$
20. $x = 2y^2 + 4y - 1$

Find the distance between the pair of points. Where appropriate, find an approximation to three decimal places.

21. $(9, 5)$ and $(6, 1)$
22. $(1, 10)$ and $(7, 2)$
23. $(0, -7)$ and $(3, -4)$
24. $(6, 2)$ and $(6, -8)$
25. $(2, 2)$ and $(-2, -2)$
26. $(5, 21)$ and $(-3, 1)$
27. $(8.6, -3.4)$ and $(-9.2, -3.4)$
28. $(5.9, 2)$ and $(3.7, -7.7)$
29. $\left(\frac{5}{7}, \frac{1}{14}\right)$ and $\left(\frac{1}{7}, \frac{11}{14}\right)$
30. $\left(0, \sqrt{7}\right)$ and $\left(\sqrt{6}, 0\right)$
31. $(-23, 10)$ and $(56, -17)$
32. $(34, -18)$ and $(-46, -38)$
33. (a, b) and $(0, 0)$
34. $(0, 0)$ and (p, q)
35. $\left(\sqrt{2}, -\sqrt{3}\right)$ and $\left(-\sqrt{7}, \sqrt{5}\right)$
36. $\left(\sqrt{8}, \sqrt{3}\right)$ and $\left(-\sqrt{5}, -\sqrt{6}\right)$
37. $(1000, -240)$ and $(-2000, 580)$
38. $(-3000, 560)$ and $(-430, -640)$

Find the midpoint of the segment with the given endpoints.

39. $(-3, 6)$ and $(2, -8)$
40. $(6, 7)$ and $(7, -9)$
41. $(8, 5)$ and $(-1, 2)$
42. $(-1, 2)$ and $(1, -3)$
43. $(-8, -5)$ and $(6, -1)$
44. $(8, -2)$ and $(-3, 4)$
45. $(-3.4, 8.1)$ and $(2.9, -8.7)$
46. $(4.1, 6.9)$ and $(5.2, -6.9)$

47. $\left(\frac{1}{6}, -\frac{3}{4}\right)$ and $\left(-\frac{1}{3}, \frac{5}{6}\right)$
48. $\left(-\frac{4}{5}, -\frac{2}{3}\right)$ and $\left(\frac{1}{8}, \frac{3}{4}\right)$
49. $\left(\sqrt{2}, -1\right)$ and $\left(\sqrt{3}, 4\right)$
50. $\left(9, 2\sqrt{3}\right)$ and $\left(-4, 5\sqrt{3}\right)$

Find an equation of the circle satisfying the given conditions.

51. Center $(0, 0)$, radius 7
52. Center $(0, 0)$, radius 4
53. Center $(-2, 7)$, radius $\sqrt{5}$
54. Center $(5, 6)$, radius $2\sqrt{3}$
55. Center $(-4, 3)$, radius $4\sqrt{3}$
56. Center $(-2, 7)$, radius $2\sqrt{5}$
57. Center $(-7, -2)$, radius $5\sqrt{2}$
58. Center $(-5, -8)$, radius $3\sqrt{2}$
59. Center $(0, 0)$, passing through $(-3, 4)$
60. Center $(3, -2)$, passing through $(11, -2)$
61. Center $(-4, 1)$, passing through $(-2, 5)$
62. Center $(-3, -3)$, passing through $(1.8, 2.6)$

Find the center and the radius of the circle. Then graph the circle.

63. $x^2 + y^2 = 36$
64. $x^2 + y^2 = 25$
65. $(x + 1)^2 + (y + 3)^2 = 4$
66. $(x - 2)^2 + (y + 3)^2 = 1$
67. $(x - 8)^2 + (y + 3)^2 = 40$
68. $(x + 5)^2 + (y - 1)^2 = 75$
69. $x^2 + y^2 = 2$
70. $x^2 + y^2 = 3$
71. $(x - 5)^2 + y^2 = \frac{1}{4}$
72. $x^2 + (y - 1)^2 = \frac{1}{25}$
73. $x^2 + y^2 + 8x - 6y - 15 = 0$
74. $x^2 + y^2 + 6x - 4y - 15 = 0$
75. $x^2 + y^2 - 8x + 2y + 13 = 0$
76. $x^2 + y^2 + 6x + 4y + 12 = 0$
77. $x^2 + y^2 - 4x = 0$
78. $x^2 + y^2 + 6x = 0$
79. $x^2 + y^2 + 10y - 75 = 0$
80. $x^2 + y^2 - 8x - 84 = 0$

81. $x^2 + y^2 + 7x - 3y - 10 = 0$

82. $x^2 + y^2 - 21x - 33y + 17 = 0$

83. $4x^2 + 4y^2 = 1$

84. $25x^2 + 25y^2 = 1$

Skill Maintenance

85. A rectangle 10 in. long and 6 in. wide is bordered by a strip of uniform width. If the perimeter of the larger rectangle is twice that of the smaller rectangle, what is the width of the border?

86. One airplane flies 60 mph faster than another. To fly a certain distance, the faster plane takes 4 hr and the slower plane takes 4 hr and 24 min. What is the distance?

Synthesis

87. ◈   Outline a procedure that would use the distance formula to determine whether three points, (x_1, y_1), (x_2, y_2), and (x_3, y_3), are collinear (lie on the same line).

88. ◈   Describe a procedure that would use the distance formula to determine whether three points, (x_1, y_1), (x_2, y_2), and (x_3, y_3), can be vertices of a right triangle.

89. Use a graph of the equation $x = y^2 - y - 6$ to approximate the solutions of each of the following equations.

a) $y^2 - y - 6 = 2$ (*Hint:* Graph $x = 2$ on the same set of axes as the graph of $x = y^2 - y - 6$.)

b) $y^2 - y - 6 = -3$

90. The horsepower of a certain kind of engine is given by the fomula

$$H = \frac{D^2 N}{2.5},$$

where N is the number of cylinders and D is the diameter, in inches, of each piston. Graph this equation, assuming that $N = 6$ (a six-cylinder engine). Let D run from 2.5 to 8.

Using the same set of axes, graph the pair of equations. Try to discover a way to obtain one graph from the other without computing points for the second graph. (*Hint:* Review Section 9.2.)

91. $y = -2x^2 + 3, \quad x = -2y^2 + 3$

92. $y = x^2 + 2x - 3, \quad x = y^2 + 2y - 3$

93. Find the point on the y-axis that is equidistant from (2, 10) and (6, 2).

94. Find the point on the x-axis that is equidistant from (−1, 3) and (−8, −4).

95. Ace Carpentry needs to cut an arch for the top of an entranceway. The arch needs to be 8 ft wide and 2 ft high. To draw the arch, the carpenters will use a stretched string with chalk attached at an end as a compass.

a) Using a coordinate system, locate the center of the circle.

b) What radius should the carpenters use to draw the arch?

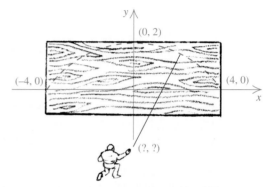

Find an equation of a circle satisfying the given conditions.

96. Center (3, −5) and tangent to (touching at one point) the y-axis

97. Center (−7, −4) and tangent to the x-axis

98. The endpoints of a diameter are (7, 3) and (−1, −3).

99. Center (−3, 5) with a circumference of 8π units

100. A ferris wheel has a radius of 24.3 ft. Assuming that the center is 30.6 ft off the ground and that the origin is below the center, as in the following figure, find an equation of the circle.

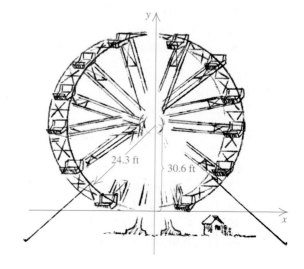

101. Prove the midpoint formula by showing that

    i) the distance from (x_1, y_1) to

$$\left(\frac{x_1 + x_2}{2}, \frac{y_1 + y_2}{2}\right)$$

    equals the distance from (x_2, y_2) to

$$\left(\frac{x_1 + x_2}{2}, \frac{y_1 + y_2}{2}\right);$$

    and

ii) the points

$$(x_1, y_1), \left(\frac{x_1 + x_2}{2}, \frac{y_1 + y_2}{2}\right),$$

and

$$(x_2, y_2)$$

lie on the same line (see Exercise 87).

10.2

Conic Sections: Ellipses and Hyperbolas

Ellipses • **Hyperbolas** • **Hyperbolas (Nonstandard Form)** •
Classifying Graphs of Equations

Ellipses

When a cone is cut at an angle, as shown, the conic section formed is an *ellipse*. You can draw an ellipse by sticking two tacks in a piece of cardboard. Then tie a string to the tacks, place a pencil as shown, and draw.

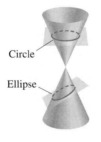

Circle

Ellipse

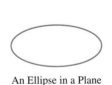

An Ellipse in a Plane

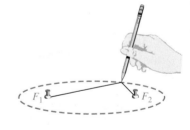

The formal mathematical definition is related to this method of drawing an ellipse. An **ellipse** is defined as the set of all points in a plane such that the *sum* of the distances from two fixed points F_1 and F_2 (called **foci;** singular, **focus**) is constant. In the figure shown, the tacks are at the foci. The midpoint of the segment F_1F_2 is the **center.** Ellipses have equations as follows. The proof is left to the exercises.

Equation of an Ellipse

The equation of an ellipse centered at the origin and parallel to an axis is

$$\frac{x^2}{a^2} + \frac{y^2}{b^2} = 1, \quad a, b > 0, \quad a \neq b. \quad \text{(Standard form)}$$

Ellipses with centers other than the origin are discussed in Exercises 59 and 60.

When graphing ellipses, it helps to first find the intercepts. If we replace x by 0, we can find the y-intercepts:

$$\frac{0^2}{a^2} + \frac{y^2}{b^2} = 1$$

$$\frac{y^2}{b^2} = 1$$

$$y^2 = b^2$$

$$y = \pm b.$$

Thus the y-intercepts are $(0, b)$ and $(0, -b)$. Similarly, the x-intercepts are $(a, 0)$ and $(-a, 0)$.

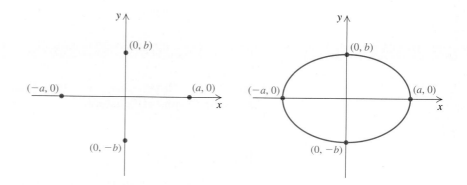

Plotting these points and filling in an oval-shaped curve, we get a graph of the ellipse. If a more precise graph is desired, we can plot more points.

For the ellipse

$$\frac{x^2}{a^2} + \frac{y^2}{b^2} = 1,$$

the x-intercepts are $(-a, 0)$ and $(a, 0)$. The y-intercepts are $(0, -b)$ and $(0, b)$.

EXAMPLE 1

Graph the ellipse $\dfrac{x^2}{4} + \dfrac{y^2}{9} = 1$.

Solution Note that

$$\frac{x^2}{4} + \frac{y^2}{9} = \frac{x^2}{2^2} + \frac{y^2}{3^2}.$$

Thus the x-intercepts are $(-2, 0)$ and $(2, 0)$, and the y-intercepts are $(0, 3)$ and $(0, -3)$. We plot these points and connect them with an oval-shaped curve. To plot some other points, we let $x = 1$ and solve for y:

$$\frac{1^2}{4} + \frac{y^2}{9} = 1$$

$$36\left(\frac{1}{4} + \frac{y^2}{9}\right) = 36 \cdot 1$$

$$36 \cdot \frac{1}{4} + 36 \cdot \frac{y^2}{9} = 36$$

$$9 + 4y^2 = 36$$

$$4y^2 = 27$$

$$y^2 = \frac{27}{4}$$

$$y = \pm\sqrt{\frac{27}{4}}$$

$$y \approx \pm 2.6.$$

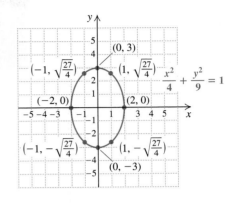

Thus, $(1, 2.6)$ and $(1, -2.6)$ can also be used to draw the graph. Similarly, the points $(-1, 2.6)$ and $(-1, -2.6)$ can also be computed and plotted. ❑

EXAMPLE 2

Graph: $4x^2 + 25y^2 = 100$.

Solution   We write the equation in standard form by multiplying on both sides by $\frac{1}{100}$:

$$\frac{1}{100}(4x^2 + 25y^2) = \frac{1}{100}(100)$$

$$\frac{1}{100}(4x^2) + \frac{1}{100}(25y^2) = 1$$

$$\frac{x^2}{25} + \frac{y^2}{4} = 1$$

$$\frac{x^2}{5^2} + \frac{y^2}{2^2} = 1.$$

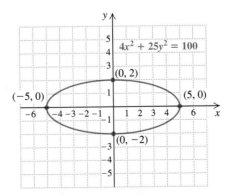

The x-intercepts are $(-5, 0)$ and $(5, 0)$, and the y-intercepts are $(0, -2)$ and $(0, 2)$. We plot the intercepts and connect them with an oval-shaped curve. Other points can also be computed and plotted. ❑

Ellipses have many applications. Earth satellites travel in elliptical orbits, and planets travel around the sun in elliptical orbits with the sun at one focus.

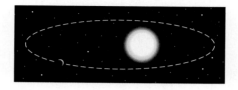

Planetary orbit   Whispering gallery

An interesting application found in some buildings is a *whispering gallery,* which is elliptical. Persons with their heads at the foci can whisper and hear each other clearly, while persons at other positions cannot hear them. This happens when sound waves emanating at one focus are reflected to the other focus, being concentrated there.

A dentist often uses a reflector light. So that the light does not hit you in the eyes, it is covered and reflected. The reflection is directed toward your mouth. One focus is at the light source and the other is at your mouth.

Hyperbolas

A **hyperbola** looks like a pair of parabolas, but the actual shapes are different. A hyperbola has two **vertices** (singular, **vertex**) and the line through the vertices is known as an **axis.** The point halfway between the vertices is called the **center.**

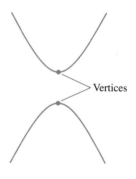

Vertices

Parabola   Hyperbola in three dimensions   Hyperbola in a plane

Equation of a Hyperbola

Hyperbolas with their centers at the origin have equations as follows:

$$\frac{x^2}{a^2} - \frac{y^2}{b^2} = 1 \qquad \text{(Axis horizontal)};$$

$$\frac{y^2}{b^2} - \frac{x^2}{a^2} = 1 \qquad \text{(Axis vertical)}.$$

> Note carefully that these equations have a 1 on the right and a minus sign between the terms.

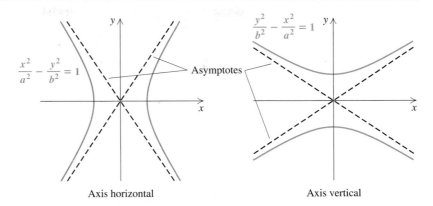

Axis horizontal   Axis vertical

Hyperbolas with horizontal or vertical axes and centers *not* at the origin are discussed in the exercise set.

To graph a hyperbola, it helps to begin by graphing the lines called **asymptotes.**

Asymptotes of a Hyperbola

For hyperbolas with equations as given in the preceding box, the asymptotes are the lines

$$y = \frac{b}{a}x \quad \text{and} \quad y = -\frac{b}{a}x.$$

As a hyperbola gets farther away from the origin, it gets closer and closer to its asymptotes. The larger $|x|$ gets, the closer the graph gets to an asymptote. The asymptotes act to "constrain" the graph of a hyperbola. Parabolas are *not* constrained by any asymptotes.

The next thing to do after sketching asymptotes is to plot vertices. Then it is easy to sketch the curve.

EXAMPLE 3

Graph: $\dfrac{x^2}{4} - \dfrac{y^2}{9} = 1$.

Solution   Note that

$$\frac{x^2}{4} - \frac{y^2}{9} = \frac{x^2}{2^2} - \frac{y^2}{3^2},$$

so $a = 2$ and $b = 3$. The asymptotes are thus

$$y = \frac{3}{2}x \quad \text{and} \quad y = -\frac{3}{2}x.$$

We sketch them, as shown on the left in the figure on the following page.

For horizontal or vertical hyperbolas centered at the origin, the vertices will also be intercepts. Since this hyperbola is horizontal, we replace y with 0 and solve for x. We see that $x^2/2^2 = 1$ when $x = \pm 2$. The intercepts are $(2, 0)$ and $(-2, 0)$. There are intercepts on only one axis. If we replace x with 0, we see that $y^2/9 = -1$ has no real-number solution.

Finally, we plot the intercepts and sketch the graph. Through each intercept, we draw a smooth curve that approaches the asymptotes closely, as shown on the right below.

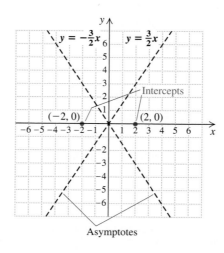

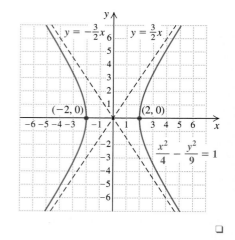

EXAMPLE 4

Graph: $\dfrac{y^2}{36} - \dfrac{x^2}{4} = 1$.

Solution Note that

$$\frac{y^2}{36} - \frac{x^2}{4} = \frac{y^2}{6^2} - \frac{x^2}{2^2} = 1.$$

> The intercept distance is found in the term without the minus sign. Here there is a y in this term, so the intercepts are on the y-axis.

The asymptotes are thus $y = \frac{6}{2}x$ and $y = -\frac{6}{2}x$, or $y = 3x$ and $y = -3x$.

With the numbers 6 and 2, we can quickly sketch a rectangle to use as a guide. Thinking of ± 2 as x-coordinates and ± 6 as y-coordinates, we form all possible ordered pairs: $(2, 6)$, $(2, -6)$, $(-2, 6)$, and $(-2, -6)$. We plot these pairs and lightly sketch a rectangle through them. The asymptotes pass through the corners (see the figure on the left below). Since the hyperbola is vertical, we graph the y-intercepts. They are $(0, 6)$ and $(0, -6)$. We now draw curves through the intercepts toward the asymptotes, as shown on the right in the figure on the following page.

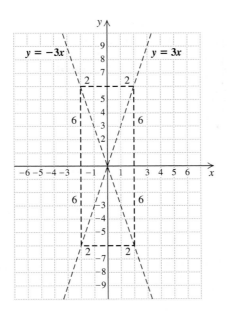 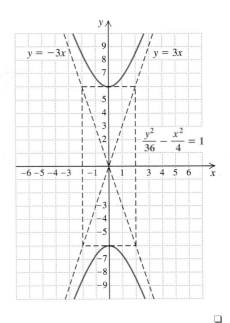

Hyperbolas (Nonstandard Form)

The equations that we have just seen for hyperbolas are the standard ones, but there are other hyperbolas. We consider some of them.

> Hyperbolas having the x- and y-axes as asymptotes have equations as follows:
>
> $$xy = c, \quad \text{where } c \text{ is a nonzero constant.}$$

EXAMPLE 5   Graph: $xy = -8$.

Solution   We first solve for y:

$$y = -\frac{8}{x}.$$

Next, we find some solutions, keeping the results in a table. Note that we cannot use 0 for x and that for large values of $|x|$, y will be close to 0. We plot the points and connect the points in the second quadrant with a smooth curve. Similarly, we connect the points in the fourth quadrant with a smooth curve. Remember that the axes serve as asymptotes.

x	y
2	−4
−2	4
4	−2
−4	2
1	−8
−1	8
8	−1
−8	1

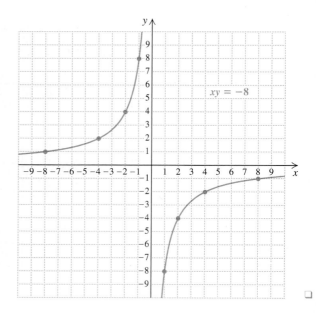

$xy = -8$

TECHNOLOGY
CONNECTION

The procedure used to draw the graph of an ellipse or a hyperbola on a grapher is similar to that used to draw a circle. For example, let's draw the graph of the hyperbola given by the equation

$$\frac{x^2}{25} - \frac{y^2}{49} = 1.$$

Solving for y gives us

$$y_1 = \frac{\sqrt{49x^2 - 1225}}{5} = \frac{7}{5}\sqrt{x^2 - 25}$$

and

$$y_2 = \frac{-\sqrt{49x^2 - 1225}}{5} = -\frac{7}{5}\sqrt{x^2 - 25}.$$

When the two pieces are drawn on the same squared axes, the result is the following:

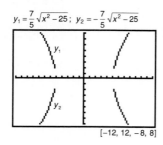

$y_1 = \frac{7}{5}\sqrt{x^2 - 25}$; $y_2 = -\frac{7}{5}\sqrt{x^2 - 25}$

$[-12, 12, -8, 8]$

Again, note the problem that the grapher has where the graph is nearly vertical.

Use a grapher to draw the graph of each of the following ellipses and hyperbolas. Use squared axes so that the shapes are not distorted.

TC1. $\dfrac{x^2}{16} + \dfrac{y^2}{60} = 1$ **TC2.** $16x^2 + 3y^2 = 64$

TC3. $\dfrac{y^2}{20} - \dfrac{x^2}{64} = 1$ **TC4.** $9x^2 - 45y^2 = 441$

Hyperbolas have many applications. A jet breaking the sound barrier creates a sonic boom whose wave front has the shape of a cone. The cone intersects the ground in one branch of a hyperbola. Some comets travel in hyperbolic orbits, and a cross section of an amphitheater may be hyperbolic in shape.

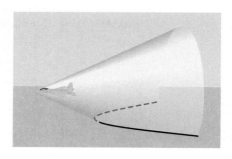

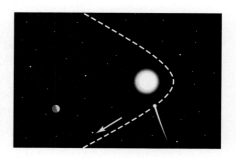

Classifying Graphs of Equations

The following is a summary of the equations and graphs of conic sections studied in this chapter.

Parabola

$y = ax^2 + bx + c, \quad a > 0$
$\quad = a(x - h)^2 + k$

$y = ax^2 + bx + c, \quad a < 0$
$\quad = a(x - h)^2 + k$

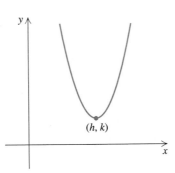

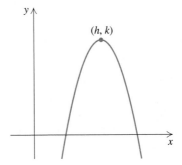

$x = ay^2 + by + c, \quad a > 0$
$\quad = a(y - k)^2 + h$

$x = ay^2 + by + c, \quad a < 0$
$\quad = a(y - k)^2 + h$

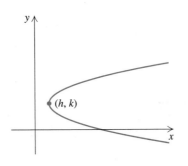

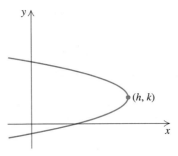

Circle

Center at the origin:

$$x^2 + y^2 = r^2$$

Center at (h, k):

$$(x - h)^2 + (y - k)^2 = r^2$$

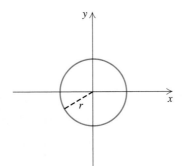

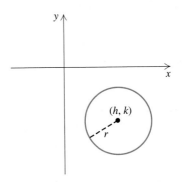

Ellipse

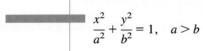

$$\frac{x^2}{a^2} + \frac{y^2}{b^2} = 1, \quad a > b$$

$$\frac{x^2}{a^2} + \frac{y^2}{b^2} = 1, \quad b > a$$

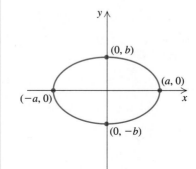

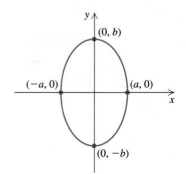

Hyperbola

$$\frac{x^2}{a^2} - \frac{y^2}{b^2} = 1$$

$$\frac{y^2}{b^2} - \frac{x^2}{a^2} = 1$$

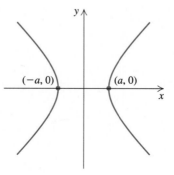

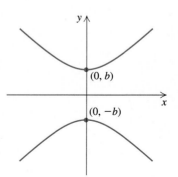

$$xy = c, \quad c > 0$$

$$xy = c, \quad c < 0$$

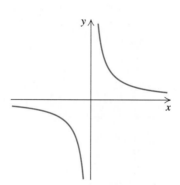

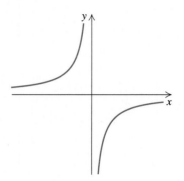

Suppose we encounter an equation that is not in one of the preceding forms. Sometimes we can find an equivalent equation that does fit one of the forms and then classify it as a circle, an ellipse, a parabola, or a hyperbola.

EXAMPLE 6

Classify the graph of the equation as a circle, an ellipse, a parabola, or a hyperbola.

a) $5x^2 = 20 - 5y^2$

b) $x + 3 + 8y = y^2$

c) $x^2 = y^2 + 4$

d) $x^2 = 16 - 4y^2$

Solution

a) We get the terms with variables on one side by adding $5y^2$:

$$5x^2 + 5y^2 = 20.$$

The fact that x and y are *both* squared tells us that we do not have a parabola. The fact that the squared terms are *added* tells us that we do not have a hyperbola. Do we have a circle? To find out, we need to get $x^2 + y^2$ by itself. We can do that by factoring the 5 out of both terms on the left and then dividing by 5 on both sides:

$$5(x^2 + y^2) = 20$$
$$x^2 + y^2 = 4$$
$$x^2 + y^2 = 2^2.$$

We can see that the graph is a circle with center at the origin and radius 2.

b) Since only one of the variables is squared, we can find the following equivalent equation:

$$x = y^2 - 8y - 3.$$

Thus the graph is a horizontal parabola that opens to the right since the coefficient of y^2, 1, is positive.

c) Both variables are squared, so we do not have a parabola. We can subtract y^2 on both sides and divide by 4 to obtain

$$\frac{x^2}{2^2} - \frac{y^2}{2^2} = 1.$$

The minus sign in this form tells us that the graph of the equation is a hyperbola.

d) Both variables are squared, so the graph is not a parabola. We obtain the following equivalent equation:

$$x^2 + 4y^2 = 16.$$

If the coefficients of the terms were the same, we would have the graph of a circle, as in part (a), but they are not. We have a plus sign between the squared terms, so the graph is not a hyperbola. We divide by 16 on both sides and obtain the equivalent equation

$$\frac{x^2}{16} + \frac{y^2}{4} = 1$$

whose graph is an ellipse.

EXERCISE SET | 10.2 |

Graph the ellipse.

1. $\dfrac{x^2}{4} + \dfrac{y^2}{1} = 1$

2. $\dfrac{x^2}{1} + \dfrac{y^2}{4} = 1$

3. $\dfrac{x^2}{16} + \dfrac{y^2}{25} = 1$

4. $\dfrac{x^2}{9} + \dfrac{y^2}{25} = 1$

5. $4x^2 + 9y^2 = 36$

6. $9x^2 + 4y^2 = 36$

7. $16x^2 + 9y^2 = 144$

8. $9x^2 + 16y^2 = 144$

9. $2x^2 + 3y^2 = 6$

10. $5x^2 + 7y^2 = 35$

11. $4x^2 + 9y^2 = 1$

12. $25x^2 + 16y^2 = 1$

13. $5x^2 + 12y^2 = 60$

14. $8x^2 + 3y^2 = 24$

Graph the hyperbola.

15. $\dfrac{x^2}{16} - \dfrac{y^2}{16} = 1$

16. $\dfrac{y^2}{9} - \dfrac{x^2}{9} = 1$

17. $\dfrac{y^2}{16} - \dfrac{x^2}{9} = 1$

18. $\dfrac{x^2}{9} - \dfrac{y^2}{4} = 1$

19. $\dfrac{x^2}{25} - \dfrac{y^2}{36} = 1$

20. $\dfrac{y^2}{9} - \dfrac{x^2}{25} = 1$

21. $x^2 - y^2 = 4$

22. $y^2 - x^2 = 25$

23. $4y^2 - 9x^2 = 36$

24. $25x^2 - 16y^2 = 400$

25. $xy = 6$

26. $xy = -4$

27. $xy = -9$

28. $xy = 3$

29. $xy = -1$

30. $xy = -2$

31. $xy = 2$

32. $xy = 1$

Classify each of the following as the equation of a circle, an ellipse, a parabola, or a hyperbola.

33. $x^2 + y^2 - 10x + 8y - 40 = 0$

34. $y + 1 = 2x^2$

35. $9x^2 - 4y^2 - 36 = 0$

36. $1 - 3y = 2y^2 - x$

37. $4x^2 + 25y^2 - 100 = 0$

38. $y^2 + x^2 = 7$

39. $x^2 + y^2 = 2x + 4y + 4$

40. $2y + 13 + x^2 = 8x - y^2$

41. $4x^2 = 64 - y^2$

42. $y = \dfrac{1}{x}$

43. $x - \dfrac{3}{y} = 0$

44. $x - 4 = y^2 + 5$

45. $y + 6x = x^2 + 6$

46. $x^2 = 16 + y^2$

47. $9y^2 = 36 + 4x^2$

48. $3x^2 + 5y^2 + x^2 = y^2 + 49$

Skill Maintenance

49. Simplify: $\sqrt[3]{125t^{15}}$.

50. Solve: $2x^2 + 10 = 0$.

51. Rationalize the denominator:
$$\dfrac{4\sqrt{2} - 5\sqrt{3}}{6\sqrt{3} - 8\sqrt{2}}.$$

52. An airplane travels 500 mi at a certain speed. A larger plane travels 1620 mi at a speed that is 320 mph faster, but takes 1 hr longer. Find the speed of each plane.

Synthesis

53. ◇ An eccentric person builds a pool table in the shape of an ellipse with a hole at one focus and a tiny dot at the other. Guests are amazed at how many bank shots the owner of the pool table makes. Explain.

54. The maximum distance of the planet Mars from the sun is 2.48×10^8 miles. The minimum distance is 3.46×10^7 miles. The sun is at one focus of the elliptical orbit. Find the distance from the sun to the other focus.

55. Let $(-c, 0)$ and $(c, 0)$ be the foci of an ellipse. Any point $P(x, y)$ is on the ellipse if the sum of the distances from the foci to P is some constant. Use $2a$ to represent this constant.

a) Show that an equation for the ellipse is given by
$$\dfrac{x^2}{a^2} + \dfrac{y^2}{a^2 - c^2} = 1.$$

b) Substitute b^2 for $a^2 - c^2$ to get standard form.

56. The Oval Office of the President of the United States is an ellipse 31 ft wide and 38 ft long. Show in a sketch precisely where the President and an adviser could sit to best use the room's acoustics. (*Hint:* See Exercise 55(b) and the discussion following Example 2.)

57. The light source in a dental lamp shines against a reflector that is shaped like a portion of an ellipse in which the light source is one focus of the ellipse. Reflected light enters a patient's mouth at the other focus of the ellipse. If the ellipse from which the reflector was formed is 2 ft wide and 6 ft long, how far should the patient's mouth be from the light source? (*Hint:* See Exercise 55(b).)

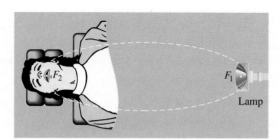

Lamp

58. Find an equation of an ellipse that has x-intercepts $(-9, 0)$ and $(9, 0)$ and y-intercepts $(0, -11)$ and $(0, 11)$.

The standard form of an ellipse parallel to an axis and centered at (h, k) is

$$\frac{(x - h)^2}{a^2} + \frac{(y - k)^2}{b^2} = 1.$$

The vertices are $(a + h, k)$, $(-a + h, k)$, $(h, b + k)$, and $(h, -b + k)$. For each of the following equations of ellipses, complete the square, if necessary, and find an equivalent equation in standard form. Find the center and the vertices. Then graph the ellipse.

59. $16x^2 + y^2 + 96x - 8y + 144 = 0$

60. $4x^2 + 25y^2 - 8x + 50y = 71$

Find an equation of a hyperbola satisfying the given conditions.

61. Having intercepts $(0, 8)$ and $(0, -8)$ and asymptotes $y = 4x$ and $y = -4x$

62. Having intercepts $(8, 0)$ and $(-8, 0)$ and asymptotes $y = 4x$ and $y = -4x$

Hyperbolas centered at (h, k), with horizontal or vertical axes, have standard equations as follows, where the asymptotes are

$$y - k = \pm \frac{b}{a}(x - h).$$

The vertices are labeled in the figures:

$$\frac{(x - h)^2}{a^2} - \frac{(y - k)^2}{b^2} = 1$$

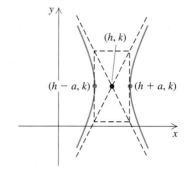

$$\frac{(y - k)^2}{b^2} - \frac{(x - h)^2}{a^2} = 1$$

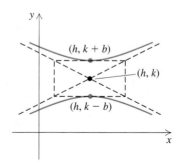

For each of the following equations of hyperbolas, complete the square, if necessary, and write in standard form. Find the center, the vertices, and the asymptotes. Then graph the hyperbola.

63. $\dfrac{(x - 2)^2}{9} - \dfrac{(y + 1)^2}{16} = 1$

64. $4x^2 - y^2 + 24x + 4y + 28 = 0$

65. $4y^2 - 25x^2 - 8y - 100x - 196 = 0$

66. $x^2 - y^2 - 2y - 4x = 6$

10.3

Nonlinear Systems of Equations

Systems Involving One Nonlinear Equation • Systems of Two Nonlinear Equations • Problem Solving

The equations that have appeared in systems of two equations have thus far all been linear. We now consider systems of two equations in which at least one equation is not linear.

Systems Involving One Nonlinear Equation

We consider a system involving an equation of a circle and an equation of a line. Let's think about the possible ways in which a circle and a line can intersect. The three possibilities are shown in the figure below. For L_1 there is no point of intersection; hence the system of equations has no real solution. For L_2 there is one point of intersection, hence one real solution. For L_3 there are two points of intersection, hence two real solutions.

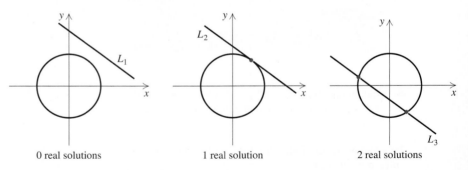

0 real solutions   1 real solution   2 real solutions

Remember that we used both *elimination* and *substitution* to solve systems of linear equations. When solving systems in which one equation is of first degree and one is of second degree, it is preferable to use the *substitution* method.

EXAMPLE 1

Solve the system

$$x^2 + y^2 = 25, \quad (1) \qquad \text{(The graph is a circle.)}$$
$$3x - 4y = 0. \quad (2) \qquad \text{(The graph is a line.)}$$

Solution   First, we solve the linear equation (2) for x:

$$x = \tfrac{4}{3}y. \quad (3)$$

Then we substitute $\tfrac{4}{3}y$ for x in Equation (1) and solve for y:

$$\left(\tfrac{4}{3}y\right)^2 + y^2 = 25$$
$$\tfrac{16}{9}y^2 + y^2 = 25$$
$$\tfrac{25}{9}y^2 = 25$$
$$y^2 = 9 \qquad \text{Multiplying by } \tfrac{9}{25} \text{ on both sides}$$
$$y = \pm 3.$$

Now we substitute these numbers for y in Equation (3) and solve for x:

for $y = 3$, $x = \frac{4}{3}(3) = 4$; for $y = -3$, $x = \frac{4}{3}(-3) = -4$.

Check: For (4, 3):

$$x^2 + y^2 = 25 \qquad\qquad 3x - 4y = 0$$

$$\overline{4^2 + 3^2 \;?\; 25} \qquad\qquad \overline{3(4) - 4(3) \;?\; 0}$$

$$16 + 9 \qquad\qquad\qquad 12 - 12$$

$$25 \;\big|\; 25 \;\; \text{TRUE} \qquad\qquad 0 \;\big|\; 0 \;\; \text{TRUE}$$

It is left to the student to confirm that $(-4, -3)$ also checks in both equations.

The pairs (4, 3) and $(-4, -3)$ check, so they are solutions. We can see the solutions in the graph. The graph of Equation (1) is a circle, and the graph of Equation (2) is a line. The graphs intersect at the points (4, 3) and $(-4, -3)$.

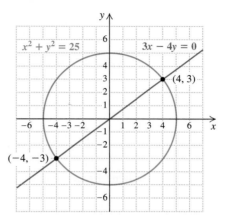

EXAMPLE 2

Solve the system

$$y + 3 = 2x, \qquad (1)$$
$$x^2 + 2xy = -1. \qquad (2)$$

Solution   First we solve the linear equation (1) for y:

$$y = 2x - 3. \qquad (3)$$

Then we substitute $2x - 3$ for y in Equation (2) and solve for x:

$$x^2 + 2x(2x - 3) = -1$$
$$x^2 + 4x^2 - 6x = -1$$
$$5x^2 - 6x + 1 = 0$$
$$(5x - 1)(x - 1) = 0 \qquad\qquad \text{Factoring}$$
$$5x - 1 = 0 \quad or \quad x - 1 = 0 \qquad \text{Using the principle of zero products}$$
$$x = \tfrac{1}{5} \quad or \qquad x = 1.$$

Now we substitute these numbers for x in Equation (3) and solve for y:

for $x = \tfrac{1}{5}$, $y = 2\left(\tfrac{1}{5}\right) - 3 = -\tfrac{13}{5}$; for $x = 1$, $y = 2(1) - 3 = -1$.

The pairs $\left(\tfrac{1}{5},\ -\tfrac{13}{5}\right)$ and $(1, -1)$ check, so they are solutions.

EXAMPLE 3

Solve the system

$$x + y = 5, \qquad (1) \qquad \text{(The graph is a line.)}$$
$$y = 3 - x^2. \qquad (2) \qquad \text{(The graph is a parabola.)}$$

Solution   We substitute $3 - x^2$ for y in the first equation:

$$x + 3 - x^2 = 5$$
$$-x^2 + x - 2 = 0 \qquad \text{Adding } -5 \text{ on both sides and rearranging}$$
$$x^2 - x + 2 = 0. \qquad \text{Multiplying by } -1 \text{ on both sides}$$

To solve this equation, we need the quadratic formula:

$$x = \frac{-b \pm \sqrt{b^2 - 4ac}}{2a}$$

$$= \frac{-(-1) \pm \sqrt{(-1)^2 - 4 \cdot 1 \cdot 2}}{2(1)}$$

$$= \frac{1 \pm \sqrt{1 - 8}}{2}$$

$$= \frac{1 \pm \sqrt{-7}}{2}$$

$$= \frac{1}{2} \pm \frac{\sqrt{7}}{2}i.$$

Solving Equation (1) for y gives us $y = 5 - x$. Substituting values for x gives

$$y = 5 - \left(\frac{1}{2} + \frac{\sqrt{7}}{2}i\right) = \frac{9}{2} - \frac{\sqrt{7}}{2}i \quad \text{and} \quad y = 5 - \left(\frac{1}{2} - \frac{\sqrt{7}}{2}i\right) = \frac{9}{2} + \frac{\sqrt{7}}{2}i.$$

The solutions are

$$\left(\frac{1}{2} + \frac{\sqrt{7}}{2}i, \frac{9}{2} - \frac{\sqrt{7}}{2}i\right) \quad \text{and} \quad \left(\frac{1}{2} - \frac{\sqrt{7}}{2}i, \frac{9}{2} + \frac{\sqrt{7}}{2}i\right).$$

There are no real-number solutions. Note in the figure below that the graphs do not intersect. Getting only nonreal solutions tells us that the graphs do not intersect.

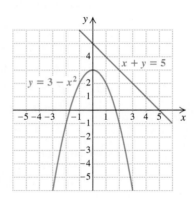

Systems of Two Nonlinear Equations

We now consider systems of two second-degree equations. Graphs of such systems can involve any two conic sections. The following figure shows some ways in which a circle and a hyperbola can intersect.

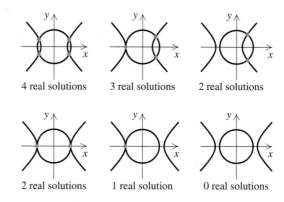

4 real solutions   3 real solutions   2 real solutions

2 real solutions   1 real solution   0 real solutions

To solve systems of two second-degree equations, we can use either the substitution method or the elimination method. The elimination method is generally better when each equation is of the form $Ax^2 + By^2 = C$. Then we can eliminate an x^2- or a y^2-term in a manner similar to the procedure we used for systems of linear equations in Chapter 3.

EXAMPLE 4

Solve the system

$$2x^2 + 5y^2 = 22, \qquad (1)$$
$$3x^2 - y^2 = -1. \qquad (2)$$

Solution   Here we multiply Equation (2) by 5 and then add the equations:

$$\begin{aligned}
2x^2 + 5y^2 &= 22 \\
15x^2 - 5y^2 &= -5 \qquad \text{\small Multiplying by 5 on both sides of Equation (2)} \\
\hline
17x^2 &= 17 \qquad \text{\small Adding} \\
x^2 &= 1 \\
x &= \pm 1.
\end{aligned}$$

If $x = 1$, $x^2 = 1$, and if $x = -1$, $x^2 = 1$, so substituting 1 or -1 for x in Equation (2), we have

$$\begin{aligned}
3 \cdot (\pm 1)^2 - y^2 &= -1 \\
3 - y^2 &= -1 \\
-y^2 &= -4 \\
y^2 &= 4 \\
y &= \pm 2.
\end{aligned}$$

Thus, if $x = 1$, $y = 2$ or $y = -2$; and if $x = -1$, $y = 2$ or $y = -2$. The possible solutions are then $(1, 2)$, $(1, -2)$, $(-1, 2)$, and $(-1, -2)$.

Check: Since $(2)^2 = 4$, $(-2)^2 = 4$, $(1)^2 = 1$, and $(-1)^2 = 1$, we can check all four pairs at once.

$$2x^2 + 5y^2 = 22$$
$$2(\pm 1)^2 + 5(\pm 2)^2 \;?\; 22$$
$$2 + 20$$
$$22 \mid 22 \text{ TRUE}$$

$$3x^2 - y^2 = -1$$
$$3(\pm 1)^2 - (\pm 2)^2 \;?\; -1$$
$$3 - 4$$
$$-1 \mid -1 \text{ TRUE}$$

The solutions are $(1, 2)$, $(1, -2)$, $(-1, 2)$, and $(-1, -2)$. □

When a product of variables is in one equation and the other is of the form $Ax^2 + By^2 = C$, we often solve for a variable in the equation with the product and then use substitution.

Solve the system

$$x^2 + 4y^2 = 20, \quad (1)$$
$$xy = 4. \quad (2)$$

Solution First we solve Equation (2) for y:

$$y = \frac{4}{x}.$$

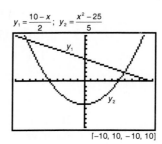

TECHNOLOGY CONNECTION

Because the algebra is often difficult, finding the solution(s) to systems of nonlinear equations provides an excellent opportunity to use a grapher. As with systems of linear equations, we simply zoom in as often as necessary and then use the Trace feature to read the coordinates of the points of intersection. Using a grapher restricts solutions to real numbers. Few graphers have the ability to find imaginary solutions.

For example, to solve the system

$$x + 2y = 10,$$
$$x^2 - 5y = 25,$$

we first solve each equation for y:

$$y_1 = \frac{10 - x}{2},$$
$$y_2 = \frac{x^2 - 25}{5}.$$

Both equations are then graphed.

$$y_1 = \frac{10 - x}{2}; \; y_2 = \frac{x^2 - 25}{5}$$

$[-10, 10, -10, 10]$

Zooming in on the points of intersection, we determine coordinates, to two decimal places, to be $(-8.43, 9.22)$ and $(5.93, 2.03)$.

Use a grapher to solve each system. Round all values to two decimal places.

TC1. $4xy - 7 = 0,$
$x - 3y - 2 = 0$

TC2. $x^2 + y^2 = 14,$
$16x + 7y^2 = 0$

Then we substitute $4/x$ for y in Equation (1) and solve for x:

$$x^2 + 4\left(\frac{4}{x}\right)^2 = 20$$

$$x^2 + \frac{64}{x^2} = 20$$

$$x^4 + 64 = 20x^2 \qquad \text{Multiplying by } x^2$$

$$x^4 - 20x^2 + 64 = 0 \qquad \text{Obtaining standard form. This equation is reducible to quadratic.}$$

$$u^2 - 20u + 64 = 0 \qquad \text{Letting } u = x^2$$

$$(u - 16)(u - 4) = 0 \qquad \text{Factoring}$$

$$u = 16 \quad or \quad u = 4. \qquad \text{Using the principle of zero products}$$

Now we substitute x^2 for u and solve these equations:

$$x^2 = 16 \quad or \quad x^2 = 4$$

$$x = \pm 4 \quad or \quad x = \pm 2.$$

Then $x = 4$ or $x = -4$ or $x = 2$ or $x = -2$. Since $y = 4/x$, if $x = 4$, $y = 1$; if $x = -4$, $y = -1$; if $x = 2$, $y = 2$; and if $x = -2$, $y = -2$. The ordered pairs $(4, 1)$, $(-4, -1)$, $(2, 2)$, and $(-2, -2)$ check. They are the solutions. ❑

Problem Solving

We now consider solving problems in which the translation is to a system of equations in which at least one is not linear.

EXAMPLE 6

For a building at a community college, an architect wants to lay out a rectangular piece of ground that has a perimeter of 204 m and an area of 2565 m². Find the dimensions of the piece of ground.

Solution

1. FAMILIARIZE. We draw a picture of the field, labeling the drawing. We let l = the length and w = the width, both in meters.

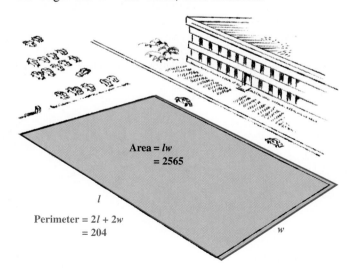

Area = lw
= 2565

l

Perimeter = $2l + 2w$
= 204

w

2. TRANSLATE.   We then have the following translation:

Perimeter: $2w + 2l = 204$;

Area: $lw = 2565$.

3. CARRY OUT.   We solve the system

$$2w + 2l = 204,$$
$$lw = 2565.$$

We solve the second equation for l and get $l = 2565/w$. Then we substitute $2565/w$ for l in the first equation and solve for w:

$$2w + 2\left(\frac{2565}{w}\right) = 204$$

$$2w^2 + 2(2565) = 204w \qquad \text{Multiplying by } w$$

$$2w^2 - 204w + 2(2565) = 0 \qquad \text{Standard form}$$

$$w^2 - 102w + 2565 = 0 \qquad \text{Multiplying by } \tfrac{1}{2}$$

> Factoring could be used instead of the quadratic formula, but the numbers are quite large.

$$w = \frac{-(-102) \pm \sqrt{(-102)^2 - 4 \cdot 1 \cdot 2565}}{2 \cdot 1}$$

$$w = \frac{102 \pm \sqrt{144}}{2} = \frac{102 \pm 12}{2}$$

$$w = 57 \quad or \quad w = 45.$$

If $w = 57$, then $l = 2565/w = 2565/57 = 45$. If $w = 45$, then $l = 2565/w = 2565/45 = 57$. Since length is usually considered to be longer than width, we have the solution $l = 57$ and $w = 45$, or $(57, 45)$.

4. CHECK.   If $l = 57$ and $w = 45$, the perimeter is $2 \cdot 57 + 2 \cdot 45$, or 204. The area is $57 \cdot 45$, or 2565. The numbers check.

5. STATE.   The length is 57 m and the width is 45 m.   ❑

EXAMPLE 7

The area of a rectangular Oriental rug is 300 ft^2, and the length of a diagonal is 25 ft. Find the dimensions of the rug.

Solution

1. FAMILIARIZE.   We draw a picture and label it. Note that there is a right triangle in the figure. We let l = the length and w = the width, both in feet.

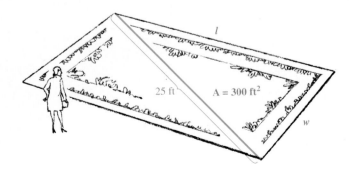

2. TRANSLATE. We translate to a system of equations:

$l^2 + w^2 = 25^2$,   Using the Pythagorean theorem

$lw = 300$.   Using the formula for the area of a rectangle

3. CARRY OUT. We solve the system

$l^2 + w^2 = 625$,

$lw = 300$

to get $(20, 15)$, $(15, 20)$, $(-20, -15)$, and $(-15, -20)$.

4. CHECK. Measurements cannot be negative and length is usually greater than width, so we check only $(20, 15)$. In the right triangle, $15^2 + 20^2 = 225 + 400 = 625$, which is 25^2. The area is $15 \cdot 20$, or 300, so we have a solution.

5. STATE. The length is 20 ft and the width is 15 ft.

EXERCISE SET | 10.3

Solve. In most cases, graphs can be used to confirm the solutions.

1. $x^2 + y^2 = 25$,
$y - x = 1$

2. $x^2 + y^2 = 100$,
$y - x = 2$

3. $4x^2 + 9y^2 = 36$,
$3y + 2x = 6$

4. $9x^2 + 4y^2 = 36$,
$3x + 2y = 6$

5. $y^2 = x + 3$,
$2y = x + 4$

6. $y = x^2$,
$3x = y + 2$

7. $x^2 - xy + 3y^2 = 27$,
$x - y = 2$

8. $2y^2 + xy + x^2 = 7$,
$x - 2y = 5$

9. $x^2 + 4y^2 = 25$,
$x + 2y = 7$

10. $y^2 - x^2 = 16$,
$2x - y = 1$

11. $x^2 - xy + 3y^2 = 5$,
$x - y = 2$

12. $m^2 + 3n^2 = 10$,
$m - n = 2$

13. $3x + y = 7$,
$4x^2 + 5y = 24$

14. $2y^2 + xy = 5$,
$4y + x = 7$

15. $a + b = 7$,
$ab = 4$

16. $p + q = -6$,
$pq = -7$

17. $2a + b = 1$,
$b = 4 - a^2$

18. $4x^2 + 9y^2 = 36$,
$x + 3y = 3$

19. $a^2 + b^2 = 89$,
$a - b = 3$

20. $xy = 4$,
$x + y = 5$

21. $x^2 + y^2 = 25$,
$y^2 = x + 5$

22. $y = x^2$,
$x = y^2$

23. $x^2 + y^2 = 9$,
$x^2 - y^2 = 9$

24. $y^2 - 4x^2 = 4$,
$4x^2 + y^2 = 4$

25. $x^2 + y^2 = 25$,
$xy = 12$

26. $x^2 - y^2 = 16$,
$x + y^2 = 4$

27. $x^2 + y^2 = 4$,
$16x^2 + 9y^2 = 144$

28. $x^2 + y^2 = 25$,
$25x^2 + 16y^2 = 400$

29. $x^2 + y^2 = 16$,
$y^2 - 2x^2 = 10$

30. $x^2 + y^2 = 14$,
$x^2 - y^2 = 4$

31. $x^2 + y^2 = 5$,
$xy = 2$

32. $x^2 + y^2 = 20$,
$xy = 8$

33. $x^2 + y^2 = 13$,
$xy = 6$

34. $x^2 + 4y^2 = 20$,
$xy = 4$

35. $3xy + x^2 = 34$,
$2xy - 3x^2 = 8$

36. $2xy + 3y^2 = 7$,
$3xy - 2y^2 = 4$

37. $xy - y^2 = 2$,
$2xy - 3y^2 = 0$

38. $4a^2 - 25b^2 = 0$,
$2a^2 - 10b^2 = 3b + 4$

39. $x^2 - y = 5$,
$x^2 + y^2 = 25$

40. $ab - b^2 = -4$,
$ab - 2b^2 = -6$

Solve.

41. A computer parts company wants to make a rectangular memory board that has a perimeter of 28 cm and a diagonal of length 10 cm. What are the dimensions of the board?

42. A bathroom tile company wants to make a new rectangular tile that has a perimeter of 6 in. and a diagonal of length $\sqrt{5}$ in. What are the dimensions of the tile?

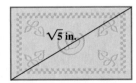

43. A rectangle has an area of 20 in^2 and a perimeter of 18 in. Find its dimensions.

44. A rectangle has an area of 2 yd^2 and a perimeter of 6 yd. Find its dimensions.

45. It will take 210 yd of fencing to enclose a rectangular field. The area of the field is 2250 yd^2. What are the dimensions of the field?

46. The diagonal of a rectangle is 1 ft longer than the length of the rectangle and 3 ft longer than twice the width. Find the dimensions of the rectangle.

47. The product of the lengths of the legs of a right triangle is 156. The hypotenuse has length $\sqrt{313}$. Find the lengths of the legs.

48. The product of two numbers is 60. The sum of their squares is 136. Find the numbers.

49. A garden contains two square peanut beds. Find the length of each bed if the sum of their areas is 832 ft^2 and the difference of their areas is 320 ft^2.

50. A certain amount of money saved for 1 yr at a certain interest rate yielded $7.50. If the principal had been $25 more and the interest rate 1% less, the interest would have been the same. Find the principal and the rate.

51. The area of a rectangle is $\sqrt{3}$ m^2, and the length of a diagonal is 2 m. Find the dimensions.

52. The area of a rectangle is $\sqrt{2}$ m^2, and the length of a diagonal is $\sqrt{3}$ m. Find the dimensions.

Skill Maintenance

Simplify.

53. $\sqrt{48}$

54. $\sqrt[4]{32a^{24}d^9}$

55. A boat travels 4 mi upstream and 4 mi back downstream. The total time for the trip is 3 hr. The speed of the stream is 2 mph. Find the speed of the boat in still water.

56. Rationalize the denominator: $\dfrac{\sqrt{x} - \sqrt{h}}{\sqrt{x} + \sqrt{h}}$.

Synthesis

57. ◈ Write a problem for a classmate to solve. Devise the problem so that a system of two non-linear equations must be solved.

58. A piece of wire 100 cm long is to be cut into two pieces and those pieces are each to be bent to make a square. The area of one square is to be 144 cm^2 greater than that of the other. How should the wire be cut?

59. Find the equation of a circle that passes through $(-2, 3)$ and $(-4, 1)$ and whose center is on the line $5x + 8y = -2$.

60. Find the equation of an ellipse centered at the origin that passes through the points $(2, -3)$ and $(1, \sqrt{13})$.

61. Four squares with sides 5 in. long are cut from the corners of a rectangular metal sheet that has an area of 340 in^2. The edges are bent up to form an open box with a volume of 350 in^3. Find the dimensions of the box.

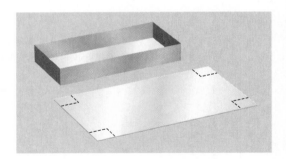

62. A company keeps records of the total revenue (money taken in) from the sale of x units of a product. It determines that total revenue R is given by

$$R = 100x - x^2.$$

The company also keeps records of the total cost of producing x units of the same product. It determines that the total cost C is given by

$$C = 20x + 1500.$$

A break-even point is a value of x for which total revenue is the same as total cost; that is, $R = C$. Find the break-even points.

Solve.

63. $p^2 + q^2 = 13,$

$\dfrac{1}{pq} = -\dfrac{1}{6}$

64. $a + b = \dfrac{5}{6},$

$\dfrac{a}{b} + \dfrac{b}{a} = \dfrac{13}{6}$

SUMMARY AND REVIEW | 10 |

IMPORTANT PROPERTIES AND FORMULAS

The Distance Formula

The distance d between any two points (x_1, y_1) and (x_2, y_2) is given by

$$d = \sqrt{(x_2 - x_1)^2 + (y_2 - y_1)^2}.$$

The Midpoint Formula

If the endpoints of a segment are (x_1, y_1) and (x_2, y_2), then the coordinates of the midpoint are

$$\left(\frac{x_1 + x_2}{2}, \frac{y_1 + y_2}{2} \right).$$

(See the summary of graphs at the end of Section 10.2.)

Parabola

Vertical with vertex at (h, k):

$$y = ax^2 + bx + c$$
$$= a(x - h)^2 + k$$

Horizontal with vertex at (h, k):

$$x = ay^2 + by + c$$
$$= a(y - k)^2 + h$$

Circle

Center at the origin:

$$x^2 + y^2 = r^2$$

Center at (h, k):

$$(x - h)^2 + (y - k)^2 = r^2$$

Ellipse

Center at the origin

Foci on x-axis: $\dfrac{x^2}{a^2} + \dfrac{y^2}{b^2} = 1, \quad a > b$ Foci on y-axis: $\dfrac{x^2}{a^2} + \dfrac{y^2}{b^2} = 1, \quad b > a$

Hyperbola

Center at the origin

Axis horizontal: $\dfrac{x^2}{a^2} - \dfrac{y^2}{b^2} = 1$ Axis vertical: $\dfrac{y^2}{b^2} - \dfrac{x^2}{a^2} = 1$

x- and y-axes asymptotes: $xy = c$

REVIEW EXERCISES

This chapter's review and test include Skill Maintenance exercises from Sections 6.5, 7.3, 7.6, and 8.2.

Find the distance between the pair of points. Where appropriate, find an approximation to three decimal places.

1. (2, 6) and (6, 6)   **2.** (−1, 1) and (−5, 4)

3. (4, 7) and (−3, −2)   **4.** (2, 3a) and (−1, a)

Find the midpoint of the segment with the given endpoints.

5. (1, 6) and (7, 6)   **6.** (−1, 1) and (−5, 4)

7. (4, 7) and (−3, −2)   **8.** (2, 3a) and (−1, a)

Find the center and the radius.

9. $(x + 2)^2 + (y - 3)^2 = 2$

10. $(x - 5)^2 + y^2 = 49$

11. $x^2 + y^2 - 6x - 2y + 1 = 0$

12. $x^2 + y^2 + 8x - 6y - 10 = 0$

13. Find an equation of the circle with center $(-4, 3)$ and radius $4\sqrt{3}$.

14. Find an equation of the circle with center $(7, -2)$ and radius $2\sqrt{5}$.

Classify the equation as a circle, an ellipse, a parabola, or a hyperbola. Then graph.

15. $4x^2 + 4y^2 = 100$ **16.** $9x^2 + 2y^2 = 18$

17. $y = -x^2 + 2x - 3$ **18.** $\dfrac{y^2}{9} - \dfrac{x^2}{4} = 1$

19. $xy = 9$ **20.** $x = y^2 + 2y - 2$

21. $xy = -3$

22. $x^2 + y^2 + 6x - 8y - 39 = 0$

Solve.

23. $x^2 - y^2 = 33,$ **24.** $x^2 - 2x + 2y^2 = 8,$
     $x + y = 11$ $2x + y = 6$

25. $x^2 - y = 3,$ **26.** $x^2 + y^2 = 25,$
     $2x - y = 3$ $x^2 - y^2 = 7$

27. $x^2 - y^2 = 3,$ **28.** $x^2 + y^2 = 18,$
     $y = x^2 - 3$ $2x + y = 3$

29. $x^2 + y^2 = 100,$ **30.** $x^2 + 2y^2 = 12,$
     $2x^2 - 3y^2 = -120$ $xy = 4$

31. A rectangle has a perimeter of 38 m and an area of 84 m². What are its dimensions?

32. Find two positive integers whose sum is 12 and the sum of whose reciprocals is $\frac{3}{8}$.

33. The perimeter of a square is 12 cm more than the perimeter of another square. Its area exceeds the area of the other by 39 cm². Find the perimeter of each square.

34. The sum of the areas of two circles is 130π ft². The difference of the circumferences is 16π ft. Find the radius of each circle.

Skill Maintenance

35. Simplify: $\sqrt[3]{81a^8b^{10}}$.

36. Solve: $x^2 + 2x + 5 = 0$.

37. Rationalize the numerator: $\dfrac{4 - \sqrt{a}}{2 + \sqrt{a}}$.

38. The speed of a moving sidewalk at an airport is 5 ft/sec. A person can walk 55 ft forward on the moving sidewalk in the same time it takes to walk 5 ft in the opposite direction. At what rate would the person walk on a nonmoving sidewalk?

Synthesis

39. ◈ Is a circle a special type of ellipse? Why or why not?

40. ◈ Explain why function notation is not used in this chapter, and list the graphs discussed for which function notation could be used.

41. Solve:
$$3x^2 + 4y^2 = 8,$$
$$x^2 - y^2 = 5.$$

42. Find the points whose distance from $(8, 0)$ and from $(-8, 0)$ is 10.

43. Find an equation of the circle that passes through $(-2, -4)$, $(5, -5)$, and $(6, 2)$.

44. Find an equation of the ellipse with the following vertices: $(-7, 0)$, $(7, 0)$, $(0, -3)$, and $(0, 3)$.

45. Find the point on the *x*-axis that is equidistant from $(-3, 4)$ and $(5, 6)$.

CHAPTER TEST | 10

Find the distance between the pair of points. If appropriate, find an approximation to three decimal places.

1. $(4, -1)$ and $(-5, 8)$ **2.** $(3, -a)$ and $(-3, a)$

Find the midpoint of the segment with the given endpoints.

3. $(4, -1)$ and $(-5, 8)$ **4.** $(3, -a)$ and $(-3, a)$

Find the center and the radius of the circle.

5. $(x + 2)^2 + (y - 3)^2 = 64$
6. $x^2 + y^2 + 4x - 6y + 4 = 0$

Classify the equation as a circle, an ellipse, a parabola, or a hyperbola. Then graph.

7. $y = x^2 - 4x - 1$
8. $x^2 + y^2 + 2x + 6y + 6 = 0$
9. $\dfrac{x^2}{9} - \dfrac{y^2}{4} = 1$ **10.** $16x^2 + 4y^2 = 64$
11. $xy = -5$ **12.** $x = -y^2 + 4y$

Solve.

13. $\dfrac{x^2}{16} + \dfrac{y^2}{9} = 1,$
$3x + 4y = 12$

14. $x^2 + y^2 = 16,$
$\dfrac{x^2}{16} - \dfrac{y^2}{9} = 1$

15. In a rational expression, the sum of the values of the numerator and the denominator is 23. The product of their values is 120. Find the values of the numerator and the denominator.

16. A rectangle with diagonal of length $5\sqrt{5}$ has an area of 22. Find the dimensions of the rectangle.

17. Two squares are such that the sum of their areas is 8 m² and the difference of their areas is 2 m². Find the length of a side of each square.

18. A rectangle has a diagonal of length 20 ft and a perimeter of 56 ft. Find the dimensions of the rectangle.

Skill Maintenance

19. Solve: $x^2 + 2x = 5$.
20. Simplify: $\sqrt[3]{48a^5b^{18}}$.
21. Rationalize the denominator: $\dfrac{4 - \sqrt{a}}{2 + \sqrt{a}}$.
22. A boat travels 6 mi upstream in the same time it takes to travel 30 mi downstream. The speed of the stream is 4 mph. Find the speed of the boat in still water.

Synthesis

23. Find an equation of the ellipse with the following vertices: $(1, 3)$, $(6, 6)$, $(11, 3)$, and $(6, 0)$.
24. Find the point on the *y*-axis that is equidistant from $(-3, -5)$ and $(4, -7)$.
25. The sum of two numbers is 36, and the product is 4. Find the sum of the reciprocals of the numbers.

Sequences, Series, and the Binomial Theorem

A student loan is in the amount of $6000. Interest is 9%, compounded annually, and the entire amount is to be repaid after 10 years. How much is to be paid back?

In Section 11.3, we will find that the amount due can be determined using the 11th term of a sequence:

$$6000, \quad 1.09 \cdot 6000,$$
$$(1.09)^2 6000, \quad (1.09)^3 6000, \ldots.$$

Dr. Jesse G. Jackson
CHIEF FINANCIAL
AID OFFICER

"I am constantly amazed at how important a role mathematics plays not just in the day-to-day operations of our office, but in the lives of our graduating students. I really never expected to use math nearly as much as I now do."

The first three sections of this chapter are devoted to *sequences* and *series*. A sequence is simply an ordered list. For example, when a baseball coach writes a batting order, a sequence is being formed. When the members of a sequence are numbers, we can find their sum. Such a sum is called a *series*.

Section 11.4 presents the *binomial theorem,* which is used to expand expressions of the form $(a + b)^n$. These expansions are themselves series.

In addition to the material in this chapter, the review and test for Chapter 11 include material from Sections 3.2, 9.4, 9.6, and 10.1.

11.1

Sequences and Series

Sequences • **Finding the General Term** • **Sums and Series** • **Sigma Notation**

Sequences

Suppose that $1000 is invested at 8%, compounded annually. The amounts to which the account will grow after 1 year, 2 years, 3 years, 4 years, and so on, are as follows:

$1080.00, $1166.40, $1259.71, $1360.49, ...

Note that we can think of this as a function that pairs 1 with the number $1080.00, 2 with the number $1166.40, 3 with the number $1259.71, 4 with the number $1360.49, and so on. A sequence is thus a *function,* where the domain is a set of consecutive positive integers beginning with 1.

If we continue computing the amounts in the account forever, we obtain an **infinite sequence,** with function values

$1080.00, $1166.40, $1259.71, $1360.49, $1469.33, $1586.87, ...

The three dots at the end indicate that the sequence goes on without stopping. If we stop after a certain number of years, we obtain a **finite sequence:**

$1080.00, $1166.40, $1259.71, $1360.49.

Sequences

An *infinite sequence* is a function having for its domain the set of positive integers: $\{1, 2, 3, 4, 5, \ldots\}$.

A *finite sequence* is a function having for its domain a set of positive integers $\{1, 2, 3, 4, 5, \ldots, n\}$, for some positive integer n.

As another example, consider the sequence given by

$$a(n) = 2^n, \quad \text{or} \quad a_n = 2^n.$$

The notation a_n means the same as $a(n)$ but is used more commonly with sequences. Some of the function values (also known as *terms* of the sequence) are as follows:

$$a_1 = 2^1 = 2,$$
$$a_2 = 2^2 = 4,$$
$$a_3 = 2^3 = 8,$$
$$a_6 = 2^6 = 64.$$

The first term of the sequence is a_1, the fifth term is a_5, and the nth term, or **general term,** is a_n. This sequence can also be denoted in the following ways:

2, 4, 8, . . . ; or

2, 4, 8, . . . , 2^n, ...   The 2^n emphasizes that the nth term of this sequence is found by raising 2 to the nth power.

EXAMPLE 1

Find the first 4 terms and the 57th term of the sequence whose general term is given by $a_n = (-1)^n/(n + 1)$.

Solution

$$a_1 = \frac{(-1)^1}{1 + 1} = -\frac{1}{2}$$

$$a_2 = \frac{(-1)^2}{2 + 1} = \frac{1}{3}$$

$$a_3 = \frac{(-1)^3}{3 + 1} = -\frac{1}{4}$$

$$a_4 = \frac{(-1)^4}{4 + 1} = \frac{1}{5}$$

$$a_{57} = \frac{(-1)^{57}}{57 + 1} = -\frac{1}{58}$$

Note that the expression $(-1)^n$ causes the signs of the terms to alternate between positive and negative, depending on whether n is even or odd.   ❏

Finding the General Term

When only the first few terms of a sequence are known, we do not know for sure what the general term is, but we can make a prediction by looking for a pattern.

EXAMPLE 2

For each sequence, predict the general term.

a) 1, 4, 9, 16, 25, . . .   **b)** $\sqrt{1}, \sqrt{2}, \sqrt{3}, \sqrt{4}, \ldots$
c) $-1, 2, -4, 8, -16, \ldots$ **d)** 2, 4, 8, . . .

Solution

a) 1, 4, 9, 16, 25, . . .
These are squares of consecutive positive integers, so the general term may be n^2.

b) $\sqrt{1}, \sqrt{2}, \sqrt{3}, \sqrt{4}, \ldots$

These are square roots of consecutive positive integers, so the general term may be $\sqrt{n}$.

c) $-1, 2, -4, 8, -16, \ldots$

These are powers of 2 with alternating signs, so the general term may be $(-1)^n[2^{n-1}]$.

d) $2, 4, 8, \ldots$

If we see the pattern of powers of 2, we will see 16 as the next term and guess 2^n for the general term. Then the sequence could be written with more terms as

$$2, 4, 8, 16, 32, 64, 128, \ldots .$$

If we see that we can get the second term by adding 2, the third term by adding 4, the next term by adding 6, and so on, we will see 14 as the next term. A general term for the sequence is then $n^2 - n + 2$, and the sequence can be written with more terms as

$$2, 4, 8, 14, 22, 32, 44, 58, \ldots .$$ ❑

Example 2(d) illustrates that with few given terms, the uncertainty about the nth term is greater.

Sums and Series

Series

> Given the infinite sequence
>
> $$a_1, a_2, a_3, a_4, \ldots, a_n, \ldots,$$
>
> the sum of the terms
>
> $$a_1 + a_2 + a_3 + \cdots + a_n + \cdots$$
>
> is called an *infinite series*. A *partial sum* is the sum of the first n terms
>
> $$a_1 + a_2 + a_3 + \cdots + a_n.$$
>
> A partial sum is also called a *finite series* and is denoted S_n.

For instance, the sequence

$$3, 5, 7, 9, \ldots, 2n + 1, \ldots$$

has the following partial sums:

$S_1 = 3,$	This is the first term of the given sequence.
$S_2 = 3 + 5 = 8,$	This is the sum of the first two terms.
$S_3 = 3 + 5 + 7 = 15,$	The sum of the first three terms
$S_4 = 3 + 5 + 7 + 9 = 24.$	The sum of the first four terms

EXAMPLE 3

For the sequence $-2, 4, -6, 8, -10, 12, -14$, find: **(a)** S_3; **(b)** S_5.

Solution

a) $S_3 = -2 + 4 + (-6) = -4$

b) $S_5 = -2 + 4 + (-6) + 8 + (-10) = -6$ ❏

Sigma Notation

When the general term of a sequence is known, the Greek letter Σ (sigma) can be used to write a series. For example, the sum of the first four terms of the sequence 3, 5, 7, 9, . . . , $2k + 1$, . . . can be named as follows, using *sigma notation*, or *summation notation:*

$$\sum_{k=1}^{4} (2k + 1).$$

This is read "the sum as k goes from 1 to 4 of $(2k + 1)$." The letter k is called the *index of summation*. Sometimes the index of summation starts at a number other than 1.

EXAMPLE 4

Find and evaluate the sum.

a) $\displaystyle\sum_{k=1}^{5} k^2$ **b)** $\displaystyle\sum_{k=1}^{4} (-1)^k(2k)$ **c)** $\displaystyle\sum_{k=0}^{3} (2^k + 5)$

Solution

a) $\displaystyle\sum_{k=1}^{5} k^2 = 1^2 + 2^2 + 3^2 + 4^2 + 5^2 = 1 + 4 + 9 + 16 + 25 = 55$

Evaluate k^2 for all integers from 1 through 5. Then add.

b) $\displaystyle\sum_{k=1}^{4} (-1)^k(2k) = (-1)^1(2 \cdot 1) + (-1)^2(2 \cdot 2) + (-1)^3(2 \cdot 3) + (-1)^4(2 \cdot 4)$

$$= -2 + 4 - 6 + 8 = 4$$

c) $\displaystyle\sum_{k=0}^{3} (2^k + 5) = (2^0 + 5) + (2^1 + 5) + (2^2 + 5) + (2^3 + 5)$

$$= 6 + 7 + 9 + 13 = 35$$ ❏

EXAMPLE 5

Write sigma notation for the sum.

a) $1 + 4 + 9 + 16 + 25$ **b)** $-1 + 3 - 5 + 7$ **c)** $3 + 9 + 27 + 81 + \cdots$

Solution

a) $1 + 4 + 9 + 16 + 25$
This is a sum of squares, $1^2 + 2^2 + 3^2 + 4^2 + 5^2$, so the general term is k^2. Sigma notation is

$$\sum_{k=1}^{5} k^2.$$

b) $-1 + 3 - 5 + 7$

Except for the alternating signs, this is the sum of the first four positive odd numbers. Note that $2k - 1$ is a formula for the kth positive odd number, and $(-1)^k = 1$ when k is even and $(-1)^k = -1$ when k is odd. The general term is thus $(-1)^k(2k - 1)$, beginning with $k = 1$. Sigma notation is

$$\sum_{k=1}^{4} (-1)^k(2k - 1).$$

To check, we can evaluate $(-1)^k(2k - 1)$ using 1, 2, 3, and 4. Then we can write the sum of the four terms.

c) $3 + 9 + 27 + 81 + \cdots$

This is a sum of powers of 3, and it is also an infinite series. We use the symbol ∞ to represent infinity and name the infinite series using sigma notation as follows:

$$\sum_{k=1}^{\infty} 3^k.$$

❑

EXERCISE SET | 11.1

In each of the following, the nth term of a sequence is given. In each case find the first 4 terms; the 10th term, a_{10}; and the 15th term, a_{15}.

1. $a_n = 3n + 1$

2. $a_n = 3n - 1$

3. $a_n = \dfrac{n}{n + 1}$

4. $a_n = n^2 + 1$

5. $a_n = n^2 - 2n$

6. $a_n = \dfrac{n^2 - 1}{n^2 + 1}$

7. $a_n = n + \dfrac{1}{n}$

8. $a_n = \left(-\dfrac{1}{2}\right)^{n-1}$

9. $a_n = (-1)^n n^2$

10. $a_n = (-1)^n(n + 3)$

11. $a_n = (-1)^{n+1}(3n - 5)$

12. $a_n = (-1)^n(n^3 - 1)$

Find the indicated term of the sequence.

13. $a_n = 4n - 7$; a_8

14. $a_n = 5n + 11$; a_9

15. $a_n = (3n + 4)(2n - 5)$; a_7

16. $a_n = (3n + 2)^2$; a_6

17. $a_n = (-1)^{n-1}(3.4n - 17.3)$; a_{12}

18. $a_n = (-2)^{n-2}(45.68 - 1.2n)$; a_{23}

19. $a_n = 5n^2(4n - 100)$; a_{11}

20. $a_n = 4n^2(11n + 31)$; a_{22}

21. $a_n = \left(1 + \dfrac{1}{n}\right)^2$; a_{20}

22. $a_n = \left(1 - \dfrac{1}{n}\right)^3$; a_{15}

23. $a_n = \log 10^n$; a_{43}

24. $a_n = \ln e^n$; a_{67}

Predict the general term, or nth term, a_n, of the sequence. Answers may vary.

25. $1, 3, 5, 7, 9, \ldots$

26. $3, 9, 27, 81, 243, \ldots$

27. $-2, 6, -18, 54, \ldots$

28. $-2, 3, 8, 13, 18, \ldots$

29. $\frac{2}{3}, \frac{3}{4}, \frac{4}{5}, \frac{5}{6}, \frac{6}{7}, \ldots$

30. $\sqrt{2}, \sqrt{4}, \sqrt{6}, \sqrt{8}, \sqrt{10}, \ldots$

31. $\sqrt{3}, 3, 3\sqrt{3}, 9, 9\sqrt{3}, \ldots$

32. $1 \cdot 2, 2 \cdot 3, 3 \cdot 4, 4 \cdot 5, \ldots$

33. $-1, -4, -7, -10, -13, \ldots$

34. $\log 1, \log 10, \log 100, \log 1000, \ldots$

Find the indicated partial sum for the sequence.

35. $1, 2, 3, 4, 5, 6, 7, \ldots ; S_7$

36. $1, -3, 5, -7, 9, -11, \ldots ; S_8$

37. $2, 4, 6, 8, \ldots ; S_5$

38. $1, \frac{1}{4}, \frac{1}{9}, \frac{1}{16}, \frac{1}{25}, \ldots ; S_5$

Rename and evaluate the sum.

39. $\displaystyle\sum_{k=1}^{5} \frac{1}{2k}$ **40.** $\displaystyle\sum_{k=1}^{6} \frac{1}{2k+1}$

41. $\displaystyle\sum_{k=0}^{5} 2^k$ **42.** $\displaystyle\sum_{k=4}^{7} \sqrt{2k-1}$

43. $\displaystyle\sum_{k=7}^{10} \log k$ **44.** $\displaystyle\sum_{k=0}^{4} \pi k$

45. $\displaystyle\sum_{k=1}^{8} \frac{k}{k+1}$ **46.** $\displaystyle\sum_{k=1}^{4} \frac{k-2}{k+3}$

47. $\displaystyle\sum_{k=1}^{5} (-1)^k$ **48.** $\displaystyle\sum_{k=1}^{5} (-1)^{k+1}$

49. $\displaystyle\sum_{k=1}^{8} (-1)^{k+1}3^k$ **50.** $\displaystyle\sum_{k=1}^{7} (-1)^k 4^{k+1}$

51. $\displaystyle\sum_{k=0}^{5} (k^2 - 2k + 3)$ **52.** $\displaystyle\sum_{k=0}^{5} (k^2 - 3k + 4)$

53. $\displaystyle\sum_{k=1}^{10} \frac{1}{k(k+1)}$ **54.** $\displaystyle\sum_{k=1}^{10} \frac{2^k}{2^k+1}$

Rewrite the sum using sigma notation.

55. $\dfrac{1}{2} + \dfrac{2}{3} + \dfrac{3}{4} + \dfrac{4}{5} + \dfrac{5}{6} + \dfrac{6}{7}$

56. $3 + 6 + 9 + 12 + 15$

57. $-2 + 4 - 8 + 16 - 32 + 64$

58. $\dfrac{1}{1^2} + \dfrac{1}{2^2} + \dfrac{1}{3^2} + \dfrac{1}{4^2} + \dfrac{1}{5^2}$

59. $4 - 9 + 16 - 25 + \cdots + (-1)^n n^2$

60. $9 - 16 + 25 + \cdots + (-1)^{n+1} n^2$

61. $5 + 10 + 15 + 20 + 25 + \cdots$

62. $7 + 14 + 21 + 28 + 35 + \cdots$

63. $\dfrac{1}{1 \cdot 2} + \dfrac{1}{2 \cdot 3} + \dfrac{1}{3 \cdot 4} + \dfrac{1}{4 \cdot 5} + \cdots$

64. $\dfrac{1}{1 \cdot 2^2} + \dfrac{1}{2 \cdot 3^2} + \dfrac{1}{3 \cdot 4^2} + \dfrac{1}{4 \cdot 5^2} + \cdots$

Skill Maintenance

Simplify.

65. $\log_3 3$ **66.** $\log_3 1$

67. $\log_3 3^7$ **68.** $\log_c c$

Synthesis

69. ◇ Explain why the equation

$$\sum_{k=1}^{n} (a_k + b_k) = \sum_{k=1}^{n} a_k + \sum_{k=1}^{n} b_k$$

is true for any positive integer n. What laws are used to justify this result?

70. ◇ **a)** Find the first few terms of the sequence $a_n = n^2 - n + 41$.
  b) What pattern do you observe?
  c) Find the 41st term. Does the pattern you found in part (b) still hold?

Find the first five terms of the sequence; then find S_5.

71. $a_n = \dfrac{1}{2^n} \log 1000^n$ **72.** $a_n = i^n, i = \sqrt{-1}$

73. $a_n = \ln (1 \cdot 2 \cdot 3 \cdots n)$

Find decimal notation, rounded to six decimal places, for the first six terms of the sequence.

74. ▦ $a_n = \sqrt{n+1} - \sqrt{n}$

75. ▦ $a_n = \left(1 + \dfrac{1}{n}\right)^n$

Some sequences are given by a *recursive definition*. The value of the first term, a_1, is given, and then we are told how to find each subsequent term from the term preceding it in the sequence. Find the first six terms of each of the following recursively defined sequences.

76. $a_1 = 1, \quad a_{n+1} = 3a_n - 2$

77. $a_1 = 0, \quad a_{n+1} = a_n^2 + 4$

78. A single cell of bacterium divides into two every 15 min. Suppose the same rate of division is maintained for 4 hr. Give a sequence that lists the number of cells after successive 15-min periods.

79. The value of an office machine is $5200. Its scrap value each year is 75% of its value the year before. Give a sequence that lists the scrap value of the machine at the start of each year for a 10-year period.

80. Katrina gets $6.20 for working in a warehouse for a publishing company. Each year she gets a $0.40 hourly raise. Give a sequence that lists Katrina's hourly salary over a 10-year period.

11.2

Arithmetic Sequences and Series

Arithmetic Sequences • Sum of the First n Terms of an
Arithmetic Sequence • Problem Solving

In this section, we concentrate on sequences and series that are said to be arithmetic (pronounced ăr′ĭth-mĕt′-ĭk).

Arithmetic Sequences

In an **arithmetic sequence,** all terms (other than the first) can be found by adding the same number to the preceding term. For example, the sequence 2, 5, 8, 11, 14, 17, . . . is arithmetic because adding 3 to any term produces the next term. In other words, the difference between any term and the preceding one is 3. Arithmetic sequences are also called *arithmetic progressions*.

Arithmetic Sequence

A sequence is *arithmetic* if there exists a number d, called the *common difference*, such that $a_{n+1} = a_n + d$ for any integer $n \geq 1$.

EXAMPLE 1

For each arithmetic sequence, identify the first term, a_1, and the common difference, d.

a) 4, 9, 14, 19, 24, . . .   **b)** 27, 20, 13, 6, −1, −8, . . .
c) 2, $2\frac{1}{2}$, 3, $3\frac{1}{2}$, . . .

Solution   To find a_1, we simply use the first term listed. To find d, we choose any term beyond the first and subtract the preceding term from it.

Sequence	First Term, a_1	Common Difference, d
a) 4, 9, 14, 19, 24, . . .	4	$5 \leftarrow 9 - 4 = 5$
b) 27, 20, 13, 6, −1, −8, . . .	27	$-7 \leftarrow 20 - 27 = -7$
c) 2, $2\frac{1}{2}$, 3, $3\frac{1}{2}$, . . .	2	$\frac{1}{2} \leftarrow 2\frac{1}{2} - 2 = \frac{1}{2}$

Here we found the common difference by subtracting a_1 from a_2. Had we subtracted a_2 from a_3 or a_3 from a_4 we would have obtained the same values for d. Thus we can check by adding d to each term in a sequence to see if we progress to the next term.

Check: **a)** $4 + 5 = 9$, $9 + 5 = 14$, $14 + 5 = 19$, $19 + 5 = 24$
b) $27 + (-7) = 20$, $20 + (-7) = 13$, $13 + (-7) = 6$,
$6 + (-7) = -1$, $-1 + (-7) = -8$
c) $2 + \frac{1}{2} = 2\frac{1}{2}$, $2\frac{1}{2} + \frac{1}{2} = 3$, $3 + \frac{1}{2} = 3\frac{1}{2}$ ❑

To find a formula for the general, or nth, term of any arithmetic sequence, we denote the common difference by d and write out the first few terms:

$a_1,$

$a_2 = a_1 + d,$

$a_3 = a_2 + d = (a_1 + d) + d = a_1 + 2d,$ Substituting for a_2

$a_4 = a_3 + d = (a_1 + 2d) + d = a_1 + 3d.$ Substituting for a_3

Note that the coefficient of d in each case is 1 less than the subscript.

Generalizing, we obtain the following formula.

Formula 1

The nth term of an arithmetic sequence is given by

$$a_n = a_1 + (n - 1)d, \quad \text{for any integer } n \geq 1.$$

EXAMPLE 2

Find the 14th term of the arithmetic sequence 4, 7, 10, 13, ...

Solution   First we note that $a_1 = 4$, $d = 3$, and $n = 14$. Then using Formula 1, we obtain

$a_n = a_1 + (n - 1)d$

$a_{14} = 4 + (14 - 1) \cdot 3 = 4 + 13 \cdot 3 = 4 + 39 = 43.$

The 14th term is 43.   ❑

EXAMPLE 3

In the sequence in Example 2, which term is 301? That is, find n if $a_n = 301$.

Solution   We substitute into Formula 1 and solve for n:

$a_n = a_1 + (n - 1)d$

$301 = 4 + (n - 1) \cdot 3$

$301 = 4 + 3n - 3$

$300 = 3n$

$100 = n.$

The term 301 is the 100th term of the sequence.   ❑

Given two terms and their places in an arithmetic sequence, we can construct the sequence.

EXAMPLE 4

The third term of an arithmetic sequence is 8, and the sixteenth term is 47. Find a_1 and d and construct the sequence.

Solution   We know that $a_3 = 8$ and $a_{16} = 47$. Thus we would have to add d thirteen times to get from 8 to 47. That is,

$8 + 13d = 47.$ a_3 and a_{16} are 13 terms apart.

Solving $8 + 13d = 47$, we obtain

$$13d = 39$$
$$d = 3.$$

We subtract d twice from a_3 to get to a_1. Thus,

$$a_1 = 8 - 2 \cdot 3 = 2. \qquad a_1 \text{ and } a_3 \text{ are 2 terms apart.}$$

The sequence is 2, 5, 8, 11, ... Note that we could have subtracted d 15 times from a_{16} in order to find a_1. ❏

In general, d should be subtracted $(n - 1)$ times from a_n in order to find a_1.

Sum of the First n Terms of an Arithmetic Sequence

When we add the terms of an arithmetic sequence, we form an **arithmetic series.** To find a formula for computing S_n when the series is arithmetic, we denote the first n terms as follows:

This is the next-to-last-term. If you add d to this term, the result is a_n.
↓

$$a_1, (a_1 + d), (a_1 + 2d), \ldots ,(a_n - 2d), (a_n - d), a_n$$

↑
This term is two terms back from the last. If you add d to this term, you get the next-to-last term, $a_n - d$.

Then S_n is given by

$$S_n = a_1 + (a_1 + d) + (a_1 + 2d) + \cdots + (a_n - 2d) + (a_n - d) + a_n. \qquad (1)$$

Reversing the order of addition, we have

$$S_n = a_n + (a_n - d) + (a_n - 2d) + \cdots + (a_1 + 2d) + (a_1 + d) + a_1. \qquad (2)$$

If we add corresponding terms of each side of Equations (1) and (2), we get

$$2S_n = [a_1 + a_n] + [(a_1 + d) + (a_n - d)] + [(a_1 + 2d) + (a_n - 2d)]$$
$$+ \cdots + [(a_n - 2d) + (a_1 + 2d)] + [(a_n - d) + (a_1 + d)] + [a_n + a_1].$$

This simplifies to

$$2S_n = [a_1 + a_n] + [a_1 + a_n] + [a_1 + a_n]$$
$$+ \cdots + [a_n + a_1] + [a_n + a_1] + [a_n + a_1].$$

Since $(a_1 + a_n)$ is being added n times, it follows that

$$2S_n = n(a_1 + a_n),$$

from which we obtain the following formula.

Formula 2

The sum of the first n terms of an arithmetic sequence is given by

$$S_n = \frac{n}{2}(a_1 + a_n).$$

EXAMPLE 5

Find the sum of the first 100 positive even numbers.

Solution The sum is

$$2 + 4 + 6 + \cdots + 198 + 200.$$

This is the sum of the first 100 terms of the arithmetic sequence for which

$$a_1 = 2, \qquad a_n = 200, \quad \text{and} \quad n = 100.$$

Substituting in the formula

$$S_n = \frac{n}{2}(a_1 + a_n),$$

we get

$$S_{100} = \frac{100}{2}(2 + 200) = 50(202) = 10{,}100.$$

Formula 2 is useful when we know a_1 and a_n, the first and last terms. When a_n is unknown, but a_1, n, and d are known, we can find S_n by using Formulas 1 and 2 together.

EXAMPLE 6

Find the sum of the first 15 terms of the arithmetic sequence 4, 7, 10, 13, ...

Solution Note that

$$a_1 = 4, \qquad d = 3, \quad \text{and} \quad n = 15.$$

Before using Formula 2, we find a_{15}:

$$a_{15} = 4 + (15 - 1)3 \qquad \text{Substituting into Formula 1}$$
$$= 4 + 14 \cdot 3 = 46.$$

Thus,

$$S_{15} = \tfrac{15}{2}(4 + 46) \qquad \text{Using Formula 2}$$
$$= \tfrac{15}{2}(50) = 375.$$

Problem Solving

For some problem-solving situations, the translation may involve sequences or series. We look at some examples.

EXAMPLE 7

Chris takes a job, starting with an hourly wage of $14.25, and is promised a raise of 15¢ per hour every 2 months for 5 years. At the end of 5 years, what will be Chris's hourly wage?

Solution

1. FAMILIARIZE. It helps to write down the hourly wage for several two-month time periods.

Beginning:	14.25,
After two months:	14.40,
After four months:	14.55,
and so on.	

What appears is a sequence of numbers: 14.25, 14.40, 14.55, ... Is it an arithmetic sequence? Yes, because we add 0.15 each time to get the next term.

We ask ourselves what we know about arithmetic sequences. The pertinent formulas are

$$a_n = a_1 + (n - 1)d$$

and

$$S_n = \frac{n}{2}(a_1 + a_n).$$

In this case, we are not looking for a sum, so it is probably the first formula that will give us our answer. We want to know the last term in a sequence. We will need to know a_1, n, and d. From our list above, we see that

$$a_1 = 14.25 \quad \text{and} \quad d = 0.15.$$

What is n? That is, how many terms are in the sequence? Each year there are 6 raises, since Chris gets a raise every 2 months. There are 5 years, so the total number of raises will be $5 \cdot 6$, or 30. There will be 31 terms: the original wage and 30 increased rates.

2. TRANSLATE. We want to find a_n for the arithmetic sequence in which $a_1 = 14.25$, $d = 0.15$, and $n = 31$.

3. CARRY OUT. Substituting in Formula 1 gives us

$$a_{31} = 14.25 + (31 - 1) \cdot 0.15$$
$$= 18.75.$$

4. CHECK. We can check the calculations. We can also calculate in a slightly different way for another check. For example, at the end of a year, there will be 6 raises, for a total raise of $0.90. At the end of 5 years, the total raise will be $5 \times \$0.90$, or $4.50. If we add that to the original wage of $14.25, we obtain $18.75. The answer checks.

5. STATE. At the end of 5 years, Chris's hourly wage will be $18.75. ❏

Example 7 is one in which the calculations or the translation could be done in a number of ways. There is often a variety of ways in which a problem can be solved. You should use the one that is best or easiest for you. In this chapter, however, we will concentrate on the use of sequences and series and their related formulas in problem solving.

EXAMPLE 8

A stack of telephone poles has 30 poles in the bottom row. There are 29 poles in the second row, 28 in the next row, and so on. How many poles are in the stack if there are 5 poles in the top row?

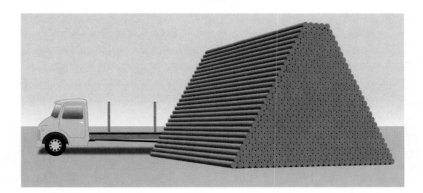

Solution

1. FAMILIARIZE.   A picture will help in this case. The following figure shows the ends of the poles and the way in which they stack. There are 30 poles on the bottom, and we see that there will be one fewer in each succeeding row. How many rows will there be?

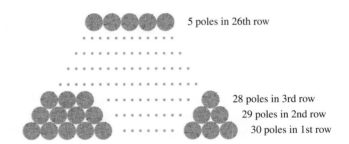

5 poles in 26th row

28 poles in 3rd row
29 poles in 2nd row
30 poles in 1st row

We go from 30 poles in a row, down to 5 poles in the top row, so there must be 26 rows.

We want the sum

$$30 + 29 + 28 + \cdots + 5.$$

Thus we have an arithmetic series. We recall, or look up if necessary, the formula

$$S_n = \frac{n}{2}(a_1 + a_n).$$

2. TRANSLATE.   We want to find the sum of the first 26 terms of an arithmetic sequence in which $a_1 = 30$ and $a_{26} = 5$.

3. CARRY OUT.   Substituting into Formula 2, we have

$$S_{26} = \frac{26}{2}(30 + 5)$$
$$= 13 \cdot 35 = 455.$$

4. CHECK. In this case, we can check the calculations by doing them again. A longer, harder way would be to do the entire addition:

$$30 + 29 + 28 + \cdots + 5.$$

5. STATE. There are 455 poles in the stack.   ❑

E X E R C I S E   S E T | 11 . 2 |

Find the first term and the common difference.

1. 2, 7, 12, 17, . . .

2. 1.06, 1.12, 1.18, 1.24, . . .

3. 7, 3, −1, −5, . . .

4. −9, −6, −3, 0, . . .

5. $\frac{3}{2}, \frac{9}{4}, 3, \frac{15}{4}, \ldots$

6. $\frac{3}{5}, \frac{1}{10}, -\frac{2}{5}, \ldots$

7. $2.12, $2.24, $2.36, $2.48, . . .

8. $214, $211, $208, $205, . . .

9. Find the 12th term of the arithmetic sequence 2, 6, 10, ...

10. Find the 11th term of the arithmetic sequence 0.07, 0.12, 0.17, ...

11. Find the 17th term of the arithmetic sequence 7, 4, 1, ...

12. Find the 14th term of the arithmetic sequence 3, $\frac{7}{3}$, $\frac{5}{3}$, ...

13. Find the 13th term of the arithmetic sequence $1200, $964.32, $728.64, ...

14. Find the 10th term of the arithmetic sequence $2345.78, $2967.54, $3589.30, ...

15. In the sequence of Exercise 9, what term is 106?

16. In the sequence of Exercise 10, what term is 1.67?

17. In the sequence of Exercise 11, what term is −296?

18. In the sequence of Exercise 12, what term is −27?

19. Find a_{17} when $a_1 = 5$ and $d = 6$.

20. Find a_{20} when $a_1 = 14$ and $d = -3$.

21. Find a_1 when $d = 4$ and $a_8 = 33$.

22. Find a_1 when $d = 8$ and $a_{11} = 26$.

23. Find n when $a_1 = 5$, $d = -3$, and $a_n = -76$.

24. Find n when $a_1 = 25$, $d = -14$, and $a_n = -507$.

25. For an arithmetic sequence in which $a_{17} = -40$ and $a_{28} = -73$, find a_1 and d. Write the first 5 terms of the sequence.

26. In an arithmetic sequence, $a_{17} = \frac{25}{3}$ and $a_{32} = \frac{95}{6}$. Find a_1 and d. Write the first 5 terms of the sequence.

27. Find the sum of the first 20 terms of the arithmetic series $5 + 8 + 11 + 14 + \cdots$.

28. Find the sum of the first 14 terms of the arithmetic series $11 + 7 + 3 + \cdots$.

29. Find the sum of the first 300 natural numbers.

30. Find the sum of the first 400 natural numbers.

31. Find the sum of the even numbers from 2 to 100, inclusive.

32. Find the sum of the odd numbers from 1 to 99, inclusive.

33. Find the sum of all multiples of 7 from 7 to 98, inclusive.

34. Find the sum of all multiples of 4 that are between 14 and 523.

35. If an arithmetic series has $a_1 = 2$ and $d = 5$, find S_{20}.

36. If an arithmetic series has $a_1 = 7$ and $d = -3$, find S_{32}.

Problem Solving

37. A gardener is making a triangular planting, with 35 plants in the front row, 31 in the second row, 27 in the third row, and so on. If the pattern is consistent, how many plants will there be in the last row? How many plants are there altogether?

38. A formation of a marching band has 14 marchers in the front row, 16 in the second row, 18 in the third row, and so on, for 25 rows. How many marchers are in the last row? How many marchers are there altogether?

39. How many poles will be in a pile of telephone poles if there are 50 in the first layer, 49 in the second, and so on, until there are 6 in the last layer?

40. If 10¢ is saved on October 1, 20¢ on October 2, 30¢ on October 3, and so on, how much is saved during October? (October has 31 days.)

41. A family saves money in an arithmetic sequence: $600 the first year, $700 the second, and so on, for 20 years. How much do they save in all (disregarding interest)?

42. Jacob saves $30 on August 1, $50 on August 2, $70 on August 3, and so on. How much will he have saved in August? (August has 31 days.)

43. Theaters are often built with more seats per row as the rows move toward the back. Suppose the main floor of a theater has 28 seats in the first row, 32 in the second, 36 in the third, and so on, for 50 rows. How many seats are on the main floor?

44. Shirley sets up an investment such that it will return $5000 the first year, $6125 the second year, $7250 the third year, and so on, for 25 years. How much in all is received from the investment?

Skill Maintenance

Convert to an exponential equation.

45. $\log_a P = k$

46. $\ln t = a$

Find an equation of the circle satisfying the given conditions.

47. Center $(0, 0)$, radius 9

48. Center $(-2, 5)$, radius $3\sqrt{2}$

Synthesis

49. ◈ The sum of the first n terms of an arithmetic sequence is given by

$$S_n = \frac{n}{2}[2a_1 + (n - 1)d].$$

Use Formulas 1 and 2 to explain how this equation is derived.

50. ◈ It is said that as a young child, the mathematician Karl F. Gauss (1777–1855) was able to compute the sum $1 + 2 + 3 + \cdots + 100$ very quickly in his head. Explain how Gauss might have done this and present a formula for the sum of the first n natural numbers.

51. Find three numbers in an arithmetic sequence such that the sum of the first and third is 10 and the product of the first and second is 15.

52. Find a formula for the sum of the first n consecutive odd numbers starting with 1:

$$1 + 3 + 5 + \cdots + (2n - 1).$$

53. ▦ In an arithmetic sequence, $a_1 = \$8760$ and $d = -\$798.23$. Find the first 10 terms of the sequence.

54. ▦ Find the sum of the first 10 terms of the sequence given in Exercise 53.

55. Prove that if p, m, and q are consecutive terms in an arithmetic sequence, then

$$m = \frac{p + q}{2}.$$

56. *Business: Straight-line depreciation.* A company buys an office machine for $5200 on January 1 of a given year. The machine is expected to last for 8 years, at the end of which time its *trade-in*, or *salvage, value* will be $1100. If the company figures the decline in value to be the same each year, then the trade-in values, after t years, $0 \leq t \leq 8$, form an arithmetic sequence given by

$$a_t = C - t\left(\frac{C - S}{N}\right),$$

where $C =$ the original cost of the item ($5200), $N =$ the years of expected life (8), and $S =$ the salvage value ($1100).

a) Find the formula for a_t for the straight-line depreciation of the office machine.

b) Find the salvage value after 0 years, 1 year, 2 years, 3 years, 4 years, 7 years, and 8 years.

11.3

Geometric Sequences and Series

Geometric Sequences • **Sum of the First n Terms of a Geometric Sequence** • **Infinite Geometric Series** • **Problem Solving**

In an arithmetic sequence, we added a certain number to each term to get the next term. With the kind of sequence we consider now, each term is *multiplied* by a certain number to get the next term. These are called *geometric sequences* or *geometric progressions*. We also consider *geometric series*.

Geometric Sequences

Consider the sequence

2, 6, 18, 54, 162, ...

If we multiply each term by 3, we obtain the next term. Sequences in which each term can be multiplied by a certain number in order to get the next term are called **geometric.** We call this multiplier the *common ratio* because it is found by dividing any term by the preceding term.

Geometric Sequence

A sequence is *geometric* if there exists a number r, called the *common ratio,* such that

$$\frac{a_{n+1}}{a_n} = r, \quad \text{or} \quad a_{n+1} = a_n \cdot r \quad \text{for any integer } n \geq 1.$$

EXAMPLE 1

For each geometric sequence, identify the common ratio.

a) 3, 6, 12, 24, 48, . . .   **b)** 3, −6, 12, −24, 48, −96, . . .
c) $5200, $3900, $2925, $2193.75, . . .   **d)** $1000, $1080, $1166.40, . . .

Solution

	Sequence	*Common Ratio*	
a)	3, 6, 12, 24, 48, . . .	2	$\frac{6}{3} = 2$, $\frac{12}{6} = 2$, and so on
b)	3, −6, 12, −24, 48, −96, . . .	−2	$\frac{-6}{3} = -2$, $\frac{12}{-6} = -2$, and so on
c)	$5200, $3900, $2925, $2193.75, . . .	0.75	$\frac{\$3900}{\$5200} = 0.75$, $\frac{\$2925}{\$3900} = 0.75$
d)	$1000, $1080, $1166.40, . . .	1.08	$\frac{\$1080}{\$1000} = 1.08$

We now find a formula for the general, or nth, term of a geometric sequence. Let a_1 be the first term and let r be the common ratio. We write out the first few terms as follows:

$a_1,$

$a_2 = a_1 r,$

$a_3 = a_2 r = (a_1 r)r = a_1 r^2,$ Substituting $a_1 r$ for a_2

$a_4 = a_3 r = (a_1 r^2)r = a_1 r^3.$ Substituting $a_1 r^2$ for a_3

Note that the exponent is 1 less than the subscript.

Generalizing, we obtain the following.

Formula 3

The nth term of a geometric sequence is given by

$$a_n = a_1 r^{n-1}, \quad \text{for any integer } n \geq 1.$$

EXAMPLE 2

Find the 7th term of the geometric sequence 4, 20, 100, ...

Solution First we note that

$$a_1 = 4 \quad \text{and} \quad n = 7.$$

To find the common ratio, we can divide any term by its predecessor, provided it has one. Since the second term is 20 and the first is 4, we get

$$r = \frac{20}{4}, \quad \text{or } 5.$$

The formula

$$a_n = a_1 r^{n-1}$$

gives us

$$a_7 = 4 \cdot 5^{7-1} = 4 \cdot 5^6 = 4 \cdot 15{,}625 = 62{,}500. \qquad \square$$

EXAMPLE 3

Find the 10th term of the geometric sequence

$$64, \ -32, \ 16, \ -8, \ \ldots .$$

Solution First we note that

$$a_1 = 64, \qquad n = 10, \quad \text{and} \quad r = \frac{-32}{64}, \quad \text{or } -\frac{1}{2}.$$

Then, using Formula 3, we have

$$a_{10} = 64 \cdot \left(-\frac{1}{2}\right)^{10-1} = 64 \cdot \left(-\frac{1}{2}\right)^9 = 2^6 \cdot \left(-\frac{1}{2^9}\right) = -\frac{1}{2^3} = -\frac{1}{8}. \qquad \square$$

Sum of the First n Terms of a Geometric Sequence

We want to find a formula for S_n when a sequence is geometric:

$$a_1, \ a_1 r, \ a_1 r^2, \ a_1 r^3, \ \ldots , \ a_1 r^{n-1}, \ \ldots .$$

The **geometric series** S_n is given by

$$S_n = a_1 + a_1r + a_1r^2 + \cdots + a_1r^{n-2} + a_1r^{n-1}. \tag{1}$$

We want to develop a formula that allows us to find this sum without a great deal of adding. If we multiply on both sides of Equation (1) by r, we have

$$rS_n = a_1r + a_1r^2 + a_1r^3 + \cdots + a_1r^{n-1} + a_1r^n. \tag{2}$$

Subtracting corresponding sides of Equation (2) from Equation (1), we see that the red terms drop out, leaving

$$S_n - rS_n = a_1 - a_1r^n,$$

or

$$S_n(1-r) = a_1(1-r^n). \qquad \text{Factoring}$$

Dividing on both sides by $1-r$ gives us the following formula.

Formula 4

> The sum of the first n terms of a geometric sequence is given by
> $$S_n = \frac{a_1(1-r^n)}{1-r}, \quad \text{for any } r \neq 1.$$

EXAMPLE 4

Find the sum of the first 7 terms of the geometric sequence 3, 15, 75, 375, ...

Solution   First we note that

$$a_1 = 3, \quad n = 7, \quad \text{and} \quad r = \frac{15}{3}, \quad \text{or } 5.$$

Then, using Formula 4, we have

$$S_7 = \frac{3(1-5^7)}{1-5} = \frac{3(1-78,125)}{-4}$$

$$= \frac{3(-78,124)}{-4}$$

$$= 58,593.$$

Infinite Geometric Series

Suppose we consider the sum of the terms of an infinite geometric sequence, such as 2, 4, 8, 16, 32, ... We get what is called an **infinite geometric series:**

$$2 + 4 + 8 + 16 + 32 + \cdots.$$

Here, as n grows larger and larger, the sum of the first n terms, S_n, becomes larger and larger without bound. There are also infinite series that get closer and closer to some specific number. Here is an example:

$$\frac{1}{2} + \frac{1}{4} + \frac{1}{8} + \frac{1}{16} + \cdots + \frac{1}{2^n} + \cdots.$$

Let's consider S_n for the first five values of n:

$$S_1 = \tfrac{1}{2} \qquad\qquad\qquad\quad = \tfrac{1}{2} = 0.5,$$
$$S_2 = \tfrac{1}{2} + \tfrac{1}{4} \qquad\qquad\quad = \tfrac{3}{4} = 0.75,$$
$$S_3 = \tfrac{1}{2} + \tfrac{1}{4} + \tfrac{1}{8} \qquad\quad = \tfrac{7}{8} = 0.875,$$
$$S_4 = \tfrac{1}{2} + \tfrac{1}{4} + \tfrac{1}{8} + \tfrac{1}{16} \qquad = \tfrac{15}{16} = 0.9375,$$
$$S_5 = \tfrac{1}{2} + \tfrac{1}{4} + \tfrac{1}{8} + \tfrac{1}{16} + \tfrac{1}{32} = \tfrac{31}{32} = 0.96875.$$

> The denominator of the sum is 2^n, where n is the subscript of S. The numerator is $2^n - 1$.

Thus, for this particular series, we have

$$S_n = \frac{2^n - 1}{2^n} = \frac{2^n}{2^n} - \frac{1}{2^n} = 1 - \frac{1}{2^n}.$$

Note that the value of S_n is less than 1 for any value of n, but as n gets larger and larger, the values of S_n get closer and closer to 1. We say that 1 is the *limit* of S_n and that 1 is the sum of this infinite geometric sequence. An infinite geometric series is denoted S_∞. It can be shown (but we will not do it here) that the sum of the terms of an infinite geometric sequence exists if and only if $|r| < 1$ (that is, the absolute value of the common ratio is less than 1).

To find a formula for the sum of an infinite geometric sequence, we first consider the sum of the first n terms:

$$S_n = \frac{a_1(1 - r^n)}{1 - r} = \frac{a_1 - a_1 r^n}{1 - r}. \qquad \text{Using the distributive law}$$

For $|r| < 1$, it follows that values of r^n get closer and closer to 0 as n gets larger. (Choose a number between -1 and 1 and check this by finding larger and larger powers on your calculator.) As r^n gets closer and closer to 0, so does $a_1 r^n$. Thus S_n gets closer and closer to $a_1/(1 - r)$.

Formula 5

When $|r| < 1$, the limit of an infinite geometric series is given by

$$S_\infty = \frac{a_1}{1 - r}.$$

EXAMPLE 5

Determine whether each infinite geometric series has a limit. If a limit exists, find it.

a) $1 + 3 + 9 + 27 + \cdots$ **b)** $-2 + 1 - \tfrac{1}{2} + \tfrac{1}{4} - \tfrac{1}{8} + \cdots$

Solution

a)   Here $r = 3$, so $|r| = |3| = 3$. Since $|r| \not< 1$, the series does *not* have a limit.
b)   Here $r = -\tfrac{1}{2}$, so $|r| = |-\tfrac{1}{2}| = \tfrac{1}{2}$. Since $|r| < 1$, the series *does* have a limit. We

find the limit by substituting into Formula 5:

$$S_\infty = \frac{-2}{1 - \left(-\frac{1}{2}\right)} = \frac{-2}{\frac{3}{2}} = -\frac{4}{3}.$$ ❑

EXAMPLE 6

Find fractional notation for 0.63636363 . . . .

Solution We can express this as

0.63 + 0.0063 + 0.000063 + · · · .

This is an infinite geometric series, where $a_1 = 0.63$ and $r = 0.01$. Since $|r| < 1$, this series has a limit:

$$S_\infty = \frac{a_1}{1 - r} = \frac{0.63}{1 - 0.01} = \frac{0.63}{0.99} = \frac{63}{99}.$$

Thus fractional notation for 0.63636363 . . . is $\frac{63}{99}$, or $\frac{7}{11}$. ❑

Problem Solving

For some problem-solving situations, the translation may involve geometric sequences or series.

EXAMPLE 7

Suppose someone offered you a job for the month of September (30 days) under the following conditions. You will be paid $0.01 for the first day, $0.02 for the second, $0.04 for the third, and so on, doubling your previous day's salary each day. How much would you earn? (Would you take the job? Make a guess before reading further.)

Solution

1. FAMILIARIZE. You earn $0.01 the first day, $0.01(2) the second day, $0.01(2)(2) the third day, and so on. The amounts form a geometric sequence with $a_1 = \$0.01$, $r = 2$, and $n = 30$.

2. TRANSLATE. The amount earned is the geometric series

$$\$0.01 + \$0.01(2) + \$0.01(2^2) + \$0.01(2^3) + \cdots + \$0.01(2^{29}),$$

where

$$a_1 = \$0.01, \quad n = 30, \quad \text{and} \quad r = 2.$$

3. CARRY OUT. Using the formula

$$S_n = \frac{a_1(1 - r^n)}{1 - r},$$

we have

$$S_{30} = \frac{\$0.01(1 - 2^{30})}{1 - 2}$$
$$= \frac{\$0.01(-1,073,741,823)}{-1} \quad \text{Using a calculator}$$
$$= \$10,737,418.23.$$

4. CHECK. The calculations can be repeated as a check.

5. STATE. The pay exceeds $10.7 million for the month. Most people would probably take the job! ❏

EXAMPLE 8

A student loan is in the amount of $6000. Interest is to be 9% compounded annually, and the entire amount is to be paid after 10 years. How much is to be paid back?

Solution

1. FAMILIARIZE. Suppose we let P represent any principal amount. At the end of one year, the amount owed will be $P + 0.09P$, or $1.09P$. That amount will be the principal for the second year. The amount owed at the end of the second year will be $1.09 \times$ New principal $= 1.09(1.09P)$, or 1.09^2P. Thus the amount owed at the beginning of successive years is as follows:

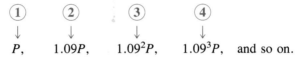

$$P, \qquad 1.09P, \qquad 1.09^2P, \qquad 1.09^3P, \quad \text{and so on.}$$

We have a geometric sequence. The amount owed at the beginning of the 11th year will be the amount owed at the end of the 10th year.

2. TRANSLATE. We have a geometric sequence with $a_1 = 6000$, $r = 1.09$, and $n = 11$. The appropriate formula is

$$a_n = a_1 r^{n-1}.$$

3. CARRY OUT. We substitute and calculate:

$$a_{11} = \$6000(1.09)^{11-1} = \$6000(1.09)^{10}$$

$\qquad \approx \$6000(2.3673637)$ Using a calculator to approximate 1.09^{10}

$\qquad \approx \$14,204.18.$ Rounded to the nearest hundredth

4. CHECK. A check, by repeating the calculations, is left to the student.

5. STATE. A total of $14,204.18 is to be paid back at the end of 10 years. ❏

EXAMPLE 9

A bungee jumper rebounds 60% of the height jumped. A bungee jump is made using a cord that stretches to 200 ft.

a) After jumping and then rebounding 9 times, how far has a bungee jumper traveled upward (the total rebound distance)?

b) Approximately how far will a jumper have traveled upward (bounced) before coming to rest?

Solution

1. FAMILIARIZE. Let's do some calculations and look for a pattern.

First fall:	200 ft
First rebound:	0.6×200, or 120 ft
Second fall:	120 ft, or 0.6×200
Second rebound:	0.6×120, or $0.6(0.6 \times 200)$, which is 72 ft
Third fall:	72 ft, or $0.6(0.6 \times 200)$
Third rebound:	0.6×72, or $0.6(0.6(0.6 \times 200))$, which is 43.2 ft

The rebound distances form a geometric sequence:

$$\begin{array}{cccc} \textcircled{1} & \textcircled{2} & \textcircled{3} & \textcircled{4} \\ \downarrow & \downarrow & \downarrow & \downarrow \\ 0.6 \times 200, & 0.6^2 \times 200, & 0.6^3 \times 200, & 0.6^4 \times 200, \dots, \end{array}$$

or

$$120, \quad 0.6 \times 120, \quad 0.6^2 \times 120, \quad 0.6^3 \times 120, \dots .$$

2. TRANSLATE.

a) The total rebound distance after 9 bounces is the sum of a geometric sequence. The first term is 120 and the common ratio is 0.6. There will be 9 terms, so we can use Formula 4:

$$S_n = \frac{a_1(1 - r^n)}{1 - r}.$$

b) Theoretically, the jumper will never stop bouncing. Realistically, the bouncing will eventually stop. We can approximate the actual distance bounced by considering an infinite number of bounces. We use Formula 5:

$$S_\infty = \frac{a_1}{1 - r}.$$

3. CARRY OUT.

a) We substitute into the formula and calculate:

$$S_9 = \frac{120[1 - (0.6)^9]}{1 - 0.6}$$

$$\approx 297. \qquad \text{Using a calculator}$$

b) We substitute and calculate:

$$S_\infty = \frac{120}{1 - 0.6} = 300.$$

4. CHECK. We can do the calculations again.

5. STATE.

a) In 9 bounces, the bungee jumper will have traveled upward a total distance of about 297 ft.

b) The jumper will travel upward about 300 ft before coming to rest. ❑

EXERCISE SET | 11.3

Find the common ratio for the geometric sequence.

1. 2, 4, 8, 16, . . .

2. 12, -4, $\frac{4}{3}$, $-\frac{4}{9}$, . . .

3. 1, -1, 1, -1, . . .

4. -5, -0.5, -0.05, -0.005, . . .

5. $\frac{1}{2}$, $-\frac{1}{4}$, $\frac{1}{8}$, $-\frac{1}{16}$, . . .

6. $\frac{2}{3}$, $-\frac{4}{3}$, $\frac{8}{3}$, $-\frac{16}{3}$, . . .

7. 75, 15, 3, $\frac{3}{5}$, . . .

8. 6.275, 0.6275, 0.06275, . . .

9. $\dfrac{1}{x}, \dfrac{1}{x^2}, \dfrac{1}{x^3}, \cdots$

10. $5, \dfrac{5m}{2}, \dfrac{5m^2}{4}, \dfrac{5m^3}{8}, \cdots$

11. $780, $858, $943.80, $1038.18, . . .

12. $5600, $5320, $5054, $4801.30, . . .

Find the indicated term for the geometric sequence.

13. 2, 4, 8, 16, . . . ; the 6th term

14. 2, −10, 50, −250, . . . ; the 9th term

15. $2, 2\sqrt{3}, 6, \ldots$; the 9th term

16. 1, −1, 1, −1, . . . ; the 57th term

17. ▦ $\dfrac{8}{243}, \dfrac{8}{81}, \dfrac{8}{27}, \cdots$; the 10th term

18. ▦ $\dfrac{7}{625}, \dfrac{-7}{125}, \dfrac{7}{25}, \cdots$; the 13th term

19. ▦ $1000, $1080, $1166.40, . . . ; the 12th term

20. ▦ $1000, $1070, $1144.90, . . . ; the 11th term

Find the *n*th, or general, term for the geometric sequence.

21. 1, 3, 9, . . . **22.** 25, 5, 1, . . .

23. 1, −1, 1, −1, . . . **24.** 2, 4, 8, . . .

25. $\dfrac{1}{x}, \dfrac{1}{x^2}, \dfrac{1}{x^3}, \cdots$ **26.** $5, \dfrac{5m}{2}, \dfrac{5m^2}{4}, \cdots$

For Exercises 27–34, use Formula 4 to find the indicated sum.

27. S_7 for the geometric series $6 + 12 + 24 + \cdots$

28. S_6 for the geometric series $16 - 8 + 4 - \cdots$

29. S_7 for the geometric series $\frac{1}{18} - \frac{1}{6} + \frac{1}{2} - \cdots$

30. S_5 for the geometric series $6 + 0.6 + 0.06 + \cdots$

31. S_8 for the series $1 + x + x^2 + x^3 + \cdots$

32. S_{10} for the series $1 + x^2 + x^4 + x^6 + \cdots$

33. ▦ S_{16} for the geometric sequence
$200, $200(1.06), $200(1.06)^2, \ldots$

34. ▦ S_{23} for the geometric sequence
$1000, $1000(1.08), $1000(1.08)^2, \ldots$

Determine whether the infinite geometric series has a limit. If a limit exists, find it.

35. $4 + 2 + 1 + \cdots$ **36.** $7 + 3 + \frac{9}{7} + \cdots$

37. $25 + 20 + 16 + \cdots$

38. $12 + 9 + \frac{27}{4} + \cdots$

39. $100 - 10 + 1 - \frac{1}{10} + \cdots$

40. $-6 + 18 - 54 + 162 - \cdots$

41. $8 + 40 + 200 + \cdots$

42. $-6 + 3 - \frac{3}{2} + \frac{3}{4} - \cdots$

43. $0.3 + 0.03 + 0.003 + \cdots$

44. $0.37 + 0.0037 + 0.000037 + \cdots$

45. $500(1.02)^{-1} + $500(1.02)^{-2} + $500(1.02)^{-3} + \cdots$

46. $1000(1.08)^{-1} + $1000(1.08)^{-2} + $1000(1.08)^{-3} + \cdots$

Find fractional notation for the infinite sum. (These are geometric series.)

47. 0.4444 . . . **48.** 9.999999 . . .

49. 0.55555 . . . **50.** 0.66666 . . .

51. 0.15151515 . . . **52.** 0.12121212 . . .

Solve. Use a calculator as needed for evaluating formulas.

53. A ping-pong ball is dropped from a height of 16 ft and always rebounds one fourth of the distance fallen. How high does it rebound the 6th time?

54. Approximate the total of the rebound heights of the ball in Exercise 53.

55. Yorktown has a current population of 100,000, and the population is increasing by 3% each year. What will the population be in 15 years?

56. How long will it take for the population of Yorktown to double? (See Exercise 55.)

57. A student borrows $1200. The loan is to be repaid in 13 years at 12% interest, compounded annually. How much will be repaid at the end of 13 years?

58. A piece of paper is 0.01 in. thick. It is folded repeatedly in such a way that its thickness is doubled each time for 20 times. How thick is the result?

59. A superball dropped from the top of the Washington Monument (556 ft high) always rebounds three fourths of the distance fallen. How far (up and down) will the ball have traveled when it hits the ground for the 6th time?

60. Approximate the total distance that the ball of Exercise 59 will have traveled when it comes to rest.

61. Suppose you accepted a job for the month of February (28 days) under the following conditions. You will be paid $0.01 the 1st day, $0.02 the 2nd, $0.04 the 3rd, and so on, doubling your previous day's salary each day. How much would you earn?

62. Leslie is saving money in a savings account for retirement. At the beginning of each year, $1000 is invested at 11%, compounded annually. How much will be in the retirement fund at the end of 40 years?

Skill Maintenance _____

Solve the system.

63. $5x - 2y = -3,$
$2x + 5y = -24$

64. $x - 2y + 3z = 4,$
$2x - y + z = -1,$
$4x + y + z = 1$

Synthesis _____

65. ◈ Write a problem for a classmate to solve. Devise the problem so that a geometric series is involved and the solution is "The total amount in the bank is

$$900(1.08)^{40},$$

or about $19,550."

66. ◈ The infinite series

$$S_\infty = 2 + \frac{1}{2} + \frac{1}{2 \cdot 3} + \frac{1}{2 \cdot 3 \cdot 4} + \frac{1}{2 \cdot 3 \cdot 4 \cdot 5}$$
$$+ \frac{1}{2 \cdot 3 \cdot 4 \cdot 5 \cdot 6} + \cdots$$

is not geometric, but it does have a sum. Using $S_1,$ $S_2, S_3, S_4, S_5,$ and $S_6,$ make a conjecture about the value of S_∞ and explain your reasoning.

67. Find the sum of the first n terms of
$$1 + x + x^2 + \cdots.$$

68. Find the sum of the first n terms of
$$x^2 - x^3 + x^4 - x^5 + \cdots.$$

69. The sides of a square are each 16 cm long. A second square is inscribed by joining the midpoints of the sides, successively. In the second square we repeat the process, inscribing a third square. If this process is continued indefinitely, what is the sum of all of the areas of all the squares? (*Hint:* Use an infinite geometric series.)

11.4

The Binomial Theorem

Binomial Expansion Using Pascal's Triangle •
Binomial Expansion Using Factorial Notation

Binomial Expansion Using Pascal's Triangle

Consider the following expanded powers of $(a + b)^n$:

$$(a + b)^0 = 1$$
$$(a + b)^1 = a + b$$
$$(a + b)^2 = a^2 + 2a^1b^1 + b^2$$
$$(a + b)^3 = a^3 + 3a^2b^1 + 3a^1b^2 + b^3$$
$$(a + b)^4 = a^4 + 4a^3b^1 + 6a^2b^2 + 4a^1b^3 + b^4$$
$$(a + b)^5 = a^5 + 5a^4b^1 + 10a^3b^2 + 10a^2b^3 + 5a^1b^4 + b^5.$$

Each expansion is a polynomial. There are some patterns to be noted:

1. There is one more term than the power of the binomial, n. That is, there are $n + 1$ terms in the expansion of $(a + b)^n$.
2. In each term, the sum of the exponents is the power to which the binomial is raised.
3. The exponents of a start with n, the power of the binomial, and decrease to 0. The last term has no factor of a. The first term has no factor of b, so powers of b start with 0 and increase to n.
4. The coefficients start at 1 and increase through certain values about "half"-way and then decrease through these same values back to 1. Let's study the coefficients further.

Suppose we want to find an expansion of $(a + b)^8$. The patterns we noticed above indicate 9 terms in the expansion:

$$a^8 + c_1a^7b + c_2a^6b^2 + c_3a^5b^3 + c_4a^4b^4 + c_5a^3b^5 + c_6a^2b^6 + c_7ab^7 + b^8.$$

How can we determine the values for the c's? We can answer this question in two different ways. The first method seems to be the easiest, but is not always. It involves writing down the coefficients in a triangular array as follows. We form what is known as **Pascal's triangle:**

```
(a + b)⁰:                    1
(a + b)¹:                 1     1
(a + b)²:              1     2     1
(a + b)³:           1     3     3     1
(a + b)⁴:        1     4     6     4     1
(a + b)⁵:     1     5    10    10     5     1
```

There are many patterns in the triangle. Find as many as you can.

Perhaps you discovered a way to write the next row of numbers, given the numbers in the row above it. There are always 1's on the outside. Each remaining number is the sum of the two numbers above:

```
                        1
                     1     1
                  1     2     1
               1     3     3     1
            1     4     6     4     1
         1     5    10    10     5     1
      1     6    15    20    15     6     1
```

We see that in the last row

the 1st and last numbers are 1;
the 2nd number is $1 + 5$, or 6;
the 3rd number is $5 + 10$, or 15;
the 4th number is $10 + 10$, or 20;
the 5th number is $10 + 5$, or 15; and
the 6th number is $5 + 1$, or 6.

Thus the expansion of $(a + b)^6$ is

$$(a + b)^6 = 1a^6 + 6a^5b + 15a^4b^2 + 20a^3b^3 + 15a^2b^4 + 6ab^5 + 1b^6.$$

To find the expansion for $(a + b)^8$, we complete two more rows of Pascal's triangle:

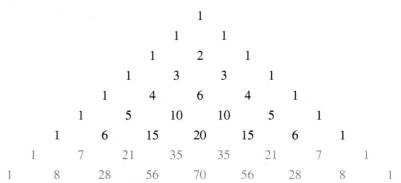

```
                        1
                    1       1
                1       2       1
            1       3       3       1
        1       4       6       4       1
    1       5       10      10      5       1
1       6       15      20      15      6       1
    1   7       21      35      35      21      7       1
1       8       28      56      70      56      28      8       1
```

Thus the expansion of $(a + b)^8$ is

$$(a + b)^8 = 1a^8 + 8a^7b + 28a^6b^2 + 56a^5b^3 + 70a^4b^4 + 56a^3b^5$$
$$+ 28a^2b^6 + 8ab^7 + 1b^8.$$

We can generalize our results as follows:

The Binomial Theorem (Form 1)

For any binomial $a + b$ and any natural number n,

$$(a + b)^n = c_0a^nb^0 + c_1a^{n-1}b^1 + c_2a^{n-2}b^2 + \cdots + c_{n-1}a^1b^{n-1} + c_na^0b^n,$$

where the numbers $c_0, c_1, c_2, \ldots, c_n$ are from the $(n + 1)$st row of Pascal's triangle.

EXAMPLE 1

Expand: $(u - v)^5$.

Solution   We note that $a = u$, $b = -v$, and $n = 5$. We use the 6th row of Pascal's triangle: 1   5   10   10   5   1.
Then we have

$(u - v)^5 = [u + (-v)]^5$

$= 1(u)^5 + 5(u)^4(-v)^1 + 10(u)^3(-v)^2 + 10(u)^2(-v)^3 + 5(u)^1(-v)^4 + 1(-v)^5$

$= u^5 - 5u^4v + 10u^3v^2 - 10u^2v^3 + 5uv^4 - v^5.$

Note that the signs of the terms alternate between $+$ and $-$. When $-v$ is raised to an odd power, the sign is $-$.   ❏

EXAMPLE 2

Expand: $\left(2t + \dfrac{3}{t}\right)^6$.

Solution   We note that $a = 2t$, $b = 3/t$, and $n = 6$. We use the 7th row of Pascal's triangle: 1   6   15   20   15   6   1.

We use the binomial theorem (Form 2) with $a = 3x$, $b = y$, and $n = 4$:

$$(3x + y)^4 = \binom{4}{0}(3x)^4 + \binom{4}{1}(3x)^3(y) + \binom{4}{2}(3x)^2(y)^2 + \binom{4}{3}(3x)(y)^3 + \binom{4}{4}(y)^4$$

$$= \frac{4!}{4!0!}3^4x^4 + \frac{4!}{3!1!}3^3x^3y + \frac{4!}{2!2!}3^2x^2y^2 + \frac{4!}{1!3!}3xy^3 + \frac{4!}{0!4!}y^4$$

$$= 81x^4 + 108x^3y + 54x^2y^2 + 12xy^3 + y^4. \qquad \text{Simplifying} \qquad \square$$

EXAMPLE 6

Expand: $(x^2 - 2y)^5$.

Solution   We have $a = x^2$, $b = -2y$, and $n = 5$:

$$(x^2 - 2y)^5 = \binom{5}{0}(x^2)^5 + \binom{5}{1}(x^2)^4(-2y) + \binom{5}{2}(x^2)^3(-2y)^2$$

$$+ \binom{5}{3}(x^2)^2(-2y)^3 + \binom{5}{4}(x^2)(-2y)^4 + \binom{5}{5}(-2y)^5$$

$$= \frac{5!}{5!0!}x^{10} + \frac{5!}{4!1!}x^8(-2y) + \frac{5!}{3!2!}x^6(-2y)^2 + \frac{5!}{2!3!}x^4(-2y)^3$$

$$+ \frac{5!}{1!4!}x^2(-2y)^4 + \frac{5!}{0!5!}(-2y)^5$$

$$= x^{10} - 10x^8y + 40x^6y^2 - 80x^4y^3 + 80x^2y^4 - 32y^5. \qquad \square$$

Note that in the binomial theorem (Form 2), $\binom{n}{0}a^nb^0$ gives us the first term, $\binom{n}{1}a^{n-1}b^1$ gives us the second term, $\binom{n}{2}a^{n-2}b^2$ gives us the third term, and so on. This can be generalized to give a method for finding a specific term without writing the entire expansion.

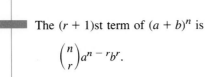

The $(r + 1)$st term of $(a + b)^n$ is

$$\binom{n}{r}a^{n-r}b^r.$$

EXAMPLE 7

Find the 5th term in the expansion of $(2x - 3y)^7$.

Solution   First, we note that $5 = 4 + 1$. Thus, $r = 4$, $a = 2x$, $b = -3y$, and $n = 7$. Then the 5th term of the expansion is

$$\binom{7}{4}(2x)^{7-4}(-3y)^4, \quad \text{or} \quad \frac{7!}{3!4!}(2x)^3(-3y)^4, \quad \text{or} \quad 22,680x^3y^4. \qquad \square$$

It is because of the binomial theorem that $\binom{n}{r}$ is called a *binomial coefficient*.

We can now explain why 0! is defined to be 1. In the binomial expansion, we want $\binom{n}{0}$ to equal 1 and we also want the definition

$$\binom{n}{r} = \frac{n!}{(n-r)!r!}$$

to hold for all whole numbers n and r. Thus we must have

$$\binom{n}{0} = \frac{n!}{(n-0)!0!} = \frac{n!}{n!0!} = 1.$$

This will be satisfied if $0!$ is defined to be 1.

EXERCISE SET | 11.4

Simplify.

1. $9!$ **2.** $10!$

3. $11!$ **4.** $12!$

5. $\dfrac{7!}{4!}$ **6.** $\dfrac{8!}{6!}$

7. $\dfrac{9!}{5!}$ **8.** $\dfrac{10!}{7!}$

9. $\binom{8}{2}$ **10.** $\binom{7}{4}$

11. $\binom{9}{6}$ **12.** $\binom{10}{7}$

13. $\binom{20}{18}$ **14.** $\binom{30}{3}$

15. $\binom{35}{2}$ **16.** $\binom{40}{38}$

Expand. Use both of the methods shown in this section.

17. $(m+n)^5$ **18.** $(a-b)^4$

19. $(x-y)^6$ **20.** $(p+q)^7$

21. $(x^2-3y)^5$ **22.** $(3c-d)^7$

23. $(3c-d)^6$ **24.** $(t^{-2}+2)^6$

25. $(x-y)^3$ **26.** $(x-y)^5$

27. $\left(\dfrac{1}{x}+y\right)^7$ **28.** $(2s-3t^2)^3$

29. $\left(a-\dfrac{2}{a}\right)^9$ **30.** $\left(2x+\dfrac{1}{x}\right)^9$

31. $(a^2+b^3)^5$ **32.** $(x^3+2)^6$

33. $\left(\sqrt{3}-t\right)^4$ **34.** $\left(\sqrt{5}+t\right)^6$

35. $(x^{-2}+x^2)^4$ **36.** $\left(\dfrac{1}{\sqrt{x}}-\sqrt{x}\right)^6$

Find the indicated term of the binomial expression.

37. 3rd, $(a+b)^6$ **38.** 6th, $(x+y)^7$

39. 12th, $(a-2)^{14}$ **40.** 11th, $(x-3)^{12}$

41. 5th, $\left(2x^3-\sqrt{y}\right)^8$ **42.** 4th, $\left(\dfrac{1}{b^2}+\dfrac{b}{3}\right)^7$

43. Middle, $(2u-3v^2)^{10}$

44. Middle two, $\left(\sqrt{x}+\sqrt{3}\right)^5$

Skill Maintenance

Solve.

45. $\log_2 x + \log_2 (x-2) = 3$

46. $\log_3 (x+2) - \log_3 (x-2) = 2$

47. $e^t = 280$

48. $\log_5 x^2 = 2$

Synthesis

49. ▣   At one point in a recent season, Darryl Strawberry of the Los Angeles Dodgers had a batting average of 0.313. Suppose he came to bat 5 times in a game. The probability of his getting exactly 3 hits is the 3rd term of the binomial expansion of $(0.313 + 0.687)^5$. Find that term and use your calculator to estimate the probability.

50. ▣   The probability that a woman will be either widowed or divorced is 85%. Suppose 8 women are interviewed. The probability that exactly 5 of them will be either widowed or divorced in their lifetime is the 6th term of the binomial expansion of $(0.15 + 0.85)^8$. Find that term and use your calculator to estimate the probability.

51. ▪ In reference to Exercise 49, the probability that Strawberry will get *at most* 3 hits is found by adding the last 4 terms of the binomial expansion of $(0.313 + 0.687)^5$. Find these terms and use your calculator to estimate the probability.

52. ▪ In reference to Exercise 50, the probability that *at least* 6 of the women will be widowed or divorced is found by adding the last three terms of the binomial expansion of $(0.15 + 0.85)^8$. Find these terms and use your calculator to estimate the probability.

53. Prove that

$$\binom{n}{r} = \binom{n}{n-r}$$

for any whole numbers n and r.

54. Find the term of
$$\left(\frac{3x^2}{2} - \frac{1}{3x}\right)^{12}$$
that does not contain x.

55. Find the middle term of $(x^2 - 6y^{3/2})^6$.

56. Find the ratio of the 4th term of
$$\left(p^2 - \frac{1}{2}p\sqrt[3]{q}\right)^5$$
to the 3rd term.

57. Find the term containing $\dfrac{1}{x^{1/6}}$ of
$$\left(\sqrt[3]{x} - \frac{1}{\sqrt{x}}\right)^7.$$

58. What is the degree of $(x^2 + 3)^4$?

SUMMARY AND REVIEW | 11 |

IMPORTANT PROPERTIES AND FORMULAS

Arithmetic sequence: $a_{n+1} = a_n + d$

nth term of an arithmetic sequence: $a_n = a_1 + (n-1)d$

Sum of the first n terms of an arithmetic sequence: $S_n = \frac{n}{2}(a_1 + a_n)$

Geometric sequence:	$a_{n+1} = a_n \cdot r$
nth term of a geometric sequence:	$a_n = a_1 r^{n-1}$
Sum of the first n terms of a geometric sequence:	$S_n = \dfrac{a_1(1 - r^n)}{1 - r}$
Limit of an infinite geometric series:	$S_\infty = \dfrac{a_1}{1 - r}, \quad |r| < 1$
Factorial notation:	$n! = n(n-1)(n-2) \cdots 3 \cdot 2 \cdot 1$
Binomial coefficient:	$\dbinom{n}{r} = \dfrac{n!}{(n-r)!r!}$
Binomial theorem:	$(a+b)^n = \dbinom{n}{0}a^n + \dbinom{n}{1}a^{n-1}b +$
	$\dbinom{n}{2}a^{n-2}b^2 + \cdots + \dbinom{n}{n}b^n$

REVIEW EXERCISES

This chapter's review and test include Skill Maintenance exercises from Sections 3.2, 9.4, 9.6, and 10.1.

Find the first 4 terms; the 8th term, a_8; and the 12th term, a_{12}.

1. $a_n = 4n - 3$

2. $a_n = \dfrac{n-1}{n^2 + 1}$

Predict the general term. Answers may vary.

3. $-2, -4, -6, -8, -10, \ldots$

4. $1, 4, 9, 16, 25, \ldots$

Rename and evaluate the sum.

5. $\displaystyle\sum_{k=1}^{5} (-2)^k$

6. $\displaystyle\sum_{k=2}^{7} (1 - 2k)$

Rewrite using sigma notation.

7. $4 + 8 + 12 + 16 + 20$

8. $\dfrac{-1}{2} + \dfrac{1}{4} + \dfrac{-1}{8} + \dfrac{1}{16} + \dfrac{-1}{32}$

9. Find the 14th term of the arithmetic sequence $-6, 1, 8, \ldots$.

10. Find d when $a_1 = 11$ and $a_{10} = 35$. Assume an arithmetic sequence.

11. Find a_1 and d when $a_{12} = 25$ and $a_{24} = 40$. Assume an arithmetic sequence.

12. Find the sum of the first 17 terms of the arithmetic series $-8 + (-11) + (-14) + \cdots$.

13. Find the sum of all the multiples of 6 from 12 to 318 inclusive.

Solve.

14. An auditorium has 31 seats in the first row, 33 seats in the second row, 35 seats in the third row, and so on, for 18 rows. How many seats are there in the 17th row?

15. In Exercise 14, how many seats are there in all 18 rows of the auditorium?

16. Find the 20th term of the geometric sequence $2, 2\sqrt{2}, 4, \ldots$.

17. Find the common ratio of the geometric sequence $2, \frac{4}{3}, \frac{8}{9}, \ldots$.

18. Find the nth term of the geometric sequence $-2, 2, -2, \ldots$.

19. Find the nth term of the geometric sequence $3, \frac{3}{4}x, \frac{3}{16}x^2, \ldots$.

20. Find S_6 for the geometric series $3 + 12 + 48 + \cdots$.

21. Find S_{12} for the geometric series
$$3x - 6x + 12x - \cdots .$$

Determine whether the infinite geometric series has a limit. If a limit exists, find it.

22. $6 + 3 + 1.5 + 0.75 + \cdots$

23. $0.04 + 0.008 + 0.0016 + \cdots$

24. $2 + (-2) + 2 + (-2) + \cdots$

25. $0.04 + 0.08 + 0.16 + 0.32 + \cdots$

26. $\$2000 + \$1900 + \$1805 + \$1714.75 + \cdots$

27. Find fractional notation for $0.555555 \ldots$.

28. Find fractional notation for $0.39393939 \ldots$.

Solve.

29. You take a job, starting with an hourly wage of $\$11.40$. You are promised a raise of 20¢ per hour every 3 months for 8 years. At the end of 8 years, what will be your hourly wage?

30. A stack of logs has 42 poles in the bottom row. There are 41 poles in the second row, 40 poles in the third row, and so on. How many poles are in the stack?

31. A student loan is in the amount of $\$10,000$. Interest is 7%, compounded annually, and the amount is to be paid off in 12 years. How much is to be paid back?

32. Find the total rebound distance of a ball, given that it is dropped from a height of 12 m and each rebound is one third of the preceding one.

Simplify.

33. $8!$

34. $\dbinom{8}{3}$

35. Find the 3rd term of $(a + b)^{20}$.

36. Expand: $(x - 2y)^4$.

Skill Maintenance

Solve.

37. $3x - \ y = 7,$
$2x + 3y = 5$

38. $\log (x + 5) - \log x = 1$

39. Express in terms of logarithms of a, b, c, and d:
$$\log \sqrt{\frac{a^6 b^8}{c^2 d^4}}.$$

40. Find the distance between the points $(9, 5)$ and $(4, -7)$.

Synthesis

41. ◈ Explain what happens to the terms of a geometric sequence with $|r| < 1$ as n gets larger.

42. ◈ Compare the two forms of the binomial theorem given in the text. Under what circumstances would one be better to use than the other?

43. Find the sum of the first n terms of the geometric series $1 - x + x^2 - x^3 + \cdots$.

44. Expand: $(x^{-3} + x^3)^5$.

CHAPTER TEST ⎸ 11

1. Find the first 5 terms and the 16th term of a sequence with general term $a_n = 6n - 5$.

2. Predict the general term of the sequence
$$\tfrac{4}{3}, \tfrac{4}{9}, \tfrac{4}{27}, \ldots .$$

3. Rename and evaluate:
$$\sum_{k=1}^{5} (3 - 2^k).$$

4. Rewrite using sigma notation:
$$1 + 8 + 27 + 64 + 125.$$

5. Find the 12th term, a_{12}, of the arithmetic sequence $9, 4, -1, \ldots$.

Assume arithmetic sequences for Questions 6 and 7.

6. Find the common difference d when $a_1 = 9$ and $a_7 = 11\tfrac{1}{4}$.

7. Find a_1 and d when $a_5 = 16$ and $a_{10} = -3$.

8. Find the sum of all the multiples of 12 from 24 to 240 inclusive.

9. Find the 6th term of the geometric sequence $72, 18, 4\tfrac{1}{2}, \ldots$.

10. Find the common ratio of the geometric sequence $22\tfrac{1}{2}, 15, 10, \ldots$.

11. Find the nth term of the geometric sequence $3, -9, 27, \ldots$.

12. Find the sum of the first 9 terms of the geometric series

$$(1 + x) + (2 + 2x) + (4 + 4x) + \cdots.$$

Determine whether the infinite geometric series has a limit. If a limit exists, find it.

13. $0.5 + 0.25 + 0.125 + \cdots$

14. $0.5 + 1 + 2 + 4 + \cdots$

15. $\$1000 + \$80 + \$6.40 + \cdots$

16. Find fractional notation for $0.85858585 \ldots$.

17. You take a job, starting with an hourly wage of $12.80. You are promised a raise of 20¢ per hour every 4 months for 8 years. At the end of 8 years, what will be your hourly wage?

18. A stack of poles has 52 poles in the bottom row. There are 51 poles in the second row, 50 poles in the third row, and so on. How many poles are in the stack?

19. A student loan is in the amount of $20,000. Interest is to be 8%, compounded annually, and the amount is to be paid off in 10 years. How much is to be paid back?

20. Find the total rebound distance of a ball that is dropped from a height of 18 m, with each rebound two thirds of the preceding one.

21. Simplify: $\begin{pmatrix} 13 \\ 11 \end{pmatrix}$.

22. Expand: $(x^2 - 3y)^5$.

23. Find the 4th term in the expansion of $(a + x)^{12}$.

Skill Maintenance

24. Solve: $4^{2x - 3} = 64$.

25. Find an equation of the circle with center $(1, -2)$ and radius $3\sqrt{3}$.

26. Solve:

$$y = 3x + 5,$$
$$2x + 5y = 8.$$

27. Express in terms of logarithms of a, b, and c:

$$\log_t \frac{c^7}{a^5 b^4}.$$

Synthesis

28. Find a formula for the sum of the first n even natural numbers:

$$2 + 4 + 6 + \cdots + 2n.$$

29. Find the sum of the first n terms of

$$1 + \frac{1}{x} + \frac{1}{x^2} + \frac{1}{x^3} + \cdots.$$

CUMULATIVE REVIEW | 1–11

Simplify.

1. $(-9x^2y^3)(5x^4y^{-7})$

2. $|-3.5 + 9.8|$

3. $2y - [3 - 4(5 - 2y) - 3y]$

4. $(10 \cdot 8 - 9 \cdot 7)^2 - 54 \div 9 - 3$

5. Evaluate

$$\frac{ab - ac}{bc}$$

when $a = -2$, $b = 3$, and $c = -4$.

Perform the indicated operations and simplify.

6. $(5a^2 - 3ab - 7b^2) - (2a^2 + 5ab + 8b^2)$

7. $(-3x^2 + 4x^3 - 5x - 1) + (9x^3 - 4x^2 + 7 - x)$

8. $(2a - 1)(3a + 5)$ **9.** $(3a^2 - 5y)^2$

10. $\dfrac{1}{x - 2} - \dfrac{4}{x^2 - 4} + \dfrac{3}{x + 2}$

11. $\dfrac{x^2 - 6x + 8}{3x + 9} \cdot \dfrac{x + 3}{x^2 - 4}$

12. $\dfrac{3x + 3y}{5x - 5y} \div \dfrac{3x^2 + 3y^2}{5x^3 - 5y^3}$

13. $\dfrac{x - \dfrac{a^2}{x}}{1 + \dfrac{a}{x}}$

Factor.

14. $4x^2 - 12x + 9$ **15.** $27a^3 - 8$

16. $a^3 + 3a^2 - ab - 3b$ **17.** $15y^4 + 33y^2 - 36$

18. For the function described by

$$f(x) = 3x^2 - 4x,$$

find $f(-2)$.

19. Divide:

$$(7x^4 - 5x^3 + x^2 - 4) \div (x - 2).$$

Solve.

20. $9(x - 1) - 3(x - 2) = 1$

21. $x^2 - 2x = 48$

22. $\dfrac{6}{x} + \dfrac{6}{x + 2} = \dfrac{5}{2}$

23. $\dfrac{7x}{x - 3} - \dfrac{21}{x} + 11 = \dfrac{63}{x^2 - 3x}$

24. $5x + 3y = 2$,
$3x + 5y = -2$

25. $x + y - z = 0$,
$3x + y + z = 6$,
$x - y + 2z = 5$

26. $\sqrt{x - 5} = 5 - \sqrt{x}$

27. $x^4 - 29x^2 + 100 = 0$

28. $x^2 + y^2 = 8$,
$x^2 - y^2 = 2$

29. 🖩 $5^x = 8$

30. $\log (x^2 - 25) - \log (x + 5) = 3$

31. $\log_4 x = -2$ **32.** $7^{2x+3} = 49$

33. $|2x - 1| \leq 5$ **34.** $7x^2 + 14 = 0$

35. $x^2 + 4x = 3$ **36.** $y^2 + 3y > 10$

Solve.

37. The perimeter of a rectangle is 34 ft. The length of a diagonal is 13 ft. Find the dimensions of the rectangle.

38. A telephone company charges $0.40 for the first minute and $0.25 for every other minute of a long-distance call placed before 5 P.M. The rates after 5 P.M. drop to $0.30 for the first minute and $0.20 for every other minute of the call. A certain call placed before 5 P.M. costs $4.20. How much would a call of the same duration placed after 5 P.M. cost?

39. Find three consecutive integers whose sum is 198.

40. A pentagon with all five sides the same size has a perimeter equal to that of an octagon in which all eight sides are the same size. One side of the pentagon is 2 less than 3 times one side of the octagon. What is the perimeter of each figure?

41. A chemist has two solutions of ammonia and water. Solution A is 6% ammonia and solution B is 2% ammonia. How many liters of each solution are needed in order to obtain 80 L of a solution that is 3.2% ammonia?

42. An airplane can fly 190 mi with the wind in the same time it takes to fly 160 mi against the wind. The speed of the wind is 30 mph. How fast can the plane fly in still air?

43. Bianca can do a certain job in 21 min. Dahlia can do the same job in 14 min. How long would it take to do the job if the two worked together?

44. The centripetal force F of an object moving in a circle varies directly as the square of the velocity v and inversely as the radius r of the circle. If $F = 8$ when $v = 1$ and $r = 10$, what is F when $v = 2$ and $r = 16$?

45. A farmer wants to fence in a rectangular area next to a river. (Note that no fence will be needed along the river.) What is the area of the largest region that can be fenced in with 100 ft of fencing?

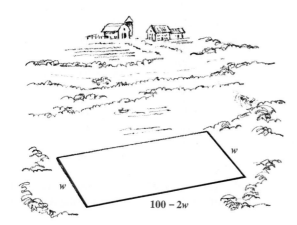

Graph.

46. $3x - y = 6$

47. $\dfrac{x^2}{25} + \dfrac{y^2}{4} = 1$

48. $y = \log_2 x$

49. $2x - 3y < -6$

50. Graph: $f(x) = -2(x - 3)^2 + 1$.
   a) Label the vertex.
   b) Draw the line of symmetry.
   c) Find the maximum or minimum value.

51. Solve $V = P - Prt$ for r.

52. Solve $I = \dfrac{R}{R + r}$ for R.

53. Find an equation of the line containing the point $(-1, 4)$ and perpendicular to the line whose equation is $3x - y = 6$.

Evaluate.

54. $\begin{vmatrix} -5 & -7 \\ 4 & 6 \end{vmatrix}$

55. $\begin{vmatrix} 2 & -1 & 1 \\ 1 & 2 & 0 \\ 3 & -1 & 1 \end{vmatrix}$

56. Multiply $(8.9 \times 10^{-17})(7.6 \times 10^4)$. Write scientific notation for the answer.

57. Multiply and simplify: $\sqrt{8x}\ \sqrt{8x^3y}$.

58. Simplify: $(25x^{4/3}y^{1/2})^{3/2}$.

59. Divide and simplify:
$$\frac{\sqrt[3]{15x}}{\sqrt[3]{3y^2}}.$$

60. Rationalize the denominator:
$$\frac{1 - \sqrt{x}}{1 + \sqrt{x}}.$$

61. Write a single radical expression:
$$\frac{\sqrt[3]{(x + 1)^5}}{\sqrt{(x + 1)^3}}.$$

62. Multiply these complex numbers:
$$(3 + 2i)(4 - 7i).$$

63. Write a quadratic equation whose solutions are $5\sqrt{2}$ and $-5\sqrt{2}$.

64. Find the center and the radius of the circle
$$x^2 + y^2 - 4x + 6y - 23 = 0.$$

65. Express as a single logarithm:
$$\tfrac{2}{3}\log_a x - \tfrac{1}{2}\log_a y + 5\log_a z.$$

66. Convert to an exponential equation: $\log_a c = 5$.

Find each of the following using a calculator or table.

67. log 5677.2

68. antilog (-4.8904)

69. ln 5677.2

70. antilog$_e$ (-4.8904)

Population growth. The Virgin Islands has an exponential growth rate of 0.7%. In 1991, the population was 99,404.

71. Write an exponential function describing the growth of the population of the Virgin Islands.

72. ▦ Predict the population for the year 2000.

73. Find the distance between the points $(-1, -5)$ and $(2, -1)$.

74. Find the 21st term of the arithmetic sequence 19, 12, 5, ...

75. Find the sum of the first 25 terms of the arithmetic series $-1 + 2 + 5 + \cdots$.

76. Find the general term of the geometric sequence 16, 4, 1, ...

77. Find the 7th term of $(a - 2b)^{10}$.

78. ▦ Find the sum of the first 9 terms of the geometric series $x + 1.5x + 2.25x + \cdots$.

79. ▦ On Mark's 9th birthday, his grandmother opened a savings account for him with $100. The account draws 6% interest, compounded annually. If Mark neither adds to nor withdraws any money from the bank, how much will be in the account on his 18th birthday?

T A B L E S

TABLE 1 Powers, Roots, and Reciprocals

n	n^2	n^3	$\sqrt{n}$	$\sqrt[3]{n}$	$\sqrt{10n}$	$\dfrac{1}{n}$	n	n^2	n^3	$\sqrt{n}$	$\sqrt[3]{n}$	$\sqrt{10n}$	$\dfrac{1}{n}$
1	1	1	1.000	1.000	3.162	1.0000	51	2,601	132,651	7.141	3.708	22.583	.0196
2	4	8	1.414	1.260	4.472	.5000	52	2,704	140,608	7.211	3.733	22.804	.0192
3	9	27	1.732	1.442	5.477	.3333	53	2,809	148,877	7.280	3.756	23.022	.0189
4	16	64	2.000	1.587	6.325	.2500	54	2,916	157,464	7.348	3.780	23.238	.0185
5	25	125	2.236	1.710	7.071	.2000	55	3,025	166,375	7.416	3.803	23.452	.0182
6	36	216	2.449	1.817	7.746	.1667	56	3,136	175,616	7.483	3.826	23.664	.0179
7	49	343	2.646	1.913	8.367	.1429	57	3,249	185,193	7.550	3.849	23.875	.0175
8	64	512	2.828	2.000	8.944	.1250	58	3,364	195,112	7.616	3.871	24.083	.0172
9	81	729	3.000	2.080	9.487	.1111	59	3,481	205,379	7.681	3.893	24.290	.0169
10	100	1,000	3.162	2.154	10.000	.1000	60	3,600	216,000	7.746	3.915	24.495	.0167
11	121	1,331	3.317	2.224	10.488	.0909	61	3,721	226,981	7.810	3.936	24.698	.0164
12	144	1,728	3.464	2.289	10.954	.0833	62	3,844	238,328	7.874	3.958	24.900	.0161
13	169	2,197	3.606	2.351	11.402	.0769	63	3,969	250,047	7.937	3.979	25.100	.0159
14	196	2,744	3.742	2.410	11.832	.0714	64	4,096	262,144	8.000	4.000	25.298	.0156
15	225	3,375	3.873	2.466	12.247	.0667	65	4,225	274,625	8.062	4.021	25.495	.0154
16	256	4,096	4.000	2.520	12.648	.0625	66	4,356	287,496	8.124	4.041	25.690	.0152
17	289	4,913	4.123	2.571	13.038	.0588	67	4,489	300,763	8.185	4.062	25.884	.0149
18	324	5,832	4.243	2.621	13.416	.0556	68	4,624	314,432	8.246	4.082	26.077	.0147
19	361	6,859	4.359	2.668	13.784	.0526	69	4,761	328,509	8.307	4.102	26.268	.0145
20	400	8,000	4.472	2.714	14.142	.0500	70	4,900	343,000	8.367	4.121	26.458	.0143
21	441	9,261	4.583	2.759	14.491	.0476	71	5,041	357,911	8.426	4.141	26.646	.0141
22	484	10,648	4.690	2.802	14.832	.0455	72	5,184	373,248	8.485	4.160	26.833	.0139
23	529	12,167	4.796	2.844	15.166	.0435	73	5,329	389,017	8.544	4.179	27.019	.0137
24	576	13,824	4.899	2.884	15.492	.0417	74	5,476	405,224	8.602	4.198	27.203	.0135
25	625	15,625	5.000	2.924	15.811	.0400	75	5,625	421,875	8.660	4.217	27.386	.0133
26	676	17,576	5.099	2.962	16.125	.0385	76	5,776	438,976	8.718	4.236	27.568	.0132
27	729	19,683	5.196	3.000	16.432	.0370	77	5,929	456,533	8.775	4.254	27.749	.0130
28	784	21,952	5.292	3.037	16.733	.0357	78	6,084	474,552	8.832	4.273	27.928	.0128
29	841	24,389	5.385	3.072	17.029	.0345	79	6,241	493,039	8.888	4.291	28.107	.0127
30	900	27,000	5.477	3.107	17.321	.0333	80	6,400	512,000	8.944	4.309	28.284	.0125
31	961	29,791	5.568	3.141	17.607	.0323	81	6,561	531,441	9.000	4.327	28.460	.0123
32	1,024	32,768	5.657	3.175	17.889	.0312	82	6,724	551,368	9.055	4.344	28.636	.0122
33	1,089	35,937	5.745	3.208	18.166	.0303	83	6,889	571,787	9.110	4.362	28.810	.0120
34	1,156	39,304	5.831	3.240	18.439	.0294	84	7,056	592,704	9.165	4.380	28.983	.0119
35	1,225	42,875	5.916	3.271	18.708	.0286	85	7,225	614,125	9.220	4.397	29.155	.0118
36	1,296	46,656	6.000	3.302	18.974	.0278	86	7,396	636,056	9.274	4.414	29.326	.0116
37	1,369	50,653	6.083	3.332	19.235	.0270	87	7,569	658,503	9.327	4.431	29.496	.0115
38	1,444	54,872	6.164	3.362	19.494	.0263	88	7,744	681,472	9.381	4.448	29.665	.0114
39	1,521	59,319	6.245	3.391	19.748	.0256	89	7,921	704,969	9.434	4.465	29.833	.0112
40	1,600	64,000	6.325	3.420	20.000	.0250	90	8,100	729,000	9.487	4.481	30.000	.0111
41	1,681	68,921	6.403	3.448	20.248	.0244	91	8,281	753,571	9.539	4.498	30.166	.0110
42	1,764	74,088	6.481	3.476	20.494	.0238	92	8,464	778,688	9.592	4.514	30.332	.0109
43	1,849	79,507	6.557	3.503	20.736	.0233	93	8,649	804,357	9.644	4.531	30.496	.0108
44	1,936	85,184	6.633	3.530	20.976	.0227	94	8,836	830,584	9.695	4.547	30.659	.0106
45	2,025	91,125	6.708	3.557	21.213	.0222	95	9,025	857,375	9.747	4.563	30.822	.0105
46	2,116	97,336	6.782	3.583	21.448	.0217	96	9,216	884,736	9.798	4.579	30.984	.0104
47	2,209	103,823	6.856	3.609	21.679	.0213	97	9,409	912,673	9.849	4.595	31.145	.0103
48	2,304	110,592	6.928	3.634	21.909	.0208	98	9,604	941,192	9.899	4.610	31.305	.0102
49	2,401	117,649	7.000	3.659	22.136	.0204	99	9,801	970,299	9.950	4.626	31.464	.0101
50	2,500	125,000	7.071	3.684	22.361	.0200	100	10,000	1,000,000	10.000	4.642	31.623	.0100

TABLE 2   Common Logarithms

x	0	1	2	3	4	5	6	7	8	9
1.0	.0000	.0043	.0086	.0128	.0170	.0212	.0253	.0294	.0334	.0374
1.1	.0414	.0453	.0492	.0531	.0569	.0607	.0645	.0682	.0719	.0755
1.2	.0792	.0828	.0864	.0899	.0934	.0969	.1004	.1038	.1072	.1106
1.3	.1139	.1173	.1206	.1239	.1271	.1303	.1335	.1367	.1399	.1430
1.4	.1461	.1492	.1523	.1553	.1584	.1614	.1644	.1673	.1703	.1732
1.5	.1761	.1790	.1818	.1847	.1875	.1903	.1931	.1959	.1987	.2014
1.6	.2041	.2068	.2095	.2122	.2148	.2175	.2201	.2227	.2253	.2279
1.7	.2304	.2330	.2355	.2380	.2405	.2430	.2455	.2480	.2504	.2529
1.8	.2553	.2577	.2601	.2625	.2648	.2672	.2695	.2718	.2742	.2765
1.9	.2788	.2810	.2833	.2856	.2878	.2900	.2923	.2945	.2967	.2989
2.0	.3010	.3032	.3054	.3075	.3096	.3118	.3139	.3160	.3181	.3201
2.1	.3222	.3243	.3263	.3284	.3304	.3324	.3345	.3365	.3385	.3404
2.2	.3424	.3444	.3464	.3483	.3502	.3522	.3541	.3560	.3579	.3598
2.3	.3617	.3636	.3655	.3674	.3692	.3711	.3729	.3747	.3766	.3784
2.4	.3802	.3820	.3838	.3856	.3874	.3892	.3909	.3927	.3945	.3962
2.5	.3979	.3997	.4014	.4031	.4048	.4065	.4082	.4099	.4116	.4133
2.6	.4150	.4166	.4183	.4200	.4216	.4232	.4249	.4265	.4281	.4298
2.7	.4314	.4330	.4346	.4362	.4378	.4393	.4409	.4425	.4440	.4456
2.8	.4472	.4487	.4502	.4518	.4533	.4548	.4564	.4579	.4594	.4609
2.9	.4624	.4639	.4654	.4669	.4683	.4698	.4713	.4728	.4742	.4757
3.0	.4771	.4786	.4800	.4814	.4829	.4843	.4857	.4871	.4886	.4900
3.1	.4914	.4928	.4942	.4955	.4969	.4983	.4997	.5011	.5024	.5038
3.2	.5051	.5065	.5079	.5092	.5105	.5119	.5132	.5145	.5159	.5172
3.3	.5185	.5198	.5211	.5224	.5237	.5250	.5263	.5276	.5289	.5307
3.4	.5315	.5328	.5340	.5353	.5366	.5378	.5391	.5403	.5416	.5428
3.5	.5441	.5453	.5465	.5478	.5490	.5502	.5514	.5527	.5539	.5551
3.6	.5563	.5575	.5587	.5599	.5611	.5623	.5635	.5647	.5658	.5670
3.7	.5682	.5694	.5705	.5717	.5729	.5740	.5752	.5763	.5775	.5786
3.8	.5798	.5809	.5821	.5832	.5843	.5855	.5866	.5877	.5888	.5899
3.9	.5911	.5922	.5933	.5944	.5955	.5966	.5977	.5988	.5999	.6010
4.0	.6021	.6031	.6042	.6053	.6064	.6075	.6085	.6096	.6107	.6117
4.1	.6128	.6138	.6149	.6160	.6170	.6180	.6191	.6201	.6212	.6222
4.2	.6232	.6243	.6253	.6263	.6274	.6284	.6294	.6304	.6314	.6325
4.3	.6335	.6345	.6355	.6365	.6375	.6385	.6395	.6405	.6415	.6425
4.4	.6435	.6444	.6454	.6464	.6474	.6484	.6493	.6503	.6513	.6522
4.5	.6532	.6542	.6551	.6561	.6571	.6580	.6590	.6599	.6609	.6618
4.6	.6628	.6637	.6646	.6656	.6665	.6675	.6684	.6693	.6702	.6712
4.7	.6721	.6730	.6739	.6749	.6758	.6767	.6776	.6785	.6794	.6803
4.8	.6812	.6821	.6830	.6839	.6848	.6857	.6866	.6875	.6884	.6893
4.9	.6902	.6911	.6920	.6928	.6937	.6946	.6955	.6964	.6972	.6981
5.0	.6990	.6998	.7007	.7016	.7024	.7033	.7042	.7050	.7059	.7067
5.1	.7076	.7084	.7093	.7101	.7110	.7118	.7126	.7135	.7143	.7152
5.2	.7160	.7168	.7177	.7185	.7193	.7202	.7210	.7218	.7226	.7235
5.3	.7243	.7251	.7259	.7267	.7275	.7284	.7292	.7300	.7308	.7316
5.4	.7324	.7332	.7340	.7348	.7356	.7364	.7372	.7380	.7388	.7396
x	0	1	2	3	4	5	6	7	8	9

TABLE 2 *(continued)*

x	0	1	2	3	4	5	6	7	8	9
5.5	.7404	.7412	.7419	.7427	.7435	.7443	.7451	.7459	.7466	.7474
5.6	.7482	.7490	.7497	.7505	.7513	.7520	.7528	.7536	.7543	.7551
5.7	.7559	.7566	.7574	.7582	.7589	.7597	.7604	.7612	.7619	.7627
5.8	.7634	.7642	.7649	.7657	.7664	.7672	.7679	.7686	.7694	.7701
5.9	.7709	.7716	.7723	.7731	.7738	.7745	.7752	.7760	.7767	.7774
6.0	.7782	.7789	.7796	.7803	.7810	.7818	.7825	.7832	.7839	.7846
6.1	.7853	.7860	.7868	.7875	.7882	.7889	.7896	.7903	.7910	.7917
6.2	.7924	.7931	.7938	.7945	.7952	.7959	.7966	.7973	.7980	.7987
6.3	.7993	.8000	.8007	.8014	.8021	.8028	.8035	.8041	.8048	.8055
6.4	.8062	.8069	.8075	.8082	.8089	.8096	.8102	.8109	.8116	.8122
6.5	.8129	.8136	.8142	.8149	.8156	.8162	.8169	.8176	.8182	.8189
6.6	.8195	.8202	.8209	.8215	.8222	.8228	.8235	.8241	.8248	.8254
6.7	.8261	.8267	.8274	.8280	.8287	.8293	.8299	.8306	.8312	.8319
6.8	.8325	.8331	.8338	.8344	.8351	.8357	.8363	.8370	.8376	.8382
6.9	.8388	.8395	.8401	.8407	.8414	.8420	.8426	.8432	.8439	.8445
7.0	.8451	.8457	.8463	.8470	.8476	.8482	.8488	.8494	.8500	.8506
7.1	.8513	.8519	.8525	.8531	.8537	.8543	.8549	.8555	.8561	.8567
7.2	.8573	.8579	.8585	.8591	.8597	.8603	.8609	.8615	.8621	.8627
7.3	.8633	.8639	.8645	.8651	.8657	.8663	.8669	.8675	.8681	.8686
7.4	.8692	.8698	.8704	.8710	.8716	.8722	.8727	.8733	.8739	.8745
7.5	.8751	.8756	.8762	.8768	.8774	.8779	.8785	.8791	.8797	.8802
7.6	.8808	.8814	.8820	.8825	.8831	.8837	.8842	.8848	.8854	.8859
7.7	.8865	.8871	.8876	.8882	.8887	.8893	.8899	.8904	.8910	.8915
7.8	.8921	.8927	.8932	.8938	.8943	.8949	.8954	.8960	.8965	.8971
7.9	.8976	.8982	.8987	.8993	.8998	.9004	.9009	.9015	.9020	.9025
8.0	.9031	.9036	.9042	.9047	.9053	.9058	.9063	.9069	.9074	.9079
8.1	.9085	.9090	.9096	.9101	.9106	.9112	.9117	.9122	.9128	.9133
8.2	.9138	.9143	.9149	.9154	.9159	.9165	.9170	.9175	.9180	.9186
8.3	.9191	.9196	.9201	.9206	.9212	.9217	.9222	.9227	.9232	.9238
8.4	.9243	.9248	.9253	.9258	.9263	.9269	.9274	.9279	.9284	.9289
8.5	.9294	.9299	.9304	.9309	.9315	.9320	.9325	.9330	.9335	.9340
8.6	.9345	.9350	.9355	.9360	.9365	.9370	.9375	.9380	.9385	.9390
8.7	.9395	.9400	.9405	.9410	.9415	.9420	.9425	.9430	.9435	.9440
8.8	.9445	.9450	.9455	.9460	.9465	.9469	.9474	.9479	.9484	.9489
8.9	.9494	.9499	.9504	.9509	.9513	.9518	.9523	.9528	.9533	.9538
9.0	.9542	.9547	.9552	.9557	.9562	.9566	.9571	.9576	.9581	.9586
9.1	.9590	.9595	.9600	.9605	.9609	.9614	.9619	.9624	.9628	.9633
9.2	.9638	.9643	.9647	.9652	.9657	.9661	.9666	.9671	.9675	.9680
9.3	.9685	.9689	.9694	.9699	.9703	.9708	.9713	.9717	.9722	.9727
9.4	.9731	.9736	.9741	.9745	.9750	.9754	.9759	.9763	.9768	.9773
9.5	.9777	.9782	.9786	.9791	.9795	.9800	.9805	.9809	.9814	.9818
9.6	.9823	.9827	.9832	.9836	.9841	.9845	.9850	.9854	.9859	.9863
9.7	.9868	.9872	.9877	.9881	.9886	.9890	.9894	.9899	.9903	.9908
9.8	.9912	.9917	.9921	.9926	.9930	.9934	.9939	.9943	.9948	.9952
9.9	.9956	.9961	.9965	.9969	.9974	.9978	.9983	.9987	.9991	.9996
x	0	1	2	3	4	5	6	7	8	9

TABLE 3 Geometric Formulas

Plane Geometry:

Rectangle
Area: $A = lw$
Perimeter: $P = 2l + 2w$

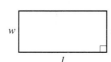

Square
Area: $A = s^2$
Perimeter: $P = 4s$

Triangle
Area: $A = \frac{1}{2}bh$

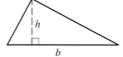

Triangle
Sum of Angle Measures:
$A + B + C = 180°$

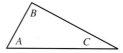

Right Triangle
Pythagorean Theorem
(Equation):
$a^2 + b^2 = c^2$

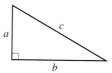

Parallelogram
Area: $A = bh$

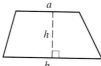

Trapezoid
Area: $A = \frac{1}{2}h(b_1 + b_2)$

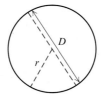

Circle
Area: $A = \pi r^2$
Circumference:
$C = \pi D = 2\pi r$
$\left(\frac{22}{7} \text{ and } 3.14 \text{ are different}\right.$
approximations for π $)$

Solid Geometry:

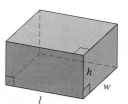

Rectangular Solid
Volume: $V = lwh$

Cube
Volume: $V = s^3$

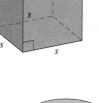

Right Circular Cylinder
Volume: $V = \pi r^2 h$
Total Surface Area:
$S = 2\pi rh + 2\pi r^2$

Right Circular Cone
Volume: $V = \frac{1}{3}\pi r^2 h$
Total Surface Area:
$S = \pi r^2 + \pi rs$
Slant Height:
$s = \sqrt{r^2 + h^2}$

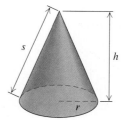

Sphere
Volume: $V = \frac{4}{3}\pi r^3$
Surface Area: $S = 4\pi r^2$

CHAPTER 1

Exercise Set 1.1, pp. 8–9

1. $n - 4$　**3.** $2x$　**5.** $0.32n$　**7.** $\frac{1}{2}x + 7$　**9.** $0.19t - 4$
11. $x - y + 5$　**13.** $xy - 4$　**15.** $0.35t + 1$　**17.** 13
19. 42　**21.** 7　**23.** 4　**25.** 25　**27.** 4
29. 3 *is* a solution.　**31.** 7 *is not* a solution.　**33.** 6 *is*
a solution.　**35.** 5 *is* a solution.　**37.** 0.4 *is not* a
solution.　**39.** $\frac{17}{3}$ *is* a solution.　**41.** $\{a, e, i, o, u\}$ or
$\{a, e, i, o, u, y\}$　**43.** $\{2, 4, 6, 8, \ldots\}$
45. $\{5, 10, 15, 20, \ldots\}$　**47.** $\{x | x$ is an odd number
between 10 and 30$\}$　**49.** $\{x | x$ is a whole number less than 5$\}$
51. $\{n | n$ is a multiple of 5 between 7 and 79$\}$　**53.** True
55. False　**57.** True　**59.** True　**61.** True
63. True　**65.** True　**67.** False　**69.** True

71. ◈　**73.** $3(x + y)$　**75.** $\dfrac{n - m}{n + m}$　**77.** $\{0\}$

79. $\{x | x$ is a real number$\}$

81.

Exercise Set 1.2, pp. 17–19

1. 7　**3.** 9　**5.** 6.2　**7.** 0　**9.** $1\frac{7}{8}$　**11.** 4.21
13. -9 is less than or equal to -1; true　**15.** -7 is
greater than 1; false　**17.** 3 is greater than or equal to
-5; true　**19.** -9 is less than -4; true　**21.** -4 is
greater than or equal to -4; true　**23.** -5 is less than
-5; false　**25.** 17　**27.** -11　**29.** -3.2　**31.** $-\frac{11}{35}$
33. -8.5　**35.** $\frac{5}{9}$　**37.** -4.5　**39.** 0　**41.** -6.4
43. -7.29　**45.** 4.8　**47.** 0　**49.** $6\frac{1}{3}$　**51.** -7
53. 2.7　**55.** -1.79　**57.** 0　**59.** 0.03　**61.** 2
63. -5　**65.** 4　**67.** -17　**69.** -3.1　**71.** $-\frac{11}{10}$

73. 2.9　**75.** 7.9　**77.** -28　**79.** 24　**81.** -21
83. -7.2　**85.** 0　**87.** 5.44　**89.** 5　**91.** -5
93. -73　**95.** 0　**97.** 7　**99.** $\frac{1}{5}$　**101.** $-\frac{1}{9}$　**103.** $\frac{3}{2}$
105. $-\frac{11}{3}$　**107.** $\frac{5}{6}$　**109.** $-\frac{6}{5}$　**111.** $\frac{8}{27}$

113. $-\frac{3}{8}$　**115.** $-\frac{7}{3}$　**117.** $\dfrac{15x}{35}$　**119.** $-\dfrac{12a}{32}$

121. $10x$　**123.** $8a$　**125.** $2x + 5y$　**127.** $-3y + 2x$

129. $\dfrac{y}{3} \cdot \dfrac{x}{2}$　**131.** $(2 \cdot 8)x$　**133.** $(x + 2y) + 5$

135. $3a + 3$　**137.** $4x - 4y$　**139.** $-10a - 15b$
141. $2ab - 2ac + 2ad$　**143.** $8(x + y)$　**145.** $9(p - 1)$
147. $7(x - 3)$　**149.** $2(x - y + z)$　**151.** $0.7x - 5$

153. -6　**155.** ◈　**157.** $\frac{2}{3}$　**159.** $\dfrac{2(x - 2)}{x + 2}$

Exercise Set 1.3, pp. 24–25

1. Equivalent　**3.** Equivalent　**5.** Equivalent
7. Not equivalent　**9.** Not equivalent　**11.** 14.6
13. 8　**15.** 18　**17.** 5　**19.** 24　**21.** 7　**23.** $9a$
25. $-3b$　**27.** $15y$　**29.** $11a$　**31.** $-8t$　**33.** $10x$
35. $8x - 8y$　**37.** $2c + 10d$　**39.** $22x + 18$
41. $2x - 33y$　**43.** 8　**45.** 21　**47.** 2　**49.** 2　**51.** $\frac{18}{5}$
53. 0　**55.** $\frac{4}{5}$　**57.** $\frac{4}{3}$　**59.** $\frac{37}{5}$　**61.** 13　**63.** $-a - 5$
65. $m + 1$　**67.** $5d - 12$　**69.** $-7x + 14$
71. $-9x + 21$　**73.** $44a - 22$　**75.** -190
77. $-12y - 145$　**79.** 2　**81.** 2　**83.** 7　**85.** 5
87. $-\frac{51}{31}$　**89.** 5　**91.** 2　**93.** $\frac{3}{5}$　**95.** $-\frac{1}{2}$　**97.** -1
99. $\varnothing$　**101.** $\mathbb{R}$　**103.** $\varnothing$　**105.** $\mathbb{R}$
107. $\{1, 2, 3, 4, 5, 6, 7, 8, 9\}$;
$\{x | x$ is a positive integer less than 10$\}$　**109.** -1.7
111. 54　**113.** 4　**115.** ◈　**117.** 0.00705
119. -4.1762　**121.** 8　**123.** $\frac{224}{29}$

Exercise Set 1.4, pp. 31–33

1. Let x and $x + 9$ be the numbers; $x + (x + 9) = 81$
3. Let t be Deirdre's time; $(5 - 3.2)t = 2.7$　**5.** Let t be

the boat's time; $(12 + 3)t = 35$ **7.** Let x, $x + 1$, and $x + 2$ be the angle measures; $x + (x + 1) + (x + 2) = 180$ **9.** Let t be the number of minutes climbing; $3500t = 29{,}000 - 8000$.   **11.** Let x be the measure of the second angle; $3x + x + (2x - 12) = 180$ **13.** Let n be the first odd number; $n + 2(n + 2) + 3(n + 4) = 70$ **15.** Let s be the length of a side in the smaller square; $4s + 4(s + 2) = 100$ **17.** Let x be the first number; $x + (3x - 6) + [\frac{2}{3}(3x - 6) + 2] = 172$ **19.** Let x be the score on the next test; $\dfrac{93 + 89 + 72 + 80 + 96 + x}{6} = 88$

21 4.1   **23.** 50.3   **25.** 320   **27.** 63   **29.** 165 and 330   **31.** 1.5 hr   **33.** 14 and 16   **35.** A 6-m piece and a 4-m piece   **37.** 96°, 32°, 52°   **39.** 20   **41.** All real numbers   **43.** ◈   **45.** 90 in²   **47.** 1,200,000

Exercise Set 1.5, pp. 42–43

1. 3^7 **3.** 5^9 **5.** a^3 **7.** $15x^6$ **9.** $21m^{13}$ **11.** $x^{10}y^{10}$ **13.** a^6 **15.** $2x^3$ **17.** m^5n^4

19. $5x^6y^4$ **21.** $-7a^3b^{10}$ **23.** $-3x^4y^6z^6$ **25.** $\dfrac{1}{6^3}$

27. $\dfrac{1}{9^5}$ **29.** $\dfrac{1}{-11}$ **31.** $\dfrac{1}{(5x)^3}$ **33.** $\dfrac{x^2}{y^3}$ **35.** $\dfrac{x^2}{y^2}$

37. x^3y^2 **39.** $\dfrac{1}{x^2y^5}$ **41.** $\dfrac{x^3}{y^5}$ **43.** $\dfrac{y^7z^4}{x^2}$ **45.** 3^{-4}

47. $(-16)^{-2}$ **49.** $\dfrac{1}{6^{-4}}$ **51.** $\dfrac{6}{x^{-2}}$ **53.** $(5y)^{-3}$

55. $\dfrac{y^{-4}}{3}$ **57.** 8^{-4}, or $\dfrac{1}{8^4}$ **59.** 8^{-6}, or $\dfrac{1}{8^6}$

61. b^{-3}, or $\dfrac{1}{b^3}$ **63.** a^3 **65.** $-28m^5n^5$

67. $-14x^{-11}$, or $\dfrac{-14}{x^{11}}$ **69.** 5^{3a} **71.** 4^5

73. 10^{-9}, or $\dfrac{1}{10^9}$ **75.** 81   **77.** a^5

79. $3ab^{-3}$, or $\dfrac{3a}{b^3}$ **81.** $-\dfrac{4}{3}x^9y^{-2}$, or $-\dfrac{4x^9}{3y^2}$

83. $\dfrac{3}{2}x^3y^{-2}$, or $\dfrac{3x^3}{2y^2}$ **85.** 10^a **87.** 4^6

89. 8^{-12}, or $\dfrac{1}{8^{12}}$ **91.** 6^{12} **93.** $27x^6y^6$

95. $\dfrac{1}{4}x^{-6}y^8$, or $\dfrac{y^8}{4x^6}$

97. $\dfrac{1}{36}a^4b^{-6}c^{-2}$, or $\dfrac{a^4}{36b^6c^2}$

99. $\dfrac{5a^4b}{2}$ **101.** $\dfrac{8x^9y^3}{27}$ **103.** 1

105. $\dfrac{4}{25}x^{-4}y^{22}$, or $\dfrac{4y^{22}}{25x^4}$ **107.** 23   **109.** 27   **111.** $-\dfrac{6}{11}$

113. $-\dfrac{65}{7}$ **115.** 117   **117.** 28   **119.** -5 **121.** $20\frac{2}{3}$

123. | 9 | + | 3 | × | 4 | + | 1 | = | ÷ | 2 | = |

125. | 8 | − | 2 | × | 3 | + | 4 | ÷ | 1 | 5 | = |

127. | 9 | − | 2 | xʸ | 8 | + | 4 | = | ÷ | 3 | = |

129. | 7 | − | 2 | ÷ | 3 | = | xʸ | 4 | − | 5 |

| ÷ | 8 | = | **131.** 19   **133.** 59, 61, and 63

135. ◈   **137.** $3a^{-x-4}$ **139.** y^{-4}, or $\dfrac{1}{y^4}$ **141.** $3a^{2+2a}$

143. $2x^{a+2}y^{b-2}$ **145.** $2^{-2a-2b+ab}$ **147.** $\frac{2}{27}$

Exercise Set 1.6, pp. 47–49

1. $w = \dfrac{A}{l}$ **3.** $I = \dfrac{W}{E}$ **5.** $r = \dfrac{d}{t}$ **7.** $l = \dfrac{V}{wh}$

9. $m = \dfrac{E}{c^2}$ **11.** $l = \dfrac{P - 2w}{2}$, or $\dfrac{P}{2} - w$

13. $a^2 = c^2 - b^2$ **15.** $r^2 = \dfrac{A}{\pi}$

17. $h = \dfrac{2}{11}W + 40$ **19.** $r^3 = \dfrac{3V}{4\pi}$

21. $h = \dfrac{2A}{b_1 + b_2}$ **23.** $m = \dfrac{rF}{v^2}$ **25.** $n = \dfrac{q_1 + q_2 + q_3}{A}$

27. $t = \dfrac{d_2 - d_1}{v}$ **29.** $d_1 = d_2 - vt$

31. $m = \dfrac{r}{1 + np}$ **33.** $a = \dfrac{y}{b - c^2}$ **35.** 12 cm

37. About 8.5 cm   **39.** \$1571.43   **41.** 9 ft   **43.** 7 years   **45.** 34   **47.** 9225 g   **49.** 72%   **51.** -58.44 **53.** ◈   **55.** 7.4 cm   **57.** 3 months

59. $a = \dfrac{2s - 2v_it}{t^2}$ **61.** $V_1 = \dfrac{T_1P_2V_2}{P_1T_2}$ **63.** $c = \dfrac{a}{x} - b$

Exercise Set 1.7, pp. 54–55

1. 4.7×10^{10} **3.** 8.63×10^{17} **5.** 1.6×10^{-8} **7.** 7×10^{-11} **9.** 4.07×10^{11} **11.** 6.03×10^{-7} **13.** 4.927×10^{11} **15.** 0.0004   **17.** 673,000,000   **19.** 0.0000000008923   **21.** 90,300,000,000   **23.** 0.00000004037   **25.** 8,007,000,000,000   **27.** 9.7×10^{-5} **29.** 1.3×10^{-11} **31.** 8.3×10^{10} **33.** 2.0×10^3 **35.** 1.51×10^{-12} **37.** 2.5×10^3 **39.** 5.0×10^{-4} **41.** 3.0×10^{11} **43.** 2.00×10^{10} **45.** 2.0×10^{-16} **47.** 2.5   **49.** 4.5×10^2 **51.** 1×10^{11} **53.** 6.5×10 **55.** 10^5, or 100,000 light

years   **57.** 3.08×10^{26} Å   **59.** 1×10^{22} cu Å
61. 0.02 m³   **63.** 5.8×10^8 miles
65. $-\frac{1}{12}$ **67.** $x - 6$ **69.** ◈
71. $8 \cdot 10^{-90}$ is larger by 7.1×10^{-90} **73.** 8
75. 8×10^{18} grains

Review Exercises: Chapter 1, pp. 59–60

1. [1.1] $\frac{x}{y} - 5$ **2.** [1.1] -25

3. [1.1] {2, 4, 6, 8, 10, 12};
{$x|x$ is an even number between 1 and 13}   **4.** [1.1] Yes
5. [1.2] 7.3   **6.** [1.2] 4.09   **7.** [1.2] 0
8. [1.2] -13.1 **9.** [1.2] $-\frac{23}{35}$ **10.** [1.2] $\frac{7}{15}$
11. [1.2] 4.01   **12.** [1.2] -11.5 **13.** [1.2] $-\frac{1}{6}$
14. [1.2] -5.4 **15.** [1.2] 6.3   **16.** [1.2] $-\frac{5}{12}$
17. [1.2] -4.8 **18.** [1.2] 5   **19.** [1.2] -9.1
20. [1.2] $-\frac{21}{4}$ **21.** [1.2] $a + 5$ **22.** [1.2] $y7$
23. [1.2] $x5 + y$, or $y + 5x$ **24.** [1.2] $4 + (a + b)$
25. [1.2] $x(y7)$ **26.** [1.2] $7m(n + 2)$ **27.** [1.3] $x + 12$
28. [1.3] $47x - 60$ **29.** [1.3] 6.6   **30.** [1.3] $\frac{27}{2}$
31. [1.3] $-\frac{4}{11}$ **32.** [1.3] ∅   **33.** [1.3] ℝ
34. [1.4] $2x - 13 = 21$ **35.** [1.4] 49
36. [1.4] 90°, 30°, 60°   **37.** [1.5] $-10a^5b^8$
38. [1.5] $4xy^6$ **39.** [1.5] 1   **40.** [1.5] 3^3, or 27

41. [1.5] 5^3a^6, or $125a^6$ **42.** [1.5] $-\frac{1}{8}a^9b^{-6}$, or $\frac{a^9}{-8b^6}$

43. [1.5] $x^{-4}y^{-6}z^8$, or $\frac{z^8}{x^4y^6}$ **44.** [1.5] $\frac{1}{16}a^{-20}b^{16}$, or $\frac{b^{16}}{16a^{20}}$

45. [1.5] $-\frac{16}{7}$ **46.** [1.5] 0   **47.** [1.6] $m = PS$

48. [1.6] $x = \frac{c}{m - r}$ **49.** [1.6] 2 cm

50. [1.7] 3.086×10^{13} **51.** [1.7] 1.03×10^{-7}
52. [1.7] 3.7×10^7 **53.** [1.7] 2.0×10^{-6}
54. [1.7] 1.4×10^4 mm³, or 1.4×10^{-5} m³
55. [1.3] ◈ To write an equation that has no solution,
begin with a simple equation that is false for any value of
x, such as $x = x + 1$. Then add or multiply by the same
quantities on both sides of the equation to construct a
more complicated equation with no solution.
56. [1.1] ◈ Examine what each letter represents to
determine whether a letter is a variable or a constant. For
example, in $d = r \cdot t$, if r is fixed at a specific speed in
the statement of a problem and d and t are not, then d
and t would be variables.
57. [1.7] 0.0000003%   **58.** [1.1], [1.5] $-\frac{23}{24}$
59. [1.4], [1.6] The 17-in. pizza is a better deal, because
it costs less per square inch.

60. [1.6] 729 cm³   **61.** [1.6] $z = y - \frac{x}{m}$

62. [1.5] $3^{-2a - 2ab - 4b}$ **63.** [1.4], [1.6] $88.\overline{3}$
64. [1.3] -39 **65.** [1.3] $-40x$

66. [1.2] $a2 + cb + cd + ad =$
$ad + a2 + cb + cd = a(d + 2) + c(b + d)$
67. [1.1] $\sqrt{5}/4$; answers may vary

Test: Chapter 1, pp. 60–61

1. [1.4] 94   **2.** [1.4] 55.5, 28.1   **3.** [1.2] -41
4. [1.2] -3.7 **5.** [1.2] -2.11 **6.** [1.2] -14.2
7. [1.2] -43.2 **8.** [1.2] $-\frac{19}{12}$ **9.** [1.2] -33.92
10. [1.2] -84.84 **11.** [1.2] $\frac{5}{49}$ **12.** [1.2] 6
13. [1.2] 0.7   **14.** [1.2] $-\frac{4}{3}$ **15.** [1.2] $8y$
16. [1.2] $11a - 27b$ **17.** [1.2] $-33c + 44d$
18. [1.2] $-24xy - 28xz$ **19.** [1.3] -2 **20.** [1.3] $\frac{15}{2}$
21. [1.4] 17, 19, 21   **22.** [1.3] $-8x - 1$

23. [1.3] $24b - 9$ **24.** [1.5] $-72x^{-10}y^{-6}$, or $\frac{-72}{x^{10}y^6}$

25. [1.5] $\frac{13}{15}$ **26.** [1.5] $\frac{1}{36}x^{-4}y^8$, or $\frac{y^8}{36x^4}$

27. [1.5] $\frac{1}{4}x^6y^{-8}$, or $\frac{x^6}{4y^8}$ **28.** [1.5] 1   **29.** [1.5] $-\frac{3}{2}$

30. [1.7] 1.8×10^5 **31.** [1.7] 2.01×10^{-7}
32. [1.7] 5.0×10^9 **33.** [1.7] 3.8×10^2

34. [1.7] 4.2×10^8 mi   **35.** [1.6] $V_2 = \frac{T_2P_1V_1}{T_1P_2}$

36. [1.5] $16^cx^{6ac}y^{2bc + 2c}$ **37.** [1.5] $-9a^3$

38. [1.5] $\frac{4}{7y^2}$

CHAPTER 2

Technology Connection, Section 2.1

TC1.
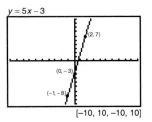

TC2.
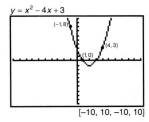

TC3.
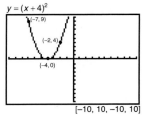

Exercise Set 2.1, pp. 70–71

1.

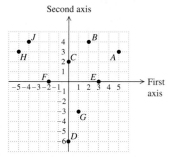

3.

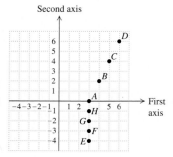

5.

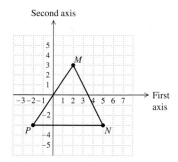

7. III **9.** II **11.** I **13.** IV **15.** Yes **17.** No
19. Yes **21.** Yes **23.** Yes **25.** No **27.** Yes
29. No

31.

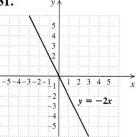

33.

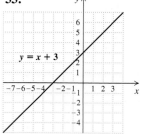

35.

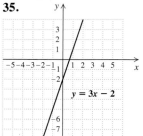

37.

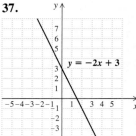

39.

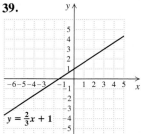

41.

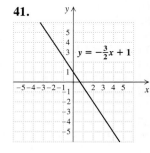

43.

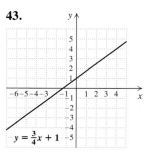

45.

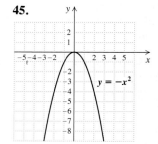

47.

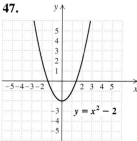

49.

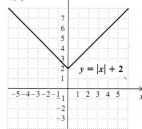

51.

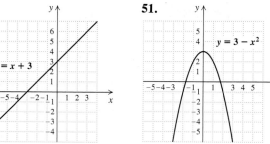

53.

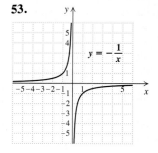

55. 26 ft   **57.** −1.4   **59.** ◈   **61.** ◈
63. (a); (d)

65. 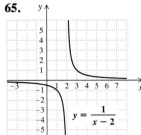   **67.**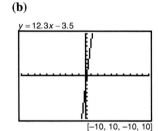

69. ◈

71. (a)   (b)

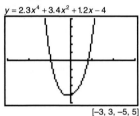

(c)

39. About 2.5 hr

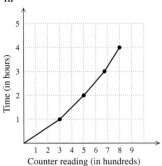

41. 64,000

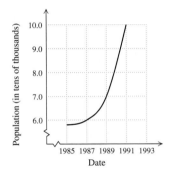

43. About $310,000

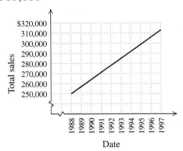

Exercise Set 2.2, pp. 76–80

1. No   **3.** Yes   **5.** Yes   **7.** No   **9.** Yes
11. Function   **13.** Function   **15.** A relation, but not a
function.   **17.** (a) 1; (b) −3; (c) −6; (d) 9; (e) $a + 3$
19. (a) 4; (b) 9; (c) 49; (d) $5t^2 + 4$; (e) $20a^2 + 4$
21. (a) 15; (b) 32; (c) 20; (d) 4; (e) $27r^2 + 6r − 1$
23. (a) $\frac{3}{5}$; (b) $\frac{1}{3}$; (c) $\frac{4}{7}$; (d) 0; (e) $\frac{x − 1}{2x − 1}$
25. $4\sqrt{3}$ cm^2 **27.** 36π in$^2 \approx 113.04$ in^2
29. 14°F   **31.** 159.48 cm   **33.** 75   **35.** 1.4 million
37. 3.5 drinks

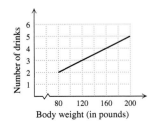

45. Yes   **47.** Yes   **49.** No   **51.** No
53. 18, 20, 22   **55.** $l = \dfrac{S − 2wh}{2h + 2w}$ **57.** ◈
59. (a) 3.1497708; (b) 55.7314683; (c) 3178.20675;
(d) 1166.70323
61. $g(x) = \frac{15}{4}x − \frac{13}{4}$ **63.** ◈   **65.** At 2 min, 40 sec and
at 5 min, 40 sec   **67.** 1 every 3 min

Exercise Set 2.3, pp. 87–89

1. Slope = 4; y-intercept = 5   **3.** Slope = −2;
y-intercept = −6   **5.** Slope = $-\frac{3}{8}$; y-intercept = −0.2
7. Slope = 0.5; y-intercept = −9   **9.** Slope = 0;
y-intercept = 7   **11.** $y = \frac{2}{3}x − 7$ **13.** $y = −4x + 2$
15. $y = -\frac{7}{9}x + 3$ **17.** $y = 5x + \frac{1}{2}$

19. Slope $= \frac{5}{2}$; y-intercept $= 1$

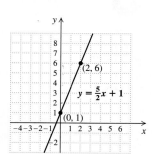

21. Slope $= -\frac{5}{2}$; y-intercept $= 4$

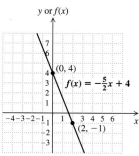

35. Slope $= -\frac{3}{4}$; y-intercept $= 3$

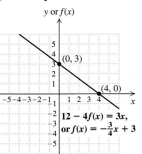

37. Slope $= 0$; y-intercept $= 4$

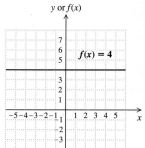

23. Slope $= 2$; y-intercept $= -5$

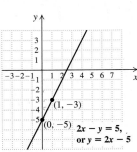

25. Slope $= \frac{1}{3}$; y-intercept $= 6$

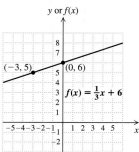

39. (a)

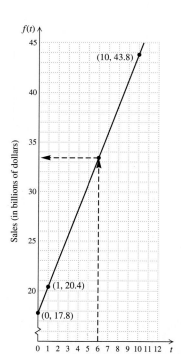

(b) $33.4 billion

27. Slope $= -\frac{2}{7}$; y-intercept $= 1$

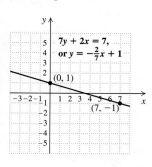

29. Slope $= -0.25$; y-intercept $= 2$

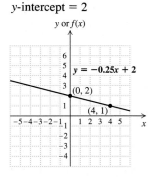

31. Slope $= \frac{4}{5}$; y-intercept $= -2$

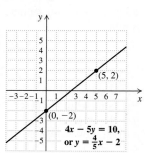

33. Slope $= \frac{5}{4}$; y-intercept $= -2$

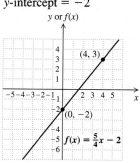

41. (a)

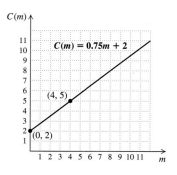

(b) About $6.00

43. (a)

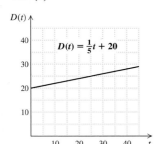

(b) About 27 quadrillion joules

45. (a)

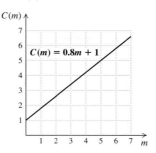

(b) About $4.25; **(c)** about 8 min

47.

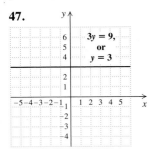

49.

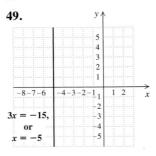

51.

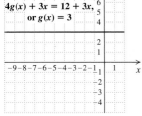

53. Linear; slope $= -\frac{3}{5}$ **55.** Linear
57. Linear; slope $= -\frac{1}{2}$ **59.** Not linear
61. Not linear   **63.** Not linear   **65.** $45x + 54$
67. ◈   **69.** Slope $= -\frac{5}{r}$; y-intercept $= \frac{p}{r}$

71. Slope $= -\frac{r}{p}$; y-intercept $= \frac{s}{p}$ **73.** Linear

75. Linear   **77.** $f(x) = \frac{47}{20}x + 3.1$
79.

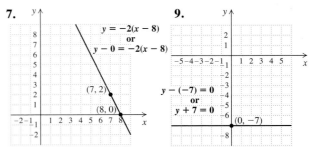

Exercise Set 2.4, pp. 94–95

For Exercises 1–15, each graph is a line through the intercepts given below.
1. $(0, -2)$, $(2, 0)$ **3.** $(0, -1)$, $\left(\frac{1}{3}, 0\right)$
5. $(0, -5)$, $(4, 0)$ **7.** $(0, -5)$, $(-1, 0)$
9. $(0, -3)$, $(5, 0)$ **11.** $(0, 3)$, $(-5, 0)$
13. $(0, -7)$, $(2.8, 0)$ **15.** $\left(0, \frac{7}{2}\right)$, $\left(\frac{7}{5}, 0\right)$ **17.** 2
19. -2 **21.** 5   **23.** $-\frac{1}{3}$ **25.** Slope is undefined.
27. 0   **29.** -1 **31.** 10 km/h   **33.** 0.6 ton per hour
35. 300 ft per min   **37.** $0.4 million per year
39. Slope is undefined.   **41.** 3   **43.** 0   **45.** $\frac{5}{7}$
47. Slope is undefined.   **49.** $\frac{1}{2}$
51. Slope is undefined.   **53.** -2 **55.** 0   **57.** 0
59. 3   **61.** 4 miles   **63.** ◈   **65.** $(50, -2)$, $(25, -1)$, $(-25, 1)$, $(-50, 2)$, answers may vary
67. $\left(-\frac{b}{m}, 0\right)$ **69. (a)** $-\frac{5c}{4b}$; **(b)** slope is undefined;
(c) $\frac{a+d}{f}$

Exercise Set 2.5, pp. 100–102

1.

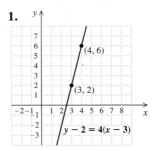

3.

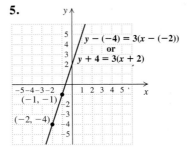

5.

7.

9.

11.

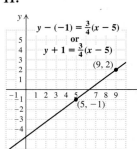

$y - (-1) = \frac{3}{4}(x - 5)$

or

$y + 1 = \frac{3}{4}(x - 5)$

(9, 2)

(5, −1)

13. $y = 5x - 13$
15. $y = -\frac{2}{3}x - \frac{13}{3}$
17. $y = -0.6x - 5.8$
19. $f(x) = \frac{1}{2}x + \frac{7}{2}$
21. $f(x) = x$
23. $f(x) = \frac{5}{2}x + 5$
25. $f(x) = \frac{1}{4}x + \frac{17}{4}$
27. $f(x) = \frac{2}{5}x$

29. $f(x) = 3x + 5$ **31. (a)** $R(t) = -0.075t + 46.8$;
(b) 41.9 sec, 41.7 sec;   **(c)** 2021
33. (a) $A(t) = 4.325t + 132.7$; **(b)** 184.6 million
35. (a) $N(t) = 1.125t + 14.5$; **(b)** 33.625 million tons
37. \$151,000; \$201,000   **39. (a)** $E(t) = 0.15t + 65$;
(b) 72.5 years   **41.** Yes   **43.** No   **45.** Yes
47. $y = -\frac{1}{2}x + \frac{17}{2}$ **49.** $y = \frac{5}{7}x - \frac{17}{7}$ **51.** $y = \frac{1}{3}x + 4$
53. $y = -\frac{3}{2}x - \frac{13}{2}$ **55.** Yes   **57.** No   **59.** $y = \frac{1}{2}x + 4$
61. $y = \frac{4}{3}x - 6$ **63.** $y = \frac{5}{2}x + 9$ **65.** $y = -\frac{5}{3}x - \frac{41}{3}$
67. \$35   **69.** $82\frac{2}{3}$ **71.** ◈   **73.** 21.1°C
75. (a) $f(x) = \frac{1}{3}x + \frac{10}{3}$; **(b)** $f(3) = \frac{13}{3}$; **(c)** $a = 290$
77. $k = -\frac{40}{9}$

Exercise Set 2.6, pp. 107–109

1. 1   **3.** 7   **5.** −29   **7.** −41   **9.** −30   **11.** 110
13. $\frac{1}{2}$ **15.** $\frac{10}{11}$ **17.** 13   **19.** $x^2 - x + 1$ **21.** 5
23. 2   **25.** 42   **27.** $-\frac{3}{4}$ **29.** $\frac{1}{6}$
31. $\{x|x$ is a real number$\}$
33. $\{x|x$ is a real number and $x \neq 2\}$
35. $\{x|x$ is a real number and $x \neq 0\}$
37. $\{x|x$ is a real number and $x \neq 1\}$
39. $\{x|x$ is a real number and $x \neq 2$ and $x \neq 4\}$
41. $\{x|x$ is a real number and $x \neq -2$ and $x \neq 4\}$
43. $\{x|x$ is a real number and $x \neq 3\}$
45. $\{x|x$ is a real number and $x \neq 4\}$
47. $\{x|x$ is a real number and $x \neq 4$ and $x \neq 5\}$
49. $\{x|x$ is a real number and $x \neq -2.5$ and $x \neq -1\}$
51. $\{x|x$ is a real number and $x \neq 0$ and $x \neq 4\}$
53. $\{x|x$ is a real number and $x \neq 1$ and $x \neq 4\}$
55. $\{x|x$ is a real number and $x \neq 2$, $x \neq 3$, and $x \neq 4\}$
57. $\{x|0 \leqslant x \leqslant 9\}$; $\{x|3 \leqslant x \leqslant 10\}$; $\{x|3 \leqslant x \leqslant 9\}$;
$\{x|3 \leqslant x \leqslant 9\}$;

59.

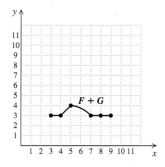

F + G

61. $\frac{7}{5}$ **63.** 86   **65.** ◈
67. $\{x|x$ is a real number and $x \neq 2$, $x \neq 3$, and $x \neq 1\}$
69. For $f + g$, $f - g$, and $f \cdot g$, domain $= \{-2, -1, 0, 1\}$;
for f/g, domain $= \{-2, 0, 1\}$
71. $\{x|x$ is a real number and $x \neq 4$, $x \neq 2$, $x \neq -2$,
and $x \neq 3\}$

73. Answers may vary. $f(x) = \dfrac{1}{x + 2}$, $g(x) = \dfrac{1}{x - 5}$

75. (a) 67 million; **(b)** 50 million; **(c)** 50 million

Review Exercises: Chapter 2, p. 111

1. [2.1] Yes   **2.** [2.1] No   **3.** [2.1] No
4. [2.1] Yes
5. [2.1]

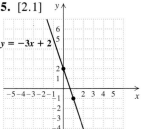

$y = -3x + 2$

6. [2.1]

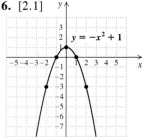

$y = -x^2 + 1$

7. [2.3]

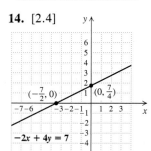

$8x + 32 = 0$

8. [2.3] Slope is −4;
y-intercept is −9
9. [2.3] Slope is $\frac{1}{3}$;
y-intercept is $-\frac{7}{6}$
10. [2.3] Yes
11. [2.3] Yes
12. [2.3] No
13. [2.3] No

14. [2.4]

$\left(-\frac{7}{2}, 0\right)$ $\left(0, \frac{7}{4}\right)$

$-2x + 4y = 7$

15. [2.4] $\frac{4}{7}$ **16.** [2.4]
Slope is undefined
17. [2.5] $y - 4 =$
$-2(x + 3)$
18. [2.5] $f(x) = \frac{4}{3}x + \frac{7}{3}$
19. [2.5] Perpendicular
20. [2.5] Parallel

21. [2.5] Parallel   **22.** [2.5] $y = \frac{3}{5}x - \frac{31}{5}$
23. [2.5] $y = -\frac{5}{3}x - \frac{5}{3}$ **24.** [2.2] −5   **25.** [2.2] −8
26. [2.6] 57   **27.** [2.6] −10   **28.** [2.6] $-\frac{7}{4}$
29. [2.2] $2a + 2b - 5$ **30.** [2.6] $\{x|x$ is a real number$\}$
31. [2.6] $\{x|x$ is a real number and $x \neq \frac{5}{2}\}$ **32.** [1.2] $\frac{2}{15}$
33. [1.5] $25a^6b^2$ **34.** [1.3] −26
35. [1.7] 5.28×10^6 **36.** [2.2] ◈ In a function, since
each member of the domain corresponds to *exactly one*
member of the range, it follows that each member of the
domain corresponds to *at least one* member of the range.

Therefore, a function is a relation. In a relation, every member of the domain corresponds to *at least one,* but not necessarily *exactly one,* member of the range. Therefore, a relation may or may not be a function. **37.** [2.4] ◈ The slope of a line is the rise between two points on the line divided by the run between those points. For a vertical line, there is no run between any two points, and division by 0 is undefined; therefore, the slope is undefined. For a horizontal line, there is no rise between any two points, so the slope is 0/run, or 0. **38.** [1.5], [2.4] -9 **39.** [2.5] $-\frac{9}{2}$

Test: Chapter 2, p. 112

1. [2.1] Yes   **2.** [2.1] No   **3.** [2.1] No   **4.** [2.1] No
5. [2.1]   **6.** [2.1]

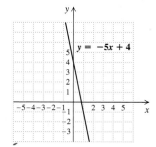

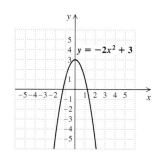

7. [2.3] Slope is 3; y-intercept is -5 **8.** [2.3] Slope is $\frac{4}{3}$; y-intercept is -3
9. [2.3]   **10.** [2.3] **(a)** Linear;
                                    (b) nonlinear; **(c)** linear
                                    11. [2.4]

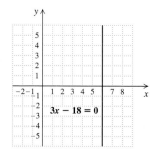

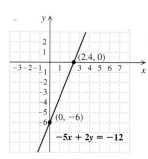

12. [2.4] $\frac{5}{8}$ **13.** [2.4] 0
14. [2.4] **(a)** Zero slope; **(b)** undefined slope
15. [2.5] $y + 4 = 4(x + 2)$ **16.** [2.5] $f(x) = -x + 2$
17. [2.5] Parallel
18. [2.5] Perpendicular   **19.** [2.5] $y = \frac{2}{5}x + \frac{16}{5}$
20. [2.5] $y = -\frac{5}{2}x - \frac{11}{2}$ **21.** [2.2], [2.6] **(a)** -4;
(b) 5; **(c)** -2; **(d)** -5 **22.** [2.5] **(a)** $C(m) = 0.3m + 25$;
(b) \$175   **23.** [1.3] All real numbers
24. [1.7] 5.28×10^{-4} **25.** [1.5] 30   **26.** [1.2] -6.8
27. [2.2], [2.4] **(a)** 30 miles; **(b)** 15 mph
28. [2.5] $s = -\frac{3}{2}r + \frac{27}{2}$

CHAPTER 3

Technology Connection, Section 3.1

TC1. $x = 1.59$, $y = 5.72$ **TC2.** $x = -0.26$, $y = 57.06$
TC3. $x = 2.08$, $y = 1.04$
TC4. $x = 0.87$, $y = -0.32$

Exercise Set 3.1, pp. 120–122

1. $x + y = -42$, $x - y = 52$
3. $x + y = 40$, $4.95x + 7.95y = 282$
5. $x + y = 90$, $x + \frac{1}{2}y = 64$
7. $x + y = 250$, $3.50x + 7y = 1347.50$
9. $2w + 2l = 228$, $w = l - 42$
11. $x + y = 40$, $2x + 3y = 89$
13. $x + y = 400$, $20x + 30y = 11{,}000$
15. $x + y = 12$, $30x + 60y = 600$ **17.** Yes   **19.** No
21. Yes   **23.** No   **25.** Yes   **27.** $(3, 1)$ **29.** $(3, 2)$
31. $(1, -5)$ **33.** $(2, 1)$ **35.** $\left(\frac{5}{2}, -2\right)$ **37.** $(3, -2)$
39. $\varnothing$ **41.** $(4, -5)$ **43.** $(3, -4)$
45. $\{(x, y) | 2x - 3y = 6\}$ **47.** All except 39   **49.** 45
51. -3 **53.** $\frac{9}{20}$ **55.** ◈   **57.** 1983 and 1990
59. 1987   **61.** Answers may vary.
(a) $x + y = 6$, $x - y = 4$; **(b)** $x + y = 1$, $2x + 2y = 3$;
(c) $x + y = 1$, $2x + 2y = 2$
63. $b = 2s$, $b - 10 = 3(s - 10)$
65. $2l + 2w = 156$, $l = 4(w - 6)$ **67.** $(3, 3)$, $(-5, 5)$
69. $(-0.39, -1.10)$ **71.** $(-0.13, 0.67)$

Exercise Set 3.2, pp. 128–129

1. $(-4, 3)$ **3.** $(-3, -15)$ **5.** $(2, -2)$ **7.** $(-2, 1)$
9. $\left(\frac{1}{2}, \frac{1}{2}\right)$ **11.** $\left(\frac{19}{8}, \frac{1}{8}\right)$ **13.** $\varnothing$ **15.** $\{(x, y) | x - 3 = y\}$
17. $(1, 2)$ **19.** $(3, 0)$ **21.** $(-1, 2)$ **23.** $\left(\frac{128}{31}, -\frac{17}{31}\right)$
25. $(6, 2)$ **27.** $\left(\frac{140}{13}, -\frac{50}{13}\right)$ **29.** $(4, 6)$
31. $\left(\frac{110}{19}, -\frac{12}{19}\right)$ **33.** $\{(x, y) | 2x + 3y = 1\}$ **35.** $\varnothing$
37. $\left(\frac{1}{2}, -\frac{1}{2}\right)$ **39.** $\left(-\frac{4}{3}, -\frac{19}{3}\right)$ **41.** $\left(\frac{1000}{11}, -\frac{1000}{11}\right)$
43. $(140, 60)$ **45.** 11%   **47.** ◈
49. $(23.118879, -12.039964)$ **51.** $\left(\dfrac{a + 2b}{7}, \dfrac{a - 5b}{7}\right)$
53. $p = 2$, $q = -\frac{1}{3}$ **55.** $\left(-\frac{1}{4}, -\frac{1}{2}\right)$

Exercise Set 3.3, pp. 137–139

1. 5, -47 **3.** 12 white, 28 printed   **5.** 38° and 52°
7. 115 children's plates, 135 adults' plates
9. Width = 36 ft, length = 78 ft
11. 31 2-point, 9 3-point
13. Lumber: 100; plywood: 300
15. 4 30-sec spots; 8 60-sec spots
17. 150 lb soybean meal, 200 lb corn meal
19. $12\frac{1}{2}$ L of Arctic Antifreeze, $7\frac{1}{2}$ L of Frost-No-More
21. \$6800 at 9%, \$8200 at 10%
23. \$12,500 at 10%, \$14,500 at 12%
25. 84 adult, 33 children   **27.** Paula is 32; Bob is 20

29. Length $= 76$ m, width $= 19$ m
31. 7 \$5 bills, 15 \$1 bills   **33.** 375 km   **35.** $1\frac{3}{4}$ hr
37. 24 mph   **39.** $x + 29$ **41.** $y = -\frac{3}{4}x - \frac{7}{2}$
43. Burl 40, son 20
45. Width $= \frac{102}{5}$ in., length $= \frac{288}{5}$ in.   **47.** $4\frac{4}{7}$ L
49. 82   **51.** First train: 36 km/h; second train: 54 km/h
53. 3 girls, 4 boys

Exercise Set 3.4, pp. 145–146

1. Yes   **3.** $(1, 2, 3)$ **5.** $(-1, 5, -2)$ **7.** $(3, 1, 2)$
9. $(-3, -4, 2)$ **11.** $(2, 4, 1)$ **13.** $(-3, 0, 4)$
15. Dependent   **17.** $\left(\frac{1}{2}, 4, -6\right)$ **19.** $\left(\frac{1}{2}, \frac{1}{3}, \frac{1}{6}\right)$
21. $\left(\frac{1}{2}, \frac{2}{3}, -\frac{5}{6}\right)$ **23.** $(15, 33, 9)$ **25.** $\left(\frac{1}{4}, -\frac{1}{2}, -\frac{1}{4}\right)$
27. $\left(\frac{98}{5}, \frac{304}{5}, \frac{498}{5}\right)$ **29.** $\varnothing$ **31.** Dependent
33. $c = \dfrac{2F + td}{t}$, or $\dfrac{2F}{t} + d$ **35.** ◈
37. $(1, -1, 2)$ **39.** $(1, -2, 4, -1)$
41. $\left(-1, \frac{1}{5}, -\frac{1}{2}\right)$ **43.** 12   **45.** $3x + 4y + 2z = 12$

Exercise Set 3.5, pp. 149–151

1. 17, 9, 79   **3.** 4, 2, -1
5. $A = 34°$, $B = 104°$, $C = 42°$
7. $A = 25°$, $B = 50°$, $C = 105°$
9. \$41.1 billion on newspaper, \$36 billion on television, \$7.7 billion on radio
11. Steak: 2; baked potato: 1; broccoli: 2
13. Oriental: 385; African-American: 200; Caucasian: 154
15. A: 1500; B: 1900; C: 2300
17. A: 900 gal/hr; B: 1300 gal/hr; C: 1500 gal/hr
19. \$45,000 at 8%; \$15,000 at 6%; \$20,000 at 9%
21. 2   **23.** $\dfrac{a^3}{b}$ **25.** 20   **27.** $180°$ **29.** 35

Exercise Set 3.6, pp. 154–155

1. $\left(\frac{3}{2}, \frac{5}{2}\right)$ **3.** $(-4, 3)$ **5.** $\left(\frac{1}{2}, \frac{3}{2}\right)$ **7.** $\left(\frac{3}{2}, -4, 3\right)$
9. $(2, -2, 1)$ **11.** $\left(4, \frac{1}{2}, -\frac{1}{2}\right)$ **13.** $(1, -3, -2, -1)$
15. 4 dimes, 30 nickels   **17.** Mix 5 lb of the \$4.05-per-pound granola with 10 lb of the \$2.70-per-pound granola.   **19.** \$400 at 7%, \$500 at 8%, \$1600 at 9%   **21.** $\frac{69}{10}$, or 6.9   **23.** -32
25. ◈   **27.** 1324

Exercise Set 3.7, p. 159

1. 3   **3.** 36   **5.** -10 **7.** -3 **9.** 5   **11.** $(2, 0)$
13. $\left(-\frac{25}{2}, -\frac{11}{2}\right)$ **15.** $\left(\frac{3}{2}, \frac{13}{14}, \frac{33}{14}\right)$ **17.** $(2, -1, 4)$
19. $(1, 2, 3)$ **21.** $\frac{333}{245}$ **23.** One piece, 20.8 ft; the other piece, 12 ft   **25.** 3   **27.** An equation of the line through (x_1, y_1) and (x_2, y_2) is

$$y - y_1 = \frac{y_2 - y_1}{x_2 - x_1}(x - x_1),$$

which is equivalent to

$$yx_2 - yx_1 - y_1x_2 + y_1x_1 = y_2x - y_2x_1 - y_1x + y_1x_1$$

or

$$y_2x_1 + y_1x - y_2x + yx_2 - yx_1 - y_1x_2 = 0. \quad (1)$$

$$\begin{vmatrix} x & y & 1 \\ x_1 & y_1 & 1 \\ x_2 & y_2 & 1 \end{vmatrix} = 0$$

is equivalent to

$$x(y_1 - y_2) - x_1(y - y_2) + x_2(y - y_1) = 0$$

or

$$xy_1 - xy_2 - x_1y + x_1y_2 + x_2y - x_2y_1 = 0. \quad (2)$$

Equations (1) and (2) are equivalent.

Exercise Set 3.8, pp. 163–165

1. (a) $P(x) = 20x - 600,000$; (b) 30,000 units
3. (a) $P(x) = 50x - 120,000$; (b) 2400 units
5. (a) $P(x) = 80x - 10,000$; (b) 125 units
7. (a) $P(x) = 40x - 75,000$; (b) 1875 units
9. (a) $P(x) = 75x - 195,000$; (b) 2600 units
11. (\$10, 1400)   **13.** (\$22, 474)   **15.** (\$50, 6250)
17. (\$10, 1070)   **19.** (a) $C(x) = 750x + 125,000$;
(b) $R(x) = 1050x$; (c) $P(x) = 300x - 125,000$; (d) \$5000
loss, \$85,000 profit; (e) 417 units   **21.** (a) $C(x) =$
$20x + 10,000$; (b) $R(x) = 100x$; (c) $P(x) = 80x - 10,000$;
(d) \$150,000 profit, \$6000 loss; (e) 125 units
23.

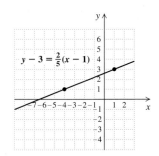

$y - 3 = \frac{2}{5}(x - 1)$

25. $\frac{5}{2}$ **27.** (\$5, 300)   **29.** ◈
31. (a) 4526; (b) \$870

Review Exercises: Chapter 3, pp. 167–168

1. [3.1] $(3, 2)$ **2.** [3.1] $(-2, 1)$ **3.** [3.2] $\left(\frac{2}{5}, -\frac{4}{5}\right)$
4. [3.2] $\varnothing$ **5.** [3.2] $\left(-\frac{11}{15}, -\frac{43}{30}\right)$ **6.** [3.2] $\left(\frac{37}{19}, \frac{53}{19}\right)$
7. [3.2] $\left(\frac{76}{17}, -\frac{2}{119}\right)$ **8.** [3.2] $(2, 2)$ **9.** [3.3] CD, \$14;
cassette, \$9   **10.** [3.3] 4 hours   **11.** [3.3] 10 liters of
Cleanse-O, 30 liters of Tingle   **12.** [3.4] $(10, 4, -8)$
13. [3.4] $\left(-\frac{7}{3}, \frac{125}{27}, \frac{20}{27}\right)$ **14.** [3.4] $(2, 0, 4)$ **15.** [3.2] $\varnothing$
16. [3.4] $\left(\frac{8}{9}, -\frac{2}{3}, \frac{10}{9}\right)$ **17.** [3.2] $\{(x, y)|3x + 4y = 6\}$

18. [3.4] $\left(2, \frac{1}{3}, -\frac{2}{3}\right)$ **19.** [3.5] $A = 90°$, $B = 67\frac{1}{2}°$, $C = 22\frac{1}{2}°$ **20.** [3.5] 641   **21.** [3.5] \$20 bills: 5; \$5 bills: 15; \$1 bills: 19   **22.** [3.6] $\left(55, -\frac{89}{2}\right)$
23. [3.6] $(-1, 1, 3)$ **24.** [3.7] 2   **25.** [3.7] 9
26. [3.7] $(6, -2)$ **27.** [3.7] $(-3, 0, 4)$
28. [3.8] (\$3, 81)
29. [3.8] **(a)** $C(x) = 175x + 35{,}000$; **(b)** $R(x) = 225x$; **(c)** $P(x) = 50x - 35{,}000$; **(d)** \$25,000 profit, \$10,000 loss;
(e) 700 beds   **30.** [1.3] $-\frac{4}{3}$ **31.** [1.6] $t = \dfrac{Q}{a - 4}$
32. [2.5] $y = -\frac{7}{3}x + 12$ **33.** [1.4] 69, 76, 83
34. [3.5] ◈ To solve a problem involving four variables, go through the Familiarize and Translate steps as usual. The resulting system of equations can be solved using the elimination method just as for three variables but likely with more steps.
35. [3.4] ◈ A system of equations can be both dependent and inconsistent if it is equivalent to a system with fewer equations that has no solution. An example is a system of three equations in three unknowns in which two of the equations represent the same plane, and the third represents a parallel plane.   **36.** [3.1] $(0, 2)$, $(1, 3)$
37. [3.5] $a = -\frac{2}{3}$, $b = -\frac{4}{3}$, $c = 3$; $f(x) = -\frac{2}{3}x^2 - \frac{4}{3}x + 3$

Test: Chapter 3, pp. 168–169

1. [3.2] $\left(3, -\frac{11}{3}\right)$ **2.** [3.2] $\left(\frac{15}{7}, -\frac{18}{7}\right)$
3. [3.2] $\left(-\frac{3}{2}, -\frac{3}{2}\right)$ **4.** [3.2] $\varnothing$ **5.** [3.3] $l = 30$, $w = 18$ **6.** [3.3] 70 two-piece, 62 three-piece
7. [3.4] Dependent   **8.** [3.4] $\left(2, -\frac{1}{2}, -1\right)$ **9.** [3.4] $\varnothing$
10. [3.4] $(0, 1, 0)$ **11.** [3.6] $\left(\frac{34}{107}, -\frac{104}{107}\right)$
12. [3.6] $(3, 1, -2)$ **13.** [3.7] 34   **14.** [3.7] 133
15. [3.7] $\left(\frac{13}{18}, \frac{7}{27}\right)$ **16.** [3.5] 3.5 hr   **17.** [3.8] (\$3, 55)
18. [3.8] **(a)** $C(x) = 40{,}000 + 45x$; **(b)** $R(x) = 80x$; **(c)** $P(x) = 35x - 40{,}000$; **(d)** \$12,500 profit;
(e) 1143 rackets   **19.** [1.4] \$35   **20.** [1.6] $a = \dfrac{P + 3b}{4}$
21. [1.3] $\frac{9}{5}$ **22.** [2.5] $y + 1 = -\frac{5}{8}(x - 5)$, or $y = -\frac{5}{8}x + \frac{17}{8}$ **23.** [2.5], [3.3] $m = 7$, $b = 10$
24. [3.5] Adults: 1651; senior citizens: 1346; children: 335

| CUMULATIVE REVIEW: 1-3

1. [1.3] 14.87   **2.** [1.3] -22 **3.** [1.3] -42.9
4. [1.3] 20   **5.** [1.3] $-\dfrac{21}{4}$ **6.** [1.3] 6   **7.** [1.3] -5
8. [1.3] $\dfrac{10}{9}$ **9.** [1.3] $-\dfrac{32}{5}$ **10.** [1.3] $\dfrac{18}{17}$
11. [1.5] x^{11} **12.** [1.5] $-\dfrac{40x}{y^5}$ **13.** [1.5] $-288x^4y^{18}$

14. [1.5] y^{10} **15.** [1.5] $-\dfrac{2a^{11}}{5b^{33}}$ **16.** [1.5] $\dfrac{81x^{36}}{256y^8}$
17. [1.7] 1.1   **18.** [1.7] 1.12×10^6
19. [1.7] 4.0×10^2 **20.** [1.7] 4.00×10^6
21. [1.6] $b = \dfrac{2A}{h} - t$ **22.** [2.1] Yes   **23.** [2.1] No
24. [2.3]

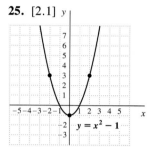

$y = -2x + 3$

25. [2.1]

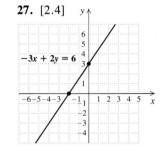

$y = x^2 - 1$

26. [2.3]

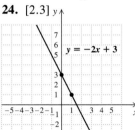

$4x + 16 = 0$

27. [2.4]

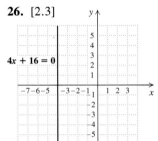

$-3x + 2y = 6$

28. [2.3] Slope is $\frac{9}{4}$; y-intercept is -3 **29.** [2.4] $\frac{4}{3}$
30. [2.5] $y = -3x - 5$ **31.** [2.5] $y = -\frac{1}{10}x + \frac{12}{5}$
32. [2.5] Parallel   **33.** [2.5] $y = -2x + 5$
34. [2.2] -31 **35.** [2.2] 3   **36.** [2.6] -43
37. [2.6] $8a^2 + 4a - 4$ **38.** [3.2] $(1, 1)$
39. [3.2] $(-2, 3)$ **40.** [3.2] $\left(-3, \frac{2}{5}\right)$
41. [3.4] $(-3, 2, -4)$ **42.** [3.4] $(0, -1, 2)$
43. [3.7] 14   **44.** [3.7] 0   **45.** [3.3] 8, 18
46. [3.3] $48\frac{8}{9}$ oz of Soakem; $71\frac{1}{9}$ oz of Rinsem
47. [1.4] 3, 5, 7   **48.** [1.4] 90
49. [3.3] $l = 10$ cm, $w = 6$ cm   **50.** [3.3] 19 nickels, 15 dimes   **51.** [3.5] \$120   **52.** [3.5] 23 wins, 33 losses, 8 ties   **53.** [3.5] 1st $= 12$, 2nd $= 59$, 3rd $= 0$
54. [1.5] $-12x^{2a}y^{b + y + 3}$ **55.** [2.5] \$151,000
56. [2.5] $m = -\frac{5}{9}$, $b = -\frac{2}{9}$

| CHAPTER 4

Exercise Set 4.1, pp. 181–183

1. No, no, no, yes   **3.** No, yes, yes, no
5. $\{x | x > 4\}$, $(4, \infty)$ **7.** $\{t | t \leqslant 6\}$, $(-\infty, 6]$

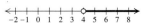

9. $\{y|y < -3\}$, $(-\infty, -3)$

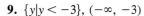

11. $\{x|x \geqslant -6\}$, $[-6, \infty)$

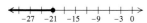

13. $\{x|x > -5\}$, or $(-5, \infty)$

15. $\{y|y < 6\}$, or $(-\infty, 6)$

17. $\{a|a \leqslant -21\}$, or $(-\infty, -21]$

19. $\{t|t \geqslant -5\}$, or $[-5, \infty)$

21. $\{y|y > -6\}$, or $(-6, \infty)$

23. $\{x|x \leqslant 9\}$, or $(-\infty, 9]$

25. $\{x|x \geqslant 3\}$ or $[3, \infty)$

27. $\{x|x < -60\}$, or $(-\infty, -60)$

29. $\{x|x \leqslant 0.9\}$, or $(-\infty, 0.9]$

31. $\left\{x|x \leqslant \frac{5}{6}\right\}$, or $\left(-\infty, \frac{5}{6}\right]$

33. $\{x|x < 6\}$, or $(-\infty, 6)$

35. $\{y|y \leqslant -3\}$, or $(-\infty, -3]$

37. $\left\{y|y > \frac{2}{3}\right\}$, or $\left(\frac{2}{3}, \infty\right)$

39. $\{x|x \geqslant 11.25\}$, or $[11.25, \infty)$

41. $\left\{x|x \leqslant \frac{1}{2}\right\}$, or $\left(-\infty, \frac{1}{2}\right]$

43. $\left\{y|y \leqslant -\frac{53}{6}\right\}$, or $\left(-\infty, -\frac{53}{6}\right]$

45. $\left\{x|x > -\frac{2}{17}\right\}$, or $\left(-\frac{2}{17}, \infty\right)$ **47.** $\left\{m|m > \frac{7}{3}\right\}$, or $\left(\frac{7}{3}, \infty\right)$

49. $\{r|r < -3\}$, or $(-\infty, -3)$ **51.** $\{x|x \geqslant 2\}$, or $[2, \infty)$

53. $\{y|y < 5\}$, or $(-\infty, 5)$ **55.** $\left\{x|x \leqslant \frac{4}{7}\right\}$, or $\left(-\infty, \frac{4}{7}\right]$

57. A score of 84 or better   **59.** Mileage $\leqslant 330$

61. All calls greater than $1\frac{1}{2}$ min

63. Gross sales greater than \$15,000

65. $\{n|n < 100 \text{ hr}\}$ **67.** \$20,000   **69.** \$1.99

71. $\{b|b > \$1450\}$ **73.** (a) $\{t|t < 1945.4°\text{F}\}$;

(b) $\{t|t < 1761.44°\text{F}\}$ **75.** (a) $\left\{x|x < 8181\frac{9}{11}\right\}$;

(b) $\left\{x|x > 8181\frac{9}{11}\right\}$

77.

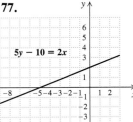

$5y - 10 = 2x$

79. 16   **81.** ◈

83. $\left\{x|x \leqslant \dfrac{2}{a-1}\right\}$

85. $\left\{y|y \geqslant \dfrac{5b + 2a}{b(a-2)}\right\}$

87. $\left\{x|x > \dfrac{4m - 2c}{d - (5c + 2m)}\right\}$

89. False; $a = 2$, $b = 3$, $c = 4$, $d = 5$; $2 - 4 = 3 - 5$

91. ◈   **93.** All real numbers

95. All real numbers except 0

Exercise Set 4.2, pp. 189–191

1. $\{6, 8\}$ **3.** $\varnothing$ **5.** $\{1, 2, 3, 4\}$

7. $1 < x < 6$; $(1, 6)$ **9.** $-7 \leqslant y \leqslant -3$; $[-7, -3]$

11. $-3 < x \leqslant 4$; $(-3, 4]$ **13.** $-6 < x \leqslant 2$; $(-6, 2]$

15. $-2 \leqslant x < 5$; $[-2, 5)$ **17.** $1 \leqslant x < 5$; $[1, 5)$

19. $\{x|-4 < x < 6\}$, or $(-4, 6)$ **21.** $\{y|-2 < y \leqslant 2\}$, or $(-2, 2]$ **23.** $\left\{x|-\frac{5}{3} \leqslant x \leqslant \frac{4}{3}\right\}$, or $\left[-\frac{5}{3}, \frac{4}{3}\right]$

25. $\{x|-1 < x \leqslant 6\}$, or $(-1, 6]$ **27.** $\left\{x|-\frac{3}{2} \leqslant x < \frac{9}{2}\right\}$, or $\left[-\frac{3}{2}, \frac{9}{2}\right)$ **29.** $\{x|10 < x \leqslant 14\}$, or $(10, 14]$

31. $\{1, 4, 5, 6, 7, 8, 11\}$ **33.** $\{1, 2, 3, 4, 5, 6, 8\}$

35. $\{4, 8, 11\}$

37.   **39.**

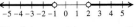

41. $\{x|x < -9 \text{ or } x > -5\}$, or $(-\infty, -9) \cup (-5, \infty)$

43. $\left\{x|x \leqslant \frac{5}{2} \text{ or } x \geqslant 11\right\}$, or $\left(-\infty, \frac{5}{2}\right] \cup [11, \infty)$

45. $\{x|x \geqslant -3\}$, or $[-3, \infty)$

47. $\left\{x|x \leqslant -\frac{5}{4} \text{ or } x > -\frac{1}{2}\right\}$, or $\left(-\infty, -\frac{5}{4}\right] \cup \left(-\frac{1}{2}, \infty\right)$

49. $\{x|x \text{ is a real number}\}$, or $\mathbb{R}$, or $(-\infty, \infty)$

51. $\{x|x < -4 \text{ or } x > 2\}$, or $(-\infty, -4) \cup (2, \infty)$

53. $\left\{x|x < \frac{79}{4} \text{ or } x > \frac{89}{4}\right\}$, or $\left(-\infty, \frac{79}{4}\right) \cup \left(\frac{89}{4}, \infty\right)$

55. $\left\{x|x \leqslant -\frac{13}{2} \text{ or } x \geqslant \frac{29}{2}\right\}$, or $\left(-\infty, -\frac{13}{2}\right] \cup \left[\frac{29}{2}, \infty\right)$

57. $\{x|x \text{ is a real number}\}$, or $\mathbb{R}$, or $(-\infty, \infty)$

59. $\left(-\frac{16}{13}, -\frac{41}{13}\right)$ **61.** $-\frac{27}{7}$ **63.** ◈

65. (a) $1945.4° \leqslant F < 4820°$; (b) $1761.44° \leqslant F < 3956°$

67. Sizes between 6 and 13   **69.** From 1973 to 1987

71. $\left\{a\middle|-\frac{3}{2}\le a\le 1\right\}$, or $\left[-\frac{3}{2},\,1\right]$;

73. $\{x|-4<x\le 1\}$, or $(-4,\,1]$;

75. $\left\{y\middle|y<-\frac{4}{3}\ or\ y>4\right\}$, or $\left(-\infty,\,-\frac{4}{3}\right)\cup(4,\,\infty)$;

75. $\left\{x\middle|-\frac{1}{8}<x<\frac{1}{2}\right\}$, or $\left(-\frac{1}{8},\,\frac{1}{2}\right)$;

77. $\left\{x\middle|-\frac{5}{4}\le x\le\frac{23}{4}\right\}$, or $\left[-\frac{5}{4},\,\frac{23}{4}\right]$

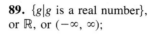

79. $\varnothing$

77. True   **79.** False   **81.** False   **83.** $\left\{x\middle|-\frac{2}{5}\le x\le 2\right\}$, or $\left[-\frac{2}{5},\,2\right]$

81. $\{x|x<-4\ or\ x>5\}$, or $(-\infty,\,-4)\cup(5,\,\infty)$;

83. $\{x|-5<x<19\}$, or $(-5,\,19)$;

Exercise Set 4.3, pp. 196–198

1. $\{-3,\,3\}$ **3.** $\varnothing$ **5.** $\{0\}$ **7.** $\{-5.5,\,5.5\}$
9. $\{-9,\,15\}$ **11.** $\left\{-\frac{1}{2},\,\frac{7}{2}\right\}$ **13.** $\left\{-\frac{3}{2},\,\frac{17}{2}\right\}$ **15.** $\varnothing$
17. $\{-11,\,11\}$ **19.** $\{-8,\,8\}$ **21.** $\left\{-\frac{11}{5},\,\frac{11}{5}\right\}$
23. $\{-7,\,8\}$ **25.** $\{-12,\,2\}$ **27.** $\{-1,\,2\}$
29. $\left\{-\frac{1}{3},\,3\right\}$ **31.** 4   **33.** 11   **35.** 33   **37.** 34
39. $\left\{-\frac{11}{2},\,\frac{3}{4}\right\}$ **41.** $\left\{-\frac{3}{2}\right\}$ **43.** $\left\{-\frac{3}{5},\,5\right\}$
45. $\mathbb{R}$ **47.** $\left\{-\frac{3}{2}\right\}$ **49.** $\left\{0,\,\frac{24}{23}\right\}$ **51.** $\left\{\frac{8}{3},\,32\right\}$
53. $\{x|-3<x<3\}$, or $(-3,\,3)$;

55. $\{x|x\le-2\ or\ x\ge 2\}$, or $(-\infty,\,-2]\cup[2,\,\infty)$;

85. $\left\{x\middle|x\le-\frac{2}{15}\ or\ x\ge\frac{14}{15}\right\}$, or $\left(-\infty,\,-\frac{2}{15}\right]\cup\left[\frac{14}{15},\,\infty\right)$;

87. $\{m|-12\le m\le 2\}$, or $[-12,\,2]$;

57. $\{t|t\le-5.5\ or\ t\ge 5.5\}$, or $(-\infty,\,-5.5]\cup[5.5,\,\infty)$;

59. $\{x|2<x<4\}$, or $(2,\,4)$;

89. $\{g|g$ is a real number$\}$, or $\mathbb{R}$, or $(-\infty,\,\infty)$;

91. $\{x|-1\le x\le 2\}$, or $[-1,\,2]$;

61. $\{x|-7\le x\le 3\}$, or $[-7,\,3]$;

93. 24,640 m^2 **95.** ◈   **97.** $\left\{x\middle|x\ge\frac{5}{2}\right\}$
99. $\{x|x\ge-5\}$ **101.** $\left\{1,\,-\frac{1}{4}\right\}$ **103.** $\varnothing$
105. $\mathbb{R}$ **107.** $|x|<3$ **109.** $|x|\ge 6$
111. $|x+3|>5$ **113.** $|p-5|\le\frac{1}{96}$, or $|5-p|\le\frac{1}{96}$, or $|p-60|\le\frac{1}{8}$, or $|60-p|\le\frac{1}{8}$ **115.** ◩

63. $\{x|x<2\ or\ x>4\}$, or $(-\infty,\,2)\cup(4,\,\infty)$;

65. $\left\{x\middle|-\frac{1}{2}\le x\le\frac{7}{2}\right\}$, or $\left[-\frac{1}{2},\,\frac{7}{2}\right]$;

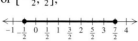

Technology Connection, Section 4.4

TC1.

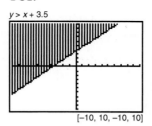

$y>x+3.5$
$[-10,\,10,\,-10,\,10]$

TC2.

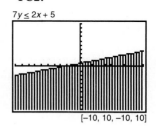

$7y\le 2x+5$
$[-10,\,10,\,-10,\,10]$

67. $\{y|y$ is a real number$\}$, or $\mathbb{R}$, or $(-\infty,\,\infty)$;

69. $\left\{x\middle|x\le-\frac{5}{4}\ or\ x\ge\frac{23}{4}\right\}$, or $\left(-\infty,\,-\frac{5}{4}\right]\cup\left[\frac{23}{4},\,\infty\right)$;

TC3.

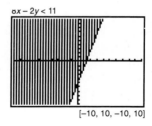

$6x-2y<11$
$[-10,\,10,\,-10,\,10]$

TC4.

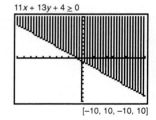

$11x+13y+4\ge 0$
$[-10,\,10,\,-10,\,10]$

71. $\{y|-9<y<15\}$, or $(-9,\,15)$;

73. $\left\{x\middle|-\frac{7}{2}\le x\le\frac{1}{2}\right\}$, or $\left[-\frac{7}{2},\,\frac{1}{2}\right]$;

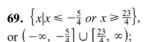

A-14   ANSWERS

Exercise Set 4.4, pp. 204–205

1. Yes

3. No

5. $y > 2x$

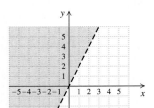

7. $y < x + 1$

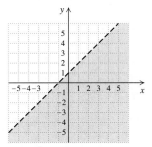

9. $y > x - 2$

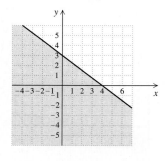

11. $x + y < 4$

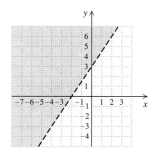

13. $3x + 4y \leq 12$

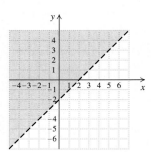

15. $2y - 3x > 6$

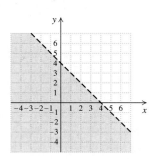

17. $3x - 2 \leq 5x + y$

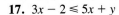
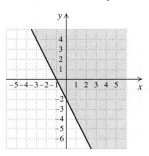

19. $x < -4$

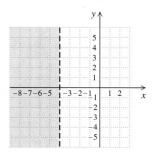

21. $y > -2$

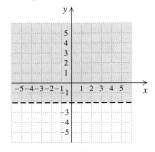

23. $-4 < y < -1$

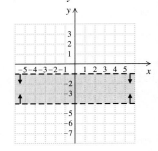

25. $-3 \leq x \leq 3$

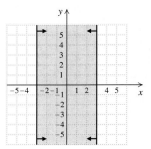

27. $0 \leq x \leq 5$

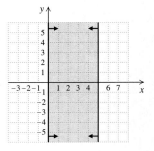

29.

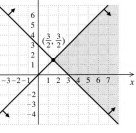

31.

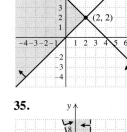

33.

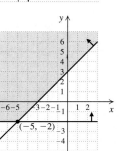

35.

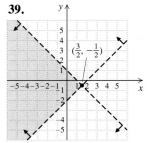

37.

39.

41.

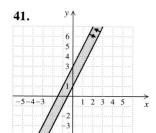

43.

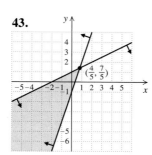

45.

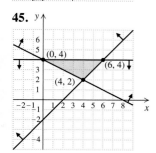

47.

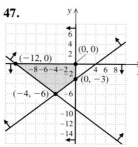

49.

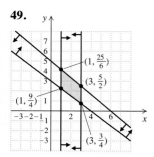

51. 15; 20   **53.** $\frac{10}{17}$
55. ◈

57.

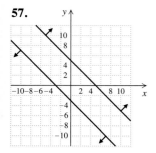

59.

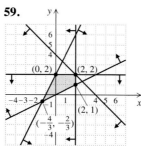

61. $0 \leq W \leq 50$,
   $0 \leq L \leq 94$

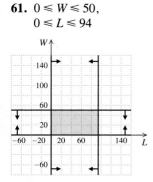

63. $35c + 75a > 1000$,
   $a \geq 0$,
   $c \geq 0$

65. 〰

Exercise Set 4.5, pp. 210–212

1. Maximum 168, when $x = 0$, $y = 6$; minimum 0, when $x = 0$, $y = 0$
3. Maximum 152, when $x = 7$, $y = 0$; minimum 32, when $x = 0$, $y = 4$
5. Maximum 50, when $x = 5$, $y = 11$; minimum 5, when $x = 0$, $y = 1$
7. Maximum income of $18 an hour occurs when 100 of each type of biscuit is made
9. Maximum score of 425 when 5 matching questions and 15 essays are done
11. 60 motorcycles and 100 bicycles
13. 100 units of lumber and 300 units of plywood
15. Maximum income of $3110 when $22,000 is invested in corporate bonds and $18,000 is invested in municipal bonds
17. Maximum profit of $2520 when 125 batches of Smello and 187.5 batches of Roppo are made
19. 30 P-1 airplanes, 10 P-2 airplanes
21. 25 chairs, 9 sofas

Review Exercises: Chapter 4, p. 214

1. [4.1] $\{x | x \leq -4\}$, or $(-\infty, -4]$;

2. [4.1] $\{x | x > 1\}$, or $(1, \infty)$;

3. [4.1] $\{a | a \leq -21\}$, or $(-\infty, -21]$;

4. [4.1] $\{y | y \geq -7\}$, or $[-7, \infty)$;

5. [4.1] $\left\{y | y > -\frac{15}{4}\right\}$, or $\left(-\frac{15}{4}, \infty\right)$;

6. [4.1] $\{y | y > -30\}$, or $(-30, \infty)$;

7. [4.1] $\{x | x > -3\}$, or $(-3, \infty)$;

8. [4.1] $\left\{y | y < -\frac{6}{5}\right\}$, or $\left(-\infty, -\frac{6}{5}\right)$;

9. [4.1] $\{x | x < -3\}$, or $(-\infty, -3)$;

10. [4.1] $\left\{y | y > -\frac{220}{23}\right\}$, or $\left(-\frac{220}{23}, \infty\right)$;

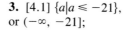

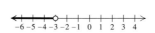

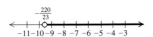

11. [4.1] $\left\{x \mid x \leqslant -\frac{5}{2}\right\}$, or $\left(-\infty, -\frac{5}{2}\right]$;

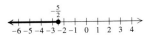

12. [4.1] Scores higher than or equal to 92
13. [4.1] $10,000

14. [4.2] $\{1, 5, 9\}$ **15.** [4.2] $\{1, 2, 3, 5, 6, 9\}$
16. [4.2] $\{x \mid -7 < x \leqslant 2\}$, or $(-7, 2]$
17. [4.2] $\left\{x \mid -\frac{5}{4} < x < \frac{5}{2}\right\}$, or $\left(-\frac{5}{4}, \frac{5}{2}\right)$
18. [4.2] $\{x \mid x < -3 \ or \ x > 1\}$, or $(-\infty, -3) \cup (1, \infty)$
19. [4.2] $\{x \mid x < -11 \ or \ x \geqslant -6\}$, or $(-\infty, -11) \cup [-6, \infty)$
20. [4.2] $\{x \mid x < -11 \ or \ x > 1\}$, or $(-\infty, -11) \cup (1, \infty)$
21. [4.2] $\{x \mid x \leqslant -6 \ or \ x \geqslant 8\}$, or $(-\infty, -6] \cup [8, \infty)$
22. [4.3] $\{6, -6\}$ **23.** [4.3] $\varnothing$ **24.** [4.3] $\{x \mid x \leqslant -3.5 \ or \ x \geqslant 3.5\}$, or $(-\infty, -3.5] \cup [3.5, \infty)$
25. [4.3] $\{-5, 9\}$ **26.** [4.3] $\left\{x \mid -\frac{17}{2} < x < \frac{7}{2}\right\}$, or $\left(-\frac{17}{2}, \frac{7}{2}\right)$
27. [4.3] $\left\{x \mid x \leqslant -\frac{11}{3} \ or \ x \geqslant \frac{19}{3}\right\}$, or $\left(-\infty, -\frac{11}{3}\right] \cup \left[\frac{19}{3}, \infty\right)$
28. [4.3] $\left\{-14, \frac{4}{3}\right\}$ **29.** [4.3] $\varnothing$
30. [4.3] $\{x \mid -12 \leqslant x \leqslant 4\}$, or $[-12, 4]$
31. [4.3] $\{x \mid x < 0 \ or \ x > 10\}$, or $(-\infty, 0) \cup (10, \infty)$
32. [4.4]   **33.** [4.4]

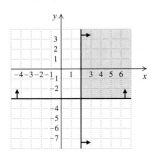   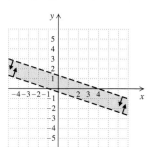

34. [4.4]

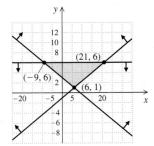

35. [4.5] Maximum 42 when $x = 7$, $y = 17$; minimum 7 when $x = 0$, $y = 3$
36. [4.5] $40,000 in municipal bonds, $20,000 in mutual funds; maximum income $6800   **37.** [3.2] $\left(0, \frac{5}{2}\right)$
38. [1.3] -22

39. [2.3]

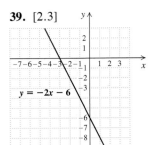

$y = -2x - 6$

40. [3.3] 38,592 ft^2

41. [4.3] ◈ The equation $|x| = p$ has two solutions when p is positive because x can be either p or $-p$. The same equation has no solution when p is negative because no number has a negative absolute value.
42. [4.4] ◈ The solution set of a system of inequalities is all numbers that make *all* the individual inequalities true. This consists of numbers that are common to all the individual solution sets, or the intersection of the graphs.
43. [4.3] $\left\{x \mid -\frac{8}{3} \leqslant x \leqslant -2\right\}$, or $\left[-\frac{8}{3}, -2\right]$
44. [4.1] False; $-4 < 3$ is true, but $(-4)^2 < 9$ is false.
45. [4.3] $|x - 1.1| \leqslant 0.03$

Test: Chapter 4, p. 215

1. [4.1] $\{x \mid x < 14\}$, or $(-\infty, 14)$
2. [4.1] $\{y \mid y > -50\}$, or $(-50, \infty)$
3. [4.1] $\{y \mid y \leqslant -2\}$, or $(-\infty, -2]$
4. [4.1] $\left\{a \mid a \leqslant \frac{11}{5}\right\}$, or $\left(-\infty, \frac{11}{5}\right]$
5. [4.1] $\{y \mid y > 1\}$, or $(1, \infty)$
6. [4.1] $\left\{x \mid x > \frac{5}{2}\right\}$, or $\left(\frac{5}{2}, \infty\right)$
7. [4.1] $\left\{x \mid x \leqslant \frac{7}{4}\right\}$, or $\left(-\infty, \frac{7}{4}\right]$
8. [4.1] More than 66.7 miles
9. [4.1] Scores higher than or equal to 93
10. [4.2] $\{3, 5\}$
11. [4.2] $\{1, 3, 5, 7, 9, 11, 13\}$
12. [4.2] $\{x \mid -1 < x < 6\}$, or $(-1, 6)$
13. [4.2] $\left\{x \mid -\frac{2}{5} < x \leqslant \frac{9}{5}\right\}$, or $\left(-\frac{2}{5}, \frac{9}{5}\right]$
14. [4.2] $\left\{x \mid x < -4 \ or \ x > -\frac{5}{2}\right\}$, or $(-\infty, -4) \cup \left(-\frac{5}{2}, \infty\right)$
15. [4.2] $\{x \mid x \leqslant -3 \ or \ x \geqslant 2\}$, or $(-\infty, -3] \cup [2, \infty)$
16. [4.2] $\{x \mid x < 3 \ or \ x > 6\}$, or $(-\infty, 3) \cup (6, \infty)$
17. [4.2] $\left\{x \mid 4 \leqslant x < \frac{15}{2}\right\}$, or $\left[4, \frac{15}{2}\right)$
18. [4.3] $\{9, -9\}$
19. [4.3] $\{x \mid x < -3 \ or \ x > 3\}$, or $(-\infty, -3) \cup (3, \infty)$
20. [4.3] $\left\{x \mid -\frac{7}{8} < x < \frac{11}{8}\right\}$, or $\left(-\frac{7}{8}, \frac{11}{8}\right)$
21. [4.3] $\left\{x \mid x \leqslant -\frac{13}{5} \ or \ x \geqslant \frac{7}{5}\right\}$, or $\left(-\infty, -\frac{13}{5}\right] \cup \left[\frac{7}{5}, \infty\right)$
22. [4.3] $\{1\}$ **23.** [4.3] $\varnothing$
24. [4.3] $\{x \mid -99 \leqslant x \leqslant 111\}$, or $[-99, 111]$

25. [4.4]

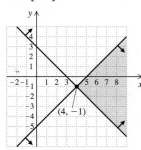

26. [4.4]

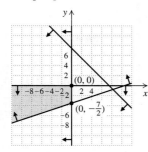

27. [4.5] Maximum 57 when $x = 6$, $y = 9$; minimum 5 when $x = 1$, $y = 0$ **28.** [4.5] 0 of A, 10 of B; 120
29. [1.3] $-\frac{33}{2}$ **30.** [3.2] $\left(-\frac{127}{8}, \frac{89}{8}\right)$
31. [2.3]

y or $f(x)$

$f(x) = \frac{3}{5}x - 3$

32. [3.3] 30-sec: 10; 60-sec: 5
33. [4.3] $\varnothing$
34. [4.2] $\left\{x \mid \frac{1}{5} < x < \frac{4}{5}\right\}$, or $\left(\frac{1}{5}, \frac{4}{5}\right)$

CHAPTER 5

Exercise Set 5.1, pp. 223–225

1. 4, 3, 2, 1, 0; 4   **3.** 3, 7, 6, 0; 7
5. 5, 6, 2, 1, 0; 6
7. $-4y^3 - 6y^2 + 7y + 23$; $-4y^3$; -4
9. $3x^7 + 5x^2 - x + 12$; $3x^7$; 3
11. $-a^7 + 8a^5 + 5a^3 - 19a^2 + a$; $-a^7$; -1
13. $12 + 4x - 5x^2 + 3x^4$
15. $3xy^3 + x^2y^2 - 9x^3y + 2x^4$
17. $-7ab + 4ax - 7ax^2 + 4x^6$
19. $P(4) = 54$, $P(0) = 2$
21. $P(-2) = -45$, $P\left(\frac{1}{3}\right) = -8\frac{19}{27}$ **23.** -18 **25.** 19
27. -12 **29.** 4   **31.** 11   **33.** About 449
35. 1024 ft   **37.** $18,750   **39.** $8375   **41.** 44.46 in²
43. $2x^2$ **45.** $a + 6$ **47.** $-6a^2b - 2b^2$
49. $9x^2 + 2xy + 15y^2$ **51.** $5x^2 + 2y^2 + 5$
53. $6a + b + c$ **55.** $-4a^2 - b^2 + 3c^2$
57. $-2x^2 + x - xy - 1$ **59.** $5x^2y - 4xy^2 + 5xy$
61. $9r^2 + 9r - 9$ **63.** $1.7x^2y - \frac{2}{15}xy + \frac{19}{12}xy^2$
65. $-(5x^3 - 7x^2 + 3x - 6)$, $-5x^3 + 7x^2 - 3x + 6$
67. $-(-12y^5 + 4ay^4 - 7by^2)$, $12y^5 - 4ay^4 + 7by^2$
69. $13x - 6$ **71.** $-4x^2 - 3x + 13$ **73.** $2a - 4b + 3c$
75. $-2x^2 + 6x$ **77.** $-4a^2 + 8ab - 5b^2$

79. $8a^2b + 16ab + 3ab^2$
81. $0.06y^4 + 0.032y^3 - 0.94y^2 + 0.93$ **83.** $x^4 - x^2 - 1$
85. $P(x) = 280x - x^2 - 5000$ **87.** $9700   **89.** $3y - 6$
91. ◈   **93.** $68x^5 - 81x^4 - 22x^3 + 52x^2 + 2x + 250$
95. $45x^5 - 8x^4 + 208x^3 - 176x^2 + 116x - 25$
97. $5x^2 - 8x$ **99.** $47x^{4a} + 40x^{3a} + 30x^{2a} + x^a + 4$
101. $x^{5b} + 4x^{4b} + x^{3b} - 6x^{2b} - 9x^b$

Exercise Set 5.2, pp. 232–233

1. $10y^3$ **3.** $-20x^3y$ **5.** $-10x^5y^6$ **7.** $6x - 2x^2$
9. $3a^2b + 3ab^2$ **11.** $15c^3d^2 - 25c^2d^3$
13. $6x^2 + x - 12$ **15.** $s^2 - 9t^2$ **17.** $x^2 - 2xy + y^2$
19. $2y^2 + 9xy - 56x^2$ **21.** $a^4 - 5a^2b^2 + 6b^4$
23. $x^3 - 64$ **25.** $x^3 + y^3$
27. $a^4 + 5a^3 - 2a^2 - 9a + 5$
29. $4a^3b^2 + 4a^3b - 10a^2b^2 - 2a^2b + 3ab^3 + 7ab^2 - 6b^3$
31. $x^2 - \frac{3}{4}x + \frac{1}{8}$ **33.** $3.25x^2 - 0.9xy - 28y^2$
35. $12x^4 - 21x^3 - 14x^2 + 35x - 10$ **37.** $a^2 + 5a + 6$
39. $y^2 + y - 6$ **41.** $x^2 + 6x + 9$ **43.** $x^2 - 4xy + 4y^2$
45. $b^2 - \frac{5}{6}b + \frac{1}{6}$ **47.** $2x^2 + 13x + 18$
49. $100a^2 - 2.4ab + 0.0144b^2$ **51.** $4x^2 - 4xy - 3y^2$
53. $4a^2 + \frac{4}{3}a + \frac{1}{9}$ **55.** $4x^6 - 12x^3y^2 + 9y^4$
57. $a^4b^4 + 2a^2b^2 + 1$ **59.** $400a^{10} - 6.4a^5b + 0.0256b^2$
61. $9x^2 - 24x + 16$ **63.** $c^2 - 4$ **65.** $4a^2 - 1$
67. $9m^2 - 4n^2$ **69.** $x^6 - y^2z^2$ **71.** $m^4 - m^2n^2$
73. $\frac{1}{4}p^8 - \frac{4}{9}q^2$ **75.** $x^4 - 1$ **77.** $a^4 - 2a^2b^2 + b^4$
79. $a^2 + 2ab + b^2 - 1$ **81.** $4x^2 + 12xy + 9y^2 - 16$
83. $A = P + 2Pi + Pi^2$ **85.** (a) $t^2 + 3t - 4$;
(b) $5h + 2ah + h^2$ **87.** 11   **89.** A: 75, B: 84, C: 63
91. ◈   **93.** $-\frac{4}{3}x^6y^{11}$ **95.** $-\frac{9}{4}r^{22}s^{14}$ **97.** z^{5n^5}
99. $16x^4 - 32x^3 + 16x^2$
101. $-a^4 - 2a^3b + 25a^2 + 2ab^3 - 25b^2 + b^4$
103. $r^8 - 2r^4s^4 + s^8$
105. $a^2 + 2ac - b^2 - 2bd + c^2 - d^2$
107. $10y^2 - 38xy + 24x^2 - 212y + 306x + 930$
109. $16x^4 + 4x^2y^2 + y^4$ **111.** $x^{4a} - y^{4b}$
113. $a^3 - b^3 + 3b^2 - 3b + 1$
115. $\frac{1}{81}x^{12} - \frac{8}{81}x^6y^4 + \frac{16}{81}y^8$ **117.** $x^{a^2 - b^2}$

Exercise Set 5.3, pp. 237–239

1. $2a(2a + 1)$ **3.** $y(y - 5)$ **5.** $y^2(y + 9)$
7. $3x^2(2 - x^2)$ **9.** $4xy(x - 3y)$ **11.** $3(y^2 - y - 3)$
13. $2a(2b - 3c + 6d)$ **15.** $5(2a^4 + 3a^2 - 5a - 6)$
17. $-3(x - 4)$ **19.** $-6(y + 12)$
21. $-2(x^2 - 2x + 6)$ **23.** $-3(y^2 - 8x)$
25. $-3y(y^2 - 4y + 5)$ **27.** $-(x^2 - 3x + 7)$
29. $-a(a^3 - 2a^2 + 13)$ **31.** (a) $h(t) = -16t(t - 5)$;
(b) $h(3) = 96$ **33.** $N(x) = \frac{1}{6}x(x^2 + 3x + 2)$
35. $P(n) = \frac{1}{2}n(n - 3)$ **37.** $R(x) = 0.4x(700 - x)$
39. $(b - 2)(a + c)$ **41.** $(x - 2)(2x + 13)$ **43.** 0
45. $(c + d)(a + b)$ **47.** $(b - 1)(b^2 + 2)$
49. $(a - 3)(a^2 - 2)$ **51.** $12x(2x^2 - 3x + 6)$

53. $x^3(x - 1)(x^2 + 1)$ **55.** $(y^2 + 3)(2y^2 + 5)$
57. 40 nickels, 26 dimes, 54 quarters **59.** ◈
61. $P(x) = x(x(x(x - 3) - 5) + 4) + 2$
63. $P(x) = x(x(x(x(5x + 0) - 3) + 4) - 5) + 1$
65. $P(x) = x(x(x(2x - 3) + 5) + 6) - 4$; $P(5) = 1026$,
$P(-2) = 60$, $P(10) = 17{,}556$ **67.** $2(y^{2a} + 3)(2y^{2a} + 5)$
69. $x^a(4x^b + 7x^{-b})$

Exercise Set 5.4, pp. 246–247

1. $(x + 5)(x + 4)$ **3.** $(t - 5)(t - 3)$ **5.** $(x - 9)(x + 3)$
7. $2(y - 4)(y - 4)$ **9.** $p(p + 9)(p - 6)$
11. $(x + 9)(x + 5)$ **13.** $(y + 9)(y - 7)$
15. $(t - 7)(t - 4)$ **17.** $(x + 5)(x - 2)$
19. $(x + 2)(x + 3)$ **21.** $(8 - x)(7 + x)$
23. $y(8 - y)(4 + y)$ **25.** $(x^2 + 16)(x^2 - 5)$
27. Not factorable using integers **29.** $(x + 9y)(x + 3y)$
31. $(x - 7y)(x - 7y)$ **33.** $(x^2 + 1)(x^2 + 49)$
35. $(x^3 - 7)(x^3 + 9)$ **37.** $(3x + 2)(x - 6)$
39. $x(3x - 5)(2x + 3)$ **41.** $(3a - 4)(a - 2)$
43. $(5\bar{y} + 2)(7y + 4)$ **45.** $2(5t - 3)(t + 1)$
47. $4(2x + 1)(x - 4)$ **49.** $x(3x - 4)(4x - 5)$
51. $x^2(7x + 1)(2x - 3)$ **53.** $(3a - 4)(a + 1)$
55. $(3x + 1)(3x + 4)$ **57.** $(1 + 12z)(3 - z)$
59. $-2(2t - 3)(2t + 5)$ **61.** $x(3x + 1)(x - 2)$
63. $(24x + 1)(x - 2)$ **65.** $3x(7x + 3)(3x + 4)$
67. $4(10x^4 + 4x^2 - 3)$ **69.** $(4a - 3b)(3a - 2b)$
71. $(2x - 3y)(x + 2y)$ **73.** $(2x - 7y)(3x - 4y)$
75. $(3x - 5y)(3x - 5y)$ **77.** $(3x^3 + 2)(2x^3 - 1)$
79. $h(0) = 224$ ft; $h(1) = 288$ ft; $h(3) = 320$ ft;
$h(4) = 288$ ft; $h(6) = 128$ ft
81. ◈ **83.** $(pq + 4)(pq + 3)$ **85.** $\left(x + \frac{4}{5}\right)\left(x - \frac{1}{5}\right)$
87. $(y - 0.1)(y + 0.5)$ **89.** $(7ab + 6)(ab + 1)$
91. $6x(x + 9)(x + 4)$ **93.** $(2x^a + 1)(2x^a - 3)$
95. $(bx + a)(dx + c)$ **97.** $a(2r + s)(r + s)$
99. $[3(x - 7) - 1][2(x - 7) + 5]$, or $(3x - 22)(2x - 9)$
101. $31, -31, 14, -14, 4, -4$ **103.** Since
$ax^2 + bx + c = (mx + r)(nx + s)$, from FOIL we know
that $a = mn$, $c = rs$, and $b = ms + rn$. If $P = ms$ and
$Q = rn$, then $b = P + Q$. Since $ac = mnrs = msrn$, we
have $ac = PQ$.

Exercise Set 5.5, pp. 251–252

1. $(y - 3)^2$ **3.** $(x + 7)^2$ **5.** $(x + 1)^2$ **7.** $2(a + 2)^2$
9. $(y - 6)^2$ **11.** $y(y - 9)^2$ **13.** $3(2a + 3)^2$
15. $2(x - 10)^2$ **17.** $(1 - 4d)^2$ **19.** $(y^2 + 4)^2$
21. $(0.5x + 0.3)^2$ **23.** $(p - q)^2$ **25.** $(a + 2b)^2$
27. $(5a - 3b)^2$ **29.** $(x^2 + y^2)^2$ **31.** $(x + 4)(x - 4)$
33. $(p + 7)(p - 7)$ **35.** $(pq + 5)(pq - 5)$
37. $6(x + y)(x - y)$ **39.** $4x(y^2 + z^2)(y + z)(y - z)$
41. $a(2a + 7)(2a - 7)$
43. $3(x^4 + y^4)(x^2 + y^2)(x + y)(x - y)$
45. $a^2(3a + 5b^2)(3a - 5b^2)$ **47.** $\left(\frac{1}{5} - x\right)\left(\frac{1}{5} + x\right)$

49. $(0.2x + 0.3y)(0.2x - 0.3y)$
51. $(m - 7)(m + 2)(m - 2)$ **53.** $(a - 2)(a + b)(a - b)$
55. $(a + b + 10)(a + b - 10)$
57. $(a + b + 3)(a + b - 3)$
59. $(r - 1 - 2s)(r - 1 + 2s)$
61. $2(5a + m + n)(5a - m - n)$
63. $(3 - a + b)(3 + a - b)$ **65.** $(2, -1, 3)$
67. $\left\{x | -\frac{4}{7} \le x \le 2\right\}$, or $\left[-\frac{4}{7}, 2\right]$ **69.** ◈
71. $x(x - 15)(x + 15)$ **73.** $3x(y - 25)^2$
75. $12(x - 3y)^2$ **77.** $-\frac{1}{54}(4r + 3s)^2$ **79.** $(4a - 3b)^2$
81. $(0.3x + 0.8)^2$ **83.** $(x^{2a} + y^b)(x^{2a} - y^b)$
85. $(5y^a - x^b + 1)(5y^a + x^b - 1)$ **87.** $3(x + 3)^2$
89. $(3x^n - 1)^2$
91. $P(a + h) - P(a) = (a + h)^2 - a^2$
$= (a + h - a)(a + h + a)$

Exercise Set 5.6, p. 255

1. $(x + 2)(x^2 - 2x + 4)$ **3.** $(y - 4)(y^2 + 4y + 16)$
5. $(w + 1)(w^2 - w + 1)$ **7.** $(2a + 1)(4a^2 - 2a + 1)$
9. $(y - 2)(y^2 + 2y + 4)$ **11.** $(2 - 3b)(4 + 6b + 9b^2)$
13. $(4y + 1)(16y^2 - 4y + 1)$
15. $(2x + 3)(4x^2 - 6x + 9)$ **17.** $(a - b)(a^2 + ab + b^2)$
19. $\left(a + \frac{1}{2}\right)\left(a^2 - \frac{1}{2}a + \frac{1}{4}\right)$ **21.** $2(y - 4)(y^2 + 4y + 16)$
23. $3(2a + 1)(4a^2 - 2a + 1)$
25. $r(s + 4)(s^2 - 4s + 16)$
27. $5(x - 2z)(x^2 + 2xz + 4z^2)$
29. $(x + 0.1)(x^2 - 0.1x + 0.01)$
31. $8(2x^2 - t^2)(4x^4 + 2x^2t^2 + t^4)$
33. $2y(y - 4)(y^2 + 4y + 16)$
35. $(z + 1)(z^2 - z + 1)(z - 1)(z^2 + z + 1)$
37. $(t^2 + 4y^2)(t^4 - 4t^2y^2 + 16y^4)$ **39.** 228 ft^2
41. $\left\{x | -\frac{33}{5} \le x \le 9\right\}$, or $\left[-\frac{33}{5}, 9\right]$ **43.** ◈
45. $(x^{2a} + y^b)(x^{4a} - x^{2a}y^b + y^{2b})$
47. $3(x^a + 2y^b)(x^{2a} - 2x^ay^b + 4y^{2b})$
49. $\frac{1}{3}\left(\frac{1}{2}xy + z\right)\left(\frac{1}{4}x^2y^2 - \frac{1}{2}xyz + z^2\right)$
51. $7\left(x + \frac{1}{2}\right)\left(x^2 - \frac{1}{2}x + \frac{1}{4}\right)$ **53.** $y(3x^2 + 3xy + y^2)$
55. $4(3a^2 + 4)$
57. $P(a + h) - P(a)$
$= (a + h)^3 - a^3$
$= (a + h - a)((a + h)^2 + a(a + h) + a^2)$
$= h(3a^2 + 3ah + h^2)$

59.

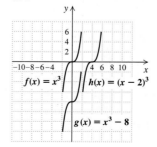

Exercise Set 5.7, pp. 258–259

1. $(x + 12)(x - 12)$ **3.** $(2x + 3)(x + 4)$
5. $3(x^2 + 2)(x^2 - 2)$ **7.** $(a + 5)^2$ **9.** $2(x - 11)(x + 6)$
11. $(3x + 5y)(3x - 5y)$
13. $(m + 1)(m^2 - m + 1)(m - 1)(m^2 + m + 1)$
15. $(x + y + 3)(x - y + 3)$
17. $2(5x - 4y)(25x^2 + 20xy + 16y^2)$
19. $(m^3 + 10)(m^3 - 2)$ **21.** $(a + d)(c - b)$
23. $(2c - d)^2$ **25.** $(2x - 7)(x^2 + 2)$
27. $2(x + 3)(x + 2)(x - 2)$
29. $2(2x + 3y)(4x^2 - 6xy + 9y^2)$ **31.** $(6y - 5)(6y + 7)$
33. $(a^4 + b^4)(a^2 + b^2)(a + b)(a - b)$
35. $ab(a + 4b)(a - 4b)$ **37.** $(b - 2)(a + c)$
39. $7a(a^3 - 2a^2 + 3a - 1)$ **41.** $(9ab + 2)(3ab + 4)$
43. $y(2y - 5)(4y^2 + 10y + 25)$
45. $(a - b - 3)(a + b + 3)$ **47.** 55 correct, 20 wrong
49. ◈   **51.** $(2x - 3)(10x + 13)$
53. $4(2a - 1)(9a + 40)$ **55.** $7\left(x - \frac{1}{2}\right)\left(x^2 + \frac{1}{2}x + \frac{1}{4}\right)$
57. $-b(3a^2 + 3ab + b^2)$
59. $(x + 1)(x - 1)(x + 7)(x - 7)$
61. $(3x^{2s} + 4y^t)(9x^{4s} - 12x^{2s}y^t + 16y^{2t})$
63. $(2x + y - r + 3s)(2x + y + r - 3s)$ **65.** $c(c^w - 1)^2$
67. $y(y^4 + 1)(y^2 + 1)(y + 1)(y - 1)$
69. $3(a + b - c - d)(a + b + c + d)$ **71.** $2m(m^2 + 3)$
73. 0

Technology Connection, Section 5.8

TC1. $x = -2.45, 1.31$ **TC2.** $x = 7.30$
TC3. No solution   **TC4.** $x = -8.98, -4.56$
TC5. $x = -1.21, 3.45, 8.98$
TC6. $x = -1.50, 3.97, 4.43$
TC7. $x = -3.68, 0.97$ **TC8.** $x = 0.51, 4.08$

Exercise Set 5.8, pp. 267–270

1. $\{-7, 4\}$ **3.** $\{4\}$ **5.** $\{6\}$ **7.** $\{-5, -4\}$
9. $\{0, -8\}$ **11.** $\{-3, 3\}$ **13.** $\{-6, 6\}$
15. $\{-5, -9\}$ **17.** $\{-9, 7\}$ **19.** $\{7, 4\}$ **21.** $\{8, -4\}$
23. $\left\{-\frac{2}{3}, -2\right\}$ **25.** $\left\{\frac{3}{4}, \frac{1}{2}\right\}$ **27.** $\{0, 6\}$ **29.** $\left\{-\frac{3}{4}, \frac{2}{3}\right\}$
31. $\{-2, 2\}$ **33.** $\left\{\frac{1}{2}, 7\right\}$ **35.** $\left\{-\frac{5}{7}, \frac{2}{3}\right\}$ **37.** $\left\{0, \frac{1}{5}\right\}$
39. $\left\{-\frac{1}{5}, \frac{1}{5}\right\}$ **41.** $\{0, 3, -2\}$ **43.** -12 or 11
45. Length is 12 cm; width is 7 cm   **47.** 3 cm
49. 3 cm   **51.** Length is 8 m; width is 3 m   **53.** 20 ft
55. The height of the antenna tower is 16 ft and the distance d is 12 ft.
57. The height is 7 m and the base is 16 m.
59. $-10, -8$, and -6; 6, 8, and 10   **61.** 41 ft
63. Length is 100 m; width is 75 m   **65.** 9 and 11
67. 7 sec   **69.** 3
71. Faster car: 54 mph; slower car: 39 mph
73. $\left\{x | x > \frac{10}{3}\right\}$, or $\left(\frac{10}{3}, \infty\right)$ **75.** $\left\{-\frac{5}{3}, 5, 4\right\}$

77. $\{2, -2\}$ **79.** $\{3, -3, -2\}$ **81.** 4 cm
83. Tugboat: 15 km/h; freighter: 8 km/h
85. $\{-1.91, 4.32\}$ **87.** $\{6.90\}$ **89.** $\{3.48\}$

Review Exercises: Chapter 5, pp. 271–272

1. [5.1] 0; -6 **2.** [5.1] 4   **3.** [5.1] 7, 11, 3, 2; 11
4. [5.1] $-5x^3$; -5
5. [5.1] $-3x^2 + 2x^3 + 3x^6y - 7x^8y^3$
6. [5.1] $-7x^8y^3 + 3x^6y + 2x^3 - 3x^2$
7. [5.1] $-x^2y - 2xy^2$ **8.** [5.1] $ab + 4 + 12ab^2$
9. [5.1] $-x^3 + 2x^2 + 5x + 2$
10. [5.1] $-3x^4 + 3x^3 - x + 16$
11. [5.1] $x^3 + 6x^2 - x - 4$
12. [5.1] $-8xy^2 + 4xy + 13x^2y$ **13.** [5.1] $9x - 7$
14. [5.1] $-2a + 6b + 7c$ **15.** [5.1] $16p^2 - 8p$
16. [5.1] $6x^2 - 7xy + 3y^2$ **17.** [5.2] $-18x^3y^4$
18. [5.2] $x^8 - x^6 + 5x^2 - 3$
19. [5.2] $8a^2b^2 + 2abc - 3c^2$ **20.** [5.2] $4x^2 - 25y^2$
21. [5.2] $4x^2 - 20xy + 25y^2$
22. [5.2] $20x^4 - 18x^3 - 47x^2 + 69x - 27$
23. [5.2] $x^4 + 8x^2y^3 + 16y^6$ **24.** [5.2] $x^3 - 125$
25. [5.2] $x^2 - \frac{1}{2}x + \frac{1}{18}$ **26.** [5.3] $x(6x + 5)$
27. [5.3] $3y^2(3y^2 - 1)$
28. [5.3] $3x(5x^3 - 6x^2 + 7x - 3)$
29. [5.4] $(a - 9)(a - 3)$ **30.** [5.4] $(3m + 2)(m + 4)$
31. [5.5] $(5x + 2)^2$ **32.** [5.7] $4(y + 2)(y - 2)$
33. [5.5] $(a + 9)(a - 9)$ **34.** [5.3] $(a + 2b)(x - y)$
35. [5.3] $(y + 2)(3y^2 - 5)$
36. [5.7]$(a^2 + 9)(a + 3)(a - 3)$
37. [5.3] $4(x^4 + x^2 + 5)$
38. [5.6] $(3x - 2)(9x^2 + 6x + 4)$
39. [5.6] $(0.4b - 0.5c)(0.16b^2 + 0.2bc + 0.25c^2)$
40. [5.7] $y(y^2 + 1)(y + 1)(y - 1)$ **41.** [5.3] $2z^6(z^2 - 8)$
42. [5.7] $2y(3x^2 - 1)(9x^4 + 3x^2 + 1)$
43. [5.6] $(1 + a)(1 - a + a^2)$ **44.** [5.5] $4(3x - 5)^2$
45. [5.4] $(3t + p)(2t + 5p)$
46. [5.7] $(x + 3)(x - 3)(x + 2)$
47. [5.5] $(a - b + 2t)(a - b - 2t)$ **48.** [5.8] $\{10\}$
49. [5.8] $\left\{\frac{2}{3}, \frac{3}{2}\right\}$ **50.** [5.8] $\left\{\frac{5}{4}, \frac{1}{2}\right\}$ **51.** [5.8] 5
52. [5.8] 3, 5, 7; $-7, -5, -3$ **53.** [5.8] 5 in. $\times$ 8 in.
54. [5.2] $a^2 + 2ah + h^2 + 3a + 3h$
55. [3.4] $(0, -2, 7)$ **56.** [4.1] $\left\{x | x > -\frac{7}{3}\right\}$, or $\left(-\frac{7}{3}, \infty\right)$
57. [4.3] $\left\{x | -\frac{4}{3} \leq x \leq 8\right\}$, or $\left[-\frac{4}{3}, 8\right]$
58. [4.3] $\left\{x | x \leq -\frac{4}{3} \text{ or } x \geq 8\right\}$, or $\left(-\infty, -\frac{4}{3}\right] \cup [8, \infty)$
59. [3.5] A: 112, B: 90, C: 85
60. [5.8] ◈ The roots of a polynomial function are the x-coordinates of the points at which the graph of the function crosses the x-axis.
61. [5.8] ◈ The principle of zero products states that if a product is equal to 0, at least one of the factors must be 0. If a product is nonzero, we cannot conclude that any one of the factors is a particular value.

62. [5.6], [5.7] $2(2x - y)(4x^2 + 2xy + y^2)(2x + y) \times$
$(4x^2 - 2xy + y^2)$ **63.** [5.6] $-2(3x^2 + 1)$
64. [5.2], [5.6] $a^3 - b^3 + 3b^2 - 3b + 1$ **65.** [5.2] z^{5n^5}
66. [5.8] $\{0, \frac{1}{8}, -\frac{1}{8}\}$

Test: Chapter 5, p. 273

1. [5.1] 4; 2 **2.** [5.2] $2ah + h^2 - 5h$ **3.** [5.1] 9
4. [5.1] $5x^5y^4 - 2x^4y - 4x^2y + 3xy^3$ **5.** [5.1] $-4a^3$
6. [5.1] $3xy + 3xy^2$ **7.** [5.1] $-3x^3 + 3x^2 - 6y - 7y^2$
8. [5.1] $7a^3 - 6a^2 + 3a - 3$
9. [5.1] $7m^3 + 2m^2n + 3mn^2 - 7n^3$ **10.** [5.1] $6a - 8b$
11. [5.1] $7x^2 - 7x + 13$ **12.** [5.1] $2y^2 + 5y + y^3$
13. [5.2] $64x^3y^3$ **14.** [5.2] $12a^2 - 4ab - 5b^2$
15. [5.2] $x^3 - 2x^2y + y^3$
16. [5.2] $-3m^4 - 13m^3 + 5m^2 + 26m - 10$
17. [5.2] $16y^2 - 72y + 81$ **18.** [5.2] $x^2 - 4y^2$
19. [5.3] $x(9x + 7)$ **20.** [5.3] $8y^2(3y + 2)$
21. [5.7] $(y + 5)(y + 2)(y - 2)$
22. [5.4] $(p - 14)(p + 2)$ **23.** [5.4] $(6m + 1)(2m + 3)$
24. [5.5] $(3y + 5)(3y - 5)$
25. [5.7] $3(r - 1)(r^2 + r + 1)$ **26.** [5.5] $(3x - 5)^2$
27. [5.5] $(z + 1 + b)(z + 1 - b)$
28. [5.5] $(x^4 + y^4)(x^2 + y^2)(x + y)(x - y)$
29. [5.5] $(y + 4 + 10t)(y + 4 - 10t)$
30. [5.7] $5(2a - b)(2a + b)$
31. [5.7] $2(4x - 1)(3x - 5)$
32. [5.7] $2ab(2a^2 + 3b^2)(4a^4 - 6a^2b^2 + 9b^4)$
33. [5.8] $\{6, -3\}$ **34.** [5.8] $\{-5, 5\}$
35. [5.8] $\{-\frac{3}{2}, -7\}$ **36.** [5.8] $l = 8$ cm; $w = 5$ cm
37. [4.3] $\{x | -6 < x < \frac{2}{3}\}$, or $\left(-6, \frac{2}{3}\right)$
38. [4.3] $\{x | x < -6 \text{ or } x > \frac{2}{3}\}$, or $(-\infty, -6) \cup \left(\frac{2}{3}, \infty\right)$
39. [4.1] $\{x | x > 1\}$, or $(1, \infty)$ **40.** [3.4] $(5, -1, -2)$
41. [3.5] 31 multiple-choice, 26 true–false, 13 fill-in
42. (a) [5.2] $x^5 + x + 1$;
(b) [5.2], [5.7] $(x^2 + x + 1)(x^3 - x^2 + 1)$
43. [5.4] $(3x^n + 4)(2x^n - 5)$

CHAPTER 6

Technology Connection, Section 6.1

TC1. $\{x | x \neq -3.50 \text{ and } x \neq 3.50\}$
TC2. $\{x | x \neq -3.10 \text{ and } x \neq 4.70\}$
TC3. $\{x | x \neq 1.76 \text{ and } x \neq 2.24\}$
TC4. $\{x | x \neq 2.17\}$

Exercise Set 6.1, pp. 283–285

1. $v(0) = \frac{2}{3}$, $v(3) = \frac{23}{6}$, $v(7) = \frac{163}{10}$
3. $r(0) = 0$, $r(4) = -184$, $r(5)$ does not exist
5. $g(0) = -\frac{9}{4}$, $g(2)$ does not exist, $g(-1) = -\frac{11}{9}$
7. $f(-3) = 0$, $f(0) = \frac{9}{5}$, $f(1)$ does not exist
9. $\{x | x$ is a real number and $x \neq 3\}$

11. $\{t | t$ is a real number and $t \neq 0$ and $t \neq 4\}$
13. $\{x | x$ is a real number and $x \neq 2$ and $x \neq -2\}$
15. $\{x | x$ is a real number and $x \neq 2$ and $x \neq 6\}$
17. $\dfrac{3x(x + 1)}{3x(x + 3)}$ **19.** $\dfrac{(t - 3)(t + 3)}{(t + 2)(t + 3)}$ **21.** $\dfrac{(x^2 - 3)(x + 6)}{(x - 6)(x + 6)}$
23. $\dfrac{3y}{5}$ **25.** $\dfrac{2}{t^4}$ **27.** $a - 3$ **29.** $\dfrac{2x - 3}{4}$ **31.** $\dfrac{y - 3}{y + 3}$
33. $\dfrac{6}{5}$ **35.** $-\dfrac{6}{5}$ **37.** $\dfrac{t + 4}{t - 4}$ **39.** $\dfrac{x + 8}{x - 4}$ **41.** $\dfrac{4 + t}{4 - t}$
43. $\dfrac{3t^3x}{5}$ **45.** $\dfrac{3x^2}{25}$ **47.** $\dfrac{y + 4}{2}$ **49.** $\dfrac{(x + 4)(x - 4)}{x(x + 3)}$
51. $\dfrac{-2}{t^4(t + 3)(t + 2)}$ **53.** $\dfrac{(x + 5)(2x + 3)}{7x}$ **55.** $c(c - 2)$
57. $\dfrac{a^2 + ab + b^2}{3(a + 2b)}$ **59.** $\dfrac{1}{2x + 3y}$ **61.** $\dfrac{4a^4}{b^4}$ **63.** $\dfrac{3}{y^5}$
65. $\dfrac{(y - 3)(y + 2)}{y^6}$ **67.** $\dfrac{2a + 1}{a + 2}$ **69.** $-x^2$
71. $\dfrac{(x + 4)(x + 2)}{3(x - 5)}$ **73.** $\dfrac{y(y^2 + 3)}{(y + 3)(y - 2)}$
75. $\dfrac{x^2 + 4x + 16}{(x + 4)^2}$ **77.** $\dfrac{(2a + b)^2}{2(a + b)}$ **79.** $\left(\dfrac{7}{2}, \dfrac{5}{2}\right)$
81. $\dfrac{16}{3}$ **83.** ◈

85.
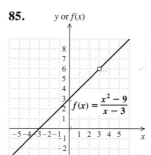

87. $\dfrac{2s}{r + 2s}$
89. $\dfrac{x - 3}{(x + 1)(x + 3)}$
91. $\dfrac{m - t}{m + t + 1}$
93. $\dfrac{x^2 + xy + y^2 + x + y}{x - y}$
95. $\dfrac{-2x}{x - 1}$
97. $\{x | x$ is a real number and $x \neq -8.70$ and $x \neq 0\}$
99. $\{x | x$ is a real number and $x \neq -1.25$ and $x \neq 5.32\}$

Exercise Set 6.2, pp. 291–293

1. $\dfrac{4}{a}$ **3.** $-\dfrac{1}{a^2b}$ **5.** 2 **7.** $\dfrac{3y + 5}{y - 2}$ **9.** $\dfrac{1}{x - 4}$
11. $\dfrac{1}{a + 3}$ **13.** $a + b$ **15.** $\dfrac{11}{x}$ **17.** $\dfrac{1}{x + 5}$
19. $\dfrac{1}{t^2 + 4}$ **21.** $\dfrac{1}{m^2 + mn + n^2}$ **23.** $\dfrac{2y^2 + 22}{(y - 5)(y + 4)}$
25. $\dfrac{3x - 1}{x + 1}$ **27.** $\dfrac{x + y}{x - y}$ **29.** $\dfrac{3x - 4}{(x - 2)(x - 1)}$

31. $\dfrac{8x+1}{(x+1)(x-1)}$ **33.** $\dfrac{2x-14}{15(x+5)}$

35. $\dfrac{-a^2+7ab-b^2}{(a-b)(a+b)}$ **37.** $\dfrac{x-5}{(x+5)(x+3)}$

39. $\dfrac{y}{(y-2)(y-3)}$ **41.** $\dfrac{3y-10}{(y+4)(y-5)}$

43. $\dfrac{3y^2-3y-29}{(y-3)(y+8)(y-4)}$ **45.** $\dfrac{2x^2-13x+7}{(x+3)(x-1)(x-3)}$

47. $\dfrac{10a-16}{(a^2-5a+4)(a^2-4)}$ **49.** $\dfrac{4t^2-2t-14}{(t+2)(t-2)}$ **51.** 0

53. $\dfrac{-3x^2-3x-4}{(x-1)(x+1)}$ **55.** $\dfrac{4t+26}{(t+2)(t+3)(t-4)(t+1)}$

57. $\dfrac{3y^6z^7}{7x^5}$ **59.** 2 rolls of dimes, 5 rolls of nickels,

5 rolls of quarters   **61.** ◈
63. $x^4(x^2+1)(x+1)(x-1)(x^2+x+1)(x^2-x+1)$
65. $8a^4,\ 8a^4b,\ 8a^4b^2,\ 8a^4b^3,\ 8a^4b^4,\ 8a^4b^5,\ 8a^4b^6,\ 8a^4b^7$

67. 420 years   **69.** $\dfrac{x^4+4x^3-2x^2}{(x+2)(x-2)(x+5)}$

71. $\dfrac{x(x+5)}{x+2}$, $x\ne 0$, $x\ne -5$, $x\ne 2$

73. $\dfrac{2y^2+3-7x^3y}{x^2y^2}$ **75.** $\dfrac{5y+23}{5-2y}$ **77.** $\dfrac{3}{x+8}$

79. $\dfrac{3t^2(t+3)}{-2t^2+13t-7}$

Exercise Set 6.3, pp. 299–301

1. $\dfrac{1+4x}{1-3x}$ **3.** $\dfrac{x^2-1}{x^2+1}$ **5.** $\dfrac{3y+4x}{4y-3x}$ **7.** $\dfrac{x+y}{x}$

9. $\dfrac{a^2(b-3)}{b^2(a-1)}$ **11.** $\dfrac{1}{a-b}$ **13.** $-\dfrac{1}{x(x+h)}$ **15.** $\dfrac{y-3}{y+5}$

17. $\dfrac{4x-7}{7x-9}$ **19.** $\dfrac{a^2-3a-6}{a^2-2a-3}$ **21.** $\dfrac{x+2}{x+3}$ **23.** $\dfrac{1}{y+3}$

25. $\dfrac{a+1}{2a+5}$ **27.** $\dfrac{-1-3x}{8-2x}$ **29.** $-y$

31. $\dfrac{2(5a^2+4a+12)}{5(a^2+6a+18)}$ **33.** $\dfrac{3x^2(x+2y)}{y(x^2+3xy+3y^2)}$

35. $\dfrac{2(x-2)(x+1)}{(2x-1)(x+3)}$ **37.** $\dfrac{-y-1}{y+8}$ **39.** $\dfrac{2(a+1)^2}{(a+3)(2a-1)}$

41. $\dfrac{x^2+21x+8}{x^2+3x-34}$ **43.** $x=\dfrac{a}{b}-y$

45. Perimeter of the first rectangle is 15; perimeter of the second rectangle is 29

47. ◈   **49.** $\dfrac{5}{6}$ **51.** $\dfrac{b-a}{ab}$ **53.** $\dfrac{3x+2}{5x+3}$

55. $\dfrac{-3(2x+1)}{x^2(x+h)^2}$ **57.** $\dfrac{1}{(1-x-h)(1-x)}$

59. $\{x|x$ is a real number and $x\ne 0$ and $x\ne -2$ and $x\ne 2\}$ **61.** ◼

Technology Connection, Section 6.4

TC1. x-coordinates of points of intersection should approximate the solutions of Examples 1–5.

Exercise Set 6.4, pp. 305–306

1. $\frac{51}{2}$ **3.** 144   **5.** -2 **7.** 5   **9.** No solution
11. 2   **13.** $-5,-1$ **15.** $2,-\frac{3}{2}$ **17.** No solution
19. 2   **21.** $\frac{17}{4}$ **23.** 11   **25.** $-\frac{10}{3}$ **27.** $\frac{3}{5}$ **29.** 5
31. $3,2$ **33.** -145 **35.** -3 **37.** $-6,5$
39. No solution   **41.** $(3x-y)(3x+y)(9x^2+y^2)$
43. 25 multiple-choice, 30 true–false, 15 fill-in
45. ◈   **47.** $\frac{1}{5}$
49. All real numbers except 1 and -1 are solutions.
51. 0.9465556   **53.** Identity   **55.** ◼

Exercise Set 6.5, pp. 313–315

1. $\frac{35}{12}$ **3.** $-3,-2$ **5.** 8 and 9, -9 and -8
7. $3\frac{3}{14}$ hr   **9.** $8\frac{4}{7}$ hr   **11.** $3\frac{9}{52}$ hr
13. 10 days for Claudia; 40 days for Jan   **15.** $1\frac{1}{5}$ hr
17. 6 hr for Jake; 12 hr for Skyler   **19.** 5 hr
21. $22\frac{1}{2}$ hr for the new machine; 45 hr for the old machine   **23.** 7 mph   **25.** $10\frac{3}{13}$ ft/sec
27. Rosanna: $3\frac{1}{3}$ mph; Simone: $5\frac{1}{3}$ mph
29. A train: 46 mph; B train: 58 mph
31. Steamboat: 30 km/h; freighter: 20 km/h
33. 9 km/h   **35.** 2 km/h   **37.** 20 mph   **39.** 20 mph
41. 6000   **43.** $-7, 11$ **45.** ◈
47. Speed of boat: 14 km/h; speed of stream: 10 km/h
49. $3\frac{3}{4}$ km/h   **51.** $49\frac{1}{2}$ hr
53. $21\frac{9}{11}$ minutes after 4:00   **55.** 48 km/h

Exercise Set 6.6, pp. 319–320

1. $6x^4-3x^2+8$ **3.** $-2a^2+4a-3$
5. y^3-2y^2+3y **7.** $-6x^5+3x^3+2x$
9. $1-ab^2-a^3b^4$ **11.** $-2pq+3p-4q$

13. $x+7$ **15.** $a-12+\dfrac{32}{a+4}$

17. $x-6+\dfrac{-7}{x-5}$ **19.** $y-5$

21. $y^2-2y-1+\dfrac{-8}{y-2}$

23. $2x^2-x+1+\dfrac{-5}{x+2}$

25. $a^2 + 4a + 15 + \dfrac{72}{a-4}$ **27.** $4x^2 - 6x + 9$

29. $x^2 + 6$ **31.** $x^3 + x^2 - 1 + \dfrac{1}{x-1}$

33. $2y^2 + 2y - 1 + \dfrac{8}{5y-2}$

35. $2x^2 - x - 9 + \dfrac{3x+12}{x^2+2}$

37. $2x^3 + x^2 - 1 + \dfrac{x+1}{x^2+1}$ **39.** 0, 5

41. 12, 13, and 14 **43.** ◈ **45.** $x^2 + 2y$

47. $x^3 + x^2y + xy^2 + y^3$ **49.** $\dfrac{14}{3}$

51. **(a)** $f(x) = 3 + \dfrac{1}{x+2}$;

(b), (c) The graph of f looks like the graph of g, shifted up 3 units. The graph of g looks like the graph of h, shifted to the left 2 units.

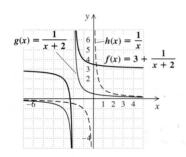

Exercise Set 6.7, pp. 323–324

1. $x^2 - x + 1$, R -4, or $x^2 - x + 1 + \dfrac{-4}{x-1}$

3. $a + 7$, R -47, or $a + 7 + \dfrac{-47}{a+4}$

5. $x^2 - 5x - 23$, R -43, or $x^2 - 5x - 23 + \dfrac{-43}{x-2}$

7. $3x^2 - 2x + 2$, R -3, or $3x^2 - 2x + 2 + \dfrac{-3}{x+3}$

9. $y^2 + 2y + 1$, R 12, or $y^2 + 2y + 1 + \dfrac{12}{y-2}$

11. $3x^3 + 9x^2 + 2x + 6$ **13.** $x^2 + 3x + 9$
15. $y^4 + y^3 + y^2 + y + 1$
17. $3x^3 + 2x^2 - 2x - 3$, R 2, or

$3x^3 + 2x^2 - 2x - 3 + \dfrac{2}{x+2}$

19. $3x^2 + 6x - 3$, R 2, or $3x^2 + 6x - 3 + \dfrac{2}{x+\frac{1}{3}}$

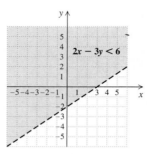

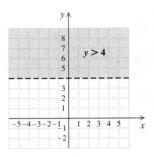

25. ◈
27. **(a)** Remainder is 0; **(b)** If $f(x) = (x-4) \cdot p(x)$ for some polynomial $p(x)$, then $f(4) = (4-4)\,p(4) = 0$; **(c)** $4^3 - 5(4)^2 + 5(4) - 4 = 0$

Exercise Set 6.8, pp. 327–328

1. $d_1 = \dfrac{d_2 W_1}{W_2}$ **3.** $t = \dfrac{2s}{v_1 + v_2}$ **5.** $r_1 = \dfrac{Rr_2}{r_2 - R}$

7. $s = \dfrac{Rg}{g-R}$ **9.** $r = \dfrac{2V - IR}{2I}$ **11.** $q = \dfrac{fp}{p-f}$

13. $r = \dfrac{nE - IR}{In}$ **15.** $H = m(t_1 - t_2)S$

17. $e = \dfrac{rE}{R+r}$ **19.** $a = \dfrac{S - Sr}{1 - r^n}$ **21.** 12

23. $5\frac{5}{9}$ ohms **25.** $t = \dfrac{uv}{v+u}$

27. $t_2 = \dfrac{d_2 - d_1}{v} + t_1$ **29.** 7%

31. $T = -\dfrac{I_f}{I_t} + 1$ **33.** $h = \dfrac{2gR^2}{V^2} - R$ **35.** 3

37.

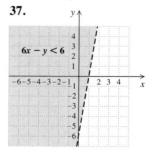

39. $(t + 2b)(t^2 - 2tb + 4b^2)$
41. 4527 miles **43.** $\{pq,\ 2pq\}$

Review Exercises: Chapter 6, pp. 330–331

1. [6.1] $\{x | x$ is a real number and $x \neq 6$ and $x \neq -1\}$
2. [6.1] **(a)** $-\frac{15}{8}$; **(b)** 0; **(c)** 12 **3.** [6.2] $48x^3$
4. [6.2] $(x+5)(x-2)(x-4)$ **5.** [6.2] $x^2 - 2x + 4$

6. [6.2] $\dfrac{1}{x-4}$ **7.** [6.1] $\dfrac{b^2 c^6 d^2}{a^5}$ **8.** [6.2] $\dfrac{3np + 4m}{18m^2 n^4 p^2}$

9. [6.1] $\dfrac{y-8}{2}$ **10.** [6.1] $\dfrac{(x-2)(x+5)}{x-5}$

11. [6.1] $\dfrac{3a-1}{a-3}$ **12.** [6.1] $\dfrac{(x^2+4x+16)(x-6)}{(x+4)(x+2)}$

13. [6.2] $\dfrac{x-3}{(x+1)(x+3)}$ **14.** [6.2] $\dfrac{x^2+11xy+y^2}{(x-y)(x+y)}$

15. [6.2] $\dfrac{2x^3+2x^2y+2xy^2-2y^3}{(x-y)(x+y)}$

16. [6.2] $\dfrac{-y}{(y+4)(y-1)}$ **17.** [6.3] $\dfrac{3}{4}$

18. [6.3] $\dfrac{a^2b^2}{2(b^2-ba+a^2)}$

19. [6.3] $\dfrac{(y+11)(y+5)}{(y-5)(y+2)}$ **20.** [6.3] $\dfrac{(14-3x)(x+3)}{2x^2+16x+6}$

21. [6.4] 2   **22.** [6.4] $\frac{28}{11}$ **23.** [6.4] 6
24. [6.4] No solution   **25.** [6.4] 3
26. [6.5] $5\frac{1}{7}$ hr   **27.** [6.5] 24 mph
28. [6.5] Motorcycle, 62 mph; car, 70 mph
29. [6.6] $4s^2-3s-2rs^2$

30. [6.6] $y^2+4y+16$ **31.** [6.6] $4x+3+\dfrac{-9x-5}{x^2+1}$

32. [6.7] $x^2+6x+20+\dfrac{54}{x-3}$

33. [6.7] $4x^2-6x+18+\dfrac{-59}{x+3}$

34. [6.8] $s=\dfrac{Rg}{g-R}$ **35.** [6.8] $m=\dfrac{H}{S(t_1-t_2)}$

36. [6.8] $c=\dfrac{b+3a}{2}$ **37.** [6.8] $t_1=\dfrac{-A}{vT}+t_2$

38. [4.4]

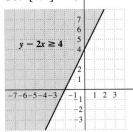

$y-2x\geq 4$

39. [4.4]

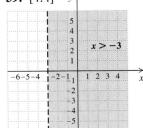

$x>-3$

40. [5.7] $(5x-2y)(25x^2+10xy+4y^2)$
41. [5.7] $(x+6)(6x-7)$ **42.** [5.8] $-6,\frac{7}{6}$
43. [2.2] 40   **44.** [6.2], [6.3], [6.4] ◈ The least common denominator was used to add and subtract rational expressions, to simplify complex rational expressions, and to solve rational equations.   **45.** [6.1], [6.4] ◈ A rational *expression* is a quotient of two polynomials. Expressions can be simplified, multiplied, or added, but they cannot be solved for a variable. A rational *equation* is an equation containing rational expressions. In a rational equation, we often can solve for a variable.   **46.** [6.4] All real numbers except 0 and 13
47. [6.3], [6.4] 45   **48.** [6.5] Anna, 56; Franz, 42

Test: Chapter 6, pp. 331–332

1. [6.1] $\dfrac{4(y-1)}{3}$ **2.** [6.1] $\dfrac{x^2-3x+9}{x+4}$

3. [6.2] $(x-4)(x+4)(x^2+4x+16)$
4. [6.2] $(x-3)(x+11)(x-9)$

5. [6.2] $\dfrac{25x+x^3}{x+5}$ **6.** [6.2] $3(a-b)$

7. [6.2] $\dfrac{a^3-a^2b+4ab+ab^2-b^3}{(a-b)(a+b)}$

8. [6.2] $\dfrac{-2(2x^2+5x+20)}{(x-4)(x+4)(x^2+4x+16)}$

9. [6.2] $\dfrac{y-4}{(y+3)(y-2)}$ **10.** [6.3] $\dfrac{5y-3x}{2y+3x}$

11. [6.3] $\dfrac{(x-9)(x-6)}{(x+6)(x-3)}$ **12.** [6.3] $\dfrac{4x^2-14x+2}{3x^2+7x-11}$

13. [6.1] $\frac{1}{10}$ **14.** [6.1] 0
15. [6.1] $\{x\,|\,x$ is a real number and $x\neq -3$ and $x\neq 4\}$
16. [6.4] 1
17. [6.4] $-\frac{21}{4}$ **18.** [6.5] $1\frac{31}{32}$ hr

19. [6.6] $\dfrac{4b^2c}{a}-\dfrac{5bc^2}{2a}+3bc$

20. [6.6] $y-14+\dfrac{-20}{y-6}$

21. [6.6] $6x^2-9+\dfrac{5x+22}{x^2+2}$

22. [6.7] $x^2+9x+40+\dfrac{153}{x-4}$

23. [6.8] $b_1=\dfrac{2A}{h}-b_2$

24. [6.5] 5 and 6; -6 and -5 **25.** [6.5] $3\frac{3}{11}$ mph
26. [6.8] 2.5 ohms   **27.** [2.2] a^2+2a-2
28. [5.7] $8(2t+3)(t-3)$ **29.** [5.8] $-\frac{3}{2},3$

30. [4.4]

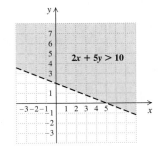

$2x+5y>10$

31. [6.4] All real numbers except 0 and 15
32. [6.2] $(1 - t^6)(1 + t^6)$
33. [6.1], [6.4] x-intercept: (11, 0); y-intercept; $\left(0, \frac{11}{4}\right)$

CUMULATIVE REVIEW: 1–6

1. [1.1], [1.2] 10   **2.** [1.7] 5.76×10^9
3. [2.3] Slope is $\frac{7}{4}$; y-intercept is $(0, -3)$
4. [2.5] $y = -\frac{10}{3}x + \frac{11}{3}$ **5.** [3.2] $(-3, 4)$
6. [3.4] $(-2, -3, 1)$ **7.** [3.3] 12 small, 15 large
8. [3.5] First number is 12, second number is $\frac{1}{2}$, third number is $7\frac{1}{2}$.
9. [3.7] 42   **10.** [5.8] $\frac{1}{4}$ **11.** [5.8] $-\frac{25}{7}, \frac{25}{7}$
12. [4.1] $\{x|x > -3\}$; or $(-3, \infty)$
13. [4.1] $\{y|y \geqslant -50\}$; or $[-50, \infty)$
14. [4.1] $\{x|x \geqslant -1\}$; or $[-1, \infty)$
15. [4.2] $\{x|-10 < x < 13\}$; or $(-10, 13)$
16. [4.2] $\{x|x < -\frac{4}{3} \text{ or } x > 6\}$; or $\left(-\infty, -\frac{4}{3}\right) \cup (6, \infty)$
17. [4.3] $\{x|x < -6.4 \text{ or } x > 6.4\}$;
or $(-\infty, -6.4) \cup (6.4, \infty)$
18. [4.3] $\left\{\frac{4}{3}, \frac{8}{3}\right\}$
19. [4.3] $\{x|-\frac{13}{4} \leqslant x \leqslant \frac{15}{4}\}$; or $\left[-\frac{13}{4}, \frac{15}{4}\right]$
20. [6.4] $-\frac{5}{3}$ **21.** [6.4] -1 **22.** [6.4] No solution
23. [6.4] $\frac{1}{3}$ **24.** [1.6] $n = \dfrac{m - 12}{3}$ **25.** [6.8] $a = \dfrac{Pb}{3 - P}$
26. [4.4]   **27.** [4.4]

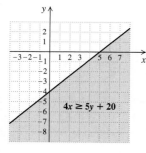
$4x \geqslant 5y + 20$

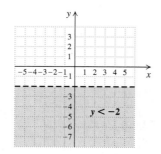
$y < -2$

28. [5.1] $-3x^3 + 9x^2 + 3x - 3$ **29.** [5.2] $-15x^4y^4$
30. [5.1] $5a + 5b - 5c$
31. [5.2] $15x^4 - x^3 - 9x^2 + 5x - 2$
32. [5.2] $4x^4 - 4x^2y + y^2$ **33.** [5.2] $4x^4 - y^2$
34. [5.1] $-m^3n^2 - m^2n^2 - 5mn^3$

35. [6.1] $\dfrac{y - 6}{2}$ **36.** [6.1] $x - 1$

37. [6.2] $\dfrac{a^2 + 7ab + b^2}{(a - b)(a + b)}$ **38.** [6.2] $\dfrac{-m^2 + 5m - 6}{(m + 1)(m - 5)}$

39. [6.2] $\dfrac{3y^2 - 2}{3y}$ **40.** [6.3] $\dfrac{y - x}{xy(x + y)}$

41. [6.6] $9x^2 - 13x + 26 + \dfrac{-50}{x + 2}$

42. [5.3] $2x^2(2x + 9)$ **43.** [5.4] $(x - 6)(x + 14)$

44. [5.5] $(4y - 9)(4y + 9)$
45. [5.6] $8(2x + 1)(4x^2 - 2x + 1)$ **46.** [5.5] $(t - 8)^2$
47. [5.7] $x^2(x - 1)(x + 1)(x^2 + 1)$
48. [5.6] $(0.3b - 0.2c)(0.09b^2 + 0.06bc + 0.04c^2)$
49. [5.4] $(4x - 1)(5x + 3)$ **50.** [5.4] $(3x + 4)(x - 7)$
51. [5.3] $(x^3 + y)(x^2 - y)$
52. [2.6], [6.1] $\{x|x$ is a real number and $x \neq 2$ and $x \neq 5\}$
53. [6.5] 1 hr   **54.** [5.8] 30 ft
55. [5.8] All such sets of even integers satisfy this condition.
56. [5.2] $x^3 - 12x^2 + 48x - 64$
57. [5.8] $-3, 3, -5, 5$
58. [4.3] $\{x|-3 \leqslant x \leqslant -1 \text{ or } 7 \leqslant x \leqslant 9\}$; or $[-3, -1] \cup [7, 9]$
59. [6.4] All real numbers except 9 and -5
60. [5.8] $0, \frac{1}{4}, -\frac{1}{4}$

CHAPTER 7

Exercise Set 7.1, pp. 342–343

1. $4, -4$ **3.** $12, -12$ **5.** $20, -20$ **7.** $7, -7$
9. $-\frac{7}{6}$ **11.** 14   **13.** $-\frac{4}{9}$ **15.** 0.3   **17.** -0.07
19. $p^2 + 4$ **21.** $\dfrac{x}{y + 4}$ **23.** $f(6) = \sqrt{20}, f(2) = 0$,
0 and -3 are not in the domain   **25.** $t(4) = -3$, -4 is not in the domain, $t(0) = -1, t(12) = -5$
27. $f(5) = \sqrt{26}, f(-5) = \sqrt{26}, f(0) = 1, f(10) = \sqrt{101}$, $f(-10) = \sqrt{101}$ **29.** $g(2) = \sqrt{17}, g(-2) = 1$,
$g(3) = 6$, -3 is not in the domain   **31.** $4|x|$ **33.** $7|c|$
35. $|a + 1|$ **37.** $|x - 2|$ **39.** $|2x + 7|$ **41.** 5
43. -1 **45.** $-\frac{2}{3}$ **47.** $|x|$ **49.** $5|a|$ **51.** 6
53. $|a + b|$ **55.** y **57.** $x - 2$ **59.** $4x$ **61.** $6b$
63. $a + 1$ **65.** $2(x + 1)$, or $2x + 2$ **67.** $3t - 2$
69. 3   **71.** $2x$ **73.** -6 **75.** $5y$ **77.** $0.2(x + 1)$
79. $2x^2$ **81.** $f(7) = 2, f(26) = 3, f(-9) = -2$,
$f(-65) = -4$ **83.** $g(19) = 2$, -13 and 1 are not in the domain, $g(84) = 3$ **85.** $\{x|x \geqslant -7\}$ **87.** $\left\{t|t \geqslant \frac{9}{2}\right\}$
89. $\{x|x \leqslant 5\}$ **91.** $\mathbb{R}$ **93.** $\left\{z|z \geqslant -\frac{3}{5}\right\}$ **95.** $\left\{t|t \geqslant \frac{5}{3}\right\}$
97. $a^9b^6c^{15}$ **99.** $x^2 - 9$ **101.** ◈   **103.** (a) 13;
(b) 15; (c) 18; (d) 20
105. A.180   **107.** A.181

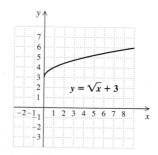
$y = \sqrt{x} + 3$

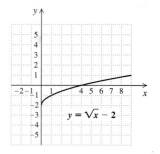
$y = \sqrt{x} - 2$

Exercise Set 7.2, pp. 347–348

1. $\sqrt[4]{x}$ **3.** 2 **5.** 3 **7.** 3 **9.** $\sqrt[3]{xyz}$ **11.** $\sqrt[5]{a^2b^2}$
13. $\sqrt[3]{a^2}$ **15.** 8 **17.** 343 **19.** 243
21. $\sqrt[4]{81^3x^3}$, or $27\sqrt[4]{x^3}$ **23.** $125x^6$
25. $\sqrt[3]{64a^4b^8}$, or $4ab^2\sqrt[3]{ab^2}$ **27.** $20^{1/3}$ **29.** $17^{1/2}$
31. $x^{3/2}$ **33.** $m^{2/5}$ **35.** $(cd)^{1/4}$ **37.** $(xy^2z)^{1/5}$
39. $(3mn)^{3/2}$ **41.** $(8x^2y)^{5/7}$
43. $\dfrac{1}{x^{1/3}}$ **45.** $\dfrac{1}{(2rs)^{3/4}}$ **47.** $10^{2/3}$ **49.** $x^{2/3}$
51. $\dfrac{5}{x^{1/4}}$ **53.** $\dfrac{3m^{1/2}}{5}$ **55.** $5^{7/8}$ **57.** $7^{1/4}$ **59.** $8.3^{7/20}$
61. $10^{6/25}$ **63.** $a^{23/12}$ **65.** $x^{2/7}$ **67.** $m^{1/6}n^{1/8}$
69. $a^4b^{3/2}$ **71.** $\sqrt[3]{a^2}$ **73.** $2y^2$ **75.** $2\sqrt[4]{2}$ **77.** $\sqrt[3]{2x}$
79. $2c^2d^3$ **81.** $\dfrac{m^2n^4}{2}$ **83.** $\sqrt[4]{r^2s}$ **85.** $3ab^3$
87. $-3, 3$ **89.** $\frac{1}{3}$ **91.** ◈ **93.** $\sqrt[10]{x^5y^3}$ **95.** $\sqrt[4]{2xy^2}$
97. (a) 1.8 m; (b) 3.1 m; (c) 1.5 m; (d) 5.3 m
99. 7.937×10^{-13} to 1

Exercise Set 7.3, pp. 354–355

1. $\sqrt{6}$ **3.** $\sqrt[3]{10}$ **5.** $\sqrt[4]{72}$ **7.** $\sqrt{30ab}$ **9.** $\sqrt[5]{18t^3}$
11. $\sqrt{x^2-a^2}$ **13.** $\sqrt[3]{0.06x^2}$ **15.** $\sqrt[4]{x^3-1}$
17. $\sqrt{\dfrac{6y}{5x}}$ **19.** $\sqrt[7]{\dfrac{5x-15}{4x+8}}$
21. $\sqrt[6]{392}$ **23.** $\sqrt[6]{4x^5}$ **25.** $\sqrt[6]{x^5-4x^4+4x^3}$
27. $\sqrt[10]{x^9y^7}$ **29.** $\sqrt[12]{x^{11}(y+1)^{10}}$ **31.** $\sqrt[15]{a^{11}b^{13}c^{14}}$
33. $3\sqrt{3}$ **35.** $3\sqrt{5}$ **37.** $2\sqrt{2}$ **39.** $2\sqrt{6}$
41. $6x^2\sqrt{5}$ **43.** $2\sqrt[3]{100}$ **45.** $-2x^2\sqrt[3]{2}$
47. $3x^2\sqrt[3]{2x^2}$ **49.** $2x^2\sqrt[3]{10x^2}$ **51.** $2\sqrt[4]{2}$ **53.** $3\sqrt[4]{10}$
55. $2a^2\sqrt[4]{6}$ **57.** $3cd\sqrt[4]{2d^2}$ **59.** $(x+y)\sqrt[3]{x+y}$
61. $20(m+n)^2\sqrt[3]{(m+n)^2}$ **63.** $-ab^2c^3\sqrt[5]{abc^2}$
65. $3\sqrt{2}$ **67.** $6\sqrt{5}$ **69.** $4\sqrt{3}$ **71.** $3\sqrt[3]{2}$
73. $5bc^2\sqrt{2b}$ **75.** $-5x^2\sqrt[3]{6x}$ **77.** $2y^3\sqrt[3]{2}$
79. $(b+3)^2$ **81.** $4a^3b\sqrt{6ab}$ **83.** $a(b+c)^2\sqrt[5]{a(b+c)}$
85. $a\sqrt[10]{ab^7}$ **87.** $2xy^2\sqrt[6]{2x^5y}$ **89.** $xy\sqrt[4]{xy^3}$
91. $a^2b^2c^2\sqrt[12]{a^2bc^2}$ **93.** $yz\sqrt[12]{x^{11}yz^5}$
95. $2^2a(b-5)^2\sqrt[6]{2a(b-5)}$ **97.** 13.416 **99.** 7.477
101. 0.454 **103.** 4.414 **105.** 156,524.7584
107. 684,251.4158 **109.** 0.0001987460691
111. 0.0007071774883 **113.** 8
115. $(2x-7)(2x+7)$ **117.** ◈ **119.** (a) 20 mph;
(b) 37.4 mph; (c) 42.4 mph **121.** 10 **123.** $x \geqslant -\frac{3}{2}$

Exercise Set 7.4, pp. 359–360

1. $\sqrt{7}$ **3.** 3 **5.** $y\sqrt{5y}$ **7.** $2\sqrt[3]{a^2b}$ **9.** $3\sqrt{2xy}$
11. $2x^2y^2$ **13.** $\sqrt{x^2+xy+y^2}$ **15.** $\sqrt[12]{a^5}$ **17.** $\sqrt[12]{x^2y}$

19. $\sqrt[10]{ab^9c^3}$ **21.** $\sqrt[20]{(3x-1)^3}$ **23.** $\dfrac{4}{5}$ **25.** $\dfrac{4}{3}$ **27.** $\dfrac{7}{y}$
29. $\dfrac{5y\sqrt{y}}{x^2}$ **31.** $\dfrac{2x\sqrt[3]{x^2}}{3y}$ **33.** $\dfrac{2a}{3}$ **35.** $\dfrac{ab^2}{c^2}\sqrt[4]{\dfrac{a}{c^2}}$
37. $\dfrac{2x}{y^2}\sqrt[5]{\dfrac{x}{y}}$ **39.** $\dfrac{xy}{z^2}\sqrt[6]{\dfrac{y^2}{z^3}}$ **41.** $6a\sqrt{6a}$ **43.** $64b^3$
45. $54a^3b\sqrt{2b}$ **47.** $3c^2d\sqrt[3]{3c^2d}$ **49.** $x\sqrt[3]{25xy^2}$
51. $x\sqrt[4]{x^2y^3}$ **53.** $4a^2\sqrt[3]{a^2b^2}$ **55.** $4xy\sqrt[4]{y^2}$, or $4xy\sqrt{y}$
57. 8 **59.** Length is 20; width is 5 **61.** ◈
63. a^3bxy^2 **65.** $2yz\sqrt{2z}$

Exercise Set 7.5, pp. 363–364

1. $8\sqrt{3}$ **3.** $3\sqrt[3]{5}$ **5.** $13\sqrt[3]{y}$ **7.** $7\sqrt{2}$ **9.** $6\sqrt[3]{3}$
11. $21\sqrt{3}$ **13.** $38\sqrt{5}$ **15.** $122\sqrt{2}$ **17.** $9\sqrt[3]{2}$
19. $(1+6a)\sqrt{5a}$ **21.** $(2-x)\sqrt[3]{3x}$ **23.** $3\sqrt{2y-2}$
25. $(x+3)\sqrt{x-1}$ **27.** $15\sqrt[3]{4}$ **29.** 0
31. $(3x^2+x)\sqrt[4]{x-1}$ **33.** $2\sqrt{6}-18$
35. $\sqrt{6}-\sqrt{10}$ **37.** $2\sqrt{15}-6\sqrt{3}$ **39.** -6
41. $3a\sqrt[3]{2}$ **43.** $x^2+\sqrt[4]{3x^3}$ **45.** 18 **47.** -3
49. -19 **51.** $a+\sqrt{3a}+\sqrt{2a}+\sqrt{6}$
53. $6+\sqrt[3]{5}-\sqrt[3]{25}$ **55.** -6
57. $2\sqrt[3]{9}-3\sqrt[3]{6}-2\sqrt[3]{4}$ **59.** $4+2\sqrt{3}$
61. $a^2+2a\sqrt{b}+b$ **63.** $4x^2-4x\sqrt[4]{y}+\sqrt{y}$
65. $m+2\sqrt{mn}+n$ **67.** $\sqrt[6]{a^5}+\sqrt[4]{a^3b}$
69. $x\sqrt[6]{xy^5}-\sqrt[15]{x^{13}y^{14}}$
71. $2m^2+m\sqrt[4]{n}+2m\sqrt[3]{n^2}+\sqrt[12]{n^{11}}$
73. $a^2-a\sqrt[3]{c}+a\sqrt[4]{b}-\sqrt[12]{b^3c^4}$
75. $\sqrt[4]{x^3}+\sqrt[10]{x^7z^2}-\sqrt[12]{x^3y^4z^4}-\sqrt[15]{x^3y^5z^8}$
77. $-\frac{19}{5}$ **79.** $-28\sqrt{3}-22\sqrt{5}$
81. $ac[(3ac+2)\sqrt{ab}-2c\sqrt[3]{ab}]$ **83.** 6
85. $14+2\sqrt{15}-6\sqrt{2}-2\sqrt{30}$ **87.** $3\sqrt[3]{3}+2\sqrt[3]{9}-8$

Exercise Set 7.6, pp. 369–370

1. $\dfrac{\sqrt{30}}{5}$ **3.** $\dfrac{\sqrt{70}}{7}$ **5.** $\dfrac{2\sqrt{15}}{5}$ **7.** $\dfrac{2\sqrt[3]{6}}{3}$
9. $\dfrac{\sqrt[3]{75ac^2}}{5c}$ **11.** $\dfrac{y\sqrt[3]{180x^2y}}{6x^2}$ **13.** $\dfrac{\sqrt[3]{x^2y^2}}{xy}$
15. $\dfrac{\sqrt{14a}}{6}$ **17.** $\dfrac{3\sqrt{5y}}{10xy}$ **19.** $\dfrac{\sqrt[3]{90xy}}{10xy^2}$ **21.** $\dfrac{7}{\sqrt{21x}}$
23. $\dfrac{2}{\sqrt{6}}$ **25.** $\dfrac{52}{3\sqrt{91}}$ **27.** $\dfrac{7}{\sqrt[3]{98}}$ **29.** $\dfrac{7x}{\sqrt{21xy}}$
31. $\dfrac{5y^2}{x\sqrt[3]{150x^2y^2}}$ **33.** $\dfrac{ab}{3\sqrt{ab}}$ **35.** $\dfrac{x^2y}{\sqrt{2xy}}$
37. $\dfrac{ab}{\sqrt[3]{5ab^2}}$ **39.** $\dfrac{x^2y}{\sqrt[3]{3x^2y}}$ **41.** $\dfrac{5(8+\sqrt{6})}{58}$

43. $-2\sqrt{7}(\sqrt{5} + \sqrt{3})$ **45.** $\dfrac{\sqrt{15} + 20 - 6\sqrt{2} - 8\sqrt{30}}{-77}$

47. $\dfrac{x - 2\sqrt{xy} + y}{x - y}$ **49.** $\dfrac{4 + 3\sqrt{6}}{2}$

51. $\dfrac{6x - 7\sqrt{xy} - 3y}{4x - 9y}$ **53.** $\dfrac{-11}{4(\sqrt{3} - 5)}$

55. $\dfrac{-22}{\sqrt{6} + 5\sqrt{2} + 5\sqrt{3} + 25}$ **57.** $\dfrac{x - y}{x + 2\sqrt{xy} + y}$

59. $\dfrac{7}{43\sqrt{2} + 66}$ **61.** $\dfrac{3 - 4x}{3 - 3\sqrt{3x} + 2x}$

63. $\dfrac{a^2 - b^2 c}{a^2 + (a - ab)\sqrt{c} - bc}$ **65.** 6 **67.** ◈

69. $\dfrac{ab + (a - b)\sqrt{a + b} - a - b}{a + b - b^2}$

71. $\dfrac{b^2 + \sqrt{b}}{b^2 + b + 1}$ **73.** $\dfrac{x - 19}{x + 10\sqrt{x + 6} + 31}$

75. $\left(\dfrac{5x + 4y - 3}{xy}\right)\sqrt{xy}$ **77.** $\dfrac{7\sqrt{3}}{39}$

Technology Connection, Section 7.7

TC1. *x*-coordinates of points of intersection should approximate the solutions of the examples.

Exercise Set 7.7, pp. 374–375

1. 2 **3.** 12 **5.** 168 **7.** No solution **9.** 3
11. 19 **13.** $\frac{80}{3}$ **15.** 4 **17.** 64 **19.** -27
21. 397 **23.** No solution **25.** 3 **27.** $\frac{1}{64}$ **29.** -6
31. 5 **33.** $-\frac{1}{4}$ **35.** 3 **37.** 9 **39.** 7 **41.** $\frac{80}{9}$
43. -1 **45.** 6, 2 **47.** No solution **49.** 10
51. 15 **53.** 2 **55.** ◈ **57.** About 208 miles
59. $-\frac{8}{9}$ **61.** $-8, 8$ **63.** $-4, 3$ **65.** 1, 8
67. $\frac{1}{36}$, 36

Exercise Set 7.8, pp. 379–382

1. $\sqrt{34}$; 5.831 **3.** $\sqrt{288}$; 16.971 **5.** 5
7. $\sqrt{31}$; 5.568 **9.** $\sqrt{12}$; 3.464 **11.** $\sqrt{n - 1}$
13. $\sqrt{325}$; 18.028 ft **15.** $\sqrt{14{,}500}$; 120.416 ft
17. 12 in. **19.** 50 ft **21.** $b = 9$; $c = 9\sqrt{2} \approx 12.728$
23. $a = 7$; $b = 7\sqrt{3} \approx 12.124$
25. $a = \dfrac{5\sqrt{3}}{3} \approx 2.887$; $c = \dfrac{10\sqrt{3}}{3} \approx 5.774$
27. $a = \dfrac{13\sqrt{2}}{2} \approx 9.192$; $b = \dfrac{13\sqrt{2}}{2} \approx 9.192$
29. $c = 24$; $b = 12\sqrt{3} \approx 20.785$ **31.** $h = 4\sqrt{3} \approx 6.928$
33. $c = 5\sqrt{2} \approx 7.071$ **35.** $\dfrac{19\sqrt{2}}{2} \approx 13.435$

37. Neither; both have the same area, 300 ft^2
39. $h = 2\sqrt{3} \approx 3.464$ ft **41.** 8 ft **43.** $(0, -4), (0, 4)$
45. 3, 8 **47.** ◈ **49.** $\sqrt{75}$ cm

Exercise Set 7.9, pp. 388–389

1. $i\sqrt{15}$, or $\sqrt{15}i$ **3.** $4i$ **5.** $-2i\sqrt{3}$, or $-2\sqrt{3}i$
7. $i\sqrt{3}$, or $\sqrt{3}i$ **9.** $9i$ **11.** $7i\sqrt{2}$, or $7\sqrt{2}i$
13. $-7i$ **15.** $4 - 2\sqrt{15}i$ **17.** $(2 + 2\sqrt{3})i$
19. $8 + i$ **21.** $9 - 5i$ **23.** $7 + 4i$ **25.** $-2 - 3i$
27. $-1 + i$ **29.** $11 + 6i$ **31.** -30 **33.** $-\sqrt{30}$
35. $-5\sqrt{6}$ **37.** $-12\sqrt{2}$ **39.** -40 **41.** 35
43. $10 + 15i$ **45.** $-12 - 21i$ **47.** $1 + 5i$
49. $18 + 14i$ **51.** $38 + 9i$ **53.** $2 - 46i$ **55.** $5 - 37i$
57. $-11 - 16i$ **59.** $13 - 47i$ **61.** $5 - 12i$
63. $-5 + 12i$ **65.** $-5 - 12i$ **67.** $-i$ **69.** 1
71. -1 **73.** i **75.** -1 **77.** $-125i$ **79.** 8
81. $1 - 26i$ **83.** 0 **85.** 0 **87.** 1 **89.** $5 - 8i$
91. $\frac{3}{2} + \frac{1}{2}i$ **93.** $\frac{3}{29} + \frac{7}{29}i$ **95.** $-\frac{7}{6}i$ **97.** $-\frac{3}{7} - \frac{8}{7}i$
99. $\frac{8}{5} + \frac{1}{5}i$ **101.** $-i$ **103.** $\frac{19}{13} - \frac{9}{13}i$ **105.** Yes
107. No **109.** 7 **111.** $-\frac{4}{3}, 7$ **113.** $\frac{250}{41} + \frac{200}{41}i$
115. 8 **117.** $\frac{3}{5} + \frac{9}{5}i$ **119.** 1

Review Exercises: Chapter 7, pp. 393–394

1. [7.1] $\frac{2}{3}$ **2.** [7.1] 0.07 **3.** [7.1] 5
4. [7.1] $\{x | x \geq \frac{7}{2}\}$, or $[\frac{7}{2}, \infty)$ **5.** [7.1] $9|a|$
6. [7.1] $|c + 8|$ **7.** [7.1] $|x - 3|$ **8.** [7.1] $|2x + 1|$
9. [7.1] -2 **10.** [7.1] $-\frac{1}{3}$ **11.** [7.1] $|x|$
12. [7.1] 3 **13.** [7.2] $(8x^6 y^2)^{4/5}$, or $8^{4/5} x^{24/5} y^{8/5}$
14. [7.2] $\sqrt[4]{(5a)^3}$ **15.** [7.2] $\sqrt[12]{x^4 y^3}$ **16.** [7.2] $-\dfrac{x^3 y^4}{2}$
17. [7.3] 15.652 **18.** [7.3] 10.583 **19.** [7.3] -0.236
20. [7.3] $3xy\sqrt{2y}$ **21.** [7.3] $3a\sqrt[3]{a^2 b^2}$
22. [7.3] $3x^2 y^2 \sqrt[6]{3xy^2}$ **23.** [7.4] $y\sqrt[3]{6}$ **24.** [7.4] $\dfrac{5\sqrt{x}}{2}$
25. [7.4] $8xy^2$ **26.** [7.4] $2a\sqrt[3]{2ab^2}$ **27.** [7.5] $30\sqrt[3]{5}$
28. [7.5] $15\sqrt{2}$ **29.** [7.5] $9 - \sqrt[3]{4}$
30. [7.5] $-43 - 2\sqrt{10}$ **31.** [7.5] $8 - 2\sqrt{7}$
32. [7.6] $\dfrac{10a\sqrt{3} - 10\sqrt{3ab}}{a - b}$
33. [7.6] $\dfrac{30a}{\sqrt{3a}(\sqrt{a} + \sqrt{b})}$, or $\dfrac{30a}{\sqrt{3}(a + \sqrt{ab})}$
34. [7.7] 4 **35.** [7.7] 13 **36.** [7.8] 9 cm
37. [7.8] 5 ft **38.** [7.9] $-2\sqrt{2}i$ **39.** [7.9] $-2 - 9i$
40. [7.9] $1 + i$ **41.** [7.9] 29 **42.** [7.9] i
43. [7.9] $9 - 12i$ **44.** [7.9] $\frac{2}{5} + \frac{3}{5}i$ **45.** [7.9] $\frac{9}{5} - \frac{12}{5}i$
46. [5.8] 12, 13, 14 **47.** [6.4] $\frac{23}{7}$ **48.** [5.8] $-\frac{9}{2}, 3$

49. [6.1] $\dfrac{x(x+2)}{(x+y)(x-3)}$

50. [7.1] An absolute-value sign must be used to simplify $\sqrt[k]{x^k}$ when k is even, since x may be negative. If x is negative while k is even, the radical expression cannot be simplified to x, since $\sqrt[k]{x^k}$ represents the principal, or positive, root. When k is odd, there is only one root, and it will be positive or negative depending on the sign of x. Thus there is no absolute-value sign when k is odd.

51. [7.9] ◈ Every real number is a complex number, but there are complex numbers that are not real. A complex number $a + bi$ is not real if $b \neq 0$.

52. [7.7] 3   **53.** [7.9] $-\frac{2}{5} + \frac{9}{10}i$

Test: Chapter 7, pp. 394–395

1. [7.1] $\frac{10}{7}$ **2.** [7.1] $\{t | t \geq -5\}$, or $[-5, \infty)$
3. [7.1] $6|y|$ **4.** [7.1] $|x + 5|$ **5.** [7.1] -2
6. [7.1] 4   **7.** [7.2] $(5xy^2)^{5/2}$ **8.** [7.2] $2ab^3 \sqrt[3]{2a^2b}$
9. [7.3] 0.00109727   **10.** [7.3] $2x^3$
11. [7.4] $\dfrac{5x\sqrt{x}}{6y^2}$ **12.** [7.4] $4a\sqrt[3]{4ab^2}$
13. [7.5] $38\sqrt{2}$ **14.** [7.5] -20
15. [7.5] $2x^2 - x\sqrt[3]{y^2} + 2xy\sqrt{y} - y^2\sqrt[6]{y}$
16. [7.6] $-\dfrac{13 + 8\sqrt{2}}{41}$ **17.** [7.7] 7
18. [7.8] Longer leg: $10\sqrt{3} \approx 17.321$ cm; hypotenuse: 20 cm   **19.** [7.9] $3\sqrt{2}i$ **20.** [7.9] $7 + 5i$
21. [7.9] -30 **22.** [7.9] $-2i$ **23.** [7.9] $-\frac{77}{50} + \frac{7}{25}i$
24. [7.9] $-1 + 5i$ **25.** [5.8] $-\frac{1}{3}, \frac{5}{2}$ **26.** [6.1] $\dfrac{x-3}{x-4}$
27. [6.4] No solution   **28.** [5.8] 16, 18 and $-18, -16$
29. [7.7] 3   **30.** [7.9] $-\frac{17}{4}i$

CHAPTER 8

Technology Connection, Section 8.1

TC1. x-intercepts should be approximations of $-3 + \sqrt{5}$ and $-3 - \sqrt{5}$ for Example 8; approximations of $(-5 + \sqrt{37})/2$ and $(-5 - \sqrt{37})/2$ for Example 11.
TC2. A grapher can only give rational-number approximations of the two irrational solutions. An *exact* solution cannot be found with a grapher.

Exercise Set 8.1, pp. 405–406

1. $\pm\sqrt{5}$ **3.** $\pm\frac{2}{5}i$ **5.** $\pm\dfrac{\sqrt{6}}{2}$ **7.** 5, -9
9. $3 \pm \sqrt{21}$ **11.** $13 \pm 2\sqrt{2}$ **13.** $7 \pm 2i$ **15.** $9 \pm \sqrt{34}$

17. $\dfrac{3 \pm \sqrt{14}}{2}$ **19.** 5, -11 **21.** 9, 5
23. $x^2 + 10x + 25$, $(x + 5)^2$ **25.** $x^2 - 8x + 16$, $(x - 4)^2$
27. $x^2 - 24x + 144$, $(x - 12)^2$
29. $x^2 + 9x + \frac{81}{4}$, $\left(x + \frac{9}{2}\right)^2$ **31.** $x^2 - 7x + \frac{49}{4}$, $\left(x - \frac{7}{2}\right)^2$
33. $x^2 + \frac{2}{3}x + \frac{1}{9}$, $\left(x + \frac{1}{3}\right)^2$ **35.** $x^2 - \frac{5}{6}x + \frac{25}{144}$, $\left(x - \frac{5}{12}\right)^2$
37. $x^2 + \frac{9}{5}x + \frac{81}{100}$, $\left(x + \frac{9}{10}\right)^2$ **39.** $-7, -1$
41. $5 \pm \sqrt{47}$ **43.** $-5, -1$ **45.** 3, 7   **47.** $-3, -2$
49. $-2 \pm \sqrt{3}$ **51.** $5 \pm \sqrt{2}$ **53.** $-3 \pm 2i$
55. $-\frac{1}{2}, 3$ **57.** $-\frac{3}{2}, -\frac{1}{2}$ **59.** $-\frac{1}{2}, \frac{2}{3}$ **61.** $-1 \pm \dfrac{\sqrt{2}}{2}$
63. $\dfrac{5 \pm \sqrt{61}}{6}$ **65.** $6\frac{1}{4}\%$ **67.** 20%   **69.** 4%
71. About 9.3 sec   **73.** About 9.5 sec
75.

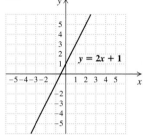

y = 2x + 1

77. 4.6   **79.** ◈   **81.** $b = 16, -16$ **83.** $0, \frac{7}{2}, -8, -\frac{10}{3}$
85. A: 15 km/h; B: 8 km/h   **87.** ◈

Exercise Set 8.2, pp. 410–411

1. $-3 \pm \sqrt{5}$ **3.** $-1, -\frac{5}{3}$ **5.** $\dfrac{1 \pm i\sqrt{3}}{2}$ **7.** $2 \pm 3i$
9. $\dfrac{-3 \pm \sqrt{41}}{2}$ **11.** $-1 \pm 2i$ **13.** 0, -1 **15.** $0, -\frac{9}{14}$
17. $\frac{2}{5}$ **19.** $-1, -2$ **21.** 5, 10   **23.** $\dfrac{13 \pm \sqrt{509}}{10}$
25. $1 \pm 2i$ **27.** $\frac{2}{3}, \frac{3}{2}$ **29.** $-2, 3$ **31.** $\frac{3}{4}, -2$ **33.** $\dfrac{1 \pm 3i}{2}$
35. $2, -1 \pm i\sqrt{3}$ **37.** 1.3, -5.3 **39.** 5.2, 0.8
41. 2.8, -1.3 **43.** 1.9, -0.3
45. 30 lb of Kenyan; 20 lb of Peruvian
47. (a) $\left(-\frac{5}{2}, 0\right), (4, 0)$; (b) $\dfrac{3 - \sqrt{73}}{4}, \dfrac{3 + \sqrt{73}}{4}$
49. 1.1753905, -0.4253905
51. $\dfrac{-1 \pm \sqrt{1 + 4\sqrt{2}}}{2}$ **53.** $\dfrac{-\sqrt{5} \pm \sqrt{5 + 4\sqrt{3}}}{2}$
55. $\sqrt{3}, \dfrac{3 - \sqrt{3}}{2}$ **57.** $\frac{1}{2}$

Exercise Set 8.3, pp. 415–416

1. 6, −4   **3.** $\pm\frac{3}{2}$ **5.** −20, 25   **7.** −1, 4
9. $\pm\sqrt{37}$ **11.** −5, 2   **13.** $\pm 2\sqrt{7}$ **15.** 10, $\frac{21}{2}$
17. 2, 3   **19.** $\frac{1}{15}$, 2   **21.** −1, −6   **23.** −4
25. First part, 12 km/h; second part, 8 km/h   **27.** 35
mph   **29.** Super-prop: 350 mph; turbo-jet: 400 mph
31. Speed out: 60 mph; speed back: 50 mph
33. 10 mph   **35.** 6 hr   **37.** 3.24 mph   **39.** 1, 5
41. $2y\sqrt[3]{9x^2}$ **43.** ◈   **45.** $4 \pm 2\sqrt{2}$ **47.** $\pm\sqrt{2}$
49. ◈

Exercise Set 8.4, pp. 419–420

1. One real   **3.** Two imaginary   **5.** Two real
7. One real   **9.** Two imaginary   **11.** Two real
13. Two real   **15.** Two real   **17.** Two real
19. Two imaginary   **21.** One real
23. $x^2 + 2x − 99 = 0$ **25.** $x^2 − 14x + 49 = 0$
27. $x^2 + 8x + 15 = 0$ **29.** $3x^2 − 14x + 8 = 0$
31. $6x^2 − 5x + 1 = 0$ **33.** $25x^2 − 20x − 12 = 0$
35. $x^2 − 4\sqrt{2}x + 6 = 0$ **37.** $x^2 + 3\sqrt{5}x + 10 = 0$
39. $x^2 + 9 = 0$ **41.** $x^2 − 10x + 29 = 0$
43. $2x^2 − 2x + 5 = 0$
45. Six 30-second commercials; six 60-second
commercials   **47.** ◈   **49.** $a = 1, b = 2, c = −3$
51. (a) $k = −\frac{3}{5}$; **(b)** $−\frac{1}{3}$ **53. (a)** $k = 9 + 9i$; **(b)** $3 + 3i$
55. $x^2 − \sqrt{3}x + 8 = 0$ **57.** $a = 8, b = 20, c = −12$
59. $k = 6$

Exercise Set 8.5, pp. 423–424

1. $\pm\sqrt{5}$ **3.** $\pm\sqrt{3}, \pm 3$ **5.** $\pm 1, \pm\frac{\sqrt{5}}{3}$ **7.** 81, 1
9. $\frac{4}{9}$ **11.** $\pm\sqrt{7}, \pm 2\sqrt{2}$ **13.** 7, −1, 5, 1   **15.** 4
17. $−\frac{1}{2}, \frac{1}{3}$ **19.** −1, 2   **21.** −27, 8   **23.** 16
25. 32, −243   **27.** 1   **29.** 6, 1   **31.** $−\frac{3}{2}$
33. $3 \pm \sqrt{10}, −1 \pm \sqrt{2}$ **35.** $3x^2\sqrt{x}$
37. $\dfrac{x^3 + x^2 + 2x + 2}{x^3 − 1}$
39. $(\sqrt{7}, 0), (−\sqrt{7}, 0), (1, 0), (−1, 0)$ **41.** ± 1.99
43. $\frac{432}{143}$ **45.** 9   **47.** −2, 1   **49.** −3, −1, 1, 4

Exercise Set 8.6, pp. 430–433

1. $k = 8; y = 8x$ **3.** $k = 3.6; y = 3.6x$
5. $k = \frac{15}{4}; y = \frac{15}{4}x$ **7.** $k = 1.6; y = 1.6x$ **9.** 6 amperes
11. $4.29   **13.** 50 kg   **15.** 3.36   **17.** $k = 60; y = \dfrac{60}{x}$
19. $k = 12; y = \dfrac{12}{x}$ **21.** $k = 36; y = \dfrac{36}{x}$

23. $k = 9; y = \dfrac{9}{x}$ **25.** $\frac{2}{9}$ ampere   **27.** 160 cm^3
29. $6\frac{2}{3}$ hours   **31.** $y = 15x^2$ **33.** $y = \dfrac{0.0015}{x^2}$
35. $y = xz$ **37.** $y = 0.3xz^2$ **39.** $y = \dfrac{4wx^2}{z}$ **41.** $y = \dfrac{xz}{5wp}$
43. 36 mph   **45.** 6.25 km   **47.** 1600 km
49. 2 mm   **51.** 8.17 mph   **53.** $y = −\frac{2}{3}x − 5$
55. $f(3) = 9$ **57.** ◈   **59.** ◈
61. y is multiplied by 8.   **63.** $\dfrac{6\sqrt{2}}{5} \approx 1.697$ m
65. Q varies directly as the square of p and inversely as
the cube of q.

Exercise Set 8.7, pp. 439–441

1.

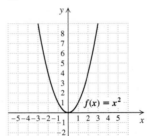

3.

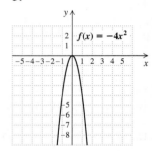

5.

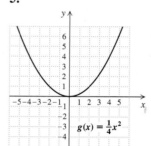

7.

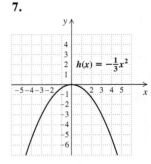

9.

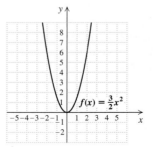

11. Vertex: $(−1, 0)$;
line of symmetry: $x = −1$

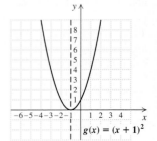

13. Vertex: (4, 0); line of symmetry: $x = 4$

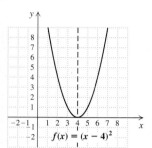

$f(x) = (x - 4)^2$

15. Vertex: (3, 0); line of symmetry: $x = 3$

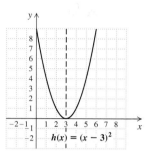

$h(x) = (x - 3)^2$

29. Vertex: $(-9, 0)$; line of symmetry: $x = -9$

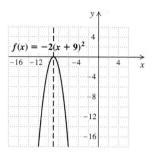

$f(x) = -2(x + 9)^2$

31. Vertex: $\left(\frac{1}{2}, 0\right)$; line of symmetry: $x = \frac{1}{2}$

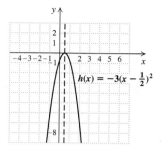

$h(x) = -3\left(x - \frac{1}{2}\right)^2$

17. Vertex: $(-4, 0)$; line of symmetry: $x = -4$

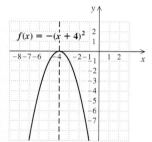

$f(x) = -(x + 4)^2$

19. Vertex: (1, 0); line of symmetry: $x = 1$

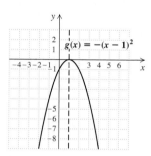

$g(x) = -(x - 1)^2$

33. Vertex: (3, 1); line of symmetry: $x = 3$; minimum: 1

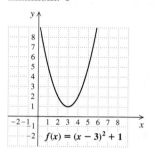

$f(x) = (x - 3)^2 + 1$

35. Vertex: $(-1, -2)$; line of symmetry: $x = -1$; minimum: -2

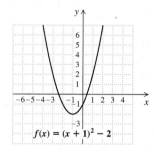

$f(x) = (x + 1)^2 - 2$

21. Vertex: (1, 0); line of symmetry: $x = 1$

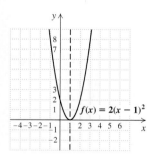

$f(x) = 2(x - 1)^2$

23. Vertex: (3, 0); line of symmetry: $x = 3$

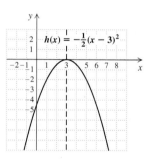

$h(x) = -\frac{1}{2}(x - 3)^2$

37. Vertex: $(-4, 1)$; line of symmetry: $x = -4$; minimum: 1

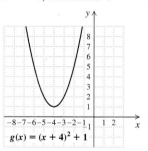

$g(x) = (x + 4)^2 + 1$

39. Vertex: (5, 2); line of symmetry: $x = 5$; minimum: 2

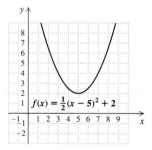

$f(x) = \frac{1}{2}(x - 5)^2 + 2$

25. Vertex: $(-1, 0)$; line of symmetry: $x = -1$

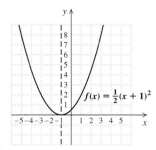

$f(x) = \frac{1}{2}(x + 1)^2$

27. Vertex: (2, 0); line of symmetry: $x = 2$

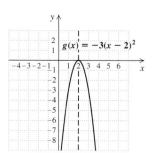

$g(x) = -3(x - 2)^2$

41. Vertex: $(1, -3)$; line of symmetry: $x = 1$; maximum: -3

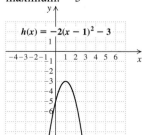

$h(x) = -2(x - 1)^2 - 3$

43. Vertex: $(-4, 1)$; line of symmetry: $x = -4$; maximum: 1

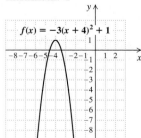

$f(x) = -3(x + 4)^2 + 1$

45. Vertex: (1, 2);
line of symmetry: $x = 1$;
maximum: 2

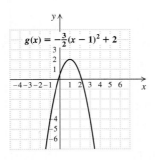

47. Vertex: (9, 5); line of symmetry: $x = 9$; minimum: 5
49. Vertex: $(-6, 11)$; line of symmetry: $x = -6$;
maximum: 11
51. Vertex: $\left(-\frac{1}{4}, -13\right)$; line of symmetry: $x = -\frac{1}{4}$;
minimum: -13
53. Vertex: $(10, -20)$; line of symmetry: $x = 10$;
maximum: -20
55. Vertex: $(-4.58, 65\pi)$; line of symmetry: $x = -4.58$;
minimum: 65π
57. $(-50, 350, 0)$ **59.** ◈ **61.** $f(x) = -2x^2 + 4$
63. $f(x) = 2(x - 6)^2$ **65.** $f(x) = -2(x - 3)^2 + 8$
67. $f(x) = 2(x + 3)^2 + 6$ **69.** $g(x) = -\frac{1}{2}(x - 1)^2 - 6$
71. **73.**

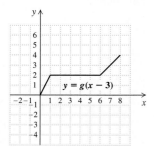

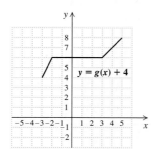

75. **77.**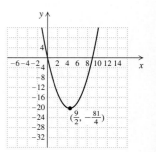

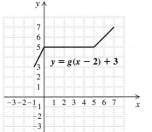

1. $f(x) = (x - 1)^2 - 4$ **3.** $g(x) = (x + 3)^2 + 4$

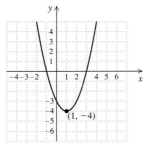

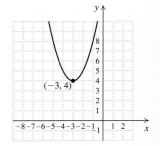

5. $f(x) = (x + 2)^2 - 5$ **7.** $h(x) = 2(x + 4)^2 - 7$

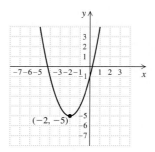

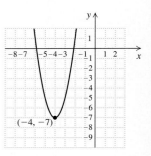

9. $f(x) = -(x - 2)^2 + 10$ **11.** $g(x) = \left(x + \frac{3}{2}\right)^2 - \frac{49}{4}$

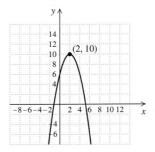

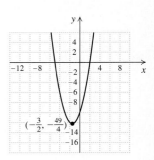

13. $f(x) = 3(x - 4)^2 + 2$ **15.** $h(x) = \left(x - \frac{9}{2}\right)^2 - \frac{81}{4}$

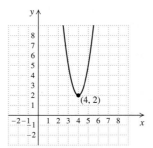

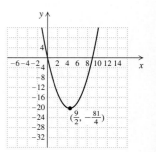

17. $f(x) = -2(x + 1)^2 - 4$

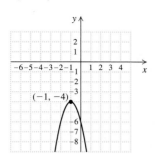

$(-1, -4)$

19. $g(x) = 2\left(x - \frac{5}{2}\right)^2 + \frac{3}{2}$

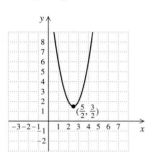

$\left(\frac{5}{2}, \frac{3}{2}\right)$

21. $f(x) = -3\left(x + \frac{1}{2}\right)^2 + \frac{7}{4}$

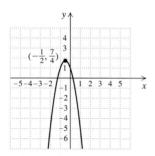

$\left(-\frac{1}{2}, \frac{7}{4}\right)$

23. $h(x) = \frac{1}{2}(x + 4)^2 - \frac{5}{3}$

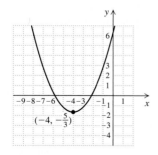

$\left(-4, -\frac{5}{3}\right)$

25. $(2 + \sqrt{3}, 0), (2 - \sqrt{3}, 0)$ **27.** $(3, 0), (-1, 0)$
29. $(4, 0), (-1, 0)$ **31.** $(2, 0), (1, 0)$

33. $\left(\dfrac{-2 + \sqrt{6}}{2}, 0\right), \left(\dfrac{-2 - \sqrt{6}}{2}, 0\right)$

35. None exists.   **37.** 5   **39.** ◈
41. (a) Minimum: -6.953660714;
(b) $(-1.056433682, 0), (2.413576539, 0)$
43. (a) 3.4, -2.4; (b) 2.3, -1.3
45. $f(x) = m\left(x - \dfrac{n}{2m}\right)^2 + \dfrac{4mp - n^2}{4m}$

47.   **49.**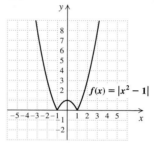

$f(x) = |x^2 - 1|$

Exercise Set 8.9, pp. 452–456

1. 19 ft by 19 ft; 361 ft^2 **3.** 64; 8 and 8
5. 121; 11 and 11   **7.** -4; 2 and -2
9. $-\frac{81}{4}$; $-\frac{9}{2}$ and $\frac{9}{2}$ **11.** $\frac{49}{4}$; $-\frac{7}{2}$ and $-\frac{7}{2}$
13. 200 ft^2; 10 ft by 20 ft   **15.** 4 ft by 4 ft
17. $f(x) = 2x^2 + 3x - 1$ **19.** $f(x) = -\frac{1}{4}x^2 + 3x - 5$
21. (a) $f(x) = -4x^2 + 40x + 2$; (b) \$98
23. (a) $A(s) = 0.05s^2 - 5.5s + 250$, where $A(s)$ is the
number of daytime accidents (for every 200 million km)
and s is the speed of travel in km/h; (b) 100
25. (a) $P(d) = \frac{1}{64}d^2 + \frac{5}{16}d + \frac{5}{2}$; (b) \$9.94
27. $P(x) = -x^2 + 192x - 5000$; a maximum profit of
\$4216 occurs when 96 units are produced and sold.

29. $s = \sqrt{\dfrac{A}{6}}$ **31.** $r = \sqrt{\dfrac{Gm_1m_2}{F}}$ **33.** $c = \sqrt{\dfrac{E}{m}}$

35. $b = \sqrt{c^2 - a^2}$ **37.** $k = \dfrac{3 + \sqrt{9 + 8N}}{2}$

39. $r = \dfrac{-\pi h + \sqrt{\pi^2 h^2 + 2\pi A}}{2\pi}$ **41.** $n = \dfrac{1 + \sqrt{1 + 8N}}{2}$

43. $w = \dfrac{-2l + \sqrt{4l^2 + 2A}}{2}$ **45.** $g = \dfrac{4\pi^2 l}{T^2}$

47. $r = -1 + \dfrac{-P_2 + \sqrt{P_2^2 + 4AP_1}}{2P_1}$

49. $v = \dfrac{c}{m}\sqrt{m^2 - m_0^2}$

51. (a) 3.9 sec; (b) 1.9 sec; (c) 79.6 m   **53.** 0.89 sec
55. 14   **57.** 2.9 sec   **59.** 2.5 m/sec   **61.** 7%
63. $2\sqrt{5}i$ **65.** Base: 19 cm, height: 19 cm; 180.5 cm^2

67. $n = \pm\sqrt{\dfrac{r^2 \pm \sqrt{r^4 + 4m^4r^2p - 4mp}}{2m}}$ **69.** \$6

71. 78.4 ft   **73.** $d = \dfrac{-\pi h + \sqrt{\pi^2h^2 + 2\pi A}}{\pi}$

75. $L = \sqrt{\dfrac{A}{2}}$

Technology Connection, Section 8.10

TC1. $\{x|-0.78 \leq x \leq 1.59\}$, or $[-0.78, 1.59]$
TC2. $\{x|x \leq -0.21 \text{ or } x \geq 2.47\}$,
or $(-\infty, -0.21] \cup [2.47, \infty)$
TC3. $\{x|x < -1.26 \text{ or } x > 2.33\}$,
or $(-\infty, -1.26) \cup (2.33, \infty)$
TC4. $\{x|x > -1.37\}$, or $(-1.37, \infty)$

Exercise Set 8.10, pp. 463–464

1. $\{x|x < -3 \text{ or } x > 5\}$, or $(-\infty, -3) \cup (5, \infty)$
3. $\{x| -1 \leq x \leq 2\}$, or $[-1, 2]$
5. $\{x| -1 < x < 2\}$, or $(-1, 2)$
7. $\{x|x \leq -3 \text{ or } x \geq 3\}$, or $(-\infty, -3] \cup [3, \infty)$
9. $\{x|x \text{ is a real number}\}$, or $(-\infty, \infty)$
11. $\{x|2 < x < 4\}$, or $(2, 4)$
13. $\{x|x < -2 \text{ or } 0 < x < 2\}$, or $(-\infty, -2) \cup (0, 2)$
15. $\{x| -3 < x < -1 \text{ or } x > 2\}$, or $(-3, -1) \cup (2, \infty)$
17. $\{x|x < -3 \text{ or } -2 < x < 1\}$, or $(-\infty, -3) \cup (-2, 1)$
19. $\{x|x < 4\}$, or $(-\infty, 4)$
21. $\{x|x < -1 \text{ or } x > 3\}$, or $(-\infty, -1) \cup (3, \infty)$
23. $\left\{x| -\frac{2}{3} \leq x < 3\right\}$, or $\left[-\frac{2}{3}, 3\right)$
25. $\left\{x|2 < x < \frac{5}{2}\right\}$, or $\left(2, \frac{5}{2}\right)$
27. $\{x|x < -1 \text{ or } 2 < x < 5\}$, or $(-\infty, -1) \cup (2, 5)$
29. $\{x|x \leq 0 \text{ or } x > 2\}$, or $(-\infty, 0] \cup (2, \infty)$
31. $\{x|x > 0\}$, or $(0, \infty)$
33. $\{x|x < -4 \text{ or } 1 < x < 3\}$, or $(-\infty, -4) \cup (1, 3)$
35. $\left\{x|0 < x < \frac{1}{2}\right\}$, or $\left(0, \frac{1}{2}\right)$ **37.** $\sqrt[15]{a^{11}b^{13}}$
39. ◈   **41.** All real numbers   **43.** $\left\{x|x < \frac{1}{4} \text{ or } x > \frac{5}{2}\right\}$
45. (a) $\{t|0 < t < 2\}$; **(b)** $\{t|t > 10\}$
47. $\{n|9 \leq n \leq 23\}$, or $[9, 23]$
49. $f(x) < 0$ for $\{x|x < -2\}$, $f(x) > 0$
for $\{x| -2 < x < 1 \text{ or } x > 1\}$
51. $f(x) < 0$ for $\{x|0 < x < 1\}$, $f(x) > 0$ for $\{x|x > 1\}$
53. $f(x) < 0$ for $\{x|x < -3 \text{ or } -2 < x < 0 \text{ or } 1 < x < 2\}$,
$f(x) > 0$ for $\{x| -3 < x < -2 \text{ or } 0 < x < 1 \text{ or } x > 2\}$

Review Exercises: Chapter 8, pp. 467–468

1. $[8.1] \pm\dfrac{\sqrt{14}}{2}$ **2.** $[8.2]\ 0, -\dfrac{5}{14}$ **3.** $[8.1]\ 3, 9$

4. $[8.2]\ \dfrac{-3 \pm i\sqrt{7}}{8}$ **5.** $[8.2]\ \dfrac{7 \pm i\sqrt{3}}{2}$ **6.** $[8.2]\ 3, 5$

7. $[8.2]\ -0.3, -3.7$ **8.** $[8.1]\ x^2 - 12x + 36; (x - 6)^2$
9. $[8.1]\ x^2 + \frac{3}{5}x + \frac{9}{100}; \left(x + \frac{3}{10}\right)^2$ **10.** $[8.1]\ 4, -2$
11. $[8.1]\ 3 \pm 2\sqrt{2}$ **12.** $[8.1]\ 10\%$ **13.** $[8.1]\ 6.7$ sec
14. $[8.3]\ 4, -2$ **15.** $[8.3]\ -5, 3$ **16.** $[8.3]\ 4 \pm 4\sqrt{2}$
17. $[8.3]\ \dfrac{1 \pm \sqrt{481}}{15}$
18. [8.3] 50 mph on first part, then 40 mph on
second part   **19.** [8.3] 6 hr   **20.** [8.4] Two real
21. [8.4] Two nonreal   **22.** $[8.4]\ 25x^2 + 10x - 3 = 0$
23. $[8.4]\ x^2 + 8x + 16 = 0$ **24.** $[8.5]\ 2, -2, 3, -3$
25. $[8.5]\ 3, -5$ **26.** $[8.5]\ \pm\sqrt{7}, \pm\sqrt{2}$
27. [8.6] 500 watts   **28.** [8.6] 64 L
29. [8.7]
30. [8.8] **(a)** Vertex:
(3, 5), line of symmetry:
$x = 3$; **(b)**

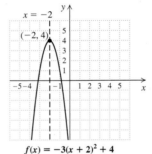

$f(x) = -3(x + 2)^2 + 4$
Maximum: 4

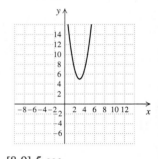

31. [8.8] (7, 0), (2, 0)   **32.** [8.9] 5 sec
33. $[8.9]\ p = \dfrac{9\pi^2}{N^2}$ **34.** $[8.9]\ T = \dfrac{1 \pm \sqrt{1 + 24A}}{6}$
35. [8.9] -121; 11 and -11
36. $[8.9]\ f(x) = -x^2 + 6x - 2$
37. $[8.10]\ \{x|-1 < x < 7\}$, or $(-1, 7)$
38. $[8.10]\ \{x|-3 < x \leq 5\}$, or $(-3, 5]$
39. [3.3] 210 kg of A; 90 kg of B   **40.** [7.7] 2
41. $[6.2]\ \dfrac{x + 1}{(x - 3)(x - 1)}$ **42.** $[7.3]\ 3t^5s\sqrt[3]{s}$
43. [8.1], [8.2], [8.8] ◈ Completing the square was used
to solve quadratic equations and to graph quadratic
functions by rewriting the function in the form
$f(x) = a(x - h)^2 + k$.
44. [8.10] ◈ From graphs of polynomial functions, it
can be seen that for outputs to change from positive to
negative, they must at some point equal zero. In solving a
polynomial inequality, we form the intervals by finding
all zeros of the related equation. The outputs cannot
change sign except at the interval boundaries. Thus if a
function has a positive output for one number in an
interval, it will be positive for *all* the numbers in the
interval.
45. [8.3] All real numbers except -13 and -7
46. $[8.4]\ h = 60, k = 60$ **47.** [8.5] 18 and 324

Test: Chapter 8, pp. 468–469

1. [8.1] $\pm\dfrac{2\sqrt{3}}{3}$ **2.** [8.2] 9, 2 **3.** [8.2] $\dfrac{-1\pm i\sqrt{3}}{2}$

4. [8.2] 0.4, −4.4 **5.** [8.3] 0, 2

6. [8.5] $\pm\sqrt{\dfrac{5+\sqrt{5}}{2}}, \pm\sqrt{\dfrac{5-\sqrt{5}}{2}}$

7. [8.1] $x^2+14x+49$; $(x+7)^2$
8. [8.1] $x^2-\frac{2}{7}x+\frac{1}{49}$; $\left(x-\frac{1}{7}\right)^2$
9. [8.1] −6, 3 **10.** [8.1] $-5\pm\sqrt{10}$
11. [8.3] 16 km/h **12.** [8.3] 2 hr
13. [8.4] Two nonreal **14.** [8.4] $x^2-4\sqrt{3}x+9=0$

15. [8.6] $\dfrac{833}{125}$, or 6.664 in^2

16. [8.7]

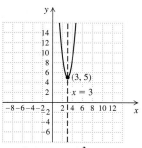

$f(x)=4(x-3)^2+5$
Minimum: 5

17. [8.8] **(a)** (−1, −8), $x=-1$;
(b)

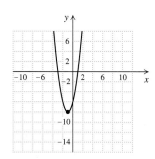

18. [8.8] (3, 0) and (−2, 0)

19. [8.9] $r=\sqrt{\dfrac{3V}{\pi}-R^2}$ **20.** [8.9] −16

21. [8.9] $f(x)=\frac{1}{5}x^2-\frac{3}{5}x$
22. [8.10] $\{x\mid-2<x<1\ or\ x>2\}$, or $(-2,1)\cup(2,\infty)$
23. [7.7] 6 **24.** [7.3] $ab\sqrt[4]{2a^2}$

25. [6.2] $\dfrac{x-8}{(x+6)(x+8)}$ **26.** [3.3] Each is 36.

27. [8.4] $\frac{1}{2}$ **28.** [8.3] All real numbers except 11 and 7

CHAPTER 9

Technology Connection, Section 9.1

TC1.

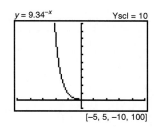

TC2.

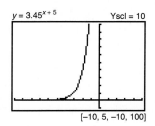

TC3.

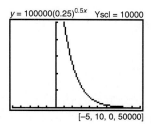

TC4.

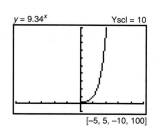

TC5.

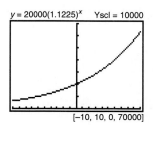

TC6.

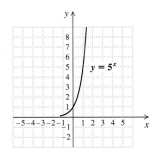

Exercise Set 9.1, pp. 478–479

1.

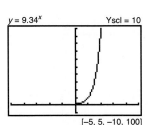

3.

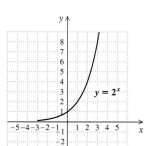

5.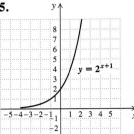

7.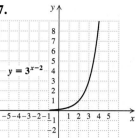

25.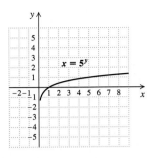

27.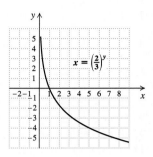

9.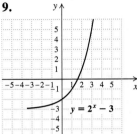

11.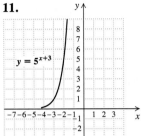

29.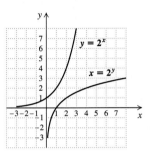

31.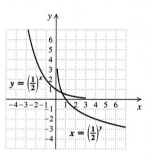

13.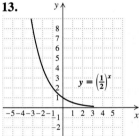

15.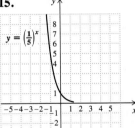

33. (a) 384,160; **(b)** 2,066,105;
(c)

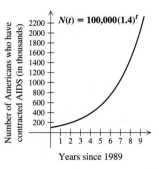

17.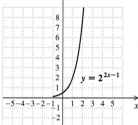

19.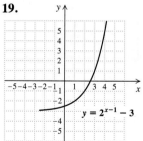

35. (a) 250,000; 166,667; 49,383; 4335;
(b)

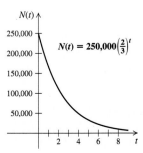

21.

23.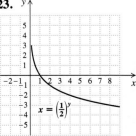

37. (a) 7.5 million, 18.4 million, 110.2 million, 661.4 million, 58,320 million, 5,142,752.7 million;
(b)

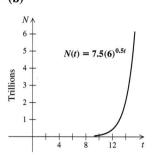

39. x^{-2}, or $\dfrac{1}{x^2}$ **41.** x^{-7}, or $\dfrac{1}{x^7}$ **43.** ◈   **45.** $\pi^{2.4}$

47.

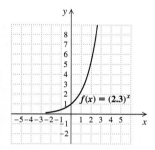

49.

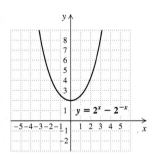

51.

53.

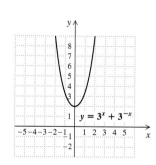

55.

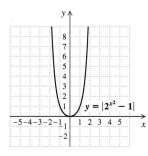

57.

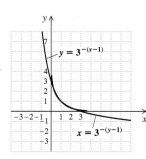

59. (a) 19 wpm, 36 wpm, 66 wpm, 115 wpm;
(b)

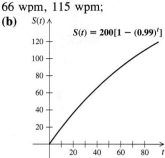

Technology Connection, Section 9.2

TC1. C   **TC2.** A   **TC3.** B   **TC4.** D

Exercise Set 9.2, pp. 489–490

1. $f \circ g(x) = 12x^2 - 12x + 5$; $g \circ f(x) = 6x^2 + 3$
3. $f \circ g(x) = \dfrac{16}{x^2} - 1$; $g \circ f(x) = \dfrac{2}{4x^2 - 1}$
5. $f \circ g(x) = x^4 - 2x^2 + 2$; $g \circ f(x) = x^4 + 2x^2$
7. $f(x) = x^2$; $g(x) = 5 - 3x$
9. $f(x) = x^5$; $g(x) = 3x^2 - 7$
11. $f(x) = \dfrac{1}{x}$; $g(x) = x - 1$

13. $f(x) = \dfrac{1}{\sqrt{x}}$; $g(x) = 7x + 2$

15. $f(x) = \dfrac{x + 1}{x - 1}$; $g(x) = x^3$ **17.** Yes   **19.** No
21. Yes   **23.** No   **25. (a)** Yes; **(b)** $f^{-1}(x) = x - 4$
27. (a) Yes; **(b)** $f^{-1}(x) = 5 - x$
29. (a) Yes; **(b)** $g^{-1}(x) = x + 5$
31. (a) Yes; **(b)** $f^{-1}(x) = \dfrac{x}{3}$

33. (a) Yes; **(b)** $g^{-1}(x) = \dfrac{x - 2}{3}$

35. No   **37. (a)** Yes; **(b)** $f^{-1}(x) = \dfrac{1}{x}$

39. (a) Yes; **(b)** $f^{-1}(x) = \dfrac{3x - 1}{2}$

41. (a) Yes; **(b)** $f^{-1}(x) = \sqrt[3]{x + 1}$
43. (a) Yes; **(b)** $g^{-1}(x) = \sqrt[3]{x} + 2$
45. (a) Yes; **(b)** $f^{-1}(x) = x^2$, $x \geq 0$
47. (a) Yes; **(b)** $f^{-1}(x) = \sqrt{\dfrac{x - 3}{2}}$

49.

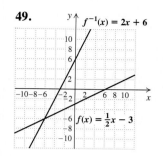

$f^{-1}(x) = 2x + 6$
$f(x) = \frac{1}{2}x - 3$

51.

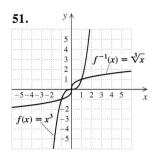

$f^{-1}(x) = \sqrt[3]{x}$
$f(x) = x^3$

53.

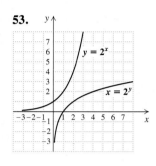

$y = 2^x$
$x = 2^y$

55.

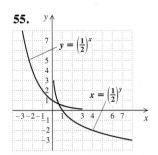

$y = \left(\frac{1}{2}\right)^x$
$x = \left(\frac{1}{2}\right)^y$

57.

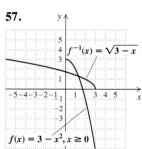

$f^{-1}(x) = \sqrt{3 - x}$
$f(x) = 3 - x^2, x \geq 0$

59. (1) $f^{-1} \circ f(x) = f^{-1}(f(x)) = f^{-1}\left(\frac{4}{5}x\right) = \frac{5}{4}\left(\frac{4}{5}x\right) = x$;

(2) $f \circ f^{-1}(x) = f(f^{-1}(x)) = f\left(\frac{5}{4}x\right) = \frac{4}{5}\left(\frac{5}{4}x\right) = x$

61. (1) $f^{-1} \circ f(x) = f^{-1}(f(x)) = f^{-1}\left(\dfrac{1-x}{x}\right)$

$= \dfrac{1}{\left(\dfrac{1-x}{x}\right) + 1} = \dfrac{1}{\dfrac{1-x+x}{x}}$

$= x$;

(2) $f \circ f^{-1}(x) = f(f^{-1}(x)) = f\left(\dfrac{1}{x+1}\right)$

$= \dfrac{1 - \left(\dfrac{1}{x+1}\right)}{\left(\dfrac{1}{x+1}\right)} = \dfrac{\dfrac{x+1-1}{x+1}}{\dfrac{1}{x+1}} = x$

63. (a) 40, 42, 46, 50; (b) $f^{-1}(x) = x - 32$;
(c) 8, 10, 14, 18

65. $y = 9x$ **67.** ◈   **69.** $g(x) = \dfrac{x}{2} + 20$

71. Inverses   **73.** Not inverses

Exercise Set 9.3, pp. 495–496

1.

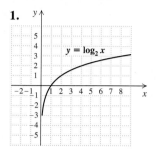

$y = \log_2 x$

3.

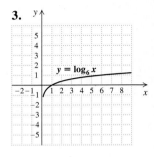

$y = \log_6 x$

5.

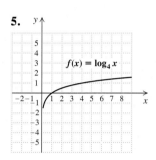

$f(x) = \log_4 x$

7.

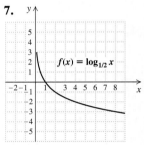

$f(x) = \log_{1/2} x$

9.

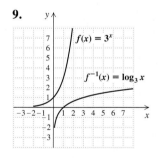

$f(x) = 3^x$
$f^{-1}(x) = \log_3 x$

11. $3 = \log_{10} 1000$
13. $-3 = \log_5 \frac{1}{125}$
15. $\frac{1}{3} = \log_8 2$
17. $0.3010 = \log_{10} 2$
19. $2 = \log_e t$
21. $t = \log_Q x$
23. $2 = \log_e 7.3891$
25. $-2 = \log_e 0.1353$
27. $3^t = 8$ **29.** $5^2 = 25$
31. $10^{-1} = 0.1$

33. $10^{0.845} = 7$ **35.** $e^{2.9957} = 20$ **37.** $t^k = Q$
39. $e^{-1.3863} = 0.25$ **41.** $r^{-x} = T$ **43.** 9   **45.** 5
47. 4   **49.** 3   **51.** 13   **53.** 1   **55.** $\frac{1}{2}$ **57.** 2
59. 2   **61.** -1 **63.** 0   **65.** 4   **67.** -2 **69.** 0
71. 1   **73.** $\frac{2}{3}$ **75.** 3   **77.** t

79. $\dfrac{x(3y - 2)}{2y + x}$ **81.** $\dfrac{1}{4096}$ **83.** $\dfrac{1}{\sqrt[3]{t^2}}$ **85.** ◈

87.

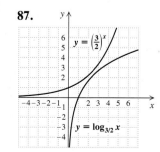

$y = \left(\frac{3}{2}\right)^x$
$y = \log_{3/2} x$

89.

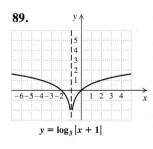

$y = \log_3 |x + 1|$

91. 25   **93.** $-\frac{7}{16}$ **95.** 3   **97.** 0   **99.** -2

Exercise Set 9.4, pp. 501–502

1. $\log_2 32 + \log_2 8$ **3.** $\log_4 64 + \log_4 16$
5. $\log_c B + \log_c x$
7. $\log_a (6 \cdot 70)$ **9.** $\log_c (K \cdot y)$ **11.** $3 \log_a x$
13. $6 \log_c y$ **15.** $-3 \log_b C$ **17.** $\log_a 67 - \log_a 5$
19. $\log_b 3 - \log_b 4$ **21.** $\log_a \frac{15}{7}$
23. $2 \log_a x + 3 \log_a y + \log_a z$
25. $\log_b x + 2 \log_b y - 3 \log_b z$
27. $\frac{1}{3}\left(4 \log_c x - 3 \log_c y - 2 \log_c z\right)$
29. $\frac{1}{4}\left(8 \log_a x + 12 \log_a y - 3 - 5 \log_a z\right)$
31. $\log_a \dfrac{\sqrt[3]{x^2}\,\sqrt{y}}{y}$ **33.** $\log_a \dfrac{2x^4}{y^3}$ **35.** $\log_a \dfrac{\sqrt{a}}{x}$
37. 2.708   **39.** 0.51   **41.** -1.609 **43.** $\frac{3}{2}$
45. 2.609   **47.** 3.218   **49.** 9   **51.** m **53.** 4
55. -7 **57.** i **59.** $\frac{3}{5} + \frac{4}{5}i$ **61.** ◈
63. $\log_a (x^6 - x^4y^2 + x^2y^4 - y^6)$
65. $\frac{1}{2} \log_a (1 - s) + \frac{1}{2} \log_a (1 + s)$ **67.** $\frac{10}{3}$ **69.** -2
71. False   **73.** False   **75.** True

Technology Connection, Section 9.5

TC1.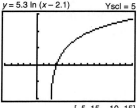

TC2.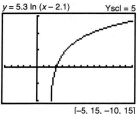

TC3.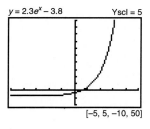

TC4.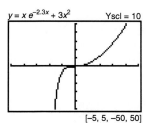

TC5.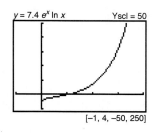

TC6.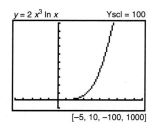

Exercise Set 9.5, p. 508

1. 0.3010   **3.** 0.8021   **5.** 1.6532   **7.** 2.6405
9. 4.1271   **11.** -1.2840 **13.** 1000   **15.** 501.1872
17. 3.0001   **19.** 0.2841   **21.** 0.0011   **23.** 0.6931
25. 4.1271   **27.** 8.3814   **29.** -5.0832 **31.** 36.7890
33. 0.0023   **35.** 1.0057   **37.** 5.8346×10^{14}
39. 7.6331   **41.** 2.5702   **43.** 3.3219   **45.** 0.6419
47. -2.3219 **49.** -2.3219 **51.** 3.5471

53. 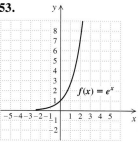   **55.**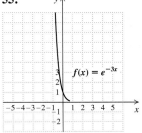

57. 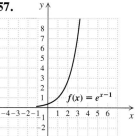   **59.**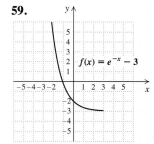

61. 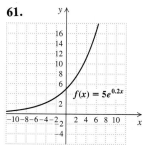   **63.**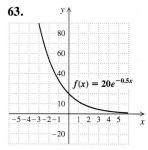

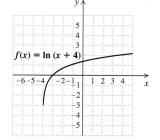   **67.**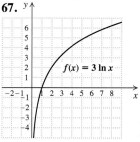

69.
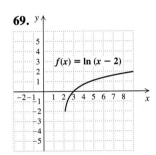

71.
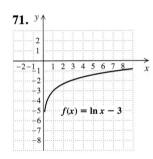

73. $\frac{5}{2}$, $-\frac{5}{2}$ **75.** 16, 256 **77.** ◆

79. $\ln M = \dfrac{\log M}{\log e}$ **81.** 52.5084 **83.** 4.9855

Technology Connection, Section 9.6

TC1. 0.38 **TC2.** -1.96 **TC3.** 0.90 **TC4.** -1.53
TC5. 3.45, 0.0001 (the function is undefined at zero)
TC6. -0.75, 0.75

Exercise Set 9.6, pp. 512–513

1. 3 **3.** 4 **5.** $\frac{5}{2}$ **7.** $\frac{3}{5}$ **9.** 3.170 **11.** 3.322
13. $\frac{5}{2}$ **15.** -3, -1 **17.** 1.404 **19.** 4.605
21. 2.303 **23.** 140.671 **25.** 2.710 **27.** 3.607
29. 5.646 **31.** 27 **33.** $\frac{1}{8}$ **35.** 10 **37.** $\frac{1}{100}$

39. $e^2 \approx 7.389$ **41.** $\dfrac{1}{e} \approx 0.368$ **43.** 66 **45.** 10

47. $\frac{1}{3}$ **49.** 3 **51.** $\frac{2}{5}$ **53.** 5

55. $\dfrac{1}{25}x^{-14/3}y^{4/3}z^{-4}$, or $\dfrac{y^{4/3}}{25\,x^{14/3}z^4}$ **57.** $c = \sqrt{\dfrac{E}{m}}$

59. ◆ **61.** -4 **63.** 2 **65.** $\pm\sqrt{34}$ **67.** $10^{100,000}$
69. 1, 100 **71.** $\frac{1}{100,000}$, 100,000 **73.** $-\frac{1}{3}$ **75.** $\frac{3}{2}$
77. -3

Exercise Set 9.7, pp. 519–522

1. (a) 5.5 years; (b) 0.8 year
3. (a) 37.7 years; (b) 11.9 years **5.** (a) 3.5 years;
(b) 13.6 years **7.** (a) $P(t) = P_0 e^{0.09t}$;
(b) \$1094.17, \$1197.22; (c) 7.7 years **9.** 69.3 years
11. (a) $P(t) = 5.2e^{0.016t}$, where t is the number of years
after 1990; (b) 6.1 billion; (c) 2017
13. (a) 68%; (b) 54%, 40%; (c)
(d) 6.9 months

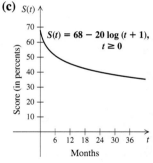

15. 6.2 **17.** 7.8 **19.** 10^{-7} moles per liter
21. 6.3×10^{-4} moles per liter **23.** 8.25 **25.** 2000
27. (a) $N(t) = e^{0.363t}$, where t is the number of years
after 1967; (b) 77,111 **29.** 1860 years **31.** 7.2 days
33. 23% per minute **35.** (a) $V(t) = 110,000e^{0.2822t}$,
where t is the number of years after 1986;
(b) \$3,251,528; (c) 2.5 years; (d) 2002

37. $\dfrac{(x-5)(x-3)}{x^2 + 5x - 6}$ **39.** ◆ **41.** ◆

Review Exercises: Chapter 9, pp. 523–525

1. [9.1]

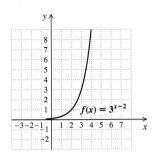

2. [9.1]

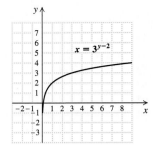

3. [9.3]

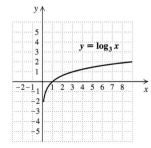

4. [9.5]

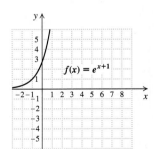

5. [9.2] $f \circ g(x) = 9x^2 - 30x + 25$, $g \circ f(x) = 3x^2 - 5$
6. [9.2] No **7.** [9.2] $f^{-1}(x) = x - 2$

8. [9.2] $g^{-1}(x) = \dfrac{7x + 3}{2}$

9. [9.2] $f^{-1}(x) = \dfrac{2 - 5x}{x}$, or $\dfrac{2}{x} - 5$

10. [9.2] $g^{-1}(x) = \dfrac{\sqrt[3]{x}}{2}$ **11.** [9.3] $4^x = 16$

12. [9.3] $10^{0.3010} = 2$ **13.** [9.3] $\left(\frac{1}{2}\right)^{-3} = 8$
14. [9.3] $16^{3/4} = 8$ **15.** [9.3] $\log_{10} 10,000 = 4$
16. [9.3] $\log_{25} 5 = \frac{1}{2}$ **17.** [9.3] $\log_7 \frac{1}{49} = -2$
18. [9.3] $\log_{2.718} 20.1 = 3$
19. [9.4] $4 \log_a x + 2 \log_a y + 3 \log_a z$
20. [9.4] $\log_a x + \log_a y - 2 \log_a z$
21. [9.4] $\frac{1}{4}(2 \log z - 3 \log x - \log y)$
22. [9.4] $2 \log_q x + \frac{1}{3} \log_q y - 4 \log_q z$

23. [9.4] $\log_a (8 \cdot 15)$, or $\log_a 120$

24. [9.4] $\log_a \frac{72}{12}$, or $\log_a 6$ **25.** [9.4] $\log \frac{a^{1/2}}{bc^2}$

26. [9.4] $\log_a \sqrt[3]{\frac{x}{y^2}}$ **27.** [9.3] 1   **28.** [9.3] 0

29. [9.4] 17   **30.** [9.4] -7 **31.** [9.4] 6.93
32. [9.4] -3.2698 **33.** [9.4] 8.7601
34. [9.4] 3.2698   **35.** [9.4] 2.54995
36. [9.4] -3.6602 **37.** [9.5] -2.2027
38. [9.5] 7.8621   **39.** [9.5] 29,798.88
40. [9.5] 0.0361   **41.** [9.5] 213.50   **42.** [9.5] -2.3065
43. [9.5] 5.5965   **44.** [9.5] 0.000000163
45. [9.5] 10.0821   **46.** [9.5] -2.6921
47. [9.5] 0.00002593   **48.** [9.5] 3.4934×10^{19}
49. [9.5] 0.4307   **50.** [9.5] 1.7097   **51.** [9.6] $\frac{1}{9}$
52. [9.6] 2   **53.** [9.6] $\frac{1}{10,000}$ **54.** [9.6] $e^2 \approx 7.3891$
55. [9.6] $\frac{7}{2}$ **56.** [9.6] 1.5266   **57.** [9.6] 2   **58.** [9.6] 7
59. [9.6] 8   **60.** [9.6] 20   **61.** [9.6] $\sqrt{43}$
62. [9.7] **(a)** 62; **(b)** 46.8; **(c)** 35 months
63. [9.7] **(a)** $k = 0.05$, $C(t) = \$4.65e^{0.05t}$;
(b) \$34.36; **(c)** 1999; **(d)** 13.9 years
64. [9.7] 4.3%   **65.** [9.7] 8.25 years
66. [9.7] 3463 years   **67.** [9.7] 6.6   **68.** [9.7] 8.3

69. [8.9] $T = \dfrac{-b \pm \sqrt{b^2 + 4aQ}}{2a}$ **70.** [8.5] ± 4, $\pm\sqrt{5}$

71. [7.9] $\dfrac{-11}{10} - \dfrac{17}{10}i$ **72.** [6.3] $\dfrac{c - 2a}{2b + 3c}$

73. [9.3] ◈ Negative numbers do not have logarithms because logarithm bases are positive, and there is no power to which a positive number can be raised to yield a negative number.   **74.** [9.6] ◈ Taking the logarithm on each side of an equation produces an equivalent equation because the logarithm function is one-to-one. If two quantities are equal, their logarithms must be equal, and if the logarithms of two quantities are equal, the quantities must be the same.
75. [9.6] e^{e^3} **76.** [9.6] -3, -1 **77.** [9.6] $\left(\frac{8}{3}, -\frac{2}{3}\right)$

Test: Chapter 9, pp. 525–526

1. [9.1]

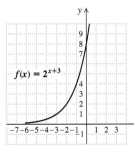

2. [9.3]

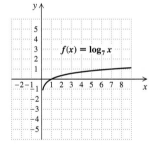

3. [9.2] $f \circ g(x) = 25x^2 - 15x + 2$, $g \circ f(x) = 5x^2 + 5x - 2$

4. [9.2] No   **5.** [9.2] $f^{-1}(x) = \dfrac{x + 3}{4}$

6. [9.2] $g^{-1}(x) = 4x + 2$ **7.** [9.2] $f^{-1}(x) = \dfrac{1 + 2x}{x - 1}$

8. [9.3] $\log_4 x = -3$ **9.** [9.3] $\log_{256} 16 = \dfrac{1}{2}$

10. [9.3] $4^2 = 16$ **11.** [9.3] $7^m = 49$

12. [9.4] $3 \log a + \dfrac{1}{2} \log b - 2 \log c$

13. [9.4] $\log_a \dfrac{x^{1/3}z^2}{y^3}$ **14.** [9.4] 23   **15.** [9.3] 1

16. [9.3] 0   **17.** [9.4] -0.544 **18.** [9.4] 0.69
19. [9.4] 1.322   **20.** [9.5] -1.9101
21. [9.5] 445,040.98   **22.** [9.5] 0.000000054777
23. [9.5] 4.0913   **24.** [9.5] -4.3949
25. [9.5] 292.19   **26.** [9.5] 1.1881   **27.** [9.6] 5
28. [9.6] 2   **29.** [9.6] 10,000   **30.** [9.6] $\frac{1}{3}$
31. [9.6] 0.0937   **32.** [9.6] $e^{1/4} \approx 1.2840$
33. [9.6] 9   **34.** [9.7] **(a)** 2.45 ft/sec; **(b)** 984,262
35. [9.7] **(a)** $P(t) = 24e^{0.012t}$, where $P(t)$ is in millions and t is in years; **(b)** 29 million, 34 million; **(c)** 2000; **(d)** 57.8 years   **36.** [9.7] 3.5%   **37.** [9.7] 9.1 years
38. [9.7] 4684 years   **39.** [9.7] 8.34   **40.** [9.7] 7.0

41. [8.5] 1, 64   **42.** [8.9] $t = \dfrac{b \pm \sqrt{b^2 + 4aS}}{2a}$

43. [7.9] 29   **44.** [6.3] $\dfrac{1}{2x}$ **45.** [9.6] 316, -309
46. [9.4] 2

CUMULATIVE REVIEW: 1–9

1. [1.1], [1.5] 2   **2.** [1.2] 6   **3.** [1.5] $\dfrac{y^{12}}{16x^8}$

4. [1.5] $\dfrac{20x^6z^2}{y}$ **5.** [1.5] $\dfrac{-y^4}{3z^5}$ **6.** [1.3] $-4x - 1$

7. [1.5] 25   **8.** [1.3] $\frac{11}{2}$ **9.** [5.8] $\frac{2}{5}$, -5
10. [3.2] $(3, -1)$ **11.** [3.4] $(1, -2, 0)$
12. [8.2] $-3 \pm \sqrt{14}$ **13.** [5.8] 5, -2 **14.** [6.4] $\frac{9}{2}$
15. [8.3] $\frac{5}{8}$ **16.** [7.7] 5   **17.** [7.7] $\frac{1}{2}$ **18.** [8.1] $\pm 5i$
19. [8.5] 9, 25   **20.** [8.5] ± 3, ± 2 **21.** [9.3] 8
22. [9.3] 7   **23.** [9.6] $\frac{3}{2}$ **24.** [9.6] $\frac{80}{9}$
25. [8.10] $\{x | x < -5 \text{ or } x > 1\}$, or $(-\infty, -5) \cup (1, \infty)$
26. [4.3] $\{x | x \leqslant -3 \text{ or } x \geqslant 6\}$, or $(-\infty, -3] \cup [6, \infty)$

27. [6.8] $a = \dfrac{Db}{b - D}$ **28.** [6.8] $q = \dfrac{pf}{p - f}$

29. [1.6] $B = \dfrac{3M - 2A}{2}$, or $B = \dfrac{3}{2}M - A$

30. [3.7] 2   **31.** [3.7] 3   **32.** [1.4] 8
33. [1.6] 36 m, 20 m

34. [3.5] A: 15°; B: 45°; C: 120°   **35.** [6.5] $5\frac{5}{11}$ hr
36. [3.3] 60 L of Swim Clean; 40 L of Pure Swim
37. [6.5] $2\frac{7}{9}$ km/h   **38.** [8.9] -49, -7 and 7
39. [6.5] Freight: 27 mph; passenger: 40 mph
40. [9.7] 78   **41.** [9.7] 67.5   **42.** [9.7] $P(t) = 430e^{0.01t}$,
where P is in millions and t is in years since 1961.
43. [9.7] 616 million, 641 million   **44.** [8.6] 18
45. [5.1] $7p^2q^3 - 2p^3q + pq - 6$
46. [5.1] $8x^2 - 11x - 1$ **47.** [5.2] $9x^4 - 12x^2y + 4y^2$
48. [5.2] $10a^2 - 9ab - 9b^2$

49. [6.1] $\dfrac{(x + 4)(x - 3)}{2(x - 1)}$ **50.** [6.3] $\dfrac{1}{x - 4}$

51. [6.1] $\dfrac{a + 2}{6}$ **52.** [6.2] $\dfrac{7x + 4}{(x + 6)(x - 6)}$

53. [5.3] $x(y - 2z + w)$
54. [5.6] $(1 - 5x)(1 + 5x + 25x^2)$
55. [5.4] $2(3x - 2y)(x + 2y)$ **56.** [5.3] $(x^3 + 7)(x - 4)$
57. [5.5] $(a - 5 + 9b)(a - 5 - 9b)$ **58.** [5.5] $2(m + 3n)^2$
59. [5.7] $(x - 2y)(x + 2y)(x^2 + 4y^2)$ **60.** [2.2] -12

61. [6.6] $x^3 - 2x^2 - 4x - 12 + \dfrac{-42}{x - 3}$

62. [1.7] 1.8×10^{-1} **63.** [1.7] 5.0×10^{13}
64. [7.4] $2y^2 \sqrt[3]{y}$ **65.** [7.3] $14xy^2 \sqrt{x}$

66. [7.2] $81a^8b \sqrt[3]{b}$ **67.** [7.6] $\dfrac{6 + \sqrt{y} - y}{4 - y}$

68. [7.4] $\sqrt[10]{\dfrac{1}{(x + 5)^3}}$ **69.** [7.9] $12 + 4\sqrt{3}i$

70. [7.9] $\frac{18}{25} + \frac{1}{25}i$ **71.** [9.2] $f^{-1}(x) = \dfrac{x - 7}{-2}$

72. [2.5] $y = -5x - 3$ **73.** [2.5] $y = \frac{1}{2}x + \frac{13}{2}$

74. [2.4]

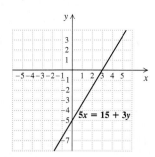

75. [8.8]

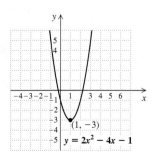

76. [9.3]

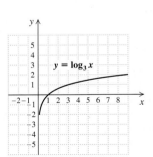

77. [9.1]

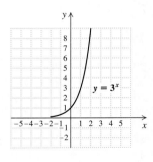

78. [4.4]

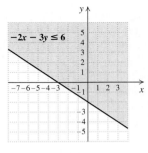

79. [8.7]

$f(x) = 2(x + 3)^2 + 1$
Minimum: 1

80. [9.4] $2 \log a + 3 \log c - \log b$

81. [9.4] $\log\left(\dfrac{x^3}{y^{1/2}z^2}\right)$ **82.** [9.3] $a^x = 5$

83. [9.3] $\log_x t = 3$ **84.** [9.5] -1.2545
85. [9.5] 278,804.64   **86.** [9.5] 2.5479
87. [9.5] 0.0253   **88.** [9.6] 0.354
89. [6.4] All real numbers except 1 and -2
90. [9.6] $\frac{1}{3}$, $\frac{10,000}{3}$ **91.** [8.3] 35 mph

CHAPTER 10

Technology Connection, Section 10.1

TC1.

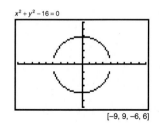

TC2.

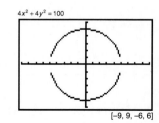

TC3.

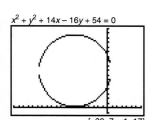

$x^2 + y^2 + 14x - 16y + 54 = 0$

$[-20, 7, -1, 17]$

TC4.

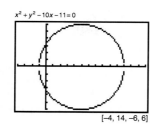

$x^2 + y^2 - 10x - 11 = 0$

$[-4, 14, -6, 6]$

17.

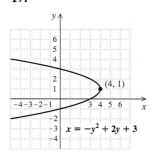

$(4, 1)$

$x = -y^2 + 2y + 3$

19.

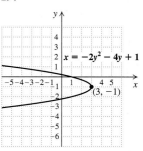

$x = -2y^2 - 4y + 1$

$(3, -1)$

Exercise Set 10.1, pp. 537–539

1.

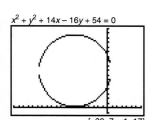

$y = x^2$

$(0, 0)$

3.

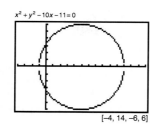

$(-3, -2)$

$x = y^2 + 4y + 1$

21. 5   **23.** $\sqrt{18} \approx 4.243$ **25.** $\sqrt{32} \approx 5.657$

27. 17.8   **29.** $\dfrac{\sqrt{41}}{7} \approx 0.915$ **31.** $\sqrt{6970} \approx 83.487$

33. $\sqrt{a^2 + b^2}$ **35.** $\sqrt{17 + 2\sqrt{14} + 2\sqrt{15}} \approx 5.677$

37. $\sqrt{9{,}672{,}400} \approx 3110.048$ **39.** $\left(-\dfrac{1}{2}, -1\right)$

41. $\left(\dfrac{7}{2}, \dfrac{7}{2}\right)$ **43.** $(-1, -3)$ **45.** $(-0.25, -0.3)$

47. $\left(-\dfrac{1}{12}, \dfrac{1}{24}\right)$ **49.** $\left(\dfrac{\sqrt{2} + \sqrt{3}}{2}, \dfrac{3}{2}\right)$

51. $x^2 + y^2 = 49$ **53.** $(x + 2)^2 + (y - 7)^2 = 5$

55. $(x + 4)^2 + (y - 3)^2 = 48$

57. $(x + 7)^2 + (y + 2)^2 = 50$ **59.** $x^2 + y^2 = 25$

61. $(x + 4)^2 + (y - 1)^2 = 20$

63. $(0, 0), 6$ **65.** $(-1, -3), 2$

5.

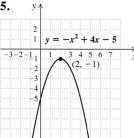

$y = -x^2 + 4x - 5$

$(2, -1)$

7.

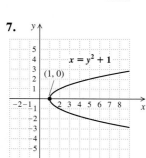

$x = y^2 + 1$

$(1, 0)$

9.

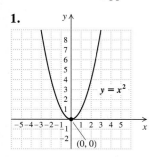

$x = -1 \cdot y^2$

$(0, 0)$

$x = -y^2 + 2y$

11.

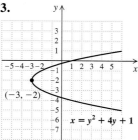

$(1, 1)$

$x = -y^2 + 2y$

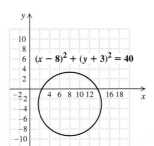

$x^2 + y^2 = 36$

$(x + 1)^2 + (y + 3)^2 = 4$

67. $(8, -3), 2\sqrt{10}$ **69.** $(0, 0), \sqrt{2}$

13.

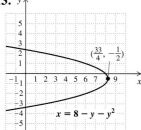

$\left(\dfrac{33}{4}, -\dfrac{1}{2}\right)$

$x = 8 - y - y^2$

15.

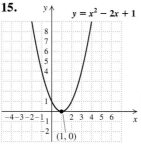

$y = x^2 - 2x + 1$

$(1, 0)$

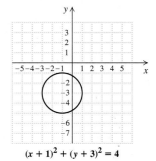

$(x - 8)^2 + (y + 3)^2 = 40$

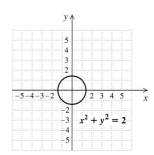

$x^2 + y^2 = 2$

71. $(5, 0)$, $\frac{1}{2}$

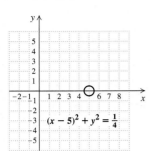

$(x - 5)^2 + y^2 = \frac{1}{4}$

73. $(-4, 3)$, $2\sqrt{10}$

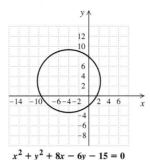

$x^2 + y^2 + 8x - 6y - 15 = 0$

91. Reflect one graph across the line $y = x$ to obtain the other.

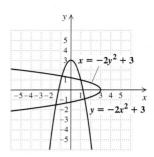

$x = -2y^2 + 3$

$y = -2x^2 + 3$

75. $(4, -1)$, 2

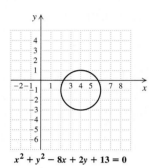

$x^2 + y^2 - 8x + 2y + 13 = 0$

77. $(2, 0)$, 2

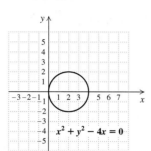

$x^2 + y^2 - 4x = 0$

93. $(0, 4)$ **95.** **(a)** $(0, -3)$; **(b)** 5 ft
97. $(x + 7)^2 + (y + 4)^2 = 16$
99. $(x + 3)^2 + (y - 5)^2 = 16$
101. Let $P_1 = (x_1, y_1)$, $P_2 = (x_2, y_2)$, and

$M = \left(\dfrac{x_1 + x_2}{2}, \dfrac{y_1 + y_2}{2} \right)$. Let $d(AB)$ denote the

distance from point A to point B.

i) $d(P_1M) = \sqrt{\left(\dfrac{x_1 + x_2}{2} - x_1 \right)^2 + \left(\dfrac{y_1 + y_2}{2} - y_1 \right)^2}$

$= \dfrac{1}{2}\sqrt{(x_2 - x_1)^2 + (y_2 - y_1)^2};$

$d(P_2M) = \sqrt{\left(\dfrac{x_1 + x_2}{2} - x_2 \right)^2 + \left(\dfrac{y_1 + y_2}{2} - y_2 \right)^2}$

$= \dfrac{1}{2}\sqrt{(x_1 - x_2)^2 + (y_1 - y_2)^2}$

$= \dfrac{1}{2}\sqrt{(x_2 - x_1)^2 + (y_2 - y_1)^2} = d(P_1M).$

ii) $d(P_1M) + d(P_2M) = \dfrac{1}{2}\sqrt{(x_2 - x_1)^2 + (y_2 - y_1)^2}$

$+ \dfrac{1}{2}\sqrt{(x_2 - x_1)^2 + (y_2 - y_1)^2}$

$= \sqrt{(x_2 - x_1)^2 + (y_2 - y_1)^2}$

$= d(P_1P_2).$

79. $(0, -5)$, 10

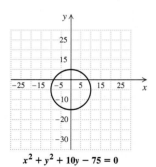

$x^2 + y^2 + 10y - 75 = 0$

81. $\left(-\dfrac{7}{2}, \dfrac{3}{2} \right)$, $\dfrac{7\sqrt{2}}{2}$

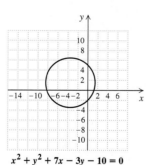

$x^2 + y^2 + 7x - 3y - 10 = 0$

83. $(0, 0)$, $\dfrac{1}{2}$

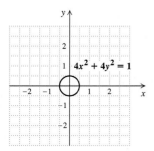

$4x^2 + 4y^2 = 1$

85. 4 in.   **87.** ◈
89. **(a)** $3.4, -2.4$;
(b) $2.3, -1.3$

Technology Connection, Section 10.2

TC1.

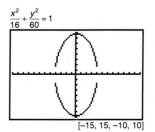

$\dfrac{x^2}{16} + \dfrac{y^2}{60} = 1$

$[-15, 15, -10, 10]$

TC2.

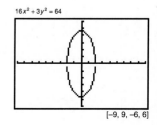

$16x^2 + 3y^2 = 64$

$[-9, 9, -6, 6]$

TC3.

$$\frac{y^2}{20} - \frac{x^2}{64} = 1$$

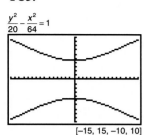

[−15, 15, −10, 10]

TC4.

$9x^2 - 45y^2 = 441$

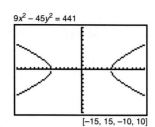

[−15, 15, −10, 10]

Exercise Set 10.2, pp. 550–551

1.

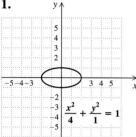

$$\frac{x^2}{4} + \frac{y^2}{1} = 1$$

3.

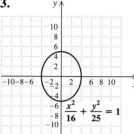

$$\frac{x^2}{16} + \frac{y^2}{25} = 1$$

5.

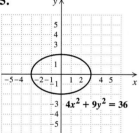

$4x^2 + 9y^2 = 36$

7.

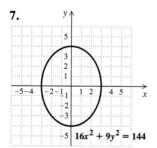

$16x^2 + 9y^2 = 144$

9.

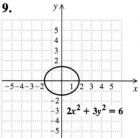

$2x^2 + 3y^2 = 6$

11.

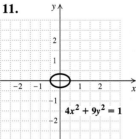

$4x^2 + 9y^2 = 1$

13.

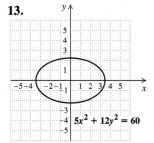

$5x^2 + 12y^2 = 60$

15.

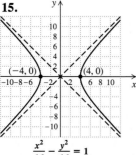

$$\frac{x^2}{16} - \frac{y^2}{16} = 1$$

17.

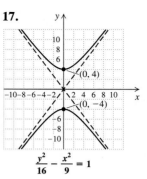

$$\frac{y^2}{16} - \frac{x^2}{9} = 1$$

19.

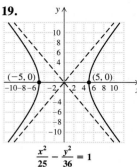

$$\frac{x^2}{25} - \frac{y^2}{36} = 1$$

21.

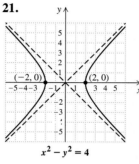

$x^2 - y^2 = 4$

23.

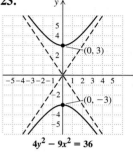

$4y^2 - 9x^2 = 36$

25.

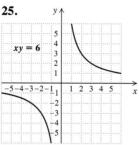

$xy = 6$

27.

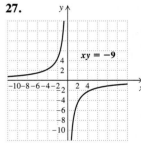

$xy = -9$

29.

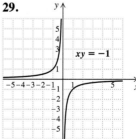

$xy = -1$

31.

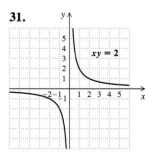

$xy = 2$

33. Circle   **35.** Hyperbola   **37.** Ellipse   **39.** Circle
41. Ellipse   **43.** Hyperbola   **45.** Parabola

47. Hyperbola   **49.** $5t^5$ **51.** $\dfrac{13 + 8\sqrt{6}}{10}$

53.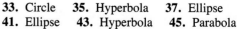

55. (a) Let $F_1 = (-c, 0)$ and $F_2 = (c, 0)$. Then the sum of the distances from the foci to P is $2a$. By the distance formula,

$$\sqrt{(x + c)^2 + y^2} + \sqrt{(x - c)^2 + y^2} = 2a, \quad \text{or}$$
$$\sqrt{(x + c)^2 + y^2} = 2a - \sqrt{(x - c)^2 + y^2}.$$

Squaring, we get

$$(x + c)^2 + y^2 = 4a^2 - 4a\sqrt{(x - c)^2 + y^2} + (x - c)^2 + y^2,$$

or $x^2 + 2cx + c^2 + y^2$

$$= 4a^2 - 4a\sqrt{(x - c)^2 + y^2} + x^2 - 2cx + c^2 + y^2.$$

Thus

$$-4a^2 + 4cx = -4a\sqrt{(x - c)^2 + y^2}$$
$$a^2 - cx = a\sqrt{(x - c)^2 + y^2}.$$

Squaring again, we get

$$a^4 - 2a^2cx + c^2x^2 = a^2(x^2 - 2cx + c^2 + y^2)$$
$$a^4 - 2a^2cx + c^2x^2 = a^2x^2 - 2a^2cx + a^2c^2 + a^2y^2,$$

or

$$x^2(a^2 - c^2) + a^2y^2 = a^2(a^2 - c^2)$$
$$\frac{x^2}{a^2} + \frac{y^2}{a^2 - c^2} = 1.$$

(b) When P is at $(0, b)$, it follows that $b^2 = a^2 - c^2$. Substituting, we have

$$\frac{x^2}{a^2} + \frac{y^2}{b^2} = 1.$$

57. 5.66 ft
59. $\dfrac{(x + 3)^2}{1} + \dfrac{(y - 4)^2}{16} = 1$; **61.** $\dfrac{y^2}{64} - \dfrac{x^2}{4} = 1$
C: $(-3, 4)$;
V: $(-2, 4)$, $(-4, 4)$,
$(-3, 8)$, $(-3, 0)$

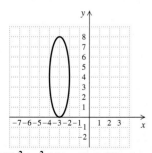

$16x^2 + y^2 + 96x - 8y + 144 = 0$

63. C: $(2, -1)$; V: $(-1, -1)$,
$(5, -1)$;
asymptotes: $y + 1 = \frac{4}{3}(x - 2)$,
$y + 1 = -\frac{4}{3}(x - 2)$;

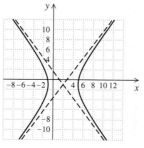

65. $\dfrac{(y - 1)^2}{25} - \dfrac{(x + 2)^2}{4} = 1$; C: $(-2, 1)$;
V: $(-2, 6)$, $(-2, -4)$;
asymptotes: $y - 1 = \frac{5}{2}(x + 2)$,
$y - 1 = -\frac{5}{2}(x + 2)$;

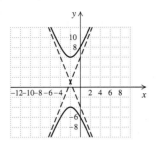

Technology Connection, Section 10.3

TC1. $(-1.50, -1.17)$; $(3.5, 0.5)$
TC2. $(-2.77, 2.52)$; $(-2.77, -2.52)$

Exercise Set 10.3, pp. 559–560

1. $(-4, -3)$, $(3, 4)$ **3.** $(0, 2)$, $(3, 0)$ **5.** $(-2, 1)$
7. $\left(\dfrac{5 + \sqrt{70}}{3}, \dfrac{-1 + \sqrt{70}}{3}\right)$, $\left(\dfrac{5 - \sqrt{70}}{3}, \dfrac{-1 - \sqrt{70}}{3}\right)$
9. $(3, 2)$, $\left(4, \frac{3}{2}\right)$ **11.** $\left(\frac{7}{3}, \frac{1}{3}\right)$, $(1, -1)$
13. $(1, 4)$, $\left(\frac{11}{4}, -\frac{5}{4}\right)$
15. $\left(\dfrac{7 + \sqrt{33}}{2}, \dfrac{7 - \sqrt{33}}{2}\right)$, $\left(\dfrac{7 - \sqrt{33}}{2}, \dfrac{7 + \sqrt{33}}{2}\right)$
17. $(3, -5)$, $(-1, 3)$ **19.** $(8, 5)$, $(-5, -8)$
21. $(4, -3)$, $(4, 3)$, $(-5, 0)$ **23.** $(-3, 0)$, $(3, 0)$
25. $(-4, -3)$, $(-3, -4)$, $(3, 4)$, $(4, 3)$
27. $\left(\dfrac{6\sqrt{21}}{7}, \dfrac{4i\sqrt{35}}{7}\right)$, $\left(\dfrac{6\sqrt{21}}{7}, -\dfrac{4i\sqrt{35}}{7}\right)$,
$\left(-\dfrac{6\sqrt{21}}{7}, \dfrac{4i\sqrt{35}}{7}\right)$, $\left(-\dfrac{6\sqrt{21}}{7}, -\dfrac{4i\sqrt{35}}{7}\right)$
29. $(-\sqrt{2}, -\sqrt{14})$, $(-\sqrt{2}, \sqrt{14})$, $(\sqrt{2}, -\sqrt{14})$, $(\sqrt{2}, \sqrt{14})$
31. $(-2, -1)$, $(-1, -2)$, $(1, 2)$, $(2, 1)$
33. $(-3, -2)$, $(-2, -3)$, $(2, 3)$, $(3, 2)$
35. $(2, 5)$, $(-2, -5)$ **37.** $(3, 2)$, $(-3, -2)$
39. $(-3, 4)$, $(3, 4)$, $(0, -5)$
41. Length: 8 cm; width: 6 cm
43. Length: 5 in.; width: 4 in.
45. Length: 75 yd; width: 30 yd   **47.** 13 and 12
49. 24 ft, 16 ft   **51.** Length: $\sqrt{3}$ m; width: 1 m
53. $4\sqrt{3}$ **55.** 3.7 mph   **57.** ◈
59. $(x + 2)^2 + (y - 1)^2 = 4$

61. 10 in. by 7 in. by 5 in.
63. $(-2, 3), (2, -3), (-3, 2), (3, -2)$

Review Exercises: Chapter 10, pp. 562–563

1. [10.1] 4   **2.** [10.1] 5   **3.** [10.1] $\sqrt{130} \approx 11.402$
4. [10.1] $\sqrt{9 + 4a^2}$ **5.** [10.1] $(4, 6)$
6. [10.1] $\left(-3, \frac{5}{2}\right)$ **7.** [10.1] $\left(\frac{1}{2}, \frac{5}{2}\right)$
8. [10.1] $\left(\frac{1}{2}, 2a\right)$ **9.** [10.1] $(-2, 3), \sqrt{2}$
10. [10.1] $(5, 0), 7$ **11.** [10.1] $(3, 1), 3$
12. [10.1] $(-4, 3), \sqrt{35}$
13. [10.1] $(x + 4)^2 + (y - 3)^2 = 48$
14. [10.1] $(x - 7)^2 + (y + 2)^2 = 20$
15. [10.2], [10.1] Circle   **16.** [10.2] Ellipse

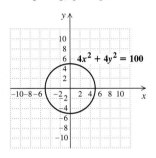

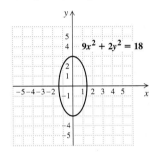

17. [10.2], [10.1]
Parabola

18. [10.2] Hyperbola

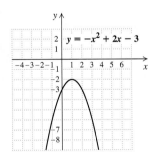

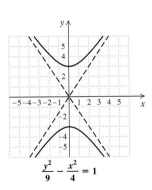

19. [10.2] Hyperbola

20. [10.2], [10.1]
Parabola

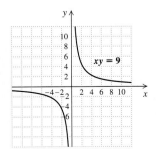

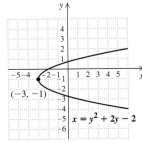

21. [10.2] Hyperbola

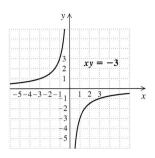

22. [10.2], [10.1] Circle

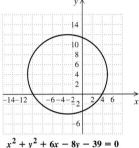

$x^2 + y^2 + 6x - 8y - 39 = 0$

23. [10.3] $(7, 4)$ **24.** [10.3] $(2, 2), \left(\frac{32}{9}, -\frac{10}{9}\right)$
25. [10.3] $(0, -3), (2, 1)$
26. [10.3] $(4, 3), (4, -3), (-4, 3), (-4, -3)$
27. [10.3] $(2, 1), \left(\sqrt{3}, 0\right), (-2, 1), \left(-\sqrt{3}, 0\right)$
28. [10.3] $(3, -3), \left(-\frac{3}{5}, \frac{21}{5}\right)$
29. [10.3] $(6, 8), (6, -8), (-6, 8), (-6, -8)$
30. [10.3] $(2, 2), (-2, -2), \left(2\sqrt{2}, \sqrt{2}\right),$
$\left(-2\sqrt{2}, -\sqrt{2}\right)$
31. [10.3] 12 m by 7 m   **32.** [10.3] 4 and 8
33. [10.3] 32 cm, 20 cm   **34.** [10.3] 3 ft, 11 ft
35. [7.3] $3a^2b^3 \sqrt[3]{3a^2b}$ **36.** [8.2] $-1 \pm 2i$
37. [7.6] $\dfrac{16 - a}{8 + 6\sqrt{a} + a}$ **38.** [6.5] 6 ft/sec

39. [10.1], [10.2] ◈ A circle is a special type of ellipse, where $a = b$.
40. [10.1], [10.2] ◈ Function notation is not used in this chapter because many of the relations are not functions. Function notation could be used for vertical parabolas and for hyperbolas that have the axes as asymptotes.
41. [10.3] $(2, i), (2, -i), (-2, i), (-2, -i)$
42. [10.1], [10.3] $(0, 6), (0, -6)$
43. [10.1], [10.3] $(x - 2)^2 + (y + 1)^2 = 25$

44. [10.2] $\dfrac{x^2}{49} + \dfrac{y^2}{9} = 1$ **45.** [10.1] $\left(\dfrac{9}{4}, 0\right)$

Test: Chapter 10, p. 563

1. [10.1] $9\sqrt{2} \approx 12.728$ **2.** [10.1] $2\sqrt{9 + a^2}$
3. [10.1] $\left(-\frac{1}{2}, \frac{7}{2}\right)$ **4.** [10.1] $(0, 0)$
5. [10.1] $(-2, 3), 8$ **6.** [10.1] $(-2, 3), 3$

7. [10.1], [10.2] Parabola

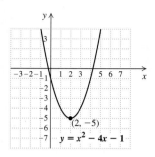

$$y = x^2 - 4x - 1$$

8. [10.1], [10.2] Circle

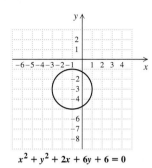

$$x^2 + y^2 + 2x + 6y + 6 = 0$$

9. [10.2] Hyperbola

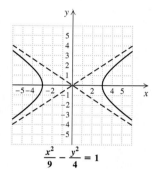

$$\frac{x^2}{9} - \frac{y^2}{4} = 1$$

10. [10.2] Ellipse

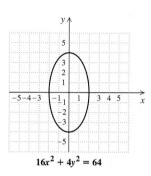

$$16x^2 + 4y^2 = 64$$

11. [10.2] Hyperbola

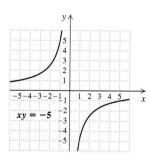

$$xy = -5$$

12. [10.1], [10.2] Parabola

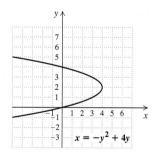

$$x = -y^2 + 4y$$

13. [10.3] (0, 3), (4, 0) **14.** [10.3] (4, 0), (−4, 0)
15. [10.3] 15 and 8 or 8 and 15 **16.** [10.3] 2 by 11
17. [10.3] $\sqrt{5}$ m, $\sqrt{3}$ m **18.** [10.3] 16 ft by 12 ft
19. [8.2] $-1 \pm \sqrt{6}$ **20.** [7.3] $2ab^6\sqrt[3]{6a^2}$
21. [7.6] $\dfrac{8 - 6\sqrt{a} + a}{4 - a}$ **22.** [6.5] 6 mph
23. [10.2] $\dfrac{(x-6)^2}{25} + \dfrac{(y-3)^2}{9} = 1$
24. [10.1] $\left(0, -\frac{31}{4}\right)$ **25.** [10.3] 9

CHAPTER 11

Exercise Set 11.1, pp. 570–571

1. 4, 7, 10, 13; 31; 46 **3.** $\frac{1}{2}, \frac{2}{3}, \frac{3}{4}, \frac{4}{5}; \frac{10}{11}; \frac{15}{16}$
5. −1, 0, 3, 8; 80; 195 **7.** 2, $2\frac{1}{2}$, $3\frac{1}{3}$, $4\frac{1}{4}$; $10\frac{1}{10}$; $15\frac{1}{15}$
9. −1, 4, −9, 16; 100; −225
11. −2, −1, 4, −7; −25; 40 **13.** 25 **15.** 225
17. −23.5 **19.** −33,880 **21.** $\frac{441}{400}$ **23.** 43
25. $2n - 1$ **27.** $(-1)^n 2(3)^{n-1}$ **29.** $\dfrac{n+1}{n+2}$ **31.** $3^{n/2}$
33. $-(3n - 2)$ **35.** 28 **37.** 30
39. $\frac{1}{2} + \frac{1}{4} + \frac{1}{6} + \frac{1}{8} + \frac{1}{10} = \frac{137}{120}$
41. $2^0 + 2^1 + 2^2 + 2^3 + 2^4 + 2^5 = 63$
43. $\log 7 + \log 8 + \log 9 + \log 10 \approx 3.7024$
45. $\dfrac{1}{2} + \dfrac{2}{3} + \dfrac{3}{4} + \dfrac{4}{5} + \dfrac{5}{6} + \dfrac{6}{7} + \dfrac{7}{8} + \dfrac{8}{9} = \dfrac{15,551}{2520}$
47. $(-1)^1 + (-1)^2 + (-1)^3 + (-1)^4 + (-1)^5 = -1$
49. $(-1)^2 3^1 + (-1)^3 3^2 + (-1)^4 3^3 + (-1)^5 3^4 +$
$(-1)^6 3^5 + (-1)^7 3^6 + (-1)^8 3^7 + (-1)^9 3^8 = -4920$
51. $3 + 2 + 3 + 6 + 11 + 18 = 43$
53. $\dfrac{1}{1 \cdot 2} + \dfrac{1}{2 \cdot 3} + \dfrac{1}{3 \cdot 4} + \dfrac{1}{4 \cdot 5} + \dfrac{1}{5 \cdot 6} + \dfrac{1}{6 \cdot 7} +$
$\dfrac{1}{7 \cdot 8} + \dfrac{1}{8 \cdot 9} + \dfrac{1}{9 \cdot 10} + \dfrac{1}{10 \cdot 11} = \dfrac{10}{11}$
55. $\displaystyle\sum_{k=1}^{6} \dfrac{k}{k+1}$ **57.** $\displaystyle\sum_{k=1}^{6} (-2)^k$ **59.** $\displaystyle\sum_{k=2}^{n} (-1)^k k^2$
61. $\displaystyle\sum_{k=1}^{\infty} 5k$ **63.** $\displaystyle\sum_{k=1}^{\infty} \dfrac{1}{k(k+1)}$ **65.** 1 **67.** 7
69. ◆ **71.** $\frac{3}{2}, \frac{3}{2}, \frac{9}{8}, \frac{3}{4}, \frac{15}{32}, \frac{171}{32}$
73. 0, ln 2, ln 6, ln 24, ln 120; ln 34,560
75. 2, 2.25, 2.370370, 2.441406, 2.488320, 2.521626
77. 0, 4, 20, 404, 163,220, 26,640,768,404
79. $5200, $3900, $2925, $2193.75, $1645.31,
$1233.98, $925.49, $694.12, $520.59, $390.44

Exercise Set 11.2, pp. 578–579

1. $a_1 = 2, d = 5$ **3.** $a_1 = 7, d = -4$ **5.** $a_1 = \frac{3}{2}, d = \frac{3}{4}$
7. $a_1 = $2.12, d = 0.12 **9.** 46 **11.** −41
13. −$1628.16 **15.** 27th **17.** 102nd **19.** 101
21. 5 **23.** 28 **25.** $a_1 = 8; d = -3; 8, 5, 2, -1, -4$
27. 670 **29.** 45,150 **31.** 2550 **33.** 735 **35.** 990
37. 3; 171 **39.** 1260 **41.** $31,000 **43.** 6300
45. $a^k = P$ **47.** $x^2 + y^2 = 81$ **49.** ◆ **51.** 3, 5, 7
53. $8760, $7961.77, $7163.54, $6365.31, $5567.08,
$4768.85, $3970.62, $3172.39, $2374.16, $1575.93
55. Let d = the common difference. Since p, m, and q form an arithmetic sequence, $m = p + d$ and $q = p + 2d$.
Then $\dfrac{p+q}{2} = \dfrac{p + (p + 2d)}{2} = p + d = m$.

Exercise Set 11.3, pp. 586–588

1. 2   **3.** −1   **5.** $-\dfrac{1}{2}$ **7.** $\dfrac{1}{5}$ **9.** $\dfrac{1}{x}$ **11.** 1.1

13. 64   **15.** 162   **17.** 648   **19.** $2331.64

21. $a_n = 3^{n-1}$ **23.** $a_n = (-1)^{n-1}$ **25.** $a_n = \dfrac{1}{x^n}$

27. 762   **29.** $\dfrac{547}{18}$

31. $\dfrac{1-x^8}{1-x}$, or $(1+x)(1+x^2)(1+x^4)$ **33.** $5134.51

35. 8   **37.** 125   **39.** $\dfrac{1000}{11}$ **41.** No   **43.** $\dfrac{1}{3}$

45. $25,000   **47.** $\dfrac{4}{9}$ **49.** $\dfrac{5}{9}$ **51.** $\dfrac{5}{33}$ **53.** $\dfrac{1}{256}$ ft

55. 155,797   **57.** $5236.19   **59.** 3100.35 ft

61. $2,684,354.55   **63.** $\left(-\dfrac{63}{29}, -\dfrac{114}{29}\right)$ **65.** ◈

67. $\dfrac{1-x^n}{1-x}$ **69.** 512 cm²

Exercise Set 11.4, pp. 594–595

1. 362,880   **3.** 39,916,800   **5.** 210   **7.** 3024
9. 28   **11.** 84   **13.** 190   **15.** 595
17. $m^5 + 5m^4n + 10m^3n^2 + 10m^2n^3 + 5mn^4 + n^5$
19. $x^6 - 6x^5y + 15x^4y^2 - 20x^3y^3 + 15x^2y^4 - 6xy^5 + y^6$
21. $x^{10} - 15x^8y + 90x^6y^2 - 270x^4y^3 + 405x^2y^4 - 243y^5$
23. $729c^6 - 1458c^5d + 1215c^4d^2 - 540c^3d^3 + 135c^2d^4 - 18cd^5 + d^6$ **25.** $x^3 - 3x^2y + 3xy^2 - y^3$
27. $x^{-7} + 7x^{-6}y + 21x^{-5}y^2 + 35x^{-4}y^3 + 35x^{-3}y^4 + 21x^{-2}y^5 + 7x^{-1}y^6 + y^7$
29. $a^9 - 18a^7 + 144a^5 - 672a^3 + 2016a - 4032a^{-1} + 5376a^{-3} - 4608a^{-5} + 2304a^{-7} - 512a^{-9}$
31. $a^{10} + 5a^8b^3 + 10a^6b^6 + 10a^4b^9 + 5a^2b^{12} + b^{15}$
33. $9 - 12\sqrt{3}t + 18t^2 - 4\sqrt{3}t^3 + t^4$
35. $x^{-8} + 4x^{-4} + 6 + 4x^4 + x^8$ **37.** $15a^4b^2$
39. $-745,472a^3$ **41.** $1120x^{12}y^2$
43. $-1,959,552u^5v^{10}$ **45.** 4   **47.** $t \approx 5.6348$

49. $\dbinom{5}{2}(0.313)^3(0.687)^2 \approx 0.145$

51. $\dbinom{5}{2}(0.313)^3(0.687)^2 + \dbinom{5}{3}(0.313)^2(0.687)^3 +$
$\dbinom{5}{4}(0.313)(0.687)^4 + \dbinom{5}{5}(0.687)^5 \approx 0.964$

53. $\dbinom{n}{n-r} = \dfrac{n!}{[n-(n-r)]!\,(n-r)!}$
$= \dfrac{n!}{r!\,(n-r)!} = \dbinom{n}{r}$

55. $-4320x^6y^{9/2}$ **57.** $-\dfrac{35}{x^{1/6}}$

Review Exercises: Chapter 11, pp. 596–597

1. [11.1] 1, 5, 9, 13; 29; 45

2. [11.1] $0, \dfrac{1}{5}, \dfrac{1}{5}, \dfrac{3}{17}; \dfrac{7}{65}; \dfrac{11}{145}$ **3.** [11.1] $a_n = -2n$
4. [11.1] $a_n = n^2$
5. [11.1] $-2 + 4 + (-8) + 16 + (-32) = -22$
6. [11.1] $-3 + (-5) + (-7) + (-9) + (-11) + (-13) = -48$

7. [11.1] $\displaystyle\sum_{k=1}^{5} 4k$ **8.** [11.1] $\displaystyle\sum_{k=1}^{5} \dfrac{1}{(-2)^k}$

9. [11.2] 85   **10.** [11.2] $\dfrac{8}{3}$
11. [11.2] $d = 1.25$, $a_1 = 11.25$ **12.** [11.2] -544
13. [11.2] 8580   **14.** [11.2] 63   **15.** [11.2] 864
16. [11.3] $1024\sqrt{2}$ **17.** [11.3] $\dfrac{2}{3}$

18. [11.3] $a_n = 2(-1)^n$ **19.** [11.3] $a_n = 3\left(\dfrac{x}{4}\right)^{n-1}$

20. [11.3] 4095   **21.** [11.3] $-4095x$ **22.** [11.3] 12
23. [11.3] 0.05   **24.** [11.3] No   **25.** [11.3] No
26. [11.3] $40,000   **27.** [11.3] $\dfrac{5}{9}$ **28.** [11.3] $\dfrac{13}{33}$
29. [11.2] $17.80   **30.** [11.2] 903
31. [11.3] $22,521.92   **32.** [11.3] 6 m
33. [11.4] 40,320   **34.** [11.4] 56   **35.** [11.4] $190a^{18}b^2$
36. [11.4] $x^4 - 8x^3y + 24x^2y^2 - 32xy^3 + 16y^4$
37. [3.2] $\left(\dfrac{26}{11}, \dfrac{1}{11}\right)$ **38.** [9.6] $\dfrac{5}{9}$
39. [9.4] $3 \log a + 4 \log b - \log c - 2 \log d$
40. [10.1] 13
41. [11.3] ◈ For a geometric sequence with $|r| < 1$, as n gets larger, the absolute value of the terms gets smaller, since $|r^n|$ gets smaller.   **42.** [11.4] ◈ The first form of the binomial theorem draws the coefficients from Pascal's triangle; the second form uses factorial notation. The second form avoids the need to compute all preceding rows of Pascal's triangle, and is generally easier to use when only one term of an expansion is needed. When several terms of an expansion are needed and n is not large (say, $n \leq 8$), it is often easier to use Pascal's triangle.

43. [11.3] $\dfrac{1-(-x)^n}{x+1}$
44. [11.4] $x^{-15} + 5x^{-9} + 10x^{-3} + 10x^3 + 5x^9 + x^{15}$

Test: Chapter 11, p. 597–598

1. [11.1] 1, 7, 13, 19, 25; 91   **2.** [11.1] $a_n = 4\left(\dfrac{1}{3}\right)^n$
3. [11.1] $1 - 1 - 5 - 13 - 29 = -47$

4. [11.1] $\displaystyle\sum_{k=1}^{5} k^3$ **5.** [11.2] -46 **6.** [11.2] $\dfrac{3}{8}$

7. [11.2] $a_1 = 31.2$; $d = -3.8$ **8.** [11.2] 2508
9. [11.3] $\dfrac{9}{128}$ **10.** [11.3] $\dfrac{2}{3}$ **11.** [11.3] $(-1)^{n+1}3^n$
12. [11.3] $511 + 511x$ **13.** [11.3] 1

14. [11.3] No   **15.** [11.3] $\dfrac{\$25,000}{23} \approx \$1086.96$

16. [11.3] $\dfrac{85}{99}$ **17.** [11.2] $17.60   **18.** [11.2] 1378

19. [11.3] $43,178.50   **20.** [11.3] 36 m   **21.** [11.4] 78
22. [11.4] $x^{10} - 15x^8y + 90x^6y^2 - 270x^4y^3 + 405x^2y^4 - 243y^5$ **23.** [11.4] $220a^9x^3$ **24.** [9.6] 3
25. [10.1] $(x-1)^2 + (y+2)^2 = 27$ **26.** [3.2] $(-1, 2)$
27. [9.4] $7 \log_t c - 5 \log_t a - 4 \log_t b$
28. [11.2] $n(n+1)$

29. [11.3] $\dfrac{1 - \left(\frac{1}{x}\right)^n}{1 - \frac{1}{x}}$, or $\dfrac{x^n - 1}{x^{n-1}(x-1)}$

CUMULATIVE REVIEW: 1–11

1. [1.5] $-45x^6y^{-4}$, or $\dfrac{-45x^6}{y^4}$ **2.** [1.2] 6.3

3. [1.3] $-3y + 17$ **4.** [1.5] 280   **5.** [1.1], [1.2] $\frac{7}{6}$
6. [5.1] $3a^2 - 8ab - 15b^2$
7. [5.1] $13x^3 - 7x^2 - 6x + 6$ **8.** [5.2] $6a^2 + 7a - 5$

9. [5.2] $9a^4 - 30a^2y + 25y^2$ **10.** [6.2] $\dfrac{4}{x+2}$

11. [6.1] $\dfrac{x-4}{3(x+2)}$ **12.** [6.1] $\dfrac{(x+y)(x^2+xy+y^2)}{x^2+y^2}$

13. [6.3] $x - a$ **14.** [5.5] $(2x-3)^2$
15. [5.6] $(3a-2)(9a^2+6a+4)$
16. [5.3] $(a^2-b)(a+3)$ **17.** [5.7] $3(y^2+3)(5y^2-4)$

18. [2.2] 20   **19.** [6.6] $7x^3 + 9x^2 + 19x + 38 + \dfrac{72}{x-2}$

20. [1.3] $\frac{2}{3}$ **21.** [5.8] 8, -6 **22.** [6.4] $-\frac{6}{5}$, 4
23. [6.4] No solution   **24.** [3.2] $(1, -1)$
25. [3.4] $(2, -1, 1)$ **26.** [7.7] 9   **27.** [8.5] ± 5, ± 2
28. [10.3] $(\sqrt{5}, \sqrt{3}), (\sqrt{5}, -\sqrt{3}), (-\sqrt{5}, \sqrt{3}),$
$(-\sqrt{5}, -\sqrt{3})$ **29.** [9.6] $\ln 8/\ln 5 \approx 1.2920$
30. [9.6] 1005   **31.** [9.3] $\frac{1}{16}$ **32.** [9.6] $-\frac{1}{2}$
33. [4.3] $\{x | -2 \le x \le 3\}$ **34.** [8.1] $\pm i\sqrt{2}$
35. [8.2] $-2 \pm \sqrt{7}$ **36.** [8.10] $\{y | y < -5 \text{ or } y > 2\}$
37. [10.3] 5 ft by 12 ft   **38.** [1.4] $3.34
39. [1.4] 65, 66, 67   **40.** [3.3] $11\frac{3}{7}$
41. [3.3] 24 L of A, 56 L of B   **42.** [6.5] 350 mph
43. [6.5] $8\frac{2}{5}$ min or 8 min, 24 sec   **44.** [8.6] 20
45. [8.9] 1250 ft^2
46. [2.4]   **47.** [10.2]

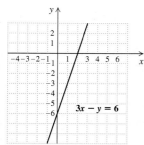

$3x - y = 6$

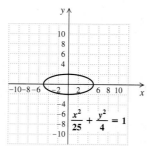

$\dfrac{x^2}{25} + \dfrac{y^2}{4} = 1$

48. [9.3]

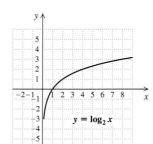

$y = \log_2 x$

49. [4.4]

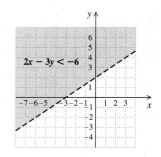

$2x - 3y < -6$

50. [8.7]

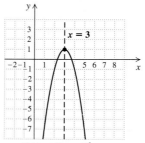

$f(x) = -2(x-3)^2 + 1$
Vertex (3, 1)
Maximum: 1

51. [1.6] $r = \dfrac{V-P}{-Pt}$ **52.** [6.8] $R = \dfrac{Ir}{1-I}$

53. [2.5] $y = -\frac{1}{3}x + \frac{11}{3}$ **54.** [3.7] -2 **55.** [3.7] -2
56. [1.7] 6.8×10^{-12} **57.** [7.3] $8x^2\sqrt{y}$

58. [7.2] $125x^2y^{3/4}$ **59.** [7.4] $\dfrac{\sqrt[3]{5xy}}{y}$

60. [7.6] $\dfrac{1 - 2\sqrt{x} + x}{1-x}$ **61.** [7.4] $\sqrt[6]{x+1}$

62. [7.9] $26 - 13i$ **63.** [8.4] $x^2 - 50 = 0$
64. [10.1] Center: $(2, -3)$; radius = 6

65. [9.4] $\log_a \dfrac{\sqrt[3]{x^2} \cdot z^5}{\sqrt{y}}$ **66.** [9.3] $a^5 = c$

67. [9.5] 3.7541   **68.** [9.5] 0.0000129
69. [9.5] 8.6442   **70.** [9.5] 0.0075
71. [9.7] $P(t) = 99{,}404e^{0.007t}$ **72.** [9.7] 105,868
73. [10.1] 5   **74.** [11.2] -121 **75.** [11.2] 875
76. [11.3] $16\left(\frac{1}{4}\right)^{n-1}$ **77.** [11.4] $13{,}440a^4b^6$
78. [11.3] $74.88671875x$ **79.** [11.3] $168.95
80. [6.4] All real numbers except 0 and -12
81. [9.6] 81   **82.** [8.6] y gets divided by 8

83. [7.9] $-\dfrac{7}{13} + \dfrac{2\sqrt{30}}{13}i$ **84.** [3.5] 84 yr

A

Abscissa, 65

Absolute value, 10, 12, 57, 191, 213
 and distance, 193, 194, 533
 equations with, 191
 inequalities with, 194
 and radical expressions, 339, 341

Absolute-value function on a grapher,
 196

Absolute-value principle, 192, 195,
 213

Addition
 associative law, 16, 57
 commutative law, 15, 57
 of complex numbers, 384
 of exponents, 34
 of functions, 103
 of polynomials, 221
 of radical expressions, 361
 of rational expressions, 285, 288
 of real numbers, 11, 56

Addition principle
 for equations, 20, 58
 for inequalities, 176, 213

Additive inverse, 11
 of polynomials, 222

Algebra of functions, 103, 110

Algebraic expression, 2
 evaluating, 4
 least common multiple, 288, 329
 translating to, 3

Algebraic logic, 41

Angles
 complementary, 114
 supplementary, 120

Angstrom (Å), 55

Antilogarithms, 503, 504

Applications, *see* Applied problems;
 Formulas; Index of
 Applications

Applied problems, 28–33, 46, 48, 49,
 52–55, 59–61, 74, 77–80, 85,
 87–89, 91, 94, 95, 97,
 100–102, 108, 112, 114, 115,
 120–122, 128–139, 146–151,
 154, 155, 159–165, 167–171,
 179, 180, 182, 183, 190, 191,
 197, 198, 205, 207, 208,
 210–212, 214, 215, 220,
 223–225, 233, 235, 237, 238,
 247, 252, 255, 258, 263, 265,
 266, 268–270, 272, 273, 284,
 292, 293, 300, 306–315, 320,
 324–328, 330–334, 343, 347,
 348, 355, 360, 375, 393, 394,
 404–406, 411, 413, 414, 416,
 419, 424, 425, 428–433,
 447–456, 464, 477, 479, 490,
 513, 514, 516, 517, 519–522,
 524–528, 538, 550, 557–560,
 562, 571, 575–579, 584, 585,
 587, 594, 595, 597–601. *See
 also the Index of Applications
 for specific applications.*

Approximating
 solutions of quadratic equations,
 410
 square roots, 352

Area, *see* Index of Applications:
 Geometric Applications

Arithmetic progressions, *see*
 Arithmetic sequences

Arithmetic sequences, 572
 common difference, 572
 nth term, 573
 sum of n terms, 575

Arithmetic series, 574

Ascending order, 219

Associative laws, 16, 57

Asymptote of an exponential function,
 474

Asymptotes of a hyperbola, 543

Average speed, 315

Axes, 64

Axis of a hyperbola, 542

B

Back-substitution, 152

Base, 34
 changing, 505

Binomial expansion
 using factorial notation, 591
 using Pascal's triangle, 588
 $(r + 1)$st term, 593

Binomial theorem, 590, 592

SELECTED KEYS OF THE GRAPHING CALCULATOR

Determines portion of curve(s) that is viewed.

Magnifies or reduces a portion of curve being viewed and can "square" the graph to reduce distortion.

The window in which graphs and mathematical symbolism appear.

Used to enter equation(s) that is to be graphed.

Used to determine coordinates of points on a curve.

Determines whether curves are dotted or connected, if curves are drawn sequentially or simultaneously, and if a grid is to appear.

Used to graph equations that were entered using the $Y=$ key.

Used to write the variable x.

Used to move cursor.

Used to insert characters in previously entered expressions.

Used to delete previously entered characters.

These keys behave similarly to the corresponding keys on a scientific calculator (see facing page).

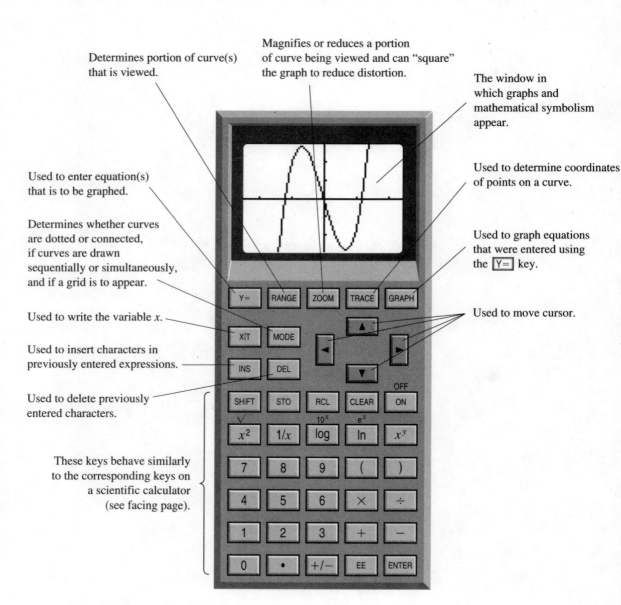

The use of a graphing calculator is optional in this text.